CELL SURFACE CARBOHYDRATES AND BIOLOGICAL RECOGNITION

PROGRESS IN CLINICAL AND BIOLOGICAL RESEARCH

Series Editors:

George J. Brewer Kurt Hirschhorn
Vincet P. Eijsvoogel Seymour S. Kety
Robert Grover Sidney Undenfriend
　　　　Jonathan W. Uhr

Vol 1: Erythrocyte Structure and Function, George J. Brewer, *Editor*
Vol 2: Preventability of Perinatal Injury, Karlis Adamsons and Howard A. Fox, *Editors*
Vol 3: Infections of the Fetus and the Newborn Infant, Saul Krugman and Anne A. Gershon, *Editors*
Vol 4: Conflicts in Childhood Cancer, Lucius F. Sinks and John O. Godden, *Editors*
Vol 5: Trace Components of Plasma: Isolation and Clinical Significance, G.A. Jamieson and T.J. Greenwalt, *Editors*
Vol 6: Prostatic Disease, H. Marberger, H. Haschek, H.K.A. Schirmer, J.A.C. Colston, and E. Witkin, *Editors*
Vol 7: Blood Pressure, Edema and Proteinuria in Pregnancy, Emanuel A. Friedman, *Editor*
Vol 8: Cell Surface Receptors, Garth L. Nicolson, Michael A. Raftery, Martin Rodbell, and C. Fred Fox, *Editors*
Vol 9: Membranes and Neoplasia: New Approaches and Strategies, Vincent T. Marchesi, *Editor*
Vol 10: Diabetes and Other Endocrine Disorders During Pregnancy and in the Newborn, Maria I. New and Robert H. Fiser, *Editors*
Vol 11: Clinical Uses of Frozen-Thawed Red Blood Cells, John A. Griep, *Editor*
Vol 12: Breast Cancer, Albert C.W. Montague, Geary L. Stonesifer, Jr., and Edward F. Lewison, *Editors*
Vol 13: The Granulocyte: Function and Clinical Utilization, Tibor J. Greenwalt and G.A. Jamieson, *Editors*
Vol 14: Zinc Metabolism: Current Aspects in Health and Disease, George J. Brewer and Ananda S. Prasad, *Editors*
Vol 15: Cellular Neurobiology, Zach Hall, Regis Kelly, and C. Fred Fox, *Editors*
Vol 16: HLA and Malignancy, Gerald P. Murphy, *Editor*
Vol 17: Cell Shape and Surface Architecture, Jean Paul Revel, Ulf Henning, and C. Fred Fox, *Editors*
Vol 18: Tay-Sachs Disease: Screening and Prevention, Michael M. Kaback, *Editor*
Vol 19: Blood Substitutes and Plasma Expanders, G.A. Jamieson and T.J. Greenwalt, *Editors*
Vol 20: Erythrocyte Membranes: Recent Clinical and Experimental Advances, Walter C. Kruckeberg, John W. Eaton, and George J. Brewer, *Editors*
Vol 21: The Red Cell, George J. Brewer, *Editor*
Vol 22: Molecular Aspects of Membrane Transport, Dale Oxender and C. Fred Fox, *Editors*
Vol 23: Cell Surface Carbohydrates and Biological Recognition, Vincent T. Marchesi, Victor Ginsburg, Phillips W. Robbins, and C. Fred Fox, *Editors*

CELL SURFACE CARBOHYDRATES AND BIOLOGICAL RECOGNITION

Proceedings of the ICN—UCLA Symposium
held at Keystone, Colorado
February 1977

Editors

VINCENT T. MARCHESI
Yale University School of Medicine
New Haven, Connecticut

VICTOR GINSBURG
National Institutes of Health
Bethesda, Maryland

PHILLIPS W. ROBBINS
Massachusetts Institute of Technology
Cambridge, Massachusetts

C. FRED FOX
University of California
Los Angeles, California

Alan R. Liss, Inc. • New York

Copyright 1978 Alan R. Liss, Inc
Address all inquiries to the publisher:
Alan R. Liss, Inc, 150 Fifth Avenue, New York, New York 10011

LC 78-417 • ISBN 0-8451-0023-8

The owner of copyright for this book hereby grants permission to users to make photocopy reproductions of any part or all of its contents for personal or internal organizational use, or for personal or internal use of specific clients. For copying beyond that permitted by Sections 107 or 108 of the US Copyright Law, this consent is given on the condition that the copier pay the stated per-copy fee through the Copyright Clearance Center, Inc, PO Box 765, Schenectady, NY 12301. The per-copy fee appears on the opening pages of the articles on pages 437–546. For all other articles in this book, the fee is listed in the most current issue of "Permissions to Photocopy" (Publisher's Fee List, distributed by CCC, Inc). This consent does not extend to other kinds of copying, such as copying for general distribution, advertising or promotional purposes, creating new collective works, or for resale.

Printed in the United States of America

NOTE

Pages 1–546 of this volume are reprinted from Journal of Supramolecular Structure, Volumes 6 and 7, 1977, and Volume 8, Number 1, 1978. The page numbers in the table of contents, author index, and subject index of this volume correspond to the page numbers at the foot of these pages.

Contents

Preface xiii

Glycosphingolipids in Membrane Architecture
Frances J. Sharom and Chris W.M. Grant 1

Conformation and Intermolecular Interactions of Carbohydrate Chains
Edwin R. Morris, David A. Rees, David Thom, and E. Jane Welsh 11

Isolation and Partial Characterization of the Heterophile Antigen of Infectious Mononucleosis From Bovine Erythrocytes
J.M. Merrick, R. Schifferle, K. Zadarlik, K. Kano, and F. Milgrom 27

Homogenous Antibodies Directed Against Human Cell Surface Antigens: I. The Mouse Spleen Fragment Culture Response to T and B Cell Lines Derived From the Same Individual
Lois A. Lampson, Ivor Royston, and Ronald Levy 43

Chemical and Immunological Studies of Cell Surfaces From Normal and Transformed Cells
W.J. Grimes, Gary A. Van Nest, and Arthur R. Kamm 51

Microbial Carbohydrate Specific Antibodies Distinguish Between Different Stages of Differentiating Mouse Cerebellum
Ekkhart Trenkner and Siddhartha Sarkar 67

Carbon-13 as a Tool for the Study of Carbohydrate Structures, Conformations and Interactions
H.A. Nunez, T.E. Walker, R. Fuentes, J. O'Connor, A. Serianni, and R. Barker 75

Distribution of a Major Surface-Associated Glycoprotein, Fibronectin, in Cultures of Adherent Cells
Deane F. Mosher, Olli Saksela, Jorma Keski-Oja, and Antti Vaheri 91

The Amino- and Carboxyl-Terminal Sequence of Bovine Rhodopsin
Paul A. Hargrave and Shao-Ling Fong 99

Contents

Sialic Acid Uptake by BHK Cells and Subsequent Incorporation Into Glycoproteins and Glycolipids
Carlos B. Hirschberg and Mary Yeh 111

Properties of a Penicillium GDP-Mannose: Glycopeptide Mannosyltransferase Solubilized With Triton X-100
J.E. Gander and Faye Fang 119

Characterization of the Fc Receptors of the Murine Leukemia L1210
Sheldon M. Cooper and Yugalkishore Sambray 131

Plants Interact With Microbial Polysaccharides
Peter Albersheim, Arthur R. Ayers, Jr., Barbara S. Valent, Jürgen Ebel, Michael Hahn, Jack Wolpert, and Russell Carlson 139

Localization of Some Glycolipid Glycosylating Enzymes in the Golgi Apparatus of Rat Kidney
Becca Fleischer 159

Distribution of Glycoconjugates in Mouse Fibroblasts With Varying Degrees of Tumorigenicity
D.J. Winterbourne and P.T. Mora 171

Interaction of Cartilage Proteoglycans With Hyaluronic Acid
Vincent C. Hascall 181

Structural Analysis of a Membrane Glycoprotein: Glycophorin A
Heinz Furthmayr 201

Glycosylation of VSV Glycoprotein Is Similar in Cystic Fibrosis, Heterozygous Carrier, and Normal Human Fibroblasts
Lawrence A. Hunt and Donald F. Summers 215

Cell Surface Carbohydrates of Preimplantation Embryos as Assessed by Lectin Binding
Anna G. Brownell 225

Biochemical Characteristics, Metabolism, and Antitumor Activity of Several Acetylated Hexosamines
R.J. Bernacki, M. Sharma, N.K. Porter, Y. Rustum, B. Paul, and W. Korytnyk 237

Role of Mannosyl Phosphoryl Polyisoprenol in Biosynthesis of Mammary Glycoproteins
Inder K. Vijay and Steven R. Fram 253

Plant Lectins Detect Age and Region Specific Differences in Cell Surface Carbohydrates and Cell Reassociation Behavior of Embryonic Mouse Cerebellar Cells
Mary E. Hatten and Richard L. Sidman 269

Cell Surface Changes Accompanying Myoblast Differentiation
L.T. Furcht, G. Wendelschafer-Crabb, and P.A. Woodbridge 279

Surface Antigens of the Embryonic Chick Myoblast: Expression on
Freshly Trypsinized Cells
Martin Friedlander and Donald A. Fischman 295

Organization and Polysaccharides of Sponge Aggregation Factor
Susie Humphreys, Tom Humphreys, and James Sano 311

Membrane Assembly: Synthesis and Intracellular Processing of the
Vesicular Stomatitis Viral Glycoprotein
Flora N. Katz, James E. Rothman, David M. Knipe, and
Harvey F. Lodish 325

Comparison of the Carbohydrate of Sindbis Virus Glycoproteins With
the Carbohydrate of Host Glycoproteins
Kenneth Keegstra and David Burke 343

The Biosynthesis of Mannolipids and Mannose-Containing Complex
Glycans by the Retina
Edward L. Kean 353

Synthesis, Secretion, and Attachment of LETS Glycoprotein in
Normal and Transformed Cells
Richard O. Hynes, Antonia T. Destree, Vivien M. Mautner,
and Iqbal U. Ali 369

The Turnover of a Tissue Specific Cell Surface Ligand Which
Inhibits Lectin Induced Capping
James McDonough and Jack Lilien 381

Glycoprotein Synthesis as a Function of Epithelial Cell Arrangement:
Biosynthesis and Release of Glycoproteins by Human Breast and
Prostate Cells in Organ Culture
Zoltán Tökés and Gerald B. Dermer 391

The Area-Code Hypothesis: The Immune System Provides Clues to
Understanding the Genetic and Molecular Basis of Cell Recognition
During Development
L. Hood, H.V. Huang, and W.J. Dreyer 407

Ganglioside Structure and Distribution: Are They Localized at the
Nerve Ending?
R.W. Ledeen 437

Cell Surface Glycosyltransferases — Do They Exist?
Wolfgang Deppert and Gernot Walter 455

Immunochemical Purification of Probe-Labeled Plasma Membrane
Proteins: An Approach to the Molecular Anatomy of the Cell
Surface
Paolo Comoglio, Guido Tarone, and Marilena Bertini 475

Contents

Enzymatic Conversion of Proteins to Glycoproteins by Lipid-Linked
Saccharides: A Study of Potential Exogenous Acceptor Proteins
Kathryn E. Kronquist and William J. Lennarz 487

Sialic Acid: A Specific Role in Hematopoietic Spleen Colony
Formation
Quentin Tonelli and Russel H. Meints 503

Dimensions and Specificities of Recognition Sites on Lectins and
Antibodies
Elvin A. Kabat 515

Glycoprotein and Protein Precursors to Plasma Membranes in
Vesicular Stomatitis Virus Infected HeLa Cells
Paul H. Atkinson 525

Receptor-Mediated Uptake of Lysosomal Enzymes
William S. Sly, Arnold Kaplan, Daniel T. Achord, Frederick E. Brot,
and C. Elliott Bell 547

A New Approach to the Structural Determination of Glycoproteins
and Polysaccharides: Anhydrous HF Solvolysis
Andrew J. Mort 553

Toward a Mechanism of Myoblast Fusion
K.A. Knudsen and A.F. Horwitz 563

Cell Surface Carbohydrate Recognition and the Viability of Erythrocytes in Circulation
David Aminoff, William C. Bell, and William G. VorderBruegge 569

Proteins Containing Reductively Aminated Disaccharides: Chemical
and Immunochemical Characterization
Gary R. Gray, Barbara A. Schwartz, and Barbara J. Kamicker 583

Comparative Biochemistry of Nucleotide-Linked Sugars
Victor Ginsburg 595

Immunochemistry of Streptococcal Group C Polysaccharide and the
Nature of its Crossreaction With the Forssman Glycolipid
John E. Coligan, Blair A. Fraser, and Thomas J. Kindt 601

Use of Common Plant Lectins for Isolation and Characterization of
Constitutive and Developmentally Regulated Cell Surface
Associated Glycoproteins of Dictyostelium discoideum
John E. Geltosky, Jasodhara Ray, and Richard A. Lerner 613

The Initial Glycosylation of the Sindbis Virus Membrane Proteins
Bartholomew M. Sefton 621

Developmentally Regulated Lectins in Cellular Slime Molds and
Embryonic Chick Tissues
Samuel H. Barondes 633

Properties of a Double Gradient Model for Retinotectal Specificity
    Richard B. Marchase and Stephen Roth 637

Affinity Labeling of a Cell Surface Receptor for Epidermal Growth Factor
    Manjusri Das, Tokichi Miyakawa, and C. Fred Fox 647

Lectin Receptors and Cell Surface Recognition
    R. Colin Hughes 657

Author Index 669

Subject Index 671

Preface

Many knowledgeable biologists would say, almost reflexly, that complex carbohydrates probably play a pivotal role in determining the specificity of many biological recognition phenomenon. Although the basis for this prejudice would certainly vary with the holder, most would cite the fact that complex carbohydrates seem to be present on all mammalian cell surfaces. It is also often pointed out that the complex carbohydrates are extremely variable in chemical composition and three-dimensional conformation, thus, these molecules could conceivably contain an infinite amount of stereochemical information. When such complex carbohydrates are experimentally modified, either by enzymes or specific chemical reagents, one often notes changes in the surface behavior of the modified cells and also marked differences in surface antigenicity. Interesting effects are also achieved by treating cells with different plant lectins. These sugar-specific proteins bind to most cell surfaces, some seem to show a special selectivity for tumor cells, and in some cases such bound lectins have the capacity to mimic the effects of hormones.

In spite of this wealth of circumstantial information linking complex carbohydrates with recognition sites, we still do not know which carbohydrates on cell membranes mediate these functions, nor how such molecules are displayed to convey their information. Apart from what is known about the chemistry of specific blood-group antigens, there is little data linking specific complex carbohydrates with specific recognition determinants. We are also extremely ignorant as to how and where complex carbohydrates are arranged three-dimensionally on the cell surface. Part of this problem stems from the fact that we do not know enough about how model carbohydrates can arrange themselves in space under physiologic conditions.

The papers in this symposium volume attempt to address many of these questions by describing results from a diverse set of biological systems.

This conference was supported in part through contracts awarded by the National Institutes of Health: National Institute of General Medical Sciences (PD-108185-7) and the National Cancer Institute (PD-204856-7). We also wish to acknowledge the continued financial sponsorship of these meetings provided by a yearly gift from ICN Pharmaceuticals, Inc. Finally, the organizers wish to express their gratitude to Fran Stusser for her efforts in bringing this conference to fruition.

Vincent T. Marchesi
Victor Ginsburg
Phillips W. Robbins

Glycosphingolipids in Membrane Architecture

Frances J. Sharom and Chris W. M. Grant

Department of Biochemistry, University of Western Ontario, London, Ontario, Canada, N6A 5C1

As part of a program to investigate the behavior and interactions of glycolipids in biological membranes we have synthesized spin-labeled derivatives of 2 families of carbohydrate-bearing ceramides (glycosphingolipids): simple neutral glycolipids and gangliosides. Galactosyl ceramide has been synthesized with the spin label at 3 different positions on the fatty acid chain. It has been studied in bilayers of various different lipids and lipid mixtures and compared to the corresponding phospholipid spin labels. Considerable similarity has been found between the behavior of galactosyl ceramide and phosphatidylcholine. These similarities include a negligible flip-flop rate, a flexibility gradient in the acyl chains, and exclusion from phosphatidylserine domains in the face of a Ca^{2+}-induced lateral phase separation. Evidence for dramatic clustering of simple neutral glycolipids has not been found. Glycosphingolipids do seem to have the capacity to increase rigidity in fluid lipid bilayers. A general procedure has been developed for covalent attachment of a nitroxide spin label to the headgroup region of complex glycolipids such as gangliosides. Studies of beef brain gangliosides labeled in this manner and incorporated into bilayers of phosphatidylcholine indicate that the headgroup oligosaccharides are in rapid, random motion as opposed to being in any way immobilized. This headgroup mobility depends very little on the fluidity or rigidity of the bilayer. However, headgroup mobility decreases, perhaps as a result of cooperative headgroup interactions, with increasing bilayer concentration of unlabeled ganglioside.

Key words: glycosphingolipids, membrane structure, flip-flop, spin label, glycolipids, gangliosides

Considerable research effort is currently being directed toward fitting individual biological membrane components into a 3-dimensional model of membrane architecture. This model has to allow not only for intricate arrangements and complex associations, but also for dramatic rearrangements and association changes. We have been approaching the

Abbreviations: EPR — electron paramagnetic resonance; GC — galactosyl ceramide; PC — phosphatidylcholine

Received March 21, 1977; accepted May 3, 1977

© 1977 Alan R. Liss, Inc., 150 Fifth Avenue, New York, NY 10011

problem of glycosphingolipid behavior and interactions largely by studying spin-labeled derivatives in lipid bilayer systems. The spin label technique has proven very useful in membrane research, giving early impetus to innovative concepts such as the fluid bilayer nature of membranes (1), slow flip-flop (2), rapid lateral diffusion (3–5), and fluidity gradients (1).

The usefulness of spectroscopic probes in studies such as those outlined here is greatly increased if they are covalently attached in some clearly understood fashion to the molecule of interest. For this reason we have synthesized spin-labeled derivatives of 2 glycolipid families: simple neutral glycosphingolipids (6) and gangliosides (7) (Fig. 1). Incorporation of such labeled glycolipids into bilayers containing various mixtures of lipids is straightforward. The fate and behavior of the spin-labeled component can then be followed via its EPR[1] spectrum.

Glycosphingolipids (lipids which have carbohydrate headgroups glycosidically linked to a sphingosine backbone) are of considerable interest as outer surface components of mammalian cells. They have been implicated in a wide variety of processes related to recognition, adhesion, and growth control. An immediate question is whether they can be fitted into current membrane models in the same way as phosphatidylcholine or whether

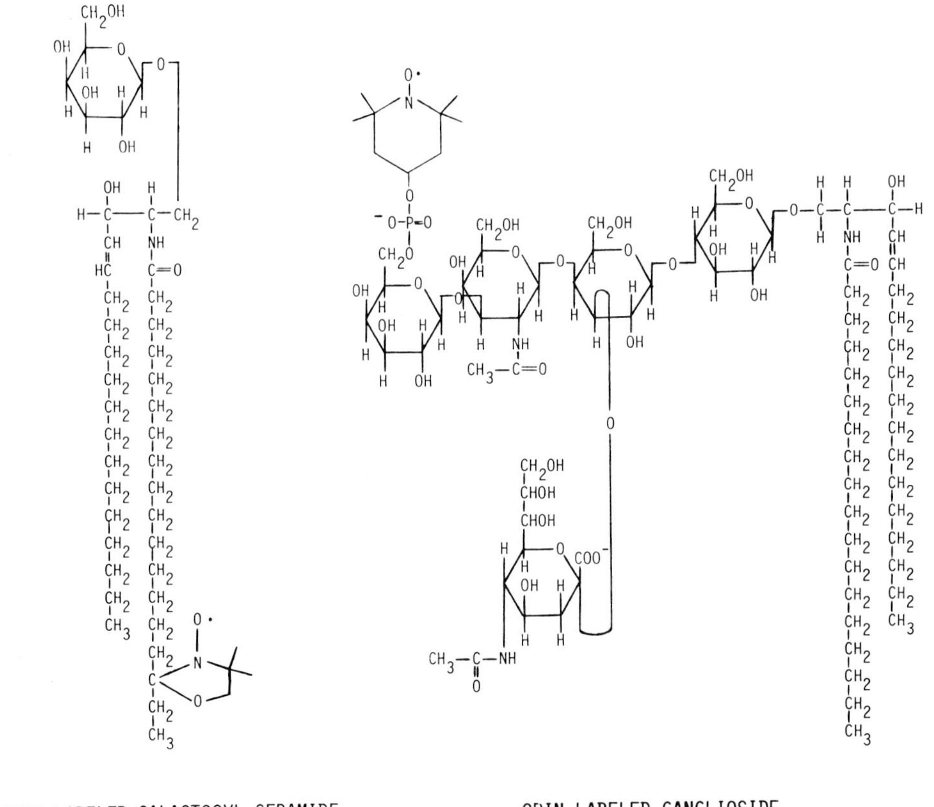

Fig. 1.

one may expect behavior peculiar to glycosphingolipids (i.e., as a result of the sphingosine backbone or the carbohydrate headgroup).

MATERIALS AND METHODS

For the preparation of the spin-labeled neutral glycolipid the fatty acid of natural GC was removed by base hydrolysis and replaced with 1 of 3 spin-labeled fatty acids (6). Corresponding phospholipid labels were prepared by acylation of palmitoyl lysophosphatidylcholine with the same 3 spin-labeled fatty acids (7). Beef brain gangliosides were labeled in the headgroup region by covalent attachment of the small spin label tempophosphate (8). This reaction relies on the formation of phosphate esters with primary alcohols (of which there is 1 per carbohydrate residue). The reagent ratios can be controlled and generally we have employed conditions which give 1 or fewer spin labels per ganglioside.

Antisera to natural GC were raised in rabbits as outlined by Alving et al. (9). The immunoglobulin fraction was purified by 3 successive ammonium sulfate precipitations and was then redissolved in isotonic saline at the original serum concentration. Antisera were kindly titred by Dr. Carl Alving of the Walter Reed Army Medical Center, Washington, DC.

Lipid sources and purification procedures were as described elsewhere (6, 8, 10). Lipid mixtures were made by dissolving appropriate amounts of each in $CHCl_3$ of $CHCl_3/CH_3OH$ and drying extensively in vacuo prior to rehydration.

The flip-flop rate of galactosyl ceramide in egg PC was determined by the basic approach of Kornberg and McConnell (2) but using GC derivatives spin labeled in the fatty acid chain. Flip-flop studies via ascorbate reduction of fatty acid labeled phospholipids have been reported by other workers (11). Single bilayer egg PC vesicles ($\sim 250°$ Å diameter) containing 4 mol % (12,3) spin-labeled GC were prepared by sonication in 50 mM phosphate buffer pH 7.0 at a concentration of 40 μmol/ml. Approximately 8 2-min bursts of sonication with a microtip probe sonicator were required (Heat Systems Ultrasonics Model W140) with ice bath cooling. The resulting clear suspension was centrifuged at 18,000 × g for 20 min and the (small) pellet of titanium discarded. We then cooled 1.1 ml of the suspension to $0°C$ and incubated with 0.1 ml of 90 mM sodium ascorbate pH 7.0 for 75 min. After this ascorbate treatment (to reduce outward-facing spin labels) the sample was run down a calibrated column of Sephadex G-25 at $0°C$ to remove excess ascorbate (ascorbate does not penetrate lipid bilayer vesicles at $0°C$). This sample was then incubated at $23°C$ to allow glycolipid flip-flop to occur. At various time intervals (0–5 hr) 100 μl aliquots were removed, chilled to $0°C$, and treated with 8 μl of 90 mM ascorbate for 75 min to reduce any label which had flip-flopped to the outer monolayer. The EPR spectrum of each aliquot was recorded at $0°C$ following this final treatment, and the intensity of the low-field peak relative to the initial intensity at time 0 (h_t/h) was measured.

In order to determine the time necessary (75 min) for complete reduction of outward-facing (12, 3) spin-labeled glycolipid the following experiment was performed. Lipid vesicles containing 5 mol % GC spin label in egg PC (lipid concentration 27 μmol/ml) were made by sonication as above. A 50 μl aliquot was removed and cooled to $0°C$ prior to addition of 5 μl of cold 250 mM sodium ascorbate pH 7.0. The sample was immediately transferred to a precooled EPR sample tube and its spectrum followed continuously at $0°C$. Monitoring the peak height ratio h_t/h as a function of time showed that after 75 min essentially all outward-facing label was reduced.

RESULTS AND DISCUSSION

The sequence of reactions used to produce (fatty acid) spin-labeled GC relies on the presence of 1 base-hydrolyzable amide ester linkage. On this basis the same reaction sequence should be capable of producing other spin-labeled glycosyl ceramides [such as glucosyl ceramide or lactosyl ceramide (12)]. A complication arises if any of the carbohydrate groups possess N-acetyl or N-glycolyl moieties which will also be subject to base hydrolysis (e.g., gangliosides, globoside). The spin label (nitroxide ring) does not interfere with the function of GC as a receptor. For instance immunoglobulins directed against natural GC dramatically agglutinate liposomes of egg PC containing 2% natural or spin-labeled GC whereas they do not affect suspensions of pure egg PC liposomes.

The technique employed to spin label gangliosides is quite general in that the requirement is only for a primary alcohol (of which there is 1 per sugar residue). Random introduction of up to 1 spin label per ganglioside headgroup adds both an extra negative charge and a new group. This should not detract too severely from extrapolation of the behavior of spin-labeled gangliosides to that of gangliosides in general since the latter are already highly negatively charged and possess a wide variety of headgroup structures.

Behavior of Neutral Glycosphingolipids

One of the first experiments we performed was to simply incorporate small amounts of spin-labeled galactosyl ceramides into fully hydrated bilayers of various phosphatidylcholines. The immediate observation was that the gross behavior of neutral glycolipids is similar to that of phospholipids (6, 10). That is, the EPR spectra (which are sensitive to mobility, orientation, spin label clustering, and environment polarity) of spin-labeled galactosyl ceramides are similar to those of the corresponding spin-labeled phosphatidylcholines. Certainly there is no dramatic clustering or immobilization of the glycolipids at concentrations of a few % in bilayers of egg PC (6). The possibility of clustering at higher concentrations of surface carbohydrate has not yet been investigated in the case of simple neutral glycolipids.

It was originally established using spin-labeled phospholipids that Ca^{2+} can induce dramatic lateral phase separations in fluid bilayers rich in phosphatidylserine (13, 14). Apparently Ca^{2+}-induced cross-linking of the headgroup carboxylic acid residues leads to formation of more rigid domains selectively enriched in phosphatidylserine which coexist with PC-enriched domains in the same bilayer (13–15). Spin labels are quite sensitive to this phenomenon because any selective pooling of the labeled component shows up clearly in spin exchange broadening of the EPR spectrum (6, 13). We have shown that GC is also subject to this Ca^{2+}-induced selective exclusion from cross-linked phosphatidylserine domains (6). In fact when Ca^{2+} is added to a suspension of liposomes containing the ternary lipid mixture, phosphatidylserine/PC/GC (40:10:1 mole ratio) the result seems to be formation of coexisting phosphatidylserine-enriched and PC/GC-enriched domains. Hence, in this case at least, a phospholipid and glycolipid have shown the same dynamic response to a lipid-based perturbation.

Order Parameters

A basic similarity between phosphatidylcholines and galactosyl ceramide is clearly seen in lipid bilayer order parameters derived from spin label data. The order parameter, S, has been described in detail by other workers (7, 16–18). Its measurement relies on spin label sensitivity to orientation in the magnetic field and the reflection of this sensitivity in the EPR spectrum. An S value approaching 1.0 for a fatty acid spin label derivative indicates that the fatty acid backbone region to which the label is attached is in general oriented so that its long axis has a time average orientation ⊥ to the plane of the membrane. S values approaching 0 indicate that the backbone region to which the label is attached is randomly oriented (note however that these statements represent a great oversimplification — see especially discussion of flexibility gradients by McConnell in Ref. 18). In general the degree of fatty acid chain orientation ⊥ to the plane of the membrane decreases towards the center of the bilayer. In the case of (12, 3) spin-labeled lipids S values may be calculated using parameters measured directly from EPR spectra of liposome suspensions. The S values for lipids labeled with the (5, 10) and (1, 14) labels are generally derived from spectra of oriented bilayers on flat surfaces. Figure 2 shows order parameter plots for bilayers of egg PC derived using PC and GC spin labels. Since the spin-labeled lipids were kept at very low concentrations (and are not subject to dramatic clustering) the S values should reflect conditions of the egg PC bilayer immediately surrounding the particular spin label used. Qualitatively the 2 sets of data are very similar. However, quantitatively the egg PC order parameters measured using glycolipid spin labels are consistently higher than those found with phospholipids. Such an observation would be consistent with a tighter, more orderly packing of phospholipids immediately adjacent to glycosphingolipids and/or with glycolipids sitting slightly higher in the membrane (perhaps due to a bulky, H-bonded headgroup) (10). Neither of these possibilities has been ruled out, but we have shown (10) that natural GC

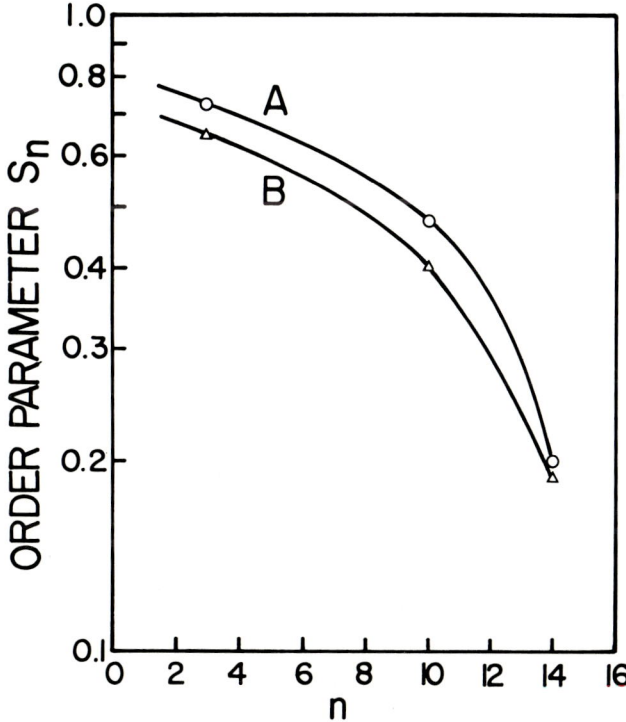

Fig. 2. The order parameter, S, plotted as a function of n, where n is the number of methylene carbons between the spin label ring and the carboxyl function of the fatty acid. The data is for fully hydrated bilayers of egg phosphatidylcholine containing 1% spin-labeled A) galactosyl ceramide and B) phosphatidylcholine. Temperature of all spectra, 23°C. [From Ref. 10 with permission from Elsevier/North-Holland Biomedical Press.]

and beef brain gangliosides confer increased order and rigidity on fluid phospholipid bilayers as would be expected in the case of tighter packing around glycolipids.

Glycolipid Flip-Flop

The flip-flop phenomenon in lipid bilayers was first measured (and found to be slow) using headgroup spin-labeled phospholipids (2). The approach used was to prepare sealed, single bilayer vesicles whose membranes contained a small amount of the spin-labeled lipid. Spin labels in the outer monolayer can be completely reduced by the addition of ascorbate at 0°C, but labels in the inner monolayer are unaffected since the bilayer is impermeable to ascorbate at this temperature. Reduced labels disappear permanently from the EPR spectrum, and peak height can be used as a measure of what fraction remains at the inner surface. Figure 3 shows that the (12, 3) GC spin label may be used for this kind of experiment (see also Ref. 11) but that incubation with cold ascorbate must be carried out for some 75 min to ensure complete reduction of labels in the outer monolayer. The amount of unreduced label (some 31%) is in agreement with that expected from theory for small vesicles (2). Figure 4 shows that if, following ascorbate treatment (see Materials and Methods), the excess ascorbate is removed and the sample incubated at 23°C, there is no appreciable flip-flop of unreduced label to the outer surface over a period of some 5 hr.

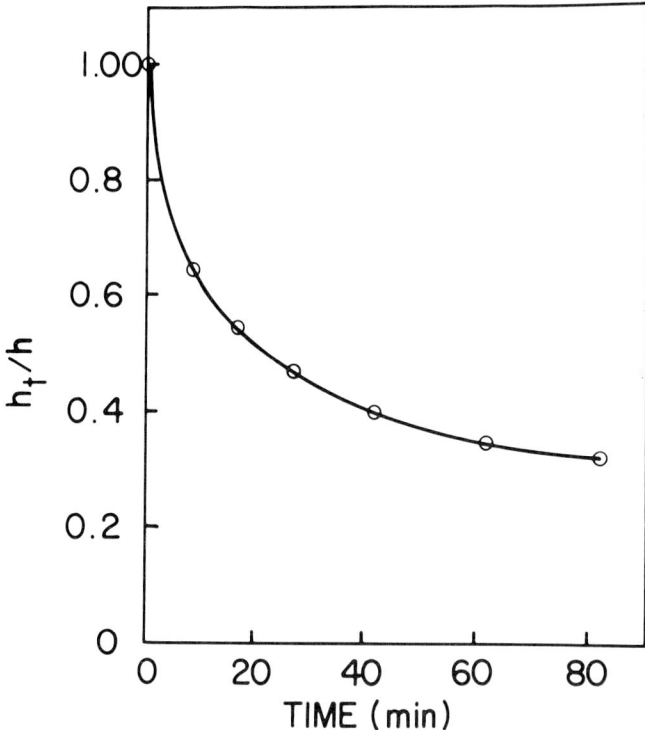

Fig. 3. Reduction kinetics for (12, 3) spin-labeled galactosyl ceramide in sealed lipid bilayer vesicles (~250 Å diameter) at 0°C. The reducing agent is ascorbate added externally and has no access to spin labels occupying the inner monolayer. Lipid mixture is egg phosphatidylcholine containing 5 mol % spin-labeled glycolipid. h_t/h is the ratio of the intensity of the low-field peak at time t to the intensity at time 0.

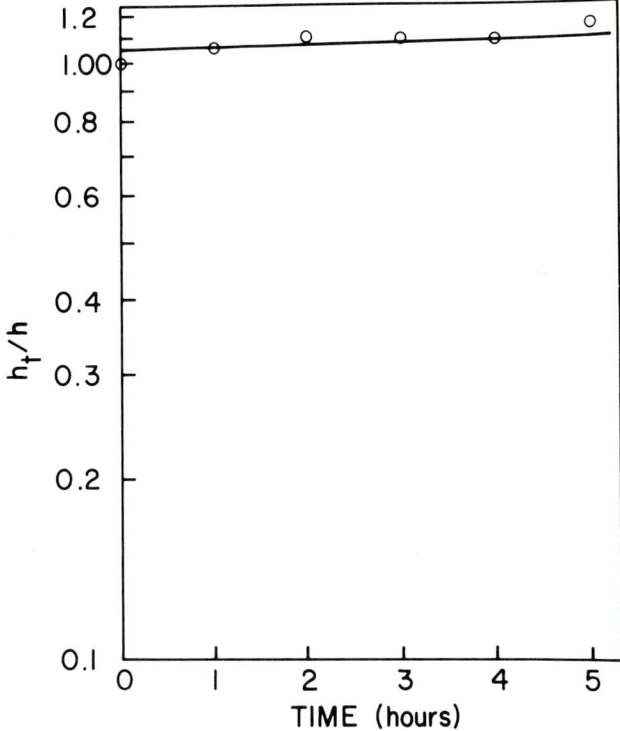

Fig. 4. Fraction of spin-labeled galactosyl ceramide which remains protected at the inner surface of sealed bilayer vesicles following an initial treatment with ascorbate at 0°C, removal of excess ascorbate, and incubation for the times indicated at 23°C. h_t/h is the ratio of the intensity of the low-field peak at time t to the intensity at time 0. An appreciable flip-flop rate would lead to a steady drop in this ratio. The lipid mixture is egg phosphatidylcholine containing 4 mol % spin-labeled glycolipid.

In preliminary experiments very similar results have been obtained with the headgroup-labeled gangliosides. In this case the reduction by ascorbate at 0°C is very fast (less than 8 min) leaving a similar proportion of protected label (28%) to that seen in the case of galactosyl ceramide. The conclusion drawn from these experiments is that in very fluid phospholipid bilayers (and hence, by interpolation, in bilayer regions of real membranes) the flip-flop of glycolipids has a half time much longer than 5 hr at 23°C. These experiments put no upper limit on the flip-flop half time.

Ganglioside Behavior and Carbohydrate Headgroup Interactions

So far we have restricted our work with spin-labeled gangliosides to beef brain mixtures labeled in the headgroup itself (see Fig. 1). Such labels should be sensitive to events involving the carbohydrate moieties. Experiments described here were carried out with samples possessing up to 1 nitroxide ring per ganglioside (i.e., 80% or more of the sugar rings are unlabeled). When such labeled ganglioside preparations are studied at 23°C in egg PC bilayers (fluid) or dipalmitoyl PC bilayers (rigid) the spectrum is that of a highly mobile nitroxide (correlation time, $\tau_c \sim 3.8 \times 10^{-10}$ sec in 10 mM phosphate buffer pH 7.0)

(Fig. 5). Hence it seems that, at least at low concentrations in the bilayer, ganglioside headgroups are not highly immobilized and the headgroup behavior is more or less independent of the physical state of the lipid bilayer. This type of labeled ganglioside is probably not very sensitive to spin-exchange broadening resulting from ganglioside clustering since the latter can occur without the nitroxides being brought into close proximity. Hence we have attempted to investigate this phenomenon and that of carbohydrate-carbohydrate headgroup interactions by monitoring spin label mobility (see below).

The existence of surface carbohydrate (whether on proteins or lipids) has been considered potentially important in mediating cell-cell adhesion (19) and association of components within a given cell (20). A possible mechanism for such an involvement would be cooperative attractive forces between carbohydrate residues such as those which are known to exist in certain periodic polysaccharides in solution (21). However, the existence of analogous forces among aperiodic oligosaccharides such as those characteristic of mammalian cell surfaces has not been demonstrated. Such interactions could lead to or stabilize glycolipid and/or glycoprotein clustering such as that which has been postulated to occur in the human erythrocyte membrane (22, 23). In aqueous solvents, the formation of H bonds to water reduces the importance of macromolecule inter- or intramolecular H bonds. But apparently the possibility of forming numerous highly directional H bonds can be very significant — presumably contributing to the extreme insolubility of compounds like cellulose, and of course to the helical nature of DNA and certain polysaccharides such as starch. This

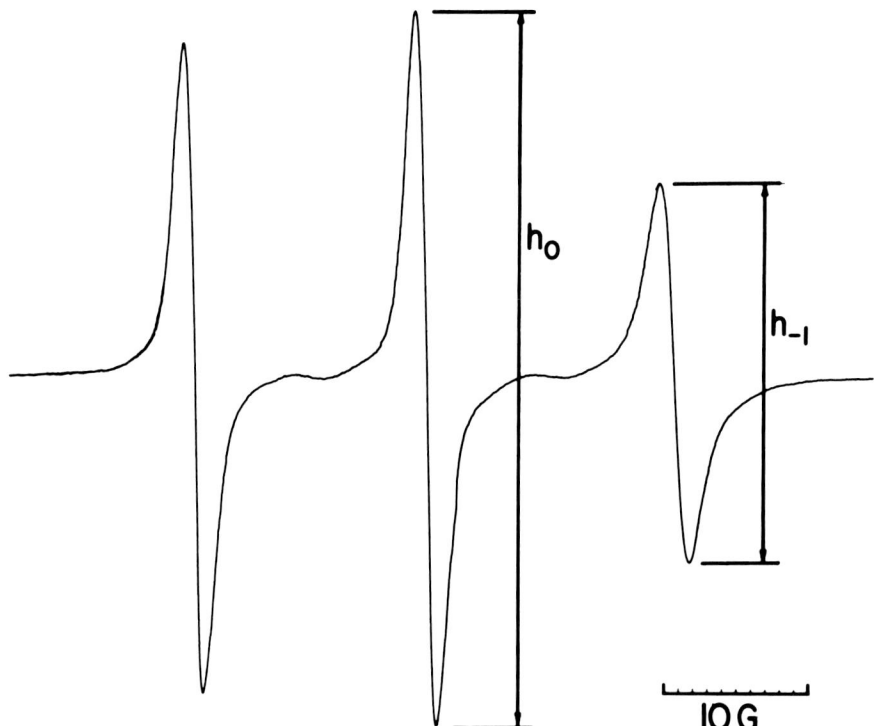

Fig. 5. Typical EPR spectrum of spin-labeled ganglioside (0.45 spin labels per ganglioside) at a concentration of 3.3 mol % in bilayers of egg phosphatidylcholine. The buffer is 10 mM phosphate pH 7.0. The peak heights used for correlation time measurements are indicated (temperature 23°C).

phenomenon should be most significant in cell surface regions where the local carbohydrate concentration is high. In general mammalian cells have much higher surface carbohydrate concentrations than that which exists in our lipid bilayers containing a few mol% ganglioside. We have used a spectral parameter related to the correlation time, τ_c, to examine the effect on ganglioside headgroup mobility of increasing surface carbohydrate concentrations (8; see Fig. 6).

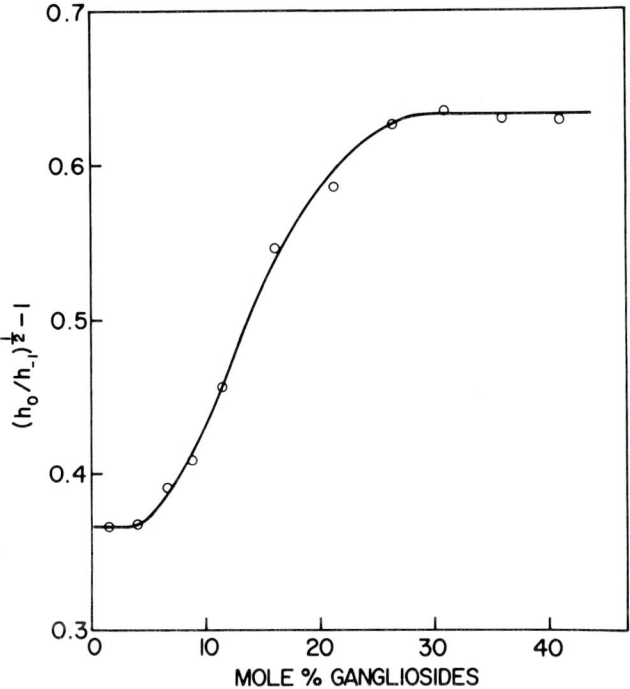

Fig. 6. Effect of increasing amounts of total ganglioside on the oligosaccharide headgroup mobility of spin-labeled gangliosides in fluid lipid vesicles at 23°C. Egg phosphatidylcholine vesicles contain 1.5 mol % spin-labeled ganglioside (0.45 spin labels per ganglioside) and variable amounts of unlabeled ganglioside. Headgroup mobility is inversely related to $[(h_0/h_{-1})^{1/2}-1]$. Vesicles were suspended in 10 mM phosphate pH 7.0.

$$\tau_c = 6.5 \times 10^{-10} \, W_0 \, [(h_0/h_{-1})^{1/2}-1]$$

Where W_0 is the linewidth of the mid-field line, and h_0 and h_{-1} are the heights of the mid- and high-field lines of the spectrum respectively (Fig. 5). This equation assumes rapid, isotropic tumbling of the nitroxide label. For the purposes of the work described here the actual value of τ_c is of little interest except insofar as changes in it may reflect changes in the interactions or mobility of the entire oligosaccharide portion of the ganglioside. In fact we often plot just the fraction, $[(h_0/h_{-1})^{1/2}-1]$ because in our system changes in W_0 are relatively small and the difficulty of accurately measuring W_0 introduces point scatter.

Figure 6 shows that headgroup immobilization increases with added ganglioside according to a sigmoidal curve*, plateauing at about 25% ganglioside in egg PC. Note that this value is for total carbohydrate which in this system is divided between the 2 surfaces, whereas in real cells all carbohydrate is concentrated at 1 surface. The interpretation of this result is not clear-cut (8), but the behavior is consistent with formation of ganglioside-enriched regions in the lipid bilayer as a result of intermolecular attractive forces between the carbohydrate headgroups.

ACKNOWLEDGMENTS

We are indebted to Professor J. R. Bolton of the Department of Chemistry for making his EPR equipment available to us. This work was aided by grants from the Medical Research Council of Canada and the Banting Research Foundation. FJS is the holder of a MRC of Canada studentship.

REFERENCES

1. McConnell HM: The Neurosciences: Second Study Program: 697, 1970.
2. Kornberg RD, McConnell HM: Biochemistry 10:1111, 1971.
3. Sackmann E, Trauble H: J Am Chem Soc 94:4492, 1972.
4. Devaux P, McConnell HM: J Am Chem Soc 94:4475, 1972.
5. Scandella CJ, Devaux P, McConnell HM: Proc Natl Acad Sci USA 69:2056, 1972.
6. Sharom FJ, Grant CWM: Biochem Biophys Res Commun 67:1501, 1975.
7. Hubbell WL, McConnell HM: J Am Chem Soc 93:314, 1971.
8. Sharom FJ, Grant CWM: Biochem Biophys Res Commun 74:1039, 1977.
9. Alving CR, Fowble JW, Joseph KC: Immunochemistry 11:475, 1974.
10. Sharom FJ, Barratt DG, Thede AE, Grant CWM: Biochim Biophys Acta 455:485, 1976.
11. Rousselet A, Guthman C, Matricon J, Bienvenue A, Devaux PF: Biochim Biophys Acta 426:357, 1976.
12. Radin NS: Methods Enzymol 28:301, 1972.
13. Ohnishi S, Ito T: Biochem Biophys Res Commun 51:132, 1973.
14. Ohnishi S, Ito T: Biochemistry 13:881, 1974.
15. Jacobson K, Papahadjopoulos D: Biochemistry 14:152, 1975.
16. Seelig J: J Am Chem Soc 92:3881, 1970.
17. McConnell HM, McFarland BG: Ann NY Acad Sci 195:207, 1972.
18. McConnell HM: In Berliner LJ (ed): "Spin Labeling, Theory and Applications." New York: Academic Press, 1976.
19. Roseman S: In Lee EYC, Smith EE (eds): "Biology and Chemistry of Eucaryotic Cell Surfaces." New York: Academic Press, 1974.
20. Nicolson GL: Biochim Biophys Acta 45:57, 1976.
21. Rees DA: In Whelan WJ (ed): "MTP International Review of Science, Biochemistry Series One," Vol. 5, "Biochemistry of Carbohydrates." London: Butterworth Scientific Publishers, 1975.
22. Ji TH, Ji I: J Mol Biol 86:129, 1974.
23. Ji TH: J Biol Chem 249:7841, 1974.

---

*When originally published (8) we were uncertain as to the shape of the curve at low ganglioside concentrations and simply drew the best straight line through these points. Further experiments have shown that the sigmoidal shape is genuine.

Conformation and Intermolecular Interactions of Carbohydrate Chains

Edwin R. Morris, David A. Rees, David Thom, and E. Jane Welsh

Unilever Research, Colworth/Welwyn Laboratory, Colworth House, Sharnbrook, Bedford, MK44 1LQ, England

For consideration of their conformations and interactions, carbohydrate chains can conveniently be divided into 3 classes on the basis of their covalent structure; namely periodic (a), interrupted periodic (b), and aperiodic (c) types. In aqueous solution carbohydrate chains often exist as highly disordered random coils. Under appropriate conditions, however, polysaccharides of types (a) and (b) can adopt a variety of ordered conformations. Physical methods, and in particular optical rotation, circular dichroism, and nuclear magnetic resonance, provide sensitive probes for the study of the mechanism and specificity of these disorder-order transitions in aqueous solution.

Intermolecular interactions between such polysaccharide chains arise from cooperative associations of long structurally regular regions which adopt the ordered conformations. For acidic polysaccharides these cooperative associations may involve alignment of extended ribbons with cations sandwiched between them. In other systems the interactions involve double helices which may then aggregate further, and geometric "matching" of different polysaccharide chains can also occur. These ordered, associated regions are generally terminated by deviations from structural regularity or by "kinks" which prevent complete aggregation of the molecules.

The complex carbohydrate chains which occur at the periphery of animal cells have very different, aperiodic structures and although their conformations are as yet poorly understood, preliminary indications are considered.

Key words: conformational analysis, polysaccharides, cooperative interactions, synergistic interactions, cooperative cation binding, spectroscopic techniques, circular dichroism, nuclear magnetic resonance, optical rotation

INTRODUCTION

Carbohydrate chains are ubiquitous components of the molecular assemblies which characterize the extracellular organization of biological tissues and we have argued (1–4), as have others, that a detailed understanding of their shapes and of their potential for intermolecular interactions is essential to the illumination of their biological roles. Recent interest has centred on complex macromolecules in which carbohydrate chains are complexed to proteins or lipids but "all carbohydrate" polymers remain important, not only in their own right but also as models on which approaches and techniques can be sharpened for application to more elaborate systems. In this discussion we shall concentrate pre-

Received April 6, 1977; accepted May 3, 1977

© 1977 Alan R. Liss, Inc., 150 Fifth Avenue, New York, NY 10011

dominantly on solution structures, since these are generally considered more relevant to biological situations, although ultimately they must be compared with the wide variety of 3-dimensional conformations defined in the solid state by x-ray diffraction (for review see Refs. 3 and 5).

In aqueous solution the favored conformations of these highly hydrated chains cannot automatically be assumed to be ordered since the conformational entropy (3, 6) arising from the continuous fluctuation of the polymers about a large number of internal linkages provides a strong drive to disordered states. Under particular circumstances, however, favorable nonbonded energy terms (hydrogen bonding, dipolar and ionic interactions, and solvent terms) can act cooperatively (3, 4, 6) to fix the macromolecules in ordered shapes. For the examples characterized to date such disorder-order transitions are almost invariably stabilized by intermolecular interactions involving alignment and cooperative association of long structurally-regular sequences of carbohydrate chains. Physical methods, and in particular circular dichroism (CD), optical rotation (OR), and nuclear magnetic resonance (NMR), provide sensitive probes of the mechanism and specificity of the interactions and we have used "simple," interacting plant, bacterial, and animal systems to characterize the subtypes of ordered structures.

For consideration of their conformations carbohydrate chains can conveniently be divided (4) into 3 classes on the basis of their covalent structure, namely periodic (a), interrupted periodic (b), and aperiodic (c) types. Naturally each class will have shared properties with other types in addition to the distinctive features illustrated here.

1. PERIODIC TYPE

These carbohydrate chains are composed of "repeating units" of sugar residues linked through identical positions and glycosidic configurations.

1. i. Homopolysaccharide Periodic Structure

For chains containing only 1 type of residue, conformational analysis confirms (7) that simple periodic sequences can generate ordered conformations which will correspond to 1 of the subclasses shown in Fig. 1. Such ordered conformations can be considered as helices and defined by their symmetry (n) and projected residue height (h).

a. Extended ribbons. These structures have $2 \leq n \leq 4$ and h close to the maximum for sugar residues (4.1–4.5 Å). Archetypes are the β-1,4 glucans: cellulose, the major skeletal component of plant cell walls, and chitin, the analogous component in fungi, insect cuticles, and crustacean shells. In ordered structures these particular examples of extended ribbons most probably adopt the "bent chain" conformation (5, 8) with alternate residues related by a 2-fold symmetry axis and hydrogen bonded between O(3) and O(5). Their skeletal function arises from their supreme ability to pack together like "planks in a timber yard," in dense, strong, microcrystalline arrays with efficient hydrogen bonding within and between the organized layers.

b. Coiled springs. These structures resembled coiled springs in various states of extension; n has a much wider variation ($2 \leq n \leq 10$) and, more significantly, h can be very small. Amylose, the α-1,4 glucan, is the supreme example. Its best-known ordered states have the spring very compressed with its geometry varying between 6, 7, and 8 residues per helical turn. Such structures do not occur in solution, where the chains have the hydrodynamic properties of a random coil (9), but for amylose they can be prepared as crystalline derivatives by addition of suitable nonpolar complexing agents. Apparently

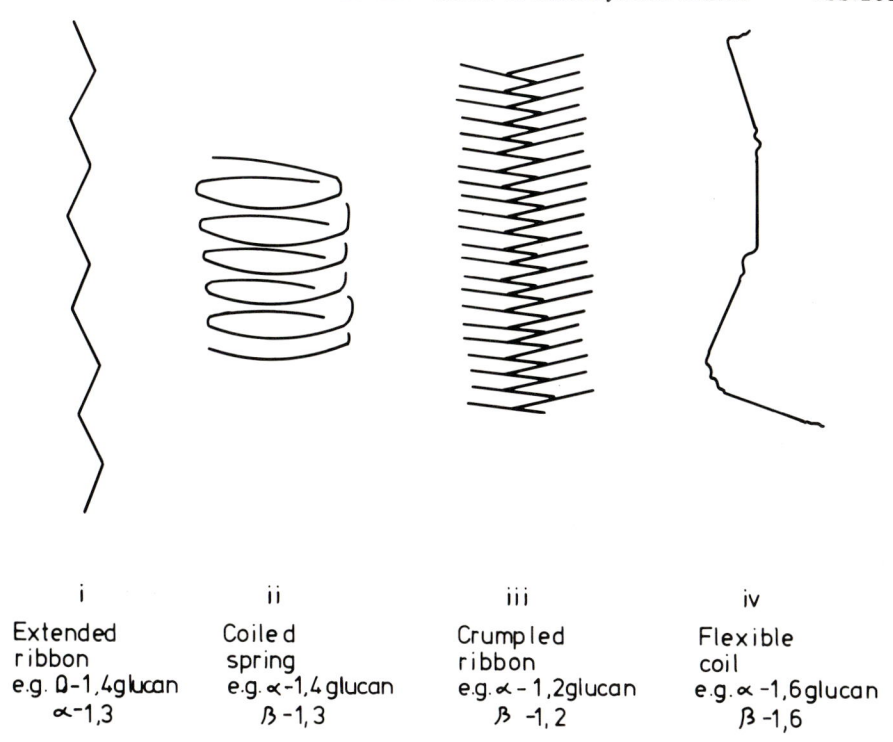

Fig. 1. Schematic representation of the subclasses of ordered conformations predicted by conformational analysis for periodic structures containing only 1 type of residue: left to right, extended ribbon, coiled spring, crumpled ribbon, and flexible coil, respectively. Such structures can be illustrated by the listed polymers of glucose.

"guest" molecules are bound in the core of the helix which can adjust to accommodate a wide variety of included molecules. The best characterized conformation, the 6-fold "V-form," is probably left-handed (10) and hydrogen bonded between O(2)–O(3) of adjacent residues and O(2)–O(6) of residues opposed on the helical surface but 6 apart in the primary sequence. Alternatively, ultimate extension is found in a potassium bromide complex considered to have n = 4 and h = 4.03 (11).

c. **Crumpled ribbons and flexible coils.** No ordered conformations of these structural types have yet been established. For crumpled ribbons extended sequences would result in multiple steric clashes, and this complex shape is seldom found in nature, although limited sequences resembling this type characterize aperiodic chains. Flexible coil sequences are characterized by an extra bond between residues, by virtue of the 1,6 linkages, which greatly enhances their conformational entropy and minimizes opportunities for ordered conformations.

1. ii. Complex Periodic Structures — Xanthan

Periodic chains can also be considerably more complex, e.g., lipopolysaccharide O-antigen chains and capsular polysaccharides of gram-negative bacteria which, in general, may contain up to 6 residues in the repeating sequence (12). A particular example is xanthan, the extracellular polysaccharide from Xanthomonas campestris, which has highly

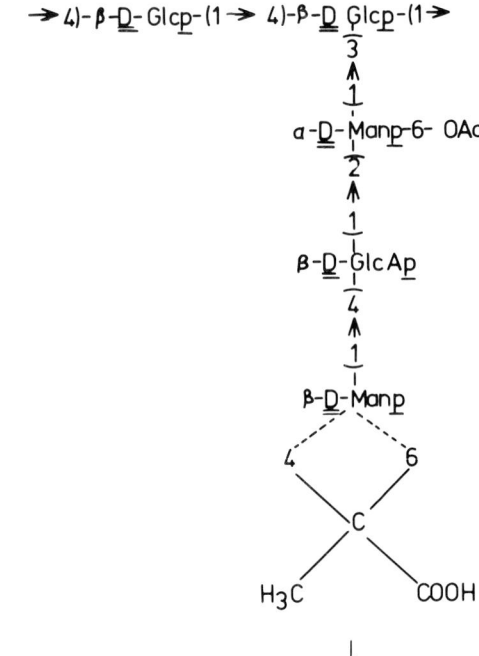

branched "repeating units" with complex trisaccharide side chains extending from alternate residues of a "cellulose" backbone (I). In the condensed state the molecule probably exists (13) as a 5-fold helix with a repeat of 47 Å: side chains align closely with the backbone and most probably stabilize the structure by favorable nonbonded interactions.

In aqueous solution, as shown in Fig. 2, viscosity, NMR, and OR all show sigmoidal temperature-dependent changes which follow very similar temperature courses (14, 15); such changes are characteristic of cooperative processes. Our interpretation is that on cooling individual chains undergo a cooperative transition (Fig. 2) from a random coil to an ordered helical conformation which may be very similar to that in the solid state. Supporting evidence is the confirmation from CD studies (14) that O-acetate chromophores on the side chains experience a more dissymmetric environment in the ordered state, as expected if side chains and backbone align. Moreover, the overall change in the observed n-π* ellipticity is negative and hence its contribution to OR must be negative. Since the observed OR transition is positive this indicates large contributions from the inaccessible CD transitions which dominate monochromatic OR. As we shall show, such effects are entirely consistent with the "locking" of glycosidic angles which accompanies such coil → helix transitions. The concentration independence of the temperature of the OR transition (15) provides supporting evidence that the change is unimolecular.

Confirming evidence for a temperature-dependent conformational restriction is provided by NMR (14). High resolution linewidths are inversely proportional to the relaxation time of the nuclei, and molecular flexibility interferes with this. For polysaccharides [^{1}H] NMR signals are normally broad and overlapping due to the restricted segmental motion of the chains. Above 85°C, however, the O-acetate and pyruvate substituents in xanthan give sharp, distinct signals. On cooling to 60°C the entire spectrum collapses confirming a transition to a rigid conformation with short relaxation times and hence signals that are

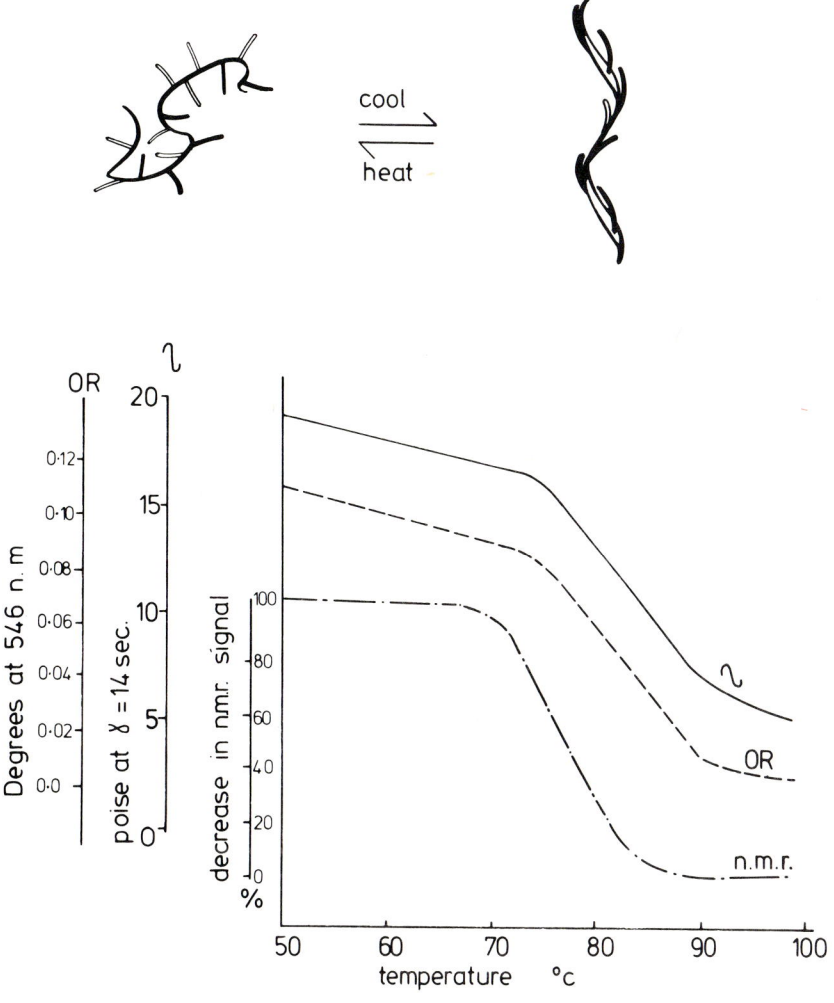

Fig. 2. Schematic representation of the suggested coil → helix transition for xanthan. The temperature dependence of the low shear viscosity (η), OR, and high resolution NMR peak areas for 1% (wt/vol) aqueous solutions of this bacterial polysaccharide is also shown.

too broad to be observed. At intermediate temperatures no significant broadening or shifting accompanies the sigmoidal decrease in peak area confirming a "two-state all or none" transition.

2. INTERRUPTED PERIODIC TYPE

These chains also contain periodic sequences capable of adopting ordered conformations, but these are separated or "interrupted" by deviations from regularity which force chains to leave ordered associations and combine with more than 1 partner. The resultant balance of ordered and "soluble" regions leads to the highly hydrated gel state which is so

2. i. Interrupted Ribbons — Alginates and Pectins

A group of acidic polysaccharides which gel by cooperative binding of divalent cations are the alginates and pectins, derived from brown algae and the cell walls and soft tissues of higher plants, respectively. For alginates (11) homopolymeric poly α-L-guluronate sequences (IIb) play the most important role in junction formation. The sequences (IIc)

II (a)   II (b)   II (c)

Alginate

which approach, albeit irregularly, alternating sequences (16) may play a minor role in gelation but probably also function with poly β-D-mannuronate sequences (IIa) as interrupting "soluble" sequences. Moreover, the relative proportion of sequence types varies with the tissue, the state of maturation, and the geographical origin of the seaweed. The available biosynthetic evidence (17) indicates that the polymer is produced as poly β-D-mannuronate and subsequently "tailored" for its biological function by incorporation of guluronate-containing sequences by an enzymatic epimerization at C(5). Conformational analysis predicts (3) that ordered conformations of both homopolymers will be extended ribbons, and this is confirmed by fiber diffraction studies (18) on such "idealized" extremes in the condensed state.

Evidence for the mechanism of gelation can be obtained from the cation binding behavior of isolated block types (19). Both homopolymeric sequences (IIa and b) show the expected statistical cation binding, but polyguluronate sequences also show a sigmoidal increase in binding affinity above a chain length of 18–20 residues. This strongly implies a cooperative binding process, confirmed by competitive ion-binding studies (20) for polyguluronate but not for other types of sequence.

Circular dichroism studies provide spectacular support for specific site-binding by polyguluronate sequences. Gelation of alginate solutions by controlled addition of divalent cations is accompanied by dramatic changes in CD (Fig. 3). The negative trough arising predominantly from contributions from polyguluronate sequences (21) is swamped by a positive transition. Moreover, the magnitide of the observed difference spectra (Fig. 2) can be correlated with the proportion of such sequences, increasing polyguluronate content giving increased difference spectra. The cation binding behavior of isolated block types confirms this view. Large CD changes are observed (22) for polyguluronate sequences with much smaller changes for alternating and insignificant changes for polymannuronate sequences. These changes in the n-π^* spectral region are consistent with the n-orbitals of polyguluronate carboxylate chromophores being specifically involved in coordination of the divalent cation.

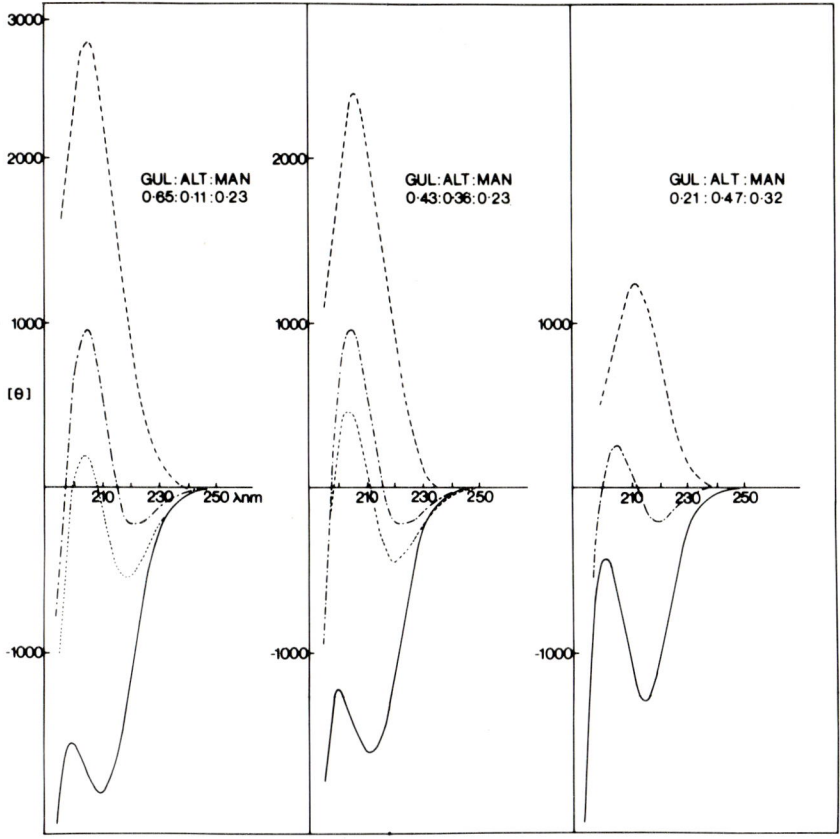

Fig. 3. Changes in CD spectra with diffusion of Ca^{++} to a final concentration of 6 mM into solutions of sodium alginates (0.1%) having the sequence compositions shown in the diagram. Spectra are shown for solutions (—) and for gels at intermediate (· · · ·) and final (· – · – ·) stages of gelation. Difference spectra (- - -) are obtained by subtraction of solution spectra from final gel spectra: $[\phi]$ = molecular ellipticity (degree – cm^2 per decimole) and λ = wavelength (nm).

The principal mechanism of alginate gelation can thus be considered in terms of an "egg-box" model (23) involving cooperative binding of cations between associated polyguluronate "ribbons" (Fig. 4). Competitive inhibition of gelation by isolated blocks and the stoichiometry of cation binding (22) strongly indicate that the predominant mechanism of association is dimerization. Physical and mathematical model building suggests an ordered structure very similar to that observed in the condensed state. Buckled ribbons pack together with cations tightly coordinated in oxygen-lined "nests" between the chains like "eggs" in an "egg-box." For individual chains (Fig. 4) each hydrophilic coordination site involves O(6) and O(5) with the glycosidic oxygen and O(2) and O(3) on the next residue in the "nonreducing" direction. Termination of polyguluronate sequences terminates particular junction zones and establishes the gel network.

Pectin, like alginate, forms firm gels with divalent cations and the "junction forming" sequences are homopolymeric, 1,4 diaxially-linked poly α-D-galacturonate sequences (IIIa; R = Na) which are geometrically very similar to polyguluronate sequences. Partial

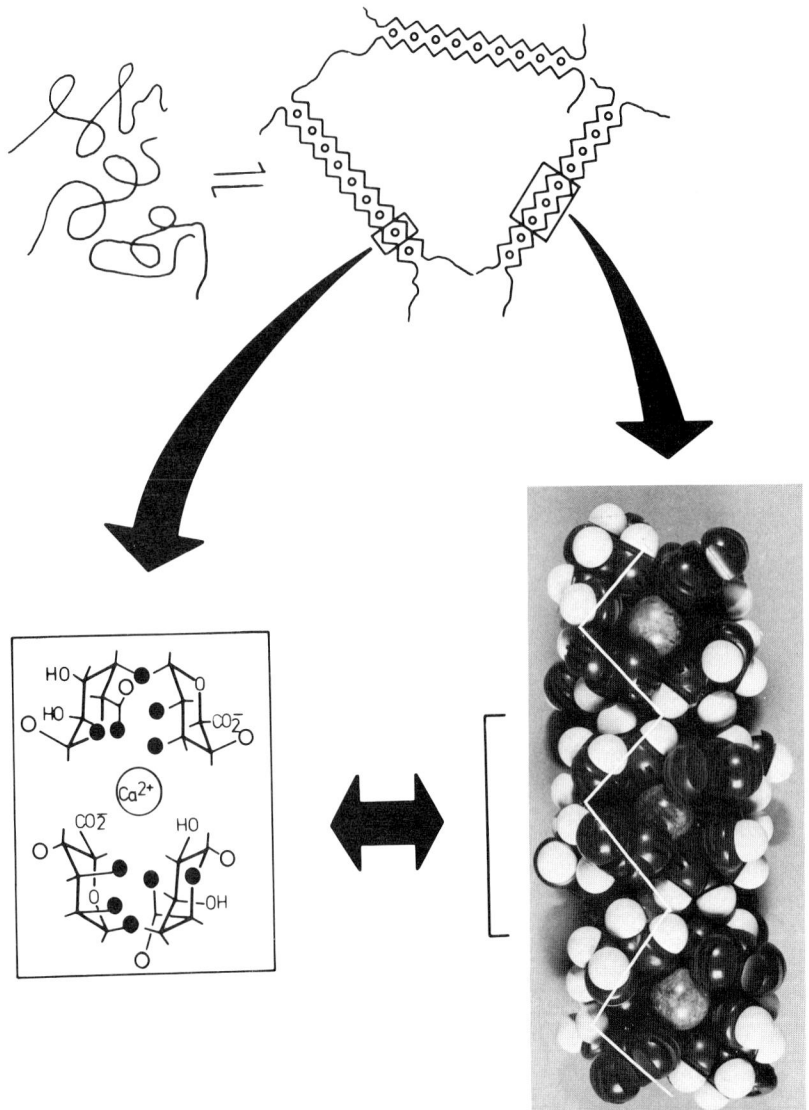

Fig. 4. Schematic representation of the "egg-box" model for the sol → gel transition in alginates. The array of calcium binding sites is shown in the space-filling model and the contour of 1 chain is traced to illustrate its buckled character. The disposition of groups coordinating the divalent cation is also shown schematically and for clarity chains have been moved apart.

methyl esterification (IIIa; R = CH$_3$) provides "soluble" regions and further complications are additional structural interruptions of severe α-1,2 linked L-rhamnose "kinks" (IIIb) and neutral sugar side chains (1, 24). The cation binding behavior of deesterified polygalacturonate sequences and low methoxy pectins are very similar to those of polyguluronate. Thus on gelation large CD charges are observed (23) in the n → π* region suggestive of an "egg-box mechanism" and polygalacturonate sequences show (20) cooperative cation binding above a chain length of 14–16 residues. As expected, highly methyl-esterified pectins show little affinity for cations and correspondingly small CD changes. They can,

[Pectin structure III(a) and III(b)]

however, gel under conditions of low pH and low water activity. The gelation predominantly involves association of extended esterified sequences (1, 23) and is accompanied by increases in the CD spectrum. Pectin can thus form ordered associations by 2 very different mechanisms, and our interpretation is that both involve stacking of extended ribbons in arrays broadly similar to those of alginate. Significant differences include termination of junction zones by a diverse and complex range of structural interruptions.

2. ii. Interrupted Helices — Carrageenans and Agarose

Algal carbohydrate chains where the gel framework involves temperature-dependent, cooperative associations in double-helical arrays are the carrageenans (κ and ι) and agarose. These are alternating copolymers, containing variable amounts of sulphate ester, and "junction forming" sequences contain 1,3-linked β-D-galactose and either 1,4-linked 3,6-anhydro-α-D- or -L-galactose residues for carrageenans (IVa) and agarose (Va) respectively (25, 26). Associated helices are terminated by "soluble kinks" (IVb and Vb) of sequences involving deviations from the regular 3,6-anhydro residues. These interruptions can be removed by chemical modification (27) to create shorter blocks or "segments" which retain the ability to adopt ordered structures but cannot gel.

[Carrageenan structure IV(a) and IV(b)]

[Agarose structure V(a) and V(b)]

For the highly sulphated ι-carrageenan the double helices have been thoroughly characterized both in solution and the solid state (28). The reversible coil → helix transition (Fig. 5) of ι-segments shows a sigmoidal increase in OR which is concentration-dependent (at constant temperature and ionic strength), corresponds to chain dimerization, and is accompanied by an exact doubling of number average and weight average molecular weights (29). The sign and magnitude of this OR shift correspond closely to that predicted (30) from the double-helix geometry in the condensed state, using semiempirical calculations of optical activity from the glycosidic angles of the polysaccharide backbone (31). Thermodynamic measurements are also consistent with double-helix formation by a "two-state all or none" mechanism and low angle x-ray scattering shows ordered rods corresponding to the expected double helical dimensions (32).

Corroborative evidence for conformational restriction is provided by [$-^{13}$C] and [$-^{1}$H] NMR measurements (33). Because of their larger chemical shifts and diminished dipolar broadening compared to [$-^{1}$H] nuclei, [$-^{13}$C] spectra provide (Fig. 5) sharp, well-resolved peaks above 80°C, i.e., for the random coil. All these peaks collapse on conversion to the rigid double helical form and concomitantly a dramatic decrease in the relaxation time (T_2) for the carbon nuclei was observed in the pulse NMR spectrometer. As for xanthan, the absence of broadening or shifting of peaks when the transition was partially completed suggested no time averaging of resonances between disordered and ordered states and hence supported the "two-state all or none" model.

For κ-carrageenan segments OR measurements again suggest (34) that double helices exist in solution, but the lower charge density due to the decrease in sulphate ester substituents leads to markedly different properties. Thus the cooperative, concentration-dependent OR transition shows distinct hysteresis in its "melting" and "setting" behavior which can be correlated with aggregation of the ordered helices. In the nonsulphated chains of agarose this aggregation is so marked that double helix formation cannot be monitored by OR, and the chains precipitate (34) from solution at the onset of the disorder → order transition. Substituted agarose segments do, however, show OR behavior indicative of double helix formation (35) and with the expected distinct hysteresis loop similar to κ-carrageenan.

In the complete carbohydrate chains the "soluble kinks" establish the 3-dimensional gel network. These gels show OR behavior which is very similar to that of the isolated segments, and their setting and melting temperatures are close to the midpoints of the observed transitions (34). Detailed calorimetric and OR analyses (32, 33) indicate that helix formation achieves 70–90% of the theoretical maximum and that each chain can evidently overcome topological constraints and take part in at least 6 independent helices. Competitive inhibition studies provide further substantiation of gelation mechanisms (36) by distinguishing between associations involving dimerization and those involving larger aggregates. As shown on Fig. 6, incorporation of isolated segments into carrageenan gels markedly decreases gel strength for ι-carrageenan whereas that of κ-carrageenan actually increases. This is the predicted result since the substitution of short segments into the double-helical junctions of ι-carrageenan must prevent the essential cross-linking whereas for κ-carrageenan gels segments should no longer do this and actually seem to produce more effective incoporation of polymer chains into aggregating juction zones and hence a stronger gel.

2. iii. Complex Interrupted Structures – Association Between Unlike Chains

Examples of this type of ordered structure are the interactions of xanthan, carrageenans,

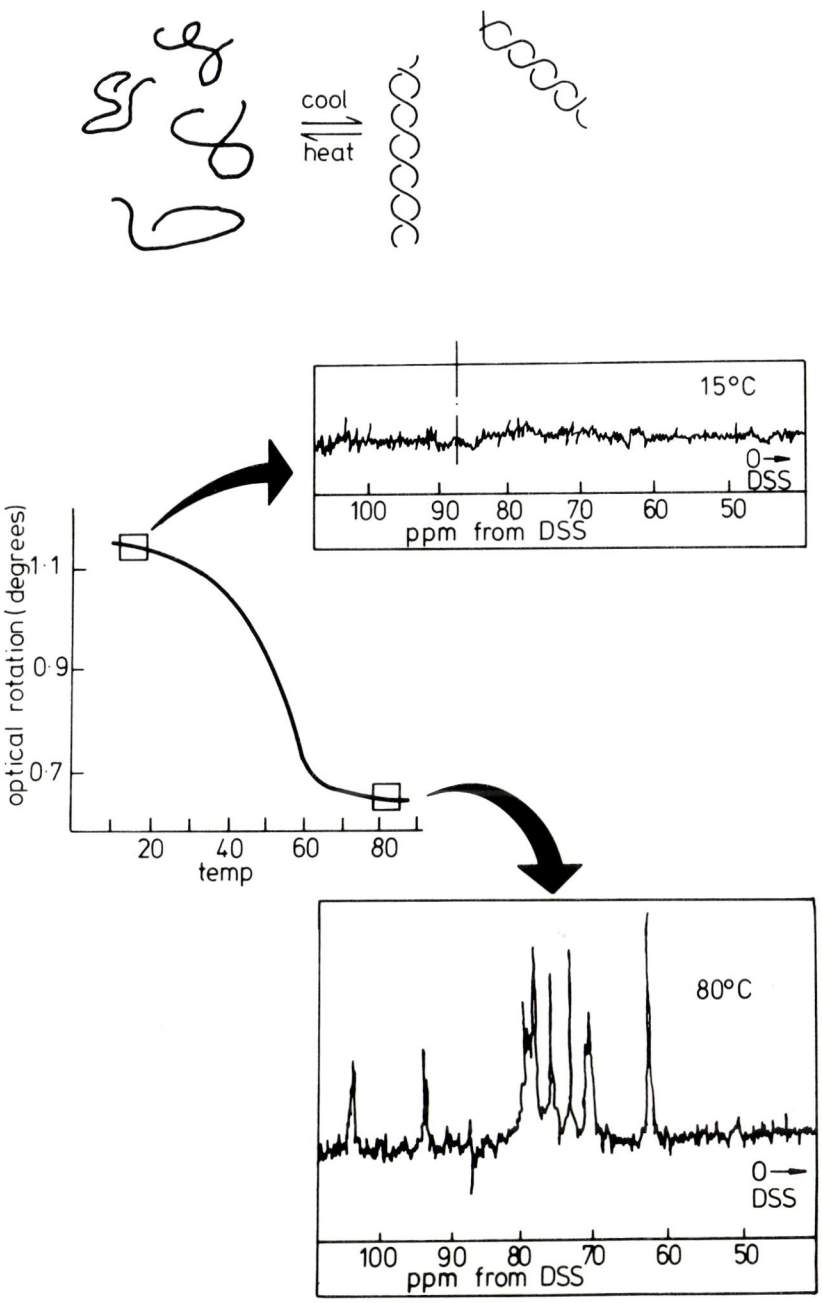

Fig. 5. Schematic representation of the coil → double helix transition of ι-carrageenan "segments." The temperature-dependent sigmoidal increase in OR and comparison of [$^{-13}$C] NMR spectra at 80°C and 15°C are also shown. The concentration was 6% (wt/vol) and 0.1 M sodium chloride.

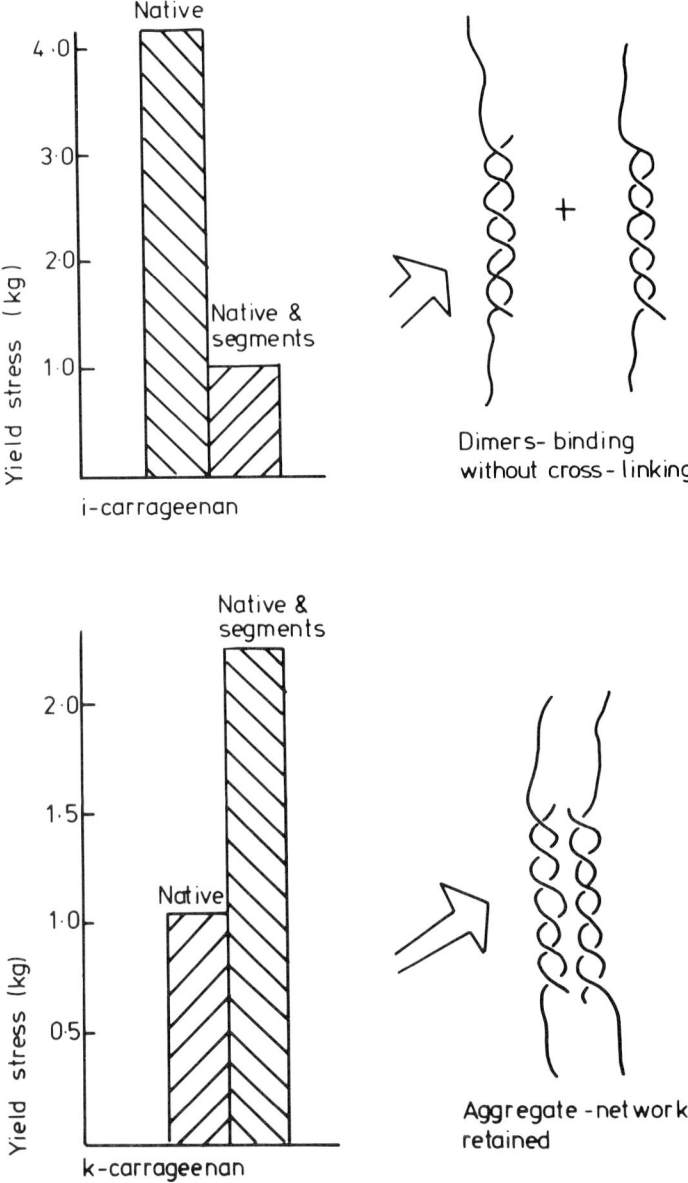

Fig. 6. Comparison of the measured gel strength for native carrageenan gels and gels formed in the presence of equal concentrations of isolated "segments": concentrations were 2.5% and 1% (wt/vol) for ι- and κ-carrageenan respectively. Proposed molecular interpretations of the experimental observations are also shown (see text).

and agar with certain plant β-1,4-linked galactomannans. Concentrations of these polymers which are incapable of gelation can be made to gel by addition of the galactomannans and isolated, nongelling "segments" show similar behavior (15, 34). Such gels show physical properties indicative of cooperative associations and are evidently cross-linked by quaternary interactions between unlike carbohydrate chains. Comparison of the gelling ability and secondary structure of different galactomannans (37) indicates that in this case junction zones involve (Fig. 7) associations between ordered helices and extended "ribbons" of "smooth" unsubstituted mannose backbone, with "hairy" galactose-substituted sequences providing complementing "soluble" sequences.

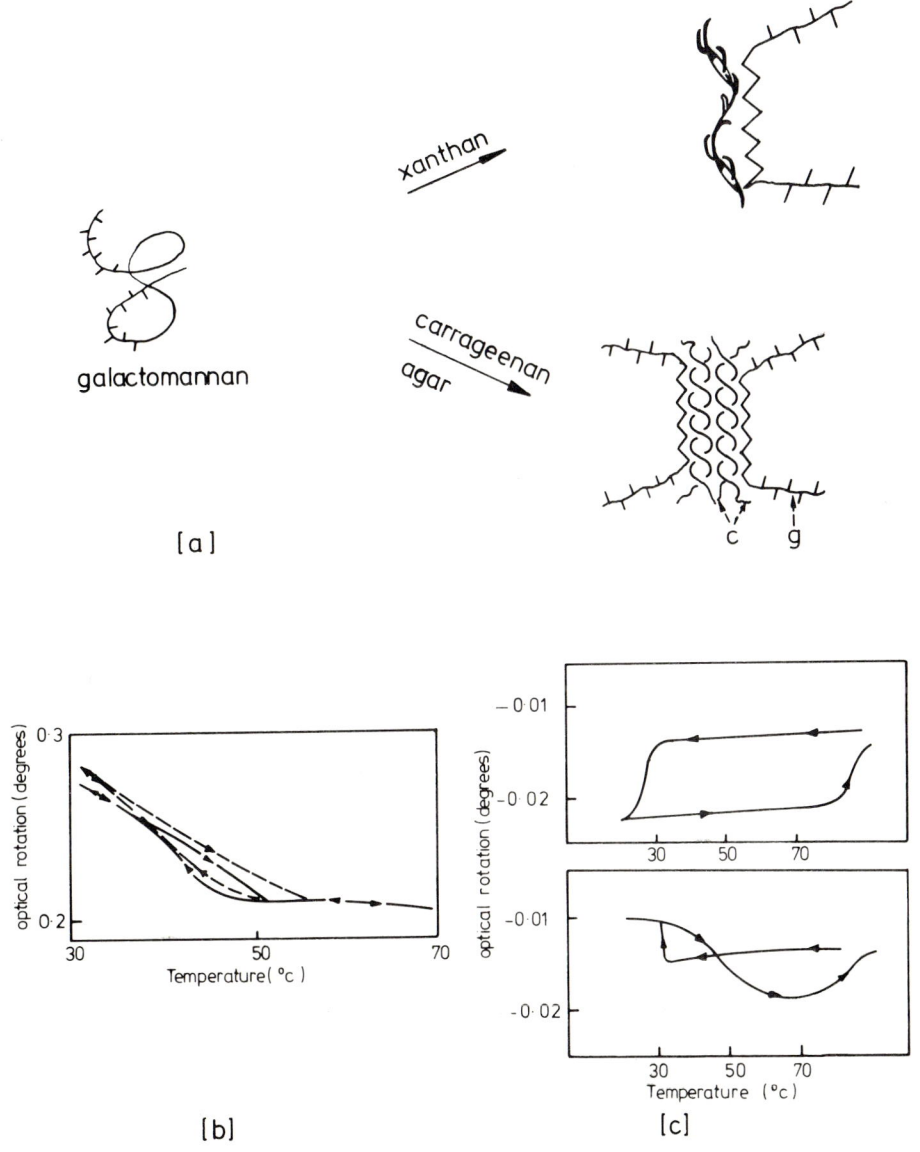

Fig. 7. a) Schematic representation of the quaternary associations between unlike carbohydrate chains: the stoichiometry of such interactions is, as yet, undetermined. Comparisons are also shown of OR variations with temperature for b) κ-carrageenan "segments" 4% (wt/vol) alone (—) and in the presence of 1% (wt/vol) galactomannan (– – –) and for c) nongelling agarose concentrations (0.05%) alone (upper graph) and with added galactomannan (0.1%; lower graph). To facilitate comparisons galactomannan-containing curves have been adjusted by subtracting expected contributions from the galactomannan alone.

On cooling, the OR transition for κ-carrageenan segments is shifted (Fig. 7) to higher temperatures, suggesting an involvement of galactomannan in the nucleation event, and likewise on heating the observed change is consistent with some stabilization of double helices by galactomannan (34). For agarose, comparison of OR transitions in the presence and absence of galactomannan shows more marked perturbations (Fig. 7). Temperature-dependent shifts on cooling again suggest some participation in helix nucleation (34) but, in addition, the transitions are opposite in sign and the "melting transition" shows a complex form when galactomannan is present. Our interpretation is that the galactomannan itself adopts an ordered shape in the junction zone and that this can be "melted out" before the agarose helix itself melts.

As previously described, disorder → order transitions in xanthan are nongelling but in the presence of galactomannans this polymer also forms (14, 15) firm gels with sharp transition points suggestive of a cooperative association between the unlike chains (Fig. 7). Such quaternary interactions may also occur (15) with other 1,4-linked chains including glucomannans, xylans, and cellulose derivatives. For carrageenans and agarose such interactions may be important in maintaining the structural integrity of plant cell walls. In the particular case of xanthan, however, the interactions may be part of a specific pathogenic "recognition process." Xanthomonas campestris is a plant pathogen which invades tissues such as cabbages, beans, and cotton through the vascular system. This intermolecular interaction suggests that the extracellular xanthan chains could serve to "recognize" particular areas of plant cell walls and to bind the bacteria to them in a single layer.

3. APERIODIC CHAINS

These carbohydrate chains, typically found covalently bound to proteins and lipids, occur at the periphery of animal cells and in secretions and have aperiodic sugar sequences. Moreover, although it is increasingly evident that such chains are often involved in cellular recognition phenomena their shapes and potential for intermolecular associations are, as yet, poorly understood. It is apparent, however, that any such ordered conformations must be very different to those previously described. The chains are characterized by short, irregularly linked sequences of up to 6 sugar units. Ordered helices or ribbons are, therefore, impossible and it is tempting to speculate that the frequent occurrence of common "core" regions with "peripheral" side chains extending from dense branch points serve to induce the aperiodic chains to adopt either stable "globular shapes" or "planar arrays" capable of aligning and interacting with ordered "backbone" structures in a manner losely analogous to that of the simpler bacterial systems (12, 13) discussed above.

Excitingly, the first steps into the detailed understanding of the conformations of these chains are now being taken. Conformational analysis (38), using predictive methods for determining protein secondary structure from primary sequence, indicates a very high probability that aperiodic chains will attach to reverse β-turns of the amino acid backbone; all 9 O-glycosidic and 19 out of 28 N-glycosidically-linked chains considered were attached to sequences of this conformational type. In addition the striking revelation of the first crystallographic evidence for spatially distinct aperiodic chains confirms that such chains can themselves adopt ordered conformations under appropriate circumstances. The crystallized polypeptide is the Fc region of an immunoglobulin IgG and, although the shape of the carbohydrate chains is not yet finalized, Huber and his co-workers (39) have shown that they adopt a distinct 3-dimensional geometry and are, indeed, linked to a β-turn of the protein backbone. As shown on Fig. 8 each carbohydrate chain does "shield" distinct, apolar regions of the polypeptide "sheet" and is also in intimate contact with a number of amino acids in the region functioning as the "hinge" in the complete antibody (Fig. 8). The uncomplexed antibody itself corresponds roughly to a symmetrical "Y" shape with little or no contact between F_{AB} and F_C domains when crystallized and showing complete disorganization of the F_C region below the hinge. Huber et al. have, therefore, proposed that the ordered F_C form corresponds to the antibody structure following complexation to antigen and this is consistent with solution studies suggesting contraction and conformational stabilization when antigen binding occurs.

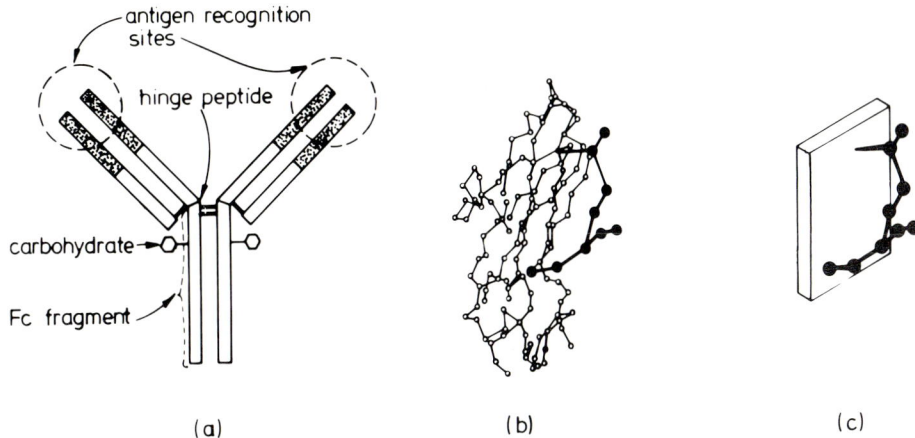

Fig. 8. Representation of antibody (IgG) molecule: a) Schematic to show relative locations of the different parts, b) contour of the carbohydrate chains (heavy lines – circles represent centers of each sugar ring) and neighbouring regions of peptide (thinner lines – circles represent C_α of each peptide unit) as deduced from X-ray diffraction analysis of F_c fragment, c) schematic of b), showing how the carbohydrate chain projects from the peptide sheet to cover 1 face.

In summary antigen binding is considered (39) to convert the antibody from a "flexible Y" to a "rigid T" shape and the function of the aperiodic carbohydrate chains appears to be in "screening" apolar protein segments and "keying in" the final association between ordered protein domains. If such conformations and interactions of these complex, aperiodic carbohydrate chains prove to be of general occurrence then the implications for molecular biology, and in particular for cellular recognition phenomenon involving multi-subunit assemblies of glycoproteins, are enormous.

NOTE ADDED IN PROOF

Additional evidence (40) consistent with the proposed role for aperiodic chains in determining the final shape and organisation of IgG antibodies is that glycoside digestion confirms the requirement of the carbohydrate moieties in recognition events mediated by the F_C region (ie through F_C receptor or complement binding) without significantly affecting their binding ability. Furthermore, differential glycosidase digestion indicated that the key sugar residues (N-acetylglucosamine and mannose) were those that make up the "core" of these complex, aperiodic carbohydrate chains.

REFERENCES

1. Rees DA: Adv Carbohydr Chem Biochem 24:267, 1967.
2. Rees DA: Biochem J 126:257, 1972.
3. Rees DA: In (Aspinall GO, ed): "Carbohydrates." MTP International Review of Science, Organic Chemistry Series One, Vol 7, London: Butterworths, 1973, p 251.
4. Rees DA: In (Whelan WJ, ed): "Biochemistry of Carbohydrates." MTP International Review of Science, Biochemistry Series One, Vol 5, London: Butterworths, 1975, p 1.
5. Marchessault RH, Sarko A: Adv Carbohydr Chem 22:421, 1967.
6. Rees DA: "Polysaccharide Shapes," Outline Series in Biology, Chapman and Hall: London, submitted for publication.
7. Rees DA, Scott WE: J Chem Soc B, 469, 1971.
8. Sundararajan PR, Marchessault RH: Can J Chem 50:792, 1972.
9. Banks W, Greenwood CT: Staerke 23:300, 1971.
10. Hyble A, Rundle RE, Williams DE: J Am Chem Soc 92:5834, 1970; French A, Zaslow B: Chem Commun 41, 1972.
11. Jackobs JJ, Bumb RR, Zaslow B: Biopolymers 6:1659, 1968.
12. Lindberg B, Svensson S:, Ref 4, p 319 (1973); Choy YM, Fehmel F, Frank N, Stirm S: J Virol 16:581, 1975; Moorhouse R, Winter WT, Arnott S: J Mol Biol (In press).
13. Moorhouse R, Walkinshaw MD, Arnott S: ACS Symposium Series, 172nd National Meeting (In press).
14. Morris ER, Rees DA, Young GA, Walkinshaw MD, Darke A: J Mol Biol 110:1, 1977.
15. Dea ICM, Morris ER, Rees DA, Welsh EJ: Carbohydr Res (In press).
16. Haug A, Larsen B: Smidsrød O, Painter T: Acta Chem Scand 23:2955, 1969; Boyd J: PhD Thesis, University College of North Wales, Bangor, 1975.
17. Haug A, Larsen B: Carbohydr Res 17:297, 1971; Madgwick J, Haug A, Larsen B: Acta Chem Scand 27:3592, 1973.
18. Atkins EDT, Niedusynski IA, Mackie W, Parker KD, Smolko EE: Biopolymers 12:1865, 1879, 1973; Mackie W: Biochem J 125:89P, 1971.
19. Kohn R, Larsen B: Acta Chem Scand 26:2455, 1972; Kohn R: Pure Appl Chem 42:371, 1975.
20. Smidsrød O, Haug A: Acta Chem Scand 26:2063, 1972.
21. Morris ER, Rees DA, Sanderson GR, Thom D: J Chem Soc, Perkin Trans 2 1418, 1975; Morris ER, Rees DA, Thom D: J Chem Soc Chem Commun 245 (1973).
22. Morris ER, Rees DA, Thom D, Young GA: (In preparation).
23. Grant GT, Morris ER, Rees DA, Smith PJC, Thom D: FEBS Lett 32:195, 1973.
24. Rees DA, Wight AW: J Chem Soc B 1366, 1971.
25. Anderson NS, Dolan TCS, Penman A, Rees DA, Mueller GP, Stancioff DJ, Stanley NF: J Chem Soc C 602, 1968; Anderson NS, Dolan TCS, Rees DA: J Chem Soc C 596, 1968.
26. Araki C, Arai K: Bull Chem Soc Jpn 40:1452, 1967.
27. Goldstein IJ, Hay GW, Lewis BA, Smith F: "Methods in Carbohydrate Chemistry." New York: Academic Press, vol 5, p 361; Rees DA: J Chem Soc 5168, 1961; 1812, 1973.
28. Anderson NS, Campbell JW, Harding NM, Rees DA, Samuel JWB: J Mol Biol 45:85, 1969; Arnott S, Scott WE, Rees DA, McNab CGA: J Mol Biol 90:253, 1974.
29. Bryce TA, Clark AH, Reid DS, Rees DA: (In preparation); Jones RA, Staples EJ, Penman A: J Chem Soc Perkin Trans 2:1608, 1973.
30. Rees DA, Scott WE, Williamson FB: Nature (London) 227:390, 1970.
31. Rees DA: J Chem Soc B 877, 1970.
32. Reid DS, Bryce TA, Clark AH, Rees DA: Faraday Discuss Chem Soc 57:230, 1974.
33. Bryce TA, McKinnon AA, Morris ER, Rees DA, Thom D: Faraday Discuss Chem Soc 57:221, 1974.
34. Dea ICM, McKinnon AA, Rees DA: J Mol Biol 68:153, 1972.
35. Dea ICM, Rees DA, Welsh EJ: (Unpublished results).
36. Morris ER, Rees DA, Young GA: (In preparation).
37. Baker CA, Whistler RL: Carbohydr Res 45:237, 1975.
38. Aubert J-P, Loucheux-Lefebvre MH: Arch Biochem Biophys 175:400, 1976; Aubert J-P, Biserte G, Loucheux-Lefebvre MH: Arch Biochem Biophys 175:410, 1976.
39. Huber R, Trends Biochem Sci 1:174, 1976; Huber R, Deisenhofer J, Colmann PM, Matsushima M, Palm W, Nature (London) 264:415, 1976.
40. Korde N, Nose M, Maramatsu T: Biochem Biophys Res Commun 75:838, 1977.

Isolation and Partial Characterization of the Heterophile Antigen of Infectious Mononucleosis From Bovine Erythrocytes

J. M. Merrick, R. Schifferle, K. Zadarlik, K. Kano, and F. Milgrom

Department of Microbiology, School of Medicine, State University of New York at Buffalo, Buffalo, New York, 14214

The heterophile antigen (Paul-Bunnell antigen, PBA) of infectious mononucleosis was isolated by extraction of an aqueous suspension of bovine erythrocyte stromata with chloroform-methanol (2:1). The upper aqueous layer contained gangliosides, PBA, and a high-molecular-weight glycoprotein. PBA and gangliosides were separated from the high-molecular-weight glycoprotein by extraction of lyophilized upper layer with chloroform-methanol solvents. Separation of PBA from gangliosides was carried out by chromatography on DEAE-cellulose with chloroform-methanol solvents. PBA appeared to be a minor glycoprotein component of the erythrocyte membrane and had both hydrophobic and hydrophilic properties. It was soluble in either organic or aqueous solvents. On SDS-polyacrylamide gel electrophoresis, it migrated as a single component that stained for protein with Coomassie blue, for carbohydrate with periodic acid-Schiff reagent, and for lipid with oil red 0; it had an apparent molecular weight of 26,000. It was composed of 62% protein with major amino acids: glutamic acid, proline, glycine, isoleucine, leucine, and threonine (158, 116, 98, 90, 85, and 82 residues per 1,000 residues, respectively). Carbohydrate content was 9.2% with major sugar constituents: sialic acid, galactosamine, and galactose. Serologic activity of PBA was destroyed by pronase but not by trypsin.

Key words: bovine erythrocytes, heterophile antigen, infectious mononucleosis, membranes, Paul-Bunnell antigen

Cellular components common to cells of different species have been recognized for many years. Forssman (1), for example, demonstrated in 1911 that the injection of guinea pig kidney suspensions into rabbits induced the formation of antibodies (Forssman antibodies) that reacted with sheep erythrocytes. Forssman antibodies have been termed heterophile antibodies, that is, antibodies which react with cells or tissues of species that are unrelated to the original species that provided the antigenic preparation used for immunization. Heterophile antigens are thus defined as antigens which stimulate the formation of or are capable of combining with heterophile antibodies. The Forssman antigen is

Abbreviations: HD — Hanganutziu-Deicher; IM — infectious mononucleosis; PAS — periodic acid-Schiff reagent; PBA — Paul-Bunnell antigen; PBS — phosphate buffered saline; SDS — sodium dodecyl sulfate.

Received March 29, 1977; accepted May 4, 1977

© 1977 Alan R. Liss, Inc., 150 Fifth Avenue, New York, NY 10011

found in a variety of mammalian species (e.g., sheep, goat, horse, guinea pig) and has been identified as a glycosphingolipid (2). A ceramide pentasaccharide with similar structures has been isolated from horse spleen and from sheep and goat erythrocytes (2, 3, 4).

Paul and Bunnell in 1932 (5) reported that sera of patients suffering from infectious mononucleosis agglutinated sheep erythrocytes. Subsequently, it was established that these heterophile antibodies which differ from Forssman antibodies, also acted on horse, goat, and bovine erythrocytes. The appearance of Paul-Bunnell antibodies in "normal" sera or in patients suffering from diseases other than IM[1] is apparently extremely rare (6). In fact, the detection of these antibodies forms the basis of a key test in the diagnosis of IM (6). Although the Epstein-Barr virus has been implicated as the causative agent of IM (7, 8), the antigenic stimulus responsible for the formation of Paul-Bunnell antibodies still remains unknown.

The heterophile antigen of IM (Paul-Bunnell antigen, PBA) has been isolated from horse, sheep, goat, and bovine erythrocytes by hot 75%- ethanol extraction of stromata (8, 9). Using similar procedures a glycoprotein possessing M and N specificity has been obtained from human erythrocytes (8). In the case of horse, sheep, and goat erythrocytes, the PBA antigenic determinants appeared to reside on a glycoprotein, that is, closely related to the human erythrocyte major sialoglycoprotein (glycophorin) both with regard to amino acid patterns and carbohydrate constituents and content. However, the bovine erythrocyte PBA differed from PBA isolated from other mammalian species in its relatively lower carbohydrate content and its resistance to proteases (8, 10). Although the overall chemical composition of the various glycoproteins appears to be similar, they are clearly distinguished from the human erythrocyte glycoprotein by their antigenic activity. The underlying structural features responsible for these antigenic differences are unknown.

In this communication we describe a fractionation procedure that permits the isolation of the neutral glycolipids, gangliosides, a high-molecular-weight glycoprotein, and PBA from bovine erythrocyte stromata. PBA has been purified and partially characterized. Its properties are of special interest since it appears to have characteristics of a hydrophilic and hydrophobic glycoprotein. In addition, we also report that the ganglioside fraction obtained from bovine erythrocyte membranes reacted with Hanganutziu-Deicher antibodies (11, 12). These antibodies, commonly referred to as "serum sickness antibodies," were initially observed by Hanganutziu and Deicher as agglutinins for sheep erythrocytes in sera of patients who received therapeutic injections of horse antitoxin. They have also been reported to be present in patients who received γ globulin fractions of goat antisera to human thymocytes (13) and in pathologic human sera of patients who had apparently never received injections of foreign species sera (14). The following differences between PBA, Forssman, and HD antigens may be noted: PBA appears on bovine and sheep erythrocytes but not on guinea pig kidney, Forssman antigen appears on sheep erythrocytes as well as guinea pig kidney but not on bovine erythrocytes, and HD antigens are found in all 3 of these tissues.

MATERIALS AND METHODS
Materials

All organic solvents were reagent grade and were redistilled before use. Trypsin treated with L-(tosylamido 2-phenyl) ethyl chloromethyl ketone was purchased from Worthington Biochemical Corp. (Freehold, New Jersey), pronase from Calbiochem (La Jolla, California), and DEAE-cellulose from Brown Co. (Berlin, New Hampshire). All other chemicals were of the highest purity commercially available.

Analytic Methods

For the determination of neutral sugars, samples were hydrolyzed in 1 N HCl for 6 h at 100°C in sealed tubes and the hydrolysates were passed through coupled columns of Dowex 50 and Dowex 1 (15). Neutral sugars were determined by means of automated borate complex anion exchange chromatography (Technicon system) at a column temperature of 45°C with the elution gradient described by Lee et al. (16). Sialic acid was estimated by the thiobarbituric acid reaction following hydrolysis in 0.1 N H_2SO_4 at 80°C for 1 h (17). Total hexose was measured by the phenol-H_2SO_4 method (18). For the determination of amino acids and amino sugars, samples were hydrolyzed in constant boiling HCl under nitrogen at 105°C for 28 h. Analyses were carried out on a Beckman 120-C analyzer. Total half-cystine was analyzed as cysteic acid and methionine as methionine sulfone after performic acid oxidation according to the method of Hirs (19). Protein was determined by the method of Lowry et al. (20); phosphate by the procedure of Bartlett (21); and sphingosine by the spectrophotometric method of Lauter and Trams (22). Fatty acids were determined after hydrolysis of 5 mg samples in 1.0 ml of 2 N KOH-methanol at 100°C for 24 h. Prior to hydrolysis, heneicosanoic acid was added as internal standard. After acidification with HCl, the hydrolysate was extracted 3 times with 1.0 ml of n-hexane. The hexane extracts were concentrated to dryness and esterification was carried out with 0.5 ml of 14% boron trifluoride in methanol in sealed tubes at 100°C for 30 min. After esterification, 0.5 ml of water was added to each tube and the contents extracted 3 times with 1.0 ml of n-hexane. The hexane extracts were concentrated to a small volume, and analysis of the fatty acid methyl esters was carried out by gas liquid chromatography (23).

Hemagglutination Inhibition Tests

These tests were carried out in plastic microtiter plates. In studies to detect PBA activity, antigen and IM serum were diluted with diluent, normal rabbit serum 1:60 in PBS (116 mM NaCl-13.6 mM sodium potassium phosphate, pH 7.0). Antigen (0.025 ml) and diluent (0.025 ml) were added to the first well and serial twofold dilutions prepared. IM serum (0.025 ml) at 4 hemagglutinating doses was added to each well and after 1 h of incubation at room temperature, 0.025 ml of a 1% suspension of sheep red blood cells in PBS was added to each well. The tubes were incubated for 2 more hours and examined for hemagglutination. HD antigenic activity was measured in the same way except HD sera were used and the diluent was PBS. One unit of hemagglutination inhibiting activity was defined as the minimum amount of material necessary to completely inhibit the agglutination of sheep erythrocytes by 4 hemagglutinating doses of IM serum (or HD serum) under standard assay conditions.

Double Diffusion Gel Precipitation Tests

These tests were performed in plastic petri dishes containing 1% agarose gel according to the procedure described by Milgrom et al. (24).

Sera

Sera of patients with IM were obtained from the Student Health Service of the State University of New York at Buffalo. Sera from patients who received γ globulin fractions of goat antisera to human thymocytes were our source of HD antibodies and were kindly supplied by Dr. B. Pirofsky of the Division of Immunology and Allergy, University of Oregon, Portland, Oregon.

SDS-Polyacrylamide Gel Electrophoresis

Samples were treated with 1% SDS in 0.005 M sodium phosphate buffer, pH 7.2, containing 2.5 M urea, 0.5% 2-mercaptoethanol, and 0.005% EDTA for 15 min at 37° and 45 min at room temperature (25). SDS-gel electrophoresis was performed in 10% gels by the method of Weber and Osborn (26). Protein was stained with Coomassie blue and carbohydrate with PAS reagent according to the procedures described by Segrest and Jackson (27). For the detection of lipid containing components, gels were fixed overnight in methanol-acetic acid-water (50:5:45), followed by staining for 24 h with a solution of oil red O prepared by mixing equal volumes of 20% trichloroacetic acid and a saturated solution of oil red O in methanol (28). Destaining was carried out with the methanol-acetic acid-water fixative solution. To compare oil red O and PAS stained gels, the latter were treated with the fixative solution until the size of the gel was similar to the gel previously treated with oil red O.

Preparation of Bovine Erythrocyte Membrane Extracts

Fresh pooled blood of cattle was obtained from a local slaughter house and collected into a sodium citrate solution (4.8% sodium citrate – 0.9% NaCl; 1.0 ml per 9.0 ml blood). The erythrocytes were washed 3 times with 3 volumes of PBS. Stromata were prepared by hemolyzing washed erythrocytes with 20–30 volumes of distilled water usually containing from 0.8 to 1.4 ml of 5% acetic acid per liter. The erythrocyte ghosts were allowed to sediment overnight at 4°C and were harvested after centrifugation at 20,000 × g for 15 min at 4°C. The ghosts were washed repeatedly with water followed by 0.9% NaCl until either the water or saline washes were colorless.

Erythrocyte glycoproteins were extracted from stromata suspensions by a modification of the method described by Kornfeld and Kornfeld (29) and Hamaguchi and Cleve (30). Aqueous suspensions of mechanically homogenized stromata (15–20 mg/ml) were stirred vigorously for 2 h with 6 volumes of a chloroform-methanol (2:1) mixture. Unless stated otherwise, all procedures were carried out at room temperature. Preparations with a final volume of less than 1.0 liter were centrifuged at 500 × g for 15 min and the aqueous layer removed and treated as described below. Preparations with a final volume greater than 1.0 liter were allowed to stand overnight. Solid debris was removed by filtration through Whatman No. 1 solvent-washed filter paper. The mixture was added to a separatory funnel and the upper aqueous layer and lower organic layer were separated. The upper layer was concentrated in vacuo in a rotary evaporator at 37°C with frequent additions of n-proponal to prevent foaming, followed by dialysis against distilled water overnight at 4°C. Further concentration to approximately the volume of the original membrane suspension was achieved by ultrafiltration with a PM-10 membrane (Amicon Company, Lexington, Massachusetts). The opalescent material was centrifuged at 38,000 × g for 40 min and the pellets were discarded. The clear supernatant fluid which contained the erythrocyte glycoproteins and gangliosides was lyophilized. The lower layer contained neutral glycolipids and was not further studied.

Separation of Gangliosides, PBA, and High Molecular Weight Glycoproteins

The lyophilized membrane extract was further resolved into a fraction which contained gangliosides, a component tentatively identified as ganglioside and PBA, and into a high-molecular-weight glycoprotein fraction. Lyophilized membrane extract (1 g) was vigorously stirred with 200 ml of chloroform-methanol-water (1:1:0.3) for 1 h. The mixture was filtered with a Büchner funnel using an aspirator and a Whatman No. 54 solvent-washed filter paper. The residue was reextracted with 200 ml chloroform-methanol-water (1:1:0.3) 3 more times and the combined extracts were concentrated under reduced pressure, dialyzed against distilled water overnight and lyophilized (ganglioside and PBA fraction). The residue remaining was further extracted sequentially, 3 times for 1 h with 100 ml of chloroform-methanol (2:1) containing 4% H_2O and 1% ammonium acetate, 3 times for 1 h with 100 ml of chloroform-methanol-water (1:1:0.3) containing 1% ammonium acetate, and 3 times for 1 h with 100 ml of chloroform-methanol-water (1:1:10) containing 1% ammonium acetate. The latter extract contained the high-molecular-weight glycoprotein. The extraction protocol is summarized in Fig. 1.

Gangliosides and PBA were resolved by chromatography on a DEAE-cellulose column prepared as described by Rouser et al. (31). DEAE-cellulose (9.6 g) in glacial acetic acid was packed in a column of 1.5 cm diameter and the column was washed successively with 6 bed volumes of methanol and 3 bed volumes of chloroform-methanol (1:1). The sample was applied directly to the column or as a suspension in chloroform-

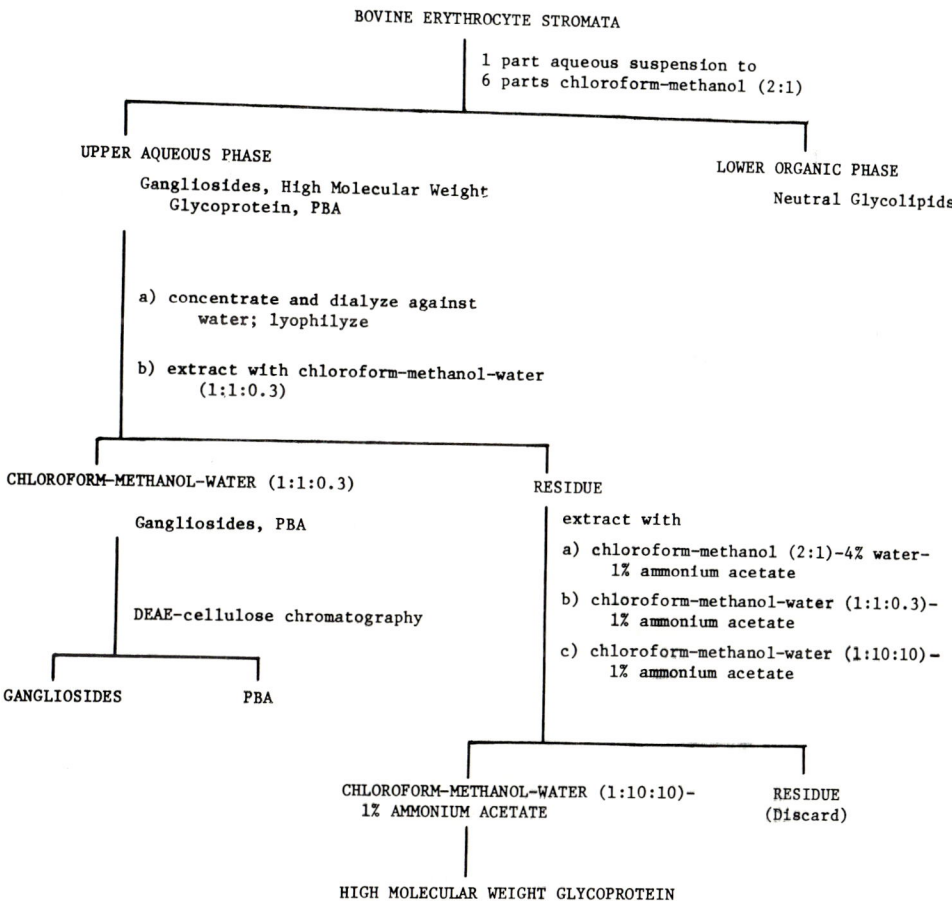

Fig. 1. Summary of fractionation procedures for the isolation of PBA, gangliosides, and high-molecular-weight glycoprotein. Bovine erythrocyte membranes were treated as diagrammed in the figure and described under Materials and Methods.

methanol (1:1) (Solvent 1). After the column was washed with 200 ml of the same solvent, gangliosides were eluted with chloroform-methanol (2:1) saturated with aqueous 28% ammonium hydroxide (Solvent 2); 10-ml fractions were collected and analyzed for hexose by the phenol-sulfuric acid method. The fractions containing hexose were combined, evaporated in vacuo, dialyzed against distilled water overnight, and lyophilized. After no further hexose containing material was detected in the fractions, the column was washed with an additional 100 ml of solvent. About 600 ml of this solvent was poured through the column. The column was then washed successively with 200 ml of chloroform-methanol (2:1) containing 4% water (Solvent 3), 200 ml of chloroform-methanol-water (1:1:0.3) (Solvent 4), and chloroform-methanol-water (1:1:0.3) containing 0.02 M ammonium acetate (Solvent 5). In the latter case 11.5 ml fractions were collected and analyzed for hexose which appeared in fractions 22 through 36. After no further hexose material was detected, the column was washed with an additional 200 ml of the same solvent. Solvent 5 eluted a component tentatively identified as a ganglioside and a small amount of PBA. PBA was next eluted from the column with a linear gradient of ammonium acetate in chloroform-methanol-water (1:1:0.3). The gradient was started with 250 ml of chloroform-methanol-water containing 0.02 M ammonium acetate in the mixing chamber and 250 ml of chloroform-methanol-water (1:1:0.3) containing 0.2 M ammonium acetate in the reservoir (Solvent 6); 10-ml fractions were collected and aliquots of selected fractions were assayed for protein by the Lowry procedure after the solvent was removed by heating at 37°C under a stream of N_2.

RESULTS

SDS-polyacrylamide gel electrophoresis of washed bovine erythrocyte membranes revealed the presence of 3 PAS staining components (Fig. 2A). These components were released into the upper aqueous layer by extraction of aqueous suspension of erythrocyte membranes with chloroform-methanol (2:1) (Fig. 2B). In a typical fractionation 1,100 mg of membrane extract was obtained from 13.18 g of stromata (Table 1). Routinely this procedure released from 60 to 80% of the sialic acid of the membrane. The more mobile PAS staining components were easily resolved from the high-molecular-weight glycoprotein by appropriate solvent extractions (Fig. 1) of lyophilized preparations of the upper aqueous phase (Fig. 2C and D). By hemagglutination inhibition tests it was readily demonstrated that PBA was in the fraction which contained the mobile PAS components (266 mg of this fraction, and 613 mg of the high-molecular-weight glycoprotein fraction were obtained from 1,100 mg of membrane extract, Table I). In addition, this fraction also contained the ganglioside components of the membrane; at least 5 gangliosides were detected on thin layer chromatograms.

The composition of the bovine erythrocyte membrane and the partially resolved components are given in Table I. The high-molecular-weight glycoprotein contained the major carbohydrate components of the erythrocyte membrane and appears to be similar to the bovine erythrocyte glycoprotein isolated by Emerson and Kornfeld (32) by extraction of ghosts with lithium 3,5-diiodosialicylate. There are however differences regarding the relative sugar content of the glycoproteins as well as the reported inability of the protein isolated by the above workers to stain with Coomassie blue after SDS-polyacrylamide gel electrophoresis. These workers also failed to detect the minor glycoprotein component of the bovine erythrocyte membrane.

PBA was resolved from gangliosides and the low-molecular-weight PAS staining

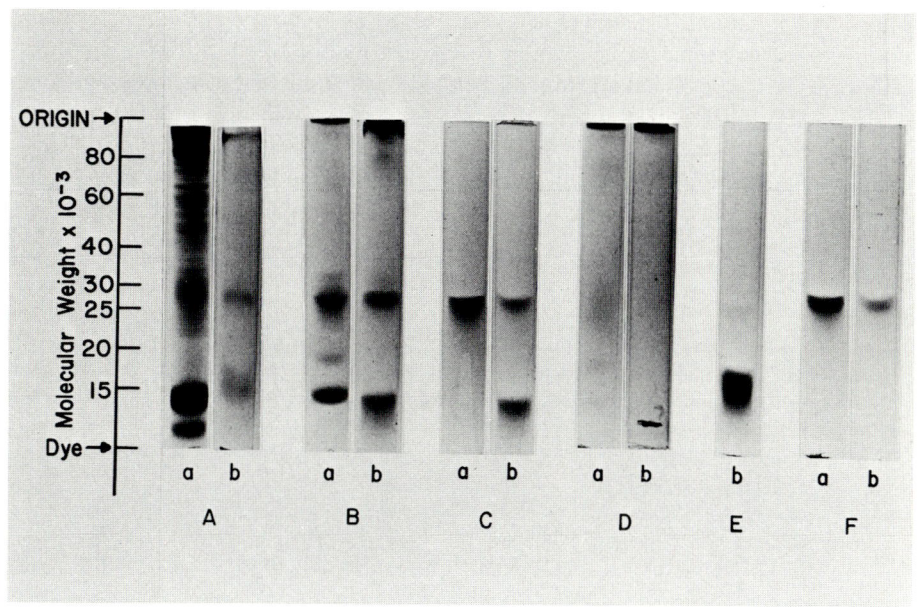

Fig. 2. SDS-polyacrylamide gel electrophoresis of bovine erythrocyte membranes and isolated components. Samples were treated with 1% SDS and subjected to electrophoresis on 10% acrylamide gels in 0.1% SDS and 0.1 M phosphate buffer, pH 7.0; a) samples stained for protein with Coomassie blue; b) samples stained for carbohydrate with periodic acid-Schiff reagent. A) washed erythrocyte membranes [a) 59 μg protein; b) 470 μg protein]; B) upper aqueous phase [a) 28 μg protein; b) 98 μg protein]; C) PBA-ganglioside fraction [a) 14 μg protein; b) 29 μg protein]; D) high-molecular-weight glycoprotein fraction [a) and b) 14 μg protein]; E) ganglioside fraction eluted from DEAE-cellulose column with solvent no. 5 [b) 14 μg protein]; F) PBA eluted from DEAE-cellulose column with solvent no. 6 [a) 13 μg protein; b) 26 μg protein].

TABLE I. Composition of Bovine Erythrocyte Membranes and Extracted Components*

			Extracted components of membrane extract	
	Erythrocyte membranes	Membrane extract[a]	Gangliosides-PBA	HGP[b]
dry weight (mg)	13,180	1,100	266	613
protein (mg)	6,063	311	85	98
fucose (μmoles)	62.2	25.3	3.3	ND[c]
mannose (μmoles)	70.4	13.2	0.9	11.8
galactose (μmoles)	1,685.7	1,122.0	116.5	1,075.2
glucose (μmoles)	179.5	55.0	33.0	18.1
sialic acid (μmoles)	361.1	217.8	38.5	159.4
N-acetylglucosamine (μmoles)	1,229.5	901.6	60.2	720.3
N-acetylgalactosamine (μmoles)	329.9	169.5	13.2	133.2
HIU[d] (1M serum)	3.0×10^7	1.66×10^7	2.55×10^7	9.8×10^4

*Extraction procedures are described in the text and summarized in Fig. 1.
[a]Membrane extract, upper aqueous phase
[b]HGP — high molecular weight glycoprotein
[c]ND — none detected
[d]HIU — hemagglutination inhibiting units

component by chromatography on DEAE-cellulose. The solvents utilized in this fractionation procedure are summarized in Table II. The distribution of HD and PBA antigenic activity in the various fractions from the column was monitored by hemagglutination inhibition tests. As shown in Table II, HD antigenic activity was principally found in the ganglioside fraction eluted with solvent system No. 2. The low-molecular-weight PAS staining component was eluted with solvent system No. 5 and did not contain significant PBA or HD antigenic acitivty. On SDS-polyacrylamide gels (Fig. 2E) this sialic acid con-

TABLE II. Separation of Gangliosides and PBA by DEAE-Cellulose Column Chromatography*

Solvent no.	Elution solvents[a]	Elution volumes ml	Substance eluted	HIU[b]-IM serum	HIU[b]-HD serum
1	C-M (1:1)	200			
2	C-M (2:1) saturated with 28% NH_4OH	600	Gangliosides	5.6×10^5	1.3×10^6
3	C-M (2:1) + 4% H_2O	200			
4	C-M-W (1:1:0.3)	200			
5	C-M-W (1:1:0.3) + 0.02 M NH_4Ac	600	Ganglioside (?)	8.0×10^4	1.1×10^4
6	C-M-W (1:1:0.3) + 0.02 M NH_4Ac to C-M-W (1:1:0.3) + 0.2 M NH_4Ac	Linear gradient 250 ml of each	PBA	1.2×10^7	3.2×10^3

*The PBA-ganglioside fraction (204 mg; 1.96×10^7 HIU-IM serum and 1.67×10^6 HIU-HD serum) was placed on a column of DEAE-cellulose (acetate) (9.6g) and the column washed with solvents in the sequence shown above.
[a]C-M-W — chloroform, methanol, water, respectively
[b]HIU — hemagglutination inhibiting units

taining component migrated in the glycolipid region and has for this reason been tentatively identified as a ganglioside. Approximately 10 mg of this component was isolated from the column. PBA antigenic activity was eluted from the DEAE-cellulose column with a linear gradient of ammonium acetate (0.02 M to 0.2 M) in chloroform-methanol-water (1:1:0.3) (Table II, Fig. 3). Approximately 42 mg of material (3.4 ng contained 1 unit of hemagglutination inhibitory activity) was obtained in this fraction from 204 mg of starting material. Chloroform was an obligatory requirement in the solvent system used for elution of PBA; methanol or aqueous ammonium acetate solutions were totally ineffective. Occasionally in some fractionations a small peak of PBA activity (about 15% of the PBA activity found in the major peak) was eluted just prior to the major peak. No difference in the components eluted in either peak were detected by serological analysis, by SDS-polyacrylamide gel electrophoresis, or in sugar or amino acid content. Thus the reasons for this elution pattern were not readily apparent. Residual PBA containing about 10% of the activity put on the column could be eluted after the gradient was completed with 200 ml of chloroform-methanol-water (1:1:0.3) containing 0.2 M ammonium acetate and 11.3 mM glacial acetic acid. Again, this material appeared identical to material eluted in the major fraction by the criteria indicated above.

To establish unequivocally the nature of the isolated antigens, double diffusion gel precipitation tests were carried out. As can be seen on Fig. 4, PBA formed a strong precipitation line with IM serum and reacted weakly with a serum containing HD antibodies. Complete removal of HD antigenic activity from PBA preparations was difficult and

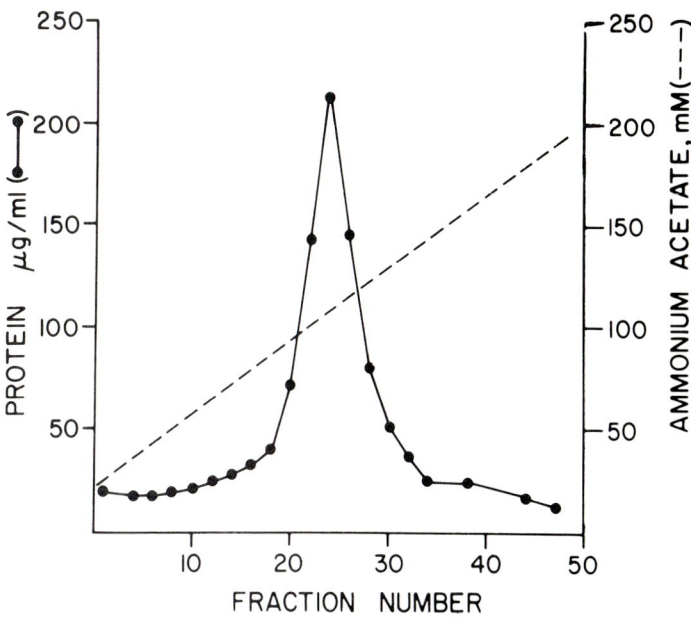

Fig. 3. DEAE-cellulose chromatography of PBA. After removal of gangliosides as described in Table II, PBA was eluted from the DEAE-cellulose column with a linear gradient of ammonium acetate (0.02–0.2 M, 250 ml each) in chloroform-methanol-water (1:1:0.3). Fractions (10 ml) were collected and selected tubes analyzed for protein.

residual activity was usually observed even in preparations that had been subjected to a second fractionation on DEAE-cellulose by the procedures described previously. With some IM sera we also observed the formation of 2 well-separated lines (Fig. 4). In earlier studies, Milgrom et al. (24) reported similar results with ultrasonic extracts of stromata prepared from trypsinized erythrocytes. Ganglioside fractions eluted with solvent No. 2 and partially purified on silica gel columns gave a strong reaction with serum containing HD antibodies and a negligible reaction with IM serum (Fig. 4). As expected, absorption of the sera with guinea pig kidney removed HD but not Paul-Bunnell antibodies.

On SDS-polyacrylamide gel electrophoresis, PBA appeared as a single glycoprotein component that stained with PAS or Coomassie blue (Fig. 2G). PBA is readily soluble in water or chloroform-methanol-water (1:1:0.3) and thus has both hydrophobic and hydrophilic properties. Its hydrophobic characteristics were also indicated by its behavior on DEAE-cellulose columns. Elution of PBA under our fractionation conditions was completely dependent on the presence of chloroform in the solvent system. These extreme hydrophobic properties suggested that PBA may be closely associated with lipid. To explore this possibility, gels were also stained with oil red O, a dye that had previously been used to stain for lipoproteins (33). Yamamoto and Lampen (33) for example, stained the membrane penicillinase phospholipoprotein of Bacillus licheniformis with oil red O. PBA was run on replicate gels, and gels stained with PAS or oil red O were compared. It was, however, necessary to reduce the size of the PAS stained gel (see Materials and Methods) so that comparisons could be made with the gel stained with oil red O. As can be seen on Fig. 5, oil red O staining did indeed detect a single component. The small differences noted be-

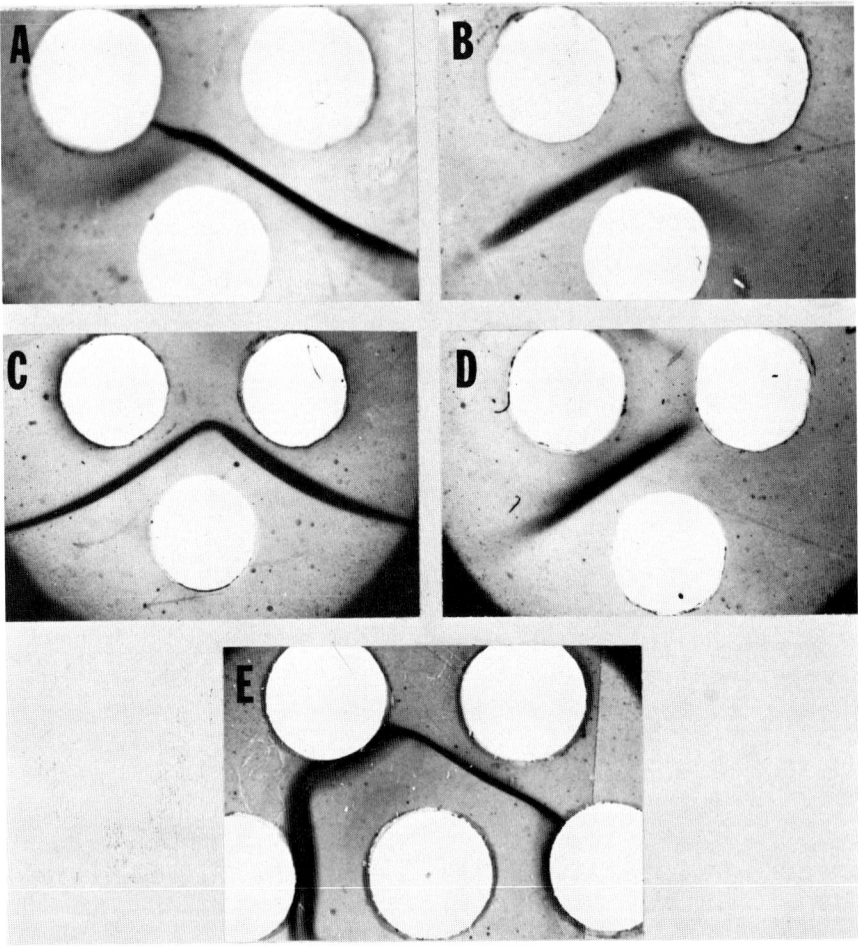

Fig. 4. Double diffusion gel precipitation studies. A) lower well: PBA, B) lower well: gangliosides. A and B) Upper left well: serum containing HD antibodies; upper right well: IM serum. C) lower well: PBA; D) lower well: gangliosides; C) upper left well: IM serum, upper right well: IM serum absorbed with guinea pig kidney; D) upper left well: HD serum, upper right well: HD serum absorbed with guinea pig kidney; E) center well: PBA; outer wells: various IM sera.

tween the position of the PAS and oil red O staining components is presumably a reflection of the differential changes in gel length that occurred during the 2 different staining procedures. These differences were also noted between oil red O and Coomassie blue staining of the bacterial membrane and exopenicillinases (33). In our studies it is unknown whether the ability of PBA to stain with oil red O is in fact due to lipid covalently attached to protein. It is also feasible that these results could have been obtained if the PBA fraction were contaminated with a lipid with mobility characteristics on SDS gels that are similar to PBA, or alternatively, staining may simply be due to the hydrophobic amino acid content of the protein.

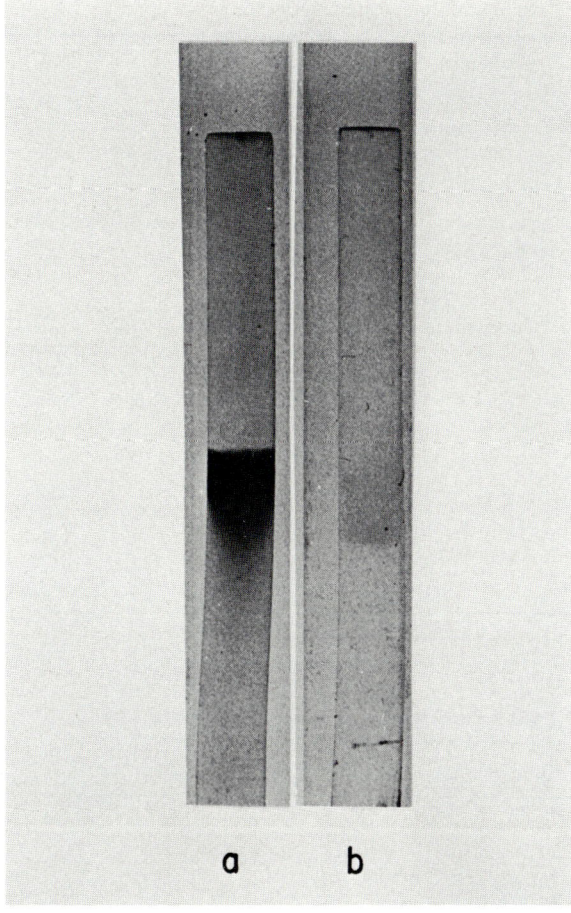

Fig. 5. SDS-polyacrylamide gel electrophoresis of PBA. Conditions as described in Fig. 2. a) Gel was stained with periodic acid-Schiff reagent; b) gel was stained with oil red O. a) 26 µg protein; b) 214 µg protein.

Table III shows the amino acid content of PBA. It can be seen that the protein contains a relatively high content of hydrophobic residues (proline, leucine, isoleucine, valine, alanine, phenylalanine). These results are in agreement with the amino acid composition of the bovine IM heterophile antigen reported by Springer (34, 35) and in somewhat less agreement with the data reported by Fletcher and Woolfolk (36). Both groups of workers extracted PBA from bovine erythrocyte stromata with hot 75% ethanol; however, Springer (34) further purified the ethanol extracted component by ethanol fractionation, gel filtration, and sucrose gradient ultracentrifugation.

The overall chemical composition of PBA is presented in Table IV. Approximately 9.2% of the antigen was analyzed as carbohydrate which differs significantly from the carbohydrate content of bovine PBA reported by Springer (35) and Fletcher and Woolfolk (36). The predominating sugar constituents were sialic acid, galactose, and N-acetylgalactosamine. The solubility of PBA in organic solvents as well as its apparent staining with oil

TABLE III. Amino Acid Composition of Bovine PBA*

Amino acid	Residues 1,000 Amino acid residues	g Amino acid 100 g Antigen
aspartic acid	59.9	4.1
threonine	81.9	4.9
serine	67.5	3.5
glutamic acid	157.7	12.0
proline	115.9	6.6
glycine	97.8	3.3
alanine	58.9	2.5
valine	57.6	3.4
1/2 cystine	–	–
methionine	27.1	2.1
isoleucine	84.5	5.6
leucine	89.9	5.9
tyrosine	–	–
phenylalanine	35.5	3.1
lysine	19.2	1.5
histidine	1.1	0.1
arginine	36.1	3.3

*Average values from 3 different samples, uncorrect for losses during hydrolysis

TABLE IV. Chemical Composition of Bovine PBA

Component	Weight %
polypeptide	62.0
sialic acid	4.3
N-acetylgalactosamine	2.5
N-acetylglucosamine	0.2
galactose	2.1
glucose	0.1
mannose	< 0.1
phosphorous	0.04
sphingosine	ND[a]
fatty acids	
16	0.05
18	0.06
18:1	0.02

[a]ND – none detected; 2.2 mg sample used for analysis

red O suggested that lipid components may be present. However, we could only detect small amounts of phosphorus and fatty acids. It still remains to be determined whether lipid components or the hydrophobic properties of the protein itself resulted in staining by oil red O.

The molecular weight of PBA estimated by SDS-polyacrylamide gel electrophoresis was 26,000 (Fig. 6). This molecular weight can only be considered tentative for the reasons discussed by Tanford and Reynolds (37) who point out that the only reliable determination of molecular weight of membrane proteins is the use of sedimentation equilibrium in SDS.

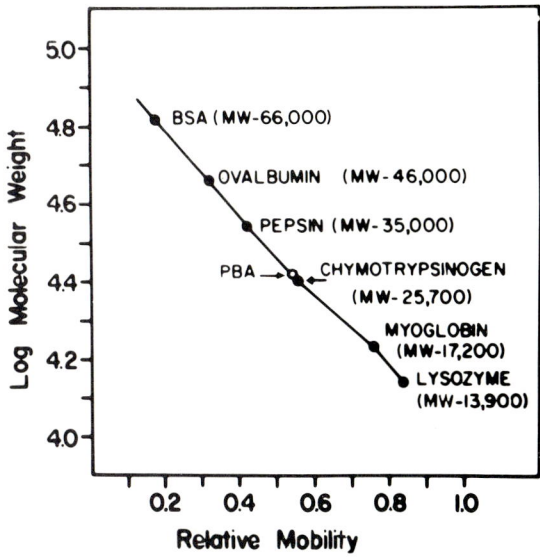

Fig. 6. Molecular weight estimation of PBA by SDS-polyacrylamide gel electrophoresis. Conditions as described in Fig. 2. The relative mobilities were calculated as the ratio of the distance moved by the protein to the distance moved by the marker dye. The protein standards used in the construction of the standard curve and their molecular weights are indicated on the graph.

Trypsin treatment of PBA failed to inactivate the antigen as determined by the hemagglutination inhibition test (Table V). SDS-polyacrylamide gel electrophoresis, however, indicated that PBA was altered by trypsin treatment and had an estimated molecular weight of 19,000. The trypsinized component stained normally with PAS but only very faintly with Coomassie blue. PBA antigenic activity was sensitive to pronase treatment (Table V) which suggests that the antigenic determinants are located on the glycoprotein.

TABLE V. Effect of Trypsin or Pronase Treatment of PBA*

Expt.	Treatment	Total HIU[a]	Activity remaining
1	PBA alone	6.4×10^4	100
	PBA + trypsin	6.4×10^4	
2	PBA alone	4.8×10^4	8
	PBA + pronase	4×10^3	

*PBA (0.20 mg) was incubated with trypsin (19.4 μg) in 0.068 M phosphate buffer, pH 7.0 (Expt. 1) or with pronase (27.5 μg) in 0.07 M Tris HCl buffer, pH 8.2 containing 1 mM $CaCl_2$ (Expt. 2). Final volume was 0.25 ml. Incubations were carried out for 24 h at 37°C in the presence of a small amount of toluene. Hemagglutination inhibiting activity was measured by the ability of PBA to inhibit the agglutination of sheep erythrocytes by infectious mononucleosis serum.
[a]HIU — hemagglutination inhibiting units

DISCUSSION

Our studies were initially undertaken in an effort to reexamine the chemical nature of the bovine IM heterophile antigen. Earlier investigations carried out by Springer (34, 35) and by Fletcher and Woolfolk (36) suggested that PBA of animal cells were glycoproteins that possessed remarkable similarity in carbohydrate and amino acid composition to the blood group M and N glycoproteins. However, the isolation procedures utilized by these workers did not exclude the possibility that their preparations were contaminated with glycolipids. The difficulty in removing ABO blood group determinants from glycoproteins isolated from human erythrocyte is well known (38).

The fractionation procedures that we have described in this communication permit the separation of glycoprotein and glycolipid constituents of the bovine erythrocyte membrane. PBA appears as a minor glycoprotein component of bovine erythrocyte membranes and has an estimated molecular weight of 26,000. Thus, our results support the conclusions reached by the above workers. The amphipathic nature of PBA is suggested by its hydrophobic and hydrophilic properties. PBA is therefore ideally suited as an integral membrane protein. The carbohydrate content of bovine PBA is significantly lower than the content reported for PBA isolated from horse, sheep, and goat erythrocytes (35, 36, 39). However interestingly, its antigenic activity is greater. At the present time the reasons for this are unknown. Sialic acid has been implicated as an important immunodeterminant since neuraminidase destroys PBA activity of sheep, horse, and goat antigens. In the case of bovine PBA, Fletcher and Woolfolk (36) reported that after neuraminidase treatment only 0.5% of the original sialic acid remained, yet PBA retained 15% of its hemagglutination inhibiting activity. In our studies, Clostridium perfringens neuraminidase (40) released 69% of the sialic even after extensive treatment with the enzyme. This resulted in about a 50% loss of activity in the standard assay. The inability of neuraminidase to completely hydrolyze sialic acid containing polymers has been observed previously (40). Cassidy et al. (40) point out that the position of attachment of sialic acid or the nature of the sugar to which it is attached influences greatly the rate of hydrolysis by neuraminidase. It is of interest to note that Hakomori and Saito (41) have described a neuraminidase resistant glycolipid which contains O-acetyl (N-glycoyl) neuraminic acid. After removal of the O-acetyl group, the sialic acid of the glycolipid was now susceptible to hydrolysis by neuraminidase.

IM is a self-limiting lymphoproliferative disease in which heterophile antibodies are characteristically produced (6). In addition, Epstein-Barr virus associated antibodies directed against viral capsid antigen, membrane antigen, and "early" antigen are also produced (42). Heterophile antibodies of the IM type are highly specific for IM and rarely appear, if at all, in other diseases. However, the role of IM heterophile antibodies in determining the self-limiting course of the disease is unknown. It is of considerable interest that Milgrom et al. (43) have reported that PBA is present on spleen cells of patients suffering from various forms of lymphoma and leukemia. These patients, however, do not produce Paul-Bunnell antibodies. These observations have prompted Milgrom et al. (43) to speculate that lack of IM heterophile antibodies may account for the malignant character of these diseases while their production in high titer accounts for the benign character of IM.

In IM the antigenic stimulus responsible for Paul-Bunnell antibody formation is not known. There are apparently few reports where PBA has been identified on human tissue during the course of the disease (44, 45). It is not known if PBA is a product of the intecting viral genome or a host cellular component that becomes uncovered or slightly modified as a result of viral infection. Our understanding of the chemical structure of PBA may provide further insight into the resolution of these possibilities.

This investigation has also demonstrated that the ganglioside fraction of bovine erythrocyte membranes reacts with HD antibodies. In humans, HD antibodies are primarily formed as a result of injection with foreign proteins, presumably containing antigenic gangliosides. However, they have also been detected in pathological human sera in patients who have not received such injections. In these cases, the antigenic stimulus remains unknown. Further studies on the identity of active ganglioside components and their antigenic determinants are underway.

ACKNOWLEDGMENTS

We are very grateful to Dr. Michael J. Levine for the aid provided in amino acid and sugar analyses, to Dr. Jack D. Klingman for his help in the fatty acid analyses, to Mr. Ian Slepian who helped in developing some of the fractionation procedures, to Dr. Ulana Loza and Mr. Walter Campbell for their assistance in the serological aspects of this study, and to Dr. Murray W. Stinson for many helpful discussions.

This investigation was supported by American Cancer Society Grant IM24B, and by Public Health Service Biomedical Research Support Grant, National Institutes of Health.

REFERENCES

1. Forssman J: Biochem Z 37:78, 1911.
2. Siddiqui B, Hakomori S: J Biol Chem 246:5766, 1971.
3. Fraser BA, Mallette MF: Immunochemistry 11:581, 1974.
4. Taketomi T, Hara A, Kawamura N, Hayashi M: J Biochem (Tokyo) 75:197, 1974.
5. Paul JR, Bunnell WW: Am J Med Sci 183:90, 1932.
6. Davidson I: In Glade PR (ed): "Infectious Mononucleosis." Philadelphia-Toronto: JB Lippincott Company, 1972, pp 19–35.
7. Henle G, Henle W, Diehl B: Proc Natl Acad Sci USA 59:94, 1968.
8. Fletcher MA, Woolfolk BJ: Biochim Biophys Acta 107:842, 1971.
9. Fletcher MA, Lo TM, Graves WR: J Immunol 117:722, 1976.
10.. Springer GF, Fletcher MA: In Heymer A, Ricken D (eds): "Organ Transplantation." Stuttgart-New York: Schattaur-Verlag, 1969, pp 35–45.
11. Hanganatziu M; Cor Soc Biol 91:1457, 1924.
12. Deicher H: Z Hyg Infecktionskr 106:561, 1926.
13. Pirofsky B, Ramirez-Mateos JC, August Z: Blood 42:385, 1973.
14. Kasukawa R, Kano K, Bloom ML, Milgrom F: Clin Exp Immunol 25:122, 1976.
15. Spiro RG: Methods Enzymol 8:3, 1966.
16. Lee YC, Johnson GS, White B, Scocca J: Anal Biochem 43:640, 1971.
17. Warren L: J Biol Chem 234:1971, 1959.
18. Dubois M, Gilles KA, Hamilton JK, Regers PA, Smith F: Anal Chem 28:350, 1956.
19. Hirs CHW: Methods Enzymol 11:197, 1967.
20. Lowry OH, Rosebrough NJ, Farr AL, Randall RL: J Biol Chem 193:265, 1951.
21. Bartlett GR: J Biol Chem 234:466, 1959.
22. Lauter CJ, Trams EG: J Lipid Res 3:136, 1932.
23. Organisciak DT, Klingman JD: Lipids 9:307, 1974.
24. Milgrom F, Loza U, Kano K: Int Arch Allerg Appl Immunol 48:82, 1975.
25. Greenberg CS, Glick MC: Biochemistry 11:3680, 1972.
26. Weber K, Osborn M: J Biol Chem 244:4406, 1969.
27. Segrest JP, Jackson RL: Methods Enzymol 28:54, 1972.
28. Bownds D, Gordon-Walker A, Gaide-Huguenin A, Robinson W: J Gen Physiol 58:225, 1971.

29. Kornfeld R, Kornfeld S: J Biol Chem 245:2536, 1970.
30. Hamaguchi H, Cleve H: Biochim Biophys Res Commun 47:459, 1972.
31. Rouser G, Kritchevsky G, Yamamoto A, Simon G, Galli C, Bauman AJ: Methods Enzymol 14:272, 1969.
32. Emerson WA, Kornfeld S: Biochemistry 15:1697, 1976.
33. Yamamoto S, Lampen JO: J Biol Chem 251:4095, 1976.
34. Springer GF: In Westphal IO, Bock HE, Grundmann E (eds): "Current Problems in Immunology." Berlin-Heidelberg-New York: Springer-Verlag, 1969, pp 47–62.
35. Springer GF: In Glade PR (ed): "Infectious Mononucleosis." Philadelphia-Toronto: JB Lippincott Company, 1972, pp 26–35.
36. Fletcher MA, Woolfolk BJ: J Immunol 107:842, 1971.
37. Tanford C, Reynolds J: Biochim Biophys Acta 457:133, 1976.
38. Marchesi VT, Furthmayer H, Tomita M: Annu Rev Biochem 45:667, 1967.
39. Fletcher MA, Lo TM, Graves WR: J Immunol 117:722, 1976.
40. Cassidy JT, Jourdian GW, Roseman S: J Biol Chem 240:3501, 1965.
41. Hakomori S, Saito T: Biochemistry 8:5082, 1969.
42. Miller G: Prog Med Virol 20:84, 1975.
43. Milgrom F, Kano K, Fjelde A, Bloom ML: Transplant Proc 7:201, 1975.
44. Andres GA, Kano K, Elwood C, Prezyna A, Sepulveda M, Milgrom F: Int Arch Allergy Appl Immunol (In press).
45. Lowenthal RM, McLauchlan SL, Tyrrell DA: Nature (London) New Biol 244:109, 1973.

Homogeneous Antibodies Directed Against Human Cell Surface Antigens: I. The Mouse Spleen Fragment Culture Response to T and B Cell Lines Derived From the Same Individual

Lois A. Lampson, Ivor Royston, and Ronald Levy

Division of Oncology, Department of Medicine, Stanford University School of Medicine, Stanford, California 94305

The use of the mouse spleen fragment culture system is extended to the production of antibodies to human lymphoblastoid cell lines. These antibodies were tested for reactivity against the immunizing cell line, and against a second cell line which had been derived from the same human blood sample. Many of the antibodies were found to discriminate between the 2 isogenic lines. These results demonstrate the potential of the mouse spleen fragment culture system to provide homogeneous reagents which detect distinguishing markers on closely related human cells.

Key words: antibody, human T cell, human B cell

Among the different techniques that have been described for the in vitro production of antibody (1–3), the mouse spleen fragment culture system (3) is of particular value because it permits one to routinely obtain the antibody product of a single clone of responding B cells. The monoclonal nature of this response (4–7) has a number of important consequences. Firstly, knowing that the responding cells are the clonal progeny of a single responding B cell allows one to infer some of the properties of that cell, such as its frequency, specificity, or requirements for triggering (8, 9). Secondly, because the antibody produced by each fragment is homogeneous with respect to its specificity, spleen fragment cultures provide a source of material for the study of the antibody molecule itself (6, 7). In both of these cases, molecules of known structure have been the antigens of choice. Finally, the homogeneous antibody may be used as a probe for the study of the antigen to which it is directed. It is only relatively recently, with the development of techniques for the production and assay of responses to complex antigens, that this final aspect of the fragment culture response is beginning to be exploited. Antigens that have been used in this context include influenza virus (10, 11), murine tumor cells (12), and, most recently, in our laboratory, human tumor cells (13) and cell lines.

Previous work in this laboratory has established the feasibility of using the mouse spleen fragment culture system to obtain monoclonal antibodies to human cell surface

Received March 29, 1977; accepted May 4, 1977

© 1977 Alan R. Liss, Inc., 150 Fifth Avenue, New York, NY 10011

antigens; of particular significance was the finding that many of the antibodies raised could differentiate between presumably related cell types, such as between chronic lymphocytic leukemias from different patients (13). In order to further explore the ability of the mouse to recognize human cell surface antigens, we have studied the mouse spleen fragment culture response to a pair of human lymphoblastoid cell lines derived from the same individual. One of these, 8392, has been characterized as a B cell line, and the other, 8402, as a T cell line (14, 15). Thus, although these 2 lines have the same genetic background, they also differ in many of their surface characteristics. The question we have asked is: How many of the mouse monoclonal antibodies produced against one of these lines will be directed against common antigens, and how many of the antibodies will be able to discriminate between the 2 cell lines?

MATERIALS AND METHODS

Cell lines. Four human lymphoblastoid cell lines, 8402, 8392, H-SB2, and SB, were used in these experiments. 8402 and 8392 are T and B cell lines derived from the same human blood sample; H-SB2 and SB are T and B cell lines derived from a second human blood sample. The derivation, characterization, and conditions for culturing these lines have been described previously (14–16).

The fragment culture system. The use of the mouse spleen fragment culture system to obtain antibodies to human cell surface antigens has been described previously (13). Briefly, for each cell line studied, (Balb/c × C57 Bl/6) F1 mice were immunized and boosted, intraperitoneally, with 10^7 human cells. From 2 to 4 weeks following the boost, $0.5\text{-}50 \times 10^6$ viable spleen cells from the immunized donors were transferred, intravenously, into lethally (1,300 rad) irradiated recipients. Within a few minutes the irradiated recipients also received an intravenous dose of 10^7 of the immunizing human cells. Within 24 h the recipient spleens were chopped into 1 mm^3 fragments, and the fragments cultured individually in Linbro plates in MEM-10% agamma horse serum-glutamine-gentamicin. Culture supernatants were harvested at 3–4 day intervals for the first month of culture, and stored at $-70°C$ until assay.

Goat anti-mouse kappa. A goat was immunized with the Fab fragments of a purified mouse anti-DNP antibody (17). The globulin fraction of the resulting goat anti-mouse Fab antiserum was purified by ammonium sulfate precipitation (at 40% saturation) and passed over an affinity column (18) of purified W3207 (a mouse $\kappa\alpha$ myeloma) protein, which was the generous gift of Ms. Carol Nottenburg. The bound goat anti-mouse κ was eluted from this column with 0.1 N acetic acid, and was radioiodinated as described previously (19) for use in the radioimmunoassay (see below).

Radioimmunoassay. The radioimmunoassay for the detection of antibody to human cell surface antigens has been described previously (13). Briefly, 25 µl of supernatant fluid was incubated with 25×10^4 glutaraldehyde-fixed target cells for 2 hr at room temperature; the cells were washed free of unbound material, and then incubated with ^{125}I-goat anti-mouse κ overnight at 4°C. The cells were then washed extensively, and the amount of label bound to the pellets was counted in a gamma counter.

Data analysis. In each experiment, 50–100 control fragments were cultured. These were fragments from the spleens of irradiated animals which had been given the stimulating dose of antigen, but no donor spleen cells. At least 12 control fluids were included in each assay, and the mean and standard deviation of the control binding activity was computed. In assessing the binding activity of experimental supernatant fluids, those whose binding

TABLE I. Antigen Dependence of the Fragment Culture Response to Human Cell Lines

Stimulator cells	Positive fragments
	Experiment 1
	Stimulator cells = 8402[a]
0	0/42 = 0%
10^5	2/96 = 2%
10^6	5/96 = 5%
10^7	3/84 = 4%
	Experiment 2
	Stimulator = 8392[b]
0	0/45 = 0%
10^5	3/90 = 3%

[a]Donor mice were primed and boosted with 10^7 8402 cells; 2×10^6 donor spleen cells were transferred to irradiated recipients, and $0-10^7$ 8402 cells were injected within a few minutes. One day later, the recipient spleens were chopped and cultured as described in Materials and Methods.
[b]Same as 1, except 8392 cells were used to immunize, boost, and stimulate.

TABLE II. Donor Cell Dependence of the Fragment Culture Response to Human Cell Lines*

	Positive fragments			
	Stimulator = 8402[a]		Stimulator = 8392[a]	
Donor cells	Experiment 1	Experiment 2	Experiment 1	Experiment 2
0	0/48 = 0%	—[b]	0/45 = 0%	—
0.5×10^6	1/96 = 1%	—	0/96 = 0%	—
2×10^6	3/84 = 4%	—	1/96 = 1%	—
$10-12 \times 10^6$	6/143 = 4%	—	3/96 = 3%	—
30×10^6	—	29/83 = 35%	—	5/96 = 5%
50×10^6	—	—	12/41 = 29%	—

*From $0-50 \times 10^6$ donor spleen cells were transferred to irradiated recipients, and 10^7 stimulator cells were injected within a few minutes. One day later, the recipient spleens were chopped and cultured as described in Materials and Methods.
[a]Different groups of donor mice were used for Experiments 1 and 2.
[b]Not done.

activity fell within 2 standard deviations of the control mean were scored as negative, and those whose binding activity fell at least 5 standard deviations from the control mean were scored as positive. Binding activities falling between these 2 cut-off points were not scored.

RESULTS

The Mouse Spleen Fragment Culture Response to the Human Lymphoblastoid Cell Lines 8402 and 8392

In order to determine the appropriate conditions of culture, mouse spleen fragment culture responses to either 8402 or 8392 were established using varying doses of donor

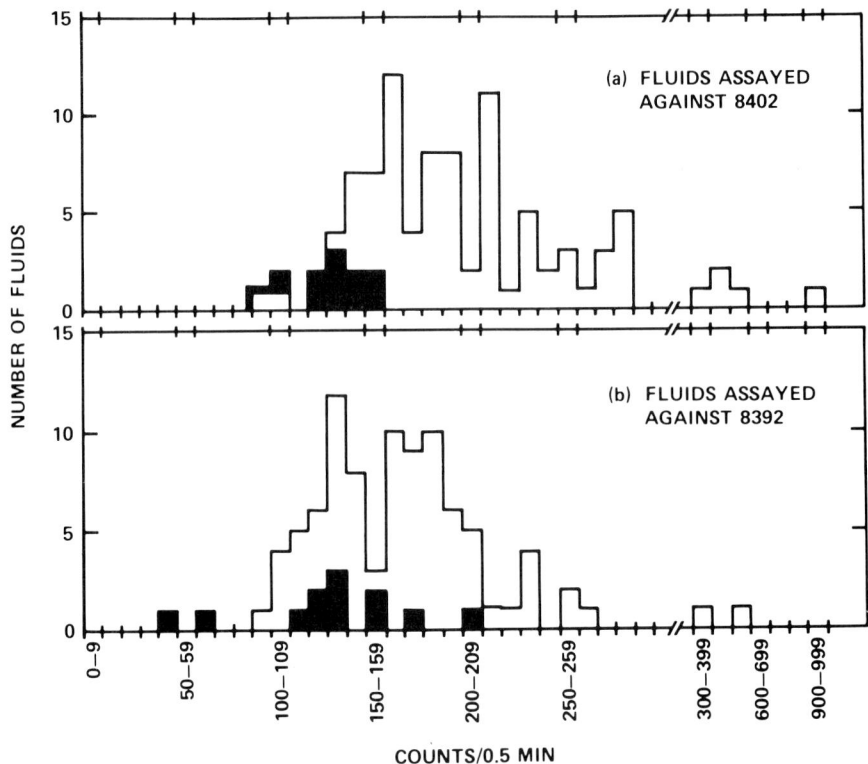

Fig. 1. Distributions of binding activities against 8402 (the immunizing cell) and its isogenic partner, 8392. Donor mice were immunized and boosted with 8402. Recipient mice received 0.5–10 × 10⁶ donor spleen cells and a stimulating dose of 8402 cells within a few minutes. Recipient spleens were chopped and cultured as described in Materials and Methods. Day 7 supernatant fluids were assayed against a) 8402 or b) 8392.

and stimulating cells. The supernatant fluids were collected on day 7 after the beginning of culture and assayed in the radioimmunoassay for antibody activity against the immunizing cell line. Table I shows the number of responding fragments obtained when 2×10^6 donor cells and from $0–10^7$ stimulator cells were given. Table I verified that the response is antigen-dependent and that 10^7 cells is a great excess of stimulating cells.

Table II shows the number of responding fragments obtained when the stimulating dose was held constant and the dose of donor cells varied over the range $0.5–50 \times 10^6$ cells. Table II confirms that the response is also donor cell-dependent. Moreover, Table II shows that at doses of up to $10–12 \times 10^6$ donor spleen cells when 8402 is the immunogen, and up to 30×10^6 donor spleen cells when 8392 is the immunogen, the cloning efficiency is low enough so that the probability of one fragment containing 2 responding B cells is low (10).

The Broad Specificity of the Response to 8402

In order to assess the broad specificity of culture fluids obtained against 8402, 90 such fluids were assayed against both 8402 and its isogenic partner, 8392. Each fluid was assayed at 3 dilutions: undiluted, 1:4, and 1:16. The results obtained with the un-

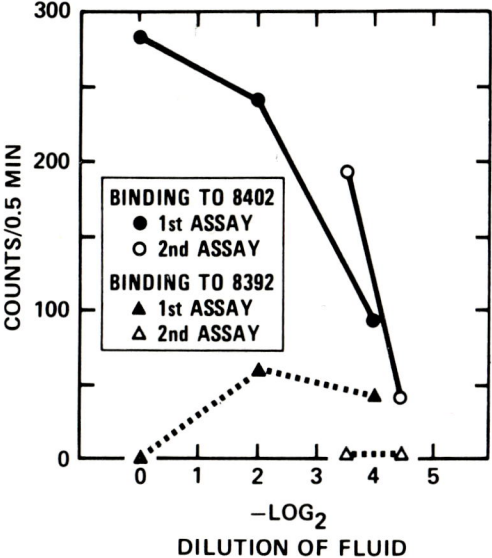

Fig. 2. Example of a discriminator culture fluid. This clone was raised against the human T cell line 8402. The supernatant fluid was positive against the immunizing cell and negative against the isogenic B cell, 8392 (see text). Circles show binding to 8402, triangles show binding to 8392; filled symbols represent first assay, open symbols represent second assay. In each case, the mean binding of 12 control fluids was subtracted from the total counts. Means and standard deviations of the control values were: first assay: 129 ± 20 (8402), 129 ± 43 (8392); second assay: 81 ± 30 (8402), 78 ± 13 (8392).

diluted fluids are shown on Fig. 1. The backgrounds seen when 12 control fluids (see Materials and Methods) were assayed against each of the 2 cell lines are also shown in Fig. 1 (shaded areas). It is apparent from Fig. 1 that, while the majority of the fluids tested bound to the immunizing T cell line (8402), when the same fluids were assayed against 8392 relatively few showed levels of binding to the target cells above background levels. A similar analysis of the data obtained with the diluted fluids gave the same results (data not shown).

In order to analyze this data in another way, individual culture fluids were classed as *discriminators* if they were positive against 8402 at at least 2 of the dilutions tested, and negative against 8392 at all 3 dilutions. (Responses were scored as positive or negative as described in Materials and Methods.) Culture fluids were classed as *nondiscriminators* if they were positive at at least 2 dilutions against both cell lines. Of the 90 culture fluids that were studied, 11 fell into either of these categories; of these 11 culture fluids, 9 were discriminators and 2 were nondiscriminators. (None of the 90 culture fluids was negative against 8402 but positive against 8392.) An example of a discriminator culture fluid is shown in Fig. 2.

The Response to 8392.

In order to determine the specificity of culture fluids obtained when a B cell was the immunizing cell, 5 culture fluids which had been raised against 8392 and originally scored as positive against that cell were reassayed against both 8392 and its isogenic partner, 8402. Using the criteria described in the preceding paragraph, all 5 of these fluids were found to be discriminators; an example of one of these is shown in Fig. 3.

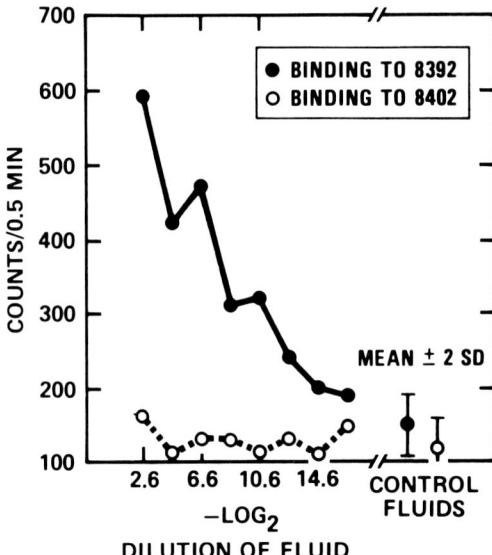

Fig. 3. Example of a discriminator culture fluid. This clone was raised against the human B cell line 8392. The supernatant fluid was positive against the immunizing cell and negative against the isogenic T cell, 8402. Filled circles show binding to 8392, open circles show binding to 8402. The mean and 2 standard deviations of the binding of 12 control fluids are also shown.

Other Responses

The results presented thus far have been taken from experiments in which the cloning efficiency was low. Even where the cloning efficiencies were much higher, so that one would expect the majority of the fragments to contain more than one responding B cell (10), there were many fragments whose fluids showed more binding activity for the immunizing cell than for its isogenic partner. In one group of experiments, 96 fluids were raised against 8402, and 77 of these (83%) were found to be positive against the immunizing cell. Of these 77 fluids, 7 showed quantitatively more binding activity against 8402 than against its isogenic partner, 8392. The binding activities of 4 of these culture fluids are illustrated in Fig. 4. (None of the 77 fragments analyzed showed more binding activity against 8392 than against 8402.)

A final group of experiments involved a second isogenic T-B pair, H-SB2 and SB. In this case, the B cell (SB) was used as the immunogen, and of 94 culture fluids that were screened for antibody activity, 92 (98%) were found to be positive against the immunizing cell. Of the 92 positive fluids, 62 showed quantitatively more activity against the immunizing B cell (SB) than against its isogenic partner (H-SB2). The binding activity of 4 of these fluids is shown in Fig. 5. (None of the fluids showed more binding activity against H-SB2 than against SB.)

DISCUSSION

Having previously shown that the mouse can make antibodies which discriminate between leukemias from different individuals (13), we wished to extend our analysis of the mouse B cell repertoire against human cells by studying the mouse spleen fragment culture

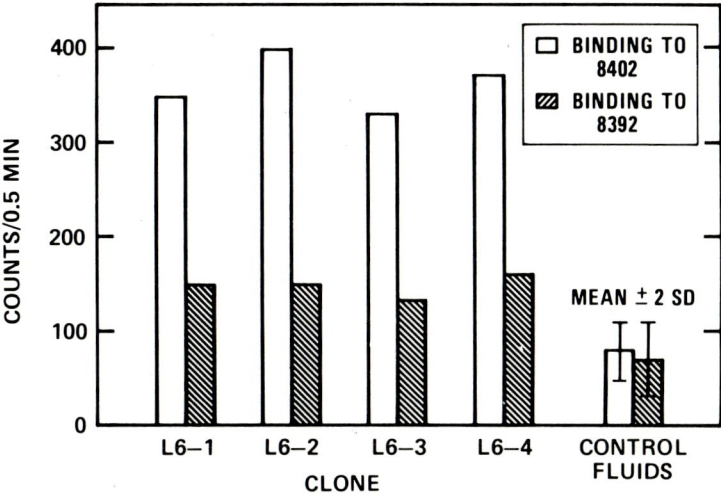

Fig. 4. Examples of culture fluids which show quantitative differences in binding to isogenic cell lines. Each clone was generated against the human T cell line 8402. The culture fluids were all positive against the immunizing cell and either negative or not scoreable against the isogenic partner, 8392. (For explanation of scoring of fluids see Materials and Methods.) In each pair, open bar shows binding to 8402, shaded bar shows binding to 8392. Mean and 2 standard deviations of the binding of 12 control fluids are also shown.

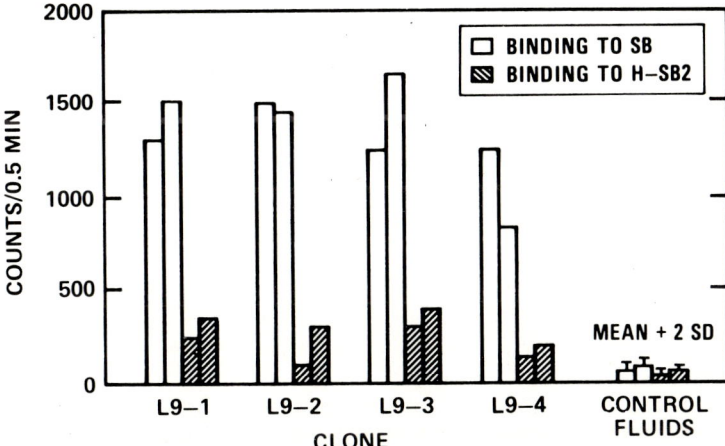

Fig. 5. Examples of culture fluids which show quantitative differences in binding to isogenic cell lines. Each clone was generated against the human B cell line SB. The culture fluids were positive against both the immunizing cell and the isogenic partner, but showed large quantitative differences in the amount of binding to the 2 cell lines. In each group of bars, open bars show binding to SB, shaded bars show binding to H-SB2. In each group, first and third bars show data from first assay, second and fourth bars show data from second assay. The mean and 2 standard deviations of the binding of 35–37 control fluids are also shown.

response to pairs of human cells with the same genetic background. In this paper, we have described conditions under which antibodies can be raised against human lymphoblastoid cell lines in the mouse spleen fragment culture system, and have presented our preliminary results showing the specificity of the responses obtained. We have shown that, at low cloning efficiency, many of the clones produce antibody which discriminates between the immunizing cell and its isogenic partner, and that, even at high cloning efficiency, many clones are found which show quantitative differences in their binding to the 2 cell lines.

We wish to stress the reciprocity of the responses that we have described. When the immunizing cell line was a human T cell line, many of the resulting culture fluids were positive for that cell line and negative for an isogenic B cell line, and none of the fluids reacted only with the B cell line. Conversely, when the immunizing cell line was a B cell line, we found many fluids which reacted only with the B cell line, and none which reacted only with an isogenic T cell line. These findings are in agreement with other reports showing that isogenic human T and B cell lines each bear distinguishing antigenic markers (15, 16, 20). The importance of the present work is that we have shown the feasibility of using the mouse spleen fragment culture system to obtain homogeneous antibodies to these markers. These antibodies may now serve as reagents for the characterization of the surface of human T and B cells, and for the identification of subpopulations of normal and transformed T and B cells.

ACKNOWLEDGMENTS

We are grateful to Dr. John Hyde for statistical advice, to Ms. Shelley Jacobs for skilled technical assistance, and to Ms. Linda Puffer for preparing the manuscript.

This work was supported by American Cancer Society Grant IM 114.

REFERENCES

1. Mishell RI, Dutton RW: J Exp Med 126:423, 1967.
2. Marbrook J: Lancet 2:1279, 1967,
3. Klinman NR: Immunochemistry 6:757, 1969.
4. Klinman NR: J Immunol 106:1345, 1971.
5. Klinman NR: J Immunol 106:1330, 1971.
6. Klinman NR: J Exp Med 136:241, 1972
7. Gearhart PJ, Sigal NH, Klinman NR: J Exp Med 141:56, 1975.
8. Klinman NR, Press JL: J Exp Med 141:1133, 1975.
9. Pierce SK, Klinman NR: J Exp Med 142:1165, 1975.
10. Gerhard W, Braciale TJ, Klinman NR: Eur J Immunol 5:720, 1975.
11. Gerhard W: J Exp Med 144:985, 1976.
12. Segal GP, Klinman NR: J Immunol 116:1539, 1976.
13. Levy R, Dilley J: J Immunol 1977 (in press).
14. Huang CC, Hou Y, Woods LK, Moore GE, Minowada J: J Natl Cancer Inst 53:655, 1974.
15. Royston I, Pitts RB, Smith RW, Graze PR: Transplant Proc Vol 7 (Suppl 1):531, 1975.
16. Royston I, Smith RW, Buell DW, Huang ES, Pagano JS: Nature (London) 251:745, 1974.
17. Haimovich J, Bergman Y, Linker-Israeli M, Haran-Ghera N: Eur J Immunol 7:226, 1977.
18. March SC, Purikh I, Cuatrecasas P: Anal Biochem 60:149, 1974.
19. Klinman NR, Aschinazi G: J Immunol 106:1338, 1971.
20. Trowbridge IS, Hyman R, Mazavskas C: Eur J Immunol 6:777, 1976.

Chemical and Immunological Studies of Cell Surfaces From Normal and Transformed Cells

W. J. Grimes, Gary A. Van Nest, and Arthur R. Kamm

Department of Biochemistry, University of Arizona, College of Medicine, Tucson, Arizona 85724

Immunological and chemical studies of cell surfaces from normal and transformed BALB/c fibroblasts have shown alterations associated with transformation. The cells studied include normal lines which do not cause tumors when injected into BALB/c mice, viral transformants, and spontaneous transformants which cause tumors that either regress or grow progressively, killing the host. The spontaneously transformed progressors include cell lines which are immunogenic and nonimmunogenic as determined by the ability of tumor excision to protect an animal from subsequent rechallenge by tumor cells. Tumor-bearing mice produce lymphocytes which are nonspecifically cytotoxic for all the normal and transformed lines. Some of the cell lines induce specific antibody formation in BALB/c hosts. Antisera have been prepared in rabbits which are specific for the transformed cell lines. These antisera can be used to determine specific surface changes on the transformed cells. Chemical studies have shown glycolipid alterations between the normal cells and some, but not all, of the transformants. Glycoproteins labeled by lactoperoxidase-^{125}I or [^{3}H]glucosamine were compared by SDS gel electrophoresis. Results from these studies do not show changes associated with malignancy. Individual glycoprotein regions from gels were treated with pronase, and the glycopeptides compared by Sephadex G-50 chromatography. Alterations in glycopeptides from several cellular glycoproteins are the only changes which appear to be associated with malignancy.

Key words: cell surface, transformed cells, glycolipids, glycoproteins

The cell surface is believed to play a central role in both growth regulation by cell-cell and cell-hormone (growth factors or chalones) interactions and cell adhesion. While it has been widely recognized that changes in cell surface components could be fundamental to cancer (for reviews see Refs. 1—3), there is no evidence that any of the reported changes detected in viral, carcinogen, or spontaneously transformed cells are primary to a transformation event. The problem has been that we lack both a biological assay for determining function and a chemical understanding of cell surface components important in cellular interactions.

The immunological response made by animals bearing tumors offers a possible approach to defining cell surface change in cancer. A number of investigators have reported evidence that a variety of both carcinogen induced and virally caused tumors

Received March 24, 1977; accepted May 16, 1977

© 1977 Alan R. Liss, Inc., 150 Fifth Avenue, New York, NY 10011

possess unique antigens (4, 5). It has been suggested that the release of these antigens (tumor specific or viral) in the form of free antigen or antigen-antibody complexes could prevent a successful rejection response (6–11).

There has been no definitive work to link results from immunological studies to chemical alterations which have been shown to accompany the transformation of a normal to a malignant cell. These reported chemical changes include altered carbohydrate compositions, a decrease in the level of complex glycolipids, loss of a 240,000 mol wt surface glycoprotein (LETS), and alterations in profiles of cellular derived glycopeptides chromatographed on molecular sieving columns (12–20). In the work reported here, we present our preliminary chemical, biological, and immunological studies of normal, spontaneously transformed, and virally transformed cells derived from BALB/c mice. We are attempting to correlate surface change with transformation, immunogenicity, and malignancy.

MATERIALS AND METHODS

Cell Lines and Cell Culture

All cell lines were derived from BALB/c mice. A_{31} is a cloned line of BALB/c 3T3 fibroblasts and was a gift from Dr. G. Todaro of NIH. c5 is a transformed cell line cloned from A_{31} as previously described (21, 22). c5T was isolated from a tumor caused by injecting c5 cells into a BALB/c mouse. MSC was derived from a Moloney strain murine sarcoma virus induced tumor in a BALB/c mouse, and was the gift of Dr. S. Russell of Scripps Clinic and Research Foundation. 3T12T cells were isolated from a tumor caused by injecting 3T12 cells (a gift from Dr. Todaro) into a BALB/c mouse. PBC cells are early passage BALB/c embryo fibroblasts (22). Cells were grown in monolayer cultures in antibiotic free Dulbecco's modified minimal essential medium supplemented with 10% fetal calf serum (Grand Island Biological Company, Berkeley, California). Routine tests for mycoplasma contamination were performed (23, 24). Cultures testing positive were not used for any experiments and were discarded.

Preparation of Lymphocytes

Spleens were excised from tumor bearing or immunized mice and placed in 5 ml Dubecco's modified minimal essential medium (DMM) in 60-mm plastic culture dishes. Lymphocytes were teased from spleens and adherent cells removed by incubating the suspension for 45 min at 37°C. Cells were then collected by sedimenting the suspension at 1,000 × g for 10 min. Erythrocytes were removed from the single cell suspension by one addition of 5 ml containing 0.75% NH_4Cl in 0.016 M Tris, pH 7.4. Lymphocytes were washed twice more in DMM and suspended in DMM-10% heat inactivated fetal calf serum at a concentration of 2×10^6 cells/ml. Such preparations retained cytotoxic reactivity for at least 24 h when stored at 25°C.

Cytotoxicity was determined by 2 methods. Following the procedure described by Hellstrom (25), target cells were plated at concentrations of 100–200 cells/well in 100 µl DMM containing 10% heat inactivated fetal calf serum. Following an overnight incubation to allow target cell attachment, 100 µl of the lymphocyte suspension was added and the plates incubated for 48 h. Lymphocytes and dead cells were removed by 2 washes with saline, and the remaining live cells stained with crystal violet and counted. In the majority of experiments, the procedure described by McKhann (26) was followed. Target cells were first plated in 35-mm tissue culture dishes at a concentration of 2×10^5 cells/dish. [^{3}H] Uridine (50 µCi at a specific activity of 29 Ci/mM) was added and the cells incubated overnight at 37°C. The labeled monolayer was then rinsed twice with sterile saline and the cells trypsinized and suspended at a concentration of 500 cells/100 μl in DMM containing 10% heat inactivated FCS. For the lymphocyte mediated killing assay, 100 μl of target cells were mixed with 100 μl of the lymphocyte suspension (usually from 1×10^4 to 2×10^5 lymphocytes) and placed in Microtest II wells. Following a 48 h incubation, the medium was removed and attached live cells rinsed twice with saline. The labeled cells were removed by 2 rinses with 200 μl of 0.5 M NaOH and placed in scintillation vials. The aqueous volume was made up to 1.0 ml by adding 0.5 ml of 20% glacial acetic acid, and radioactivity was determined using a counting solution containing 960 ml Triton X-114, 2,900 ml xylene, and 125 ml Liquiflor (New England Nuclear Corporation, Boston, Massachusetts).

Antisera

New Zealand white rabbits were inoculated subcutaneously with freeze-thaw disrupted c5T or 3T12T cells. The inoculations were at multiple sites on a bimonthly schedule over an 18-month period. The rabbits were bled by venous puncture and sera prepared in standard fashion. The sera were heat inactivated at 56°C for 30 min and stored as 0.5 ml aliquots at −70°C.

Indirect Immunofluorescence Assay

Surface specificities of rabbit antisera were determined by the indirect immunofluorescence assay as modified from Baldwin and Barker (28). Basically, monolayers of fibroblasts were rinsed and harvested by incubation for 10 min with 10^{-2} M EDTA/ solution. A (NaCl, 8 g/liter; KCl, 0.2 g/liter; Na_2HPO_4, 1.12 g/liter; KH_2PO_4, 0.2 g/liter; pH 7.4). The harvested cells were washed 3 times in Dulbecco's modified minimal essential medium (DMM) by centrifugation (300 × g, 5 min) and the final pellet readjusted to 10^6 cells/ml. The 10^6 cells were pelleted in 1.5-ml plastic conical centrifuge tubes and then resuspended in 50 μl of the appropriate heat inactivated antiserum for a 30-min incubation at 37°C. These cells were then washed 3 times with 0.2 ml DMM and resuspended in 50 μl of fluorescein isothiocyanate conjugated goat antirabbit globulin (Grand Island Biological Company) which had previously been absorbed against the target cell lines and diluted 1:3 with DMM to remove nonspecific cell surface fluorescence. Incubation was for 30 min at 37°C after which the cells were washed twice with DMM and finally resuspended in a glycerol: Solution A (50:50 vol/vol) solution. Cells were examined on a Nikon fluorescence microscope using the appropriate filters and the dark field condenser.

Dye Exclusion Cytotoxicity

This technique was performed precisely as outlined by Hilgers et al. (27) using a noncytotoxic complement source from New Zealand white rabbits.

Radioactive Labeling

Cell surface proteins were labeled by the lactoperoxidase-^{125}I method as described (16). Approximately 10^6 cells were plated on 60-mm culture dishes and incubated for 24 h in normal growth medium. The monolayers were then washed 3 times with phosphate buffered saline (PBS; 0.1 M NaCl, 2.7 mM KCl, 15.3 mM Na_2HPO_4, 1.5 mM KH_2PO_4, 0.9 mM $CaCl_2$, 0.5 mM $MgCl_2$, pH 7.2), and labeled for 10 min at room temperature in 0.5 ml/plate of PBS containing 5 mM glucose, 20 μg/ml lactoperoxidase (Calbiochem, La Jolla, California), 0.1 U/ml glucose oxidase (Calbiochem), and 250 μCi/ml $Na^{125}I$ (17.0

Ci/mg). The reaction was stopped by adding PBS containing 1.47 mg/ml cold NaI and the monolayers were washed 3 times with the same solution. The cells were then scraped from the plates into PBS containing 2 mM phenylmethylsulfonyl fluoride and pelleted by centrifugation at 2,300 rpm. The labeled pellets were stored at $-70°$. In some experiments the cells were subjected to mild trypsin treatment immediately after labeling. Such cells were washed as above and incubated for 10 min at room temperature with 10 μg/ml trypsin in a buffer consisting of 0.046 M Tris, 0.015 M $CaCl_2$, 0.088 M NaCl, pH 8.1. Control plates were incubated in the buffer alone. Control and trypsinized cells were then harvested and stored as above. Labeling of glycoproteins or glycolipids with [^{14}C]- or [^{3}H]glucosamine was performed by incubating cell monolayers for 24–48 h in DMM containing 10% fetal calf serum and either 5 μCi/ml [^{3}H]glucosamine (10.74 Ci/mmole) or 2 μCi/ml [^{14}C]-glucosamine (237.7 mCi/mmole). The labeled cells were washed 3 times with Solution A (calcium and magnesium free PBS), and removed from the flask with one round of freeze thawing in Solution A containing 2 mM phenylmethylsulfonyl fluoride. The cells were then homogenized (Dounce), centrifuged at 1,300 rpm for 5 min to remove nuclei and large pieces, and the crude mixture of labeled membrane fragments remaining in suspension was pelleted by centrifugation at 100,000 $\times$ g for 90 min. The pellet was collected and stored at $-70°C$.

SDS Polyacrylamide Gel Electrophoresis

Labeled membrane samples were dissolved in a solution containing 20% sucrose, 1% sodium dodecyl sulfate (SDS), 0.05% dithiothreitol, 0.054 M Tris, pH 6.1. The samples were boiled in the above solution and applied to either slab or tube gels. The complete electrophoresis system was as described by Neville (29) and consisted of the following: upper reservoir buffer: 0.04 M boric acid, 0.041 M Tris, 0.1% SDS, pH 8.64; lower reservoir buffer: 0.031 N HCl, 0.424 M Tris, pH 9.19; stacking gel: 3% acrylamide in 0.027 M H_2SO_4, 0.054 M Tris, 5% glycerol; resolving gel: 7.5% acrylamide in lower reservoir buffer containing 10% glycerol. The ratio of acrylamide to N,N'-methylene-bisacrylamide was 25:1 in all gels. Stacking and resolving gels were polymerized by the addition of 0.08% ammonium persulfate and 0.05% TEMED. Slab gels were dried and radioactivity was detected by autoradiography on Kodak RP/R-14 film. Tube gels were sliced into 1-mm sections and glycoproteins eluted by shaking in 1% SDS overnight. Radioactivity was determined in a Beckman liquid scintillation counter in a mixture containing 2,900 ml xylene, 960 ml Triton X-114, and 125 ml Liquiflor (New England Nuclear Corporation).

Pronase Digestion

Labeled membrane preparations or SDS gel sections containing labeled glycoproteins were incubated in a solution of 0.1 M Tris, 20 mM $CaCl_2$, pH 7.8, and pronase (Calbiochem) for 5–8 days at $37°C$. Routinely, 100 μg pronase was included on day 0 and 50 μg added every other day. A few drops of toluene were added to inhibit bacterial growth.

Gel Chromatography

Pronase glycopeptides were analyzed on a Sephadex G-50 column (1.5 $\times$ 90 cm). Samples were applied and eluted with 0.05 M potassium phosphate, 0.02% sodium azide, pH 7.2. Two-milliliter fractions were collected, and radioactivity was determined in a Beckman liquid scintillation counter.

RESULTS

The biological properties of the cell lines are summarized in the data of Table I. A_{31} is a clone of BALB/c 3T3 cells, and has normal properties including a low saturation density, the inability to grow in soft agar, lack of agglutinability by Concanavalin A, and failure to cause tumors when injected into either BALB/c or athymic nude mice. PBC are low passage BALB/c embryo fibroblasts. 3T12T cells are prepared from tumors caused by injecting 3T12 cells into BALB/c mice as described (21, 22). BALB/c mice receiving injections of from 10^3 to 5×10^5 3T12T cells will be killed by the tumor within 30 days. MSC cells are derived from a tumor caused by inoculating BALB/c mice with the Moloney pseudotype of murine sarcoma virus. MSC cells have been shown to produce virus, and can cause tumors which grow progressively or regress depending on several parameters, including the initial cell dose injected into animals (30). c5 is a spontaneously transformed clone derived from A_{31} cells and causes tumors which regress in immunocompetent BALB/c mice. Injecting c5 into nude mice leads to the development of progressively growing tumors that kill the mice within 60 days. c5T is a transformed cell line derived from a tumor caused by injecting c5 cells into BALB/c mice (prior to regression). Mice receiving 5×10^6 c5T cells will be killed by their tumor within 60 days. The 3T12T line can be classified as nonimmunogenic, since injecting irradiated tumor cells or tumor excision does not protect mice from subsequent challenge by 3T12T cells. MSC and c5T are immunogenic since the same procedures protect animals from rechallenge. Interestingly, mice which have had c5 tumors that regress are protected from subsequent rechallenge by c5T cells, indicating shared antigens between c5 and c5T cells. The immunogenicity of c5, c5T, and MSC cells is paralleled by their stimulation of antibody production in tumor bearing mice.

Immunological Studies

We studied lymphocyte cytotoxicity as a possible assay for specific cell surface alterations on tumor cells. Lymphocytes prepared from spleens and lymph nodes of mice injected 10 to 20 days previously with tumor cells were cytotoxic for cultured cells. Cytotoxic lymphocytes could be detected in mice bearing tumors caused by inoculating any of the transformed cell lines. However, while we could demonstrate killing, the lymphocytes from tumor bearers did not show any specificity (Table II). Animals bearing tumors caused by either c5T or 3T12T cells had lymphocytes capable of killing A_{31}, c5T, or 3T12T cells. In other experiments (data not shown), similar preparations of lymphocytes were cytotoxic for PBC and even Swiss 3T3 cells. Similar results were found in 10 separate preparations of lymphocytes from tumor bearing animals. Varying the ratios of lymphocytes to target cells did not change the observed pattern of a lack of specificity for killing by lymphocytes.

Kall and Hellstrom (31) recently published experiments describing methods for sensitizing lymphocytes from normal BALB/c mice in vitro. Following their procedures, we incubated lymphocytes from BALB/c mice on a monolayer of mitomycin C treated A_{31}, 3T12T, or MSC cells. The results from these experiments are shown in the data of Table III. Lymphocytes cocultivated with A_{31} cells were not activated for killing either A_{31} or any of the transformed lines. In contrast, activation of lymphocytes against 3T12T cells resulted in the production of lymphocytes which were cytotoxic for 3T12T cells and the other normal and transformed lines. Activation against virus producing MSC cells resulted

TABLE I. Biological Properties of BALB/c Cell Lines

Cell line	Growth in methocel	Con A agg. (500 μg/ml)	Tumorigenicity in nude mice	Tumorigenicity in BALB/c	BALB/c tumor bearer fate	Immunogenicity (protection from re-challenge after immunization)	Antibody formation in tumor bearers
PBC	−	−	−	−			
A31	−	−	−	−			
MSC	+	+	+	+	Regressed	+	+
c5	+	+	+	+	Regressed	+	+
c5T	+	+	+	+	Killed	+	+
3T12T	+	+	+	+	Killed	−	−

Procedures for determining the ability of cell lines to grow in methocel and the technique for measuring Concanavalin A agglutinability have been described (22). Tumorigenicity was determined by the subcutaneous injection of up to 5×10^6 cells into 3-week-old BALB/c mice or athymic nude mice. "Regressed" means that a vascularized tumor was produced which disappeared over a 1-month period. Such tumors failed to return in the next 6 months to 1 year. "Killed" means that the mice were killed by the tumor, usually within 1–2 months following the injection of 5×10^3 to 5×10^6 cells. Mice were immunized either by 4 weekly injections of 5×10^6 irradiated (5,000 rads) tumor cells or by excising tumors which had reached 1–2 cm in diameter. "Antibody" circulating in tumor bearer serum was determined as described in Materials and Methods.

TABLE II. Specificity of Cell Mediated Cytotoxicity by Lymphocytes From Mice With Tumors Caused by c5T and 3T12T

Lymphocyte source	Target cells	CPM remaining	% Cytotoxicity[a]	
None	A_{31}	21,695 ± 1,071		
Control	A_{31}	11,037 ± 1,396	—	
c5T bearer	A_{31}	7,499 ± 689	32	< 0.01[b]
3T12T bearer	A_{31}	5,840 ± 473	42	< 0.01
None	c5T	9,898 ± 779		
Control	c5T	4,727 ± 527	—	
c5T bearer	c5T	3,134 ± 481	34	< 0.01
3T12T bearer	c5T	2,321 ± 270	51	< 0.01
None	3T12T	54,244 ± 4,709		
Control	3T12T	31,846 ± 2,415	—	
c5T bearer	3T12T	18,702 ± 1,818	41	< 0.01
3T12T bearer	3T12T	16,902 ± 1,835	47	< 0.01

[a] $\dfrac{\text{CPM test lymphocytes}}{(1 - \text{CPM control lymphocytes})} \times 100$

[b] Calculated by the Student T test

Lymphocyte preparation and cell mediated killing determined by the [^{3}H] uridine assay were performed as described in Materials and Methods. Lymphocytes were prepared from tumor bearing animals while the tumors were from 1 to 2 cm in diameter (usually days 18–26).

TABLE III. In Vitro Lymphocyte Sensitization and Cytotoxicity

Sensitizing monolayer	Target[a] cell	No. surviving cells	% Cytotoxicity	
A_{31}	A_{31}	48 ± 11	—	
3T12T	A_{31}	18 ± 6	66	< 0.01
MSC	A_{31}	45 ± 5	15	N.S.
A_{31}	3T12T	83 ± 4	—	
3T12T	3T12T	8 ± 1	90	< 0.01
MSC	3T12T	72 ± 12	13	N.S.
A_{31}	MSC	24 ± 5	—	
3T12T	MSC	14 ± 2	42	< 0.01
MSC	MSC	13 ± 2	46	< 0.01

[a] 1×10^4 lymphocytes added per target well

In vitro activation. Cells for lymphocyte activation were incubated at 3×10^6 cells/ml in Dulbecco's modified minimal essential media (DMM) containing 20 μg of mitomycin C/ml for 30 min at 37°C. The cells were rinsed 2 times in DMM and 1×10^6 cells added to a 60-mm plate (20 cm² growing area). Lymphocytes were produced from spleens as described in Materials and Methods, and suspended at a concentration of $1-3 \times 10^7$ cells/5 ml in sensitizing medium RPMI 1640, with 15% FCS as described (31). Five milliliters of cell suspension was added and the plates incubated for 6 days at 37°C. The activated lymphocytes were washed in DMM containing 10% fetal calf serum and cell mediated cytotoxicity determined.

Using in vitro activated lymphocytes, 1×10^4 cells were added to each well containing target cells as described in Materials and Methods. The number of surviving cells was counted for sight wells, and the average is shown ± the standard deviation. Lymphocytes activated by culture on mitomycin C treated A_{31} cells did not show any cytotoxicity relative to experiments where either no lymphocytes or control lymphocytes were added to target cells. The number of surviving cells from experiments where mitomycin C treated A_{31} cells are incubated with lymphocytes are thus used as a control in calculations of lymphocyte cytotoxicity.

in lymphocytes with greatest cytotoxicity for MSC cells. The process of activation appears to be able to detect some basic difference between A_{31} and 3T12T cells, but once activated the lymphocytes show no specificity for the target cells they will kill.

On numerous occasions, we tested for the ability of sera from tumor bearers to block the killing of target cells by either in vitro activated or tumor bearer lymphocytes. Results from these experiments were uniformally negative. c5T or 3T12T bearer sera added to the lymphocyte killing assay did not affect target cell survival.

In another immunological approach to studies of cell surface change, we inoculated New Zealand white rabbits with freeze-thaw disrupted 3T12T or c5T cells. After absorption against A_{31}, sera from rabbits injected with 3T12T could be shown to react with transformed lines (c5, c5T, or 3T12T) only. Using antisera from rabbits injected with c5T cells, we were able to prepare antisera specific for the immunogenic (c5 and c5T) transformed cells (Table IV). We have subsequently shown that the antisera specifically precipitates a unique surface glycoprotein found only on the surfaces of c5 and c5T cells.

Chemical Studies

Glycolipids labeled in cells incubated with [^{3}H]glucosamine were compared by thin layer chromatography. The percent of label in the various glycolipid species are summarized in the data of Fig. 1. Growing or confluent A_{31} and the malignant nonimmunogenic 3T12T cells are very similar, with the majority of the label comigrating with G_{M2} and G_{D1a}. Very little globoside or G_{M3} is detected in extracts from either 3T12T or A_{31} cells. The immunogenic lines show glycolipid compositions which are altered relative to A_{31} or 3T12T cells. MSC cells show increased G_{M3} and globoside and decreased G_{M2} and G_{D1a}. c5 and c5T are very similar and show increased G_{m3} and decreased G_{M2} compared with A_{31}. The results from glucosamine labeling are similar to earlier studies where ganglioside compositions were determined by measuring sialic acid in various glycolipid species in normal and transformed BALB/c cells (32). The surface labeling patterns of proteins on A_{31}, PBC, and 3T12T cells labeled by lactoperoxidase-^{125}I were indistinguishable when preparations were resolved by SDS gel electrophoresis. Figure 2 shows ^{125}I profiles for the 3 spontaneously transformed cell lines before and after mild trypsinization (10 μg/ml for 10 min at room temperature). c5 shows nearly identical patterns before and after trypsin treatment, as does c5T. Virtually no labeling is seen in the LETS region, while a very heavily labeled band of about 105,000 mol wt appears in both these cell lines. 3T12T is strikingly different from the other 2 transformed lines, with the heaviest label appearing in a band of 240,000 mol wt, analagous to the LETS glycoprotein. This band is sensitive to mild trypsinization, as has been reported for the LETS glycoprotein. 3T12T also shows much greater labeling of several bands of about 165,000 (which are not trypsin sensitive) and about 135,000 mol wt (which are trypsin sensitive) than do c5 and c5T. There is no labeling of a 105,000 mol wt band in 3T12T.

Membrane Glycoproteins

Crude membrane glycoproteins were labeled with [^{3}H]glucosamine and prepared as described in Materials and Methods. The labeled preparations were analyzed by SDS-polyacrylamide gel electrophoresis with the results shown on Fig. 3. Three labeled regions were found in all the cell lines tested: region I, about 220,000–250,000; region II, about 105,000–165,000; and region III, appearing about 70,000 mol wt. There were virtually no differences in region I patterns between any of the cell lines. Region II labeling patterns

TABLE IV. Indirect Immunofluorescence Using Anti-c5T Rabbit Antisera

Serum type	Target cell line				
	A_{31}	3T12T	c5T	c5	MSC
control serum	−	−	−	−	−
immune serum	+	+	+	+	+
A_{31} absorbed immune serum	−	−	+	+	−

Antiserum was raised in New Zealand white (NZW) rabbits against the c5T cell line as outlined in Materials and Methods. 0.5 ml of this crude antiserum (termed "immune serum") was absorbed at 4°C for 8 h against the 80,000 × g pellet (crude membrane fraction) of 5×10^7 homogenized A_{31} cells. The crude membranes were then pelleted and the absorption repeated for a total of 3 times. The resultant antiserum was termed "A_{31} absorbed immune serum." "Control serum" was obtained from nonimmunized NZW rabbits. The cell surface specificities of the above sera were determined by the indirect immunofluorescence assay as outlined in Materials and Methods. Negative surface fluorescence ("−") means that none of the cells showed positive surface fluorescence. Positive surface fluorescence ("+" means that 95–100% of the cells demonstrated ring or cap fluorescence.

vary between the cell lines tested. A_{31} and 3T12T cells appear to possess at least 2 glycoproteins in this region, IIa and IIb. c5, c5T, and MSC show these same glycoproteins as well as an additional glycoprotein, IIc, of about 105,000 mol wt. Region III again appears similar among all the lines. This consists of a broad peak that actually separates into a doublet on some gels. MSC also shows labeling of a group of bands between regions II and III. In comparing each region it was found that A_{31} always shows relatively lower labeling of region II and III than do the transformed lines.

Glycopeptides

We next compared glycopeptides from these cell lines. [^{14}C] and/or [^{3}H] glucosamine labeled membrane fragments were digested exhaustively with pronase and the resulting glycopeptides analyzed using Sephadex G-50 chromatography. Figure 4 shows the cochromatogram of [^{3}H] glucosamine labeled A_{31} and [^{14}C] glucosamine labeled 3T12T membranes after pronase digestion. It can be seen that 3T12T is enriched in higher molecular weight glycopeptide species compared to A_{31}, confirming the earlier work of Meezan et al. and Buck et al. (18, 19). If the A_{31}-3T12T glycopeptide mixture was treated with neuraminidase prior to Sephadex G-50 chromatography, the elution patterns of glycopeptides from the 2 cell lines appeared very similar. 3T12T glycopeptides also showed an enrichment in faster eluting species when compared to PBC glycopeptides. Similar results were obtained when c5T derived glycopeptides were compared with A_{31} cells. There were no differences detected between glycopeptides of growing and confluent A_{31} cells.

The individual glycoprotein regions were separated from [^{3}H] glucosamine labeled A_{31} and [^{14}C] glucosamine labeled 3T12T membranes by electrophoresis in a large (1.5 × 15 cm) tube gel. Portions of the gel corresponding to regions I, II, and III were separated, digested exhaustively with pronase, and analyzed on Sephadex G-50. The results are presented on Fig. 5. In both cell lines, the glycopeptides from region I appear to be smallest in size, those from region II are of intermediate size, and those of region III are largest. 3T12T region I and II glycopeptides appear larger than A_{31} region I and II glycopeptides,

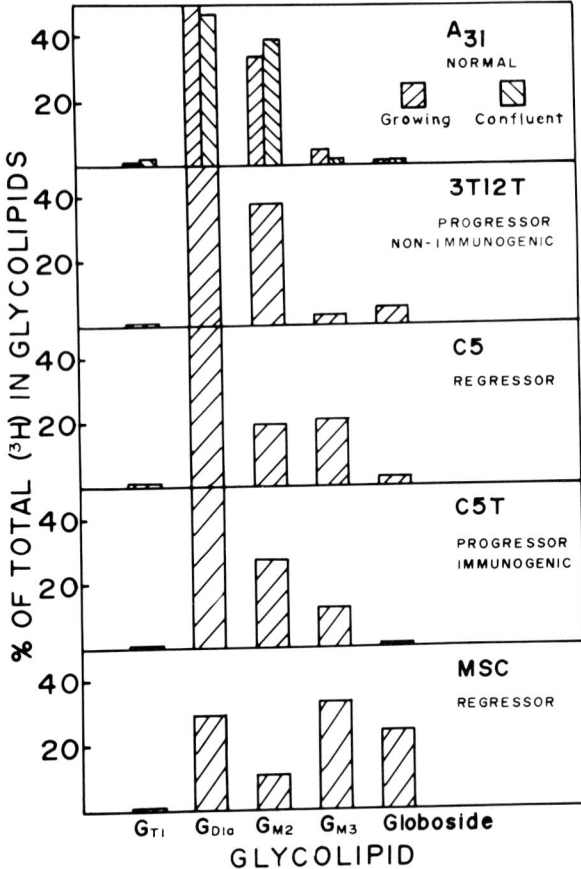

Fig. 1. Cells were labeled and membranes prepared as described in Materials and Methods. Labeled membrane fragments were extracted 3 times with $ChCl_3$:MeOH (2:1) and the extracts were pooled and dried. Radioactive samples as well as authentic glycolipid standards were applied to silica gel G-25 20 × 20 cm × 0.25 mm plates (Brinkman Instruments, Westbury, New York) and developed with chloroform:methanol:0.25% $CaCl_2$ (60:35:8). The plates were then dried and standard glycolipids were detected by reaction with iodine vapor. Labeled glycolipids were detected by cutting the developed TLC plate into 0.5-cm strips which were then placed into scintillation vials for radioactivity determinations.

respectively. Region III glycopeptides from both cell lines have similar elution profiles. There is a multiple cause for the enrichment of large glycopeptides in pronase digests of crude 3T12T membrane preparations. 3T12T has several glycoproteins (regions I and II) that produce larger glycopeptides than do the corresponding glycoproteins from A_{31}, and 3T12T shows consistently more glucosamine incorporation into glycoproteins that contain the largest glycopeptides from both cell lines (region III). If the separated region I, II, and III glycopeptides are treated with neuraminidase prior to Sephadex G-50 chromatography, the migration of all 3 regions is shifted to a smaller molecular weight. The size order of region III > region II > region I remains, however, after neuraminidase treatment. In addition, 3T12T region I and II glycopeptides remain larger than A_{31} regions I and II glycopeptides after neuraminidase treatment.

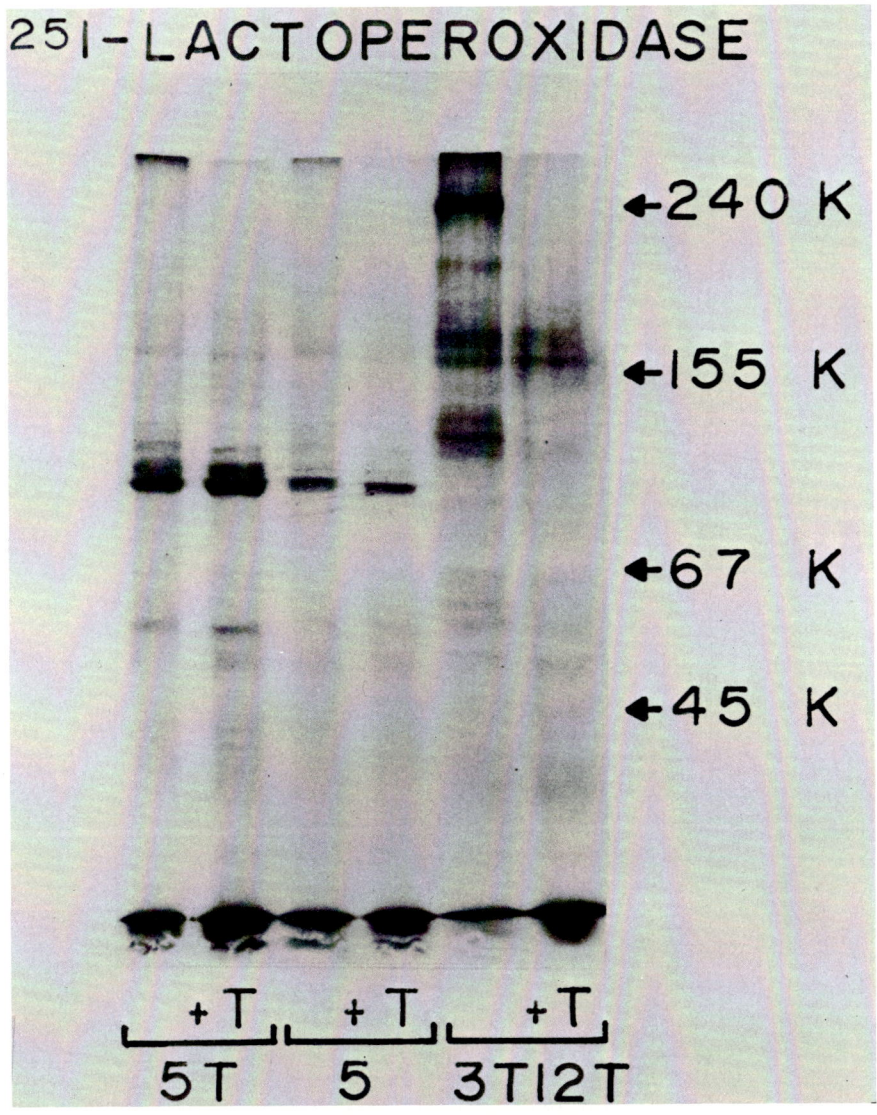

Fig. 2. SDS-polyacrylamide gel electrophoresis of ^{125}I labeled membranes from c5T, c5, and 3T12T. Approximately 50,000 cpm of each labeled cell material was applied to a 7.5% slab gel and electrophoresed at 150 V for approximately 2 h. The gel was dried and exposed on x-ray film. Molecular weight markers used include: spectrin (large band), 240,000; goat IgG, 155,000; bovine serum albumin, 67,000; hen egg albumin, 45,000. "+T" means that the ^{125}I labeled monolayer was treated with 10 μg/ml trypsin for 10 min at room temperature prior to harvest.

DISCUSSION

We want to emphasize the necessity of studying transformed cells which have well-defined tumorigenic and malignant properties. Spontaneous transformants also appear to be an important addition to the more usual experiments which included only viral or

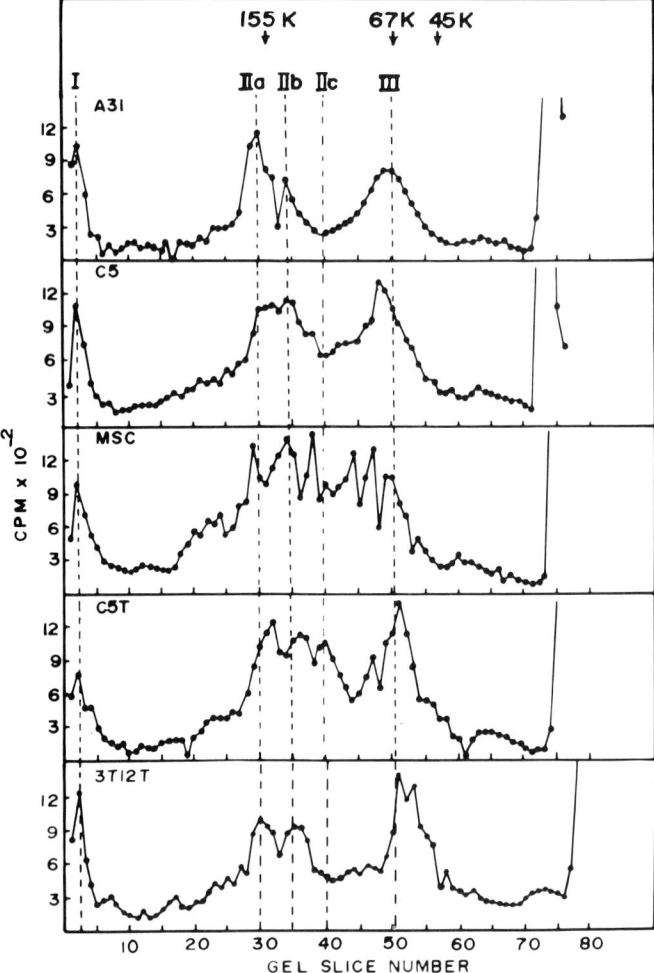

Fig. 3. SDS-polyacrylamide gel electrophoresis of [^{3}H]glucosamine labeled glycoproteins. 60,000 cpm of each labeled membrane preparation was applied to 7.5% gels in 6.0 × 120 mm glass tubes. Gels were electrophoresed at 3 mA per tube for approximately 3 h. Gels were rimmed from the glass tubes, frozen, and cut into 1-mm slices. Slices were shaken overnight in 2 ml 1% SDS to elute glycoproteins and the eluted material was counted in a Beckman liquid scintillation counter. Molecular weight markers used include: goat IgG, 155,000; bovine serum albumin, 67,000; hen egg albumin, 45,000.

carcinogen transformed cells. For example, numerous studies have shown that virally transformed cell lines have lost the ability to synthesize some of the complex glycolipids (2). In contrast, malignant spontaneous transformants frequently do not have these alterations (33). That carcinogens lead to numerous secondary changes is also not a new concept. Carcinogen transformed cells are known to be antigenically distinct, while spontaneous transformants are frequently nonimmunogenic (34–36).

The immunological response by tumor bearers offers an opportunity to detect and therefore assay for unique tumor cell antigens. The Hellstroms and others studying a number of animal model systems have developed the concept that tumor bearers make an

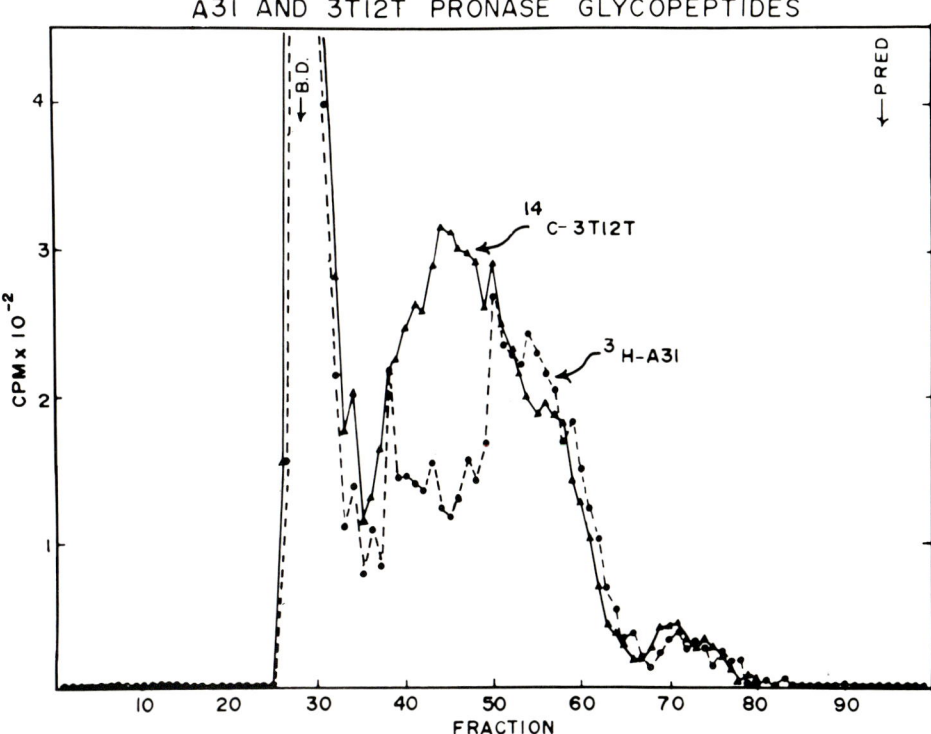

Fig. 4. Sephadex G-50 chromatography of A_{31} (●) and 3T12T (▲) pronase glycopeptides. [^{3}H] glucosamine labeled A_{31} and [^{14}C] glucosamine labeled 3T12T membrane preparations containing equal amounts of radioactivity were mixed and digested for 5 days with pronase. An aliquot of the digested material was then applied to a Sephadex G-50 column (90 × 1.5 cm). B.D.) blue dextran 2000; p. red) phenol red.

immunological response, but that progression or regression of the tumor depends on whether antigens derived from tumor cells can block lymphocyte mediated cytotoxicity (4, 10). These studies have generally used either viral or carcinogen transformed cells. The situation with human cancers is not clear. Hellstrom originally reported specific cell mediated immunity to human neoplasms of various histological types (36, 37). However, the specificity of lymphocytes from humans for their respective tumors has been questioned (38). Recently, a number of investigators have discussed problems in using microcytotoxicity assays for studies of human neoplasia (39–41).

Results from our studies of lymphocyte mediated cytotoxicity in mice bearing spontaneously transformed immunogenic and nonimmunogenic tumor cells support the concept that an immunological response can be made against progressively growing tumors. However, we were unable to demonstrate the presence of serum blocking factors. The lack of specificity for the cell precluded the use of cell mediated immunity to detect unique tumor antigens on the progressor spontaneous transformants. It is interesting to note that an immunological response can be measured even in animals bearing nonimmunogenic tumor cells. The use of antibodies both from tumor bearers and immunized mice along with specific antisera prepared from rabbits offers the best opportunity for an immunological assay capable of detecting cell surface change. Previously, rabbit antisera directed

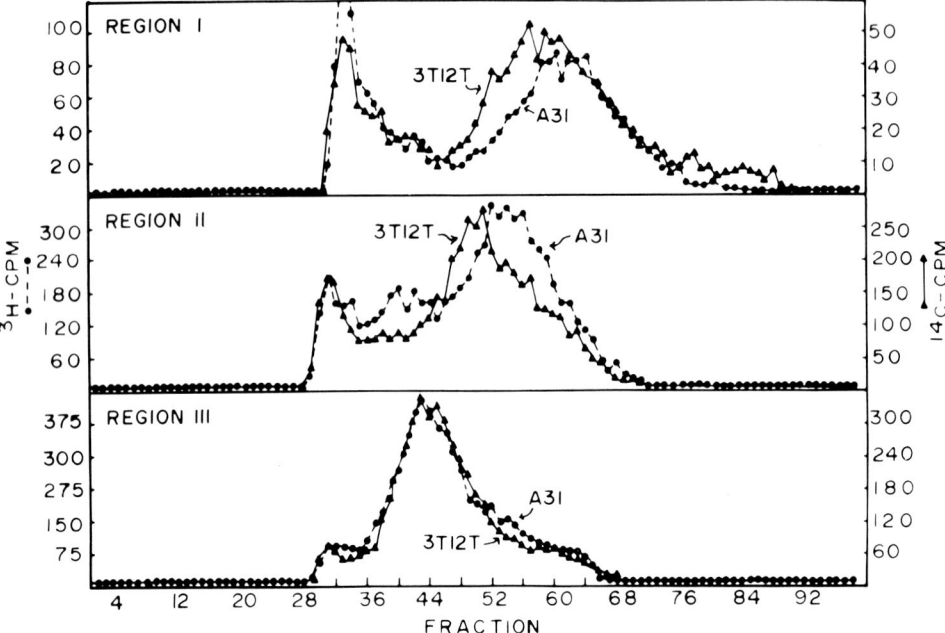

Fig. 5. Sephadex G-50 chromatography of A_{31} (●) and 3T12T (▲) pronase glycopeptides of isolated glycoprotein regions. [^{3}H]glucosamine labeled A_{31} and [^{14}C]glucosamine labeled 3T12T membrane preparations were mixed and coelectrophoresed on a 7.5% tube gel (1.5 × 15 cm). Part of the gel was sliced and counted and portions of the remaining gel corresponding to glycoprotein regions I, II, and III (see text) were excised and incubated for 8 days with pronase. The digested material from each region was then eluted from the gels, lyophilized, and applied to a Sephadex G-50 column (90 × 1.5 cm).

against human leukemic cells were used successfully to isolate a unique tumor cell antigen (42). The rabbit antiserum which reacts specifically against c5 and c5T cells is of special interest. We have now shown that the unique glycoprotein migrating in the IIc region is specifically precipitated by the alloantisera. This provides a one step purification for the unique antigen shared by c5 and c5T.

The chemical studies revealed both interesting similarities and differences between the normal and transformed cells. The immunogenic lines showed the simplified glycolipid patterns characteristic of virally transformed cells. In contrast, the most malignant line 3T12T had glycolipid compositions similar to normal cells. It is important to note that our studies only compared glucosamine labeled glycolipids. These methods would not have detected differences in fucolipids or slow migrating glycolipids (43). Similarly, the immunogenic lines c5 and c5T lacked the 240,000 mol wt ^{125}I labeled surface glycoprotein. The nonimmunogenic 3T12T progressor has levels of the LETS protein which are similar to normal cells. The glycoprotein patterns of A_{31} and 3T12T as detected by SDS-polyacrylamide gel electrophoresis were also very similar, with no new major bands appearing. Thus many of the typical transformation specific changes detected in membrane proteins, glycoproteins, and glycolipids of virally transformed cells appear to be secondary in nature and not necessary for expression of the transformed phenotype. It is interesting that the lack of substantial surface alterations in 3T12T parallels the lack of immunogenicity. It appears that a tumor cell that has very similar surface features to normal cells can be more successful in overcoming defenses of the host.

The only consistent difference detected between 3T12T and A_{31} cells was the presence of glycopeptides of higher molecular weight in the transformed cells. These results confirm earlier work using viral transformants (18, 19). Several isolated glycoprotein regions have glycopeptides of different molecular weight. The smaller glycoproteins tend to have higher molecular weight glycopeptides. These differences are retained even after neuraminidase treatment. These results confirm those of Sakiyama and Burge (44) indicating that a number of different kinds of carbohydrate structures are likely to be found on cellular glycoproteins. Glycopeptides prepared from regions I and II of 3T12T glycoproteins are both of higher molecular weight than the corresponding regions from A_{31} cells. It is of interest that the 240,000 mol wt glycoprotein while present in both the normal A_{31} and the spontaneously transformed 3T12T yields glycopeptides which are characteristically altered in the transformed cells. The presumed alteration is therefore occurring on more than one kind of carbohydrate structure. These differences remain after neuraminidase treatment, showing that the change probably involves carbohydrates other than sialic acid.

Further interpretation requires that we compile details of structures of complex polysaccharides of the normal and transformed cells. It is interesting in our system that lack of detectable change in glycolipids and I^{125}-labeled surface proteins is found in the nonimmunogenic progressor. However, 3T12T cells do show alterations in glycopeptides. These results are consistant with recent studies by Smets and co-workers where glycopeptide size profiles were also found to be correlated with malignancy (45).

ACKNOWLEDGMENTS

This work was supported by ACS Grant BC131 and NCI Grant CA12753.

REFERENCES

1. Hakomori S: Biochim Biophys Acta 417:55, 1975.
2. Nicholson G: Biochim Biophys Acta 458:1, 1976.
3. Talmadge K, Burger M: In Whelan W (ed): "Biochemistry of Carbohydrates." Baltimore: University Park Press, 1975, p 43.
4. Hellstrom K, Hellstrom I: Adv Immunol 18:209, 1974.
5. Currie G: Biochim Biophys Acta 458:135, 1975.
6. Coggin J, Ambrose K, Dierlam P, Anderson N: Cancer Res 34:2092, 1974.
7. Alexander P: Cancer Res 34:2077, 1974.
8. Currie G, Alexander P: Br J Cancer 29:72, 1974.
9. Epstein L, Knight R: Br J Cancer 31:499, 1975.
10. Baldwin R, Price M, Robins R: Nature (London) New Biol 238:185, 1972.
11. Hellstrom I, Hellstrom K: Int J Cancer 4:587, 1969.
12. Grimes W: Biochemistry 9:5083, 1970.
13. Grimes W, Greegor S: Cancer Res 36:3905, 1976.
14. Hynes R: Proc Natl Acad Sci USA 70:3180, 1973.
15. Hogg N: Proc Natl Acad Sci USA 71:489, 1974.
16. Hynes R, Humphreys K: J Cell Biol 62:438, 1974.
17. Gahmberg C, Hakomori S: Proc Natl Acad Sci USA 70:3329, 1973.
18. Meezan E, Wu H, Black P, Robbins P: Biochemistry 8:2518, 1969.
19. Buck C, Glick M, Warren L: Biochemistry 9:4567, 1970.
20. Ogata S, Muramatsu T, Kobata A: Nature (London) 259:580, 1976.
21. Grimes WJ: In Baserga B, Clarkson B (eds): "Control of Proliferation in Animal Cells." New York: Cold Spring Harbor Laboratory, 1974, p 517.
22. Van Nest G, Grimes W: Cancer Res 34:1408, 1974.

23. Levine EM: Exp Cell Res 74:99, 1972.
24. Peden K: Experimentia 31:1111, 1975.
25. Hellstrom I, Hellstrom K: In Bloom B, Glade P (eds): "In Vitro Methods in Cell Mediated Immunity." New York: Academic Press.
26. McKhann C, Cleveland P, Busk M: Natl Cancer Inst Monogr 37:37, 1973.
27. Hilgers J, Haverman J, Nusse R, van Blitterswijk W, Cleton F, Hageman P, Van Nie R, Calafat J: J Natl Cancer Inst 54:1323, 1975.
28. Baldwin R, Barker C: Br J Cancer 21:793, 1967.
29. Neville P: J Biol Chem 246:6238, 1971.
30. Russell S, Cochrane C: Int J Cancer 13:54, 1974.
31. Kall M, Hellstrom I: J Immunol 114:1083, 1975.
32. Fishman P, Brady R, Aaronson S: Biochemistry 15:201, 1976.
33. Sakiyama H, Robbins P: Fed Proc Fed Am Soc Exp Biol 32:86, 1973.
34. Embleton M, Heidelberger C: Int J Cancer 9:8, 1972.
35. Basombris M, Phren R: Int J Cancer 10:1, 1972.
36. Baldwin R, Embleton M: Int J Cancer 13:433, 1974.
37. Hellstrom I, Hellstrom K, Sjogren H, Warner G: Int J Cancer 7:1, 1971.
38. Takasugi M, Mickey M, Tarasaki P: J Natl Cancer Inst 53:1527, 1974.
39. Herberman R, Oldham R: J Natl Cancer Inst 55:749, 1975.
40. Heppner G, Henry E, Stolback L, Cummings F, McDonough E, Calabrisi P: Cancer Res 35:1931, 1975.
41. Baldwin R: J Natl Cancer Inst 55:745, 1975.
42. Billing R, Terasaki P: J Natl Cancer Inst 53:1635, 1974.
43. Itaya K, Hakomori S, Klein A: Proc Natl Acad Sci USA 73:1568, 1976.
44. Sakiyama H, Burge B: Biochemistry 11:366, 1972.
45. Smets L, van Beek W, van Rosig H: Int J Cancer 18:462, 1976.

Microbial Carbohydrate Specific Antibodies Distinguish Between Different Stages of Differentiating Mouse Cerebellum

Ekkhart Trenkner

Department of Neuroscience, Children's Hospital Medical Center, Boston, Massachusetts 02115; and Department of Neuropathology, Harvard Medical School, Boston, Massachusetts 02115

Siddhartha Sarkar

Division of Medical Genetics, Department of Medicine M-013, School of Medicine, University of California, San Diego, La Jolla, California 92093

High titered anticarbohydrate antibodies were used to identify cell surface carbohydrates during different stages in histogenesis of mouse cerebellum in a micro tissue-culture system which mimics selected features of in vivo cerebellum development. Blockage of fiber formation within the first few days in vitro and inhibition of cell migrations by carbohydrate-specific antibodies served as an assay system for possible contributions of surface carbohydrates to the behavior of developing cerebellar cells. Microbial strains were selected on the basis of carbohydrate structures of their cell wall antigens, and anticarbohydrate antibodies were raised against treated whole bacteria and yeast in rabbits. We found that antibodies to mannan were active at all stages of development tested (embryonic day 13, E13; the day of birth, P0; and postnatal day 7, P7). Antibodies to sialic acids prepared against strains B and C of Neisseria meningitidis distinguish different subterminal structures: anti-B reacted with E13 and P0 cerebellar cells, and anti-C mostly with cells older than P7. Anti-fetuin antibody recognized E13 and P0 but not P7 cell populations. Pneumococcus C strain R36A-specific antibodies were effective only after coating cells to C type carbohydrate before application of the antibody. The results demonstrate that antimicrobial carbohydrate antibodies cross-react with mammalian cell surface carbohydrate structures and therefore can be used as a powerful tool in tissue culture to analyze those structures which might control cell behaviors pertinent to cerebellar development.

Key words: mouse cerebellum, development, surface carbohydrates, antibodies

The recognition sites of external signals acting on cell surfaces or cell-cell interactions are frequently found to be associated with membrane carbohydrates (1–6). The characterization of mammalian cell surface carbohydrates responsible for cell recognition is therefore important in order to understand fundamental features in development, such as cell migration of cells into a tissue structure.

Received March 31, 1977; accepted May 23, 1977.

© 1977 Alan R. Liss, Inc., 150 Fifth Avenue, New York, NY 10011

Antibodies are unique tools to detect specific determinants present in a mixture with many other structurally unrelated molecules (7) provided that a pure immunogen is used for immunization. Therefore, in order to identify cell surface carbohydrates with immunological techniques hapten-specific antibodies are required. Relatively low titered carbohydrate-specific antibodies can be obtained either by immunizing rabbits with carbohydrate determinants coupled to protein carrier molecules or isolating natural antibodies from serum using affinity chromatography (8). In this study a group of microbial strains were selected on the basis of sialic acid and mannose determinants in the carbohydrate structures of their cell wall antigens to raise high titered antibody in rabbit.

Developing mouse cerebellum was chosen to assay the activity of carbohydrate-specific antibody. The developmental stages of cerebellum are well characterized (9), and it is composed of a relatively limited number of neuronal and glial cell types in a highly stereotyped geometric pattern (10, 11). In order to analyze and manipulate cell surfaces during the onset of development, a micro tissue-culture system was designed which mimics certain features of in vivo development, such as synapse and growth cone formation, migration of granule cells, and integration of large neuronal cells, interneurons, and differentiated granule cells into cell aggregates (12, 13). The alteration of these events, for example, blocking of fiber formation and inhibition of migrating cells by carbohydrate-specific antibodies, were used in this study to explore the contribution of surface carbohydrates in cerebellar differentiation.

MATERIALS AND METHODS

Preparation of Microcultures

Preparation of single cell suspensions was performed as described by Barkley et al. (14). Cerebellum from P0 or P7 C57BL/6J mice were washed 3 times in Ca^{2+} and Mg^{2+} Tyrode solution (CMF) after dissection and incubated in 1% trypsin in CMF at room temperature for 8 min or 14 min respectively. Trypsin was removed by washing the tissue 3 times in CMF. Tissue was transferred into Eagle's Glucose (EG) [Basal medium Eagle's (Gibco) supplemented with 0.25% glucose] containing 0.05% DNase and triturated with fire-polished long-tip Pasteur pipettes with decreasing pore sizes. The cell suspension was centrifuged at 600 rpm for 8 min at 4°C. The pellet was resuspended in culture medium [EG, supplemented with glutamine (2 mM), penicillin (25 units/ml), streptomycin (25 units/ml), and horse serum (10%) (Gibco)].

Tissue Culture Condition

Single cell suspensions were plated in microtiter plates (Falcon 3034) at a concentration of $5-7 \times 10^4$ cells per 10 µl medium per well and incubated at 35°C in a moist chamber containing 5% CO_2. When antibodies were applied $5-7 \times 10^4$ cells were added to microwells in 5 µl of medium before 5 µl of various antibody dilutions in culture medium were added.

In order to reverse antibody inhibited culture, antibody containing medium was replaced by culture medium after 1–2 days in vitro. Cultures were incubated for several days. After 2, 3, and 4 days the number of fibers growing out of reaggregates and the number of fibers connecting cell aggregates were determined in order to measure antibody reactivity.

TABLE I. Inhibition of Fiber Formation in Microcultures of Mouse Cerebellar Cells

Antibody	E13	P0 A	P0 B	P7 A	P7 B
no antibody	0	140	136	268	249
normal rabbit serum	0	126	135	252	249
rabbit anti-C	±	4	112	0	239
rabbit anti-B	+++[a]	0	118	262	249
rabbit antifetuin	++	0	148	205	223
rabbit anti-S. Thompson	not done	27	106	16	252
rabbit anti-A-1	not done	12	116	8	231
rabbit anti-A-5	++	23	130	0	216

[a] +++ indicates 95% inhibition of fiber formation

TABLE II. Inhibition of Fiber Formation in Microcultures of Mouse Cerebellum Before and After Coating Cells With Pneumococcus C-Carbohydrate

Antibody	Coated P0	Coated P7	Control P0	Control P7
mouse anti-R36A	23	14	106	252
Balb/C myeloma S107	36	86	115	246

Each number in Tables I and II represents the number of fibers extending freely from reaggregates or interconnecting reaggregates (12, 13) in 6 microwells after 3 days in culture. E13 cells were cultured following the procedure of Hatten and Sidman (23). P0 cells are from newborn and P7 from 7-day-old mice. Row A in Table I represents the number of fibers after cells were treated with various antibodies listed; row B gives the number of fibers grown after antibodies were removed from the medium. Antibody preabsorbed with adult liver and spleen of C57BL/6J mice is uniformly inactive.

Preparation of Antibody

Antibodies containing specificity for polysaccharides rich in sialic acids of Neisseria strains were prepared in rabbits following the schedule described by the Center for Disease Control (15). Mouse antipneumococcus C strain R36A (anti-R36A) were raised in C57BL/6J. Mouse myeloma protein S107 was a gift from Dr. Melvin Cohn. Antibody to fetal calf glycoprotein, fetuin, was prepared as described by Spiro (16). Antibodies were heat inactivated (56°C, 30 min) and absorbed 3 times with adult live and spleen of C57BL/6J (30 min at room temperature) in order to remove unspecific rabbit antimouse reactivity.

The antibody to yeast mannan was prepared by immunizing rabbits with S. cerevisiae mutant X 2180 – 1A-5 mannan (anti-A-5), X 2180 – 1A-1 (anti-A-1), and Salmonella Thompson (anti-S. Thompson) (17).

RESULTS

Within each microwell a reproducible series of events leads to a stable pattern of cells interacting with each other. The 3 major components, overlapping in time and probably interdependent are 1) reaggregation of cells (2–16 h after plating), 2) interconnection of reaggregates via sheets of migrating granule cells on the surface of epithelial cells and processes (axonal, dendritic, and glial) oriented in parallel and 3) via straight cables composed of mainly axon-like processes with granule cells migrating along their outer surfaces (12, 13).

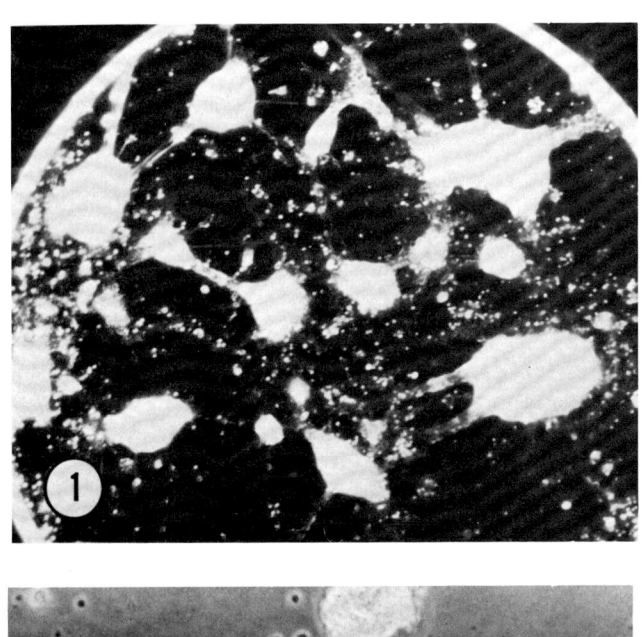

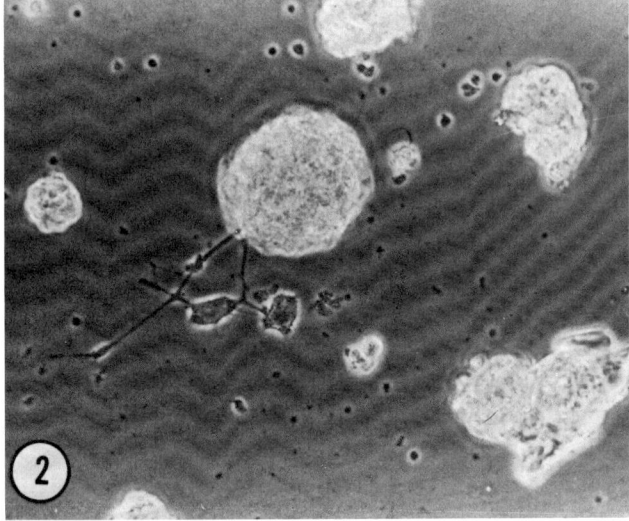

Figs. 1–4. 7×10^4 7-day-old cerebellum cells were cultured for 72 h in 10 μl of Basal Eagle's Medium in Falcon microtiter plates, in the absence of anticarbohydrate antibodies (Fig. 1, 250 ×) or with anti-Neisseria meningitis type C antibody 1:600 diluted (Fig. 2, 250 ×). The formation of fibers connecting reaggregates is blocked in the presence of antibodies. Scanning electronmicrographs demonstrate that the surface of cells and fibers is smooth under normal conditions (Fig. 3, 2,500 ×). Cell surfaces treated with antibody appear ruffled and uncoordinated filopodia and fibers are covering the surfaces of reaggregates (Fig. 4, 5,000 ×).

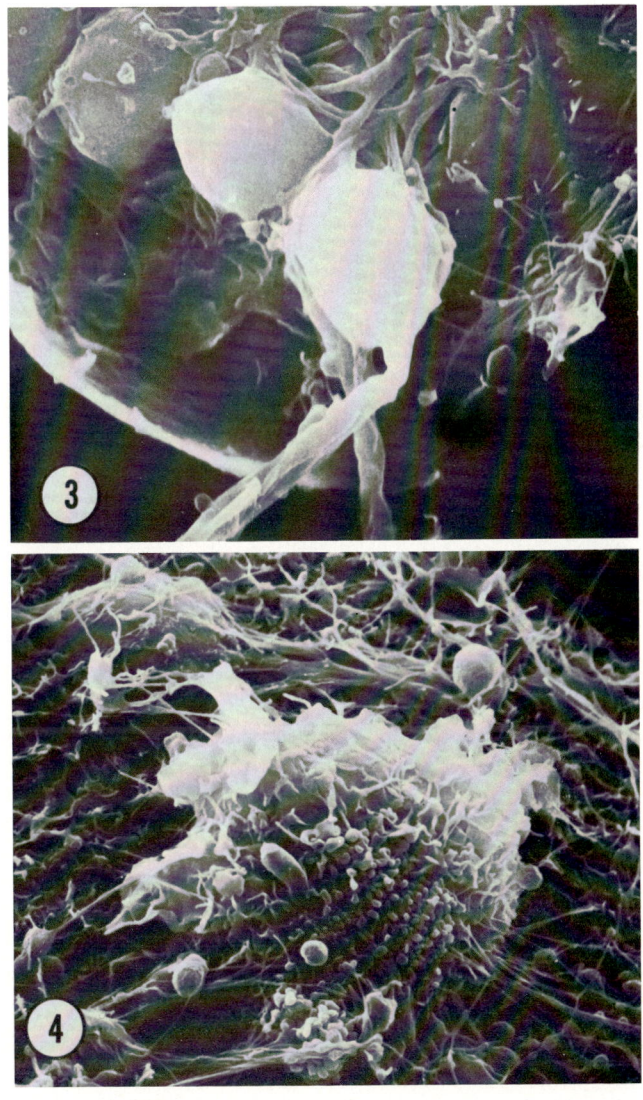

As demonstrated in this study microbial carbohydrate-specific antibodies cross-react with cerebellar cell surfaces in vitro at different stages of development (Tables I, II), as demonstrated by inhibition of fiber outgrowth (Figs. 1–4). Antibody specific for sialic acid containing determinants in Neisseria meningitidis strains was prepared by immunizing rabbits with sialic acid containing immunodeterminants in their cell wall antigen. Both groups B and C meningococcal polysaccharides were shown to be pure homopolymers of sialic acid. The 2 polysaccharides are, however, noncrossreactive (18) and differ chemically. Acetyl determination indicated that the C-polysaccharide contains both N- and O-acetyl groups, whereas the B-polysaccharide contains only N-acetyl groups. In addition the B-polysaccharide probably consists of α-ketosidic linkages whereas the C-polysaccharide might have β-ketosidic linkages (19).

As shown in Table I, row A rabbit anti-C- and -B-polysaccharide antibodies react with cerebellar cells at different stages in development. Anti-C antibodies inhibit fiber outgrowth in both P0 and P7 cerebellar cultures but not in E13 cultures. In contrast, fiber outgrowth was inhibited in E13 and P0 but not in P7 mice by anti-B and by antifetuin serum.

The mannan chemotype consists of repeated units of either α-1, 6- (mutant $\times$ 2185 1-A-5), α-1,3- (mutant $\times$ 2185 1-A-1) (20) or of α-1,2-linked mannose (S. Thompson). Antibodies raised against these strains did not cross-react with each other. When applied to cultures of P0 and P7 cerebellar cells these antibodies did not differ in their reactivity and no distinction of developmental stages could be detected (Table I).

Pneumococcus C strain R36A injected into C57BL/6J mice gives rise to antibodies which recognize phosphorylcholine (PC) (21). These antibodies did not react with cells of any stage in development (Table II). This lack of activity was particularly interesting, since cerebellar cell membrane is likely to contain relatively large amounts of PC (22). However, when pneumococcal carbohydrate was artificially inserted into the cell surface by preincubating cells for 1 h with palmitylated derivative of the antigen (21) antipneumococcal mouse antibody (IgG) or PC-specific myeloma protein (IgA) inhibited fiber outgrowth (Table II).

The inhibition due to antibody can be reversed by replacing antibody-containing medium with normal medium not later than day 2 in culture and assayed day 3 and 4 in vitro (Table I, row B).

The reactivity of antibodies was dependent upon antibody concentration added to the cultures (Fig. 5).

The results described above suggest that antimicrobial carbohydrate antibodies react with mammalian cell surface carbohydrates and distinguish prenatal from 7-day-old mouse cerebellar cells.

DISCUSSION

The interaction of cells at different times during histogenesis appears to determine a distinct sequence of events in cell migration and integration of cells into a tissue structure. These interactions are considered to be based on changes in cell surface carbohydrate structures. In order to characterize cell surface changes during development, mouse cerebellum cells were treated in microcultures with various antibodies raised against microbial carbohydrates. As demonstrated in Table I, fiber formation and presumably cell migration were blocked by these antibodies. In particular cerebellum cells of E13 and P0 express surfaces which cross-react with antibodies prepared against polysaccharides or carbohydrates containing predominantly N-acetylated sialic acid moieties (anti-N. meningitidis type B, antifetuin) whereas P7 cells react with antibody specific to both derivatives of neuraminic acids

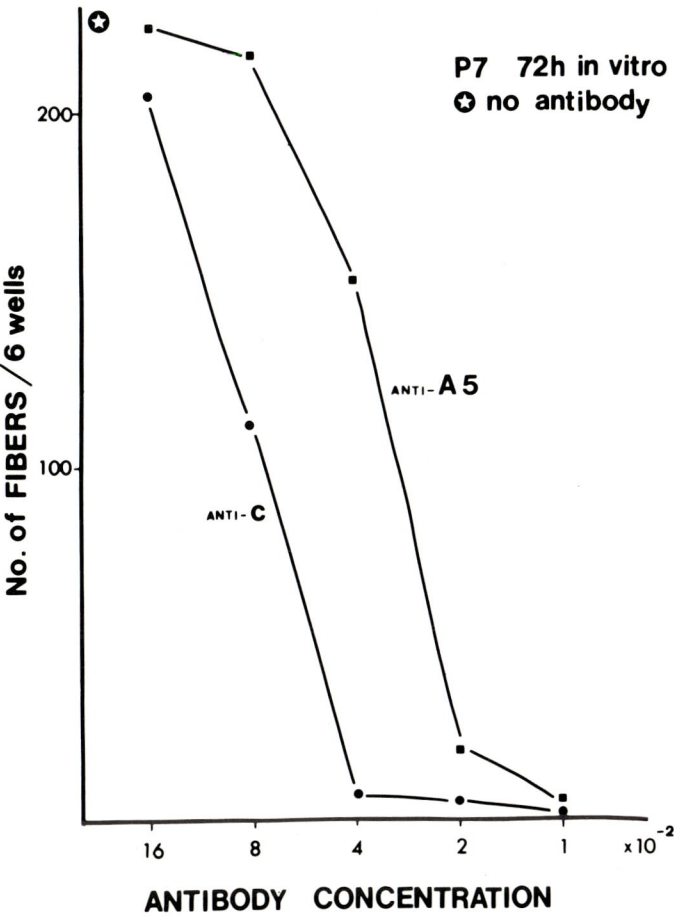

Fig. 5. Inhibition of fiber formation in cerebellar cultures is dependent upon the concentration of antibodies. Anti-C and anti-A-5 antibodies were added to P7 cells at the indicated concentrations 1 h after plating. Fibers were counted after 72 h in vitro. ★) Cultures without antibody.

(N-ac and O-ac) (anti-C). Yeast mannan antibodies directed against α-1,3- and α-1,6-linked mannose units and bacterial α-1,2-mannose unit did not discriminate between different stages of development in vitro.

These results raise several questions. First, the extent of similarity between the structures of cross-reacting sialic acid antigens of cerebellar cell surface and bacterial antigens still remains to be determined. The involvement of, for example, mammalian membrane sialic acid structures in many aspects of cellular recognition may make it difficult to ascertain the target structure(s) involved in this particular assay. Recent experiments by Hatten and Sidman (23) using different lectins as probes of surface carbohydrates demonstrated similar results, suggesting that the recognition of sugar specificities by antibodies is very similar to lectin specificities.

Second, what is the mechanism which leads to inhibition of fiber formation? The inhibition is reversible when antibodies are removed not later than 2 days after addition. Incubation of cultures with antibodies longer than 2 days leads to degeneration of predominantly neuronal cell population. It is doubtful that the inhibition caused by binding of antibodies to the surface carbohydrates or to artificially inserted molecules (Table II) is due to specific recognition of structures responsible for fiber formation (for example, growth cones); rather it is due to a pleiotropic effect caused by a nonspecific agglutination of carbohydrate-bearing surface proteins. The intensely folded cell surface in the presence of antibodies compared to the smooth surface of control cells (Figs. 3, 4) suggests that the fluidity of surface molecules is reduced or blocked by antigen-antibody reaction.

The rearrangement of surface antigens may be required to form the appropriate growth cone surface. In order to block fiber outgrowth, antibodies had to be added before fiber formation was initiated. When antibodies were added while the cell pattern was forming (1–2 days in culture) pattern formation was arrested. Under light microscopy, inhibition caused by antibodies at first appears to be restricted to the formation of fibers and connecting patterns, including the migration of granule cells but not the formation of reaggregates. Whether these results account for specific interaction of anticarbohydrate antibodies with, for example, migrating cells, still remains to be determined.

ACKNOWLEDGMENTS

E. T. is supported by a Fellowship of Deutsche Forschungsgemeinschaft. S. S. is supported by Grant 74152 of the World Health Organization.

REFERENCES

1. Roth P, McGuire EJ, Roseman S: J Cell Biol 51:525, 1971.
2. Gottschalk A, Buddecke E: In Gottschalk A (ed): "Glycoproteins." New York: American Elsevier, BBA Library Series, 1972, vol 5A, pp 529–564.
3. Cuatrecasas P: Annu Rev Biochem 43:169, 1974.
4. Sela B, Wang J, Edelman GM: Proc Natl Acad Sci USA 72:1127, 1975.
5. Ceccarini C, Muramaton T, Tsang T, Atkinson PA: Proc Natl Acad Sci USA 72:3139, 1975.
6. Mullin BR, Fishman PH, Lee G, Aloj SM, Ledley FI, Winard RJ, Kohn LD, Brady RD: Proc Natl Acad Sci USA 73:842, 1976.
7. Staub AM: In Springer GF (ed): "Polysaccharides in Biology." 5th Conf Josiah Macy Jr Foundation, New Jersey: Madison Printing Company, 1959, pp 139–214.
8. Sela B, Edelman GM: J Exp Med 145:443, 1977.
9. Miale IL, Sidman RL: Exp Neurol 4:277, 1961.
10. Rakic P: In Seeman P, Brown GM (ed): "Frontiers in Neurology and Neuroscience Research." Toronto: 1974, pp 112–132.
11. Sidman RL: In Moscona AA (ed): "The Cell Surface in Development." New York: John Wiley and Sons, 1974, pp 221–253.
12. Trenkner E, Hatten ME, Sidman RL: Proc Soc Neurosci 2:1028, 1977.
13. Trenkner E, Sidman RL: J Cell Biol (Submitted).
14. Barkley DS, Rakic LL, Chaffee JK, Wong DL: J cell Physiol 81:271, 1973.
15. Center for Disease Control, US Dept of Health, Atlanta, Georgia.
16. Spiro RG: Adv Protein Chem 27:349, 1973.
17. Sarkar S: J Reprod Med 13:93, 1974.
18. Robinson JA, Apicella MA: Infec Immunol 1:8, 1970.
19. Lin T-Y, Gotschlich EC, Dune FT, Jonsson EK: J Biol Chem 246:4703, 1971.
20. Raschke WC, Kern KA, Antalis C, Ballou CE: J Biol Chem 248: 4660, 1973.
21. Sher A, Cohn M: Eur J Immunol 2:319, 1972.
22. Folch J: J Biol Chem 174:439, 1948.
23. Hatten ME, Sidman RL: J Supramol Struct 7:1977 (in press).

Carbon-13 as a Tool for the Study of Carbohydrate Structures, Conformations and Interactions

H. A. Nunez, T. E. Walker, R. Fuentes, J. O'Connor, A. Serianni, and R. Barker

Department of Biochemistry, Michigan State University, East Lansing, Michigan 48824

The application of ^{13}C-NMR spectroscopy to problems involving the structures and interactions of carbohydrates is described. Both ^{13}C-enriched and natural abundance compounds were used and some advantages of the use of the stable isotope are described. Carbon-carbon and carbon-proton coupling constants obtained from 1-^{13}C enriched carbohydrates were employed in the assignment of their chemical shifts and to establish solution conformation. In all cases studied thus far, C-3 couples to C-1 only in the β-anomers while C-5 couples to C-1 only in the α-anomers. C-6 and C-2 always couple to C-1 in both anomeric species. The alkaline degradation of glucose [1-^{13}C] to saccharinic acids was followed by ^{13}C-NMR. The conversion of glucose [1-^{13}C] to fructose-1,6-bisphosphate [1,6-^{13}C] by enzymes of the glycolytic pathway was shown as an example of the use of ^{13}C-enriched carbohydrates to elucidate biochemical pathways. In a large number of glycosyl phosphates the ^{31}P to H-1 and ^{31}P to C-2 coupling constants demonstrate that in the preferred conformation the phosphate group lies between the O-5 and the H-1 of the pyranose ring. The influence of paramagnetic Mn^{2+} ions on the proton decoupled ^{13}C-NMR spectra of uridine diphosphate N-acetylglucosamine indicates that the Mn^{2+} interacts strongly with the pyrophosphate moiety and with the carbonyl groups of the uracil and N-acetyl groups.

Key words: ^{13}C-enriched carbohydrates, glycosyl phosphates, ^{13}C-NMR, carbohydrate conformations

Since carbon-13 has a low natural abundance of 1.1% and a small magnetic moment it is 6,000 times more difficult than hydrogen to detect by nuclear magnetic resonance methods. Recent technical developments (1) have greatly facilitated the observation of ^{13}C in natural abundance samples, and the use of isotopic enrichment with ^{13}C can further improve observation so that carbon magnetic resonance spectroscopy (^{13}C-NMR) rivals proton magnetic resonance spectroscopy (^{1}H-NMR) as a tool for application to chemical and biochemical problems (2–5). Spectrometers utilizing pulsed Fourier transformation, broad-band decoupling of protons, and computers for accumulation of time-averaged data

Received April 4, 1977; accepted May 26, 1977

T. E. Walker's present address is Los Alamos Scientific Laboratory, University of California, Los Alamos, New Mexico 87545

© 1977 Alan R. Liss, Inc., 150 Fifth Avenue, New York, NY 10011

make it possible to obtain a spectrum of a 0.5 M solution in a few minutes and that of a 10^{-3} M solution in a few hours. With isotopic enrichment, spectra can be obtained using 2 ml of a 10^{-2} M solution with a single scan (typically < 2 sec) at 14.1 kilogauss, the magnetic field used in the least expensive commercial spectrometers. The use of higher magnetic fields and larger samples make the technique applicable to systems with compounds at those concentration levels encountered in biological systems (5, 6).

In this article we present examples of some of the ways in which ^{13}C-NMR can be applied to studies of carbohydrates of biological importance. We describe applications using enriched compounds and compounds at natural abundance. The applications include: the assignment of chemical shifts to specific carbons of the common carbohydrates; the evaluation of the factors that determine the intensity of spin-spin coupling between nuclei within a monosaccharide; the effects of structural modifications and pH changes on the spectra of carbohydrates such as glycosyl phosphates; the effects of interactions between carbohydrates and metal ions; and the use of enriched compounds to follow chemical and biochemical conversions.

^{13}C-ENRICHED CARBOHYDRATES: STRUCTURE AND CONFORMATION

Hexoses containing ^{13}C at 90% enrichment in C-1 can be prepared from the corresponding pentoses by modifications of the methods used to prepare ^{14}C-labeled compounds (7, 8). The ^{13}C-NMR spectrum D-galactose [1-^{13}C] is shown on Fig. 1A. In this spectrum only 2 strong resonances are observed due to the α and β forms of the sugar. The intensities of the 2 resonances are roughly proportional to the ratio of the 2 forms in the sample. The positions of these 2 resonances unequivocally establish the chemical shifts for C-1 in the α and β forms of D-galactopyranose. The spectrum of the C-2, C-3, C-4, C-5, and C-6 carbons in D-galactose [1-^{13}C], shown on Fig. 1B, can be compared to the same region of the spectrum of unenriched D-galactose on Fig. 1C. Several significant differences exist between the spectrum of the enriched and unenriched compound. The resonances due to C-2 in the latter have virtually disappeared. The small residual resonances are due to the 10% of the enriched sample that has ^{12}C at C-1. Ninety percent of the molecules in the enriched sample give a spectrum in which the signal due to C-2 is split into 2 resonances due to spin-spin coupling to the ^{13}C at C-1. This splitting permits the unequivocal identification of the resonance due to C-2. In addition, much smaller splittings are observed in the signals for C-3 in the β form and C-5 in the α form. Both anomers show splitting of the C-6 resonance. Splittings of this type are characteristic and can be used to identify the resonances due to specific carbons.

Unequivocal assignment of ^{13}C chemical shifts in unenriched carbohydrates is often difficult. This is illustrated by incorrect assignments made by early workers (9, 10). The use of ^{13}C- and ^{2}H-enriched derivatives has permitted correct assignments to be made (11–14). These assignments are given in Table I. They can be accepted as correct and are principally due to the extensive investigations of Gorin and co-workers (11, 12, 14).

Correct assignment of chemical shifts to specific carbons is an essential prerequisite to all studies of structure and interaction. Without correct assignments the molecular meaning of observed spectral changes will be misinterpreted.

When the chemical shifts of individual carbons of monosaccharides are known it is possible to examine more complex structures such as oligosaccharides and to assign anomeric configuration, linkage position, and conformation about the glycosidic bond. The ^{13}C-NMR spectra of galactose, N-acetylglucosamine, and N-acetyllactosamine are shown on Fig. 2. The position of the linkage in the N-acetylglucosamine moiety of the

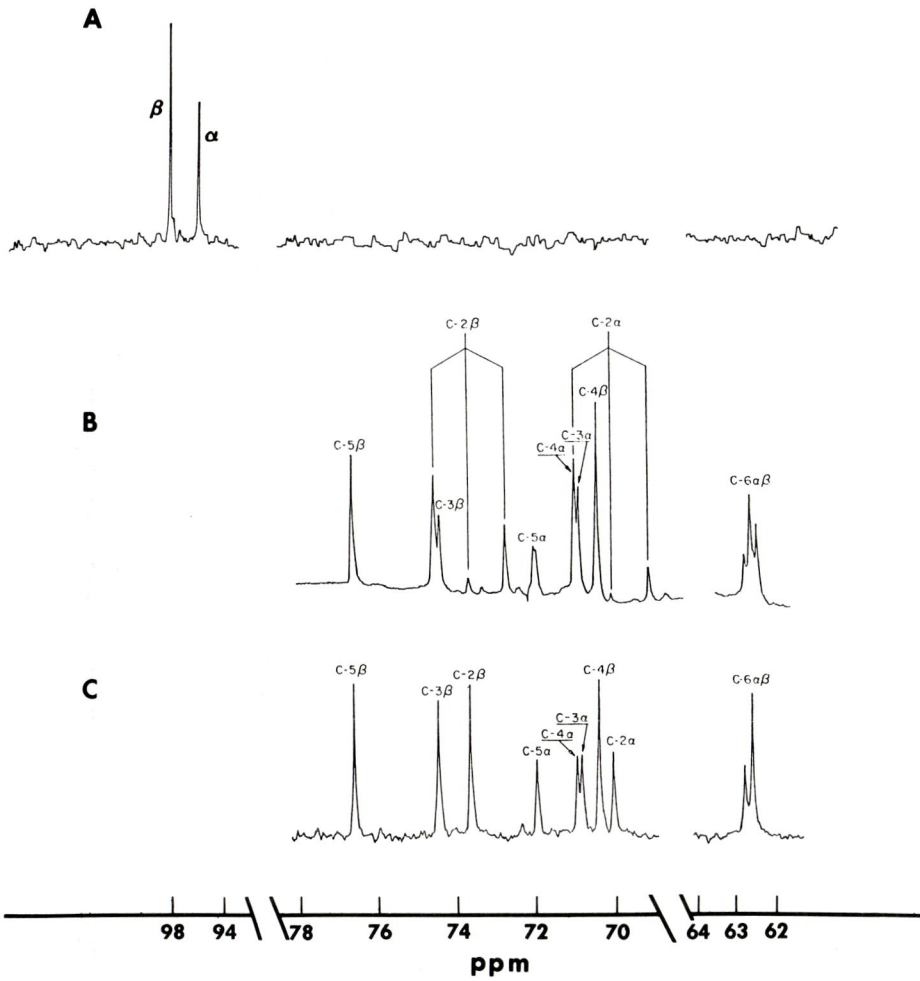

Fig. 1. A) The proton-decoupled, 25 MHz, ^{13}C-NMR spectrum of 0.2 M α and β mixture of 90%-enriched D-galactopyranose [1-^{13}C] obtained in one scan. B) The natural abundance peaks C-2 to C-6 of the enriched compound after 5,000 scans at 2.05 sec per scan. C) The C-2 to C-6 region of the ^{13}C-NMR spectrum of natural abundance α- and β-D-galactopyranose. For interpretation of the spectra see text.

latter is apparent from the shift of the resonance due to C-4. The configuration of the glycosidic linkage is easily established from the chemical shift of the C-1 of D-galactose which is characteristic of the β anomer. The conformation about the glycosidic bond was established using D-galactose [1-^{13}C] to form UDP galactose [1-^{13}C] and from it, enzymatically, N-acetyllactosamine with ^{13}C at C-1 of the galactosyl moiety (15). As described below, coupling between C-1 and C-5 of 1-^{13}C-enriched monosaccharides depends on the conformation of the substituents at C-1. Assuming that this dependence holds for the coupling between C-1 of galactose and the C-4 of N-acetylglucosamine through the glycosidic oxygen, the small coupling constant (less than 1 Hz) indicates that the preferred

TABLE I. ¹³C-NMR Resonance Assignments

	C-1	C-2	C-3	C-4	C-5	C-6	OMe
Glucopyranoses[a]							
α-D-glucose	92.7	72.14[b]	73.4	70.4	72.10	61.3	
β-D-glucose	96.5	74.8	76.4	70.3	76.6	61.5	
Methyl α-D-glucopyranoside	100.3	72.5[b]	74.2[b]	70.6	72.7	61.7	56.2
Methyl β-D-glucopyranoside	104.3	74.2	76.9	70.8	76.9	61.9	58.3
6-Deoxy-α-D-glucose	93.1	72.9	73.6	76.4	68.6	18.0	
6-Deoxy-β-D-glucose	96.8	75.6	76.6	76.1	73.0	18.0	
Methyl 6-deoxy-α-D-glucopyranoside	100.3	72.6	73.9	76.2	68.7	17.6	56.2
Methyl 6-deoxy-β-D-glucopyranoside	104.3	74.5	76.7	76.2	73.0	17.8	58.3
Methyl α-D-glucopyranosiduronic acid	100.7	71.9	73.8	72.5	71.9	d	56.7
Methyl β-D-glucopyranosiduronic acid	104.3	73.8	76.5	72.3	75.6	d	58.5
Methyl (methyl α-D-glucopyranosid) uronate	100.8	71.94	73.7	72.4	71.87	d	56.8
Methyl (methyl β-D-glucopyranosid) uronate	104.6	73.7	76.3	72.4	75.7	d	58.7
Mannopyranoses[a]							
α-D-Mannose	95.0	71.7	71.3	68.0	73.4	62.1	
β-D-Mannose	94.6	72.3	74.1	67.8	77.2	62.1	
Methyl α-D-mannopyranoside	101.9	71.2[c]	71.8[c]	68.0	73.7	62.1	55.9
Methyl β-D-mannopyranoside[d]	102.0	71.4	74.2	68.1	77.3	62.1	d
α-L-rhamnose	95.0	71.9	71.1	73.3	69.4	18.0	
β-L-rhamnose	94.6	72.4	73.8	72.9	73.1	18.0	
Methyl α-L-rhamnopyranoside	101.9	71.0	71.3	73.1	69.4	17.7	55.8
Galactopyranoses[a]							
α-D-galactose	93.6	69.8[b,c]	70.56[b]	70.63[c]	71.7	62.5	
β-D-galactose	97.7	73.3	74.2	70.1	76.3	62.3	
Methyl α-D-galactopyranoside	100.5	69.4[c]	70.6[c]	70.4	71.8	62.3	56.3
Methyl β-D-galactopyranoside	104.9	71.8	73.9	69.8	76.2	62.1	58.3
α-D-fucose	93.3	69.2[b]	70.4[b]	73.0	67.4	16.7	
β-D-fucose	97.3	72.8	74.0	72.5	71.9	16.7	
Methyl α-D-fucopyranoside	100.5	69.0	70.6	72.9	67.5	16.5	56.3
Methyl β-D-fucopyranoside	104.8	71.5	74.1	72.4	71.9	16.5	58.3
Allopyranoses[a]							
α-D-allose	93.4	67.6	72.3	66.7	67.5	61.3	
β-D-allose	94.0	71.9²	71.8	67.4	74.2[b]	61.8	

	C-1	C-2	C-3	C-4	C-5	C-6	OCH$_3$-1
Amino sugars[f]							
α-D-glucosamine·HCl	90.7	56.0	71.2	71.2	73.1	62.0	
β-D-glucosamine·HCl	94.3	58.5	73.6	71.2	77.6	62.0	
α-D-mannosamine·HCl	91.8	56.0	68.3	67.6	73.1	61.7	
β-D-mannosamine·HCl	92.4	57.1	70.9	67.4	77.4	61.8	
Pentopyranoses[a]							
α-D-xylose	93.3	72.5[b]	73.9[b]	70.4	62.1		
β-D-xylose	97.6	75.1	76.9	70.3	66.3		
Methyl α-D-xylopyranoside	100.6	72.3	74.3	70.4	62.0		56.0
Methyl β-D-xylopyranoside	105.1	74.0	76.9	70.4	66.3		58.3
α-L-arabinose[a,b]	97.8	73.0	73.5	69.6	67.5		
β-L-arabinose[a,b]	93.7	69.6	69.8	69.8	63.6		
Methyl α-L-arabinopyranoside	105.1	71.8	73.4	69.4	67.3		58.1
Methyl β-L-arabinopyranoside	101.0	69.4	69.92	69.96	63.8		56.3
Furanosides[g]							
Methyl α-D-galactofuranoside	103.1	77.4	75.5	82.3	73.7	63.4	56.1
Methyl β-D-galactofuranoside	109.2	81.9	77.8	84.0	72.0	63.9	56.1
Methyl α-D-arabinofuranoside	109.3	81.9	77.5	84.9	62.4		56.1
Methyl β-D-arabinofuranoside	103.2	77.5	75.7	83.1	64.2		56.3
Methyl α-D-lyxofuranoside	109.1	77.0	72.0	81.3	61.2		56.9
Methyl β-D-lyxofuranoside	103.2	72.9	70.7	81.9	62.4		56.5
Methyl α-D-xylofuranoside	103.0	77.7	76.0	79.3	61.5		56.6
Methyl β-D-xylofuranoside	109.6	80.9	76.0	83.5	62.1		56.2
Methyl α-D-ribofuranoside	104.2	72.1	70.8	85.5	62.6		56.5
Methyl β-D-ribofuranoside	109.0	75.3	71.9	83.9	63.9		56.3

[a] Assignments made by Gorin and Mazurek (14). Spectra were obtained on D$_2$O solutions at 33°C and are referenced to external TMS = 0 ppm.
[b] Assignments reversed from that reported by Dorman and Roberts (9).
[c] Assignments reversed from that reported by Perlin et al. (10).
[d] The position of this resonance was not reported.
[e] Assignments by Walker et al. (13) and the shifts corrected to 33°C to fit with the shifts reported by Gorin and Mazurek (12).
[f] Assignments made by Walker et al. (13). Spectra were obtained on H$_2$O solutions at ambient probe temperature (~40°C) and are referenced to external TMS = 0 ppm.
[g] Assignments made by Gorin and Mazurek (12). Spectra were obtained on D$_2$O solutions at 33°C and are referenced to external TMS = 0 ppm.

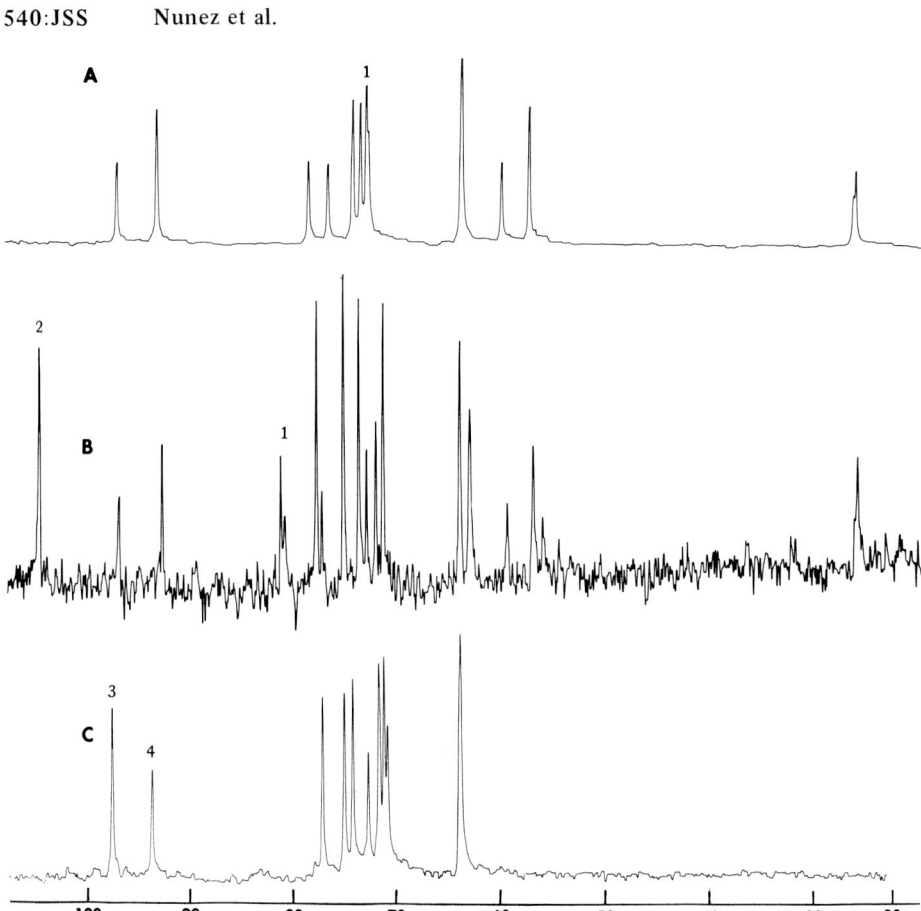

Fig. 2. A) The natural abundance proton-decoupled ^{13}C-NMR spectrum of α- and β-D-acetylglucosamine (15.08 MHz). B) The natural abundance spectrum of α- and β-N-acetyllactosamine. C) The natural abundance spectrum of α- and β-D-galactose. The peaks at 71 ppm in A and the peaks at 80 ppm in B are due to C-4; both are labelled 1. In both spectra the peak is a doublet due to the presence of α- and β-anomers. The peak at 104.1 ppm in B, labelled 2, is assigned to the β-D-galactopyranosyl linkage. The peaks at 98 and 93.8 ppm, labelled 3 and 4 in C, are the β- and α- carbon-1 resonances of D-galactose, respectively.

conformer has the C-4 of N-acetylglucosamine trans to the C-2 of the galactosyl moiety. Smith and his co-workers (16, 17) have applied ^{13}C-NMR spectroscopy to the elucidation of several important polysaccahride structures. Perlin (18), Lemieux (19), and Gorin (20) have each made important contributions to this field.

As the spectra on Fig. 1 demonstrate, ^{13}C enrichment at a specific carbon (C-1) permits the coupling between that carbon and other carbons within the structure to be observed. When this coupling involves a single bond, such as between C-1 and C-2, it does not vary greatly with structural changes and at present no attempt has been made to interpret the changes observed in carbohydrates. Coupling involving 2 or more bonds, however, seems to be sensitive to conformational changes and to depend on the geometry

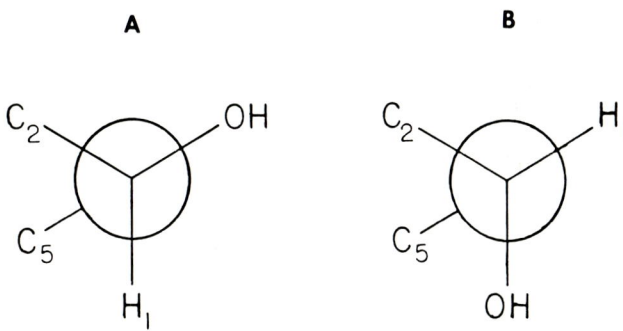

Fig. 3. The Newman projection along the C-1 to O-5 bond for the $^4C_1$ conformation of: A) β-D-glucopyranose and B) α-D-glucopyranose.

and nature of the substituent groups (21–23). Thus, when coupling between C-1 and C-5 in the hexoses is considered, the positions of the substituent groups on C-1 appear to influence the magnitude of the interaction. A "dihedral angle rule" has been proposed (24) which states that C or O substituents gauche to the coupled atom make a negative contribution to coupling (i.e., that coupling is present but has a negative sign). On the other hand, a trans substituent makes a positive contribution. When one gauche and one trans substituent are present, as on Fig. 3A, no coupling is observed since the 2 contributions cancel. When 2 gauche substituents are present, as on Fig. 3B, coupling is observed and the sign of the coupling should be negative. Although the sign of coupling has not been determined, all carbohydrates observed to date show coupling between C-1 and C-5 the α-D forms, which have the arrangement shown on Fig. 3B in the $^4C_1$ chair conformation. The data are presented in Table II.

A similar dependence of the magnitude of coupling on the position of substituents is shown in the case of coupling between C-1 and H-2 in relationship to the oxygens attached to C-1. Thus, in the cases studied, coupling is large (~ 5 Hz) only when H-2 is gauche to both oxygens at C-1. This is the case in sugars with H-2 axial and 1-OH equatorial as in β-D-glucopyranose (Table II). In the case of β-D-allose this coupling has been shown to be negative (24).

Empirical rules such as the one discussed above are helpful if used with caution. They are probably not universally applicable as can be seen by examining coupling between C-1 and C-3 in the D-hexopyranoses. Here the β-D- forms should show significant coupling if the "rule" concerning gauche and trans substituents holds. Clearly, it does not since only α-D-anomers show coupling. Although the lack of universality complicates the interpretation of coupling constants, there is ample evidence that changes in geometry affect the magnitude of coupling. Also, the values of the coupling constants in unperturbed systems are essential for studies where perturbations can be expected, as in binding to metals or proteins and in the conversion to derivatives such as phosphate esters.

TABLE II. Carbon-13 Coupling Constants of Labeled Monosaccharides

Compound	Coupling constant (Hz)[a]					
	C-1,2	C-1,3	C-1,5	C-1,6	C-1, H-1	C-1, H-2
α-D-[1-^{13}C]Glucose	46.0	–	~1.8	*	169.8	–
β-D-[1-^{13}C]Glucose	46.0	~3.5	–	*	161.2	5.5
Methyl α-D-[1-^{13}C] Glucopyranoside	46.4	–	~1.7	3.2	*	–
Methyl β-D-[1-^{13}C] Glucopyranoside	46.8	~4.1	–	4.3	*	4.1 ± 1.0
α-D-[1-^{13}C]Mannose	46.8	–	1.7	*	170.4	–
β-D-[1-^{13}C]Mannose	42.4	4.3	–	*	160.7	b
Methyl α-D-[1-^{13}C] Mannopyranoside	47.0	–	2.3	3.0	*	*
Methyl β-D-[1-^{13}C] Mannopyranoside	43.8	3.4	–	4.0	*	*
α-D-[1-^{13}C]Galactose	46.6	–	2.1	*	168.6	–
β-D-[1-^{13}C]Galactose	46.0	3.7	–	*	162.7	5.7
α-L-[1-^{13}C]Fucose	45.8	–	2.3	3.5	*	*
β-L-[1-^{13}C]Fucose	46.0	4.1	–	3.5	*	*

[a]Those couplings designated by * were not measured; those designated by – refer to no observable coupling.
[b]A broadening of ~1.6 Hz was observed in the proton coupled spectrum which could be due to C-1–H-2 coupling.

^{13}C-ENRICHED CARBOHYDRATES: CHEMICAL AND BIOCHEMICAL CONVERSIONS

In certain applications, ^{13}C has a great advantage over ^{14}C in permitting the evaluation of the transformations that occur during a chemical reaction or in a biochemical conversion. This is illustrated on Fig. 4 where the conversion of D-glucose to saccharinic acids under anaerobic conditions in 2.4 M NaOH is followed using D glucose [1-^{13}C]. The starting compound is 94% ^{13}C enriched at C-1, and other carbons are at natural abundance. As the reaction proceeds an intermediate substance is formed in which the enriched carbon resonates at a position characteristic of the 1-CH_2OH group of fructose. With time the ^{13}C label is distributed into 3 resonances characteristic of carboxylate anions and one resonance characteristic of –CH_3 groups. The time-course of the transformation is readily followed since a spectrum with good signal-to-noise characteristics can be obtained in 5 to 10 min. The changes in the location of the ^{13}C label with time are shown on Fig. 5. The data of Fig. 5 were obtained in the time taken to carry out the transformation. To obtain the same data using ^{14}C would have required extensive work-up at each time of sampling and chemical degradation to locate the isotope. The advantage of ^{13}C in this circumstance is clear.

Similar studies can be made of enzymatically catalyzed conversions. An example is presented on Fig. 6 which shows the conversion of glucose [1-^{13}C] to fructose-6-phosphate [1-^{13}C], to fructose-1,6-bisphosphate [1-^{13}C], and finally to fructose-1,6-phosphate [1,6-^{13}C] using the enzymes of the glycolytic pathway. The conversion of the latter back to glucose [1,6-^{13}C] can be followed using the appropriate phosphatases.

Other examples of the use of ^{13}C-enriched compounds to elucidate biochemical

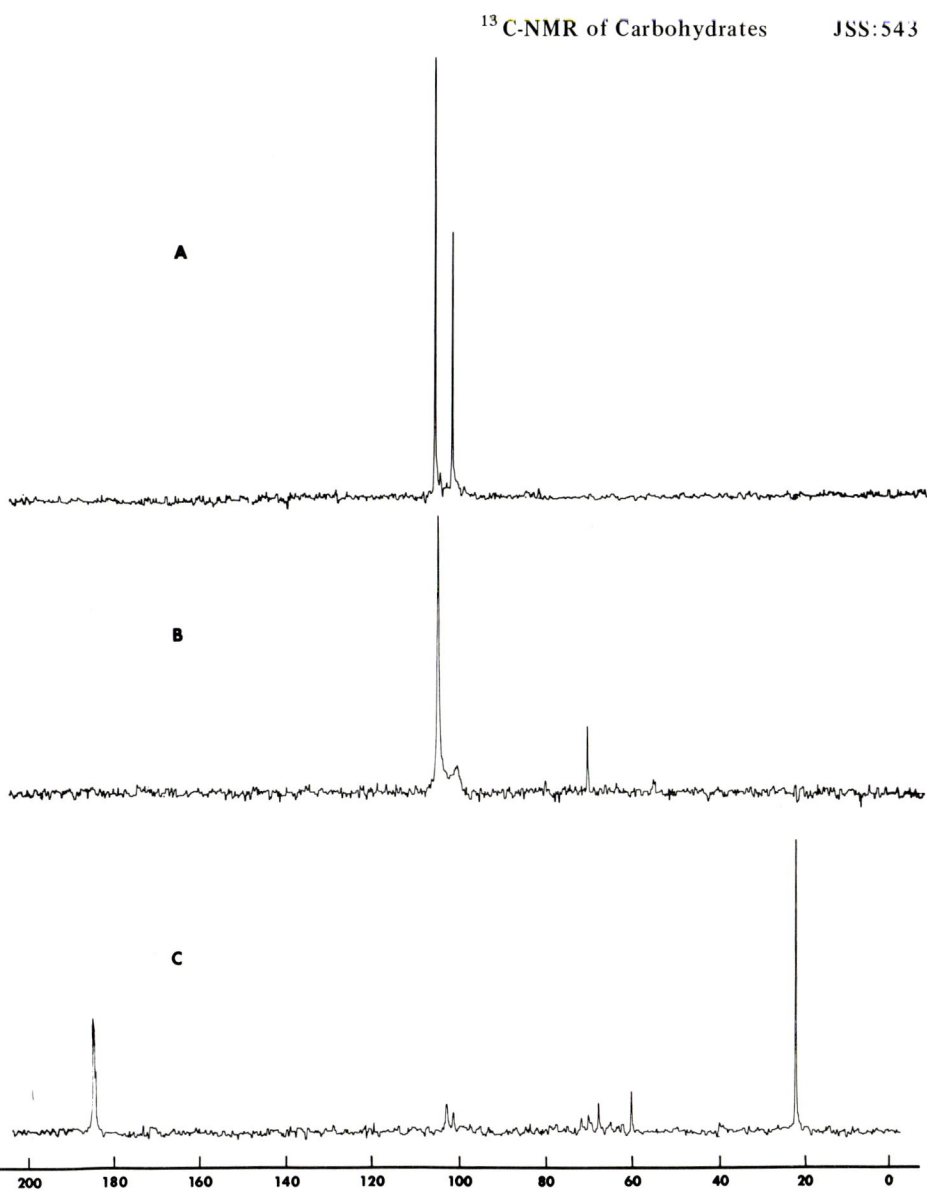

Fig. 4. The rearrangement of 0.1 M D-glucose [1-^{13}C] in 2.4 M aqueous sodium hydroxide at 37°C. A) β- and α-D-glucopyranose [1-^{13}C] in water at 37°C. B) D-glucose in 2.4 M aqueous sodium hydroxide at 37°C after 14 min of reaction, the signal for the α form is greatly broadened and the 1-CH$_2$OH of fructose is apparent. C) Reaction B after 5.5 h, resonances of 3 carboxylic acids and a methyl carbon are predominant. Small peaks due to a small proportion of D-mannose and some scrambling can be seen.

pathways are abundant. For example, the incorporation of enriched glucose into living Candida utilis cells has been followed (25). This and other examples are given in several reviews (4, 26–28).

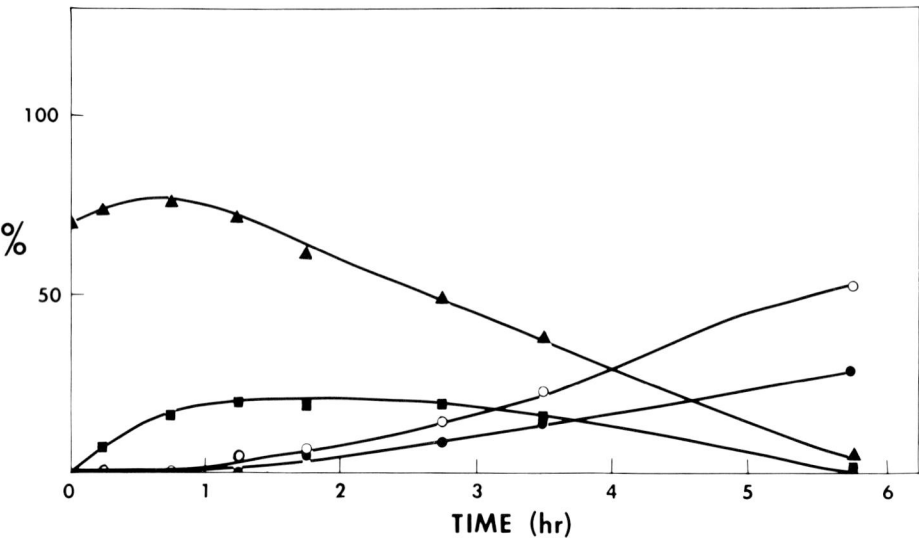

Fig. 5. The rearrangement of D-glucose [1-^{13}C] in 2.4 M aqueous sodium hydroxide at 37°C. Curves show the percentage of the total ^{13}C from the enriched carbon appearing in: ▲) β-D-glucopyranose C-1, ■) fructose C-1, ○) methyl group of saccharinic acids, ●) carboxyl group of saccharinic acids.

NATURAL ABUNDANCE CARBOHYDRATES: GLYCOSYL PHOSPHATES.

The glycosyl phosphates are important intermediates in the formation and degradation of oligo- and polysaccharides. In addition, they are structural elements in nucleoside diphosphate sugars in some polysaccharides and in glycosyl phosphate dolichol derivatives. The conformation of these glycosyl phosphates in solution has been established by a combination of proton (29) and carbon magnetic resonance spectroscopy.

Phosphorus (^{31}P) has spin equal to ½ and couples to carbon and hydrogen in the same way that these couple to each other, with the magnitude of the coupling dependent on conformation and number of bonds involved. Lee and Sarma (29) examined the proton spectra of a number of α-D-glycosyl phosphates and were able to conclude that the phosphorus atom resided in the angle subtended by the O-5 and H-1 or in the angle between H-1 and C-2 (Fig. 7). Examination of the carbon spectra of the same compounds has permitted us to conclude that the phosphorus resides between the O-5 and H-1 in all of the α-D-glycosyl phosphates examined (Table III). This conclusion arises from the magnitude of the coupling between C-2 and P (~ 8 Hz), characteristic of a trans relationship for these 2 atoms around the glycosidic bond.

The β-anomers of the glycosyl phosphates show coupling between C-2 and P of 5–7 Hz, which also can be characteristic of a trans arrangement about the glycosidic bond. This geometry is supported by the $^3J_{P-H_1}$ coupling constant of 7–9 Hz (Table III).

Examination of α-D-galactopyranosyl phosphate [1-^{13}C] indicates that the coupling and thus the conformation of the pyranosyl ring is not significantly influenced by the

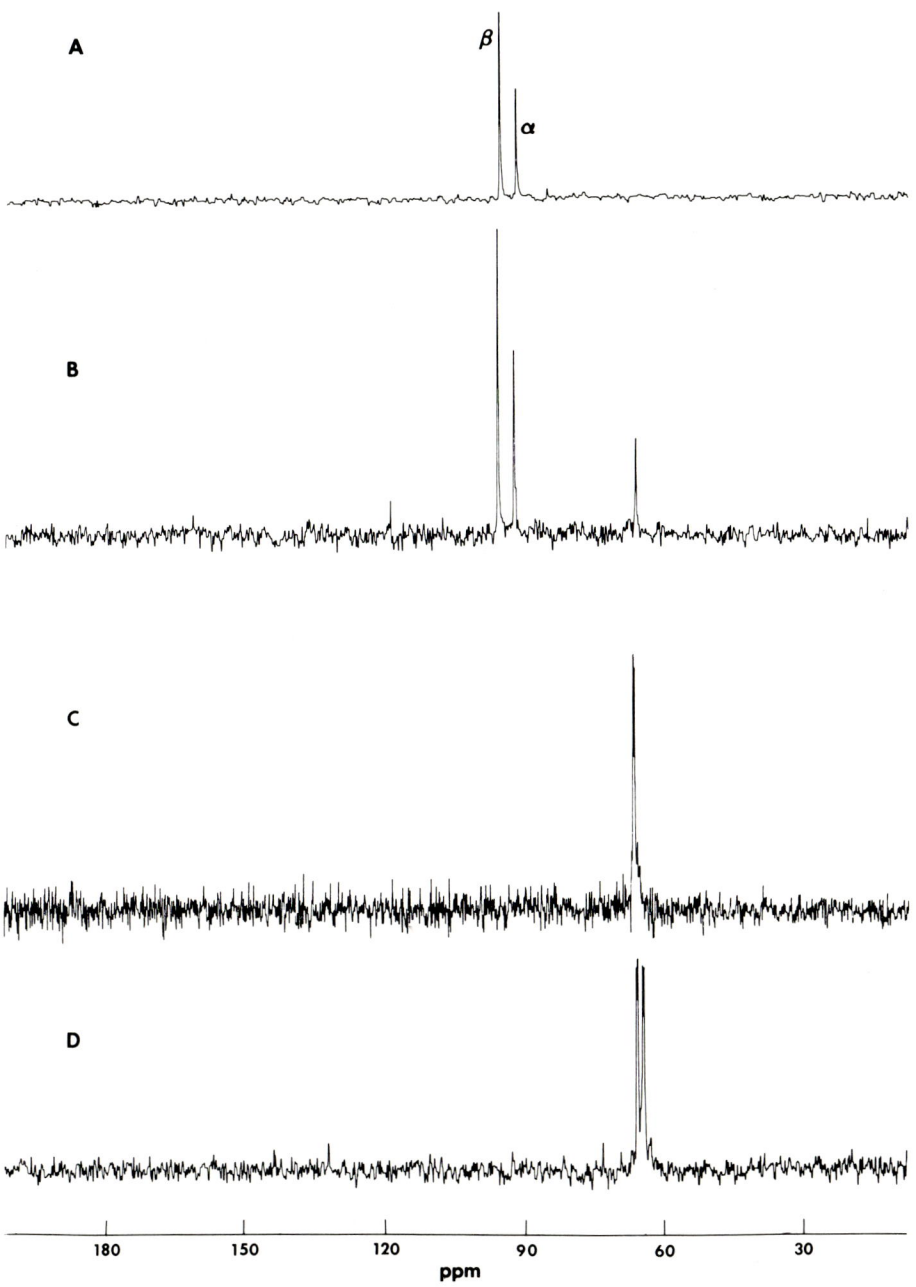

Fig. 6. The 15.08 MHz carbon-13 NMR spectra of glycolytic products obtained from an in vitro enzymatic conversion of D-glucopyranose [^{13}C-1] at an initial concentration of 50 mM and pH 7.4. ATP and MgCl$_2$ were added in 1:1 cofactor/substrate ratios. Spectra C and D show splitting due to carbon-phosphorus coupling. A) D-glucopyranose [^{13}C-1]. B) D-glucose-6-phosphate (97.4 ppm and 93.6 ppm) and D-fructose-6-phosphate [^{13}C-1] (64.2 ppm); C) D-fructose-1,6-bisphosphate [^{13}C-1]. D) D-fructose-1,6-bisphosphate [^{13}C-1,6].

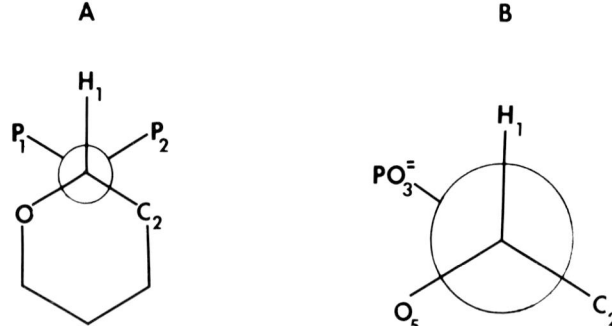

Fig. 7. The conformation of the phosphate group in α-D-glycosyl phosphates. A) The β-D glycosyl phosphate viewed from above. P_1 and P_2 represent possible orientation of phosphorus as determined by ^{1}H-NMR (29) where $^3J_{P-H-1}$ = 7.5 Hz. The ^{13}C-NMR spectra show $^3J_{P-O-C-C}$ = 8 Hz indicating that the conformation is that represented by P_1. B) A Newman projection along the glycosidic bond showing the preferred orientation of the phosphate group.

TABLE III. Phosphorus Coupling Constants for Glycosyl Phosphates

Compound	Coupling constant (Hz)[a]		
	P-C_1	P-C_2	P-H_1
α-D-glucopyranosyl-1-phosphate	5.3	7.5	7.2[b]
β-D-glucopyranosyl-1-phosphate	4.4	6.6	7.7
α-D-galactopyranosyl-1-phosphate	5.3	7.4	7.2[b]
β-D-galactopyranosyl-1-phosphate	3.7	5.1	7.6
α-D-mannopyranosyl-1-phosphate	5.3	7.5	7.7–8.6[b]
β-D-mannopyranosyl-1-phosphate	2.9	5.1	8.8
α-L-fucopyranosyl-1-phosphate	5.9	6.6	7.2
β-L-fucopyranosyl-1-phosphate	4.4	5.1	7.5

[a] All measurements were done in D_2O at pD 7.4.
[b] Taken from Lee and Sarma (29).

phosphate group. The magnitude of the coupling constants $^2J_{C-1,C-5}$ and $^3J_{C-1,C-6}$ are similar to those for α-D-galactopyranose.

The state of ionization of the phosphate group of a glycosyl phosphate markedly affects the chemical shifts of most carbons. These shifts are large enough to change the relative positions of resonances in some cases. The spectra of β-L-fucopyranosyl phosphate at pH 8 and pH 4 are shown on Fig. 8 to illustrate this point. In general, the coupling of phosphorus to C-2 is greater than coupling to C-1, a characteristic of phosphorus-carbon coupling. The resonances of those carbons which shift significantly with pH can be plotted

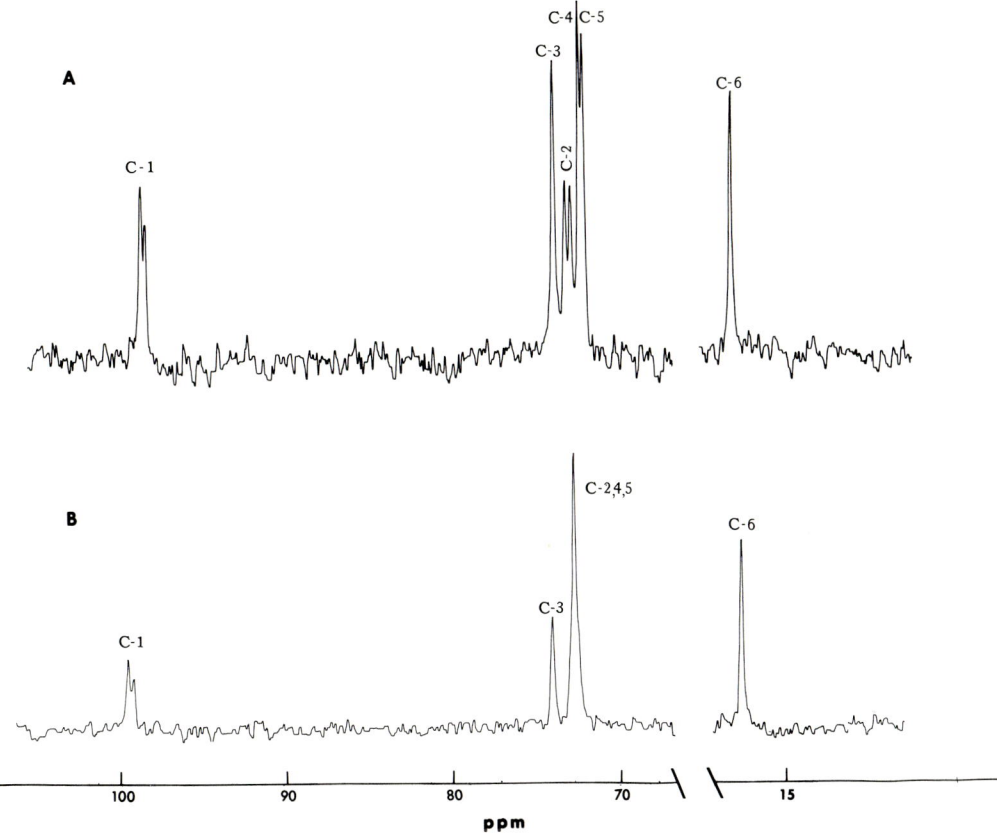

Fig. 8. The effect of pH on the 15.08 MHz proton decoupled ^{13}C-NMR spectrum of β-L-fucopyranosyl-1-phosphate in water. A) At pH 8, all 6 of the carbon resonances are resolved. The C-1 and C-2 resonances are split by the phosphorus atom through $^2J_{P-O-C}$ and $^3J_{P-O-C-C}$ coupling. B) At pH 4, the C-1, C-3 and C-6 resonances are shifted downfield about 0.5, 0.2, and 0.1 ppm respectively. C-4 and C-5 are shifted downfield slightly and C-2 is shifted upfield to the same chemical shift causing one broadened peak to appear.

as a function of pH to give titration curves that accurately reflect the ionization state of the phosphate group and allow estimation of its pK_a value. The values of the coupling constants between P and C-1 or C-2 change with pH, but at present no pattern can be discerned in these changes.

NATURAL ABUNDANCE CARBOHYDRATES: INTERACTIONS WITH METAL IONS.

Nuclear magnetic resonance spectroscopy provides a powerful tool for examining the interactions that occur between proteins and their ligands. This potential has been realized in several cases (30–34) and it is possible under ideal conditions to create a 3-dimensional map of the ligands bound to a protein (35, 36). This mapping requires the

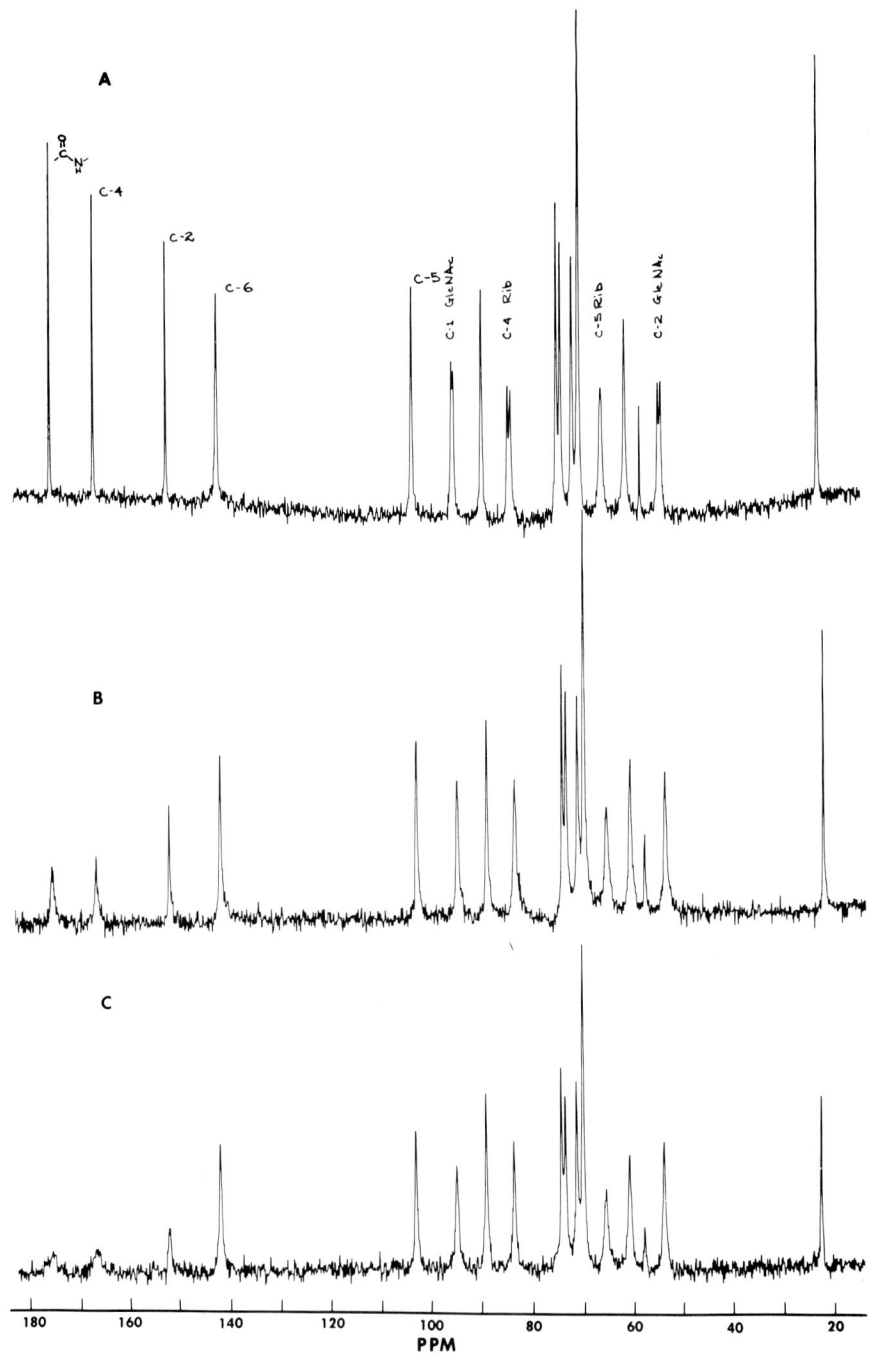

Fig. 9. The effect of Mn^{+2} ions on the proton-decoupled, ^{13}C-NMR spectra of 0.6 M UDP N-acetylglucosamine in D_2O (pD 7.0) and 37°C. A) Metal free solution. B) 5×10^{-4} M Mn^{2+}. C) 1×10^{-3} M Mn^{2+}. The spectra are discussed in the text.

presence of a paramagnetic ion, which acts to increase the rate of relaxation of magnetically susceptible nuclei, such as ^{13}C and ^{1}H, which come within its domain. The effect of the close association of a paramagnetic ion such as Mn^{2+} with a specific carbon is to broaden the resonance due to that carbon and in the extreme cause it to become unobservable.

The effect of Mn^{2+} on UDP N-acetylglucosamine is shown on Fig. 9. Very low concentrations of Mn^{2+} cause the resonances dur to the carbonyl groups of the acetyl and uracil moieties to broaden. In addition, the splitting due to P-C coupling observable in the resonances due to C-4 and C-5 of ribose and C-1 and C-2 of N-acetylglucosamine is abolished at 5×10^{-4} M Mn^{2+}. In contrast, resonances due to other carbons are affected little at 1×10^{-3} M Mn^{2+}. These observations indicate that Mn^{2+} interacts strongly with pyrophosphate moiety and with the carbonyl groups in uracil and of the acetyl group. Although no precise structure for the UDP N-acetylglucosamine/Mn^{2+} complex can be deduced from these data, they do permit the conclusions made above. Precise evaluation of the relaxation times of the carbons in the presence and absence of Mn^{2+}, together with knowledge of the reaction stoichiometry, would permit a refined model for the complex to be developed. Such studies are in progress. Since similar complexes are probably involved in biochemical conversions, and are certainly involved in Mn^{2+}-catalyzed chemical degradations (37), the evaluation of their structures is of value.

CONCLUSIONS

A number of applications of NMR spectroscopy have been described which show the kinds of information that can be obtained about carbohydrate structures, conformations, and interactions. The value of ^{13}C enrichment as a tool in structure elucidation and in following chemical and biochemical conversions has been demonstrated. The applications of these approaches to the study of more complex systems is inevitable and requires the kind of base-line studies described here.

ACKNOWLEDGMENTS

This work was supported in part by USPHS Grant GM 21731. The ^{13}C-enriched cyanide was provided by Los Alamos Scientific Laboratory.

REFERENCES

1. Farrar TC, Becker GD: "Pulse and Fourier Transform NMR." New York: Academic Press, 1971.
2. James TL: "Nuclear Magnetic Resonance in Biochemistry." New York: Academic Press, 1975.
3. Dwek RA: "NMR in Biochemistry, Applications to Enzyme Systems." London: Clarendon Press, 1973.
4. McInnes AG, Walter JA, Wright JLC, Vining LC: In Levy GC (ed): "Topics in Carbon-13 NMR Spectroscopy." New York: Wiley-Interscience, 1974, vol 1, chap 3.
5. Komoroski RA, Peat JR, Levy GC: In "Topics in Carbon-13 NMR Spectroscopy." New York: Wiley-Interscience, 1974, vol 1, chap 4.
6. Norton RS, Clouse AO, Addleman R, Allerhand A: J Am Chem Soc 99:79, 1977.
7. Mowery DF: In Whistler RL, Wolgrow ML (eds): "Methods in Carbohydrate Chemistry." New York: Academic Press, 1963, vol 2, pp 328.
8. Murray A, Williams DL: "Organic Syntheses with Isotopes." New York: Interscience Publishers, 1958.
9. Dorman DE, Roberts JD: J Am Chem Soc 92:1355, 1970.
10. Perlin AS, Casu B, Koch HJ: Can J Chem 48:2596, 1970.
11. Gorin PAJ, Mazurek M: Can J Chem 53:1212, 1975.
12. Gorin PAJ, Mazurek M: Carbohydr Res 48:171, 1976.

13. Walker TE, London RG, Whaley TW, Barker R, Matwiyoff NA: J Am Chem Soc 98:5807, 1076.
14. Gorin PAJ: Can J Chem 52:458, 1974.
15. Nunez HA, Barker R: Fed Proc Fed Am Soc Exp Biol 34:2007, 1975.
16. Bunole DR, Smith ICP, Jennings HJ: J Biol Chem 249:2275, 1974.
17. Bhattacharsee AK, Jennings HJ, Kenny CP, Martin A, Smith ICP: J Biol Chem 250:1926, 1975.
18. Perlin SA, Cyr N, Ritchie GS, Parfondry A: Carbonhydr Res 37:C1–C4, 1974.
19. Lemieux RU, Bundle DR, Baker DA: J Am Chem Soc 97:4076, 1975.
20. Gorin PAJ: Can J Chem 51:2375, 1973.
21. Barfield M, Burfitt J, Doddrell D: J Am Chem Soc 97:2631, 1975.
22. Ellis PD, Ditchfield R: In Levy GC (ed): "Topics in Carbon-13 NMR Spectroscopy." New York: Wiley-Interscience, 1976, vol 2, chap 8.
23. Coxon B: In Whistler RL, BeMiller JN (eds): "Methods in Carbohydrate Chemistry." New York: Academic Press, 1972, vol 6, pp 513–539.
24. Schwarcz JA, Perlin AS: Can J Chem 50:3667, 1972.
25. Eakin RT, Morgan LO, Gregg CT, Matwiyoff NA: FEBS Lett 28:259, 1972.
26. Matwiyoff NA, Ott DG: Science 181:1125, 1973.
27. Feeny J: In Pain RH, Smith BJ (eds): "New Techniques in Biophysics and Cell Biology." New York: Wiley-Interscience, 1975, vol 2, chap 8.
28. Anet FA, Levy GC: Science 180:141, 1973.
29. Lee C, Sarma RH: Biochemistry 15:697, 1976.
30. Fung CH, Gupta RK, Mildvan AS: Biochemistry 15:85, 1976.
31. Fung CH, Feldmann RJ, Mildvan AS: Biochemistry 15:76, 1976.
32. Fung CH, Mildvan AS, Allerhand A, Komoroski R, Scrutton MC: Biochemistry 12:620, 1973.
33. Miziorro AM, Mildvan AS: J Biol Chem 249:2743, 1974.
34. Sloan DL, Loeb LA, Mildvan AS, Feldman RJ: J Biol Chem 250:8913, 1975.
35. James TL: "Nuclear Magnetic Resonance in Biochemistry." New York: Academic Press, 1975, chap.6.
36. Brewer CF, Sternlicht H, Marcus DM, Grollman AP: Proc Natl Acad Sci USA 70:1007, 1973.
37. Nunez HA, Barker R: Biochemistry 14:3843, 1975.

Distribution of a Major Surface-Associated Glycoprotein, Fibronectin, in Cultures of Adherent Cells

Deane F. Mosher

Department of Virology, University of Helsinki, Helsinki, Finland, SF-00290, and Department of Medicine, University of Wisconsin, Madison, Wisconsin 53706

Olli Saksela, Jorma Keski-Oja, and Antti Vaheri

Department of Virology, University of Helsinki, Helsinki, Finland, SF-00290

Fibronectin was present in media and cell layers of cultures of adherent cells from human skin, kidney, lung, chest wall, liver, and heart. Cell-surface fibronectin, visualized by immunofluorescence, was in dense fibrillar (cultures from lung), discrete fibrillar (e.g., cultures from skin), or punctate (some cultures from kidney) structures. The subunit sizes of cell-surface fibronectin and fibronectin soluble in medium appeared identical in sodium dodecyl sulfate-polyacrylamide gels. To explain the polymorphism of cell-surface fibronectin, there must be chemical differences among the fibronectins synthesized by different cell strains or factors in the cell layer which influence fibronectin binding and aggregation.

Key words: binding, fibroblasts, fibronectin, immunofluorescence, receptor, secretion

Fibronectin is a glycoprotein composed of large subunits and is present in the blood, basal lamina, and blood vessels of vertebrates (1, 2). It is synthesized in fibroblast cultures, where it is found in conditioned medium, on cell surfaces (large, external, transformation sensitive protein), and intracellularly (3). Fibronectin is generally missing from the surfaces of transformed cultured cells (reviewed in Refs. 3 and 4). We estimate the subunit size of fibronectin circulating in plasma (cold-insoluble globulin) to be 2.0×10^5 daltons (5), 2.0×10^4 daltons less than the size of the subunit of cell-surface fibronectin (6, 7). However, as described below, the subunit sizes of fibronectin secreted or shed by fibroblasts into the medium and fibronectin associated with fibroblast cell surfaces appear identical. The factors that determine the distribution of fibronectin in cell cultures or within the body are not understood.

As one approach to this problem, we studied the distribution of fibronectin in cultures of a variety of adherent cells. We found that, although the amounts and subunit sizes of fibronectin were similar in cultures of adherent cells from different tissues, the arrangements of cell-surface fibronectin, visualized by immunofluorescence, were tissue specific.

Received March 29, 1977; accepted May 23, 1977

© 1977 Alan R. Liss, Inc., 150 Fifth Avenue, New York, NY 10011

MATERIALS AND METHODS

The materials and methods for lactoperoxidase-catalyzed iodination of cell-surface proteins (8), separation and detection of proteins in sodium dodecyl sulfate-polyacrylamide gels (8), radioimmunoassay of fibronectin in conditioned media and cell extracts (9; D. F. Mosher and A. Vaheri, submitted for publication), indirect immunofluorescent detection of fibronectin in fixed coverslip cultures (9; D. F. Mosher and A. Vaheri, submitted for publication), and double antibody immunoprecipitation of fibronectin intrinsically labeled with [^{35}S]-1-methionine (D. F. Mosher, O. Saksela, and A. Vaheri, submitted for publication) are described elsewhere. The antifibronectins used in the various double antibody procedures were monospecific for human plasma fibronectin (by double immunodiffusion and immunoelectrophoresis of plasma) and were adsorbed with fetal calf serum to remove cross-reactions with bovine fibronectin in the growth media (which contained 10% fetal calf serum).

Primary cultures of human embryonic tissues were prepared by standard techniques. The cells grew to confluency in 3–4 days and were subcultured (1:2 split) twice weekly. The state of fibronectin was monitored over the first few passages by radioimmunoassay of media and cell extracts and indirect immunofluorescence of cell layers. Strains of adult skin fibroblasts were established locally. Other established strains and lines of the human cells described in Table I were obtained from the American Type Culture Collection, Rockville, Maryland.

RESULTS

Newly established adherent cell strains from human embryonic skin, heart, chest wall, lung, kidney, and liver were compared to one another and to established strains from adult skin and embryonic lung and lines of rhabdomyosarcoma cells and SV-40 transformed embryonic lung cells (Table I).

All nontransformed strains had a 2.2×10^5 dalton external polypeptide which was prominently labeled by lactoperoxidase-catalyzed iodination. Extracts of 12 newly-established strains were analyzed, and the patterns of stainable polypeptide in sodium dodecyl sulfate-polyacrylamide slab gels were remarkably similar. In particular, extracts all contained a 2.2×10^5 dalton polypeptide which was bracketed by polypeptides of 2.0×10^5 and 2.5×10^5 daltons (6). The 2.2×10^5 dalton iodinated bands and protein-staining bands were of higher molecular weight (dimers and multimers) when the extracts were analyzed by electrophoresis without prior reduction (6).

All cultures had fibronectin in media and associated with cell layers. The newly established strains were robust producers of fibronectin (Table I) and secreted or shed more fibronectin into the medium than older established strains. The older established strains, in turn, secreted or shed more fibronectin into the medium than transformed lines (Table I).

The patterns of cell-associated fibronectin, visualized by indirect immunofluorescence, varied among strains derived from different tissues (Fig. 1). Cells from skin had copious amounts of patchy intracellular fibronectin in a perinuclear distribution as well as discrete surface fibrils (Fig. 1a). Cells from chest wall and liver looked much like the skin cells.

TABLE I. Fibronectin Content of Cultures 24 h After Subculture*

Cells	Passage	Protein in cell layer	Fibronectin in medium	Fibronectin in cell layer
		(μg)	(μg)	(μg)
Embryonic				
skin	6	200	14.0	1.0
kidney	6	168	10.3	0.7
heart	6	167	27.3	0.7
lung	6	274	25.5	1.8
chest wall	6	233	41.5	0.9
Mean ± SD		208 ± 41	23.7 ± 11.0	1.0 ± 0.4
ES adult skin	16	103	2.1	0.32
JKO adult skin	18	95	5.0	0.40
WI-38 embryonic lung	22	104	1.9	0.42
MRC-5 embryonic lung	19	156	6.5	0.49
Mean ± SD		115 ± 24	3.9 ± 2.0	0.40 ± 0.07
RD rhabdomyosarcoma	41	161	2.0	0.18
HT 1080 rhabdomyosarcoma	18	187	3.4	0.29
VAH transformed lung	58	149	0.7	0.12
VA13 transformed lung	132	168	1.5	0.21
RSA transformed lung	16	186	1.0	0.20
Mean ± SD		170 ± 15	1.72 ± 0.95	0.20 ± 0.05

*Values represent the mean of duplicate 20-cm^2 dishes (each containing 5 ml growth medium) sampled 24 h after subculture (1:2 split of a previously confluent culture). The differences among early-passage strains, older established strains, and transformed lines persisted through the 6th day in culture (D. F. Mosher and A. Vaheri, submitted for publication).

Heart cells had more fusiform fibrils. Cultures from lung looked similar to cultures from skin until the cultures reached confluency, and a dense network of fine extracellular fibronectin fibrils appeared (Fig. 1b). Some cultures from kidney consisted of a homogeneous population of large, round, flat cells on which fibronectin was present in punctate structures (Fig. 1c). Most of the punctate structures appeared to be in the plane of focus between the cell and substratum. Several types of fibronectin-containing cells were in secondary cultures of human kidney (Fig. 1d). In the lower left of Fig. 1d is a large round cell with both punctate and fibrillar immunofluorescence. In the upper left of Fig. 1d is a portion of an elongated cell with fibronectin fibrils oriented along the cell body. At the right of Fig. 1d are 2 cells with perinuclear (probably intracellular) fluorescence and small amounts of fibrillar fluorescence.

The older established strains of adult skin and embryonic lung fibroblasts had patterns of surface fibronectin identical to the patterns seen in the newly established strains of embryonic skin and lung, respectively. Cell-surface fibronectin was missing from the surfaces of the rhabdomyosarcoma cells and SV-40 transformed embryonic lung cells (9), and only intracellular fibronectin was seen.

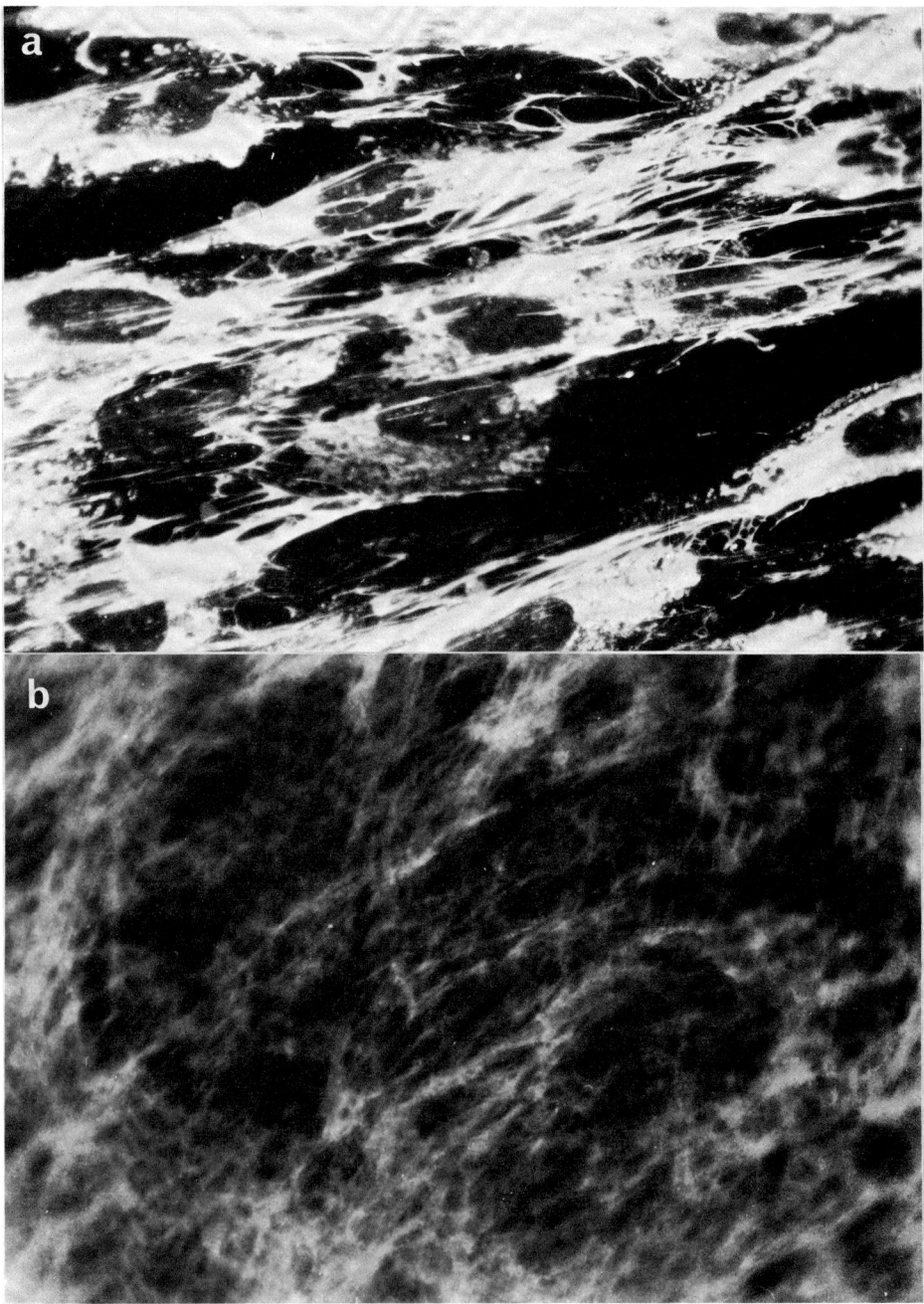

Fig. 1. Fibronectin immunofluorescence of different human embryonic cell strains. Coverslip cultures were studied 3 days after subculture. Cells were fixed with formaldehyde-acetone and stained by indirect immunofluorescence: a) embryonic skin, 4th passage; b) embryonic lung, 4th passage; c) embryonic kidney, 4th passage; and d) embryonic kidney, secondary culture. Magnification × 512.

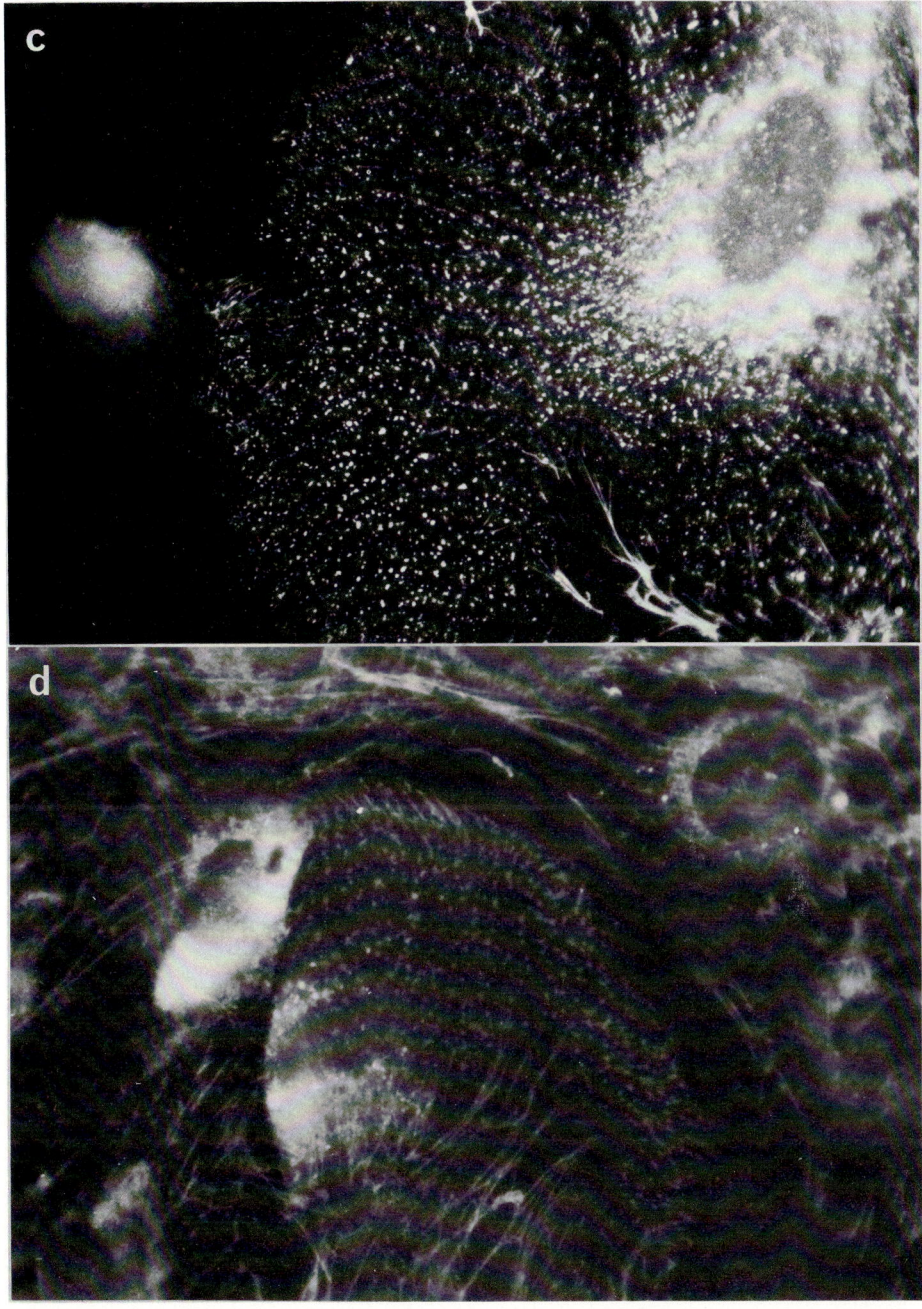

DISCUSSION

The amounts of fibronectin in culture media of the newly established strains was much greater than the amounts of fibronectin in the cell layers (Table I). Fibronectin in the media is probably in equilibrium with a portion of the fibronectin on cell surfaces. Yamada and Weston demonstrated that fibronectin (cell-surface protein) extracted from cell surfaces with low concentrations of urea could reassociate with cell surfaces (10). Hynes et al. reported that fibronectin in medium was capable of associating with cell surfaces (11). We have performed many experiments trying to demonstrate molecular weight differences between fibronectin in the medium and cell-surface fibronectin. The fibronectins were metabolically labeled with $[^{35}S]$-1-methionine, isolated by immunoprecipitation, and analyzed by sodium dodecyl sulfate-polyacrylamide slab gels combined with autoradiography. In every case, the molecular weights of fibronectins from the 2 sources were the same (12; our unpublished results).

A large fraction of cell-surface fibronectin, however, is apparently bound together by disulfide bonds (6, 13) or other types of covalent bonds (14). Most of the fibronectin-positive fibrils seen by immunofluorescence appear ultrastructurally as aggregates which are either tenuously associated with the cell surface or clearly part of the pericellular matrix (K. Hedman, A. Vaheri, and J. Wartiovaara, submitted for publication). These aggregates, if bound together by covalent bonds, could only be broken apart by proteases.

The immunofluorescence studies indicate that the distribution of fibronectin on cell surfaces varies among cell strains in a tissue-specific manner. The studies raise the question: How do cells control the distribution of cell-surface fibronectin? We have considered 4 possibilities: i) Specific surface receptors for fibronectin are arranged differently in different cell strains. ii) There are chemical differences among fibronectins synthesized by different cell strains which cause the fibronectins to bind and aggregate differently. iii) Distribution of fibronectin is influenced by other macromolecules secreted by the various strains. iv) Much of the fibronectin in the cell layer is deposited as the cells move, and cells move in a tissue-specific manner. At present, we favor the possibility of specific fibronectin receptors since it accounts for the fact that cells with different patterns of cell-surface fibronectin can exist in the same culture dish (Fig. 1d). Binding to receptors, however, would be expected to primarily influence the distribution of reversibly-bound fibronectin. The formation of fibronectin aggregates may be influenced by chemical differences among fibronectins, by interactions with other molecules of the pericellular matrix, and by cell movement. Analysis of the factors that influence the binding of fibronectin to normal cells may lead to an understanding of why most transformed cells, which do secrete or shed fibronectin (9; Table I), fail to bind it to their surfaces.

ACKNOWLEDGMENTS

This work was supported by grants awarded by the National Cancer Institute, DHEW (CA 17373), the Finnish Medical Research Council, the Finnish Cancer Foundation, and the the Sigrid Juselius Foundation.

REFERENCES

1. Linder E, Vaheri A, Ruoslahti E, Wartiovaara J: J Exp Med 142:41, 1975.
2. Ruoslahti E, Vaheri A: J Exp Med 141:497, 1975.
3. Vaheri A, Ruoslahti E, Linder E, Wartiovaara J, Keski-Oja J, Kuusela P, Saksela O: J Supramol Struct 4:63, 1976.

4. Hynes RO: Biochim Biophys Acta 458:73, 1976.
5. Mosher DF: J Biol Chem 250:6614, 1975.
6. Keski-Oja J, Mosher DF, Vaheri A: Biochem Biophys Res Commun 74:699, 1977.
7. Mosher DF: Biochim Biophys Acta (In press).
8. Keski-Oja J, Mosher DF, Vaheri A: Cell 9:29, 1976.
9. Vaheri A, Ruoslahti E: J Exp Med 142:530, 1975.
10. Yamada KM, Weston JA: Cell 5:75, 1975.
11. Hynes RO, Mautner U, Ali IU, Destree AT: J Supramol Struct (Suppl)1:36, 1977.
12. Keski-Oja J, Vaheri A, Ruoslahti E: Int J Cancer 17:261, 1976.
13. Hynes RO, Destree AT, Mautner V: In Marchesi V (ed): "Membranes and Neoplasia: New Approaches and Strategies." New York: Alan R Liss, 1977, p 189.
14. Keski-Oja J: FEBS Lett 71:325, 1976.

The Amino- and Carboxyl-Terminal Sequence of Bovine Rhodopsin

Paul A. Hargrave and Shao-Ling Fong

School of Medicine and Department of Chemistry and Biochemistry, Southern Illinois University, Carbondale, Illinois 62901

The amino terminus of bovine rhodopsin is blocked and has the sequence x-Met-Asn(CHO)-Gly-Thr-Glu-Gly-Pro-Asn-Phe-Tyr-Val-Pro-Phe-Ser-Asn(CHO)-Lys-Thr-Gly-Val-Val-Arg, where CHO represents sites of carbohydrate attachment. The carboxyl-terminal sequence of rhodopsin is Val-Ser-Lys-Thr-Glu-Thr-Ser-Gln-Val-Ala-Pro-Ala. Upon short-term digestion of rod outer segment (ROS) membranes with thermolysin, opsin ($\sim$ 35,000 daltons) is converted to a membrane-bound fragment O' ($\sim$ 30,500 daltons) and 2 peptides containing 12 amino acids are released from the carboxyl terminus of rhodopsin into the supernatant. Upon long-term digestion of ROS with thermolysin, opsin and O' are replaced by the membrane-bound fragments F_1 ($\sim$25,000 daltons), and F_2 ($\sim$9,500 daltons). When ^{32}P-ROS are digested, F_2 carries the ^{32}P. Both O' and F_1 contain the amino-terminal glycopeptide.

Key words: rhodopsin, rod cell membrane, limited proteolysis, phosphorylation site, amino-terminal, carboxyl-terminal, carbohydrate attachment

Rhodopsin is the photoreceptor protein of rod cells which are the specialized cells of the vertebrate retina responsible for black-and-white and dim-light vision. Rod outer segment membranes (ROS) contain about 50% by weight of both protein and lipid, and most of the protein (85 ± 5%) is rhodopsin (1). Rhodopsin is a glycoprotein of molecular weight 35,000–40,000 daltons (2–4). It contains 11-cis retinal in Schiff base linkage, which is important in its role as a light receptor (5). Several sulfhydryl groups in the protein are available to different chemical modification reagents (6) and cysteinyl residues have been suggested to be important in rhodopsin's cellular function (7). Rhodopsin becomes phosphorylated by a kinase in a light-dependent reaction, the role of which may be to modulate photoreceptor sensitivity (8). The only reported structural studies on rhodopsin to date are the sequence of a 9-amino acid glycopeptide (9) and the composition of the retinyl-lysine site (10). The use of proteolytic enzymes to probe the orientation of rhodopsin in its membrane environment (11) has been employed to map the location of these important functional sites of the protein in large proteolytic fragments (12). Pober and Stryer (12) have shown that thermolysin cleaves rhodopsin in the disk membrane into

Abbreviations: PAGE – polyacrylamide gel electrophoresis; PAS – periodic acid Schiff; ROS – rod outer segment; SDS – sodium dodecyl sulfate

Received March 7, 1977; accepted May 13, 1977

© 1977 Alan R. Liss, Inc., 150 Fifth Avenue, New York, NY 10011

2 large fragments: F_1, which contains carbohydrate, and F_2, which contains the retinyl-binding site.

This report summarizes our identification and sequence of the amino-terminal region of rhodopsin (13) and presents the sequence of the carboxyl-terminal tryptic peptide. We also present the current status of our efforts to localize these regions, and the phosphorylation site, in large fragments from rhodopsin.

MATERIALS AND METHODS

Preparation of Rhodopsin

Rod outer segment membranes were prepared under dim red light from frozen dark-adapted bovine retinas (Geo. Hormel and Co., Austin, Minn.) by the method of Papermaster and Dreyer (1). The membranes banding at the sucrose 1.11–1.13 g/ml interface were harvested and stored frozen at $-20°C$. In order to prepare chromatographically pure rhodopsin, ROS membranes were dissolved in 50 mM Tris acetate buffer (pH 7.8, 1 mM in Na_2EDTA) containing 5% Ammonyx L0 detergent. Agarose gel filtration was performed as described previously (14).

The Amino-Terminal Tryptic Peptide of Rhodopsin

All materials and methods concerned with the preparation of the amino-terminal tryptic peptide of rhodopsin have been previously described (13).

The Carboxyl-Terminal Region of Rhodopsin

Rhodopsin for hydrazinolysis was reduced and aminoethylated (15), dialyzed, and lyophylized. The dry residue was delipidated with chloroform:methanol (3:2) and dried over P_2O_5 under oil pump vacuum in an Abderhalden dryer ($64°C$). Protein content of the dry material was determined by amino acid analysis on acid-hydrolyzed weighed samples. Hydrazinolysis was performed at $85°C$ for varying times according to the procedure of Braun and Schroeder (16). Amino acids were separated from hydrazides by chromatography on Amberlite CG-50 resin (16). Norleucine was employed as an internal standard.

Peptide T2 from rhodopsin was obtained by chromatography on P6 polyacrylamide gel in 5% acetic acid (13). It was obtained in pure form by rechromatography on the P6 column. Homogeneity was verified by paper chromatography in butanol:acetic acid:water (17) and by electrophoresis at pH 1.65 (18). Two-hundred nanomoles of T2 was linked to triethylenetetramine resin and subjected to the automated Edman degradation on a Sequemat Model 12 solid phase sequencer (Sequemat, Inc., Watertown, Mass.) by the method of Laursen (19). PTH-amino acids were produced by the method of Tarr (20) and identified both by chromatography on polyamide sheets (21) and on silica gel using chloroform:ethanol (98:2) (22). One-thousand nanomoles of peptide T2 (670 nmol/ml) was digested with 600 μg thermolysin (CalBiochem) in 50 mM N-ethylmorpholine acetate (pH 8.0) at $40°C$ for 2.5 h. The digestion mixture was applied to electrophoresis paper and subjected to 7 V/cm for 30 min followed by 36 V/cm for 3 h in formic acid, pH 1.65. Peptides were visualized with dilute ninhydrin spray, cut out, and eluted with 5% formic acid (17). Peptide sequences were determined by the dansyl-Edman method (23).

The carboxyl-terminal tryptic peptide of rhodopsin was prepared by a method designed to be selective for such a peptide (24). Briefly, the method involves reduction and aminoethylation of the protein followed by esterification of all carboxyl groups of the protein with glycinamide. The modified rhodopsin is then digested with 10% by weight of trypsin, followed by carboxypeptidase B. The peptide mixture, in 8 M urea, is chromatographed on Dowex 1 resin at pH 11, and the peptide-containing eluant desalted by chromatography on P2 polyacrylamide gel. The purity of the carboxyl-terminal peptide was assessed by chromatography on a 1.0 × 215 cm column of P6 resin (100–200 mesh) equilibrated in 100 mM NH_4HCO_3. The peptide-containing eluant fractions from the column were further examined by preparative paper electrophoresis (as described above for peptide T2). Peptides were visualized with fluorescamine (25) and eluted from the paper as before.

Limited Proteolysis of ROS Membranes

Thermolysin digestion of ROS was performed according to Pober and Stryer (12) based on the procedure of Saari (11). Digestion was terminated either by addition of Na_2EDTA (to 10 mM) or by addition of o-phenanthroline (to 1 mM). Samples of ROS were prepared for electrophoresis by dissolving in an SDS-cocktail (1) and were submitted to SDS-PAGE (26). Gels were scanned using a Varian Techtron 635 spectrophotometer attachment. Molecular weights were estimated graphically based on the mobility of 7 standard proteins.

Membranes from digestion experiments were pelleted by centrifugation (40,000 × g, 20 min), resuspended (10 mM Tris-acetate, pH 6.9, 5 mM in Na_2EDTA) and the centrifugation repeated. The supernatants were further analyzed by amino acid analysis and by preparative paper electrophoresis. Membranes were dissolved by stirring overnight in the dark in buffer containing Ammonyx L0. The detergent extract was clarified by centrifugation (40,000 × g, 30 min). The extract was made 1 mM in $MgCl_2$, $MnCl_2$, and $CaCl_2$ prior to loading on a column of Concanavalin A-Sepharose (Pharmacia). The column was eluted with 50 mM Tris-acetate buffer (pH 6.9, 0.3% in Ammonyx L0, 1 mM in $MgCl_2$, $MnCl_2$, and $CaCl_2$), followed by the same buffer containing 500 mM α-methyl glucoside (Sigma). Protein-containing eluates were pooled and concentrated by ultrafiltration. Following dialysis to reduce the detergent concentration, samples were digested with trypsin (13). Digests were made 50 mM in pyridine, titrated to pH 2.4 with acetic acid, and chromatographed on a 0.9 × 6.0 cm column of AG50WX8 resin in the same buffer. Additional peptides were eluted with a 200 mM pyridine acetate buffer (pH 3.1). Peptides were further separated on a preparative scale by paper electrophoresis.

Phosphorylation of ROS membranes was performed by the procedure of Kühn et al. (27) as modified by McDowell and Kühn (28). Gels to be sliced were fixed in solutions lacking Coomassie Blue (26). One-millimeter slices, prepared using a Mickle gel slicer, were dissolved by heating with 0.5 ml 30% H_2O_2 in capped vials at 50°C for 15 h (29). Radioactive counting was performed using a Triton X114-xylene scintillation cocktail (30).

RESULTS

Rhodopsin and its Tryptic Glycopeptide Lacks a Free Amino Terminus

When the tryptic peptides of aminoethyl-rhodopsin are subjected to cation-exchange chromatography at low pH, several acidic peptides are not retained by the column and are eluted in the early fractions from the column (Fig. 1, inset). Three peptides have been

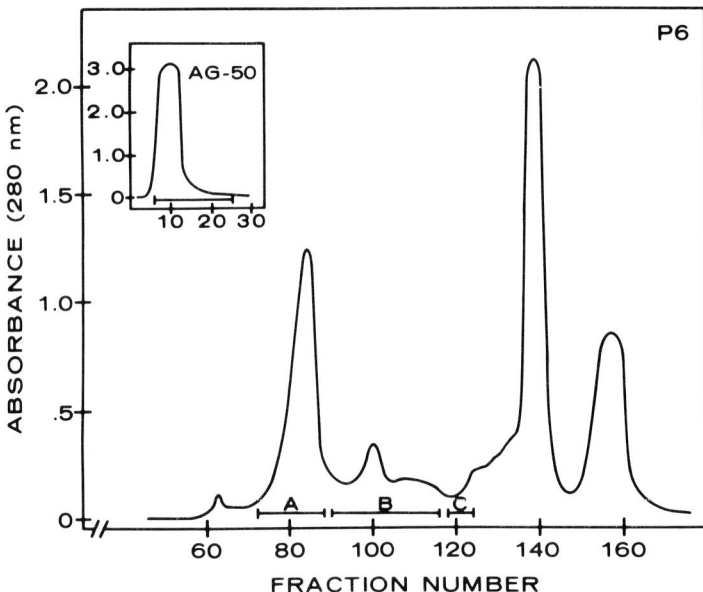

Fig. 1. Chromatography of the soluble tryptic peptides of aminoethyl rhodopsin. Inset) Soluble peptides from a tryptic digest of aminoethyl rhodopsin were chromatographed on a 0.9 × 60 cm column of AG50WX8 resin in 50 mM pyridine acetate buffer (pH 2.4) at 60 ml/h. 3.0 ml fractions were collected. Column frations which were found to contain glucosamine by amino acid analysis were pooled as indicated. Figure) Pooled glucosamine-containing fractions from chromatography on AG50 (inset) were concentrated and applied to 1.0 × 200 cm column of Biogel P6 (100–200 mesh) equilibrated in 5% acetic acid. 1.0 ml fractions were collected and their $A_{280\,nm}$ determined. Peptides were detected by paper electrophoresis of hydrolyzed aliquots. No peptide material was found in the $A_{280\,nm}$-absorbing peaks following pool C.

separated from this unbound material by gel filtration (Fig. 1). From pool A was obtained a 16-amino acid glycopeptide, T1 (Table I). Peptide T1 does not have a free amino terminus, as demonstrated by dansylation and by the Edman reaction.

Dansylation of rhodopsin prior to or following the Edman reaction shows no amino-terminal amino acid for the protein. Thus the amino-terminal blocked peptide T1 is a likely candidate for the amino-terminal peptide of rhodopsin. A method specific for isolating only the amino-terminal peptide from a protein (by tryptic cleavage at arginyl residues) produced a single peptide, T1', from rhodopsin (Table I). This peptide (T1') contained peptide T1 in its sequence. The primary sequence of this amino-terminal region of rhodopsin has been determined (13):

x-Met-Asn(CHO)-Gly-Thr-Glu-Gly-Pro-Asn-Phe-Tyr-Val-Pro-Phe-Ser-Asn(CHO)-Lys-Thr-Gly-Val-Val-Arg.

The nature of the amino-terminal blocking group is currently under investigation.

Identification of the Carboxyl-Terminal Region of Rhodopsin

When hydrazinolysis is performed on chromatographically-pure rhodopsin, alanine is identified as the carboxyl-terminal amino acid (Table II). A tryptic peptide

TABLE I. Amino Acid Analysis of Rhodopsin Peptides

Amino acid	T1[a]	T1'[a,b]	T2	T2-Th1[c]	T2-Th2[d]	T2'[e]	Th-196
Asx	3.0 (3)	3.0 (3)					
Thr	1.0	2.0 (2)	1.8 (2)		1.7 (2)	1.6 (2)	1.8 (2)
Ser	0.99 (1)	1.1 (1)	1.1 (1)		0.94 (1)	0.98 (1)	1.8 (2)
Glx	1.1 (1)	1.3 (1)	2.1 (2)		2.0 (2)	2.1 (2)	2.1 (2)
Pro	2.1 (2)	N.Q. (2)	1.1 (1)	1.0 (1)		0.97 (1)	
Gly	2.0 (2)	2.8 (3)				2.2 (2)	
Ala			2.0 (2)	2.0 (2)		2.0 (2)	
Val	0.95 (1)	3.1 (3)	1.2 (1)	0.73 (1)		1.0 (1)	1.0 (1)
Met	0.28 (1)[f]	0.60 (1)					
Tyr	0.90 (1)	N.Q.[g] (1)					
Phe	1.9 (1)	1.7 (2)					
Lys	0.89 (1)	1.1 (1)					1.0 (1)
GlcN	3.5	1.9					
% yield[h]	35	23	37	70	60	15	–

Numbers which are underlined are values used for normalization. Numbers in parentheses are integral values for the moles of amino acid present per mole of peptide. N.Q.) not quantitated but present.
[a]Hydrolyzed in the presence of phenol
[b]72-h hydrolysis
[c]Prepared by paper electrophoresis; mobility 53 cm
[d]Prepared by paper electrophoresis; mobility 41 cm
[e]Prepared by paper electrophoresis; mobility 23.6 cm
[f]Met = 0.88 when hydrolyzed in absence of phenol
[g]Could not be quantitated due to coelution with glucosamine in this analysis. Detected by paper electrophoresis.
[h]Percent yields are based on the starting material for that peptide, and do not take into account aliquots removed for analytical purposes.

TABLE II. Hydrazinolysis of Rhodopsin and its Carboxyl-Terminal Peptide

	Rhodopsin moles amino acid/mole rhodopsin		Peptide T2 moles amino acid/mole peptide	
Amino acid	60 h	80 h	60 h	100 h
Thr		0.10		
Ser		0.31		
Gly	0.07	0.20		
Ala	0.40	0.83	0.43	0.60

All yields are normalized to 100 % recovery of the norleucine internal standard.

(T2) which lacks a basic amino acid in its composition, was prepared from pool C, Fig. 1. It contains alanine (Table I) and it has an alanine carboxyl terminus as shown by hydrazinolysis (Table II). Upon digestion of T2 with thermolysin, 2 peptides were produced in good yield [T2-Th1 and T2-Th2 (Table I)]. Their sequences were determined by the dansyl-Edman technique to be Val-Ala-Pro-Ala and Thr-Glx-Thr-Ser-Glx respectively. Peptide T2, when analyzed by the solid-phase Edman method, gave the sequence

$$\text{Thr-Glu-Thr-Ser-Gln-Val-Ala-Pro-Ala.}$$

The carboxyl-terminal tryptic peptide of rhodopsin was then selectively prepared from glycinamide-modified rhodopsin. The purified peptide, T2', has the amino acid composition of T2 and 2 moles of glycine (Table I). One round of the dansyl-Edman reaction showed its amino-terminal sequence to be Thr-Glx.

Since both T2 and T2' are produced by a tryptic cleavage, the amino acid preceding the amino-terminal threonine must be basic. An overlapping peptide (Th-196) which extends this sequence has been prepared from a thermolytic digest of aminoethyl-rhodopsin (P. A. Hargrave, C. V. Barber, R. W. Siemens, K. A. Woodin and W. J. Dreyer, unpublished). Its composition is given in Table I. Its sequence, obtained by the dansyl-Edman method is

$$\text{Val-Ser-Lys-Thr-Glx-Thr-Ser-(Glx).}$$

Proteolytic Digestion of Rhodopsin in ROS Membranes

The action of thermolysin on ROS membranes resulted in the production of 3 large membrane-bound fragments (Exp. 1, Fig. 2). Opsin (O, ~ 35,000 daltons) is first converted to a smaller fragment O' (~ 30,500 daltons) in a reaction which is virtually complete in 5 min. In a separate experiment (Exp. 2), photographs of gels show the same conversion of O to O' (Fig. 3). O' is then converted to membrane-bound fragments F_1 (~ 25,000 daltons) and F_2 (~ 9,500 daltons). Opsin, O', and F_1 all stain with the PAS reagent, but F_2 does not (data not shown).

The supernatant from the 5-min digestion of ROS (Exp. 2), and its control, were further examined by amino acid analysis and by preparative paper electrophoresis. In the experimental sample, 6.5% of the 56.8 mg of ROS protein was solubilized, and in the control 0.55%. By paper electrophoresis 2 peptides from the digestion supernatant were detected with ninhydrin spray: Th-S1 (48-cm mobility) and Th-S2 (40.5-cm mobility). No additional peptides were subsequently detected using peptide-bond spray, and no peptides were found in the supernatant from control membranes incubated in the absence of thermolysin. Peptide Th-S1 had the composition of peptide Th-196 and also showed the amino-terminal sequence Val-Ser-Lys. Peptide Th-S2 had the composition of peptide T2-Th1 and also showed the amino-terminal sequence Val-Ala-Pro. More peptides were detected in the supernatant from the 15-h digestion, but they have not yet been completely characterized. Glucosamine is absent from their amino acid composition.

The protein composition of the membrane was also examined. Undigested ROS membranes, membranes after 5 min digestion (Exp. 2), and membranes after 15 h digestion (Exp. 1), were dissolved in Ammonyx L0 and chromatographed on Concanavalin A-Sepharose in the dark. The nature of the material prepared (rhodopsin, O', and F_1-F_2) was verified by SDS-PAGE. Following trypsin digestion the most acidic tryptic peptides were obtained by ion-exchange chromatography and further separated by preparative paper electrophoresis. All 3 protein samples were found to contain the amino-terminal tryptic peptide T1 (mobility 11.5–12 cm).

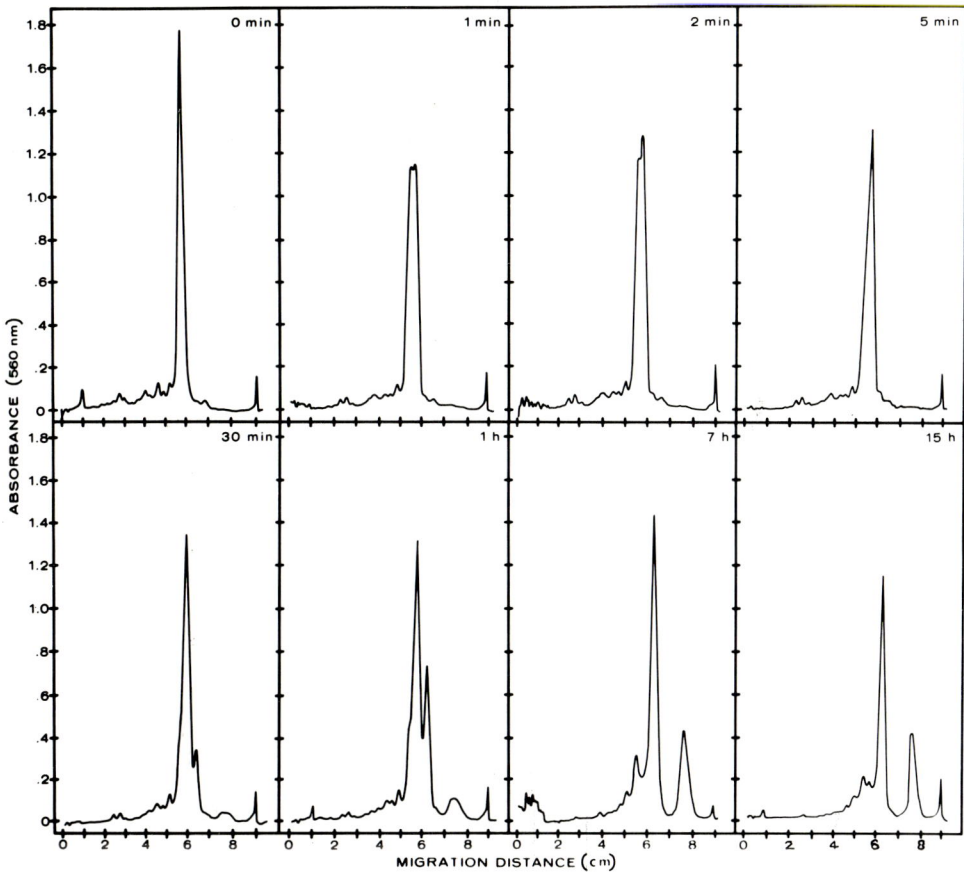

Fig. 2. Thermolysin digestion of ROS membranes: spectrophotometer tracings of stained SDS-PAGE gels. Gels were scanned at 560 nm using a Varian Techtron 635 spectrophotometer with gel scanner attachment. The spike in the tracings at 9 mm shows a nichrome wire which was inserted to mark the position of the tracking dye.

When rhodopsin, O′, or the F_1-F_2 complex of rhodopsin was loaded on the Concanavalin A-Sepharose column and eluted from it in the dark, protein recovery of greater than 90% was obtained. Pober and Stryer have shown that illuminating the F_1-F_2 complex on the column causes the dissociation of F_2, and F_1 may then be prepared by elution with α-methyl glucoside (12). However, when we reloaded the purified F_1-F_2 complex and attempted to prepare quantities of purified F_1 and F_2, poor yields were obtained. Figure 4 shows the results of an experiment in which an extract of ROS (digested to produce F_1-F_2) was chromatographed on Concanavalin A-Sepharose. Of the 26.9 mg protein loaded, 1.6 mg emerged in the unbound fraction (pool A), 1.1 mg was eluted following light exposure (pool B), and 0.95 mg was eluted with α-methyl glucoside (pool C). Examination by SDS-PAGE showed F_2 in pool B, the F_1 (and its aggregates) in pool C, as expected. Their amino acid analyses are in substantial agreement with those reported by Pober and Stryer (12).

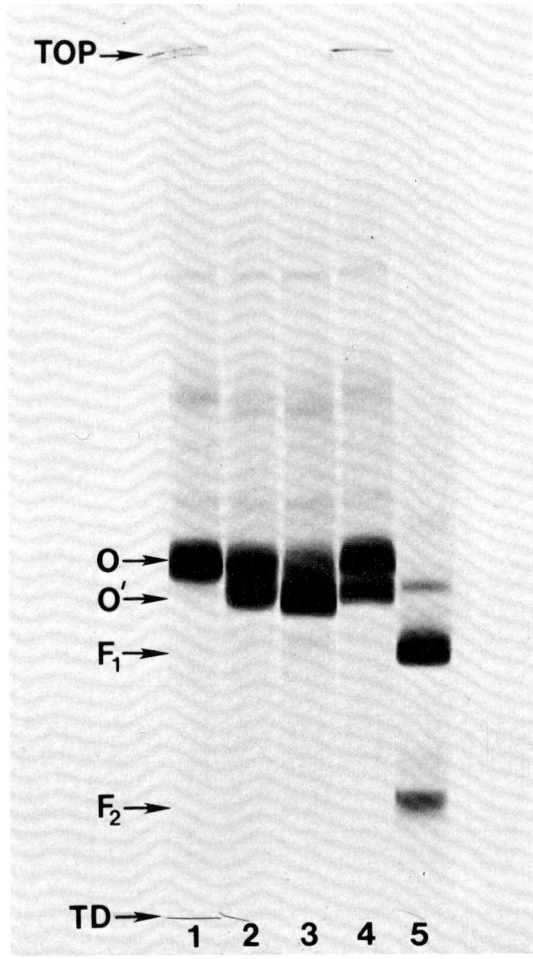

Fig. 3. Thermolysin digestion of ROS membranes: photograph of Coomassie Blue-stained SDS-PAGE gels. Gels 1–4 are from Exp. 2, a 5-min thermolysin digestion experiment in which gel 1 = 0 time, gel 2 = 1 min, gel 3 = 5 min, gel 4 = a mixture of 0-time and 5-min. Gel 5 is from a 15-h digestion of ^{32}P-ROS (Exp. 3). O = opsin, O′, F_1, and F_2 are digestion products of opsin. The nichrome wire at the base of the gels shows the original position of the tracking dye (TD).

The Location of the Phosphorylation Site(s)

Light-exposed phosphorylated ROS were submitted to SDS-PAGE before and after thermolysin digestion (Exp. 3, Fig. 5). All ^{32}P which was incorporated into ROS membrane proteins is incorporated into rhodopsin. Following proteolysis, ^{32}P migrates with the F_2 fragments and not with the large F_1. Heterogeneity in the F_2 region of the gel is shown by the radioactivity profile. In some digests of ROS we have observed a closely spaced Coomassie Blue-staining doublet in the F_2 region.

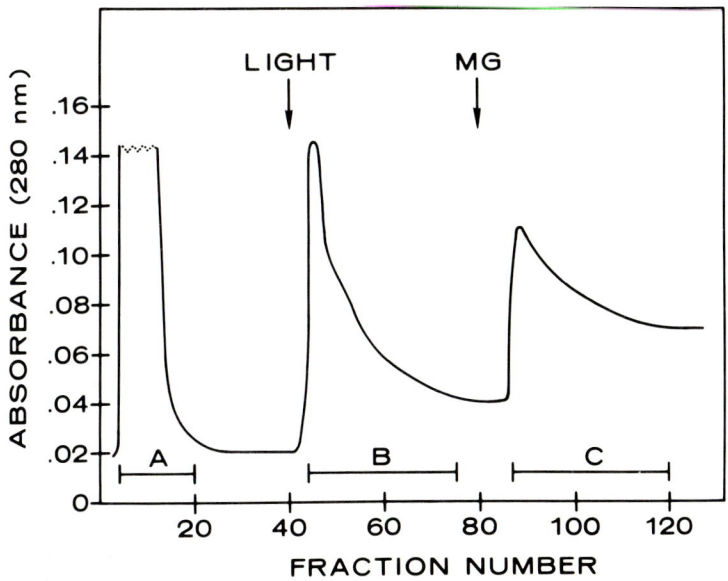

Fig. 4. Preparation of rhodopsin fragments F_1 and F_2 by chromatography on Concanavalin A-Sepharose. Thermolysin-digested ROS were solubilized in buffer containing Ammonyx L0 (see Methods) and chromatographed on a 0.9 × 17.5 cm column of Concanavalin A-Sepharose. Column flow of 12 ml/h was interrupted for 1 h during exposure to light. Flow was reduced to 6 ml/h during elution with α-methyl glucoside (MG). Fractions A, B, and C were pooled as shown.

DISCUSSION

Our results have shown that the amino-terminus of rhodopsin is blocked and that the blocked peptide T1 is the amino-terminus of rhodopsin (13). Peptide T1 is hydrophilic in composition and contains 2 sites at which carbohydrate is attached. The single carbohydrate site previously reported had been thought to account for all of rhodopsin's carbohydrate (9). This region of the rhodopsin molecule would be expected to be exposed at the membrane surface.

In contrast to a previous study in which hydrazinolysis was reported to yield no carboxyl-terminal amino acid for rhodopsin (31), we find alanine. Variable submolar amounts of glycine and serine were also produced from rhodopsin, and from lysozyme which we examined as a control. This has been reported to occur with other proteins (16). Due to its lack of a basic amino acid, the tryptic peptide T2 would be expected to be the carboxyl-terminal peptide of rhodopsin. Like rhodopsin, its carboxyl-terminal is alanine. The carboxyl-terminal status of T2 is further confirmed by its isolation using a method specific for preparing carboxyl-terminal peptides from proteins. The thermolytic peptide Th-196 contains the tryptic cleavage site and extends our sequence in the carboxyl-terminal region. This region of the molecule is hydrophilic in composition and might be expected to be exposed to an aqueous environment.

The first event in the limited thermolytic digestion of rhodopsin in ROS has been shown to be production of a membrane-bound fragment O' which has an apparent molecular weight of 30,500 daltons. This is accompanied by a release of 2 peptides into the

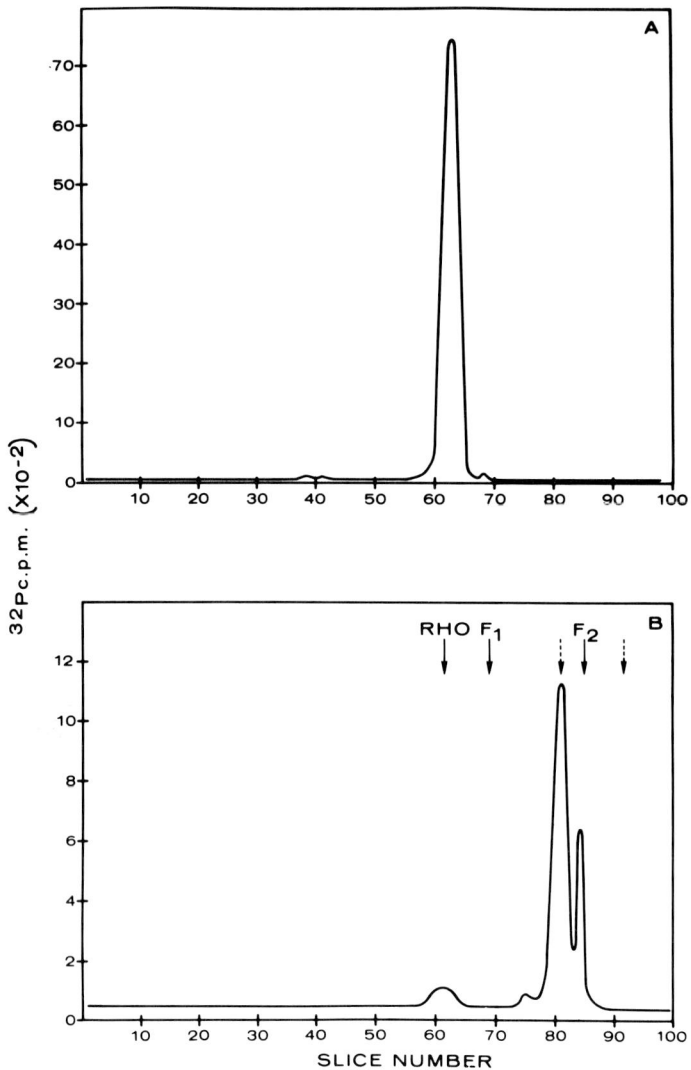

Fig. 5. ^{32}P distribution in ROS and thermolysin-digested ROS. Unstained SDS-PAGE gels of ^{32}P-ROS (A) and thermolysin-digested ^{32}P-ROS (B) were sliced in 1-mm pieces and their radioactivity determined. In B, solid arrows show the positions of rhodopsin and its F_1 and F_2 fragments in a stained duplicate gel. Dashed arrows show the positions of 2 very weakly staining bands.

supernatant. These 2 thermolytic peptides originate from the carboxyl-terminal region of rhodopsin and account for 1,200 daltons molecular weight. No glucosamine-containing peptides are released into the supernatant, and the amino-terminal tryptic peptide can be prepared from the purified O′ fragment. Additional thermolytic cleavage may have occurred at the carboxyl terminal of rhodopsin to produce small undetected membrane-bound fragments. However, the apparent molecular weights of rhodopsin and its fragments as determined by SDS-PAGE probably do not adequately reflect their true molecular weights.

The rapid production of an intermediate which may be comparable to O′ has been observed by the action of subtilisin on ROS (32). Since these authors used membranes of native sidedness, this would imply that the carboxyl-terminal region of rhodopsin is located at the extradiskal surface of the disk membrane. It is probable that the amino-terminal region of rhodopsin is located at the intradiskal membrane surface: we have shown that the amino-terminal sequence has sites of carbohydrate attachment in rhodopsin, and carbohydrate has been cytochemically localized only at the inner disk surface (33).

Following the rapid conversion of opsin to O', O' is converted to F_1 and F_2 over a period of hours. We have made no attempt to quantitate this conversion, but Saari (J. Saari, personal communication) has found F_1 and F_2 to appear concurrently and in approximately equal molar quantities. We have observed the appearance of several additional peptides in the supernatant during this conversion. No glucosamine-containing peptides are released. F_1 still stains with PAS reagent, and the amino-terminal peptide T1 is still present in the F_1-F_2 complex. Thus, the soluble peptides must come either from the region of internal cleavage which produces F_1 and F_2 from O', or from the caboxyl-terminal region of F_2.

The retinyl site of rhodopsin has been shown to be located in the F_2 fragment (12, 32). We have located the phosphorylation site(s) also in F_2 [as has Saari (personal communication)]. F_2, by its staining and ^{32}P profile, does not appear to represent a unique single fragment but is probably a small family of overlapping peptides from the carboxyl-terminal region of rhodopsin.

It would be desirable to prepare fragments F_1 and F_2 in quantity in order to further investigate their primary structure. Although we were able to elute rhodopsin and its F_1-F_2 noncovalent complex from the Concanavalin-A Sepharose column in the dark in 90% yield, very low yields were obtained for F_1 and F_2 following light exposure on the column. Our experimental conditions were not identical to those of Pober and Stryer (12), however. Better preparative scale yields might be obtained by use of their lightly-substituted Concanavalin A-agarose gel and their detergent-buffer conditions.

ACKNOWLEDGMENTS

I would like to thank Ms. Donna R. Dempsey for her capable technical assistance, and Ms. Pat Tindall for performing many of the amino acid analyses. This work was supported in part by the SIU School of Medicine and NIH Grant EY 1275.

REFERENCES

1. Papermaster DS, Dreyer WD: Biochemistry 13:2438, 1974.
2. Hubbard R: J Gen Physiol 37:381, 1954.
3. Daemen FJM, de Grip WJ, Jansen PAA: Biochim Biophys Acta 271:419, 1972.
4. Lewis MS, Krieg LC, Kirk WD: Exp Eye Res 18:29, 1974.
5. Wald G: Science 162:230, 1968.
6. de Grip WJ, Bonting SL, Daemen FJM: Biochim Biophys Acta 396:104, 1975.
7. Wald G, Brown P, Gibbons I: J Opt Soc Am 53:20, 1963.
8. Kühn H, Bader S: Biochim Biophys Acta 428:13, 1976.
9. Heller J, Lawrence MA: Biochemistry 9:864, 1970.
10. Bownds D: Nature 216:1178, 1967.
11. Saari JC: J Cell Biol 63:480, 1974.
12. Pober JS, Stryer L: J Mol Biol 95:477, 1975.
13. Hargrave PA: Biochim Biophys Acta 492:83, 1977.
14. Hargrave PA: Vision Res 16:1013, 1976.

15. Cole RD: In Hirs CHW (ed): "Methods in Enzymology." New York: Academic Press, 1967, vol 11, p 315.
16. Braun V, Schroeder WA: Arch Biochem Biophys 118:241, 1967.
17. Bennett JC: In Hirs CHW (ed): "Methods in Enzymology." New York: Academic Press, 1967, vol 11, p 330.
18. Dreyer WJ, Bynum E: In Hirs CHW (ed): "Methods in Enzymology." New York: Academic Press, 1967, vol 11, p 32.
19. Laursen RA, Bonner AG, Horn MJ: In Perham RN (ed): "Instrumentation in Amino Acid Sequence Analysis." Academic Press, 1975, p 73.
20. Tarr GE: Anal Biochem 63:361, 1975.
21. Summers MR, Smythers GW, Oroszlan S: Anal Biochem 53:624, 1973.
22. Horn MA: Personal communication.
23. Gray WR, Smith JF: Anal Biochem 33:36, 1970.
24. Hargrave PA, Wold F: Int J Peptide Protein Res 5:85, 1973.
25. Mendez E, Lai CY: Anal Biochem 65:281, 1975.
26. Fairbanks G, Steck TL, Wallach DFH: Biochemistry 10:2606, 1971.
27. Kühn H, Cook JH, Dreyer WJ: Biochemistry 12:2495, 1973.
28. McDowell JH, Kühn H: Biochemistry (In press).
29. Richards IC: Personal communication.
30. Anderson LE, McClure WO: Anal Biochem 51:173, 1973.
31. Heller J: Biochemistry 7:2906, 1968.
32. Gaw JE, Dratz EA, Vandenberg CA, Slaughter SE, Kelton EA, Lipschultz MJ, Schwartz S: Biophys J 17:80a, 1977.
33. Rölich P: Nature (London) 263:789, 1976.

Sialic Acid Uptake by BHK Cells and Subsequent Incorporation Into Glycoproteins and Glycolipids

Carlos B. Hirschberg and Mary Yeh

Edward A. Doisy Department of Biochemistry, St. Louis University School of Medicine, St. Louis, Missouri 63104

BHK cells can be grown in the presence of growth medium to which radiolabeled sialic acid has been added. After 24 h, 85% of the radioactivity in the cells is covalently bound to glycoproteins and glycolipids. No metabolism of the radiolabeled sialic acid could be detected.

Key words: sialic acid uptake, sialoglycoproteins, sialoglycolipids

We recently reported that BHK cells grown in monolayers were permeable to sialic acid (1). It was also observed that after a 1-h incubation of the cells with sialic acid approximately 18% of the sialic acid that was associated with the cells was insoluble in phosphotungstic acid. This led us to speculate that a portion of the sialic acid became covalently bound to glycoproteins and glycolipids subsequent to its entry into the cell. In addition no metabolism of the radiolabeled sialic acid could be detected.

These observations stimulated a study to determine whether BHK cells could be grown for relatively long periods (24 h) in the presence of radiolabeled sialic acid to specifically label the cellular sialoglycoproteins and sialoglycolipids (2).

MATERIALS AND METHODS

Cells

BHK 21/13 cells were a gift of Dr. M. Green (St. Louis University). Cells were grown in Dulbecco's modification of Eagle's minimal essential medium (MEM) containing in addition 4 times the normal concentration of amino acids and vitamins (3). Cells were free of mycoplasma based on their level of uridine phosphorylase activity (4) and were counted in a Celloscope particle counter.

[^{3}H]-Sialic Acid

N-acetyl-[^{3}H]-neuraminic acid was synthesized enzymatically from [G-^{3}H]-N-acetyl D-mannosamine as previously described (1). The specific activity was 2.3 Ci/mmol. Radiochemical purity was at least 99.5% based on the criteria previously described (1).

Received March 22, 1977; accepted May 24, 1977

© 1977 Alan R. Liss, Inc., 150 Fifth Avenue, New York, NY 10011

Time Course of Sialic Acid Uptake

Cells (3.9×10^5/35-mm plate) were grown in growth medium (1.3 ml; MEM containing 10% fetal calf serum) to which 12.5 μCi of [^{3}H]-sialic acid was added. At different times the medium was removed and the cells were rinsed 7 times with 2 ml of solution A (0.8% NaCl, 0.05% KCl, 0.001 M KPO$_4$, pH 7.4). Cold (4°C) phosphotungstic acid (PTA; 0.5 ml; 1% in 0.5 N HCl) was then added to each plate and the mixture was allowed to remain on ice for 15 min. The cells were then scraped with a rubber "policeman" into a 12-ml conical glass tube, and the plate was rinsed with an additional 0.5 ml of PTA. After centrifugation in a clinical centrifuge at maximum speed for 5 min the supernatant was removed and saved. The pellet was washed twice, each time with 0.5 ml of PTA. The pellet was dissolved in protosol (New England Nuclear Corporation) and the radioactivity in it and each supernatant fraction was determined as previously described (1). The cell number was determined at each time point in a parallel grown plate. After 32 h there were 2.0×10^6 cells/plate.

Identification of the Radioactivity Within Cells After 24 Hours Labeling

7.5×10^5 cells/35-mm plate were grown in growth medium (1.3 ml) to which [^{3}H]-sialic acid (12.5 μCi) had been added. After 24 h (9.4×10^5 cells/plate) the medium was removed and the cells were rinsed 7 times with 2 ml solution A. The 7th wash contained 200 dpm. Two milliliters of cold PTA was then added to the plate and allowed to remain on ice for 15 min. Cells were then scraped into a conical glass tube and the plate rinsed with an additional 1 ml of cold PTA. After centrifugation the supernatant (3,000 dpm) was removed and the pellet was washed with 1 ml of PTA. This supernatant contained 310 dpm. The pellet was washed with an additional 1 ml of PTA and after centrifugation this supernatant contained no radioactivity. The pellet was then washed twice, each time with 2 ml of water.

The pellet was suspended in H$_2$SO$_4$ (0.1 N; 0.75 ml) and heated for 1 h at 80°C with occasional stirring. After centrifugation at top speed in a clinical centrifuge, the supernatant was removed and saved. The pellet was washed with 0.25 ml of 0.1 N H$_2$SO$_4$ and after centrifugation the 2 supernatants were combined (42,000 dpm). The pellet was dissolved in protosol and the radioactivity (4,040 dpm) was determined as previously described (1). The supernatant was diluted 10-fold with water and applied to a Dowex-formate column. At least 90% of the radioactivity was characterized as sialic acid based on its behavior on this column and by ascending paper chromatography (1). The radioactivity which was soluble in PTA was applied to a Dowex-formate column after a 30-fold dilution with water. At least 90% behaved as sialic acid.

Biogel P-10 Chromatography

A PTA-insoluble pellet (700 μg of protein) was dissolved in 0.4 ml of sodium dodecyl sulfate (1% in 0.1 M phosphate buffer, pH 7.0; 0.5 ml) by boiling for 5 min. An aliquot of the sample (150 μl) was then applied to a Biogel P-10 column (13×1.0 cm) packed in the same buffer. We collected 0.5-ml fractions. For pronase treatment, 300 μg of pronase were added to an aliquot of the solubilized sample and the mixture was incubated at 37°C for 72 h. Every 12 h, an additional 300 μg of pronase was added to the reaction mixture.

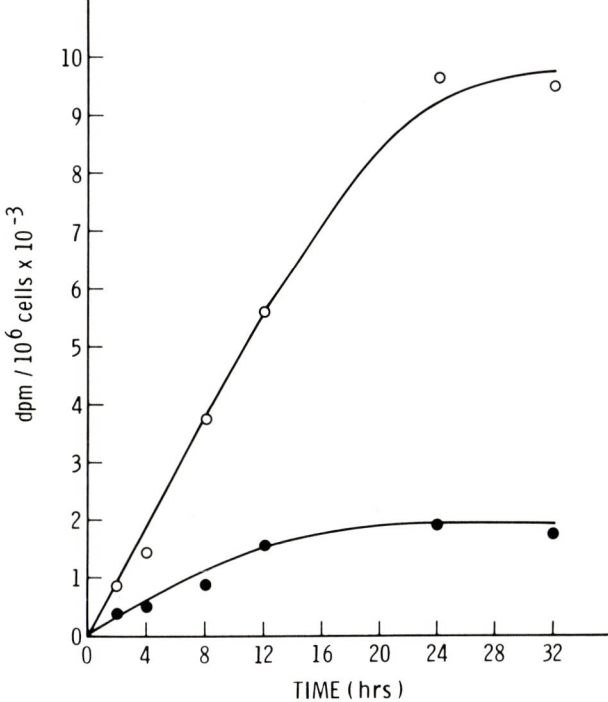

Fig. 1. Time course of [^{3}H]-sialic acid incorporation into BHK cells. Cells (3.9 × 10^5/35-mm plate) were grown in the presence of 12.5 μCi of [^{3}H]-sialic acid. The cells were washed and phosphotungstic acid (1% in 0.5 N HCl) soluble (●---●) and insoluble (○---○) radioactivity was determined as described under Methods.

SDS-Acrylamide Gel Electrophoresis

A PTA-insoluble sample (0.25 mg of protein) was solubilized in 50 μl of a mixture containing 0.05 M Na_2CO_3, 0.05 M dithiothreitol, 2% (wt/vol) sodium dodecyl sulfate, 12% (wt/vol) sucrose and 0.04% (wt/vol) bromophenol blue. Electrophoretic separation and subsequent staining, drying, and counting of gel slices (2 mm each) was performed as previously described (5).

Lipid Extraction

A PTA-insoluble pellet, from cells labeled for 24 h with radiolabeled sialic acid was dried under N_2. The pellet was extracted twice with 2 ml of chloroform-methanol 2:1. An aliquot was applied to a thin layer plate and developed in chloroform-methanol-water, 60:35:8, as previously described (6). Standard hematoside was obtained by growing cells with 1-[^{14}C]-palmitate (7).

RESULTS

Kinetics of [^{3}H]-Sialic Acid Incorporation Into Cells

Figure 1 shows that the PTA-insoluble radioactivity increased linearly for about 16 h

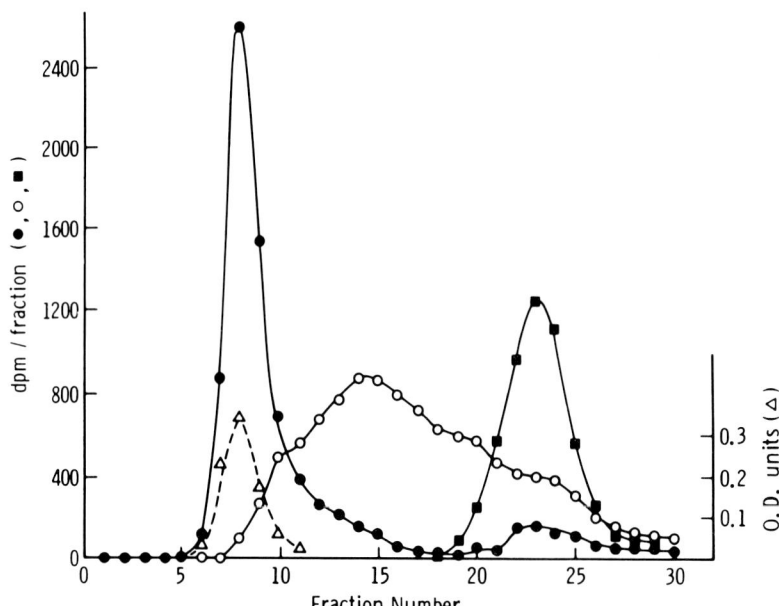

Fig. 2. Biogel P-10 chromatography. A phosphotungstic acid-insoluble pellet obtained from BHK cells which had been grown for 24 h in the presence of radiolabeled sialic acid was solubilized in 1% sodium dodecyl sulfate. An aliquot was applied to a Biogel P-10 column as described under Methods. Another aliquot was first treated with pronase (as described under Methods) and applied to the column. ●---●) phosphotungstic acid-insoluble, sodium dodecyl sulfate-soluble radioactivity; ○---○) the previous material incubated with pronase before being applied to the column; ■) standard [^{3}H]-sialic acid; △---△) blue dextran.

and became constant after 24 h. The soluble radioactivity had a similar profile. Most of the radioactivity within the cells was insoluble in phosphotungstic acid after 2 h incubation with 85% being so after 24 h.

Total radioactivity within cells was defined as that radioactivity which remained within the cells after removing the radioactive growth medium from the monolayer, followed by 7 washes with buffer. The radioactivity in the last 2 washes was constant and was never more than 3% of the total radioactivity that remained with the cells. An 8th wash with buffer to which 1 mM nonradiolabeled sialic acid had been added did not remove any additional radioactivity, suggesting that soluble radiolabeled sialic acid was not loosely bound to the cell surface. An additional correction was done to eliminate trapped radioactivity or insufficient washes as a source of error by subtracting from all time points the radioactivity obtained after a very short (less than 5 sec) incubation. This value was negligible for all time points longer than 2 h.

Identification of the Radiolabel Within the Cells

The phosphotungstic acid-soluble fraction of cells that had been grown for 24 h in the presence of radiolabeled sialic acid was analyzed for any metabolism of the radiolabel. At least 90% of the radioactivity was characterized as sialic acid based on its behavior on Dowex-formate chromatography (1). This suggests that if any metabolism of the radiolabel did occur, it must have been very small. It is possible the phosphotungstic acid-

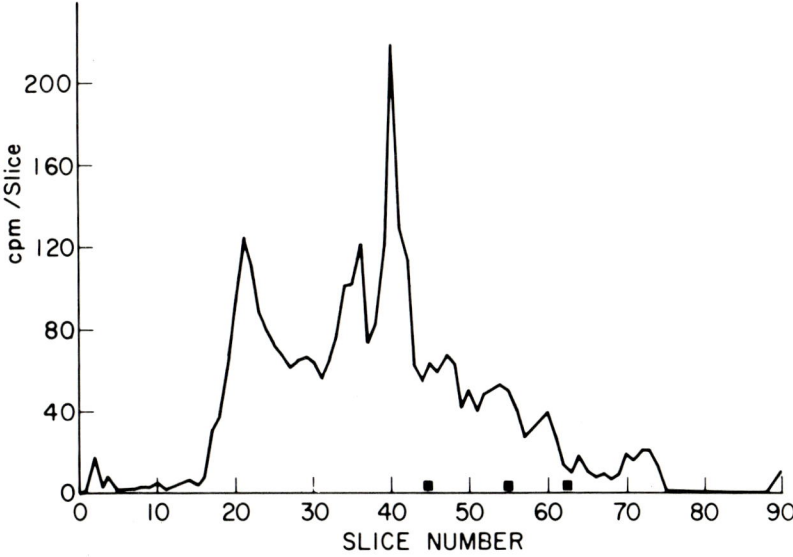

Fig. 3. SDS-acrylamide gel electrophoresis of a phosphotungstic acid-insoluble pellet of BHK cells which had been grown for 24 h in the presence of [^{3}H]-sialic acid. Slices 1–12 were the stacking gel. The dye marker ran in slice 88. The migration of the following standards is shown: bovine serum albumin, 68.5×10^3 daltons; ovalbumin, 43×10^3 daltons; and DNase, 31×10^3 daltons.

soluble radioactivity was originally present not only as free sialic acid but in another sialic acid-containing compound since the HCl, as part of the PTA reagent, may have cleaved a labile bond of such a compound yielding free radiolabeled sialic acid. Incubation of the PTA-insoluble pellet (obtained from cells grown in the presence of [^{3}H]-sialic acid for 24 h), with 0.1 N H_2SO_4 at 80°C for 60 min, solubilized over 90% of the radioactivity. These are the conditions which cleave the glycosidic linkage between sialic acid and other sugars (8). No solubilization of the radioactivity occurred if prior to heating the pellet was incubated for 1 h in cold acid.

Characterization of the Phosphotungstic Acid Insoluble Pellet

When a pellet (from cells that had been grown for 24 h in the presence of radiolabeled sialic acid) was solubilized in 1% sodium dodecyl sulfate (SDS) and subsequently applied to a Biogel P-10 column, all the radioactivity eluted with the void volume (Fig. 2). Pronase digestion of the material caused a shift of the radioactivity profile with most of the radioactivity now being retained by the column although not as long as standard free sialic acid (Fig. 2). SDS-acrylamide gel electrophoresis of the SDS-solubilized (PTA-insoluble pellet) showed comigration of the radioactivity with numerous protein bands throughout the gel (Fig. 3). Free sialic acid runs slightly behind the dye marker in this system. Extraction of the dried, PTA-insoluble pellet with chloroform-methanol 2:1 and subsequent thin layer chromatography of the extract, indicated that approximately 5% of the radioactivity, that was originally in the pellet, comigrated with hematoside. No radioactivity remained at the origin of the plate, strongly suggesting that the sample did not contain free sialic acid.

DISCUSSION

This paper describes the specific labeling of sialoglycoproteins and sialoglycolipids of BHK cells grown in the presence of radiolabeled sialic acid. The major drawback of the method is the relatively low efficiency of uptake of the radiolabel. Radiolabeled glucosamine has been extensively used in other studies to label cellular sialoglycoproteins and sialoglycolipids (9–11). However, this procedure is nonspecific towards sialoglycoproteins, particularly if the labeling is done for long periods. Radiolabeled N-acetylmannosamine has been shown by Harms et al. (12) to be a more specific precursor for labeling sialic acid moieties although significant metabolism to other compounds was also observed. A chemical procedure which has been successfully used to label sialoglycoproteins of red blood cells (13) has (as yet) not been reported to be applicable to other cells.

One important aspect of this study is the observation that no significant metabolism of the radiolabel in sialic acid (both soluble and in glycoproteins and glycolipids) could be observed, even after relatively long labeling, such as 24 h. The possibility that the observed sialic acid binding is not covalent but rather of a very tight, nonspecific nature seems unlikely for the following reasons: a) the PTA-insoluble radioactivity is solubilized by treatment with acid under conditions which are known to cleave sialic acid covalently bound to glycoproteins and glycolipids (8) (0.1 N H_2SO_4 80°, 60 min); cold sulfuric acid did not solubilize the radioactivity; b) pronase treatment of the PTA-insoluble pellet causes a shift of the radioactivity to a profile (on Biogel P-10) expected for a heterogeneous mixture of glycopeptides, rather than free sialic acid which could be expected if sialic acid were only tightly bound; c) the mobility of the chloroform-methanol soluble radioactivity on thin layer chromatography in a solvent system where free sialic acid does not migrate, is also incompatible with a noncovalent linkage of sialic acid (in this case to glycolipids); d) the radioactivity profile on SDS-acrylamide gels argues against noncovalent binding of free sialic acid to one particular protein. The fact that this behavior of sialic acid has also been observed with very different cells, mammalian and avian, established cell lines and secondary chick embryo fibroblasts and more recently in liver and kidney of mice (unpublished results) strongly suggests that this is a general phenomenon occurring in a variety of eukaryotic cells.

Evidence that most of the covalently bound radiolabeled sialic acid is in sialoglycoproteins comes from several experiments. One is the insolubility of the radioactivity in a solvent which is known to precipitate glycoproteins (1% PTA in 0.5 N HCl); another is the result of the pronase treatment of the PTA-insoluble pellet and the subsequent behavior of the radioactivity on Biogel P-10. The resulting profile is consistent with this being a heterogeneous mixture of glycopeptides of smaller size than the non-pronase-treated glycoproteins. In addition the radioactivity profile on SDS-acrylamide gel electrophoresis was similar to the labeling patterns obtained in other studies where cells were grown in the presence of radiolabeled sugars and the proteins subsequently separated by the above procedure (14–16).

Although analyses of the PTA-soluble material have shown radiolabeled sialic acid as the sole product when cells are grown in the presence of radiolabeled sialic acid, we cannot completely rule out the possibility that cleavage by a sialic acid aldolase (17) followed by resynthesis may have occurred. However, for such a mechanism to take place one would have to postulate a very low concentration of the intermediate radiolabeled N-acetylmannosamine. An unequivocal answer to this problem can be obtained by using a mixture of [1-^{14}C]- and [4,5,6,7,8,9-^{3}H]-N-acetyl neuraminic acid as precursor; such an experiment is currently in progress.

The physiological significance of both the uptake and the subsequent incorporation of sialic acid into cells is not clear. A recent study (18) has shown the existence of neuraminidase activity in serum. This enzyme could act upon surrounding sialoglycoproteins and sialoglycolipids liberating free sialic acid which may then enter the cells. However, at this time such a mechanism is only speculative.

ACKNOWLEDGMENTS

We thank Dr. A. Lambowitz for help with the SDS-gel electrophoresis. This work was supported by National Cancer Institute Grant R01-CA17015.

NOTE ADDED IN PROOF

Recent results with a mixture of 1-[^{14}C]-sialic acid and 4,5,6,7,8,9-[^{3}H]-sialic acid strongly suggest that there is no cleavage of the sialic acid backbone in the above described experiments.

REFERENCES

1. Hirschberg CB, Goodman SR, Green C: Biochemistry 15:3591, 1976.
2. Hirschberg CB, Watson D, Yeh M, Carey D: J Supramol Struct (Suppl 1):20, 1977.
3. Sakiyama H, Gross S, Robbins PW: Proc Natl Acad Sci USA 69:872, 1972.
4. Levine EM: Exp Cell Res 74:99, 1972.
5. Lambowitz AM, Chua NH, Luck D: J Mol Biol 107:223, 1976.
6. Hirschberg CB, Robbins PW: Virology 61:602, 1974.
7. Hirschberg CB, Wolf BA, Robbins PW: J Cell Physiol 85:31, 1975.
8. Spiro RG: In Neufield EF, Ginsburg V (eds): "Methods in Enzymology." New York: Academic Press 1966, vol 8, p 14.
9. Kohn P, Winzler RJ, Hoffman RC: J Biol Chem 237:304, 1962.
10. Wu HC, Meezan E, Black PH, Robbins PW: Biochemistry 8:2509, 1969.
11. Buck CA, Glick MC, Warren L: Biochemistry 9:4567, 1970.
12. Harms E, Kreisel W, Morris HP, Reutter W: Eur J Biochem 32:254, 1974.
13. Blumenfeld OO, Gallop PM, Liao TH: Biochem Biophys Res Commun 48:242, 1972.
14. Hawtrey AO, Burden TS, Robertson G: Nature (London) 252:58, 1974.
15. Hunt LA, Summers DF: J Virol 20:646, 1976.
16. Kaluza G: J Virol 19:1, 1976.
17. Brunetti P, Jourdian GW, Roseman S: J Biol Chem 237:2447, 1962.
18. Schauer R, Veh RW, Wember M, Buscher HP: Hoppe-Seyler's Z Physiol Chem 357:559, 1976.

Properties of a Penicillium GDP-Mannose: Glycopeptide Mannosyltransferase Solubilized With Triton X-100

J. E. Gander and Faye Fang

Department of Biochemistry, College of Biological Sciences, University of Minnesota, St. Paul, Minnesota 55108

Membranes from Penicillium charlesii were separated into 6 fractions by sucrose density gradient ultracentrifugation. The least dense fraction ($\rho = 1.1$ g cm^{-3}) contained GDP-mannose:glycopeptide mannosyltransferases that transferred [^{14}C] mannose onto mannopyranosyl-(seryl/threonyl)-polypeptide and phosphogalactomannan regions of peptidophosphogalactomannan. Approximately 90% of the [^{14}C] mannose incorporated was isolated as mannobiose following treatment of peptidophosphogalactomannan with 0.5 N NaOH. The remainder was located in phosphogalactomannan. About 10% of the membrane-bound mannosyltransferase activity was solubilized with 1% Triton X-100. The soluble mannosyltransferase activity was purified by affinity chromatography on peptidophosphogalactomannan-Sepharose 4B and ammonium sulfate fractionation. Mannose incorporation was shown to be a function of the concentration of added acceptor. No incorporation occurred in the absence of added acceptor or when MgCl$_2$ was substituted for MnCl$_2$. Peptidophosphogalactomannan, phosphogalactomannan, phosphomannan, and mannan, each obtained by appropriate treatment of peptidophosphogalactomannan from P. charlesii, served as mannosyl acceptors. In contrast, α-mannosidase treated peptidophosphogalactomannan did not serve an acceptor of mannosyl residues. Up to 70% of the mannose from GDP-mannose was transferred to added acceptor. Treatment of [^{14}C] mannosyl-labeled peptidophosphogalactomannan with 0.5 N NaOH released 90% of the [^{14}C] mannose as phosphogalactomannan and the remainder was released as mannobiose. [^{14}C] Mannose-labeled phosphogalactomannan was subjected to acetolysis. Mannobiose was the major [^{14}C]-labeled product isolated. Significant quantities of [^{14}C] mannose were isolated also. These results show that soluble mannosyltransferase catalyzes the formation of (1–6)-linked mannosyl residues as well as the transfer of a mannosyl residue to a (1–6)-linked mannosyl residue in the phosphogalactomannan. The specificity of the enzyme is shown by its inability to catalyze mannosyl transfer to α-mannosidase treated peptidophosphogalactomannan, or to incorporate more than 2 mannosyl residues onto the phosphogalactomannan region. Presumably the second mannosyl residue is attached by a (1–2) linkage as the mannan contains only (1–6)- and (1–2)-linked mannosyl residues (Gander et al: J Biol Chem 249:2063, 1974). No evidence was obtained for the participation of a lipid-linked mannosyl-containing intermediate in this system.

Key words: mannosyltransferase, glycopeptide, GDP-mannose, Penicillium, phosphomannan, galactofuranosyl

Abbreviations: GDP-mannose – guanosine-5'-(α-D-mannopyranosyl pyrophosphate)

Received March 21, 1977; accepted May 26, 1977.

© 1977 Alan R. Liss, Inc., 150 Fifth Avenue, New York, NY 10011

Penicillium charlesii secretes a novel glycopeptide (peptidophosphogalactomannan) which contains a phosphogalactomannan region and 10–12 mannose-containing low-molecular-weight saccharides each attached through seryl/threonyl residues to a polypeptide of approximately 30 amino acyl residues (1, 2). The phosphogalactomannan contains approximately 90 mannopyranosyl residues, 10 phosphodiester residues, and variable quantities of 5-O-β-D-galactofuranosyl residues located in 10 galactan chains attached by (1→3) glycosidic linkage to the mannan (1). Ethanolamine (2) and N,N'-dimethylethanolamine (3) are attached to the phosphogalactomannan presumably through the phosphodiester residues. The glycopeptide is likely derived from a membrane-bound lipopeptidophosphogalactomannan (4) and not from the cell walls of the Penicillium (5). The function of the glycopeptide or its precursor(s) is unknown.

We have shown previously that a membrane fraction (1.1 g cm^{-3}) from P. charlesii contains GDP-D-mannose:glycopeptide mannosyltransferase which catalyzes the incorporation of D-mannose from guanosine-5'-(α-D-mannopyranosyl pyrophosphate), (GDP-mannose), onto both the phosphogalactomannan and mannosyl but not mannobiosyl or mannotriosyl residues of the glycopeptide (6). This preparation requires Mn^{2+}; Mg^{2+}, Ca^{2+}, Co^{2+}, or Ni^{2+} will not substitute for Mn^{2+}. Following treatment of the glycopeptide with 0.5 N NaOH, approximately 90% of the mannose incorporated was located in mannobiose, and the remainder was located in phosphogalactomannan.

Guanosine-5'-(α-D-mannopyranosyl pyrophosphate) is the mannosyl donor in the biosynthesis of mannans and phosphomannans located in cell-membrane and cell wall glycoproteins of several species of yeasts and fungi (7–13). Sharma et al. (14) showed that in Saccharomyces cerevisiae the mannosylation of seryl/threonyl residues of endogenous membrane bound protein was accomplished with mannosyl-1-phosphoryl dolichol as the mannosyl donor and that GDP-D-mannose served as the direct mannosyl donor to endogenous membrane bound mannosyl-(seryl/threonyl-protein) residues forming a mannobiosyl and larger oligosaccharide. Beyond this, the role of the lipid-mannosyl carriers have not been well delineated in fungal systems.

This paper reports the solubilization and partial purification of membrane-bound mannosyltransferase which catalyzes the incorporation of mannose from GDP-mannose into the phosphogalactomannan region of peptidophosphogalactomannan.

METHODS

Materials and Methods

Stock cultures of Penicillium charlesii G. Smith ATCC 1887 were maintained as described previously (15). GDP-D-[^{14}C] mannose was purchased from Amersham/Searle Corporation. GDP-mannose was obtained from Calbiochem. Other chemicals and reagents used were reagent grade.

Purified peptidophosphogalactomannan was obtained from cultures of P. charlesii as was described previously (1), and phosphogalactomannan, phosphomannan, and mannan were obtained from peptidophosphogalactomannan by treatment with 0.5 N NaOH, and treatment with 0.01 N HCl for 1.5 or 4 h, respectively (6).

Acetolysis of phosphogalactomannan was carried out for 18 h at 37°C as described by Stewart et al. (16). Total carbohydrate was determined by the phenol sulfuric acid procedure (17). Protein was determined by the procedure of Lowry et al. (18) after precipitation of the protein with trichloroacetic acid.

^{14}C was determined by liquid scintillation counting of 0.5-cm paper strips cut from chromatograms or of 0.5 ml of aqueous samples as described previously (6).

Paper chromatography was used to separate the reaction products (6).

Lipid-soluble [^{14}C]mannose-containing substances were determined by extracting the reaction mixture with CHCl$_3$:methanol (2:1, vol/vol) followed by removal of the solvent and chromatography of the residue on Whatman 3MM paper with 95% ethanol:1 M ammonium acetate pH 7.5 (7.5:3, vol/vol) as the developing solvent.

Preparation of Membrane Bound and Soluble Mannosyltransferases

Membrane-bound mannosyltransferase from P. charlesii mycelia was obtained from membranes ($\rho = 1.1$ g cm^{-3}) as described previously (6). The mannosyltransferase(s) activity was solubilized by treating the membranes in 0.05 M Tris-HCl 1% Triton X-100, pH 7.5/5 mM 2-mercaptoethanol for 30 min at 0–4°C, followed by centrifuging the mixture at 120,000 × g for 30 min in a Beckman L2-65B ultracentrifuge. The supernatant solution obtained from this treatment was made 25 mM in MnCl$_2$ and the solution passed onto a peptidophosphogalactomannan-Sepharose 4B affinity adsorbant (19). The adsorbant was washed with buffers as described in the text. Protein in the active fraction was concentrated by adding solid ammonium sulfate to 30% saturation and the protein was dissolved in Tris-HCl, pH 7.5 containing 1% Triton X-100 and 5 mM 2-mercaptoethanol.

RESULTS

Distribution of [^{14}C] Mannose in Peptidophosphogalactomannan

We reported previously that membranes ($\rho = 1.1$ g cm^{-3}) from P. charlesii incorporate mannose from GDP-mannose into added peptidophosphogalactomannan as well as into endogenous acceptors (6). Membrane-bound mannosyltransferase was incubated with GDP-D-[^{14}C]mannose, MnCl$_2$, and peptidophosphogalactomannan for 2 h, and the peptidophosphogalactomannan was isolated (1), was treated with 0.5 N NaOH to β-eliminate the saccharides from the polypeptide, and was chromatographed to separate the phosphogalactomannan (fractions 1–4) from mannobiose (fractions 56–62) (Fig. 1a). Approximately 90% of the ^{14}C was located in the mannobiosyl fraction. The experiment was repeated with soluble mannosyltransferase substituted for the membrane-bound enzyme (Fig. 1b). About 10% of the mannosyltransferase was solubilized in 1% Triton X-100 and most of this activity was directed toward mannosylation of phosphogalactomannan. Most of the mannosyltransferase activity remained in the residue following treatment with Triton X-100 (Tonn and Gander, unpublished).

Experiments were conducted to determine if [^{14}C]mannose was incorporated into substances soluble in CHCl$_3$:methanol (2:1, vol/vol). Reaction mixtures containing soluble mannosyltransferase and approximately 6 × 10^4 cpm in GDP-[^{14}C]mannose were incubated for varying intervals with, or without, added peptidophosphogalactomannan and with either MnCl$_2$ or MgCl$_2$. The reaction mixtures were extracted with CHCl$_3$:methanol, the solvent was removed, and the residues were solubilized and chromatographed on paper. The radioactivity in the CHCl$_3$:methanol phase was variable and never exceeded 300 cpm. The distribution of radioactivity on a chromatogram showed 2 ^{14}C-containing CHCl$_3$:methanol-soluble substances which migrate like GDP-mannose and mannose (not shown). They likely represent GDP-mannose and mannose carried over from the aqueous phase. The quantity in the CHCl$_3$:methanol phase is insignificant. Significant quantities of mannose and mannose-1-phosphate were found in the aqueous phase along with unreacted

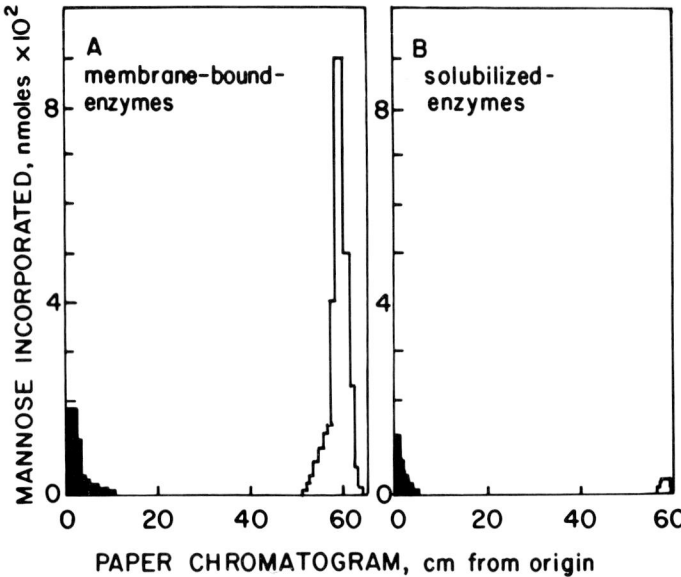

Fig. 1. Distribution of [^{14}C] mannose-containing saccharides on paper chromatograms after treating the peptidophosphogalactomannan with 0.5 N NaOH. The complete system, in a final volume of 60 μl, contained Tris-maleate, pH 7.0, in 5 mM 2-mercaptoethanol; GDP-D-[^{14}C] mannose, 5 nmole; MnCl$_2$, 1.5 μmole; 2-mercaptoethanol, 0.3 μmole; peptidophosphogalactomannan, 3 mg; and membrane preparation 150 μg of protein (A) or 49 μg of protein from membranes treated with 1% Triton X-100 (B). The reaction mixture was incubated for 2 h, the reaction was stopped by heating at 100°C for 2 min, and it was dialyzed against distilled deionized water to remove low-molecular-weight substances. The dialyzed reaction mixture was treated with 0.5 N NaOH (1) and was concentrated and applied to Whatman 3MM paper and chromatographed for approximately 16 h in 95% ethanol, 1 M ammonium acetate, pH 7.5 (7.5:3, vol/vol). The chromatograms were sectioned and ^{14}C determined as described in Materials and Methods.

GDP-mannose and immobile [^{14}C] polymer. We have shown previously that more than 70% of ^{14}C from the polymer is recovered as peptidophosphogalactomannan (6) in a system known to incorporate mannose into endogenous acceptors as well as into peptidophosphogalactomannan.

Purification of Mannosyltransferase(s)

The requirement for Mn^{2+} as a cofactor in the peptidophosphogalactomannan: mannosyltransferase catalyzed reaction suggested that the enzyme might require Mn^{2+} as a cofactor in binding the glycopeptide. The Triton X-100 solubilized mannosyltransferase preparation containing 25 mM MnCl$_2$ was passed onto a peptidophosphogalactomannan-Sepharose 4B affinity column and the adsorbant washed successively with 1) 0.05 M Tris-maleate, pH 7.5, buffer containing 25 mM MnCl$_2$ and 2-mercaptoethanol; 2) 0.05 M Tris-maleate, pH 7.5, buffer containing 0.1% Triton X-100 and 5 mM 2-mercaptoethanol; and 3) 0.05 M Tris-maleate buffer, pH 7.5, containing 0.1% Triton X-100, 5 mM 2-mercaptoethanol and 0.5 M NaCl (Fig. 2). Mannosyltransferase was released from the adsorbant with an eluent containing no MnCl$_2$. The protein in fractions 32–42 was concentrated by

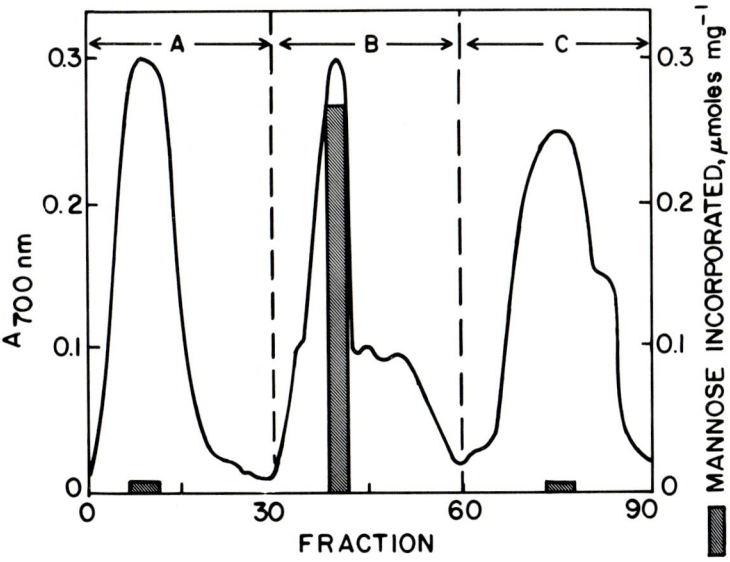

Fig. 2. Partial purification of soluble-mannosyltransferase by affinity chromatography on peptidophosphogalactomannan-Sepharose 4B. Membranes were prepared as described in Materials and Methods section and were extracted with 1% Triton X-100. The supernatant obtained following centrifuging the mixture at 120,000 × g for 30 min was made 25 mM in $MnCl_2$ and applied to a column containing peptidophosphogalactomannan-Sepharose 4B. The adsorbent was washed with 0.05 M Tris-maleate buffer, pH 7.5, containing 5 mM 2-mercaptoethanol and 25 mM $MnCl_2$ (A), 0.05 M Tris-maleate buffer, pH 7.5, containing 5 mM 2-mercaptoethanol (B), and 0.05 M Tris-maleate buffer, pH 7.5, containing 0.5 M NaCl and 5 mM 2-mercaptoethanol (C). Each fraction contained approximately 1 ml. An aliquot was removed and the protein was precipitated with trichloroacetic acid and the protein was determined by the Lowry procedure (18) (——). An aliquot was removed from the fraction containing the greatest quantity of protein from each of the 3 solvents and mannosyltransferase activity was determined (▨) as described in Fig. 1 except the reaction mixture was not dialyzed or treated with NaOH. The nmole of mannose incorporated was calculated from ^{14}C remaining at the origin of the chromatogram.

the addition of solid ammonium sulfate to 30% saturation and the residue was redissolved in Tris-maleate buffer, pH 7.5, containing 1% Triton X-100 and 5 mM 2-mercaptoethanol.

Influence of the Concentration of Mannosyltransferase, GDP-mannose, and $MnCl_2$ on Mannosyltransferase Activity

The activity of the partially purified mannosyltransferase is linear over a range of protein concentrations (not shown). Optimum mannosyltransferase activity is observed over a range of 25–32 mM $MnCl_2$ (Fig. 3). Negligible mannosyltransferase activity was observed if $MnCl_2$ is omitted or if $MnCl_2$ was replaced with $MgCl_2$ (Fig. 4). The requirement for Mn^{2+} was not decreased by the addition of 5, 10, or 20 mM $MgCl_2$. Some inhibition of mannose incorporation was observed upon the addition of $MgCl_2$ to reaction mixtures containing 10 or 20 mM $MnCl_2$ (Table I).

Mannosyltransferase activity increased with increasing GDP-mannose concentration (not shown). Half maximal activity was observed at about 0.6 mM GDP-mannose.

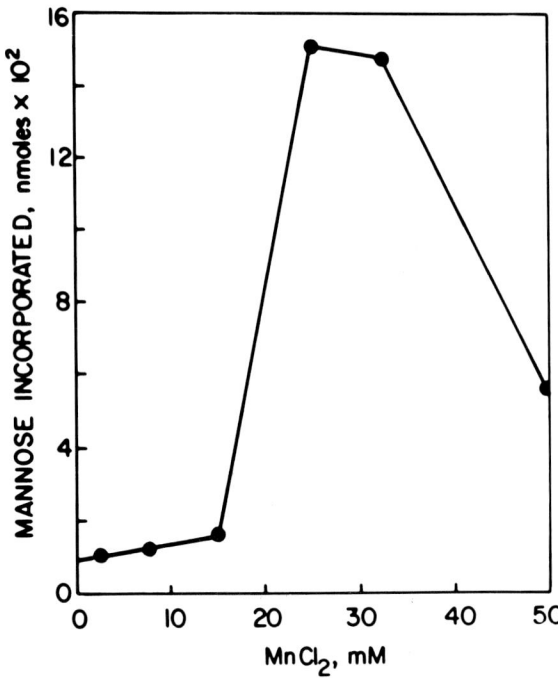

Fig. 3. Influence of MnCl$_2$ concentration on mannosyltransferase activity of the partially purified enzyme preparation. Mannosyltransferase was obtained from affinity chromatography as described in Fig. 3. The protein was concentrated by precipitation with 30% saturated ammonium sulfate and 18 µg of protein was used in each assay which contained all reaction components described in Fig. 1, except the final MnCl$_2$ concentration was as indicated on the abscissa. The mannosyltransferase was assayed as described in Fig. 3.

TABLE I. Influence of MgCl$_2$ on Mannosyltransferase Activity in the Presence of MnCl$_2$

MnCl$_2$ concentration, mM	MgCl$_2$ concentration, mM			
	0	5	10	20
	mannosyltransferase activity, nmol/mg protein			
0	0.4	0.1	0.1	0.3
5	2.1	1.3	1.8	1.4
10	3.1	1.8	2.1	2.1
20	2.7	2.9	1.8	2.2
25	2.6	—	—	—

The reaction conditions were those described in Fig. 1B except the concentration of divalent metal ion(s) were as shown above. Each reaction mixture contained 77 µg of protein.

Influence of Mannosyl Acceptor Concentration of Mannosyltransferase Activity

Peptidophosphogalactomannan was degraded stepwise to its various constituents: phosphogalactomannan, phosphomannan, and mannan. In addition, peptidophosphogalactomannan was treated with jack bean exo-α-mannosidase which degrades the oligosaccharides

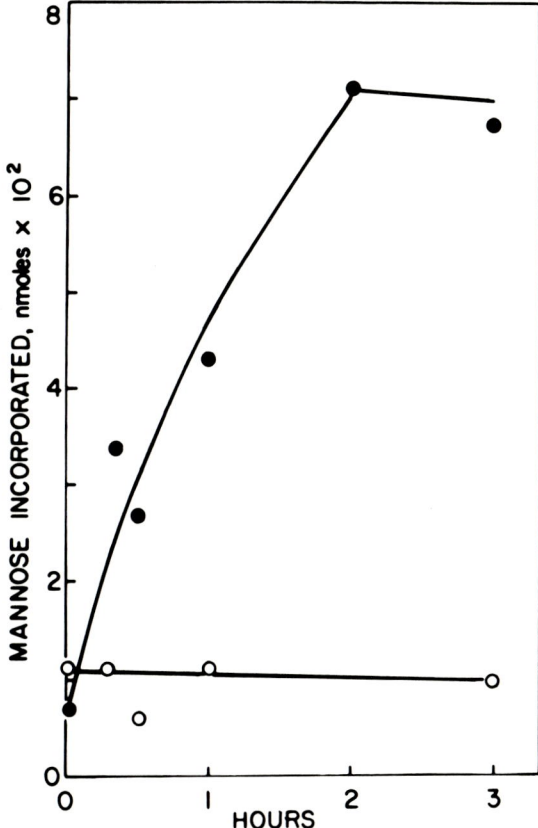

Fig. 4. Time course of mannose incorporation into peptidophosphogalactomannan in the presence of MnCl$_2$ (●) or MgCl$_2$ (○). Partially purified mannosyltransferase (17 μg) was incubated in reaction mixtures as described in Fig. 1. The reactions were stopped by heating for 2 min at 100°C after the times indicated on the abscissa and the remainder of the assay was conducted as described in Fig. 3. A separate series of reactions were conducted in which 25 mM MgCl$_2$ was substituted for MnCl$_2$.

attached to the polypeptide to mannosyl-(seryl/threonyl)-polypeptide and degrades the phosphogalactomannan region from the nonreducing terminal mannosyl residue to the first phosphomannosyl residue. All the mannan-containing polymers, except the exo-α-mannosidase treated polymer, served as mannosyl acceptors (Fig. 5) using the solubilized mannosyltransferase. Removal of the polypeptide from the glycopeptide appears to improve the polymer as a potential mannosyl acceptor from GDP-mannose. Removal of the galactan chains also improves the mannan as an acceptor when used at 10 mg 60 μl^{-1}. The long galactan chains may interfere because of increased polymer-polymer interaction with increasing concentration of mannosyl acceptor.

Acetolysis of [^{14}C] Mannose-containing Phosphogalactomannan

A reaction mixture containing phosphogalactomannan as the mannosyl acceptor was incubated 2h, and after removal of the low-molecular-weight substances by dialysis the phosphogalactomannan was isolated by precipitation as its borate-cetyltrimethylammonium

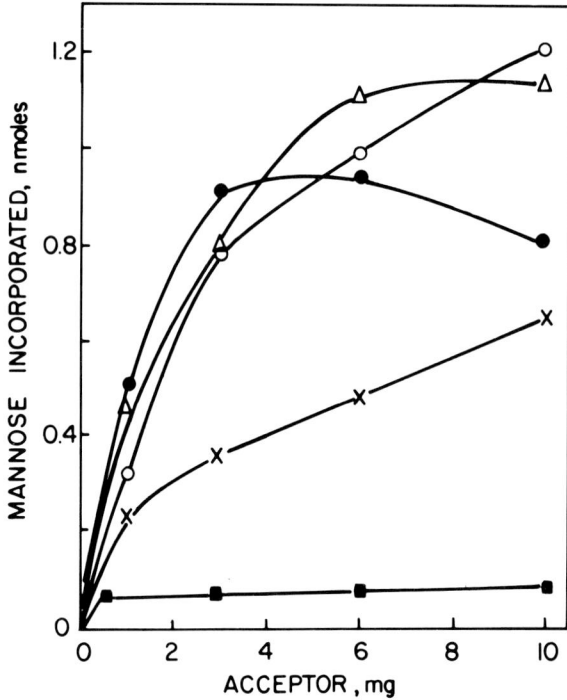

Fig. 5. Influence of acceptor concentration on mannosyltransferase activity. Crude soluble mannosyltransferase (26 μg) was incubated in reaction mixtures as described in Fig. 1, except the concentration of the acceptor was as shown in the abscissa. The reaction was stopped and assayed as described in Fig. 3. Peptidophosphogalactomannan (X), phosphogalactomannan (●), phosphomannan (○), mannan (△), and α-mannosidase-treated peptidophosphogalactomannan (■) were tested as mannosyl acceptors.

complex, followed by chromatography on DEAE-cellulose-borate (1). The fractions containing the polymer were dialyzed extensively and the samples were freeze-dried. The phosphogalactomannan was subjected to acetolysis for 18 h (16), and the products were deacetylated and fractionated on a Bio-Gel P-2 column (1). [^{14}C]Mannose was located primarily in mannobiose with lesser quantities of ^{14}C in mannose (Fig. 6). There were negligible quantities of ^{14}C in other carbohydrate-containing fractions. The large quantity of carbohydrate in the monosaccharide fraction comes from galactose which, being in the furanosyl form, is cleaved to its monomer. This figure shows that mannosyl residues were incorporated into the phosphogalactomannan in a manner to form (1–6)-linked mannosyl and mannobiosyl residues.

DISCUSSION

GDP-mannose:glycopeptide mannosyltransferase(s) that catalyze the transfer of mannosyl residues to the phosphogalactomannan region of peptidophosphogalactomannan is easily solubilized by Triton X-100. This contrasts sharply with the lack of solubility in a number of detergents of the mannosyltransferase that catalyzes the transfer of mannosyl

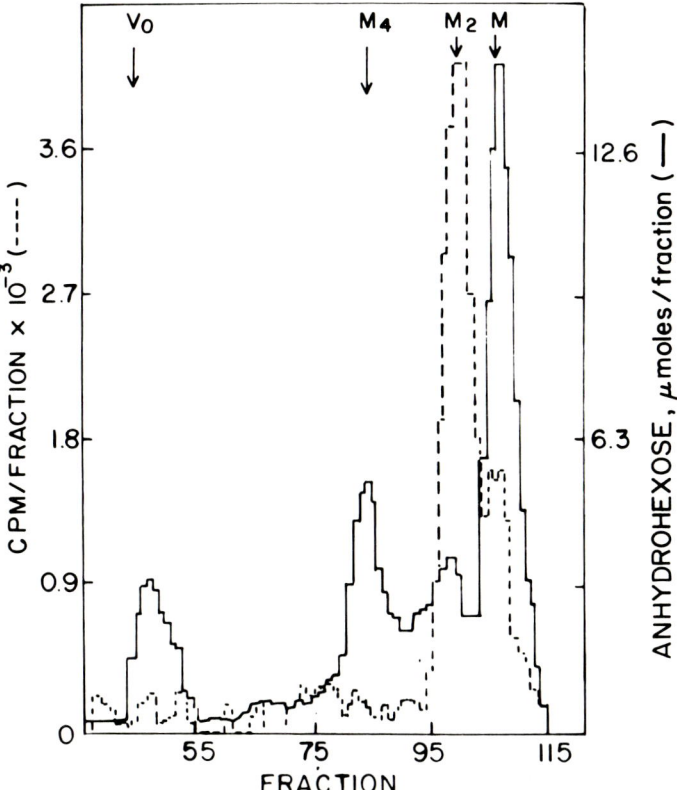

Fig. 6. Location of [^{14}C] mannosyl residues derived from [^{14}C] phosphogalactomannan following acetolysis. A reaction mixture containing 30 mg of phosphogalactomannan, 125 μg of crude mannosyltransferase, and other components at the concentrations given in Fig. 1 was incubated in a total volume of 600 μl for 2 h and the reaction was stopped by heating for 2 min at 100°C. The reaction mixture was dialyzed against distilled H_2O, and the phosphogalactomannan was isolated as described previously (1). Twenty milligrams of phosphogalactomannan was obtained. The polymer was subjected to acetolysis (1, 16), the products were deacetylated and fractionated on Bio-Gel P-2 (minus 400 mesh). The anhydrohexose (——) and ^{14}C (---) were determined in each fraction. The void volume, v_0, and the position of elution of mannose, M; mannobiose, M_2; mannotetraose, M_4 are given on the figure.

residues to mannosyl-(seryl/threonyl)-polypeptide acceptor (Tonn and Gander, unpublished). Thus, the soluble mannosyltransferase(s) are bound to the membranes relatively weakly.

The soluble mannosyltransferase was partially purified by affinity chromatography using peptidophosphogalactomannan as the ligand in a medium containing 25 mM $MnCl_2$. Removal of $MnCl_2$ from the buffered medium released the enzyme (Fig. 2). This shows that the mannosyltransferase has little affinity for peptidophosphogalactomannan in the absence of Mn^{2+}. It is of particular interest to note that the mannan, resulting after removal of phosphoryl and galactofuranosyl residues, was an excellent mannosyl acceptor. This suggests that the role of Mn^{2+} is to modify the enzyme so that it binds substrate rather than the alternative possibility that Mn^{2+} alters the conformation of the mannan by complexing with the mono-oxy anions of phosphomannan since Mn^{2+} binding to mannan would be weak at best.

The optimum concentration of $MnCl_2$ was approximately 20 mM which is several-fold greater than that usually found in physiological systems. However, maleate from the buffer most likely complexed a major portion of the divalent cation which resulted in the requirement for an aphysiological concentration of $MnCl_2$. Furthermore, it is apparent from Fig. 3 that $MgCl_2$ will not replace $MnCl_2$, nor will $MgCl_2$ reduce the quantity of $MnCl_2$ required for optimum mannosyltransferase activity (Table I). Other divalent cations (Fe^{2+}, Co^{2+}, Ca^{2+}, and Ni^{2+}) have been shown previously (6) to provide minimal or no activation of mannosyltransferase activity in the absence of Mn^{2+}.

We have no evidence for the participation of a lipid-linked mannose-containing substance as an intermediate in the incorporation of mannosyl residues into either the mannosyl-(seryl/threonyl)-polypeptide (6) or the phosphogalactomannan. The insignificant quantity of ^{14}C in the chloroform:methanol phase, following extraction of the reaction mixture containing crude, soluble mannosyltransferase with $CHCl_3$:methanol (2:1, vol/vol), could be accounted for as GDP-mannose and mannose that were carried through the extraction procedure. ^{14}C was found in only GDP-mannose and the polysaccharide following paper chromatography when the partially purified mannosyltransferase was used (not shown). Thus the solubilized crude enzyme preparation degraded GDP-mannose to mannose-1-phosphate and mannose. The time course of mannose incorporation was linear during the earliest times. However, this evidence is, at best, indirect in excluding a lipid-linked intermediate.

The acetolysis data show that the soluble crude mannosyltransferase(s) catalyze the transfer of mannosyl residues from GDP-mannose to either mannosyl-(1–6)-mannan or to mannan to give mannobiosyl-(1–6)-mannan and mannosyl-(1–6)-mannan, respectively, Currently we do not know if the mannosyl residues are added at the nonreducing terminal end of the mannan or if they branch from some internal mannosyl residue.

The concentration of peptidophosphogalactomannan used was approximately 8-fold larger than the concentration of GDP-mannose in the routine assay. We calculate that only 1 in 40 peptidophosphogalactomannan molecules in the system were mannosylated. If the polymer is released from the enzyme after each transfer of a mannosyl residue then it is improbable that any 1 mannan would have received more than 1 mannosyl residue in this system. We do not have information concerning the heterogeneity of the nonreducing terminal region of the phosphogalactomannan. For instance, the peptidophosphogalactomannan may be composed of molecules containing (1–6)-linked mannosyl, mannobiosyl, mannotriosyl, mannotetraosyl, and phosphomannotetraosyl residues at the nonreducing terminus.

ACKNOWLEDGMENTS

This work was supported in part by Research Grant GM 19978 from the General Medical Sciences division of the National Institutes of Health, United States Public Health Service, and by the University of Minnesota Agricultural Experiment Station, Scientific Journal Series No. 9842, Agricultural Experiment Station, University of Minnesota, St. Paul, Minnesota.

REFERENCES

1. Gander JE, Jentoft NH, Drewes LR, Rick PD: J Biol Chem 249:2063, 1974.
2. Rick PD, Drewes LR, Gander JE: J Biol Chem 249:2073, 1974.
3. Gander JE: Exp Mycol 1:1, 1977.
4. Beachy J; Fed Proc Fed Am Soc Exp J Biol 35:1642, 1976.
5. Gander JE, Fang F: Biochem Biophys Res Commun 71:719, 1976.
6. Gander JE, Drewes LR, Fang F, Lui AJ: J Biol Chem 252:2187, 1977.
7. Mayer RM: Biochim Biophys Acta 252:39, 1971.
8. Bretthauer RK, Kozak LP, Irwin WE: Biochem Biophys Res Commun 37:820, 1969.
9. Raizada MK, Kloepfer HG, Schutzbach JS, Ankel H: J Biol Chem 249:6080, 1974.
10. Raizada MK, Schutzbach JS, Ankel H: J Biol Chem 250:3310, 1975.
11. Behrens NH, Cabib E: J Biol Chem 243:502, 1968.
12. Nakajima T, Ballou CE: Proc Natl Acad Sci USA 72:3912, 1975.
13. Samuel O, Nordin JH: Biochem Biophys Res Commun 45:1376, 1971.
14. Sharma CB, Babczinski P, Lehle L, Tanner W: Eur J Biochem 46:35, 1974.
15. Preston JF, Gander JE: Arch Biochem Biophys 124:504, 1968.
16. Stewart TS, Mendershausen PB, Ballou CE: Biochemistry 7:1843, 1968.
17. Dubois M, Gilles KA, Hamilton JK, Rebers PA, Smith F: Anal Chem 28:350, 1956.
18. Lowry OH, Rosebrough NJ, Farr AL, Randall RJ: J Biol Chem 193:265, 1951.
19. Rietschel-Berst M, Jentoft NH, Rick PD, Pletcher C, Fang F, Gander JE: J Biol Chem 252:3219, 1977.

Characterization of the Fc Receptors of the Murine Leukemia L1210

Sheldon M. Cooper and Yugalkishore Sambray

Clinical Immunology and Rheumatic Disease Section, Department of Medicine, University of Southern California School of Medicine, Los Angeles, California, 90033

A glycoprotein extract prepared from the plasma membranes of L1210 cells was passed over columns of Sepharose 4B to which either heat-aggregated human IgG or F(ab')$_2$ fragments had been coupled. The intact IgG column bound 35.7% of the applied counts, whereas the F(ab')$_2$ columns bound 2.8%. The bound glycoproteins were eluted with citrate buffer (pH 3.2) and analyzed by sodium dodecyl sulfate-polyacrylamide gel electrophoresis. Three peaks with apparent molecular weights of 65,000, 45,000, and 28,000 daltons were identified and purified by electroelution from polyacrylamide gels. The isolated proteins were able to bind to the same subclasses of mouse IgG myeloma proteins as the intact L1210 cells, indicating that these molecules are related to L1210 surface Fc receptors. Amino acid analyses of the 3 proteins were markedly similar suggesting that the observed molecular heterogeneity might be due to carbohydrate differences. Neuraminidase digestion of the isolated proteins resulted in mobility shifts on polyacrylamide gel electrophoresis which were consistent with the interpretation that either the isolated proteins have considerably different sialic acid contents, or that removal of the sialic acid results in disaggregation of an Fc receptor molecule.

Key words: Fc receptors, membrane glycoproteins, mouse leukemia

Surface receptors for the Fc portion of the IgG molecule have been identified on a wide variety of cells (1). Cells bearing Fc receptors are identified by their ability to bind heat-aggregated IgG or antigen-antibody complexes, or to form rosettes with IgG coated erythrocytes (2, 3). It has been suggested that Fc receptors identified by aggregated IgG or immune complexes are identical (4), but a recent report demonstrated that a single cell line might possess separate Fc receptors with different IgG subclass affinities and different binding properties for aggregated versus monomeric IgG (5). The molecular identity of Fc receptors remains uncertain largely due to the lack of structural studies concerning these membrane components.

Abbreviations: PBS – phosphate-buffered saline; SDS – sodium dodecyl sulfate; PAGE – polyacrylamide gel electrophoresis; GP – glycoprotein.

Received March 23, 1977; accepted June 21, 1977

© 1977 Alan R. Liss, Inc., 150 Fifth Avenue, New York, NY 10011

L1210 is a carcinogen-induced leukemia derived from DBA/2 mice that lacks surface immunoglobulin but bears an Fc receptor. L1210 is apparently a bone-marrow derived (B) lymphocyte since it also carries a B cell alloantigen (6). We have previously shown that redistribution of L1210 Fc receptors results in the selective release from the cell surface of a 45,000 dalton Fc-binding protein (7). In this report we demonstrate that 3 proteins which bind to the Fc region of aggregated IgG can be isolated from the plasma membranes of L1210 cells. Biochemical studies indicate that the observed molecular heterogeneity of the isolated molecules may be due to differences in their carbohydrate content or to aggregation of a basic subunit.

MATERIALS AND METHODS

L1210 cells were grown in stationary culture in RPMI 1640 supplemented with 10% fetal calf serum (Flow Laboratories, Rockville, Maryland). Five $\times$ 10^{10} cells were labeled with ^{125}I (Radiochemical Centre, Amersham, England) by the lactoperoxidase technique described by Marchalonis et al. (8). The cells were swelled in hypotonic buffer consisting of 2 mM $NaHCO_3$, 0.2 mM $CaCl_2$ and 5 mM $MgCl_2$ (pH 6.8), and ruptured in a tight fitting Dounce homogenizer. The plasma membranes of the cells were prepared by a partition of the cell lysates in a 2-phase system of polyethylene glycol and dextran according to the method of Hourani et al. (9).

A glycoprotein extract of the plasma membrane preparation was obtained by the lithium diiodosalicylate technique described by Marchesi and Andrews (10). Briefly, the plasma membrane fraction was vigorously resuspended in 10 ml of 0.05 M Tris-HCl buffer (pH 7.4). Recrystallized lithium diiodosalicylate (Eastman, Rochester, New York) was added to a final concentration of 0.3 M and the solution stirred vigorously at room temperature for 15 min. Two volumes of cold distilled water were added and the solution stirred for an additional 10 min. The sample was centrifuged at 48,000 $\times$ g for 40 min and to the supernatant was added an equal volume of freshly prepared 50% phenol. After stirring for 30 min the aqueous and phenol phases were separated by centrifugation at 4,000 $\times$ g for 1 h in a swinging bucket rotor. The aqueous phase was carefully removed and extensively dialyzed against 4 changes of distilled water over a period of 72 h. In some experiments the glycoprotein extract was labeled with ^{125}I by the chloramine T method (11).

Purified human IgG (Miles Laboratories, Kankakee, Illinois) was heat aggregated at a concentration of 40 mg/ml in phosphate-buffered saline at 63°C for 15 min. An $F(ab')_2$ fraction of human IgG was prepared by pepsin digestion (12), and heat aggregated at 63°C for 30 min, at which point the solution became faintly turbid. The aggregated proteins were coupled to activated Sepharose 4B (Pharmacia, Piscataway, New Jersey) as previously described (7).

The labeled glycoprotein extract, dissolved in PBS[1] was passed over columns of Sepharose 4B coupled to either aggregated human IgG or $F(ab')_2$ fragments. The columns were washed with PBS until the radioactivity in the eluate was equal to the background radioactivity and the percentage of radioactivity bound to each column was calculated. The bound radioactivity was eluted with 0.1 M citrate buffer, pH 3.2, and analyzed by SDS-PAGE as previously described (7). The radioactive peaks were isolated by electroelution from polyacrylamide gels (13).

Columns containing Sepharose 4B coupled to nonaggregated mouse myeloma proteins of different IgG subclass (Bionetics Laboratories, Kensington, Maryland) were prepared and the isolated glycoproteins were subjected to affinity chromatography on the different columns. After the columns were washed with PBS the percentage of applied counts bound to each column was calculated. The ability of L1210 to bind the same nonaggregated myeloma proteins was determined by indirect immunofluorescence with the use of fluorescein isothiocyanate conjugated goat antimouse immunoglobulin (Antibodies, Inc., Davis, California).

The isolated glycoproteins were dissolved in 25 μl of 50 mM acetate buffer (pH 5.5) containing 20 mg/ml of NaCl and 1 mg/ml of $CaCl_2$ and incubated at 37°C for 1 and 2 h with 5 μl (2.1 international units) of neuraminidase (Type V, Sigma Chemical Company, St. Louis, Missouri). Reactions were stopped by the addition of 9 M urea to bring the total urea concentration in the sample to 6 M, and the digested and undigested samples reanalyzed by SDS-PAGE. Ovalbumin was incubated for 2 h under identical conditions and analyzed by SDS-PAGE in order to detect any proteolytic activity in the enzyme preparation. In some control tubes 5 μl of diisopropylfluorophosphate was also added to the incubation mixture.

Fc receptor activity in the glycoprotein extract was determined by incubating varying amounts of the labeled extract with either untreated sheep erythrocytes or sheep erythrocytes which had been incubated with a subagglutinating titer of rabbit IgG antisheep red cell antibody (Cordis Laboratories, Miami, Florida). Fc receptor activity represents the amount of radioactivity bound by the IgG coated erythrocytes minus the radioactivity bound by the untreated erythrocytes.

RESULTS

The glycoprotein extract prepared from 5×10^{10} L1210 cells contained approximately 8 mg of protein as determined by the method of Lowry et al. (14). The extract appeared to have Fc receptor activity since erythrocytes coated with IgG antibody bound 32% of the applied radioactivity while uncoated erythrocytes bound only 11%. In addition, when increasing amounts of the GP extract were added to a mixture of L1210 cells and IgG-coated erythrocytes we noted a progressive decrease in the number of EA rosette-forming cells, indicating a competitive inhibition between the cell-bound Fc receptors and the solubilized Fc receptors in the GP extract.

The GP extract was subjected to affinity chromatography on columns of Sepharose 4B to which either heat-aggregated human IgG or F(ab')$_2$ fragments had been covalently coupled. After washing with PBS the intact IgG column bound 35.7% of the applied labeled preparation, whereas the F(ab')$_2$ column bound only 2.8% (Table I). Approximately 75% of the radioactive material bound to the IgG column was eluted with citrate buffer.

TABLE I. Binding of Glycoprotein Extract to Sepharose 4B Columns Coupled to Heat-Aggregated IgG or F(ab')$_2$ Fragments

	% cpm Bound	
	Sepharose 4B+ aggregated IgG	Sepharose 4B+ aggregated F(ab')$_2$
Elution buffer[a]		
PBS, 0.01 M, pH 7.2	35.7	2.8
Citrate buffer, 0.1 M, pH 3.2	8.4	1.6

[a]Equal counts of glycoprotein extract were applied to the columns and eluted with PBS until the cpm in the eluate was equal to background. After the % cpm bound was calculated, the columns were then eluted with citrate buffer, pH 3.2.

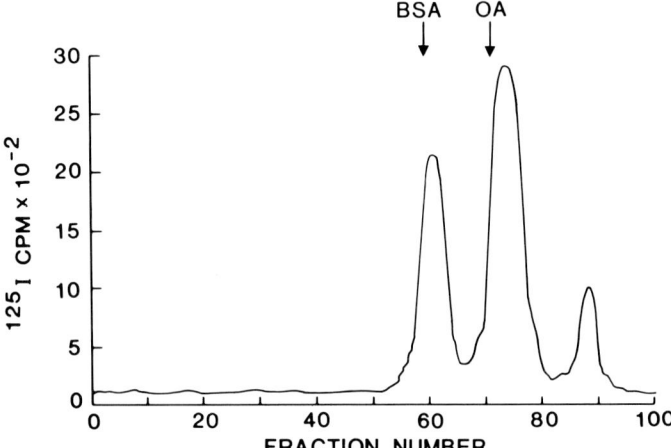

Fig. 1. SDS-PAGE analysis of the citrate eluate from the Sepharose 4B-heat aggregated human IgG column. The 5% polyacrylamide gels were sliced after electrophoresis. BSA (bovine serum albumin) and OA (ovalbumin) refer to the mobility of the marker proteins. The citrate eluate from the $F(ab')_2$ column revealed no peaks.

TABLE II. Similar IgG Subclass Affinities of Intact L1210 Cells and Isolated Fc Binding Glycoproteins

Sample	Mouse myeloma proteins		
	IgG1	IgG2a	IgG2b
L1210 cells[a]	++	++++	+
FI[b]	26.2	41.7	22.6
FII[b]	17.0	56.8	5.3
FIII[b]	10.8	19.3	10.3

[a]L1210 cells were incubated with one of the mouse myeloma proteins, washed, and then stained with fluorescein conjugated antimouse Ig. The intensity of the fluorescence was graded from + (barely visible) to ++++ (most intense).
[b]Equal numbers of counts of FI, FII, and FIII were passed over Sepharose 4B columns to which one of the myeloma proteins had been coupled. The columns were washed with phosphate buffered saline until the cpm in the eluates were equal to background. The numbers refer to the percentage of counts bound by each column.

The citrate eluates of both columns were analyzed by SDS-PAGE (Fig. 1). No distinct components were detected in the eluate from the $F(ab')_2$ column. As shown in Fig. 1, 3 distinct peaks, designated FI, FII, and FIII, were found in the eluate from the intact IgG column. This pattern was not altered when the proteolytic inhibitor, diisopropylfluorophosphate, was present throughout the plasma membrane and glycoprotein extraction procedure. FI, FII, and FIII were purified by electroelution from SDS-polyacrylamide gels (13) and each fraction contained less than 8% cross-contamination from the other fractions.

The ability of mouse myeloma proteins of different IgG subclasses to bind to intact L1210 cells was tested by indirect immunofluorescence. As judged by the intensity of the fluorescence, Table II shows that L1210 binds mouse IgG2a > IgG1 > IgG2b. Columns of Sepharose 4B coupled to the same myeloma proteins were prepared and the percentage of labeled FI, FII, and FIII bound by these columns determined (Table II). All 3 fractions revealed the same binding pattern for the mouse myeloma proteins as did the intact cells, although the actual percentage of counts bound varied considerably.

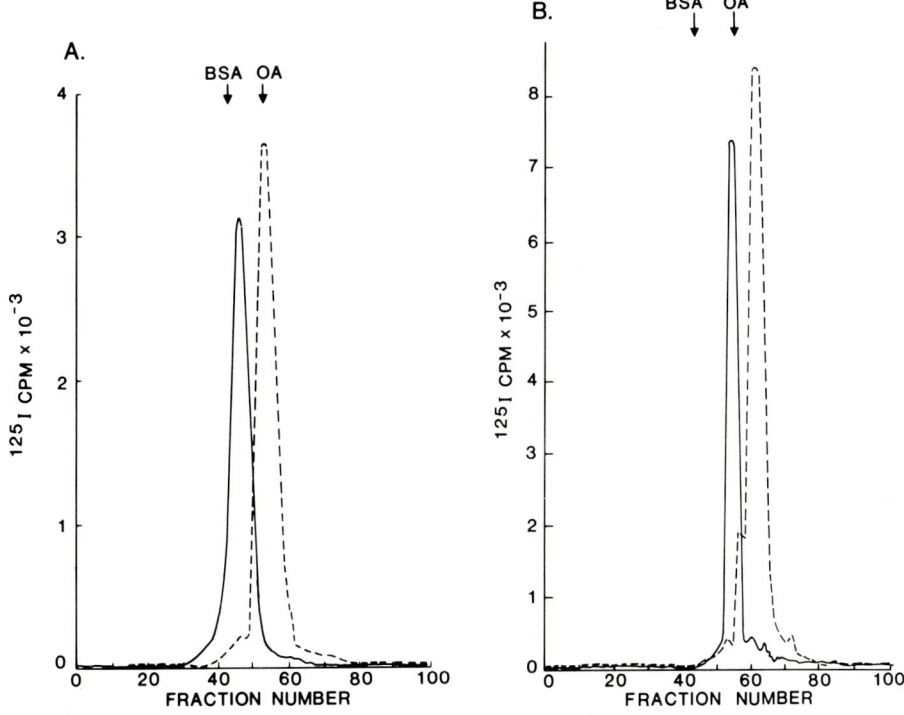

Fig. 2. Shift in mobility on 7.5% polyacrylamide gels of FI and FII after incubation with neuraminidase for 1 h. A) FI untreated (——), and after treatment with neuraminidase (---). B) FII untreated (——), and after treatment with neuraminidase (---). Incubation for 2 h did not reveal any further changes in mobility.

Reduction of FI, FII, and FIII with 2-mercaptoethanol revealed no change in molecular weight indicating that each consisted of a single polypeptide chain. Furthermore, amino acid analyses of FI, FII, and FIII showed remarkable similarity suggesting that the differences between the 3 fractions might reside primarily in the carbohydrate portions of the molecules. To explore this possibility, FI, FII, and FIII were digested for 1 h with neuraminidase and analyzed by SDS-PAGE (Fig. 2). Figure 2A shows that neuraminidase digestion of FI results in a mobility shift such that FI now migrates in the same position as FII. Upon neuraminidase digestion FII has a shift in mobility to a position intermediate between FII and FIII (Fig. 2B). No change in mobility was noted in FIII after neuraminidase digestion. Incubation with neuraminidase for 2 h did not result in any further shifts in mobility of FI and FII. Incubation of ovalbumin with neuraminidase for 2 h did not result in any degradation and the addition of diisopropylfluorophosphate to the incubation mixture did not alter the results.

DISCUSSION

These studies indicate that 3 proteins which bind to the Fc region of aggregated IgG can be isolated from the plasma membrane of the murine leukemia L1210. Since these 3 proteins exhibit the same spectrum of IgG subclass affinity as the intact cell it demonstrates that the isolated molecules are related to the Fc receptors of the cell. However at this point we do not know if each of the fractions represents a separate and distinct Fc receptor or if they are structurally related to each other.

Biochemical analyses of the isolated molecules indicate that they are composed of single polypeptide chains of similar amino acid composition, yet they differ considerably in their apparent molecular weights. Preliminary carbohydrate analyses of FI and FII have confirmed their glycoprotein nature, however FIII has not been subjected to such analysis due to insufficient material.

The data from the neuraminidase digestion suggests that the apparent molecular weight differences may be due in part to differences in the sialic acid moieties of the molecules. It is unlikely, however, that the large differences in molecular weight can be solely ascribed to differences in sialic acid content. Another interpretation of the observed heterogeneity may be that during the isolation procedure on SDS gels aggregates are formed. It has been reported that reversible aggregation of glycophorins occurs when these molecules are treated with organic solvents or run on SDS gels (15). If the heterogeneity of the Fc receptors on SDS gels is due to aggregate formation, it is possible that the neuraminidase digestion may result in disaggregation and the observed mobility shifts.

We recently reported that a 45,000 dalton Fc-binding protein is released from the surface of L1210 cells after redistribution of the cell Fc receptors by aggregated IgG and anti-IgG (7). The relationship of this molecule to the proteins described in this study is still to be determined, but it is of interest that of the 3 components isolated, it was found that FII, with an apparent molecular weight of 45,000 daltons, was the major component.

Rask and co-workers have isolated from crude membrane fractions of mouse spleen cells 3 polypeptides with molecular weights of 65,000, 18,000, and 15,000 daltons which had affinity for aggregated human IgG (16). They proposed that proteolytic degradation of the largest component gives rise to the observed molecular heterogeneity. The finding by Walker that a single cell line may possess separate Fc receptors for aggregated or monomeric IgG of different subclasses strongly suggests that Fc receptors will be found to exhibit molecular heterogeneity (5). We have preliminary data which indicates that L1210 possesses a large, complex Fc receptor which binds to IgG which is bound to antigen, while the 3 proteins described in this study bind to native IgG (17). Although the relationship of these proteins to each other has yet to be determined, it is possible that their arrangement and linkage on the cell surface may account for different binding specificities.

ACKNOWLEDGMENTS

The authors wish to thank Carey Rowan for preparing the manuscript. This work was supported in part by National Institutes of Health Grant AM 18445 and Cancer Center Grant CA14089 from the National Cancer Institute.

REFERENCES

1. Kerbel RS, Davies AJ: Cell 3:105, 1974.
2. Paraskevas F, Lee ST, Orr KB, Israels LG: J Immunol 108:1319, 1972.
3. Dickler HB, Kunkel HG: J Exp Med 136:191, 1972.
4. Dickler HB: J Exp Med 140:508, 1974.
5. Walker WS: J Immunol 116:911, 1976.
6. Freund JG, Ahmed A, Budd RE, Dorf ME, Sell KW, Vanier WE, Humphreys RE: J Immunol 117:1903, 1976.
7. Cooper SM, Sambray Y: J Immunol 117:511, 1976.
8. Marchalonis JJ, Cone RE, Santer V: Biochem J 124:921, 1971.
9. Hourani BT, Chace NH, Pincus JH: Biochem Biophys Acta 328:520, 1973.
10. Marchesi VT, Andrews EP: Science 174:1247, 1971.

11. Hunter WM, Greenwood RC: Biochem J 89:114, 1963.
12. Stanworth DR, Turner MW: In Weir DM (ed): "Handbook of Experimental Immunology, Immunochemistry." 2nd Ed. Oxford: Blackwell Scientific Publications, 1973, pp 10.16–10.20.
13. Benya PD, Padilla SR, Nimni ME: Biochemistry 16:865, 1977.
14. Lowry OH, Rosebrough NJ, Farr AL, Randall RJ, J Biol Chem 193:265, 1951.
15. Furthmayr H, Tomita M, Marchesi VT: Biochem Biophys Res Commun 65:113, 1975.
16. Rask L, Klareskog L, Ostberg L, Peterson PA: Nature 257:231, 1975.
17. Cooper SM, Sambray Y: Nature (In press).

Plants Interact With Microbial Polysaccharides

Peter Albersheim, Arthur R. Ayers, Jr., Barbara S. Valent, Jürgen Ebel, Michael Hahn, Jack Wolpert, and Russell Carlson

Department of Chemistry, University of Colorado, Boulder, Colorado 80309

Plants are resistant to almost all of the microorganisms with which they come in contact. In response to invasion by a fungus, bacterium, or a virus, many plants produce low molecular weight compounds, phytoalexins, which inhibit the growth of microorganisms. Phytoalexins are produced whether or not the invading microorganism is a pathogen. The production of phytoalexins appears to be a widespread mechanism by which plants attempt to defend themselves against pests. Molecules of microbial origin which trigger phytoalexin accumulation in plants are called elicitors. Structural polysaccharides from the mycelial walls of several fungi elicit phytoalexin accumulation in plants. Approximately 10 ng of the polysaccharide elicits the accumulation in plants of more than sufficient amounts of phytoalexin to stop the growth of microorganisms in vitro. The best characterized elicitors have been demonstrated to be β-1,3-glucans with branches to the 6 position of some of the glucosyl residues. Oligosaccharides, produced by partial acid hydrolysis of the mycelial wall glucans, are exceptionally active elicitors. The smallest oligosaccharide which is still an effective elicitor is composed of about 8 sugar residues.

Bacteria also elicit phytoalexin accumulation in plants, but the Rhizobium symbionts of legumes presumably have a mechanism which allows them to avoid either eliciting phytoalexin accumulation or the effects of the phytoalexins if they are accumulated. The lectins of legumes bind to the lipopolysaccharides of their symbiont, but not of their non-symbiont, Rhizobium. It is not known whether the lectin-lipopolysaccharide interaction is involved with the establishment of symbiosis. However, evidence will be presented that suggests that lectins are, in fact, enzymes capable of modifying the structures of the lipopolysaccharides of their symbiont, but not of their non-symbiont, Rhizobium. It will also be shown that the lipopolysaccharides isolated from different Rhizobium species and from different strains of individual Rhizobium species have different sugar compositions. Thus, the different strains of a single Rhizobium species are as different from one another as the different species of Salmonella and other gram-negative bacteria. This conclusion is substantiated by experiments demonstrating that antibodies to the lipopolysaccharide from a single Rhizobium strain can differentiate that strain from other strains of the same species as well as from other Rhizobium species. The role in symbiosis of the strain-specific O-antigens is unknown.

Key words: plants, polysaccharides, elicitors, phytoalexins, Rhizobium, nitrogen-fixation

Received April 11, 1977; accepted April 20, 1977

© 1977 Alan R. Liss, Inc., 150 Fifth Avenue, New York, NY 10011

I. ELICITORS OF PHYTOALEXIN ACCUMULATION IN PLANTS

A. Introduction

Plants are exposed to attack by an immense array of microorganisms, and yet plants are resistant to almost all of these potential pests. A microorganism which is a pathogen of a plant is a special case; pathogens have developed the ability to overcome the plant's defenses. A microorganism which can form a symbiotic relationship with a plant is also a rare case, and again the symbiont must overcome the plant's defense mechanisms. Our research group has been trying to understand how plants resist infection by the vast majority of microorganisms with which they come in contact.

B. Phytoalexins and Elicitors

Plants do not have an immune system similar to that of animals. Instead of antibodies, etc., plants produce low-molecular-weight compounds, phytoalexins, which inhibit the growth of microorganisms. Phytoalexins are produced in response to invasion by a fungus, a bacterium, or a virus. The production of phytoalexins appears to be a widespread mechanism by which plants attempt to defend themselves against pests (1–3). The molecules of microbial origin which trigger phytoalexin accumulation in plants have been called elicitors (4). Plants recognize and respond to elicitors as foreign molecules. It is highly improbable that plants have evolved separate recognition systems for every bacterial species and strain and every fungal race and every virus that plants are exposed to. Thus, elicitors are likely to be molecules common to many microbes and, in fact, the only elicitor to be well characterized and the one to be described in this paper is a fungal polysaccharide, a polysaccharide which is a structural component of the mycelial walls of many fungi.

Most plants produce several structurally related phytoalexins. The most studied phytoalexin of soybeans is glyceollin (5). Lyne et al. (6) have characterized 2 additional soybean phytoalexins which are structural isomers of glyceollin and which appear to have similar antibiotic characteristics. Glyceollin is a phenylpropanoid derivative, and thus its synthesis is probably initiated from phenylalanine via the reaction catalyzed by phenylalanine ammonia lyase (Fig. 1).

Steven Thomas in our laboratory has been studying the effect of glyceollin on a variety of microorganisms. Glyceollin is a static agent rather than a toxic agent, a trait which seems to be common to many, if not all, phytoalexins. Thomas has found that glyceollin will stop the growth of a Gram-negative bacterium, Pseudomonas glycinea, a Gram-positive bacterium, Bacillus subtilis, and baker's yeast, Saccharomyces cerevisiae. Interestingly, it requires about 25 μg/ml of glyceollin to inhibit by 50% the growth of all three of these different organisms (Fig. 2). Thus, it appears that a plant's phytoalexins can potentially protect the plant from a broad spectrum of microorganisms.

C. Assays of Elicitor Activity

Resistance of soybean (Glycine max var. L.) seedlings to Phytophthora megasperma var. sojae (Pms), the causal agent of root and stem rot, is in part due to the accumulation of glyceollin at the site of infection (7). The accumulation of glyceollin is triggered by infection but also by elicitors which are polysaccharides purified from Pms. The ability of Pms elicitor to stimulate glyceollin accumulation in soybean tissues was used as the basis for biological assays of elicitor activity. Bioassays were developed and characterized using the cotyledons (seed leaves) and the hypocotyls (upper stems) of soybean seedlings (8). A

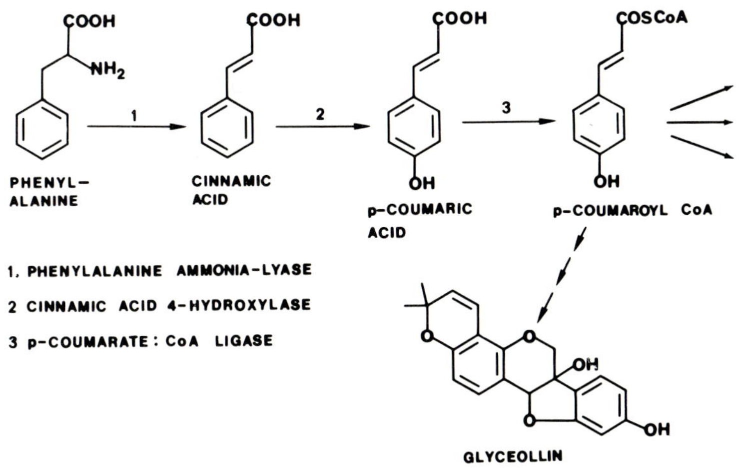

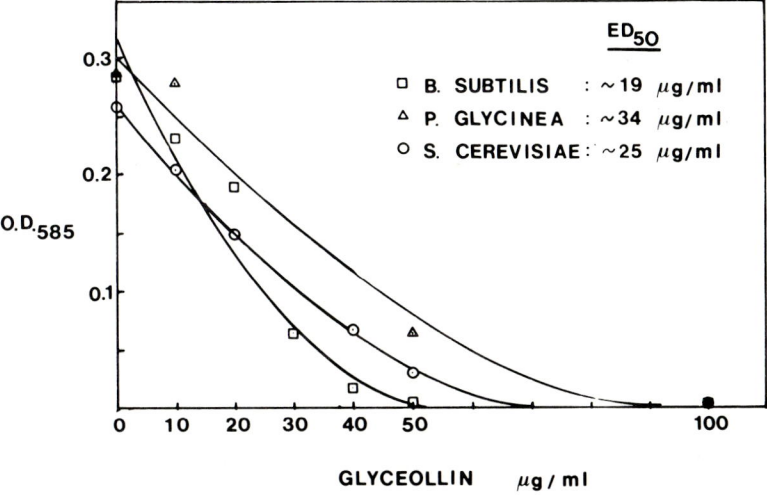

Fig. 1. Presumed biosynthetic pathway of glyceollin. Glyceollin is a phytoalexin produced by soybeans. Phytoalexins are capable of stopping the growth of microorganisms.

Fig. 2. The effect of glyceollin on the growth [optical density (O.D.) at 258 nm] of Bacillus subtilis, Pseudomonas glycinea, and Saccharomyces cerevisiae.

third bioassay was developed using suspension-cultured cells of soybeans (9). In all 3 assays, the production of glyceollin in plants is proportional to the amount of elicitor applied. This is illustrated for the hypocotyl assay in Fig. 3. In this assay, the glyceollin is

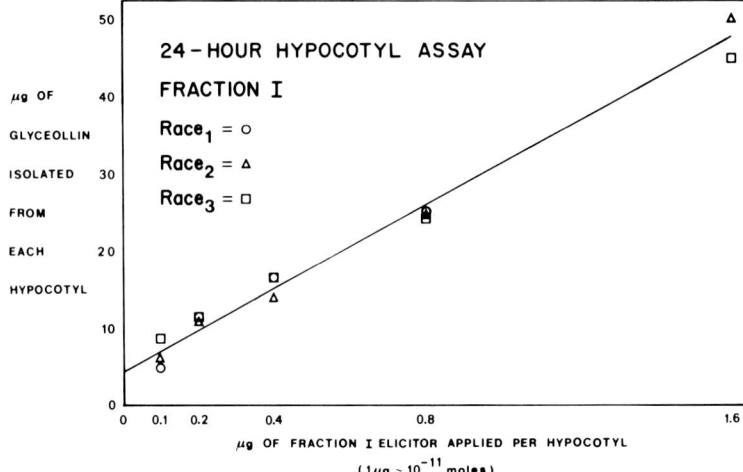

Fig. 3. The elicitors isolated from Phytophthora megasperma var. sojae races 1, 2, and 3 have equal abilities to stimulate glyceollin accumulation in the hypocotyls of 5-day-old soybean seedlings (Glycine max L. cu. Harosoy 63). Harosoy 63 is incompatible with (resistant to) Pms races 1 and 2 and compatible with (susceptible to) Pms race 3.

extracted from the hypocotyls by ethanol and then separated from contaminating compounds by thin layer chromatography. The amount of the ultraviolet absorbing glyceollin on the thin layer plates is quantitated by scraping the glyceollin from the silica gel, and measuring its absorbance at 285 nm. The identity of the glyceollin was confirmed by co-chromatography with standard glyceollin in several solvents on thin layer plates as well as by combined gas chromatography and mass spectrometry of the acetate derivative.

The cotyledon assay was used for purification of the elicitor as the cotyledon assay is less laborious than the hypocotyl assay. The cotyledon assay is based on the fact that when water droplets containing elicitor are placed on the cut surface of cotyledons, some of the glyceollin that is synthesized diffuses into the water droplets. These droplets are diluted and their absorbance at 285 nm is measured. The absorbance of this solution is reasonably proportional to the amount of glyceollin in the solution (8) even though other 285 nm absorbing compounds are also present.

D. The Effect of Pms Elicitor on Soybean Tissues

Elicitors, when introduced into flasks containing suspension-cultured soybean cells, have a dramatic effect on the cells (9). Within a few hours, the cells turn light brown. At the same time, the activity in the cells of at least one of the enzymes involved in the synthesis of glyceollin, phenylalanine ammonia lyase, is greatly increased (Figs. 1 and 4). The increase in activity of the phenylalanine ammonia lyase precedes the accumulation of glyceollin both in the cells and in the culture medium (Fig. 4). The growth of the suspension-cultured cells, as measured by fresh weight, stops upon addition of the elicitor (Fig. 5b). The cells also stop taking up ions from the media, which is another indication of the lack of growth by these cells (Fig. 5c).

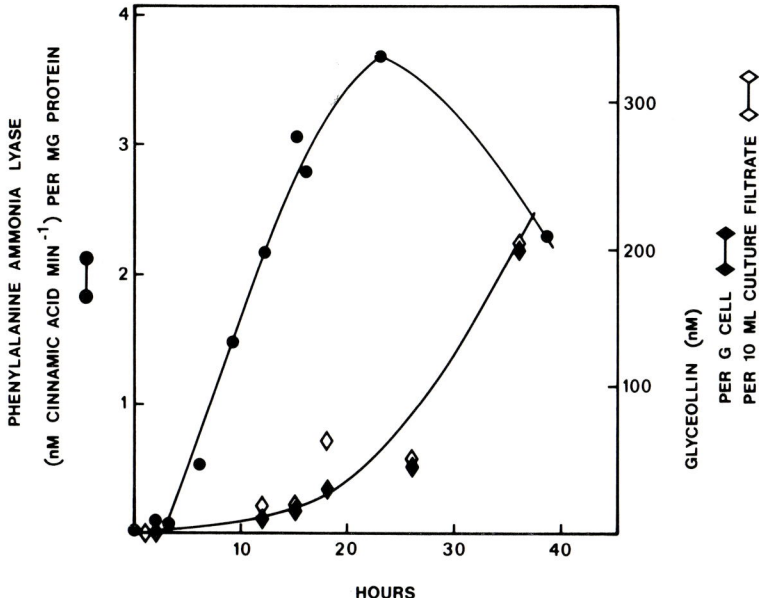

Fig. 4. The addition of 1 μg/ml of Pms elicitor to suspension-cultured soybean cells results in an increase, after about 5 hr, in the activity of phenylalanine ammonia lyase and the accumulation, after about 10 hr, of glyceollin in both the cells and in the culture fluid. Phenylalanine ammonia lyase is thought to be involved in the synthesis of glyceollin (Fig. 1).

Soybean tissues are sensitive to extremely small amounts of Pms elicitor. It is impressive to observe the effects on the growing suspension-cultured soybean cells caused by the addition of submicromolar quantities of the polysaccharide elicitor even though the cells are growing in the presence of 50 mM sucrose. The data of Fig. 3 illustrates the sensitivity of the soybean hypocotyls to the presence of the elicitor. About 10^{-12} moles of elicitor applied to a single hypocotyl stimulates quantities of glyceollin sufficient to prevent the growth of Pms and other microorganisms in vitro.

E. The Chemical Nature of the Pms Elicitor

The evidence that demonstrated that the elicitor is a polysaccharide included the fact that the elicitor is stable to autoclaving at 121°C for several hours, lacks affinity for both anion and cation exchange resins, is completely stable to treatment by pronase, and in size heterogeneous. The elicitor was first found in the fluid of old cultures of Pms and was probably released into the culture fluid by autolysis of some of the mycelia. It was later demonstrated that elicitor-active molecules with the same properties as the culture fluid elicitor could be isolated from the mycelial walls of Pms by a heat treatment similar to that used to solubilize the surface antigens from the cell walls of S. cerevisiae (10). It has now been demonstrated that the elicitor-active wall-released molecules are β-glucans. Methylation analysis of the purified elicitor has demonstrated that this glucan is largely a 3-linked polymer with glucosyl branches to about 1 out of every 3 of the backbone glucosyl residues (Table I, column 1).

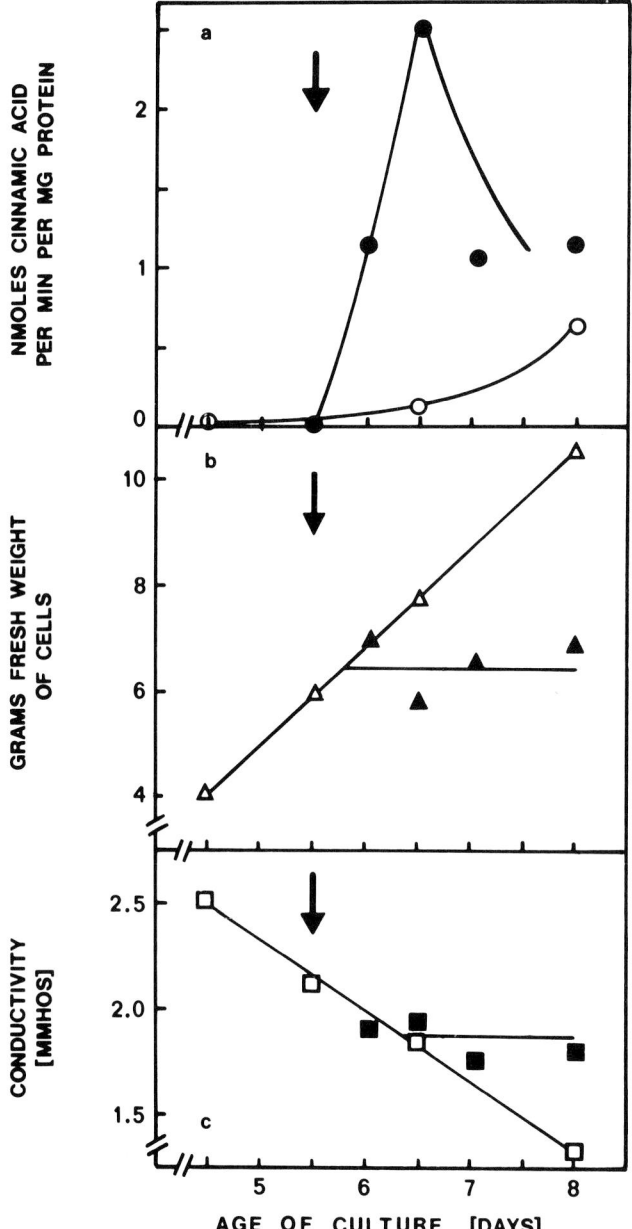

Fig. 5. The activity of phenylalanine ammonia lyase (a) increases following the addition (arrow) of 1 μg/ml of Pms elicitor to log phase suspension-cultured soybean cells. Although the elicitor does not kill the soybean cells, it does stop the growth of the cells (b) and stops the uptake by the cells of ions (primarily nitrate) from the culture media (c).

The mycelial wall-released elicitor is heterogeneous in size with an average molecular weight of approximately 100,000. An exo-β-1,3-glucanase isolated from Euglena gracilis (11) hydrolyzes approximately 90% of the glucan, leaving a more highly branched fragment (Table I, column 2). The E. gracilis exoglucanase hydrolyzes the polymer from the

TABLE I. The Glucosyl Linkage Composition of Phytophthora Megasperma var. Sojae (Pms) and Yeast Elicitors

Glucosyl linkage	Pms elicitor (before exo-)	Pms elicitor (after exo-)	Yeast elicitor
	%	%	%
Terminal	17	28	23
3-linked	54	26	11
6-linked	4	10	40
4-linked	2	6	9
3,6-linked	23	30	17

The Pms elicitor composition is given both before and after exposure of the elicitor to an exo-β-1,3-glucanase isolated from Euglena gracilis. The terminal glucosyl residues are linked only through C-1. The 3-linked, 4-linked, and 6-linked glucosyl residues are linked to other glucosyl residues through C-1 and C-3, C-4, or C-6, respectively. The 3,6-linked glucosyl residues are branch points in the glucan, being linked to other glucosyl residues through C-1, C-3, and C-6.

nonreducing end and is capable of hydrolyzing the glycosidic bond of 3-linked glucosyl residues that have other glucosyl residues attached to the 6 position. The product of the exoglucanase-degraded mycelial wall-released elicitor is still size heterogenous but has an average molecular weight of approximately 10,000. This highly branched glucan fragment retains as much activity as the undergraded elicitor.

The E. gracilis enzyme hydrolyzes β-glucans which is evidence that the Pms mycelial wall glucan is a β-linked polymer. Optical rotation and NMR studies have confirmed that the glucan is β-linked. This is not surprising as other Phytophthora cell walls have a quantitatively dominant component which is a β-3-linked glucan with some branches to C-6 (12). Indeed, it appears that as much as 60% of the mycelial wall of the Pms is composed of this polymer.

Partial acid hydrolysis of the Pms cell walls releases a series of oligosaccharides which can be partially resolved by Bio-Gel P-2 gel permeation chromatography (Fig 6). The smallest oligosaccharides do not contain detectable elicitor activity, although oligosaccharides containing as few as 7 or 8 glucosyl residues are active elicitors. The area of the P-2 column containing the smallest elicitor-active oligosaccharide has been fractionated by high pressure liquid chromatography into about 5 components. Sufficient quantities of these oligomers are now being produced to permit biological and structural characterization.

Periodate treatment of the wall-released elicitor confirms the polysaccharide nature of the active component and demonstrates the essential role of a branched oligosaccharide having terminal glycosyl residues. Exposing the elicitor to periodate eliminates almost all of the elicitor activity (Fig. 7). On the other hand, if the periodate-degraded polymers are reduced with sodium borohydride and then are subjected to mild acid hydrolysis, a considerable portion of the elicitor activity is regained (Fig. 7). Since the 3- and 3,6-linked glucosyl residues (Table I) lack vicinyl hydroxyls and are, therefore, resistant to periodate degradation, it seems likely that the periodate has destroyed the elicitor activity by modifying the terminal, 6-linked, and/or 4-linked glucosyl residues of the elicitor (Table I). Recovery of elicitor activity after partial acid hydrolysis of the periodate-treated elicitor (Fig. 7) suggests that new terminal glycosyl residues have been exposed and are able to provide the proper structure of an active elicitor. The requirement for a branched oligosaccharide is supported by our observation that 3-linked glucans which lack branches to

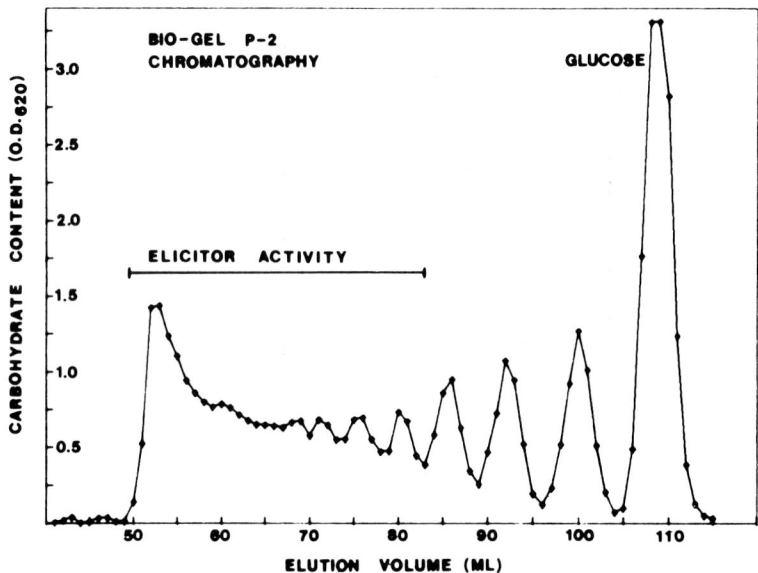

Fig. 6. Elicitor-active oligosaccharide fragments of Pms mycelial walls are generated by partial acid hydrolysis. The fragments are fractionated according to size by gel permeation chromatography. The smallest fragments possessing elicitor activity consist of 7 or 8 glycosyl residues.

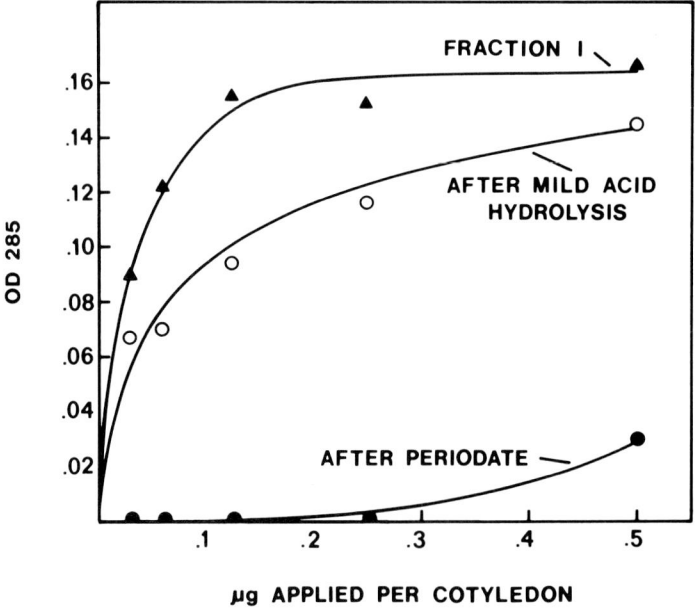

Fig. 7. Periodate (15 mM sodium metaperiodate at 20°C for 36 hr) destroys most of the Pms elicitor (Fraction I from reference 10) activity. A significant portion of this activity is recovered following reduction with sodium borohydride and mild acid hydrolysis (0.1 N H_2SO_4 at 25°C for 48 hr). The optical density at 285 nm is a measure of the amount of glyceollin accumulated in the cotyledon assay.

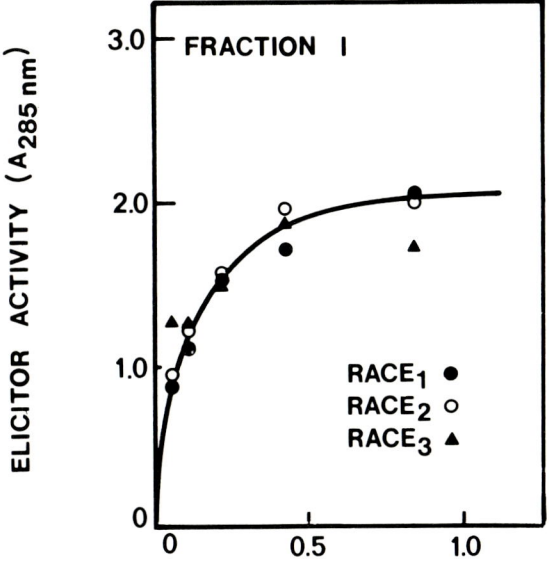

Fig. 8. The ability of the elicitors isolated from Pms races 1, 2, and 3 to stimulate glyceollin accumulation in the cotyledon assay.

C-6 or have only a single branched C-6 glucosyl residue, such as laminarin, have little or no elicitor activity (< 0.001 of the Pms elicitor).

F. Elicitors Are Widespread in Nature and Lack Species Specificity

Elicitors have been purified from 3 differentially pathogenic races of Pms and these elicitors appear structurally identical (7). Data shown on Figs. 3 and 8 demonstrate that the elicitors from the 3 Pms races have equivalent abilities to stimulate the accumulation of glyceollin soybean hypocotyls and cotyledons.

The fact that both infective and noninfective races of Pms stimulate in soybeans the accumulation of inhibitory glyceollin levels demonstrates that the accumulation of glyceollin at the observed rate is not sufficient to prevent the growth of the pathogen in susceptible cultures of soybean. This is further demonstrated in Fig. 9 in which soybeans have been exposed in the hypocotyl assay to either an incompatible race of Pms, a compatible race of Pms, or the elicitor. It can be seen that in all 3 cases it requires approximately 9 hr before inhibitory levels of glyceollin are accumulated in the plant. But the rate of glyceollin accumulation is the same whether the hypocotyls are infected with a compatible race of the pathogen, which will kill the plants within 24 hr, or an incompatible race of the pathogen, which cannot kill the plants. The fact that the purified elicitor causes the hypocotyls to produce glyceollin at the same rate as the live mycelia of both the compatible and incompatible fungi suggests that the elicitor is the mycelial component that the plant recognizes.

Soybean plants can be protected from the compatible race of the fungus by applying the elicitor to the hypocotyls 6 hr before inoculation of the hypocotyls with the mycelia of the compatible race. The mechanism by which the compatible race of the fungus avoids the inhibitory effect of glyceollin is unknown at the present time.

Soybean plants have evolved the ability to recognize and respond to the structural β-glucan of Phytophthora mycelial walls. Similar β-glucans are found in the walls of a wide range of fungi. One fungus containing such β-glucans is brewer's yeast, S. cerevisiae, a nonpathogen of plants. An elicitor has not been purified from a commercially available extract of brewer's yeast (from Difco). The 80% ethanol insoluble fraction of the yeast extract contains a very active elicitor of glyceollin accumulation in soybeans. Most of the polysaccharide in this 80% ethanol insoluble fraction is a mannan (Table II). However, yeast extract does contain small amounts of a glucan. The glucan can be almost completely separated from the mannan by binding the mannan to an affinity column consisting of Concanavalin A covalently attached to sepharose (Table II). Both the purified mannan and glucan remain contaminated by small amounts of arabinogalactan. The ribose which contaminates the 80% ethanol insoluble fraction is removed on a DEAE-cellulose column.

The elicitor activity of the crude 80% ethanol yeast extract precipitate resides in the glucan component (Fig. 10). The small amount of residual activity remaining in the mannan fraction can be attributed to the minor contamination of this fraction by glucan. The glucan is composed of the same glucosyl linkages found in the Pms elicitor and is most similar to the Pms elicitor after degradation of the Pms elicitor by the exoglucanase (Table I). The same quantities of the yeast and Pms elicitor are required to stimulate glyceollin accumulation in soybeans.

Our laboratory has obtained other evidence that the elicitor-phytoalexin story is a general one. For example, the Pms elicitor stimulates suspension-cultured cells of sycamore and parsley to produce large amounts of phenylalanine ammonia lyase activity. In addition, we have some evidence that the Pms elicitor stimulates Phaseolus vulgaris, the true bean, to make its phytoalexins. In addition, a wall glucan from Colletotrichum lindemuthianum (13), a pathogen of P. vulgaris, stimulates soybeans to produce glyceollin. And, finally, the Pms elicitor stimulates potato tubers to accumulate their phytoalexins (Mark Wade and Peter Albersheim, unpublished results).

Thus, elicitors appear to be general in nature, and diverse plants are able to respond to a single elicitor. Elicitors may therefore provide a new way of protecting plants against their pests, for elicitors may activate the plant's own defense mechanism and thereby eliminate some of the need for spraying agricultural crops with poisonous pesticides.

II. THE LECTINS OF LEGUMES INTERACT WITH THE LIPOPOLYSACCHARIDES OF NITROGEN-FIXING RHIZOBIUM SYMBIONTS

A. Rhizobium-Lectin Interactions are Specific

Bacteria, like fungi, elicit phytoalexin accumulation in plants. Therefore, the Rhizobium symbionts of legumes must either avoid eliciting phytoalexins in their hosts, metabolize the phytoalexins when they are synthesized, or be immune to the effects of the phytoalexins. In any case, molecules exist in both the Rhizobium and their legume hosts which react in such a way that a successful symbiosis can develop. These reactions do not proceed to a successful conclusion when a Rhizobium interacts with a nonhost legume. For example, Rhizobium japonicum forms a symbiotic relationship with soybeans and must avoid being rejected by the microbial defense mechanism of soybeans. But R. japonicum does not form a symbiotic relationship with other legumes such as peas. On the other hand, R. leguminosarum forms a symbiotic relationship with peas but not with soy-

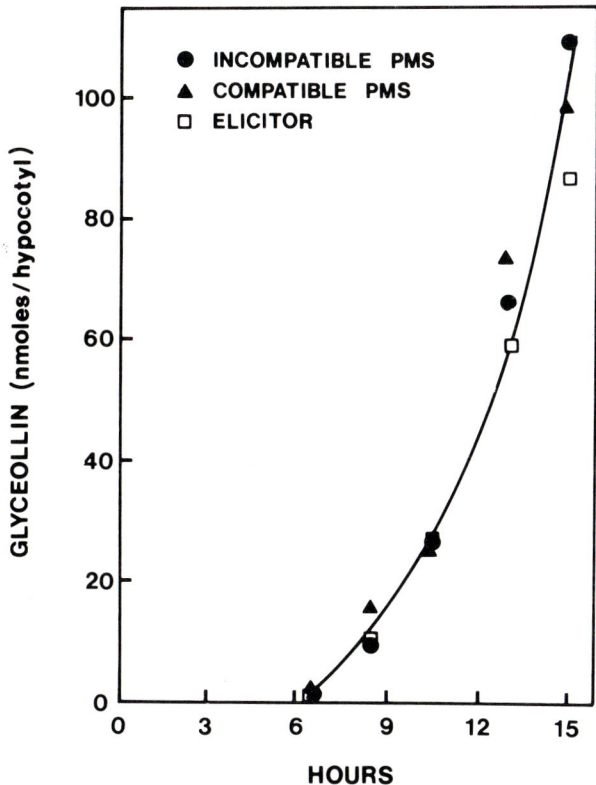

Fig. 9. Hypocotyls of soybean cultivar Harosoy 63 are stimulated to accumulate glyceollin at the same rate by incompatible Pms race 1, by compatible Pms race 3, and by Pms elicitor (isolated from either a compatible or an incompatible race).

TABLE II. The Neutral Sugar Composition of the Elicitor Obtained From Commercial Brewer's Yeast Extract (Difco)*

	Mannose	Glucose	Ribose	Arabinose	Glactose
Crude (80% EtOH Ppt)	80	8	9	2	1
Purified mannan	92	2	0	4	2
Purified glucan	2	95	0	2	1

*The values represent the % of total neutral sugar present in either the 80% ethanol precipitate of the yeast extract or in the mannan-rich fractions obtained from the 80% ethanol precipitate.

beans, and so forth. Our laboratory has been attempting to identify the molecular interactions which determine whether a particular Rhizobium-legume interaction will lead to a successful symbiosis.

B. The Lectins of Legumes Bind to the Cell Surface of Their Symbiont but not to the Cell Surface of Their Nonsymbiont Rhizobium

The lectins of legumes interact with the cell surfaces of their symbiont Rhizobium. Bohlool and Schmidt (14) were the first to provide evidence for this phenomenon. These

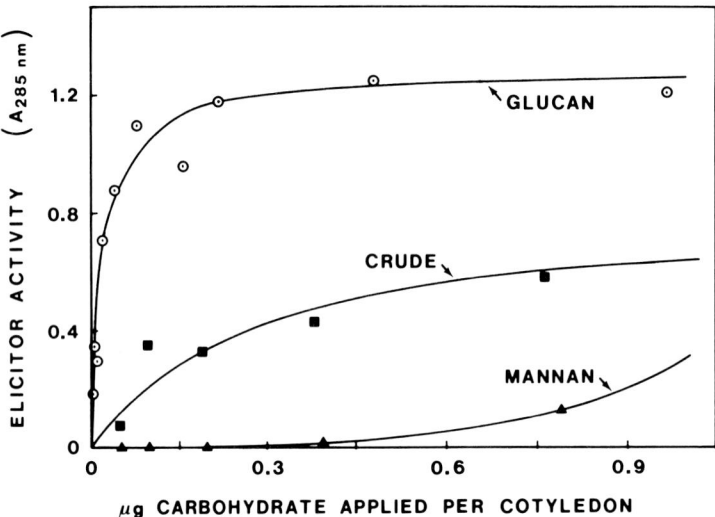

Fig. 10. The ability of the 80% ethanol precipitate of "crude" yeast extract to stimulate glyceollin accumulation in soybean cotyledons is compared with the same abilities of glucan-rich and mannan-rich fractions isolated from the "crude" yeast extract.

workers extracted soybean seed lectin and conjugated it with a fluorescent dye. They found that this fluorescence-labeled soybean lectin preparation was bound to all but 3 of 25 strains of R. japonicum, the symbiont of soybeans. They found, too, that the fluorescent-labeled lectin did not bind to any of 23 other strains representative of Rhizobium species which do not nodulate soybeans. The laboratories of W. Dietz Bauer at the Charles F. Kettering Research Institute and Jack Paxton at the University of Illinois have confirmed this basic observation (unpublished personal communication). These laboratories have found that more highly purified soybean lectin binds to about half of the symbiont R. japonicum strains, but they found no binding of soybean lectin to Rhizobium strains which do not nodulate soybeans.

C. The Lectins of Legumes Interact With the Lipopolysaccharides of Their Symbiont but Not With the Lipopolysaccharides of Their Nonsymbiont Rhizobium

The dominant surface antigens of the Gram-negative Rhizobium are the O antigens of the lipopolysaccharides. Our laboratory has asked whether legume seed lectins interact with the Rhizobium lipopolysaccharides. Affinity-purified lectins from the seeds of 4 legumes were covalently attached to Agarose by cyanogen bromide coupling. The lipopolysaccharides were purified from the 4 Rhizobium symbionts of the legumes from which the lectins were purified. The exopolysaccharides were also purified from each of the 4 Rhizobium species. The lipopolysaccharides and the exopolysaccharides were passed separately through columns containing the legume lectins bound to Agarose. It was then determined whether any of the lipopolysaccharides or exopolysaccharides were retained by the lectin columns. The lipopolysaccharides isolated from nonsymbiont Rhizobium do not bind to the lectin columns. Neither do the exopolysaccharides bind to the lectin columns, regardless of whether the exopolysaccharides were synthesized by symbiont or nonsymbiont Rhizobium. The inability of pea lectin to bind lipopolysaccharides and exopolysaccharides produced by R. japonicum, R. phaseoli, and R. spp. is illustrated by the open circles and dotted lines of Fig. 11. The pea lectin column also does not bind the exopolysaccharide of

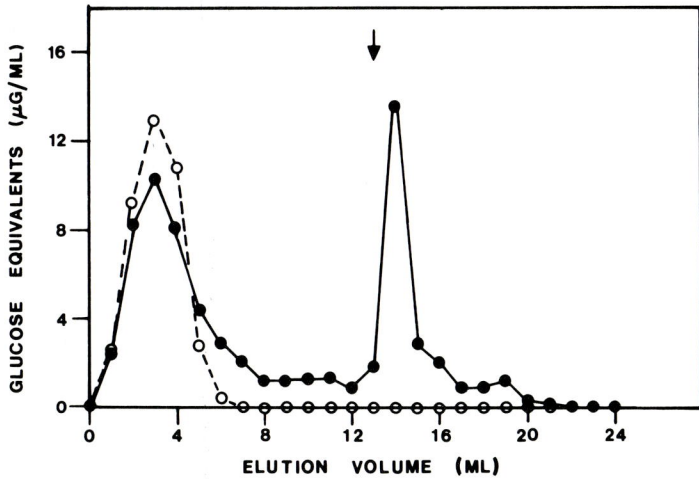

Fig. 11. The exopolysaccharides of Rhizobium japonicum, R. phaseoli, R. spp. and R. leguminosarum and the lipopolysaccharides of the first 3 of these Rhizobium species pass directly through an affinity column composed of pea lectin covalently attached to Agarose (open circles, dashed line). The lipopolysaccharide of R. leguminosarum interacts with the pea lectin (filled circles, solid line); a portion of the lipopolysaccharide is released when the column is washed (arrow) with pH 3 buffer. Peas are the symbiont host of R. leguminosarum. The polysaccharides are detected as glucose equivalents by the anthrone assay.

R. leguminosarum. But the pea lectin column does bind at least a portion of th lipopolysaccharide produced by R. leguminosarum (filled circles of Fig. 11). Thus, since R. leguminosarum is the symbiont of peas, the pea lectin binds partially the lipopolysaccharide of its symbiont Rhizobium but does not bind the lipopolysaccharides of 3 nonsymbiont Rhizobium. The data of Table III summarize the results obtained for the interaction of the 4 lipopolysaccharides and the 4 lectin columns (15). In every case, the lipopolysaccharides interact with their host-legume lectin, but not with the lectins from non-host legumes.

The binding between the lipopolysaccharides and the lectin appears to be relatively weak; continued washing of the columns with pH 7 buffer eventually washed off most of the bound lipopolysaccharide. However, as illustrated in Fig. 11, the material remaining on the column after limited washing with pH 7 buffer is immediately washed off the column with pH 3 buffer, a treatment which characteristically negates lectin binding.

D. Are Lectins Enzymes?

We are not certain why the lipopolysaccharides do not bind completely to their host-legume lectins. The amount of lipopolysaccharide which binds to the lectin column depends on the time the lipopolysaccharide interacts with the column. Indeed, there appears to be an optimal time for interaction; as illustrated on Fig. 12, more lipopolysaccharide binds to the column when allowed to react with the column for 100 min rather than for the few minutes required for the lipopolysaccharide to pass directly through the 3-ml bed volume affinity column. On the other hand, the amount of lipopolysaccharide which binds decreases again after even longer interaction times. This and other results suggest that a chemical reaction may be occurring on the lectin columns. Perhaps the

TABLE III. Interaction of Rhizobium Lipopolysaccharide (LPS) and Exopolysaccharide With Immobilized Lectins*

Rhizobium species and component	Source of immobilized lectin			
	Soybean	Pea	Red kidney bean	Jack bean
R. japonicum LPS	35	0	0	0
R. leguminosarum LPS	0	36	0	0
R. phaseoli LPS	0	0	5–23	0
R. spp. LPS	0	0	0	25
Exopolysaccharide from each of the above species	0	0	0	0

*The data represents the percentage of the anthrone positive lipopolysaccharides material which did not pass directly through the affinity column and which was eluted at pH 3.0 (15).

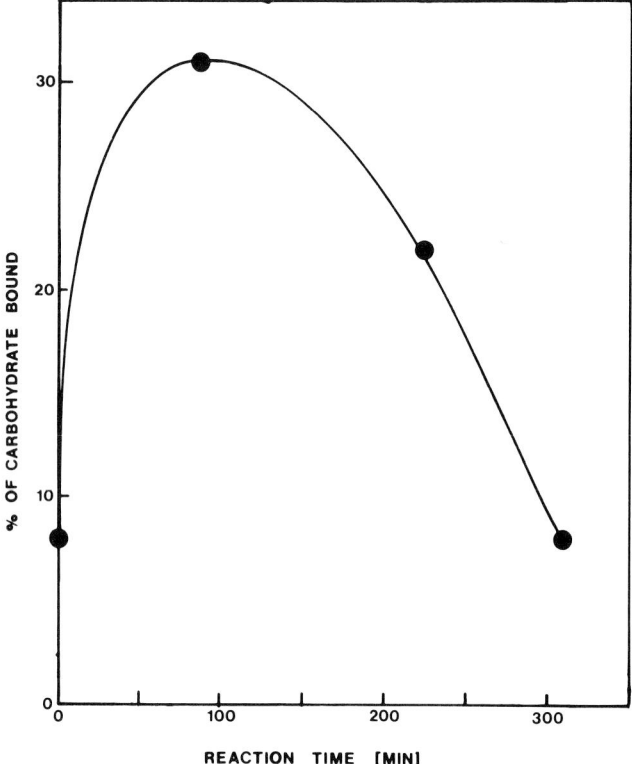

Fig. 12. The degree of binding of the Rhizobium lipopolysaccharides to their host-legume lectin columns (Fig. 11 and Table III) depends on the length of time the lipopolysaccharide interacts with the lectin column.

lectins or contaminants of the lectins might be interacting with the lipopolysaccharides of their symbiont Rhizobium in such a way that the lectins are actually altering the lipopolysaccharides. In other words, some component of the lectin columns might be catalyzing a reaction with the lipopolysaccharides.

We have tried very hard, without total success, to demonstrate whether or not lectins are enzymes. The pea lectin column does appear to alter the size of the lipopolysaccharide from R. leguminosarum (the pea symbiont), but the pea lectin does not alter the size of the lipopolysaccharides from several other Rhizobium species. The phytohemagglutinin (kidney bean lectin) column specifically alters the size of the R. phaseoli (the kidney bean symbiont) lipopolysaccharide. However, it is not safe to assume that the apparent change in size of the lipopolysaccharides is an enzyme-catalyzed alteration as the observed reaction rate was slow and the apparent size of lipopolysaccharides is highly dependent on the degree of micelle formation. Perhaps the size change resulted from a breakdown of the micelles.

Additional experiments were carried out and provided some evidence that the structure of the R. phaseoli lipopolysaccharide was being altered by interaction with the phytohemagglutinin column. These experiments depended on the fact that the untreated lipopolysaccharide of R. phaseoli binds completely to DEAE-cellulose. After interaction with the phytohemagglutinin column, approximately 50% of the carbohydrate of the lipopolysaccharide from R. phaseoli no longer binds to the DEAE-cellulose column. However, this result could also stem from a change in the degree of micelle formation of the lipopolysaccharide. And again, the rate of the observed reaction was slow. These reactions could only be demonstrated using the large amount of lectins which are covalently attached to the affinity columns. Smaller amounts of soluble lectins did not give reproducible results.

It may not be the lectins which are catalyzing the apparent reactions, but even if other proteins in the lectin preparations are responsible, these proteins must retain the observed host-symbiont specificity, for the putative catalytic factors are only capable of altering the lipopolysaccharide of the Rhizobium which are symbionts of the legumes from which the catalytic factors were isolated.

The possibility that lectins are, in fact, enzymes did not occur to us before our studies of the interactions between lectins and lipopolysaccharides. However, it might have occurred to us had we been aware of the work of Scheid and Choppin (16) who demonstrated that the larger of the 2 coat glycoproteins of paramyxovirus SV5 and the Newcastle disease virus are both a neuraminidase and a hemagglutinin. We would have been even less surprised had we been aware of the papers of Rood and Wilkinson (17, 18) who obtained a great deal of evidence that a protein secreted by Clostridium perfringens is both a sialidase and a hemagglutinin. It seems probable that lectins agglutinate animal cells because they are interacting with pseudo-substrates or perhaps one of a pair of substrates. Any enzyme involved in carbohydrate metabolism and which has 2 active sites can act as a lectin. We would imagine that under certain conditions, most if not all glycosyl transferases which have at least 2 active sites, would act as lectins. It might be that the observed relatively poor binding constants of lectins for their haptens is simply a reflection that the correct substrates have not been identified. Our biased opinion is that lectins will be found to serve many functions in plants and that many lectins are enzymes. Much more evidence will have to be obtained before this suggestion can be considered to be proven. Nevertheless, those using lectins in their studies should consider the possibility that lectins have catalytic activity.

E. The Lipopolysaccharides of Each Species of Rhizobium and Each Strain of Each Species Are Structurally Unique

If the lectins of legumes interact with the lipopolysaccharides of symbiont Rhizobium and not with the lipopolysaccharides of nonsymbiont Rhizobium, then different Rhizobium species must have different lipopolysaccharides. Carbohydrate analysis of the lipopolysaccharides purified from several Rhizobium species and from 2 or more strains of each of the species demonstrates that not only are the lipopolysaccharides of the different Rhizobium species structurally unique, but the different strains of the various Rhizobium species are structurally unique (Table IV).

TABLE IV. The Lipopolysaccharide Sugar Composition of Several Strains of Each of Three Rhizobium Species*

	R. phaseoli Strain		R. leguminosarum Strain			R. japonicum Strain		
	127K17	127K24	128C53	3HOQ1	128C63	61A23	61A135	R54a
Glucose	+	+	+	+	+	+	+	+
Galactose	+	+	+	+	+	+	+	+
Mannose	+	+	+	+	+	+	+	+
Uronic acids	+	+	+	+	+	+	+	+
KDO	+	+	+	+	+	±	±	±
Fucose	+	+	+	+	+	+	−	+
Rhamnose	−	−	+	+	−	+	−	+
2-O-Me-6-dodeoxyhexose	−	+	−	−	−	−	−	−
3-N-me-amino-3,6-dodeoxyhexose	+	−	−	−	+	−	−	−
2-Amino-2,6-dodeoxyhexose (1)	+	−	−	−	−	−	−	±
2-Amino-2,6-dodeoxyhexose (2)	−	−	−	−	−	−	−	±
Di-O-me-6-deoxyhexose	−	−	−	−	+	−	−	−
6-O-Me-hexose	−	−	−	−	−	+	−	−

\+ = sugar present in the lipopolysaccharide.
± = sugar may be present
− = sugar is not present

The lipopolysaccharides of the Rhizobium strains that we have studied are composed of sugars which are typical of the sugars which compose the lipopolysaccharides of other Gram-negative bacteria (19). Not only do these lipopolysaccharides have a diversity of sugars, but the lipopolysaccharides are characterized by the presence of rather unusual sugars such as 3-O-methyl-3-amino-3, 6-dideoxyhexoses or 2-O-methyl-6-deoxyhexose. As in other Gram-negative lipopolysaccharides, the Rhizobium lipopolysaccharides also contain 2-keto-3-deoxyoctanoic acid, a characteristic carbohydrate of the core region of Gram-negative lipopolysaccharides.

Certain sugars are found in all of the Rhizobium lipopolysaccharides examined to date. Some of all of these may be part of the core regions of the lipopolysaccharides. Other sugars are found in only one or a few different lipopolysaccharides, and these may be characterisitic of the O-antigen regions of the lipopolysaccharides. One quite common component of the Rhizobium lipopolysaccharides, not frequently reported for other Gram-negative bacteria, is galacturonic acid.

Immunological studies also indicate that the lipopolysaccharides of different strains of a Rhizobium species have different structures. The lipopolysaccharides of 2 R. leguminosarum strains have the same chamical compositions, or, in other words, the same chemotypes. But even these strains can be differentiated immunologically. For example, the lipopolysaccharide from R. phaseoli strain 127K17 was used to raise antibodies in a rabbit. These antibodies, when placed in the center well of a semimicro double diffusion plate form a precipitin band when run against the lipopolysaccharide or whole cells of

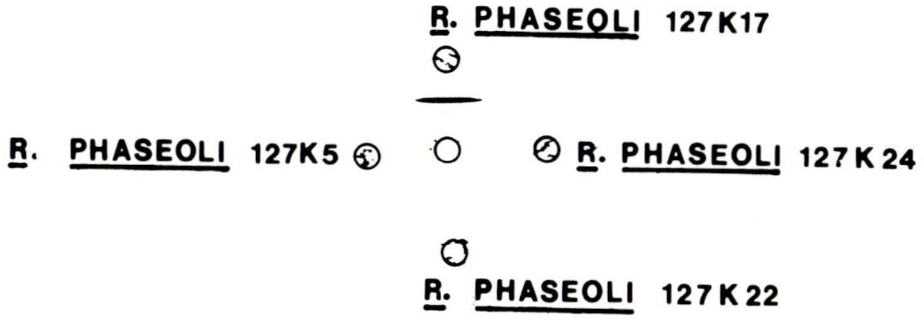

R. PHASEOLI 127K17

R. PHASEOLI 127K5   R. PHASEOLI 127K24

R. PHASEOLI 127K22

Fig. 13. Rabbit antibodies raised against the lipopolysaccharide isolated from R. phaseoli strain 127K17 were placed in the center well of this semimicro double diffusion gel. The outer wells contained frozen-dried cells (or isolated lipopolysaccharides, not shown) of the R. phaseoli strains indicated. A precipitin band only forms between the antibodies and the R. phaseoli cells of the same strain from which the lipopolysaccharide was isolated and used to raise the antibodies.

R. phaseoli 127K17 (Fig. 13). But the antibodies to R. phaseoli 127K17 do not react with whole cells of the lipopolysaccharides of the 3 other R. phaseoli strains illustrated in Fig. 13 or 3 other R. phaseoli strains not present in Fig. 13. In addition, the antibodies raised against the lipopolysaccharides of R. japonicum 61A23 react with the lipopolysaccharides of that R. japonicum strain but not with 2 other strains of R. japonicum nor with several other Rhizobium species. The specificity of the lipopolysaccharide antibodies is equally apparent in agglutination assays. Thus, the lipopolysaccharides of different Rhizobium strains have unique structures.

Since the lectins of legumes interact with the lipopolysaccharides of their symbiont Rhizobium, the legumes must interact with the different lipopolysaccharides originating from the different strains of a single Rhizobium species. If this is true, the lectins may be interacting with a common portion of the O antigens of these lipopolysaccharides or else the lectins may be interacting with the core regions of the lipopolysaccharides of their symbiont Rhizobium. We have obtained some evidence that the core regions of different Rhizobium species are in fact structurally different. This would suggest that the Rhizobium species might actually be considered to represent different genera and that the different strains of a Rhizobium species deserve species rank. Regardless of the proper classification of the Rhizobium, it is clear that the lipopolysaccharides contain enough structural information to account for the specificity of host-symbiont selection.

ACKNOWLEDGMENTS

This work has been supported in part by grants from the Energy Research and Development Administration (EY-76-S-02-1426), the Herman Frasch Foundation, the United States Department of Agriculture (616-15-73), the Rockefeller Foundation (RFGAAS 7510), and the National Science Foundation (PCM75-13897 A01).

REFERENCES

1. Ingham JL: Bot Rev 38:343, 1972.
2. Kuć J: Annu Rev Phytopathol 10:207, 1972.
3. Deverall BJ: Proc R Soc London, Ser B 181:233, 1972.
4. Keen NT, Partridge JE, Zaki AI: Phytopathology 62:768, 1972.
5. Burden RS, Bailey JA: Phytochemistry 14:1389, 1975.
6. Lyne RL, Mulheirn LJ, Leworthy DP: J Chem Soc, Chem Commun, p 497, 1976.
7. Ayers AR, Valent B, Ebel J, Albersheim P: Plant Physiol 57:766, 1976.

8. Ayers AR, Ebel J, Finelli F, Berger N, Albersheim P: Plant Physiol 57:751, 1976.
9. Ebel J, Ayers AR, Albersheim P: Plant Physiol 57:775, 1976.
10. Ayers AR, Ebel J, Valent B, Albersheim P: Plant Physiol 57:760, 1976.
11. Barras DR, Stone BA: Biochim Biophys Acta 191:342, 1969.
12. Bartnicki-Garcia S: J Gen Microbiol 42:57, 1966.
13. Anderson-Prouty AJ, Albersheim P: Plant Physiol 56:286, 1975.
14. Bohlool BB, Schmidt EL: Science 185:269, 1974.
15. Wolpert JS, Albersheim P: Biochem Biophys Res Commun 70:729, 1976.
16. Scheid A, Choppin PW: Virology 62:125, 1974.
17. Rood JI, Wilkinson RG: J Bacteriol 123:419, 1975.
18. Rood JI, Wilkinson RG: J Bacteriol 126:845, 1976.
19. Lüderitz O, Staub AM, Westphal O: Bacteriol Rev 30:193, 1966.

Localization of Some Glycolipid Glycosylating Enzymes in the Golgi Apparatus of Rat Kidney

Becca Fleischer

Department of Molecular Biology, Vanderbilt University, Nashville, Tennessee 37235

Cell fractions from rat kidney were isolated and studied for their ability to synthesize several possible intermediates in the biosynthesis of sulfatides and gangliosides. The enzymes studied include UDP-Gal:ceramide galactosyltransferase, UDP-Gal:glucosylceramide galactosyltransferase, UDP-Gal:galactosylceramide galactosyltransferase, and CMP-NAN:lactosylceramide sialyltransferase activities. The initial glycosylation of ceramide was found to be present in all of the kidney cell fractions studied. The remaining glycosylating enzymes were largely localized in the Golgi apparatus of kidney. Thus, in addition to modifying glycoproteins for secretion, the Golgi apparatus in kidney is involved in the modification of a number of glycolipids which are destined to form cell membrane components.

Key words: Golgi, glycolipid biosynthesis, glycosyltransferases, kidney cell fractions

Golgi apparatus has been isolated from liver (1–3) and shown to be the main locus in the hepatocyte for the addition of galactose and sialic acid to secreted glycoproteins (4–6). These sugars are added stepwise to the nonreducing end of the carbohydrate chains of the glucoprotein by the action of specific galactosyl or sialyltransferases which use nucleotide sugars as glycosyl donors (7). Analogous reactions have been shown to be involved in the biosynthesis of glycosphingolipids (Fig. 1) (7).

The biosynthesis of glycosphingolipids other than gangliosides has been reviewed by Morell and Braun (8). Ganglioside biosynthesis has recently been reviewed by Fishman and Brady (9). The reactions catalyzed by enzyme 1 (Fig. 1) have been shown to be present in kidney (10) and brain (11) microsomes. The synthesis of lactosylceramide (enzyme 4) has been described in spleen (12), kidney (13), and brain (14). A galactosyltransferase which adds a second galactose to galactosylceramide has also been shown to be present in rat kidney (15). The sialyltransferase which forms sialyllactosylceramide (GM_3) from lacto-

Abbreviations: CER – ceramide; Gal – galactose; Glu – glucose; UDP-Gal – uridine diphosphogalactose; UDP-Glu – uridine diphosphoglucose; PAPS – 3′-phosphoadenosine-5′-phosphosulfate; CMP-NAM – cytidine monophosphoryl-N-acetylneuraminic acid; GM_3 – sialyllactosylceramide, GD_3 – disialolactosylceramide.

Received May 23, 1977; accepted June 14, 1977

© 1977 Alan R. Liss, Inc., 150 Fifth Avenue, New York, NY 10011

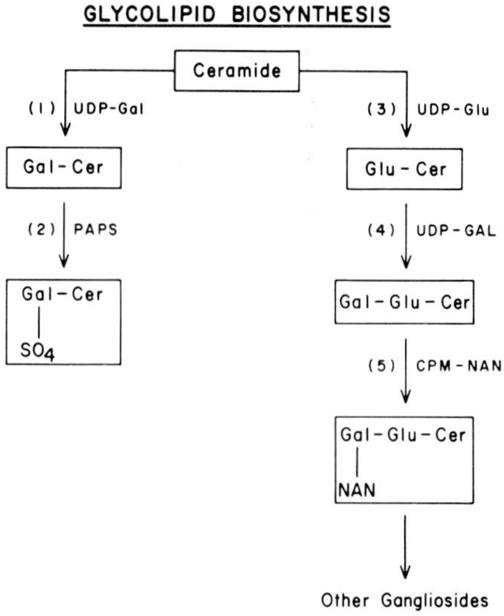

GLYCOLIPID BIOSYNTHESIS

```
                    Ceramide
          ┌────────────┴────────────┐
     (1) │ UDP-Gal             (3) │ UDP-Glu
          ▼                          ▼
       Gal-Cer                    Glu-Cer
          │                          │
     (2) │ PAPS                (4) │ UDP-GAL
          ▼                          ▼
       Gal-Cer                   Gal-Glu-Cer
          │                          │
         SO₄                    (5) │ CPM-NAN
                                     ▼
                                Gal-Glu-Cer
                                     │
                                    NAN
                                     │
                                     ▼
                              Other Gangliosides
```

Fig. 1. Two possible pathways in the biosynthesis of sulfatide and gangliosides.

sylceramide (enzyme 5) was first described in embryonic chicken brain (16). GM_3 is a precursor of more complex gangliosides in this system (16). All of the transferase activities necessary to synthesize gangliosides from ceramide have also been shown to be present in mammary gland (17). The subcellular localization to these transferases in kidney, spleen, and mammary gland are not known. In brain, a number of glycolipid glycosyltransferases have been shown to be present in synaptosome fractions (18). Keenan et al. reported enzyme 5 to be highly enriched in Golgi apparatus isolated from rat liver, although activity was also found in isolated endoplasmic reticulum fractions (19).

In order to study the possible role of the Golgi apparatus in the biosynthesis of glycosphingolipids, we developed procedures for the isolation of Golgi apparatus from rat kidney (20), a tissue rich in glycolipids. By comparing the isolated Golgi fraction with isolated plasma membranes, endoplasmic reticulum, mitochondria, and nuclei, we showed that cerebroside sulfotransferase (reaction 2, Fig. 1) was localized mainly in Golgi membranes in kidney (21). This finding brought to light a new function for the Golgi apparatus, that is, that Golgi modifies not only mucopolysaccharides and glycoproteins, but glycolipids as well. Glycolipids are not normally secreted but are destined for use in cell membranes.

In this paper we report further studies aimed at localizing some of the other galactosyl- and sialyltransferases involved in the biosynthesis of sulfatide and gangliosides in kidney. These include UDP-Gal:ceramide galactosyltransferase (enzyme 1, Fig. 1), UDP-Gal: glucosylceramide galactosyltransferase (enzyme 4) and CMP-NAN:lactosylceramide sialyltransferase (enzyme 5).

METHODS

Isolation of Cell Fractions

Male Holtzman rats, 300–350 g, fed ad libitum were used throughout. The isolation of cell fractions from rat kidneys (9) and livers (2, 22) have been described previously.

Purity of the fractions was estimated using marker enzymes as detailed previously (2, 21, 22).

Enzyme Assays

UDP-Gal was obtained from Calbiochem, (Los Angeles, California) UDP-[^{14}C]-Gal uniformly labelled in the galactose moiety was obtained from New England Nuclear Corporation (Boston, Massachusetts). It was diluted with carrier to a specific activity of about 1 mCi per mmole before use. CMP-[^{14}C]-NAN(sialic 4-^{14}C), about 1 mCi/mmole, was purchased from New England Nuclear Corporation and used without additional carrier. Hydroxy fatty acid (HFA) ceramides, a natural mixture from bovine brain, and glucosyl and galactosyl ceramides were purchased from Applied Science Laboratories Inc. (State College, Pennsylvania). N-Lignoceroyl-DL-dihydrolactocerebroside was purchased from Miles Laboratories. Inc. (Elkhart, Indiana). Human serum transferrin was obtained from Sigma Chemical Company (St. Louis, Missouri). It was desialylated by treatment with 0.05 N sulfuric acid for 1 h at 80°C followed by dialysis as described by Spiro and Spiro (23). Authentic sialyllactosylceramide (GM_3) was the kind gift of Dr. Charles Sweeley, Department of Biochemistry, Michigan State University, East Lansing, Michigan. Cardiolipin was purchased from Sylvania Chemical Company, Orange, New Jersey.

UDP-Gal:ceramide galactosyltransferase (enzyme 1) was assayed as described by E. Costantino-Ceccarine and P. Morell (10). The assay mixture contained in order of addition, 100 μg protein (lyophilized); HFA ceramides (added in benzene and evaporated) 25 μg; Tris-HCl, pH 7.8, 75 μmoles; EDTA, 0.15 μmoles; dithiothreitol, 0.15 μmoles; and UDP-[^{14}C]-Gal 0.25 μmoles; in a total volume of 130 μl.

UDP-Gal:cerebroside galactosyltransferase (enzyme 4) assays contained, in order of addition, sodium cacodylate, pH 6.6, 40 μmoles; $MnCl_2$, 0.2 μmoles; cutscum, 750 μg; protein, 50 μg; glycosyl or galactosyl ceramide, 50 μg suspended first at 10 mg/ml by homogenization in 1% (wt/wt) cutscum in water; UDP-[^{14}C]-Gal, 0.10 μmoles; in a total volume of 100 μl.

CMP-NAN:lactosylceramide sialyltransferase (enzyme 5) was assayed in a mixture containing, in order of addition, imidazole·HCl, pH 6.5, 80 μmoles; a mixture of Triton CF-54:Tween 80 (2:1, wt/wt) 200 μg; protein, 20 μg; N-lignoceroyl dihydrolactocerebroside, 50 μg, suspended first at 10 mg/ml by homogenization in 1% (w/v) Triton CF-57:Tween 80 (2:1, wt/wt) in water; and CMP-[^{14}C]-NAN, 0.10 μg; in a total volume of 100 μl. For cell fractions with low levels of activity, up to 100 μg protein was used, but the ratio of detergent/protein was kept constant.

All incubations were for 1 h at 37°C. Reaction was stopped by addition of 3.0 ml chloroform:methanol (2:1, vol/vol). Radioactive cerebrosides were recovered and counted as described previously for assay of sulfatide formation (21). For measurement of GM_3 formation, the chloroform:methanol extracts were washed once after addition of 0.6 ml 2% (wt/vol) $CaCl_2$ plus 0.3 ml 1 N formic acid to form 2 phases (24). The lower phase was then washed 3 more times with an "upper phase" made by mixing the appropriate volumes of chloroform, methanol, water, $CaCl_2$, and formic acid. The final lower phase was dried and counted as described previously (21).

UDP-Gal:N-acetylglucosamine galactosyltransferase of the fractions was assayed essentially as described previously (22) except that the final concentration of N-acetylglucosamine used was 2 mM since higher concentrations are slightly inhibitory. CMP-NAN: glycoprotein sialyltransferase was assayed by a modification of the method of Schachter et al. (4). The assay mixture contained, in order of addition, sodium 2-(N-morpholino)

TABLE I. Specific Activities* of Glycoprotein Galactosyltransferase and Sialyltransferase Enzymes in Subcellular Fractions of Rat Liver and Kidney

Fraction	Galactosyltransferase		Sialyltransferase	
	Liver	Kidney	Liver	Kidney
Homogenate	6	7	7	2
Nuclei	2	3	2	1
Rough microsomes	12	4	8	2
Smooth microsomes	41	38	33	9
Golgi	764	650	550	135
Mitochondria	0	2	0	1
Plasma membranes	5	1	23	3
Supernatant	0	0	0	0

*Specific activities expressed as nmoles galactose of N-acetylneuraminic acid transferred per hour per mg protein at 37°C.

ethanesulfonic acid (Mes) pH 5.7, 5 μmoles; protein, 20–100 μg; Triton X-100, 300 μg; desialylated human serum transferrin, 750 μg; CMP-[^{14}C]-NAN, 20 nmoles; in a total volume of 55 μl. After incubation at 37°C for 1 h, 0.02 ml of 0.3 M EDTA was added and the assay mixture placed on ice. Aliquots were spotted on Whatman No. 3 filter paper disks, dried, and protein-bound radioactivity determined as described by Mans and Novelli (25).

RESULTS

Table I summarizes the distribution in the liver and kidney subcellular fractions of galactosyltransferase and sialyltransferase activities involved in glycoprotein biosynthesis. Both activities appear to be concentrated mainly in the Golgi apparatus fraction in liver and in kidney. The low levels of activity found in other cell fractions probably reflect contamination of these fractions with Golgi membranes (21, 22).

A number of glycolipid galactosyltransferases are involved in the synthesis of glycolipids. We have measured 2 of these activities in our fractions. They are enzymes 1 and 4 (Fig. 1). The results are summarized in Table II. The activities are present in kidney but are very low in liver. Both are membrane bound. Enzyme 1 differs from enzyme 4 and from glycoprotein galactosyltransferases in that it does not appear to be localized in Golgi membranes but is distributed in all the membranous fractions studied. In contrast to the enzyme from rat brain (26, 27), we found that enzyme 1 from kidney was not stimulated by added lecithin, in the presence or absence of Triton X-100. Enzyme 4 which converts glucosylceramide to lactosylceramide, a precursor of gangliosides, appears to be a Golgi enzyme. In the course of studying the specificity of this reaction, we found that galactosylceramide was a better acceptor for galactose than glucosylceramide in this assay system (Table II). The distribution of this activity in kidney cell fractions was similar to that of enzyme 4. It is not clear whether these activities are due to the same or to different enzymes. The activity with galactosylceramide as substrate is stimulated greatly by the addition of manganese (Fig. 2). Magnesium will not substitute for manganese in this reaction. Addition of detergents also stimulates the activity up to a point (Fig. 3), after which some detergents become inhibitory.

TABLE II. Specific Activity* of Some Glycolipid Galactosyltransferases in Rat Kidney and Liver Subcellular Fractions

Fraction	Ceramide[a] (1)↓UDP-Gal Gal-Cer		Gal-Cer ↓ UDP-Gal $(Gal)_2$Cer		Glu-Cer (4) ↓ UDP-Gal Gal-Glu-Cer
	Kidney	Liver	Kidney	Liver	Kidney
Homogenate	0.110	0.015	0.2	0.03	0.1
Rough microsomes	0.048	0.01	0.2	0.03	0.2
Smooth microsomes	0.087	0.03	0.2	0.13	0.4
Golgi	0.113	0.03	18.0	0.14	8.0
Mitochondria	0.030	0.01	0.1	0.01	0.1
Plasma membranes	0.057	0.01	0.3	0.09	0.2
Supernatant	0.001	–	0.0	–	0.0

*Expressed as nmoles galactose transferred per hour per mg protein at 37°C.
[a]Ceramides containing hydroxy fatty acids only were used as substrate (10).

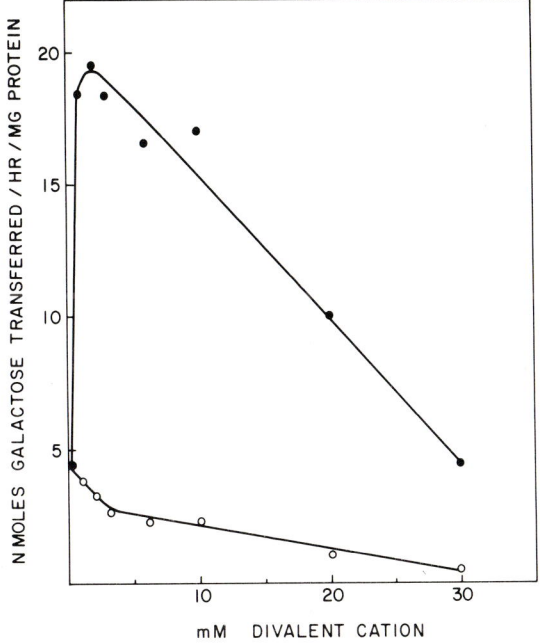

Fig. 2. Effect of varying concentrations of divalent cations in the form of their chloride salts on UDP-Gal:galactosylceramide galactosyltransferase activity of rat kidney Golgi apparatus. ●—●) Mn^{2+}; ○—○) Mg^{2+}. Other constituents added as described under "Methods". The amount of protein used was 30 µg.

We next investigated the ability of kidney Golgi fractions to synthesize GM_3 (enzyme 5). It was reported by Keenan et al. (19) that added cardiolipin stimulates the formation of GM_3 by rat liver Golgi. We therefore first studied the effect of added cardiolipin on enzyme 5 activity of rat kidney Golgi. The results are summarized in Fig. 4. Addition of cardiolipin was slightly stimulatory at pH above 6.5, while at pH 6.4 cardiolipin

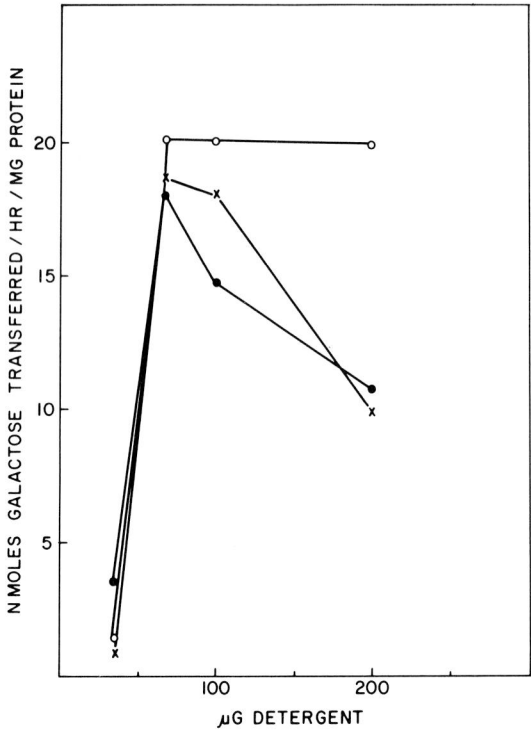

Fig. 3. Effect of varying detergent concentration and type on UDP-Gal:galactosylceramide galactosyltransferase activity of rat kidney Golgi apparatus. The final volume of the assay mixture was 100 μl and contained 20 μg protein. Other constituents added as described under "Methods." ○—○) Triton CF-54:Tween 80, 2:1 wt/wt; x—x) Triton X-100; ●—●) cutscum.

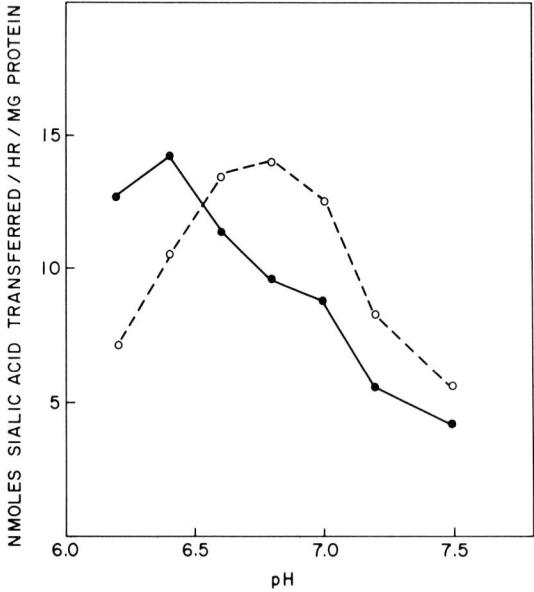

Fig. 4. Effect of varying pH of the assay mixture on CMP-NAN:lactosylceramide sialyltransferase activity of rat kidney Golgi apparatus. 40 μg protein and 10 mM Mg^{2+} were used throughout. Other constituents as described under "Methods." ●—●) Without added cardiolipin; ○—○) 7.3 μg cardiolipin were added to each assay before addition of the enzyme.

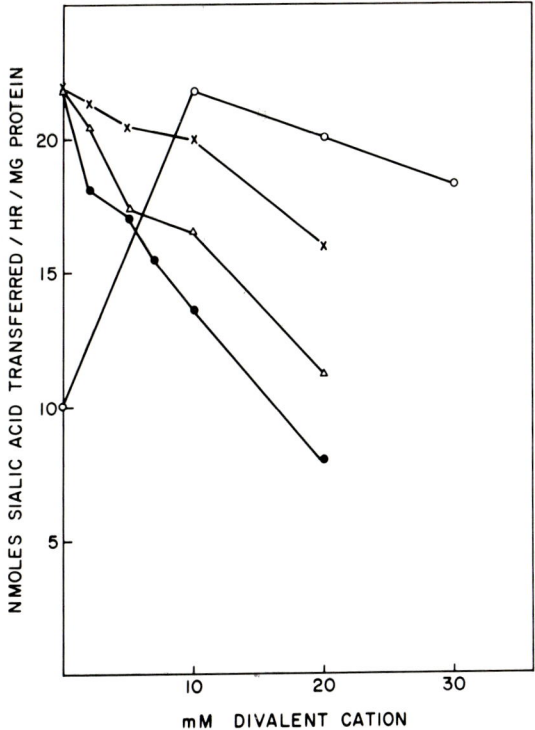

Fig. 5. Effect of divalent cations in the form of their chloride salts on CMP-NAN:lactosylceramide sialyltransferase activity of rat kidney Golgi apparatus. 40 µg protein were used throughout. x—x) Mg^{2+}; •—•) Ca^{2+}; o—o) Mg^{2+} plus 7 µg cardiolipin, pH 6.6. All but the latter were run at pH 6.4.

was inhibitory. At pH 6.6 in the presence of cardiolipin, 10 mM magnesium was necessary for maximum activity (Fig. 5). At pH 6.4 in the absence of added cardiolipin, however, added magnesium was not necessary and was slightly inhibitory at high concentrations.

Addition of cardiolipin to the assay was also reported to change the nature of the products formed when rat liver Golgi was used as a source of enzyme (19). In the absence of cardiolipin, the main product was found by these authors to be GM_3, but when cardiolipin was added, the products were a mixture of GM_3 and disialolactosylceramide (GD_3). We investigated this possibility by incubating either rat liver or rat kidney Golgi in the presence and absence of cardiolipin at optimum pH and magnesium concentrations in each case. After removal of nonlipid contaminants, the products were chromatographed on thin-layer plates. As shown in Fig. 6, one main radioactive product was obtained using either Golgi preparation in the presence or absence of added cardiolipin. This product comigrated with a standard preparation of GM_3. A small second peak was detected using some preparations of liver Golgi. The concentration of this product was not enhanced by added cardiolipin, and it did not migrate as GD_3, that is, about midway between the origin and GM_3 (28).

The distribution of enzyme 5 activity was determined in purified cell fractions from liver and kidney in the absence of added cardiolipin. The results are summarized in Table III. The highest specific activity of the enzyme was found in the Golgi fractions from both tissues. The low levels of activity found in the other cell fractions are most likely due to contamination of these fractions with Golgi apparatus.

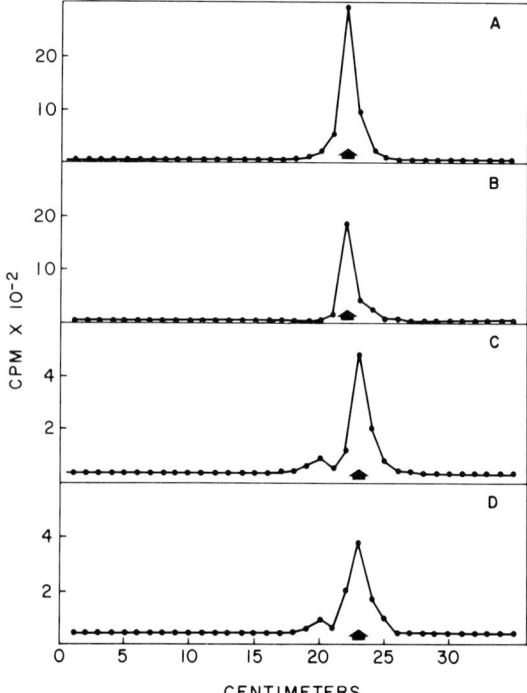

Fig. 6. Thin-layer chromatography of products of CMP-NAN: lactosylceramide sialyltransferase activity of rat kidney or liver Golgi apparatus. About 30 μg protein were used in each assay. Other constituents were as described under "Methods." After washing to remove nonlipid components, the chloroform:methanol, 2:1 (v/v) extracts were spotted on silica Gel G plates and developed 2 times with chloroform:methanol:2.5 N NaOH (65:45:9) (39). The arrows indicate the migration of standard GM_3. A) Rat kidney Golgi. B) Rat kidney Golgi plus 10 mM Mg^{2+} and 7 μg cardiolipin. C) Rat liver Golgi. D) Rat liver Golgi plus 10 mM Mg^{2+} and 7 μg cardiolipin.

TABLE III. Specific Activity* of Glycolipid Sialyltransferase in Rat Kidney and Liver Subcellular Fractions

Fraction	Gal-Glu-Cer (5) ↓ CMP-NAM NAN-Gal-Glu-Cer	
	Kidney	Liver
Homogenate	0.5	0.07
Rough microsomes	0.9	0.07
Smooth microsomes	2.4	0.70
Golgi	61.4	12.5
Mitochondria	1.2	0.0
Plasma membranes	1.2	0.2
Nuclei	0.1	0.06
Supernatant	0.0	0.0

*Specific activity expressed as nmoles N-acetylneuramic acid transferred per hour per mg protein at 37°C.

DISCUSSION

In addition to modifying glycoproteins for secretion, the Golgi apparatus appears to be involved in the modification of glycolipids as well. The initial glycosylation of ceramide to form galactosylceramide, however, does not appear to be exclusively a Golgi enzyme but is present in all of the membrane fractions from kidney. This is in sharp contrast to the sulfotransferase which converts galactosylceramide to sulfatide which we have shown previously is mainly localized in the Golgi apparatus in kidney (21). UDP-Gal:ceramide galactosyltransferase activity has also been demonstrated at a relatively high level in myelin from brain of young rats (27), indicating that in neuronal tissue that is actively forming myelin, plasma membranes can carry out this function.

Golgi apparatus from rat kidney is the main locus of galactosyltransferases which convert glucosylceramide to lactosylceramide or galactosylceramide to digalactosylceramide. These enzymes do not appear to be present in liver Golgi, and are therefore distinct from protein galactosyltransferase. UDP-Gal:ceramide galactosyltransferase also is very low in liver. This probably explains why liver, as a tissue, is low in all glycolipids.

Our studies illustrate further that Golgi apparatus from different tissues can vary considerably in its function. The most consistent function we have found in Golgi is UDP-Gal:N-acetylglucosamine galactosyltransferase, the enzyme involved in transferring the penultimate galactose to secreted glycoproteins (29). This enzyme has been shown to be present in Golgi from liver (1, 2, 4), kidney (21), pancreas (1), lung (30), thyroid (31), mammary gland (32), and testes (33). The enzyme also appears in a soluble form in milk (32), and serum (34). A slower-moving isozyme not found in normal serum has been shown to be present in sera of patients with carcinomas (35).

The sialotransferase which forms sialyllactosylceramide from lactosylceramide is present in both kidney and liver Golgi, although the activity is much higher in kidney than in liver. In both liver and kidney, this activity appears to be concentrated mainly in the Golgi apparatus and is not present to any appreciable extent in other cell fractions. In our assays using purified Golgi preparations from either liver or kidney, we could not detect the synthesis of disialyllactosylceramide in the presence or absence of added cardiolipin. This is in contrast to the results of Keenan et al. (19). Since we used a synthetic, saturated lactosylceramide as substrate while Keenan et al. used lactosylceramide isolated from milk fat globule membranes, it is possible the enzyme transferring the second sialic acid would not work with our substrate. Another possibility which we have not explored is that the extraction procedure we have used to isolate the products of the reaction discriminates between the mono- and disialyllactosylceramides and that we have only recovered the monosialylceramide.

In the absence of added cardiolipin, we found no stimulation of enzyme 5 by added Mg^{2+}. It has been shown recently by Kemp and Stoolmiller, however, that treatment with EDTA inhibits enzyme 5 activity in neuroblastoma cells and that the activity is restored by addition of 1 mM Mg^{2+} or 2 mM Mn^{2+} (36). Our enzyme preparations were not treated with EDTA and thus may contain enough bound Mg^{2+} to satisfy the requirements of this enzyme.

In contrast to neuronal tissues, GM_3 and GD_3 are the major gangliosides of kidney (37). GM_3 appears to be the precursor of the more complex gangliosides of brain (16). GM_3, the product of enzyme 5 action (Fig. 1), is the major ganglioside present in newborn rat kidney cells in tissue culture. Changes in levels of this ganglioside during viral transformation have been shown to be related to changes in the level of enzyme 5 activity (37) in the transformed cells. The induction of enzyme 5 in HeLa cells by treatment with butyrate also markedly alters the morphology of these cells and inhibits their growth (38).

The biological role of gangliosides is as yet poorly understood. Evidence that these lipids, as plasma membrane components, are involved in determining cell morphology, and may act also as cell surface receptors for viruses and hormones is accumulating (9). It would appear that in kidney, the Golgi apparatus plays a major role in the biosynthesis of these components.

ACKNOWLEDGMENTS

The author would like to thank Mr. Robert Siegel and Ms. Kathryn Dewey for their able technical assistance. This study was aided by NIH Grants AM 14632 and AM 17223.

NOTE ADDED IN PROOF

When GD_3 (the kind gift of Dr. Robert Ledeen, Albert Einstein College of Medicine, Yeshiva University, Bronx, N.Y.) was added to the assay mixture for enzyme 5, it was not recovered in the chloroform layer after washing to remove non-lipid contaminants. Chromatography of the original reaction mixture as described in Fig. 6, however, resulted in separation of CMP-NAN, GD_3, and GM_3. When this was done for assays containing rat liver Golgi with or without added Mg^{2+} and cardiolipin, two radioactive peaks were observed which comigrated with authentic GM_3 and GD_3. In the presence of added Mg^{2+} and cardiolipin, the amount of GD_3 increased while the amount of GM_3 remained the same. This confirms the observations of Keenan et al. (19).

REFERENCES

1. Fleischer B, Fleischer S, Ozawa H: J Cell Biol 43:59, 1969.
2. Fleischer B, Fleischer S: Biochim Biophys Acta 219:301, 1970.
3. Morré DJ, Hamilton RL, Mollenhauer HH, Mahley RW, Cunningham WP, Cheetham RD, LeQuire VS: J Cell Biol 44:484, 1970.
4. Schachter H, Jabbal I, Hudgen RL, Pinteric L, McGuire EJ, Roseman S: J Biol Chem 245:1090, 1970.
5. Schachter H: In Smellie RMS, Beeley JG (eds): "The Metabolism and Function of Glycoproteins." London: The Biochemical Society, 1974, p 57.
6. Fleischer B: In Harmon RE (ed): "Cellsurface Carbohydrates." ACS Monograph. New York: Academic Press (In press).
7. Roseman S: Chem Phys Lipids 5:270, 1970.
8. Morell P, Braun P: J Lipid Res 13:293, 1972.
9. Fishman PH, Brady RO: Science 194:906, 1976.
10. Costantino-Ceccarini E, Morell P: J Biol Chem 248:8240, 1973.
11. Morell P, Radin NS: Biochemistry 8:506, 1969.
12. Hildebrand J, Hauser G: J Biol Chem 244:5170, 1969.
13. Coles L, Gray GM: Biochem Biophys Res Commun 38:520, 1970.
14. Basu S, Kaufman B, Roseman S: J Biol Chem 243:5802, 1968.
15. Martensson E, Öhman R, Graves M, Svennerholm L: J Biol Chem 249:4132, 1974.
16. Steigerwald JC, Basu S, Kaufman B, Roseman S: J Biol Chem 250:6727, 1975.
17. Keenan TW: Biochim Biophys Acta 337:255, 1974.
18. Den H, Kaufman B, McGuire EJ, Roseman S: J Biol Chem 250:739, 1975.
19. Keenan TW, Morré DJ, Basu S: J Biol Chem 249:310, 1974.
20. Fleischer B, Zambrano F: Biochem Biophys Res Commun 52:951, 1973.
21. Fleischer B, Zambrano F: J Biol Chem 249:5995, 1974.

22. Fleischer B: Methods Enzymol 31:180, 1974.
23. Spiro RG, Spiro MJ: J Biol Chem 240:997, 1965.
24. Fishman PH, Simmons JL, Brady RO, Freese E: Biochem Biophys Res Commun 59:292, 1974.
25. Mans RJ, Novelli GD: Arch Biochem Biophys 94:48, 1961.
26. Neskovic NM, Sarlieve LL, Mandel P: Biochim Biophys Acta 334:309, 1974.
27. Costantino-Ceccarini E, Suzuki K: Arch Biochem Biophys 167:646, 1975.
28. Yu RK, Ledeen RW, Eng LF: J Neurochem 28:169, 1974.
29. Brew K, Vanaman TC, Hill RL: Proc Natl Acad Sci USA 59:491, 1968.
30. Fleischer B: Unpublished observations.
31. Chabaud O, Bouchilloux S, Ronin C, Ferrand M: Biochemie 56:119, 1974.
32. Coffey RG, Reithel FJ: Biochem J 109:177, 1968.
33. Cunningham WP, Mollenhauer HH, Nyquist SE: J Cell Biol 51:273, 1971.
34. Podolsky DK, Weiser MM: Biochem Biophys Res Commun 65:545, 1975.
35. Weiser MM, Podolsky DK, Isselbacher KJ: Proc Natl Acad Sci USA 73:1319, 1976.
36. Kemp SF, Stoolmiller AC: J Neurochem 27:723, 1976.
37. Puro K: Biochim Biophys Acta 189:401, 1969.
38. Brady RO, Fishman PH: Biochim Biophys Acta 355:121, 1974.
39. Esselman WJ, Laine RA, Sweeley CC: Methods Enzymol 28:140, 1972.

Distribution of Glycoconjugates in Mouse Fibroblasts With Varying Degrees of Tumorigenicity

D. J. Winterbourne and P. T. Mora

Macromolecular Biology Section, National Cancer Institute, National Institutes of Health, Bethesda, Maryland 20014

Analysis of glucosamine labeled glycoconjugates in cultured cells has been made comparing 2 clones and the parent embryonic mouse cell line. Hyaluronic acid, heparan sulphate, and chondroitin sulphate as well as a complex mixture of glycopeptides were found in the medium, the trypsinate, and the trypsinized cells, although the distribution was not uniform. The 3 cell lines had very similar in vitro growth properties, including their plating efficiency in viscous medium. However, the tumorigenicity of the cells, determined in syngeneic mice, was found to differ. All 3 cell lines were found to have similar glycoconjugate distributions, although a slight relative increase in labeled hyaluronic acid was found in the more tumorigenic mass cell line than either of the clones. The possible significance of this increase is discussed.

Key words: glycoconjugate, glycosaminoglycan, hyaluronic acid, transformation, tumorigenicity

The glycoconjugate metabolism of transformed cells, especially that of material located at the surface, has frequently been reported to differ from that of control cells. A high-molecular-weight glycoprotein on the surface of many cells appears to be almost absent in most virally transformed cells (1). Examination of the molecular weight of glycopeptides obtained from the membranes of paired cell lines has established the presence of increased levels of the larger glycopeptides in transformed cells (2, 3). Glycolipids, particularly the gangliosides series, have been claimed to exist predominantly as the simpler molecules after transformation by RNA and DNA viruses (4). Changes have also been found in glycosaminoglycan metabolism. In SV40 transformed cells, chondroitin sulphate biosynthesis has been reported to be decreased (5-9), while hyaluronic acid synthesis in some cases has been found to increase after transformation (10, 11). Histological evidence also indicated an increase in acid mucopolysaccharides at the surface of transformed cells (12, 13). Exceptions to these generalizations have been reported, and in the case of GAG metabolism a review describing the often contradictory reports of quantitative alterations and the effects of GAGs on cell growth and tumorigenicity has been published (14). The lack of general agreement is perhaps to be expected when the wide variety of cell types, transforming

Abbreviations: GAG – glycosaminoglycan; DEAE – diethylaminoethyl
Received April 5, 1977; accepted June 6, 1977.

© 1977 Alan R. Liss, Inc., 150 Fifth Avenue, New York, NY 10011

agents, and analytical methods are considered. The nonuniform use of the term "transformation" (15) only serves further to confuse the issue, especially since its relationship to tumorigenicity is uncertain (16).

In this study, a technique which allows a fairly complete quantitation of glycoconjugate production in the cultures, was applied to a family of closely related cell lines. The tumorigenicity of these cell lines was simultaneously determined in syngeneic mice. Basically the technique consists of growing a pair of cell lines in medium containing ^{3}H- or ^{14}C-labeled glucosamine, and after mixing and digesting with papain, analyzing the macromolecular products by ion exchange. Three culture compartments have been examined: the medium, material released from the cell surface by trypsin, and the trypsinized viable cells. In order to compare cells under as nearly identical conditions as possible the radioactive precursors in the medium were at the same concentration and the cultures were labeled when at the same cell density during exponential growth. Five flasks of each cell line were labeled separately and combined before analysis to reduce the effect of any variability in the cell growth.

In this report we present the results of a comparison between 2 clones and the more tumorigenic mass cell line from which they were isolated. In each experiment, a clone was labeled with [^{14}C] glucosamine and the parent cell line was labeled with [^{3}H] glucosamine. The corresponding compartments from the clone and the parent cells were combined at the earliest possible time in the analytical procedure. The advantage of this experimental design is that differences observed between the amounts of each isotope in any isolated fraction will be due only to differences in the metabolism of the 2 cell lines being compared, and not to artifacts (such as incomplete recovery of a certain component) incurred during the analysis.

MATERIALS AND METHODS

The cell line 201 was an early frozen stock of the cells which gave rise to T AL/N (17). In these experiments it was used 27 subcultures after isolation from the AL/N mouse embryo. The mass cell line was cloned by placing 1 μl drops of a very dilute cell suspension in Falcon No. 3040 Microtest II trays. Medium was added only to wells which by microscopic determination contained single cells. In this way 2 clones were isolated from 201 cells after 26 (clone 210 C) and 23 (clone 216 C) subcultures; these were analyzed 28 and 3 subcultures, respectively, after cloning. Cells were cultured in 25cm^2 Falcon tissue culture flasks in the Dulbecco-Vogt modification of Eagle's medium supplemented with 10% heat-treated fetal calf serum, without antibiotics in an atmosphere of 95% air, 5% carbon dioxide. The cells were not contaminated by mycoplasma as determined by culture methods.

For each cell line 15 replicate cultures were prepared with 2×10^5 cells per flask. Ten flasks were used to establish a growth curve and the remaining 5 flasks used for the labeling experiment. One half of the medium was changed daily and the radioactive precursors were added when the paired cell lines were at approximately the same cell density in the mid-logarithmic phase of growth. Specific radioactivities of the precursors were adjusted to give a final concentration in the medium of approximately 5 μM for both isotopes. D-[6-^{3}H(N)] glucosamine HCl in the medium was at 2.5 μCi/ml and D-[1-^{14}C]-glucosamine HCl at 0.25 μCi/ml. After 24-h labeling the medium was removed and the cell sheet washed 4 times with phosphate-buffered saline, pH 7.4. Washes and media from the

5 flasks of the 2 cell lines to be compared were pooled. The cells were trypsinized in 0.9 ml of 0.01% (wt/vol) 3 times crystallized trypsin in Tris-buffered saline at pH 7.7 for 30 min at 37°C, after which the reaction was terminated by addition of 0.1 ml of 1 mg/ml soybean trypsin inhibitor. An aliquot of the suspension was taken to enumerate the cells using a Coulter counter. The 2 suspensions were mixed, centrifuged at 200 × g for 15 min and the cell pellet and the trypsinate were separated. From this point, the compartments were analyzed as doubly labeled mixtures. Gangliosides were isolated from the cell pellet as previously described (17), but using only a single extraction and removing the cell residue by filtration through Celite. The medium and trypsinate were equilibrated with papain digestion buffer (0.2 M sodium citrate, 0.5 M sodium chloride, 5 mM cysteine, 1 mM ethylenediaminetetraacetic acid, pH 5.5) by extensive dialysis, and the cell residue was suspended in the same buffer. All 3 culture compartments (medium, trypsinate, and cell residue) were digested with activated 2 times crystallized papain at 0.5 mg/ml for 4 h at 60°C. The reaction was terminated by cooling in ice and adding trichloroacetic acid to a final concentration of 10% (wt/vol). After 1 h at 4°C the precipitate was removed by filtration, the filtrate was neutralized with sodium hydroxide and dialyzed to equilibrium against 5 changes of 10 volumes of 10 mM Tris HCl, pH 8.4.

Each compartment was analyzed by anion exchange on a 1 × 4 cm column of Whatman DE52, equilibrated with 10 mM Tris HCl, pH 8.4. After applying the sample, the column was washed with 7 bed volumes of start buffer followed by a 300 ml linear gradient of 0–0.6 M sodium chloride. Fractions of 4 ml were collected at a flow rate of 36 ml hr^{-1} cm^{-1} and the conductivity was measured in every fourth fraction. Radioactivity in 0.5 ml aliquots was determined by liquid scintillation counting in Aquasol. Quench correction, double label evaluation, and graph plotting were performed by a computer program (LSCP).

Tumorigenicity was determined by intramuscular inoculation of tenfold dilutions of trypsinized cells into the hind leg of syngeneic AL/N mice. The mice were examined weekly for 18 weeks. Tumors occurred in 6–14 weeks. All tumors were lethal. Growth in semisolid medium was as described by Risser and Pollack (18). All enzymes were obtained from Worthington Biochemical Corporation (Freehold, New Jersey) and chemicals were analytical grade. Radiochemicals and scintillation fluid were obtained from New England Nuclear Corporation (Boston, Massachusetts).

RESULTS

The growth properties of the 3 cell lines are shown in Table I. It can be seen that the clones are at least 100-fold less tumorigenic than the parent cell line 201. It should be noticed that there is very little difference between the cell lines in their in vitro growth properties, with the exception of 210 C which has a lower saturation density than the other 2 cell lines.

The fourth wash of the cells before trypsinization contained less than 0.005% of the radioactivity in the medium and less than 5% of the radioactivity subsequently released by trypsin. Under the trypsinization conditions chosen, the amount of macromolecular glucosamine label released reached a plateau after 30 min, while the plasma membrane remained sufficiently intact to exclude Trypan blue in greater than 90% of the cells. Therefore, in agreement with other reports (3, 9, 19, 20) the trypsinate represents material released from the surface of cells. Radioactivity in the medium is unlikely to be due to glucosamine spuriously bound to serum proteins (21), because in the absence of cells only

TABLE I. In Vitro and In Vivo Growth Properties of the Cell Lines

Cell line	Monolayer growth		Plating efficiency in methyl cellulose (%)	Tumorigenicity TD_{50}[a]
	Doubling time (h)	Saturation density (10^4 cells/cm^2)		
201	13	54	0.3	$10^{2.6}$
210 C	14	18	0.3	$10^{6.4}$
216 C	13	40	0.2	$10^{4.8}$

[a]TD_{50} represents the number of cells which produced tumors in 50% of the animals injected (10 animals at each dose). Plating efficiency in methyl cellulose represents the number of colonies greater than 0.1 mm (about 60 cells), as a percentage of the cells plated.

0.3% of the radioactivity was found in the high-molecular-weight fraction of papain digested medium. Also the majority of the radioactivity secreted by the cells was found to elute as hyaluronic acid from ion exchange columns (cf. Fig. 1a, b).

Elution profiles from the ion exchange column generally showed 7 radioactive peaks which eluted at highly reproducible sodium chloride concentrations. The first 4 peaks were considered to be glycopeptides due to their charge, molecular weight, and carbohydrate composition (22). The fifth peak comigrated with a commercial preparation of hyaluronic acid, was not labeled by radioactive sulphate, and was depolymerized by testicular hyaluronidase (22). This peak was therefore designated hyaluronic acid. Peaks 6 and 7 were both labeled by radioactive sulphate, while only peak 7 was degraded by testicular hyaluronidase. The principal amino sugars in these 2 peaks were glucosamine and galactosamine, respectively (22). These results are consistent with the designations of heparin sulphate (peak 6) and chondroitin sulphate (peak 7), results which have been confirmed by chondroitinase ABC and AC digestion and by nitrous acid degradation (manuscript in preparation).

The majority of the radioactivity in the medium was due to hyaluronic acid for all 3 cell lines (Fig. 1a, b). In this compartment the fourth glycopeptide peak could not readily be identified. Radioactivity in the trypsinate was more uniformly distributed and all 7 peaks could be identified (Fig. 2a, b). Minor changes in the elution positions of peaks 6 and 7 (most noticeable in Fig. 2a) could be detected between the clones and the mass cell line, with the latter consistently showing a greater overlap between the 2 peaks (Fig. 2a, b). These changes possibly reflect slight differences in degrees of sulphation. In contrast to the other 2 compartments the trypsinized cells contained little radioactivity in the GAGs, the majority eluting in the second glycopeptide peak immediately after the sodium chloride gradient was started (Fig. 3a, b).

The similarity of the profiles of the cell lines is apparent from the figures. To determine if any quantitative differences existed, the amount of radioactivity in the peaks was summed and then expressed as the percentage of radioactivity applied to the cultures incorporated into the peak per 10^6 cells. The most prominent quantitative differences occurred between the clone 210 C and its parent, 201 (Table II). This clone incorporated substantially more radioactivity per cell into all fractions except hyaluronic acid throughout the culture. The other clone (216 C) incorporated very similar amounts of radioactivity to the parent cell line in all fractions, except into hyaluronic acid, in which the incorporation was reduced. Although some peaks were not well resolved and therefore the summa-

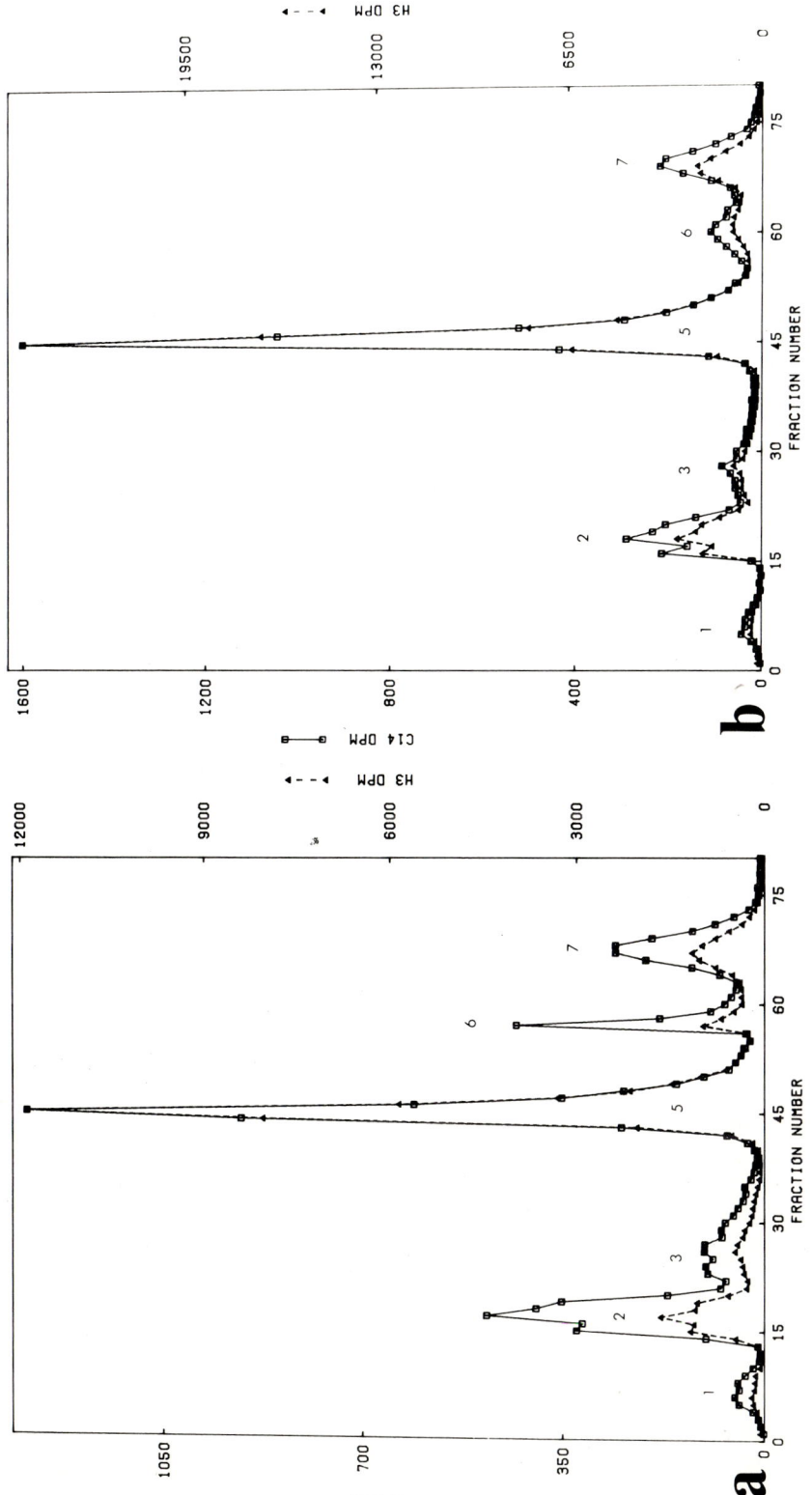

Fig. 1. Anion exchange elution profiles of papain digested media. a) Comparison of media from 216 C cells (^{14}C DPM) and 201 cells (^{3}H DPM). b) Comparison of media from 210 C cells (^{14}C DPM) and 201 cells (^{3}H DPM). Media from cells grown separately in the presence of radioactive glucosamine were pooled, digested with papain, and the high-molecular-weight material analyzed on a column of DEAE-cellulose. The gradient started at fraction 12 in a) and 13 in b). In all figures the data is plotted with the maximum DPM value at the same height for both isotopes. Recovery of applied radioactivity in the 6 elutions shown was 98.5 ± 4.1%.

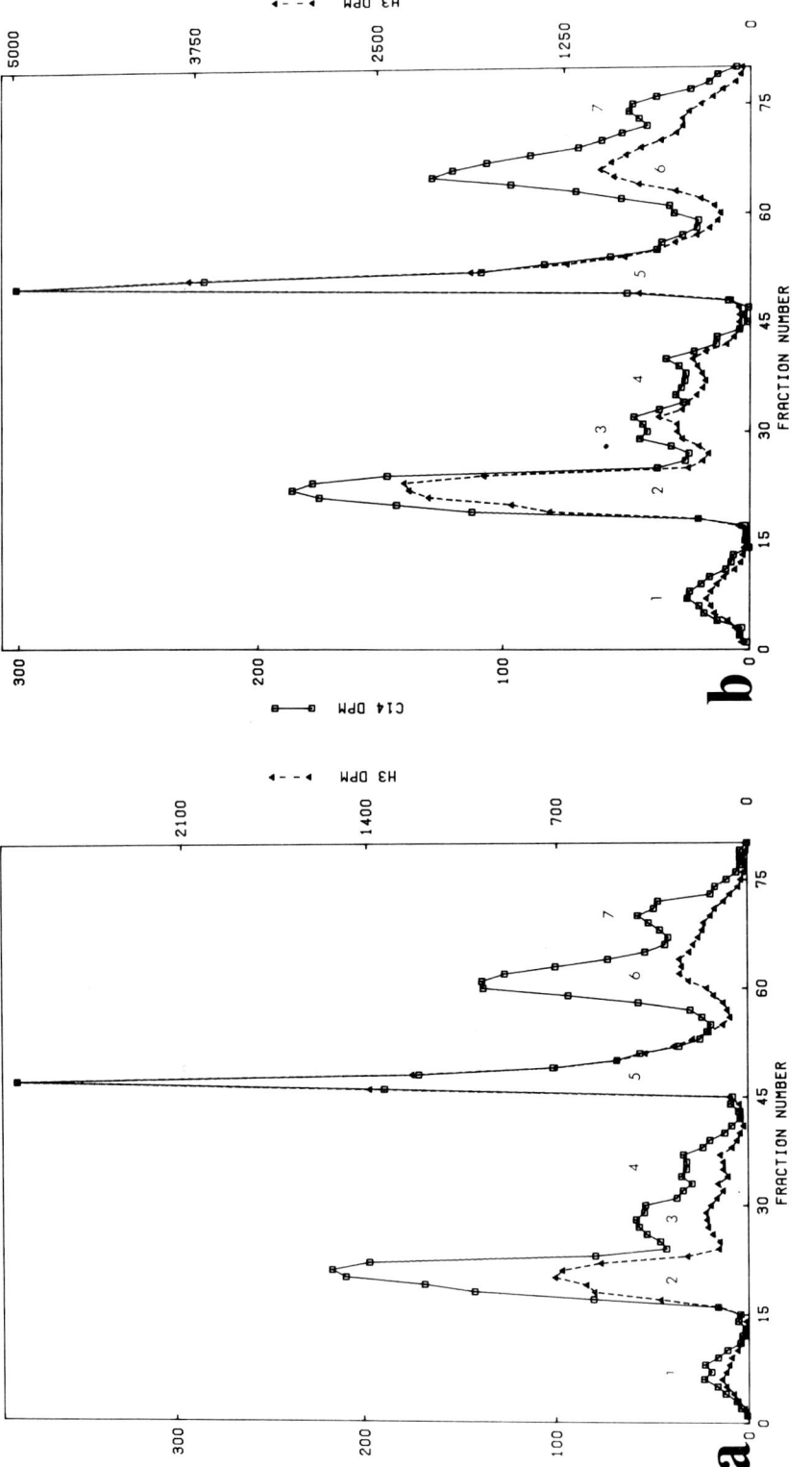

Fig. 2. Anion exchange elution profiles of papain digested trypsinates. a) Comparison of trypsinates from 210 C cells (^{14}C DPM) and 201 cells (^{3}H DPM). b) Comparison of trypsinates from 216 C cells (^{14}C DPM) and 201 cells (^{3}H DPM). Glucosamine labeled material released from the cell surface by trypsin was pooled, digested with papain, and the high-molecular-weight material analyzed on a column of DEAE-cellulose. The gradient started at fraction 15 in a) and 17 in b).

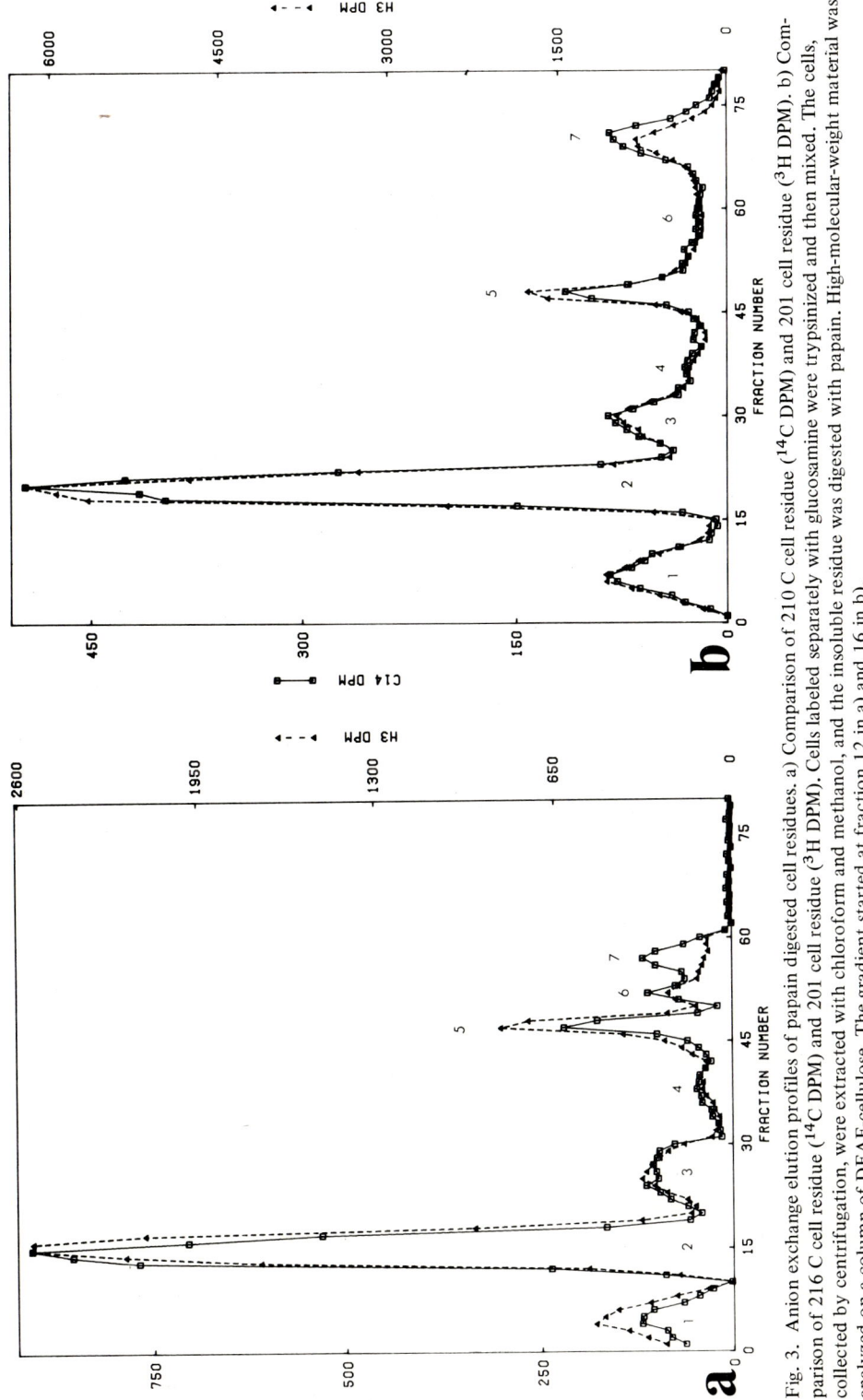

Fig. 3. Anion exchange elution profiles of papain digested cell residues. a) Comparison of 216 C cell residue (^{14}C DPM) and 201 cell residue (^{3}H DPM). b) Comparison of 210 C cell residue (^{14}C DPM) and 201 cell residue (^{3}H DPM). The cells, collected by centrifugation, were extracted with chloroform and methanol, and the insoluble residue was digested with papain. High-molecular-weight material was analyzed on a column of DEAE-cellulose. The gradient started at fraction 12 in a) and 16 in b).

TABLE II. Distribution of the Radioactive Glycoconjugates in the 3 Cell Lines*

Culture compartment	Cell line	Radioactivity incorporated per fraction			
		Glycopepties	Hyaluronic acid	Heparin sulphate	Chondroitin sulphate
Medium	210 C	66	77	18	27
	201	37	110	11	20
Trypsinate	210 C	37	21	17	7
	201	17	22	5	3
Cell	210 C	120	10	6	8
	201	59	7	3	2
Medium	216 C	32	66	12	17
	201	29	92	11	15
Trypsinate	216 C	25	13	13	4
	201	26	19	10	3
Cell	216 C	52	7	3	8
	201	61	9	3	7

*Data from the figures are expressed as $10^{-3} \times \%$ of applied radioactivity incorporated per 10^6 cells. The glucosamine precursors were present at 5 µM in all cases and at the end of the labeling period the cell densities were respectively 4×10^4 and 2×10^4 cells/cm² for 210 C and 201 and in the second experiment, 5×10^4 and 4×10^4 cells/cm² for 216 C and 201.

TABLE III. Ratio of the Radioactivity Incorporated by the Clones Compared to the Parent Cell Line*

Culture compartment	Ratio	Ratio of radioactivity			
		Glycopeptides	Hyaluronic acid	Heparin sulphate	Chondroitin sulphate
Medium	210 C:201	1.8	0.7	1.6	1.4
Trypsinate	210 C:201	2.2	1.0	3.4	2.3
Cell	210 C:201	2.0	1.4	2.0	4.0
Whole culture	210 C:201	2.0	0.8	2.2	1.7
Medium	216 C:201	1.1	0.7	1.1	1.1
Trypsinate	216 C:201	1.0	0.7	1.3	1.3
Cell	216 C:201	0.9	0.8	1.0	1.1
Whole culture	216 C:201	0.9	0.7	1.2	1.2

*The ratio seen in the whole culture is not a mean of the other ratios, due to the nonuniform distribution of radioactivity through the culture. The differences between the ratios for hyaluronic acid and those in the rest of the culture is significant ($P < 0.025$ for 210 C:201 and $P < 0.0025$ for 216 C:201, by Student's t-test).

tion may be subject to error, yet the reproducibility of these results is satisfactory as can be seen by comparing the 2 sets of values for 201 in Table II.

When the data was expressed as a ratio of the radioactivity incorporated a statistically significant decrease in the amount of radioactive hyaluronic acid relative to all the other components was found between both clones and the parent cell line (Table III).

The distribution of glycoconjugates throughout the culture can be seen in Table II. This was very similar for all 3 cell lines. Chondroitin sulphate, in agreement with other studies (5, 6, 8, 11), was found predominantly in the medium. A substantial portion of the heparin sulphate was associated with the cell surface, a location first reported by Kraemer (19). Hyaluronic acid was mostly secreted in agreement with earlier reports (11, 23), and the cells retained a major portion of the glycopeptides even after trypsinization.

DISCUSSION

The difficulty in obtaining a biological system for studying the biochemical events which may lead to tumor induction in vivo, and the lack of a general correlation between culture growth properties and tumorigenicity has recently been reviewed (16). In the cells described here there is no correlation between tumorigenicity and plating efficiency in viscous medium, in fact the highly tumorigenic cell line 201 has a very low plating efficiency (Table I), under conditions where other related cells, such as an SV 40 transformed analog of the T AL/N cells (SV AL/N) gave a minimum of 10% plating efficiency (24).

The data presented here show marked similarity in the ion exchange elution profiles of papain digested glycoconjugates from 3 related cell lines that differ in their tumorigenicity. In fact the only detected difference that correlates with higher tumorigenicity is a slightly higher labeling of hyaluronic acid relative to the other glycoconjugates. This difference is most noticeable between 210 C and 201, the cells whose tumorigenicity differs by the largest amount.

Slightly increased quantities of hyaluronic acid were found in the medium of Swiss 3T6 and 3T3 (23). Although the tumorigenicity of these cells is not known, cells isolated in a similar way from Balb/c mice showed 3T3 to have the lowest tumorigenicity (25). After transformation of 3T3 cells by DNA viruses, the amount of hyaluronic acid in the medium was greatly reduced (23). Results conflicting with this early observation have been obtained. After SV40 transformation, 3T3 cells produced larger quantities of an aggregation factor, shown to be hyaluronic acid (26). Also the radioactivity incorporated into hyaluronic acid from glucose represented a greater proportion of the total GAG label in SV3T3 than in 3T3, when the cells were at high densities (9). However, this report of a relative increase in hyaluronic acid may be due to the decreased synthesis of sulphated GAGs (5, 6, 8). Increased hyaluronic acid production after transformation by SV40 has been reported for monkey (7) and hamster (11) cells and we have unpublished data for a T-antigen positive subclone of 210 C.

Increased hyaluronic acid synthesis seems to correlate well with Rous sarcoma virus (RSV) transformation of avian cells (10, 27) and tumors induced in chicken by this virus are known to synthesize large quantities of hyaluronic acid (28). Rakusanova (29) confirmed these results with avian cells, but found that mammalian cells transformed by RSV either showed no change or produced less hyaluronic acid than normal or spontaneously transformed fibroblasts. Decreased hyaluronic acid synthesis was found in a melanoma when compared to iris melanocytes (30) and no change was noticed when comparing normal and malignant human mesothelial cells (31).

The mechanism by which increased secretion of hyaluronic acid could benefit tumor growth is not clear. The possibility that mucopolysaccharides can mask antigens has been discussed (14, 32, 33) and the exposure of lectin-binding sites by hyaluronidase was reported (34); furthermore, hyaluronic acid has been found to block in vitro stimulation of lymphocytes by phytohemagglutinin (35). Another mechanism could involve the production of an appropriate extracellular matrix in which the cells can proliferate (16). Boone et al. (36) found that Balb/c 3T3 cells readily produced tumors when implanted attached to a solid substrate. Hyaluronic acid has been shown to be a major component of the substrate-attached material thought to be involved in cell-to-substrate and cell-to-cell adhesion (37), and an increase in its synthesis was found to be an early cellular response to stimulation by epidermal growth factor (38).

Whether the increased production of hyaluronic acid by more tumorigenic cells is a general phenomenon, is being explored in other matched cell lines from this and similar series. Changes in molecular weight, both of hyaluronic acid and the other glycoconjugates, have not been excluded and we are examining this possibility also.

ACKNOWLEDGMENTS

We would like to thank Vivian McFarland for cloning the cell lines and for her help in the tumorigenicity determinations.

REFERENCES

1. Hynes RO: Biochim Biophys Acta 458:73, 1976.
2. Meezan E, Wu HC, Black PH, Robbins PW: Biochemistry 8:2518, 1969
3. Buck CA, Fuhrer JP, Soslau G, Warren L: J Biol Chem 249:1541, 1974.
4. Brady RO, Fishman PH: Biochim Biophys Acta 355:121, 1974.
5. Saito H, Uzman BG: Biochem Biophys Res Commun 43:723, 1971
6. Goggins JF, Johnson GS, Pastan I: J Biol Chem 247:5759, 1972.
7. Makita A, Shimojo H: Biochim Biophys Acta 304:571, 1973.
8. Roblin R, Albert SO, Gelb NA, Black PH: Biochemistry 14:347, 1975.
9. Cohn RH, Cassiman J, Bernfield MR: J Cell Biol 71:280, 1976.
10. Ishimoto N, Temin HM, Strominger JL: J Biol Chem 241:2052, 1966.
11. Satoh C, Duff R, Rapp F, Davidson EA: Proc Natl Acad Sci USA 70:54, 1973.
12. Martinez-Palomo A, Braislovsky C, Bernhard W: Cancer Res 29:925, 1969.
13. Morgan HR: J Virol 2:1133, 1968.
14. Nigam VN, Cantero A: Adv Cancer Res 16:1, 1972.
15. Mora PT: In Mora PT (ed): "Cell Surfaces and Malignancy." Fogarty International Center Proceedings No. 24, Washington, DC: US Government Printing Office, 1974, p 249.
16. Ponten J: Biochim Biophys Acta 458:397, 1976.
17. Mora PT, Brady RO, Bradley RM, McFarland VW: Proc Natl Acad Sci USA 63:1290, 1969.
18. Risser R, Pollack R: Virology 59:477, 1974.
19. Kraemer PM: Biochemistry 10:1437, 1971.
20. Underhill CB, Keller JM: Biochem Biophys Res Commun 63:448, 1975.
21. Herrmann M: Anal Biochem 59:293, 1974.
22. Winterbourne D: D. Phil. Thesis, University of Oxford, 1976.
23. Hamerman D, Todaro GJ, Green M: Biochim Biophys Acta 101:343, 1965.
24. McFarland VW, Mora PT, Schultz A, Pancake S: J Cell Physiol 85:101, 1975.
25. Aaronson SA, Todaro GJ: Science 162:1024, 1968.
26. Pessac D, Defendi V: Science 175:898, 1972.
27. Bader JP: J Virol 10:267, 1972.
28. Erichson S, Eng J, Morgan HR: J Exp Med 114:435, 1961.
29. Rakusanova T: Folia Biol (Prague) 15:87, 1969.
30. Satoh C, Banks J, Horst P, Kreider JW, Davidson EA: Biochemistry 13:1233, 1974.
31. Castor CW, Naylor B: Lab Invest 20:437, 1969.
32. Codington JF: In "Cellular Membranes and Tumor Cell Behavior." Baltimore: Williams and Wilkins, 1975, p 399.
33. Lippman M: Nature (London) 219:33, 1968.
34. Burger MM, Martin GS: Nature (London) New Biol 237:9, 1972.
35. Darzynkiewicz Z, Balazs EA: Exp Cell Res 66:113, 1971.
36. Boone CW, Takeichi N, Paranjpe M, Gilden R: Cancer Res 36:1626, 1976.
37. Terry AH, Culp LA: Biochemistry 13:414, 1974.
38. Lembach KJ: J Cell Physiol 89:277, 1976.

Interaction of Cartilage Proteoglycans With Hyaluronic Acid

Vincent C. Hascall

Laboratory of Biochemistry, National Institute of Dental Research, National Institutes of Health, Bethesda, Maryland 20014

Most proteoglycans are present in hyaline cartilage matrices as aggregates with as many as 100 molecules, each with average molecular weight of about 2×10^6, bound through specific, noncovalent interactions to individual strands of hyaluronic acid (HA). The interactions with HA are mediated by the HA-binding region of the core protein, which is located at one end of each of the interactive proteoglycans. A fragment of the core protein, average molecular weight of about 6×10^4, which contains the HA-binding site, can be isolated in an active form from trypsin digests of proteoglycan aggregates. The "active" HA-binding site in this preparation interacts strongly with HA-10 but weakly with HA-8, (oligomers of HA derived from partial digests of HA with testicular hyaluronidase); HA-9 derived from β-glucuronidase digestion of HA-10 also interacts strongly. No polysaccharide other than HA has been found to interact. Christner, Brown, and Dziewiatkowski (personal communication) modified the carboxyls on glucuronic acid groups in a mixture of HA-10 to HA-30, and they found that the interaction with proteoglycan no longer occurred if about 60% of the total carboxyls were a) methyl esterified, b) reduced to glucose, or c) substituted with glycine in amide linkage. Saponification of the methyl esters restored activity. Dansylation of lysine residues in the HA-binding region preparation abolished binding activity. However, when the dansylation reaction was done in the presence of HA, the HA-binding activity was protected. Acetylation of the same residues did not abolish binding activity but did prevent subsequent inactivation by dansylation. Hardingham, Ewins, and Muir (Biochem J 157:127–143, 1976) studied the effect of various amino acid modifiers on the interaction of intact proteoglycans with HA and showed that reaction of arginine residues with low concentrations of 2,3-butanedione was particularly effective in destroying binding. In sum, the data above suggests that the HA-binding region a) contains accessible arginine residues necessary for activity, b) contains lysine residues near the binding site which, when substituted with bulky groups such as dansyl, but not acetyl, sterically block interaction, and c) requires a length of HA with at least 4.5 repeat disaccharides containing 3, and possibly 4, unmodified glucuronic acid carboxyls for interaction. The possible relevance of proteoglycan-hyaluronic acid interaction to the observations that hyaluronic acid specifically inhibits proteoglycan synthesis by cultured chondrocytes is discussed.

Key words: proteoglycans, cartilage, hyaluronic acid

I. STRUCTURE AND FUNCTION OF CARTILAGE PROTEOGLYCANS

Proteoglycans are structural components of the extracellular matrix of hyaline cartilages. These macromolecules consist of a core protein structure to which a large number of chondroitin sulfate (CS) and keratan sulfate (KS) chains are covalently attached.

Received March 29, 1977; accepted June 9, 1977.

© 1977 Alan R. Liss, Inc., 150 Fifth Avenue, New York, NY 10011

The average proteoglycan molecule from bovine nasal cartilage, for example, has a molecular weight of approximately 2.5×10^6 (1, 2). It contains a core protein of about 200,000 molecular weight with about 100 chondroitin sulfate chains, each with an average molecular weight of 20,000, and 30–60 keratan sulfate chains, each with molecular weight of 4,000–8,000, distributed along specialized regions of the core protein (Fig. 1, Refs. 1–5). A large proportion of the proteoglycans have a portion of protein, the hyaluronic acid (HA)-binding region, located at one end of the core (6–8). (See section II below.) About 65% of the keratan sulfate chains are localized on another portion of the core, the KS-enriched region, adjacent to the HA-binding region, while more than 90% of the chondroitin sulfate chains are attached to the CS-enriched region, located further from the HA-binding region (Fig. 1, Refs. 5, 9). The family of proteoglycan molecules in bovine nasal cartilage is, however, widely polydisperse with molecular weights ranging from a few hundred thousand to more than 4 million (1). This range of molecular weights is primarily the result of a variation in the number of chondroitin sulfate chains bound to each core protein (1, 9–11); it is, moreover, correlated with changes in the overall lengths of proteoglycan cores observed in electron microscopic studies of proteoglycans spread on thin films of cytochrome c (12, 13), and also with amino acid and hexosamine changes which suggest that polypeptide in the CS-enriched region is longer when it contains more chondroitin sulfate chains (9–11).

In the molecular architecture of a proteoglycan molecule, the polysaccharide chains are constrained by being attached at one end to the core protein. For this reason, the intact macromolecules have hydrodynamic properties which are different from those of

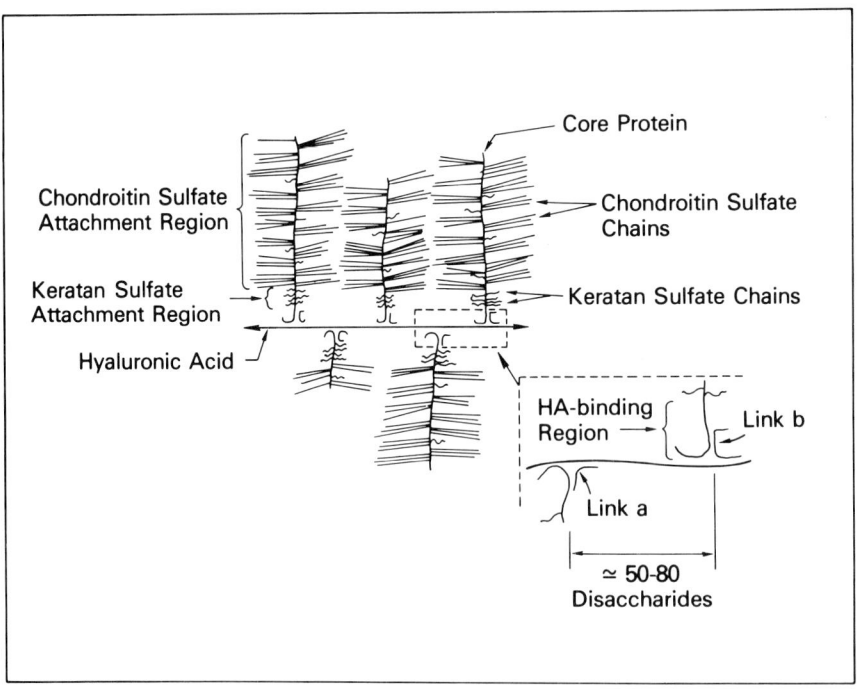

Fig. 1. Schematic model for the structure of cartilage proteoglycans and the interactions involved in the formation of aggregates.

the isolated polysaccharide chains. The highly anionic chains extend out from the core and, hence, the macromolecules occupy large molecular domains which encompass large amounts of solvent per mass of proteoglycan. The high limiting viscosity values for proteoglycans indicate that the molecules occupy solution volumes 30–50 times their dry weight (1, 2) and, for proteoglycans isolated from bovine nasal septa, have radii of gyration of almost 600 Å in 0.5 M guanidinium chloride (2). This property of intact proteoglycans is critical for the structure of cartilages. For example, the proteoglycans occupy large volumes in the extracellular matrix and provide a network for the retention of large amounts of solvent. Further, the proteoglycans can be compressed reversibly by displacing solvent from the molecular domain at the expense of increasing intramolecular interactions.

This is shown by the results of the experiment depicted in Fig. 2. Monomer (PGS) and aggregate (PGC) solutions were pelleted in the ultracentrifuge and the equilibrium packed volumes measured at different centrifugal speeds (14). For the lowest speeds, values measured at the beginning and end of the experiment coincided, indicating that the molecules were reversibly compressible. The data also indicate that under these experimental conditions, the limiting packed solute volume (extrapolating $1/\omega^2$ to zero) is between 25–30 ml/g. The concentration of proteoglycans in most hyaline cartilages is such that the largest volume they could possibly occupy in vivo would be far less than these values, and they must therefore occupy smaller domains in the tissue than their extended conformations occupy in solutions.

These physical chemical properties of the proteoglycans provide cartilages with their essential properties of resiliency and stiffness. The potential consequences for cartilages of impaired proteoglycan function have been demonstrated in several systems. Chemical (15, 16) and morphological (16) studies of cartilages in the mutant nanomelic chicken have revealed a virtual absence of cartilage specific proteoglycans, but normal amounts of type II collagen, in cartilage matrices. As a consequence, the chondrocytes are much closer together in the matrix than is normal (Fig. 3). The deficiency causes severe skeletal deformities and is lethal, usually before hatching. The classic experiments of Thomas (17) showed that intravenous injections of crude papain into young rabbits caused depolymerization of proteoglycans in cartilage matrices. This lead to a loss of tissue function, most notably in the collapse of the ear cartilages. The proteoglycans in the epiphyseal cartilages of brachiomorphic mice have been shown to be significantly undersulfated in comparison with epiphyseal cartilage from normal mice (18). Because of the reduction in charge density, there would be a decrease in the intramolecular interactions in the proteoglycan molecules with a concomitant reduction in the sizes of their molecular domains. This may account for the smaller size observed for the epiphyses in these animals and the striking foreshortening of their limbs. Increased proteolysis of proteoglycans in many osteoarthritic lesions leads to increased cartilage penetrability and loss of cartilage resiliency, (see Ref. 19 for a recent review).

Although some of the proteoglycans are apparently present in the extracellular matrices of cartilages as individual molecules, most (60–85%) are present as very large aggregates in most cartilaginous tissues (7, 12, 20–24). The function of these aggregate structures for the tissue is not known, but they may be important for immobilizing the proteoglycans and providing a more ordered structure within the matrix. The central filament of each aggregate is provided by a strand of hyaluronic acid of variable length (Fig. 1) as was first suggested by the work of Hardingham and Muir (6) (see section III below). Interactive proteoglycan monomers bind at intervals along the hyaluronic acid by means of a highly specific interaction mediated by the HA-binding region of the core protein (8).

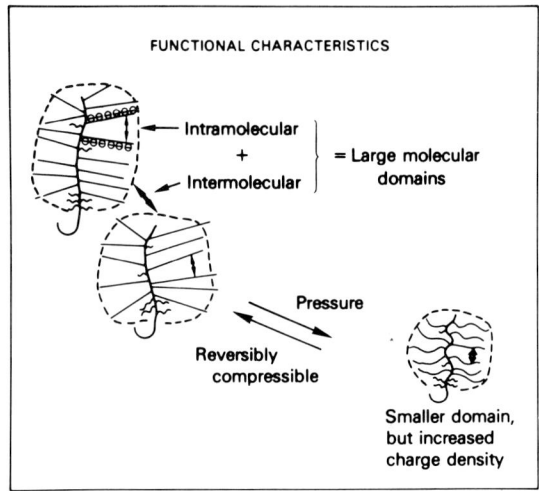

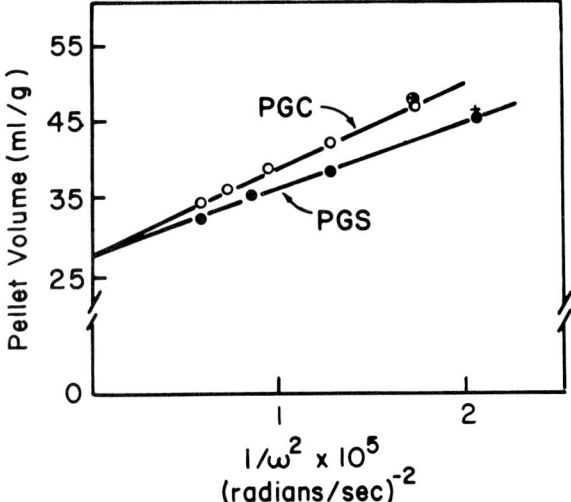

Fig. 2. a) Schematic model illustrating the functional characteristics of the intact cartilage proteoglycans. b) Compressibility of proteoglycans in centrifugal fields. Monomer (PGS) and aggregate (PGC) solutions, 2.4 mg/ml and 3.0 mg/ml respectively, in 0.5 M guanidinium chloride, 0.05 M MES, pH 5.8, were centrifuged in an analytical ultracentrifuge at different speeds. The equilibrium packed volumes of the solute were measured for each speed and are plotted against the reciprocal of the angular momentum squared, $(1/\omega^2)$. The plus symbols indicate values for the packed volumes at the lowest speeds which were observed at the end of the experiment after the other speeds were studied. The fact that they coincide with the first measurements made at those speeds indicates that the compressibilities of the solutes were reversible. The packed volumes observed for a preparation of papain digested monomer in a similar experiment were much smaller and difficult to measure reliably indicating that the hydrodynamic volumes of the intact proteoglycans are larger than those of equivalent amounts of the free glycosaminoglycan chains. Data taken from Ref. 14.

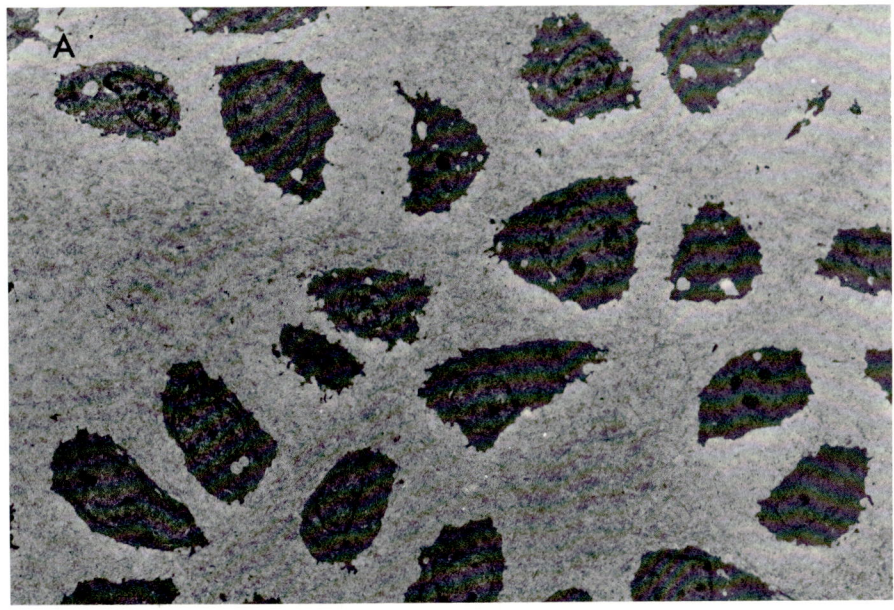

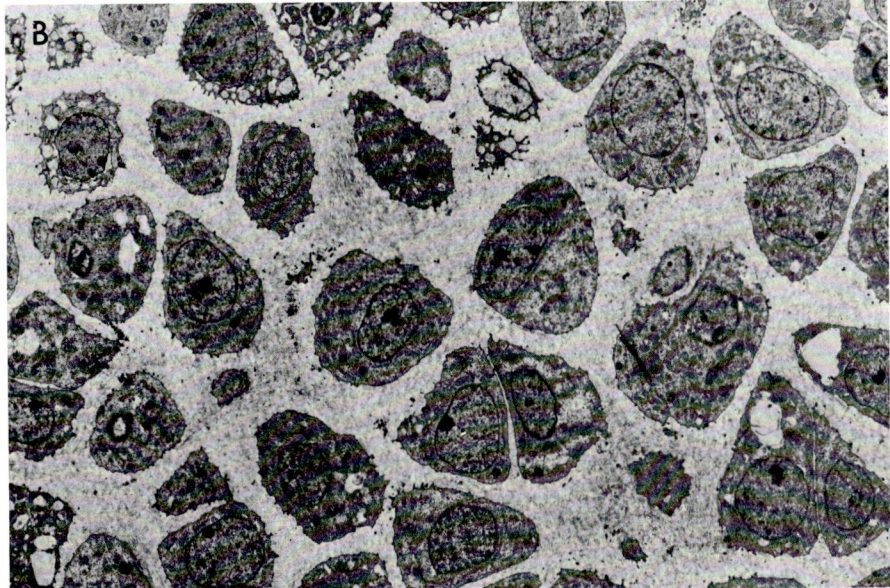

Fig. 3. Low power electron microscope pictures of sections of sternal cartilage from A) normal chick embryos and B) nanomelic chick embryos. Printed with the kind permission of Dr. Jack Pennypacker; see Ref. 16.

The average sizes of the aggregates appear to depend upon the length of the hyaluronic acid filament and can vary for different cartilages (7). Aggregates with over 100 monomers have been observed in the electron microscope (12). Two small proteins (molecular weights of 42,000 and 50,000), referred to as link a and link b respectively, are present in the structure (7, 22, 25, 26). They are capable of interacting with hyaluronic acid in the absence of proteoglycans (8), and they probably interact directly in some manner with the HA-binding region protein (Fig. 1).

II. ISOLATION OF HA-BINDING REGION PROTEIN

While interactive proteoglycan molecules will bind to hyaluronic acid in the absence of the link proteins (6), the presence of these low-molecular-weight components in the structure of the aggregate provides additional stability to the interactions (26). This enhanced stability was critical for the successful isolation of a functional HA-binding region protein preparation from intact proteoglycan aggregates. Heinegard and Hascall (8) observed that a high-molecular-weight complex consisting of hyaluronic acid and 2 associated proteins could be recovered from aggregate preparations which were digested first with chondroitinase and then trypsin. When the same experiment was done with reconstituted aggregates which contained either ^{3}H-acetylated monomer proteoglycans or ^{3}H-acetylated link protein components, it was shown that the larger of the 2 proteins associated with the hyaluronic acid was derived from the protein core of proteoglycan molecules while the smaller was derived from the link proteins. However, when a mixture of purified monomer proteoglycans with hyaluronic acid was treated with trypsin, no protein component remained bound to the hyaluronic acid. The HA-binding region preparation recovered from the trypsin digest of aggregate contained some bound keratan sulfate which accounted for the wide, continuous range of apparent molecular weights, average of about 90,000, that was observed for this preparation on sodium dodecyl sulfate-polyacrylamide gels in the presence or absence of sulfhydryl reducing agents (8). The actual size of the protein moiety in the HA-binding region appears to be smaller. Mild papain digestion of aggregates yielded a fragment from the HA-binding region of about 65,000 which contained little or no keratan sulfate (7) and trypsin digestion of aggregates isolated from the Swarm rat chondrosarcoma, which contain no keratan sulfate, yielded a HA-binding region preparation with molecular weight of about 67,000 (23, 27). More recently it has been shown that the 2 link proteins in aggregates are related to each other, with the larger link b molecules containing a glycopeptide extension which is absent from the link a molecules. This was shown by the results of the experiment depicted in Fig. 4. Sodium dodecyl sulfate-polyacrylamide electrophoretic gels were prepared from identical aliquots of aggregate samples which had been treated either with chondroitinase ABC alone or with the chondroitinase and then trypsin. After chondroitinase treatment alone, both the link b and link a molecules were present. After the subsequent trypsin treatment, the link b molecules electrophoresed with the link a molecules. This was indicated by the absence of the link b band and the proportional increase in staining intensity of the link a band. Further, Baker and Caterson (28) have provided evidence that a proportion of the link b molecules are converted to link a after treatment with sulfhydryl reducing reagents and that the extra portion of polypeptide on link b contains carbohydrate. It may be that link a is the end product of a sequence of proteolytic modifications that take place during the normal metabolism of proteoglycan aggregates. It is of interest that only link a is present in proteoglycan aggregates

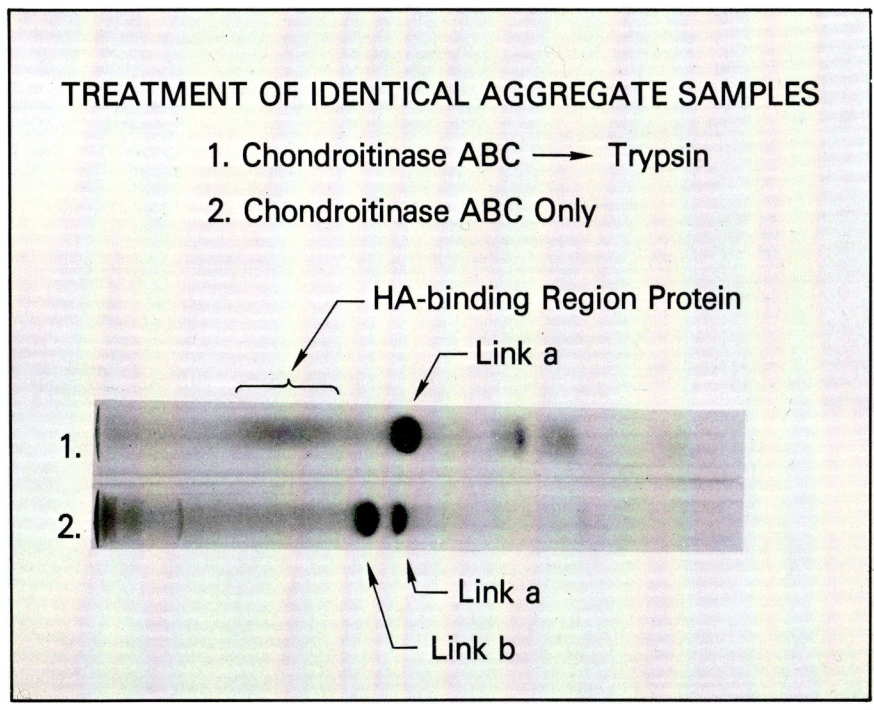

Fig. 4. Samples were treated with the enzymes indicated above as described elsewhere (8). After dialyses and lyophilization, identical quantities (about 25μg) were electrophoresed on 7% polyacrylamide gels and stained with Coomassie Blue (7).

isolated from a transplantable rat chondrosarcoma which contains high levels of proteolytic activity (23).

The above results suggest that interactions between a) the HA-binding region protein and hyaluronic acid, b) the link protein and hyaluronic acid, and c) the HA-binding region protein and the link protein, are all involved in the organization of the aggregate structure, Fig. 1.

A procedure adapted from the results of Heinegard and Hascall (8) can be used to purify the HA-binding region protein and the link protein (Ref. 5, Fig. 5). Trypsin digests of aggregate preparations are fractionated in an associative density gradient. The chondroitin sulfate peptides have high buoyant densities and are recovered in the bottom of the gradient, whereas the complex of hyaluronic acid and associated proteins has a much lower buoyant density and is recovered in the top of the gradient. The top fraction is chromatographed on Sepharose 2B to purify the complex from enzyme and peptide fragments. The individual components of the complex can subsequently be separated by chromatography on Sephadex G 200 in the presence of a dissociative solvent, 4 M guanidinium chloride. The high-molecular-weight hyaluronic acid elutes near or partially in the void volume while the HA-binding protein and the link protein are resolved separately as partially included peaks (Fig. 5). The purified HA-binding-region protein fraction, which still contains some bound keratan sulfate, is soluble in associative solvents and was shown to interact efficiently with hyaluronic acid (8) (see section IV below). The purified link

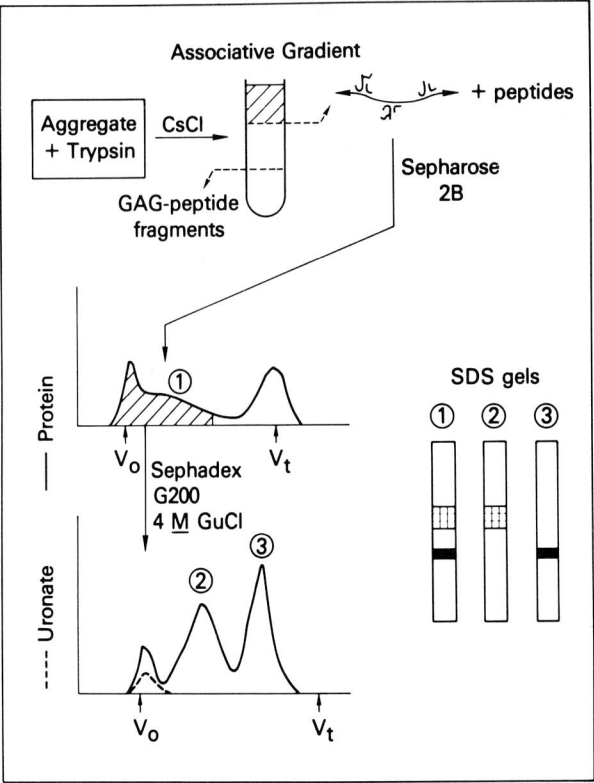

Fig. 5. Schematic outline of the procedure presently used to isolate the HA-binding-region protein and the link a protein. The elution profile on Sepharose 2B is used with kind permission of Dr. Dick Heinegard. See Ref. 8.

protein was shown to be capable of interacting with hyaluronic acid (8); however, by itself, the preparation is insoluble in most associative solvents. In recent experiments, however, it has been shown that the link protein, purified by the procedure outlined in Fig. 5, can be kept soluble in a solvent including 1 M LiCl and 0.15 M guanidinium chloride (R. A. Gelman and V. C. Hascall, unpublished observations). In this solvent the molecules still interact effectively with hyaluronic acid. Thus, it should now be possible to determine the specificity of the interaction between link protein and hyaluronic acid in a similar manner as for the HA-binding-region protein discussed below.

III. MODIFICATION OF HYALURONIC ACID

In a series of experiments that were crucial for determining the mechanism of proteoglycan aggregation, Hardingham and Muir (6) showed that in associative solvents, a large proportion of the monomer proteoglycans isolated from pig laryngeal cartilage interacted specifically with small amounts of hyaluronic acid, a glycosaminolgycan with a repeating disaccharide structure of $[(1\rightarrow 4)\text{-}\beta\text{-glucuronosyl-}(1\rightarrow 3)\text{-}\beta\text{- N-acetylglucosaminosyl}]_n$ (see Figs. 9 and 10 below). The viscosities of mixtures of the proteoglycan fraction with a preparation of umbilical cord hyaluronic acid of average molecular weight 500,000 increased up to a maximum in mixtures with slightly less than 1% (wt/wt) hyaluronic acid. These results indicated that the effective hydrodynamic sizes of the proteoglycans increased in the presence of hyaluronic acid, suggesting that aggregation was occurring. This was verified by chromatography of mixtures on Sepharose 2B; more than 50% of the proteoglycan molecules shifted from the broad, included peak observed in the absence of hyaluronic acid, into the column void volume in mixtures with 0.5–1.0% hyaluronic acid. Calculations using approximate molecular weights suggested that at maximum saturation of available hyaluronic acid, bound proteoglycans were an average of about 120 HA-monosaccharide units (60 repeat disaccharides) apart. Other polyelectrolytes, dextran sulfate, chondroitin sulfate, sodium alginate, and DNA, did not interact with proteoglycans. Further, the proteoglycan interaction with hyaluronic acid was reversed when the solution pH was lowered to pH 3–4 and when the solution concentration of guanidinium chloride was raised to 1–2 M. Since these conditions had been shown to dissociate intact proteoglycan aggregates (21, 22), Hardingham and Muir (6, 20) suggested that hyaluronic acid might be an important component of the aggregation mechanism.

Hascall and Heinegard (29) subsequently purified a series of HA-oligomers, with multiples of the repeat disaccharide, from partial digests of hyaluronic acid with testicular hyaluronidase. These oligomers were tested for their ability to interact with intact proteoglycans isolated from bovine nasal cartilage (29) and pig laryngeal cartilage (30). A competitive viscosity assay such as that indicated in Fig. 6 was used (30). When 1% (wt/wt) of a hyaluronic acid preparation was added to a monomer proteoglycan solution, the viscosity of the mixture increased to a higher, plateau value within 45 min (Fig. 7a). At that time, aliquots with various concentrations of HA-oligomers were added to equivalent mixtures of proteoglycan with hyaluronic acid, and changes in the solution viscosities

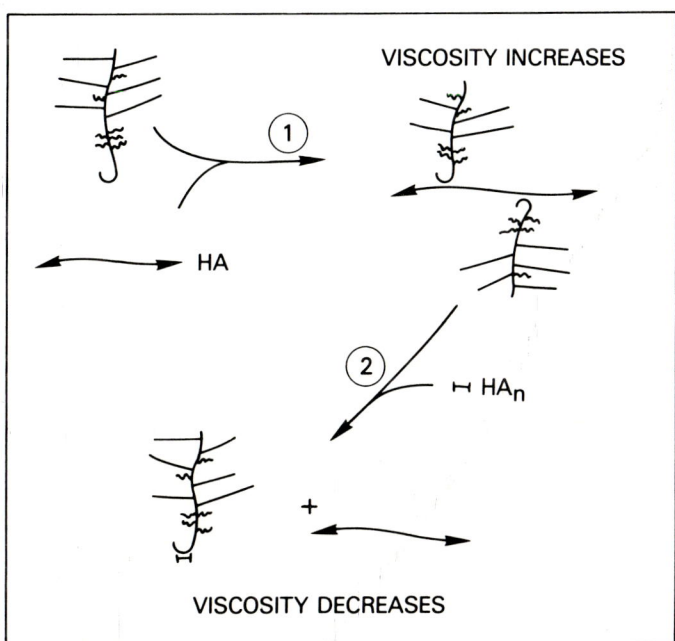

Fig. 6. Schematic outline of the procedure used to test HA-oligomers of various sizes for their ability to compete with intact hyaluronic acid for binding with proteoglycans.

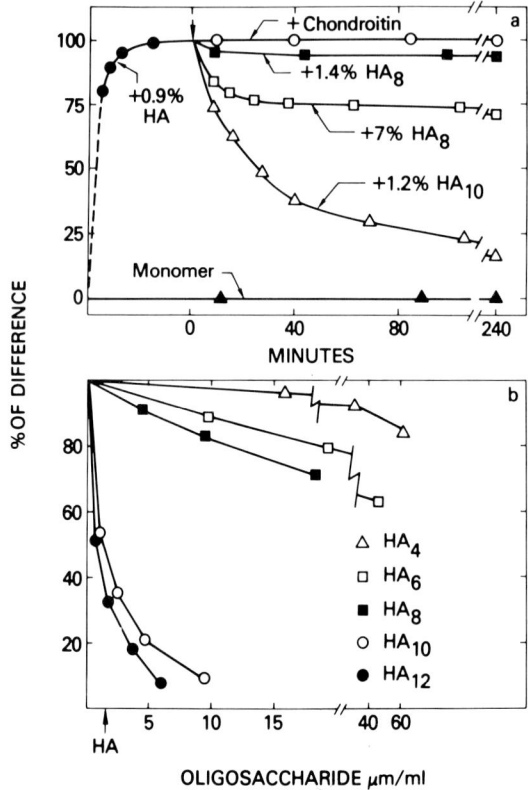

Fig. 7. The results of experiments a) from Ref. 29 and b) from Ref. 30 (with the kind permission of Drs. Timothy Hardingham and Helen Muir) which indicate the strong interactivity for binding of HA-oligomers greater than or equal to HA_{10}. See text for details.

were observed. When the HA-oligomers were capable of displacing the bound proteoglycans from intact hyaluronic acid molecules by competing for the binding site, the solution viscosities decreased. After several hours, equilibrium viscosity values intermediate between the proteoglycan solutions with and without 0.9% hyaluronic acid were observed. These values provided a measure of the equilibrium amounts of proteoglycan bound either to the HA-oligomers or to the intact molecules of hyaluronic acid. The experiments showed a striking difference between the effectiveness of binding of the HA_8 and HA_{10} oligomers (4 and 5 repeat disaccharides, respectively). A 1.2% concentration of HA_{10} decreased the viscosity to a much lower equilibrium value, about 17% of the difference, than did a much higher, 7%, concentration of HA_8 (Fig. 7a). When different concentrations of the oligomers were tested by Hardingham and Muir (30), the equilibrium viscosity levels of equimolar mixtures of HA-oligomers with intact hyaluronic acid were only 40% of the difference for HA_{10} but greater than 95% for HA_8 (Fig. 7b). Chondroitin, a structural analogue of hyaluronic acid in which the N-acetylglucosamine residues are replaced by N-acetylgalactosamine, did not displace proteoglycans from hyaluronic acid at all, and therefore did not interact with the binding site (Fig. 7a). A related polysaccharide, an unsulfated N-acetylated intermediate in the biosynthesis of heparin (31), also is not able to interact with proteoglycans (Heinegard, personal communication). This polysaccharide contains the same sugar moieties, N-acetyglucosamine and glucuronic acid, as hyaluronic acid but the glycosidic linkages are different [(1→4)-α-glucuronosyl-(1→4)-β-N-acetylglucosaminosyl]$_n$ from those in hyaluronic acid. While oligomers of HA_{10} or greater are effective in displacing proteoglycan monomers from hyaluronic acid in mixtures, they are ineffective in displacing proteoglycans from intact aggregates under similar experimental conditions (29). This provides further evidence that the link proteins add stability to the aggregates.

These experiments suggested that the highly specific interaction between the proteoglycan molecules and hyaluronic acid was mediated by a portion of the core protein in the interactive proteoglycan molecules. Evidence for this was provided when it was shown that proteoglycan core molecules, in which the chondroitin sulfate chains were enzymatically removed by chondroitinase digestion, still interacted with hyaluronic acid with the same specificity (29). Mixtures of this core preparation with different amounts of hyaluronic acid were chromatographed on Sepharose 2B in an associative solvent. The elution profiles indicated that more than 70% of the core molecules were able to bind to hyaluronic acid (Fig. 8). Further, at saturating levels of core to hyaluronic acid, each bound core molecule sterically occupied about 24 HA-monosaccharide units, (12 repeat disaccharides). This was in contrast to intact proteoglycans where the available data, including the stoichiometry of mixtures of proteoglycan with hyaluronic acid (6), the proportion of hyaluronic acid in aggregate preparations (7, 20), and electron microscopy (12), indicate that intact proteoglycan molecules are located a minimum of 80–120 HA-monosaccharide units apart. In many cases, therefore, the packing density of intact proteoglycans along hyaluronic acid in aggregates, appears to be limited by the exclusion volume of the monomers, which is primarily a function of the number and length of the attached chondroitin sulfate chains. Additional experiments in which different HA-oligomers were added to mixtures of core molecules in the presence of intact hyaluronic acid revealed that HA_{10} but not HA_8 was effective in displacing core molecules (Fig. 8), and that HA_9, an oligomer derived from β-glucuronidase digestion of HA_{10}, was almost as effective as HA_{10} (29).

Subsequent experiments indicated that an intact polypeptide fraction, the HA-binding-region protein, could be recovered from the core structure of interactive proteoglycan molecules as discussed in section II above. All of the molecules in the HA-binding region preparation were able to interact with hyaluronic acid (8). As with the core molecules, the minimum packing distance of the HA-binding-region proteins at saturation of available hyaluronic acid was about 20–24 HA-monosaccharides apart. This suggests that the keratan sulfate and the bulk of the polysaccharide attachment region protein which are present in the core, but not in the HA-binding region molecules, do not sterically interfere with the molecular packing along the hyaluronic acid molecule.

Recently, Christner, Brown, and Dziewiatkowski (32) used several chemical procedures to modify variable proportions of the carboxyl groups on glucuronic acid residues in a mixture of HA-oligomers (from HA_{10} to HA_{30}) derived from partial digests of hyaluronic acid with testicular hyaluronidase (see Fig. 9). Subsequently, the modified HA-oligomers were tested for competition with intact hyaluronic acid in an equilibrium viscosity assay similar to that described above (Fig. 6). Initially, essentially all of the carboxyl groups in a sample of the HA-oligomers were methylesterified by reacting them with diazomethane. Samples were then partially saponified to generate a series of oligomers in which the proportion of methylglucuronate to unsubstituted glucuronate groups varied. Other samples of the completely esterified preparation were treated to different extents

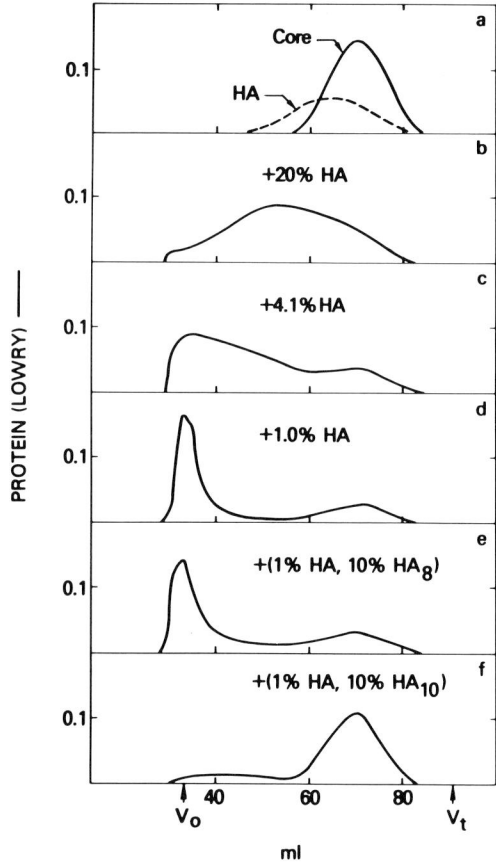

Fig. 8. A series of experiments in which the interaction of proteoglycan core molecules (isolated from chondroitinase ABC digests of monomer proteoglycans) with hyaluronic acid and HA-oligomers was tested. The dashed curve a) indicates the elution profile determined in a separate experiment of the intact hyaluronic acid sample used in all subsequent experiments (b–f). Taken from Ref. 29. See text for details.

with sodium borohydride to reduce a proportion of the methylglucuronate residues to glucose. After these samples were then completely saponified, a second series of oligomers was generated in which the proportion of glucose to glucuronate groups varied. A third set of modifications on a separate sample of the HA-oligomers involved the addition of variable amounts of the methyl ester of glycine by carbodiimide facilitated amide linkage to glucuronic acid carboxyl residues. The methyl esters on the glycines were subsequently saponified in some cases (Fig. 9). All of these series of substitutions gave similar results in the competition assay. When the proportion of free carboxylate groups to substituted or modified moieties was less than about 40% of the total, the modified oligomers showed little or no competition with intact molecules of hyaluronic acid for binding to proteoglycan and, therefore, were unable to interact with the HA-binding site. As the proportion of the unmodified glucuronate carboxyl groups increased toward 100%, competition increased approximately linearly to greater than 90% of that exhibited by the original, unsubstituted HA-oligomer sample. A statistical analysis of the results suggested that effective interaction with the HA-binding site required (probably at least 3 of

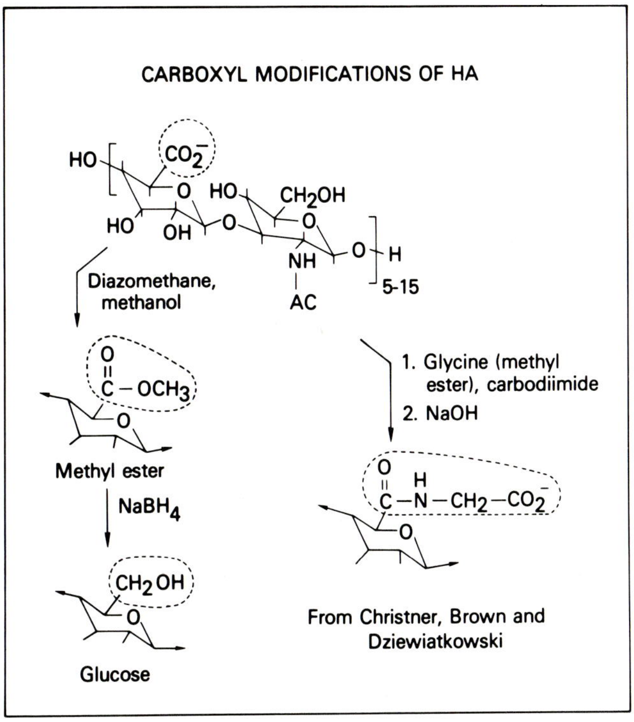

Fig. 9. Schematic outline of the reactions used to modify the carboxyl groups of HA-oligomers, Ref. 32.

the 4) unmodified carboxyl groups on the glucuronic acid residues in that portion of hyaluronic acid which is bound in the active site.

The experiments above indicate that the HA_8 oligomer derived from partial testicular hyaluronidase digestion of hyaluronic acid does not interact strongly with the HA-binding site of interactive proteoglycan molecules. The presence of an additional N-acetylglucosamine residue at the nonreducing end to give the HA_9 oligomer increases the strength of the interaction greatly (Fig. 10). It remains to be determined if the iso-HA_8-oligomer, in which glucuronic acid is at the reducing end and which could be prepared from partial digests of hyaluronic acid with leech hyaluronidase, is sufficient for strong binding. This iso-HA_8-oligomer, then, would be the minimum possible length of hyaluronic acid required for strong interaction, indicating that the active binding site in the protein core extends over at least 4 repeat disaccharides with N-acetylglucosamine at the nonreducing end (Fig. 10). At least 1, and probably more, of the N-acetylglucosamine residues are required in the active site interaction since chondroitin, in which the N-acetylglucosamines are replaced by N-acetylgalactosamines, does not interact. Finally, it is likely that as many as 3 of the 4 carboxyl residues of the glucuronic acid moieties in the active site must have unsubstituted carboxyl groups. It is also possible that protonation of the carboxyl groups is the explanation for the reversal of the interaction between proteoglycans and hyaluronic acid observed at solution pH values between 3 and 4 (6, 20, 22), because this is the region for the dissociation constants of these groups in hyaluronic acid (33).

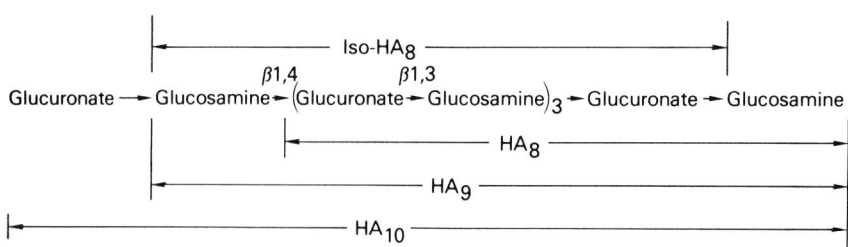

Fig. 10. Structure of various HA-oligomers.

The work of Highsmith et al. (34) on the interaction of hyaluronic acid with bovine testicular hyaluronidase has shown that the active site of this enzyme has 5 subsites for hyalobiuronate residues, the repeat disaccharide of hyaluronic acid. Thus, the enzyme interacts strongly with HA_{10} and less strongly with shorter oligosaccharides, producing primarily HA_8-, HA_6-, and HA_4- oligomers under conditions of complete digestion. The interaction of hyaluronic acid with hyaluronidase is unlike that of the interaction with the HA-binding region of the proteoglycan in 2 major respects; first the hyaluronidase hydrolyzes the hyaluronic acid, and second, it interacts with and digests chondroitin-4- and -6-sulfates as well.

IV. MODIFICATION OF THE HA-BINDING REGION PROTEIN

Sajdera and Hascall (21) showed that reduction and alkylation of disulfide groups in monomer proteoglycans prevented subsequent aggregate formation. This suggested that the conformation of the protein portion of the proteoglycan molecule was critical for the aggregation mechanism. Subsequently, it was shown that reduced and alkylated proteoglycan monomers would not interact with hyaluronic acid (6) and that reoxidation of reduced, but not alkylated, monomers restored almost 90% of the interactivity (10). Hardingham, Ewin, and Muir (10) used a variety of reagents to modify specific amino acid residues within the protein portion of proteoglycan monomers. The modified samples were then tested for their ability to interact with hyaluronic acid. The results from this study are summarized in Table I. The interaction with hyaluronic acid was inhibited after the proteoglycan preparation was treated with sufficient concentrations of reagents which modify primarily a) lysine groups (acetic anhydride, 2-methylmaleic anhydride), b) arginine groups (2,3-butanedione), or c) tryptophan groups (2-nitrophenylsulfonyl chloride). Treatment of proteoglycans with 0.055 M and 0.55 M 2-methylmaleic anhydride caused the loss of 67% and 100% of their binding capacity respectively. In the latter case most of the 2-methylmaleyl groups could be removed by treating the substituted proteoglycans with dilute acid with recovery of 60% of the binding activity. Up to two-thirds of the arginine residues in the proteoglycan could be modified by reaction with 2,3-butanedione without altering the interaction with hyaluronic acid; but with further arginine modification the interaction was almost completely abolished (Table I). Modification of about one third of the tryptophan with 2-nitrophenylsulfonyl chloride also eliminated most of the binding activity. Fluorescence measurements on the proteoglycan in an associative solvent suggested that most of the tryptophans were located in

TABLE I. Modified From Hardingham, Ewins, and Muir (10)

Effect of amino acid modification reactions on the interaction of monomer proteoglycans with hyaluronic acid

Reagent	mmolar	% Inhibition
acetic anhydride	5.3	9
	53.0	15
	530	100
2-methylmaleic anhydride	5.5	0
	55.0	67
	550	100
butane-2,3-dione	5.8	0
	12.0	95
	58.0	100

relatively hydrophobic regions. Interaction with hyaluronic acid did not appreciably alter the spectrum, suggesting that the tryptophan residues do not form direct subsite interactions stabilizing the internal, presumbaly globular, portion of the HA-binding region protein. This suggestion was supported by the observations that the fluorescence spectra of denatured proteoglycans in 0.1% sodium dodecylsulfate and of reduced and alkylated proteoglycans were shifted toward longer wavelengths, indicative of a configurational change in which some of the tryptophans moved into more polar environments.

The above results suggest that the tertiary structure of the HA-binding region protein which is essential for effective interaction with hyaluronic acid requires the presence of certain unmodified lysine, arginine, and tryptophan residues. Heinegard and Hascall (35) have recently studied the effects of modifications of amino groups in the purified HA-binding region protein on subsequent interaction with hyaluronic acid in more detail. The profiles in Fig. 11a indicate the elution position on Sepharose 6B of the HA-binding region preparation alone (dashed line) and in the presence of a high-molecular-weight hyaluronic acid in a separate experiment (solid line). The result indicates that all of the molecules in the HA-binding region preparation are capable of interacting with hyaluronic acid. Dansylation of the HA-binding region protein in the presence of cycloheptaamylose-dansyl (36) resulted in 80–100% loss of binding activity in several experiments (Fig. 11b). On the other hand, ^{3}H-acetylation of the HA-binding region protein with an effective concentration of about 5 mM acetic anhydride gave a loss of interaction of only about 20% (Fig. 11c), a result which is consistent with the observations by Hardingham et al. (10) discussed above (Table I). When the fraction of the ^{3}H-acetylated sample which retained binding activity, (Fig. 11c) was reisolated by chromatography on Sephadex G 200 in a dissociative solvent to remove hyaluronic acid and then dansylated, only a small proportion (about 10%) of the molecules were inactivated (Fig. 11d). In a separate experiment, the HA-binding-region protein was ^{3}H-acetylated in the presence of hyaluronic acid and the interactive molecules then reisolated. In this case, the amount of radioactivity incorporated into the HA-binding-region protein was only about 50% of that incorporated into the molecules that were ^{3}H-acetylated in the absence of hyaluronic acid (Fig. 11c) which suggests that many reactive groups, primarily the ϵ-amino groups of lysine residues, are partially protected from acetylation when bound to hyaluronic acid.

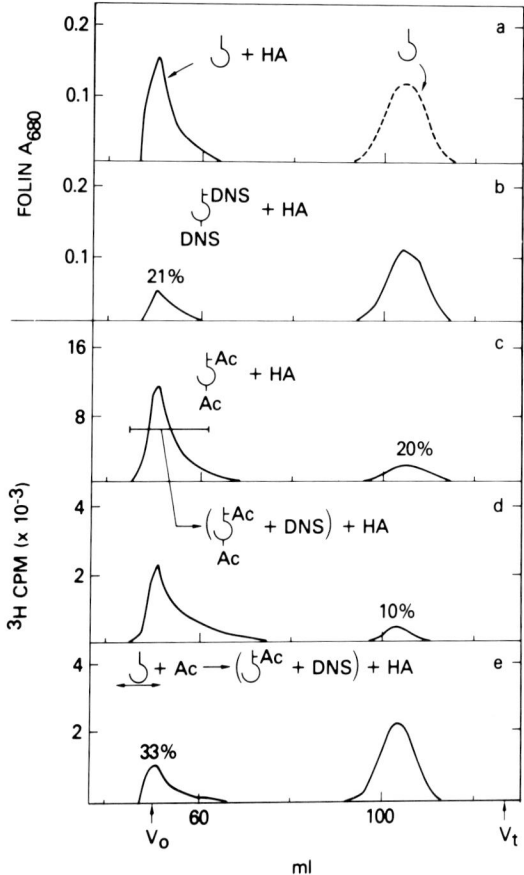

Fig. 11. Elution profiles on Sepharose 6B of aliquots of the HA-binding protein preparation which have been modified in various ways as indicated in the figure and text. The dashed line in (a) indicates the elution profile of the HA-binding region preparation alone in a separate experiment. The symbols "Ac" and "DNS" refer to acetylation and dansylation of reactive sties, respectively. The double headed arrow in (e) schematically represents a molecule of hyaluronic acid.

This was substantiated when it was shown that most of these interactive molecules could be inactivated by subsequent dansylation (Fig. 11e). These results suggest that there are lysine residues near the HA-binding site which are partially protected from substitution when hyaluronic acid is present, which can be substituted with acetyl groups without interfering with the interaction with hyaluronic acid, and which will prevent interaction with hyaluronic acid if substituted with the bulkier dansyl group.

A separate experiment was designed to see if portions of the polypeptide which are close to the HA-binding site could be selectively labeled. A sample of the HA-binding region protein was ^{3}H-dansylated in the presence of hyaluronic acid. Then, the hyaluronic acid was removed, and the sample was dansylated with unlabeled reagent. A separate sample was treated in the reverse order, dansylating first with unlabeled and then with ^{3}H-labeled reagent using the same protocol (Fig. 12). After purification, the 2 samples were treated with trypsin and the larger peptides, those excluded from Sephadex G-25, were subsequently fractionated on Sephadex G-50 (Fig. 12). While the elution profiles

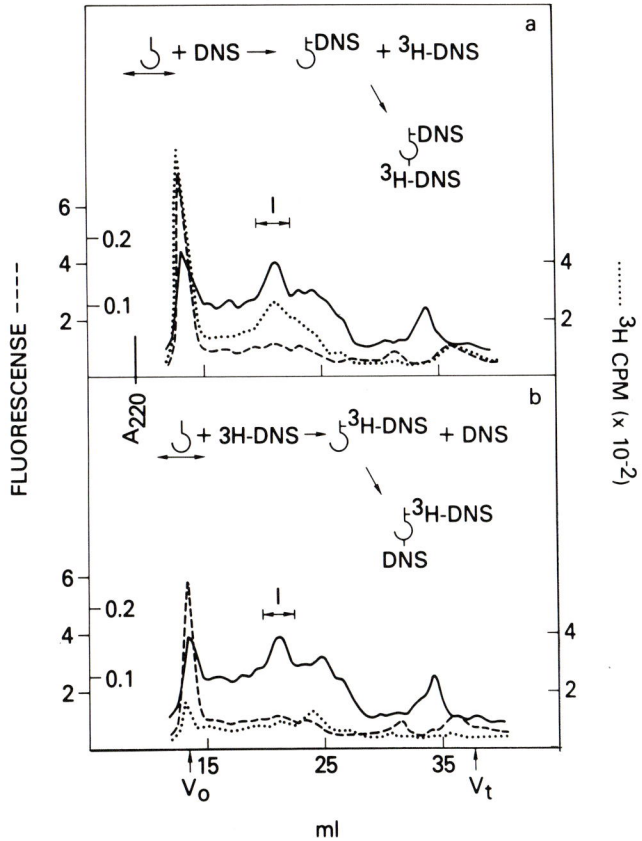

Fig. 12. Sephadex G-50 elution profiles of tryptic peptides that were isolated from dansylated HA-binding region protein samples prepared as indicated schematically in the figures. See text for details.

of absorbance at 280 nm and of fluorescence were essentially the same for each sample, the profiles of radioactivity differed. An included peak, indicated by I in Fig. 12, contained much more radioactivity when the ^{3}H-dansylation step was done after removing hyaluronic acid. Portions of this peptide(s) then, may be located in or near the HA-binding site. Similar experimental strategies should, in the future, provide many more details about the primary, secondary, and tertiary structure of the HA-binding region protein and how these provide an interaction site with hyaluronic acid. Indeed, it is fascinating to observe that proteoglycan research has reached a crossroads where many exciting new developments will depend more upon the skillful application of the research tools of protein chemistry rather than upon the more traditional methods of polysaccharide chemistry which have been critical for progress in the past.

V. THE EFFECT OF HYALURONIC ACID ON GLYCOSAMINOGLYCAN SYNTHESIS BY CHONDROCYTES

Hyaluronic acid appears to be critically involved in the process of development and differentiation of such tissues as chick cornea (37), chick vertebral column (38), and regenerating newt limb (39). (For a recent review, see Ref. 40.) In each case, there is a correlation first between cell migration and the synthesis and accumulation of hyaluronic acid, and second between the subsequent enzymatic degradation of hyaluronic acid by newly synthesized hyaluronidases, the cessation of cell migration, and initiation of cell differentiation. This is a particularly pronounced effect for the differentiation of cartilaginous tissues. Toole and his collaborators tested the effects of hyaluronic acid on differentiation of stage 26 chick embryo somite cells (41). When these cells are dispersed by mild trypsin digestion and cultured on petri dishes at high initial plating density, they normally attach, divide, and undergo chondrogenesis to form nodules, in which mounds of chondrocytes are interspersed in a typical cartilaginous extracellular matrix. The presence of exogenous hyaluronic acid in the medium at concentrations from 1 ng to 500 μg detectably inhibited cell aggregation and the formation of nodules. The inhibition was not observed with a variety of other biological polyanions, including the chondroitin sulfates, heparin, and nucleic acids (40). HA-oligosaccharides, primarily tetrasaccharides, recovered from testicular hyaluronidase digests of hyaluronic acid, were also effective inhibitors of nodule formation. Monosaccharides, on the other hand had no effect (40). The treatment with hyaluronic acid did not cause any detectable changes in cell survival or proliferation, nor in the synthesis of either type I or type II collagen; [^{35}S]sulfate incorporation into polysaccharide, however, was inhibited.

Solursh et al. (42) subsequently studied the effect of hyaluronic acid on fully differentiated chick embryo chondrocytes grown in culture. The cultures were tested for glycosaminoglycan synthesis in the presence or absence of hyaluronic acid in a medium without serum. This was necessary since the serum used normally contained appreciable amounts of hyaluronic acid. At a concentration of hyaluronic acid of 200 μg/ml, [^{35}S]sulfate incorporation was inhibited by 50% over 6–24-h incubation times. The effect was also observed with HA-oligomers, although no data regarding their size distribution was provided. No differences between control cultures and cultures treated with hyaluronic acid were observed for leucine or thymidine incorporation into macromolecules, for cellular uptake of sulfate, for collagen synthesis, or for matrix turnover. The inhibition seemed to be associated primarily with the accumulation of proteoglycans associated with the cell layer matrix. However, the chondrocytes, derived in this case from clones of differentiated cells, were able to form nodules in the presence of hyaluronic acid, although there was appreciably less extracellular matrix around the cells. Again, other polyanions were incapable of eliciting this effect.

Wiebkin et al. (43) at the same time described similar experiments for suspensions of chondrocytes derived from trypsin/collagenase digests of adult pig laryngeal cartilage. Again, as little as 0.005 μg hyaluronic acid/ml in serum-free medium was capable of significantly decreasing [^{35}S]sulfate incorporation into glycosaminoglycans by cells which had been grown in culture for several days. The inhibitory effect of hyaluronic acid was not blocked by treating the cells with chondroitinase, but was blocked if the cells were treated with trypsin. The cells recovered their ability to respond to hyaluronic acid after 6 h or longer postincubation without trypsin. Hyaluronic acid, labeled with [^{14}C]acetate, was taken up at the surfaces of the cells when added to the chondrocytes and could subsequently be released by mild trypsin digestion. Hyaluronic acid samples which were first interacted with saturating amounts of proteoglycans did not inhibit glycosaminoglycan synthesis by the cells. When HA-oligomers from testicular hyaluronidase digests were fractionated on Sephadex G-25, the included fraction, which would contain primarily tetrasaccharides and hexasaccharides, did not inhibit chondrocyte synthesis of ^{35}S-labeled glycosaminoglycans, but the excluded fraction, with longer HA-oligomers, did. Finally, other cell types which synthesize hyaluronic acid, fibroblasts and synovial cells, did not alter ^{35}S-labeled glycosaminoglycan synthesis when hyaluronic acid was added to cultures under similar experimental conditions.

More recently Handley and Lowther (44) reported experiments in which the effect of hyaluronic acid on glycosaminoglycan synthesis by chick embryo chondrocytes was investigated in the presence of a β-xyloside. These compounds, as well as xylose (45), act as exogenous acceptors for the synthesis of glycosaminoglycan chains, such as chondroitin sulfate (46–48), which are normally linked to the protein core of proteoglycans by a glycosidic bond between serine hydroxyls and xylose moieties (49). The cellular biosynthetic apparatus for glycosaminoglycan synthesis in chondrocytes diverts a large proportion of synthetic activity from the endogenous core protein to this exogenous substrate. In their experiments, Handley and Lowther (44) showed that total biosynthesis of ^{35}S-labeled glycosaminoglycans in the presence of the β-xyloside was the same independent of whether hyaluronic acid was present or not. It was suggested, then that the inhibition of synthesis in the presence of hyaluronic acid without added β-xyloside indicates that the specific effect of the hyaluronic acid is to initiate a series of intracellular events through intracellular effectors which inhibit either the synthesis of the normal proteoglycan core protein acceptor or the activity of the xylosyl transferase which is necessary to initiate the synthesis of each chondroitin sulfate chain on the protein.

Several proposals were made on the basis of the experiments described above: a) Hyaluronic acid can specifically inhibit the synthesis and accumulation of glycosaminoglycans by chondrocytes and, therefore, may be an extracellular regulator of cellular processes involved in proteoglycan synthesis. b) Chondrocytes, unlike other connective tissue cells, probably contain cell surface receptors specific for the interaction with hyaluronic acid. c) The specificity of the receptors for hyaluronic acid may have similarities with those of the matrix molecules which interact specifically with hyaluronic acid, namely the HA-binding region protein of proteoglycans and the link proteins. d) Such an interaction between an extracellular matrix component and the chondrocyte may provide control mechanisms for elaborating, maintaining, or modifying cartilage matrices.

The experiments discussed in this presentation suggest that many facets of the structure and function as well as of the differentiation and development of cartilage tissues involve specific interactions between proteins and hyaluronic acid. Undoubtedly, many new insights about this connective tissue will be uncovered as investigators discover more details about these interactions.

REFERENCES

1. Hascall VC, Sajdera SW: J Biol Chem 245:4920, 1970.
2. Pasternack SG, Veis A, Breen M: J Biol Chem 249:2206, 1974.
3. Hascall VC, Riolo RL: J Biol Chem 247:4529, 1972.
4. Hascall VC, Heinegard D: Arch Biochem Biophys 165:427, 1974.
5. Heinegard D, Axelsson I: J Biol Chem 252:1971, 1977.
6. Hardingham TE, Muir H: Biochim Biophys Acta 279:401, 1972.
7. Hascall VC, Heinegard D: J Biol Chem 249:4232, 1974.
8. Heinegard D, Hascall VC: J Biol Chem 249:4250, 1974.
9. Heinegard D: J Biol Chem 252:1980, 1977.
10. Hardingham TE, Ewins RJF, Muir H: Biochem J 157:127, 1976.
11. Rosenberg L, Wolfenstein-Todel C, Margolis R, Pal S, Strider W: J Biol Chem 251:6439, 1976.
12. Rosenberg L, Hellmann W, Kleinschmidt AK: J Biol Chem 250:1877, 1975.

13. Thyberg J, Lohmander S, Heinegard D: Biochem J 151:157, 1975.
14. Hascall VC: Doctoral dissertation, The Rockefeller University, New York, 1969.
15. Mathews MB: Nature (London) 213:1255, 1967.
16. Pennypacker JP, Goetinck PF: Dev Biol 50:35, 1976.
17. Thomas L: J Exp Med 104:245, 1956.
18. Orkin RW, Pratt RM, Martin GR: Dev Biol 50:82, 1976.
19. Howell DS, Sapolsky AI, Pita JC, Woessner JF: Semin Arthritis Rheum 5:365, 1976.
20. Hardingham TE, Muir H: Biochem Soc Trans 1:282, 1973.
21. Sajdera SW, Hascall VC: J Biol Chem 244:77, 1969.
22. Hascall VC, Sajdera SW: J Biol Chem 244:2384, 1969.
23. Oegema TR, Hascall VC, Dziewiatkowski DD: J Biol Chem 250:6151, 1975.
24. Hascall VC, Oegema TR, Brown M, Caplan AI: J Biol Chem 251:3511, 1976.
25. Keiser H, Shulman HJ, Sandson JI: Biochem J 126:163, 1972.
26. Gregory JD: Biochem J 133:383, 1973.
27. Faltz LL, Hascall VC, Heinegard D, Piez KA: Fed Proc 35:100, 1976.
28. Baker JR, Caterson BC: Biochem Biophys Res Commun 77:1, 1977.
29. Hascall VC, Heinegard D: J Biol Chem 249:4242, 1974.
30. Hardingham TE, Muir H: Biochem J 135:905, 1973.
31. Hook M, Lindahl U, Hallen A, Backstrom G: J Biol Chem 250:6065, 1975.
32. Christner J, Brown M, Dziewiatkowski DD: Biochem J. In Press.
33. Laurent TC: In Balazs EA (ed): "Chemistry and Molecular Biology of the Extracellular Matrix." New York and London: Academic Press, 1970, pp 703–732.
34. Highsmith S, Garvin JH Jr, Chipman DM: J Biol Chem 250:7473,1975.
35. Heinegard D, Hascall VC: (Submitted).
36. Kinoshita I, Iinuma F, Tsuji A: Anal Biochem 61:632, 1974.
37. Toole BP, Trelstad RL: Dev Biol 26:28, 1971.
38. Toole BP: Dev Biol 29:321, 1972.
39. Toole BP, Gross J: Dev Biol 25:57, 1971.
40. Toole BP: In Barondes SH (ed): "Neural Recognition." New York:Plenum Publishing, 1976, pp 275–329.
41. Toole BP, Jackson G, Gross J: Proc Natl Acad Sci USA 69:1384, 1972.
42. Solursh M, Vaerewyck SA, Reiter RS: Dev Biol 41:233, 1974.
43. Wiebkin OW, Hardingham TE, Muir H: In Slavkin HC, Greulich RC (eds): "Extracellular Matrix Influences on Gene Expression." New York: Academic Press, 1975, pp 209–223.
44. Handley CJ, Lowther DA: Biochim Biophys Acta 444:69, 1976.
45. Brett MJ, Robinson HC: Proc Aust Biochem Soc 4:92, 1971.
46. Okayama M, Kimata K, Suzuki S: J Biochem 74:1069, 1973.
47. Schwartz NB, Ho P-L, Dorfman A: Biochem Biophys Res Commun 71:851, 1976.
48. Robinson HC, Brett MJ, Tralaffan PJ, Lowther DA, Okayama M: Biochem J 148:25, 1975.
49. Roden L: In Balazs EA (ed): "Chemistry and Molecular Biology of the Intercellular Matrix." New York and London: Academic Press, 1970, pp 797–821.

Structural Analysis of a Membrane Glycoprotein: Glycophorin A

Heinz Furthmayr

Department of Pathology, Yale University, New Haven, Connecticut 06510

Glycophorin A is the major sialoglycoprotein of the human erythrocyte membrane. Structural studies indicate that this molecule is made up of 3 domains composed of 2 hydrophilic segments which are separated by a region of 22 nonpolar amino acids. The N-terminal half of the molecule contains all the carbohydrate associated with this protein.

Glycophorin A forms high-molecular-weight complexes which can be dissociated only under certain conditions. The site of subunit interaction is located within the hydrophobic segment, which serves both to mediate protein-protein and protein-lipid interactions within the bilayer membrane. Glycophorin A spans the membrane presumably as a dimeric complex with the carboxyterminal ends extending into the cytoplasm of the red cell. The transmembrane nature of the polypeptide chains finds strong support from the use of specific antibody-ferritin conjugates applied to thin sections of fixed and frozen intact cells.

Preliminary information on the analysis of human red cell variants which may lack some or all of the sialoglycopeptides are consistent with the presence in normal cells of a second sialoglycoprotein, provisionally labeled glycophorin B.

Key words: erythrocyte, plasma membrane, glycoproteins, amino acid sequence

The human erythrocyte membrane has been studied for many years to probe questions of membrane structure and function. A fairly well established model for this membrane system has been elaborated (1, 2). It seems that the bulk of the membrane proteins are located at the cytoplasmic side of the lipid bilayer and are linked in unknown ways to each other and to the membrane (3, 4). Only a few major proteins are associated with the membrane in a more intimate way by interacting with the hydrocarbon interior of the bilayer. These are considered to be major transmembrane proteins (5, 6). In addition to these there seem to be other proteins embedded in the membrane and protruding from the surface, which are not easily detectable by current standard methods because of the small number of copies per cell (7, 8).

Most of the data in suport of this general model have been obtained by analytical techniques such as proteolytic treatment and radiolabeling of intact cells and "leaky" or "sealed" ghosts (1, 2). Detailed structural information is largely missing since only a few of the major protein components have been isolated and purified enough to permit such studies. Particular difficulties have been encountered with the class of membrane proteins

Received May 10, 1977; accepted May 20, 1977.

which associate directly with lipid. Only recently have techniques become available which allow their isolation and separation from other membrane components (2, 6, 9).

Glycophorin A obtained from human erythrocyte membranes is the first membrane glycoprotein for which the primary structure is known (10). This knowledge not only makes it possible to test and corroborate earlier data on its orientation in the membrane, but also to study the genetics of its expression on the cell surface and to design experiments on its role in differentiation and in the mature cells.

SIALOGLYCOPROTEINS AS VIRUS RECEPTORS AND BLOOD GROUP SUBSTANCES

In the late 1940s Winzler and his colleagues isolated material from human red cell membranes by hot phenol extraction, chloroform-methanol treatment, and ultracentrifugation, which was able to inhibit myxovirus hemagglutination (11). Treatment with neuraminidase and tryptic digestion destroyed the activity suggesting that the virus receptor was a sialoglycoprotein. It was realized early that the isolated molecule tended to aggregate in aqueous buffers and that treatment with detergents, organic solvents, or acids produced smaller proteins with a molecular weight estimated at 30,000 daltons (12).

This viral receptor substance apparently is related to molecules which carry MN blood group activity (13). The MN antigens were detected by Landsteiner and Levine in 1927 by using antisera obtained by immunization of rabbits with red blood cells (14) and subsequently by plant lectins, such as that from Vicia graminea, which possesses anti-N activity (15). The first insight into the chemical nature of the MN antigens came from observations by Springer, that neuraminidase from influenza virus and from Vibrio cholerae abolishes the MN activities in human erythrocyte in addition to destroying their influenza virus receptor activity (16). Treatment of intact cells with proteolytic enzymes releases a considerable amount of the cell surface carbohydrates in the form of glycopeptides which also results in the loss of MN determinants from the cell. It was concluded that sialic acid residues serve not only as receptor sites for influenza virus, but are also essential for MN antigens, and that these activities are linked to a glycoprotein. Isolation of the MN substances resulted in proteins of very high molecular weight and studies on the antigenic structures were severely hampered by the fact that activity depended largely on the state of aggregation, being higher for the highest molecular weight aggregates. Small glycopeptides have only low activity and isolated oligosaccharides completely lack M or N activity (16). However the purity and homogeneity of these isolated products have never been rigorously established. Subsequently other methods for the isolation of these sialoglycoproteins were developed such as the use of aqueous pyridine, chloroform/methanol, detergents, affinity chromatography, and combinations of the various methods (2). The yields were usually low and the purity of the preparations was difficult to assess. In addition, a wide range of molecular weights was reported. The use of sodium dodecyl sulfate gel (SDS-gel) polyacrylamide electrophoresis added still more confusion, since a major sialoglycopeptide was found to have an apparent molecular weight of about 85,000, but the migration of the protein varied with the acrylamide concentration (17). In addition, more complex gel patterns were obtained when other electrophoresis systems were employed (18, 19).

Fairbanks et al. in their gel electrophoresis system described 4 bands for red cell membranes which were stainable with periodic acid-Schiff's reagent (20). The staining property on the gel of these peptides depends largely on the presence of sialic acid, but chemical labeling of sialic acid in intact cells or membranes and other labeling techniques confirmed these results (21). The sialoglycoproteins bands were termed PAS-1, PAS-2, PAS-3, and PAS-4. This complex pattern on SDS-gels raises several questions: What is the chemical nature of these glycopeptides, do individual bands correspond to unique glycopeptides, and what is their relationship to each other?

GLYCOPHORIN A AND GLYCOPHORIN B

Isolation of the sialoglycoprotein fraction from human red cell membranes gives a preparation with a gel pattern after PAS-staining identical to that of the total membrane (Fig. 1). Further fractionation by gel filtration in detergents (22) resolved 3 protein containing peaks (Fig. 2), and 2 of these (peaks B and C) apparently contain the same peptide (glycophorin B), which is different from the protein component in the main peak A (glycophorin A). Although both peptides appear to be related to each other and contain approximately the same amount of carbohydrate and similar, yet distinct, amino acid compositions, gel filtration and SDS-gel analysis showed that glycophorin A corresponded to the PAS-1 and glycophorin B to the PAS-3 band while the peptides eluted from the PAS-2 position appeared to be a mixture of both proteins (23). Surprisingly, reelectrophoresis of the isolated peptides did not result in single bands but revealed a tendency to aggregate to multiple-molecular-weight forms (Fig. 3). Similar observations have been made by other groups (24). Glycophorin A and glycophorin B isolated by gel filtration also do not appear as homogeneous peptides on SDS-gels (22, Fig. 4). Glycophorin B isolated from both regions of the chromatogram, although apparently homogeneous (unpublished), gives at least 4

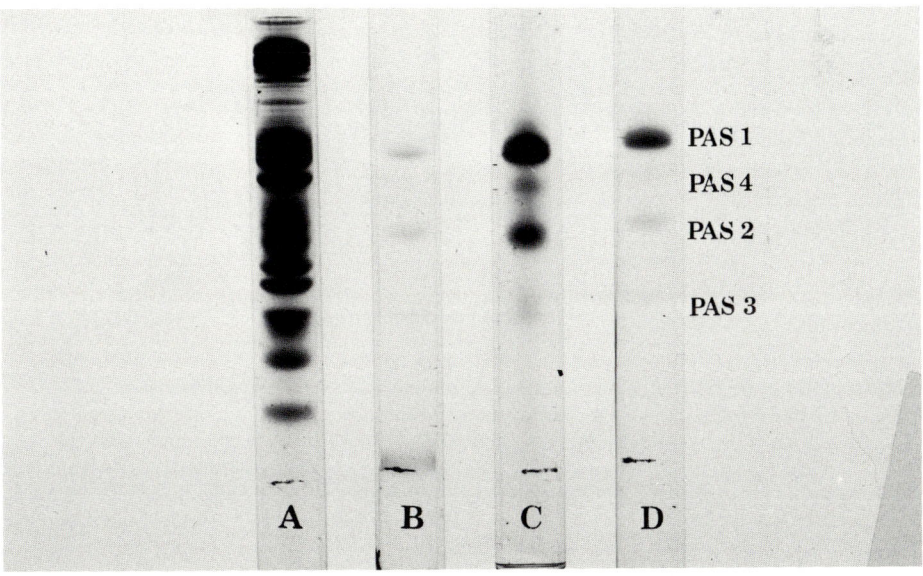

Fig. 1. Sodium dodecyl sulfate-polyacrylamide gel electrophoresis (20) of human red cell membranes (A,B) or the isolated sialoglycoprotein fraction (C,D), stained with Coomassie blue (A,C) or periodic acid-Schiff's reagent (B,D).

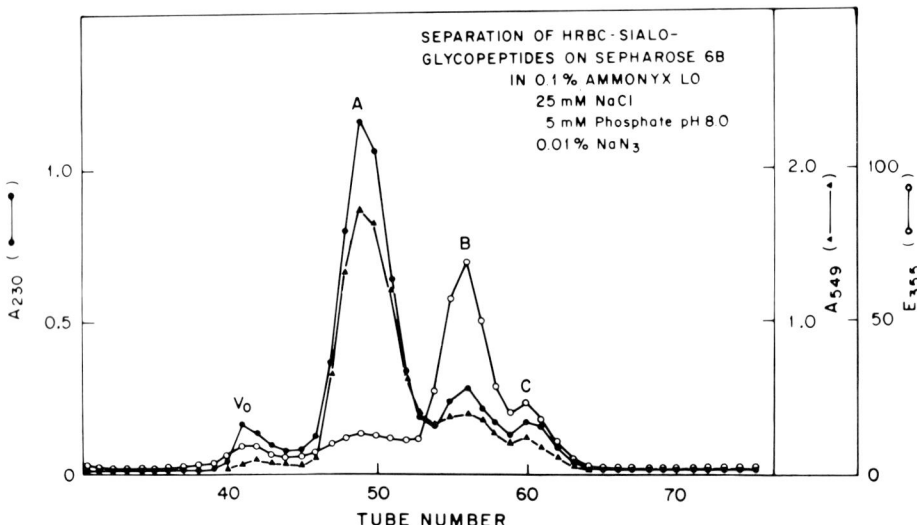

Fig. 2. Gel filtration of Biorad A 1.5 in detergent of human erythrocyte membrane sialoglycoproteins isolated by the LIS-phenol procedure. The effluent was monitored for protein (●–●–●), sialic acid (▲▲▲), and tryptophan fluorescence (excitation 290 nm, emission 355 nm, ○○○). Peak A contains glycophorin A and peaks B and C, glycophorin B. Vo is the void volume. (Reprinted with permission from Biochem Biophys Res Commun 65:113–121, 1975.)

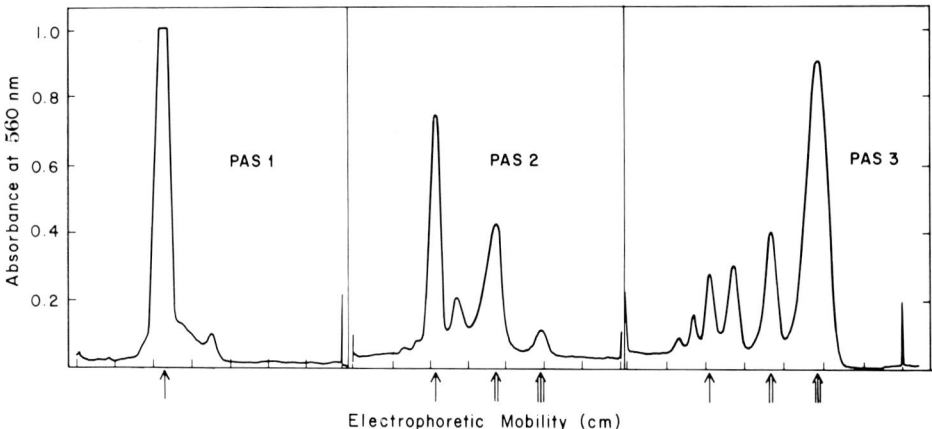

Fig. 3. Sodium dodecyl sulfate-polyacrylamide gel electrophoresis (20) of sialoglycopeptides isolated from regions of the original unfixed and unstained SDS-polyacrylamide gels corresponding to PAS-1, PAS-2, and PAS-3 (cf. Fig. 1C, D). Gel slices containing individual glycopeptides were cut from the tube gels, which had been run under standard conditions, with a razor blade and were crushed into small pieces with a spatula. After extraction of the protein by incubation overnight at room temperature in excess buffer, containing 50 mM ammonium bicarbonate and 0.05% sodium dodecyl sulfate, the gel particles were removed by centrifugation. The extract was extensively dialyzed against distilled water and lyophilized. After an additional extraction with 95% ethanol at ice temperature, the protein precipitate was dissolved in water and lyophilized again. Gel electrophoresis was done on 20 to 50 μg aliquots of the isolated proteins. The gels were stained with periodic acid-Schiff's reagent and scanned at 560 nm. The arrows indicate the positions of PAS-1 (↑), PAS-2 (↑↑), and PAS-3 (↑↑↑) on a control gel run in parallel (not shown).

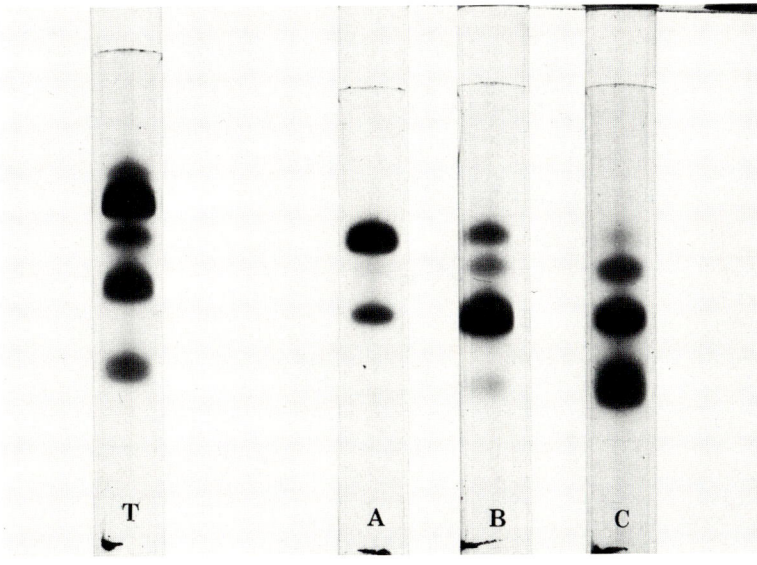

Fig. 4. Sodium dodecyl sulfate-polyacrylamide gel electrophoresis of fractions A, B, and C from Fig. 2, after extensive dialysis against distilled water and lyophilization. The gels were stained with periodic acid-Schiff's reagent. Gel T is obtained for the sialoglycoprotein mixture before fractionation.

bands comigrating with the PAS-1, -2, -3, -4 bands of the original unseparated mixture. To explain this complexity, it is assumed that these sialoglycopeptides have various preferred states of association in the presence of detergents (or in the membrane), which are not disrupted by weak detergents or even SDS under certain conditions.

Glycophorin A has been shown to be homogeneous (see structural studies discussed below) and yet it migrates in both the PAS-1 and PAS-2 position on SDS-polyacrylamide gels. The structural information available for glycophorin A allows us to interpret some of the complexities mentioned above and to exclude heterogeneity of the protein as a likely basis for these results.

When glycoprotein samples are heated in SDS before electrophoretic separation, conversion of the PAS-1 form into the PAS-2 form can be demonstrated (Fig. 5). This conversion is dependent on protein and detergent concentration, temperature, time of incubation, and furthermore is reversible (25–27, 52). At least for glycophorin A, the gel patterns are thus entirely dependent on a variety of conditions, which predictably will influence the amount of the low- or high-molecular-weight forms observed for any given experiment. These observations provided the basis for the idea that glycophorin A is an oligomeric molecule composed of low-molecular-weight subunits which can undergo reversible dissociation and reassociation reactions. Although the simplest explanation proposes dissociation of the dimeric form PAS-1 into monomers under these conditions, the absolute number of subunits in the undissociated complex is not known. Molecular weight estimates for glycoproteins on SDS-gels are not reliable and do not allow the determination of true molecular weights for the various products observed on the gels (28). In sedimentation equilibrium studies, however, a single molecular weight of 29,000 has been determined for the subunit of glycophorin A (29) in the presence of SDS, which correlates well with data

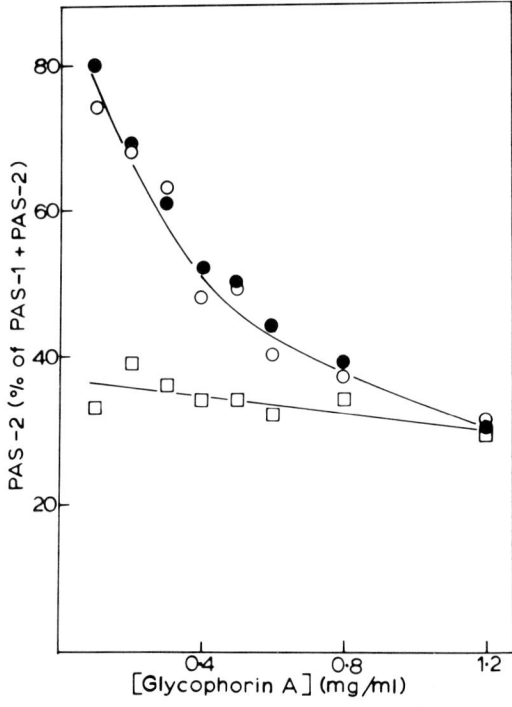

Fig. 5. Effect of incubation conditions on the distribution of glycopeptides migrating in the position of PAS-1 or PAS-2 on sodium dodecyl sulfate-polyacrylamide gels. Glycophorin A was dissolved in gel loading buffer and heated at 80°C for 15 min before electrophoresis. The gels were stained with periodic acid-Schiff's reagent, scanned at 560 nm and the peak areas determined by triangulation. The amount of PAS-2, expressed as percentage of the total PAS-1 and PAS-2, is plotted against the concentration of glycoprotein in the sample loaded onto the gel. ○) Protein diluted before heating to the concentrations shown; 50 μl of sample per gel. ●) Protein diluted before heating; 100 μl of sample per gel. □) Protein heated at 1.25 mg/ml and then diluted to the concentrations shown; 50 μl of sample per gel. (Reprinted with permission from Biochemistry 15:1448–1454, 1976. Copyright by the American Chemical Society.)

calculated from the amino acid and carbohydrate contents. The finding of a single molecular-weight species is clearly in contrast with the conclusions drawn above and the reasons for this discrepancy are not clear at present.

STRUCTURE OF GLYCOPHORIN A

The entire amino acid sequence of glycophorin A is now known and is given in Fig. 6 (30, 31). The polypeptide chain contains 131 amino acid residues and as suggested previously (32), these are arranged into 3 domains. An amino-terminal segment is the carbohydrate-containing portion rich in hydroxy amino acids, to which 15 oligosaccharide units are attached O-glycosidically and which contains one asparagine, to which a more complex carbohydrate chain is linked N-glycosidically (for a more extensive discussion of the carbohydrate structure see recent review, Ref. 2). Residues 73 through 95 do not contain any charged side chains, but have instead a high number of hydrophobic residues in addition to 3 hydroxy amino acids and most of the glycyl residues found in the poly-

HUMAN ERYTHROCYTE GLYCOPHORIN A

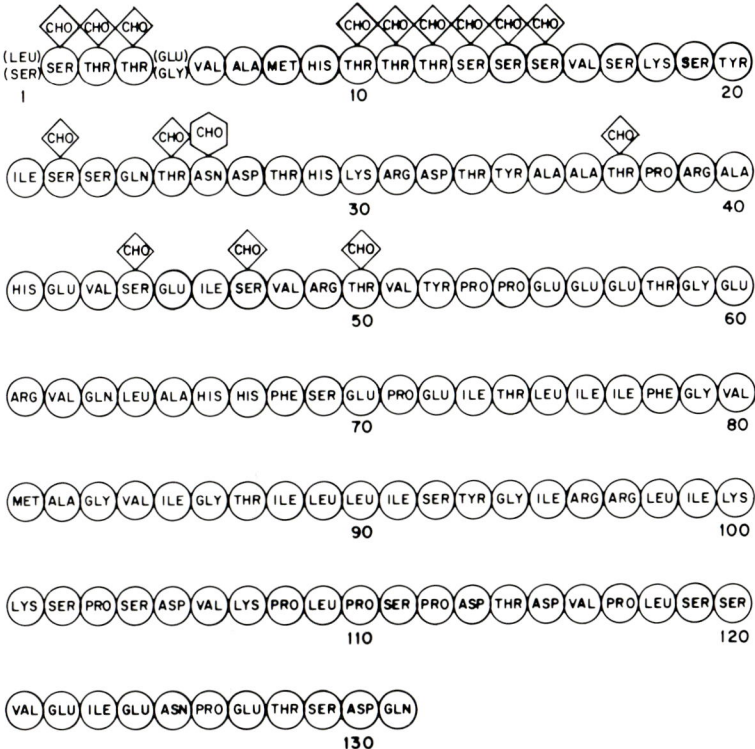

Fig. 6. Amino acid sequence of glycophorin A. Boxes above certain residues indicate attachment of oligosaccharides at these sites.

peptide chain. This region conveys properties to the protein and a tryptic fragment derived from it which made its characterization fairly difficult (31). The hydrophobic region is probably responsible for aggregation of the intact molecule in aqueous salt solutions, for its binding of high amounts of SDS, and for its association with lipids (10). The C-terminal sequence of 36 amino acid residues contains a high number of imino acids and predominantly acid amino acids.

With the exception of 2 positions of the amino acid sequence (positions 1 and 5) no evidence has been obtained suggesting additional heterogeneity of the polypeptide structure. In most of these studies pooled blood from several individuals has been used. It is conceivable that products of different genes, e.g. genes coding for the proteins which carry M or N specificities, are distinguished by a few important differences in their amino acid sequences. The heterogeneity described above could be due to such different gene products or differences could have been overlooked due to technical difficulties in the sequence determinations of multiple glycosylated peptides. Preliminary studies on peptides isolated from glycophorin A obtained from individual donors, however, do not indicate that this is the case, since no differences were seen in amino acid analysis of purified peptides. Likewise, there was no indication for heterogeneity in the nonglycosylated carboxy-terminal portion of glycophorin A (33).

In all glycophorin A preparations studied, whether from pooled or individual samples, a different type of heterogeneity however, was apparent (Fig. 7). Proteolytic cleavage using trypsin or chymotrypsin was incomplete in about one-third to one-half of the polypeptide chains at a few sites resulting in multiple peptides from regions identical in amino acid sequence, but which were different with respect to carbohydrate content (34). Cleavage at these insensitive peptide bonds in the larger peptide fragments (e.g., T1 or CH1 in Fig. 7) was only obtained after removal of terminal sialic acid residues, suggesting either heterogeneity of attachment sites of oligosaccharide chains or interaction of terminal oligosaccharide structures with the polypeptide backbone, or both. Both possibilities would have the effect of preventing cleavage by these enzymes at certain sites. It is possible that heterogeneity of this type, namely variability in the distribution as well as length of the oligosaccharide at various locations along the polypeptide backbone, determine characteristic antigenic features.

THE INTRAMEMBRANOUS SEGMENT OF GLYCOPHORIN A

Molecules which are amphiphilic, with an ionic part soluble in water and a second part repelled from water as a result of hydrophobicity, will be forced to adopt unique orientations and to adopt certain organized structures. This is the case in biological membranes which can form spontaneously and in which individual lipids are arranged in a bilayer structure. It is clear, that some of the membrane proteins interact with the hydrocarbon interior of the lipid bilayer, but the structural basis for this interaction is largely unknown.

Some membrane proteins have been found to contain a peptide or peptide region which is very much enriched in hydrophobic amino acid residues. Such hydrophobic peptides have been isolated from cytochrome b_5 (35), the coat protein of filamentous bacteriophages (36), the glycoproteins of an arborvirus (37), and glycophorin A. Amino acid sequence data are available only for some of these and even less is known about the conformation of these particular regions (38, 39). However, there are some suggestions that an α-helical structure is the preferred mode of organization of hydrophobic polypeptide regions in the apolar lipid environment (40, 41), although other arrangements have been postulated (38, 42).

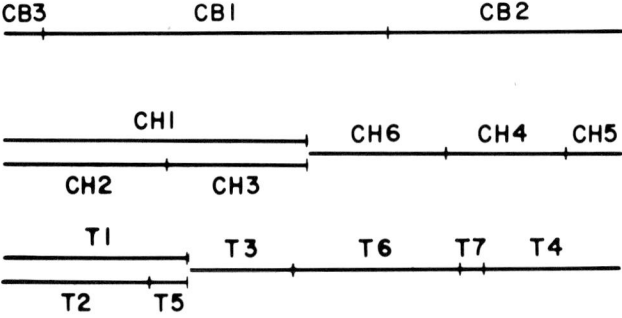

Fig. 7. Schematic representation of cyanogen bromide (CB), tryptic (T), and chymotryptic (CH) peptides of glycophorin A, drawn to scale. Total length of the polypeptide chain is 131 amino acid residues. The amino-terminal end of the molecule is to the left.

As mentioned above, glycophorin A apparently forms relatively stable complexes which can dissociate and reassociate in the presence of detergents. Two different observations suggest a role for the hydrophobic region of the glycophorin A polypeptide chain in the interactions between subunits. The first indication came from studies on the reversibility of the effect described above (26), namely that dissociated subunits can reassociate to form the original PAS-1 complex. If glycophorin A subunits can reassociate even in sodium dodecyl sulfate, then smaller peptides may inhibit complex formation of the larger subunits. Binding studies with peptides prepared from labeled glycophorin A in fact demonstrated that a tryptic peptide which contains the entire hydrophobic segment, not only binds to the smaller molecular weight form PAS-2, but also prevents reassociation of larger subunits. Binding was only observed after dissociation of the original glycophorin A complex by heat treatment, but not to the undissociated form PAS-1 (Fig. 8). Under these conditions a hybrid molecule was generated, which is composed of large, intact glycophorin A polypeptides and small hydrophobic peptides. This hybrid structure can be distinguished by SDS-gel electrophoresis from the glycophorin A complex PAS-1, but not from the dissociated form PAS-2.

In agreement with the role of the hydrophobic region for the interaction between subunits are studies on the chemical modification of glycophorin A. Alkylation with iodoacetic acid under conditions which selectively modify methionyl side chains, will result in complete conversion to the low-molecular-weight form. Modification of methionine 81 located within the hydrophobic amino acid sequence, is essential. Sodium dodecyl sulfate will prevent alkylation of this amino acid residue and thus conversion, but will not

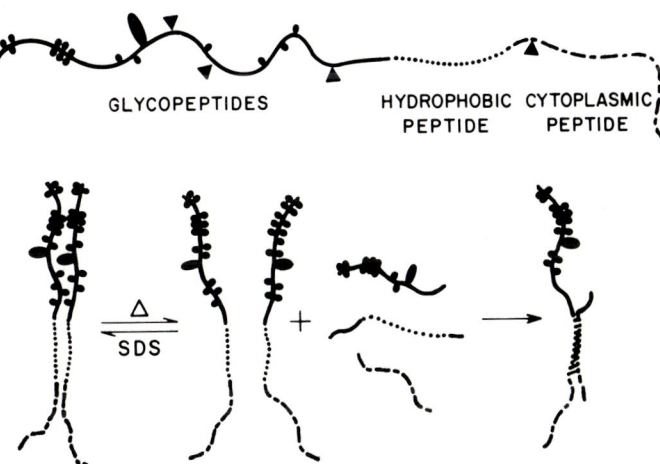

Fig. 8. Model of the glycophorin A subunit and its interaction with the hydrophobic tryptic peptide T6. The model is derived from the observation that incubation of glycophorin A together with a tryptic peptide derived from the hydrophobic segment of glycophorin A (residues 62 through 96) in sodium dodecyl sulfate solutions at 100°C generates hybrid molecules, which are composed of intact glycophorin A subunits and the short hydrophobic peptide, when analyzed by sodium dodecyl sulfate-polyacrylamide electrophoresis. Binding of the hydrophobic peptide to the glycophorin A complex (migrating in the PAS-1 position) is not observed, when incubation was done at 37°C, suggesting that dissociation of the complex is required for binding. ▲) Tryptic cleavage sites; •) location of oligosaccharide units. (Reprinted with permission from Biochemistry 15:1137–1144, 1976. Copyright by the American Chemical Society.)

protect against alkylation of the only other methionyl group in position 8 of the amino acid sequence (Fig. 9).

More information is clearly needed to describe the interaction of this part of glycophorin A with neighboring polypeptides and with membrane lipids. If the favored conformation in the bilayer is an α-helical structure, it is conceivable that separate parts of the helix are in association with the hydrocarbon chains of fatty acids and with other subunits or proteins, and that the stability of these structures is determined to a large extent by the primary structure.

It is difficult at present to extrapolate from these experiments and to make predictions about the stability of glycophorin A complexes within the natural membrane environment. An intriguing possibility exists that glycoprotein complexes are not preformed, but are generated transiently as a result of interaction with a ligand at the exterior or interior of the cell. The other alternative that some ligands require multivalent associations with receptor complexes with each subunit contributing to binding is equally attractive.

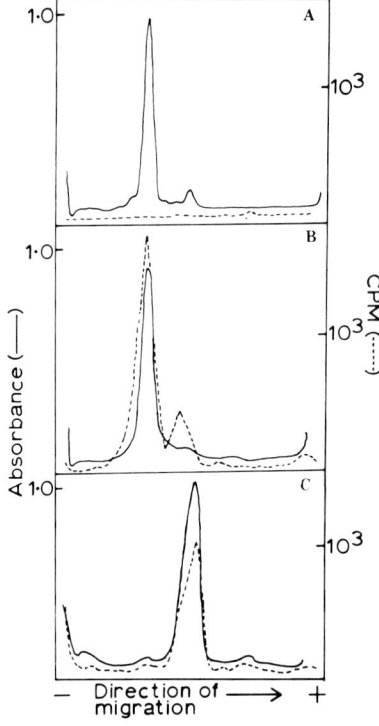

Fig. 9. Sodium dodecyl sulfate-polyacrylamide gel electrophoresis of unmodified (A) and carboxymethylated glycophorin A (B,C). B) Glycophorin A was carboxymethylated with [1-^{14}C] iodoacetic acid for 4 h at 37°C in the presence of 2% sodium dodecyl sulfate; C) glycophorin A was carboxymethylated as in (B) with unlabeled reagent and recarboxymethylated with [1-^{14}C] iodoacetic acid in the presence of 7 M guanidine hydrochloride. For further explanations see text. (Reprinted with permission from Biochemistry 15:1448–1454, 1976. Copyright by the American Chemical Society.)

ORIENTATION OF GLYCOPHORIN A IN THE MEMBRANE

The transmembrane orientation of the major sialoglycoprotein has been demonstrated repeatedly in many laboratories primarily on the basis of differential labeling of molecules in intact cells versus leaky ghost membranes (2, 43). Objections have been raised to this approach for various reasons. Enzyme-catalyzed iodination experiments have produced variable results, and some laboratories have failed to label the C-terminal segment of glycophorin at the cytoplasmic side of the membrane. Recently it has been possible to analyze this question in a more direct way for intact cells. Antibodies specific for a small region on the cytoplasmic portion of glycophorin A (Fig. 10) were coupled to ferritin and these conjugates were applied to frozen thin sections of fixed intact erythrocytes (44, Fig. 11). The electron-dense ferritin particles are found inside the cytoplasmic density regularly distributed throughout the circumference of the cell and at equal distances from the membrane. This distribution seems to reflect the linear mode of insertion of the peptide portion of glycophorin into the membrane. Since the interaction between glycophorin A and the membrane depends upon and is fixed by the hydrophobic region, one may expect that the C-terminal region has a similar fixed location within the cell.

VARIANT RED CELLS

The antigenic determinants of the MNSs and possibly Uu blood group systems are located on the sialoglycoproteins, which can be isolated from human erythrocyte membranes (45, 46). Although the antibody reagents and lectins, which are being used for blood typing have proven to be reliable, when applied carefully, they do not seem to provide markers specific enough for individual glycoproteins. Preliminary studies on rare

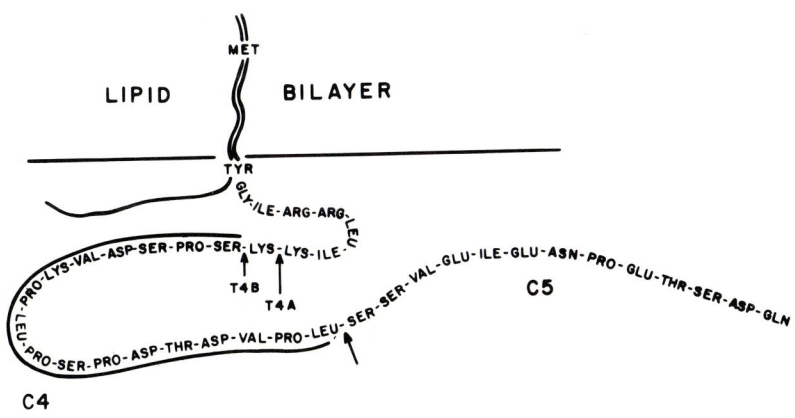

Fig. 10. The C-terminal cytoplasmic region of glycophorin A. The solid line above 17 residues within the C-terminal amino acid sequence denotes the maximum size of an antigenic determinant to which rabbit antibodies are directed. Arrows denote cleavage sites for trypsin (T) and chymotrypsin (C) to yield the fragments indicated.

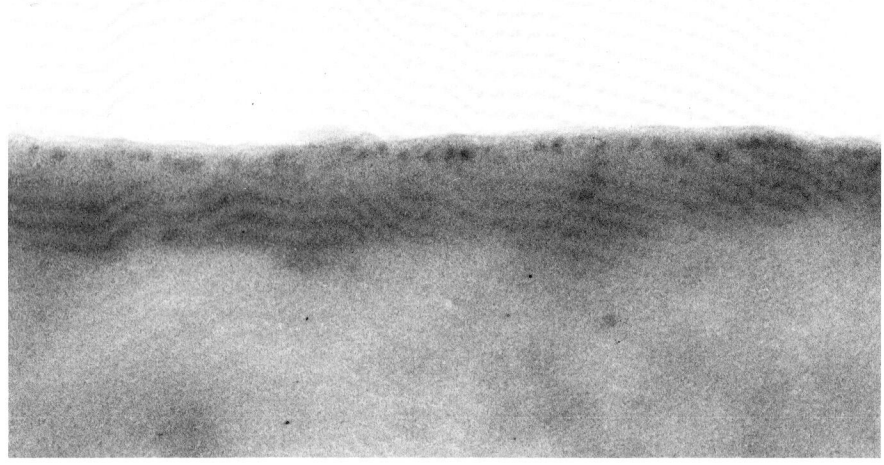

Fig. 11. Frozen thin sections of intact human red blood cells incubated with ferritin-antibody conjugates, specific for the C-terminal antigenic determinant indicated in Fig. 10 (magnification 249,900 ×). Electron-dense ferritin particles are seen at the cytoplasmic side of the membrane over the cytoplasmic density, but not at the exterior of the cell.

blood types which lack some of these determinants have shown that there may be a correlation between glycophorin A and the MN active material on the one hand and glycophorin B (or PAS-3) and Ss activity on the other hand (47, 48). En (a−) individuals who do not express MN determinants, seem to completely lack glycophorin A (49−51). On the other hand, anti-N reagents such as human or rabbit antibodies or the lectin from Vicia graminea will also react with the isolated PAS-3 glycopeptides or even with N-negative cells after trypsinization (15). Individuals who do not express Ss lack this N receptor and apparently also glycophorin B (48). There are numerous other examples of rare blood types, in which the normal expression of MNSs antigens is affected. The structure of these antigenic determinants is not known, but there is evidence for participation of terminal sialic acid residues and other carbohydrates. In view of the involvement of carbohydrate structures in antigenic activity and the known changes in sialic acid content and other carbohydrates in these variant cells, it becomes essential to develop methods which allow the quantitative determination of the amount of these glycoproteins per cell. Analysis of membranes by SDS-gel electrophoresis in combination with PAS staining of the gels or cell-surface labeling may not be tools reliable enough to study the presence or absence of these membrane proteins. It would be unwise to draw too many conclusions with respect to the genetic relationships and expression of these different proteins until such data are available (15). These deletion types, if proven, are however of considerable interest and will receive attention in attempts to study the function of glycophorins in differentiation, cell maturation, and on their effect on the mature cell.

ACKNOWLEDGMENTS

This work was supported by Public Health Service grant 5 R01 GM-19929, Membrane Center Grant P17-GM 21714, a grant from the American Cancer Society, ACS-BC 102C, and Anna Fuller Fund No. 470. The dedicated technical assistance of Mr. John Albert for obtaining frozen thin sections and electronmicroscopy is appreciated.

REFERENCES

1. Steck TL: J Cell Biol 62:1, 1974.
2. Marchesi VT, Furthmayr H, Tomita M: Annu Rev Biochem 45:667, 1976.
3. Kant JA, Steck TL: J Biol Chem 248:8457, 1973.
4. Wang K, Richards FM: J Biol Chem 249:8005, 1974.
5. Yu J, Fischman DA, Steck TL: J Supramol Struct 1:233, 1973.
6. Furthmayr H, Kahane I, Marchesi VT: J Membr Biol 26:173, 1976.
7. Steck TL, Dawson G: J Biol Chem 249:2135, 1974.
8. Gahmberg CG, Hakomori S: J Biol Chem 248:4311, 1973.
9. Yu J, Steck TL: J Biol Chem 250:9170, 1975.
10. Furthmayr H: In Cuatrecasas P, Greaves MF (eds): "Receptors and Recognition." London: Chapman and Hall, Series A, Vol 3; 103, 1977.
11. Kathan RH, Winzler RJ, Johnson CA: J Exp Med 113:37, 1961.
12. Morawiecki A: Biochim Biophys Acta 83:339, 1964.
13. Springer GF, Nagai Y, Tegtmeyer H: Biochemistry 5:3254, 1966.
14. Landsteiner K, Levine P: J Exp Med 47:757, 1928.
15. Dahr W, Uhlenbruck G, Knott W: J Immunogenet 2:87, 1975.
16. Springer GF, Ansell NJ: Proc Natl Acad Sci USA 44:182, 1958.
17. Segrest JP, Jackson RL, Andrews EP, Marchesi VT: Biochem Biophys Res Commun 44:390, 1971.
18. Mueller TJ, Dow AW, Morrison M: Biochem Biophys Res Commun 72:94, 1976.
19. Dahr W, Uhlenbruch G, Janssen E, Schmalisch R: Blut 32:171, 1976.
20. Fairbanks G, Steck TL, Wallach DFM: Biochemistry 10:2606, 1971.
21. Blumenfeld OO, Gallop PM, Liao TH: Biochem Biophys Res Commun 48:242, 1972.
22. Furthmayr H, Tomita M, Marchesi VT: Biochem Biophys Res Commun 65:113, 1975.
23. Furthmayr H: Unpublished results.
24. Janado M, Azuma J, Onodera K: J Biochem (Tokyo) 74:881, 1973.
25. Marton LSG, Garvin JE: Biochem Biophys Res Commun 52:1457, 1973.
26. Furthmayr H, Marchesi VT: Biochemistry 15:1137, 1976.
27. Silverberg M, Furthmayr H, Marchesi VT: Biochemistry 15:1448, 1976.
28. Fish WW: Methods Membr Biol 4:189, 1975.
29. Grefrath SP, Reynolds JA: Proc Natl Acad Sci USA 71:3913, 1974.
30. Tomita M, Marchesi VT: Proc Natl Acad Sci USA 72:2964, 1975.
31. Furthmayr H, Galardy RE, Tomita M, Marchesi VT: Arch Biochem Biophys, In press.
32. Marchesi VT, Tillack TW, Jackson RL, Segrest JP, Scott RE: Proc Natl Acad Sci USA 69:1445, 1972.
33. Tomita M, Furthmayr H, Marchesi VT: In preparation.
34. Tomita M, Furthmayr H, Marchesi VT: In preparation.
35. Spatz L, Strittmatter P: Proc Natl Acad Sci USA 68:1042, 1971.
36. Nakashima Y, Wiseman RL, Konigsberg W, Marvin DA: Nature (London) 253:68, 1975.
37. Uterman G, Simons K: J Mol Biol 85:569, 1974.
38. Corcoran D, Strittmatter P: Fed Proc (Abstract) 3296, 1977.
39. Asbeck F, Beyreuther K, Köhler H, Von Wettstein G, Braunitzer G: Hoppe Seyler's Z Physiol Chem 350:1047, 1969.
40. Henderson R, Unwin PNT: Nature (London) 257:28, 1975.
41. Schulte TH, Marchesi VT: In preparation.

42. Green NM, Flanagan MT: Biochem J 153:729, 1976.
43. Carraway KL: Biochim Biophys Acta 415:379, 1975.
44. Cotmore SF, Furthmayr H, Marchesi VT: J Mol Biol 113:539, 1977.
45. Hamaguchi H, Cleve H: Biochim Biophys Acta 278:271, 1972.
46. Fukuda M, Osawa T: J Biol Chem 248:5100, 1973.
47. Tanner MJA, Anstee DJ: Biochem J 153:271, 1976.
48. Dahr W, Uhlenbruck G, Issitt PD, Allen FH: J Immunogenet 2:249, 1975.
49. Dahr W, Uhlenbruck G, Leikola J, Wagstaff W, Landfried K: J Immunogenet 3:329, 1976.
50. Gahmberg CG, Myllyla G, Leikola J, Pirkola A, Nordling S: J Biol Chem 251:6108, 1976.
51. Furthmayr H: In preparation.
52. Tuech JK, Morrison M: Biochem Biophys Res Commun 59:352, 1974.

Glycosylation of VSV Glycoprotein Is Similar in Cystic Fibrosis, Heterozygous Carrier, and Normal Human Fibroblasts

Lawrence A. Hunt and Donald F. Summers

Department of Microbiology, University of Utah College of Medicine, Salt Lake City, Utah 84132

The single envelope glycoprotein of vesicular stomatitis virus was used as a specific probe of glycosyltransferase activities in fibroblasts from two cystic fibrosis patients, an obligate heterozygous carrier and a normal individual. Gel filtration of pronase-digested glycopeptides from both purified virions and infected cell-associated VSV glycoprotein which had been labeled with [^{3}H]glucosamine did not reveal any significant differences in the glycosylation patterns between the different cell cultures. All 4 cell lines were apparently able to synthesize the mannose- and glucosamine-containing core structure and branch chains terminating in sialic acid which are characteristic of asparagine-linked carbohydrate side chains in cellular glycoproteins. Analysis of tryptic glycopeptides by anion-exchange chromatography indicated that the same 2 major sites on the virus polypeptide were recognized and glycosylated in all 4 VSV-infected cell cultures. These studies suggest that the basic biochemical defect(s) in cystic fibrosis is not an absence or deficiency in enzymes responsible for the biosynthesis of complex carbohydrate side chains.

Key words: virus glycoprotein, cystic fibrosis, glycosyltransferases

Cystic fibrosis is the most common lethal genetic disease among Caucasian children, and is presumed to be transmitted as an autosomal recessive trait (1, 2). This disease is characterized by chronic pulmonary disease and pancreatic insufficiency, both of which may be secondary to a general abnormality in mucous secretions. The basic biochemical defect(s) of cystic fibrosis is unknown. Although the abnormalities are thought by a number of investigators to be expressed in all body tissues including skin fibroblasts in tissue culture (1, 2), there are no reproducibly detected abnormal phenotypes characteristic of cystic fibrosis fibroblasts. The phenomenon of elevated sweat electrolytes is utilized as a diagnostic test for cystic fibrosis, but this "sweat test" is not always reliable or easily interpreted. At present there is no methodology for either detection of "heterozygote carriers" or for prenatal diagnosis of cystic fibrosis.

Abbreviations: VSV – vesicular stomatitis virus; C.F. – cystic fibrosis

Lawrence A. Hunt is presently with the Department of Microbiology, University of Kansas School of Medicine, Kansas City, Kansas 66103
Received March 29, 1977; accepted May 25, 1977.

© 1977 Alan R. Liss, Inc., 150 Fifth Avenue, New York, NY 10011

Because of possible abnormalities in mucous secretion and cell membrane function, a defect in glycoprotein metabolism has been suggested as a possible site of the biochemical defect(s) (1,3). There have been several conflicting reports concerning differences in glycoprotein glycosyltransferases activities between normal and cystic fibrosis tissue (3–5), but there is little definitive evidence for differences in glycoprotein metabolism between normal and cystic fibrosis cells.

We have used the envelope glycoprotein of vesicular stomatitis virus, a lipid-containing animal virus, as a specific in vivo probe of glycosyltransferase activities in normal, "heterozygous carrier," and "homozygous" cystic fibrosis fibroblast cells. Fibroblasts were used even though the expression of the cystic fibrosis defect(s) in these cells is questionable because: i) they can be easily grown in tissue culture, and ii) possible differences observed between normal and cystic fibrosis fibroblasts could potentially be used in a prenatal detection assay. VSV matures by budding through the host cell membranes which have been modified by the insertion of the virus glycoprotein (G) and a nonglycosylated matrix protein (M) into the lipid bilayer composed of host cell lipids and glycolipids (6–12). The virus RNA genome codes for the G polypeptide chain, but host enzymes are responsible for the glycosylation of the virus glycoprotein. The carbohydrate moiety of the G protein has been well characterized for VSV grown in a number of tissue culture lines, and consists of 2 major N-glycosidically linked oligosaccharide side chains per glycoprotein, with the following structure (13–17):

$$\begin{array}{c}\text{fucose}\\|\ (\pm)\end{array}$$
$$\text{ASPN} - \text{GlcNAc} - \text{GlcNAc} - (\text{mannose})_{3-4} \begin{cases} \text{GlcNAc} - \text{gal} \pm \text{sialic acid} \\ \text{GlcNAc} - \text{gal} \pm \text{sialic acid} \\ \text{GlcNAc} - \text{gal} \pm \text{sialic acid} \end{cases}$$

METHODS

Human fibroblast cell lines were obtained from the Human Genetic Mutant Cell Repository at the Institute for Medical Research in Camden, New Jersey: "normal" (GM123), "obligate heterozygous carrier" (GM849), and "homozygous" cystic fibrosis (GM770 and GM1012). Cells were grown in monolayer culture in Eagle's minimal essential medium (MEM) supplemented with 10% fetal calf serum, nonessential amino acids, glutamine, and penicillin/streptomycin (all from Flow Laboratories, Rockville, Maryland).

Stock preparations of VSV (Indiana serotype) were grown in HeLa S3 suspension cultures, purified, and assayed as described previously (18, 19). Confluent cultures of human skin fibroblasts were infected with 10 plaque forming units of VSV per cell. VSV-infected cells were radiolabeled from 4 to 16 h postinfection with 10 μCi/ml D-[6-^{3}H] glucosamine (5–15 Ci/mmole), 10 μCi/ml D-[1-^{3}H] galactose (5–10 Ci/mmole) or 1 μCi/ml D-[1-^{14}C] glucosamine (45–55 mCi/mmole) (all from New England Nuclear Corporation, Boston, Massachusetts) in MEM with 2% serum and one third the normal concentration of unlabeled glucose. Alternatively, cultures were labeled with 10 μCi/ml [^{35}S] methionine (New England Nuclear Corporation; 100 Ci/mmole) in MEM with 2% serum but lacking unlabeled methionine. Virus was harvested from the supernatant medium and purified by equilibrium and velocity sedimentation in preformed sucrose gradients (18).

Radioactive proteins from infected cell membranes (12) or purified virions were analyzed by sodium dodecyl sulfate (SDS)-polyacrylamide gel electrophoresis (11), and dried slab gels were subjected to fluorography (20).

Radioactive sugar-labeled glycoprotein from purified virions or infected cell membranes was digested with pronase (B grade, Calbiochem) and chromatographed on Bio Gel P-4 (Bio Rad Laboratories) as previously described (12). Alternatively, virus glycoprotein was digested with trypsin (TPCK-treated; Worthington) and analyzed by anion exchange chromatography on DEAE-cellulose (Whatman DE-52) (16). Prior to exoglycosidase treatment, pronase glycopeptides were chromatographed on Sephadex G15/G50 (Pharmacia) to remove undigested protein and salts present in the pronase digest mixture. Pronase- or trypsin-digested glycopeptides were treated with neuraminidase (Clostridium perfringens) as described earlier (12).

RESULTS

Analysis of purified virions and the membrane fraction of VSV-infected cells by SDS-polyacrylamide gel electrophoresis indicated that the VSV G protein was the only major glycoprotein which was synthesized in infected cells and assembled into mature virions (Fig. 1), as expected from the shut-off of host protein synthesis by VSV (18). Slight differences in the mobility of G protein from VSV grown in different cell lines (compare C.F. (GM1012) to "carrier" (GM849) and C. F. (GM770)] can be attributed to heterogeneity in the amount of terminal sialic acid (J. Robertson, personal communication). The nonglycosylated VSV proteins had electrophoretic mobilities identical to those of the [^{35}S]methionine-labeled proteins in VSV grown on HeLa cells, based upon the pattern of methionine-labeled proteins from VSV grown on cell line GM849 (Fig. 1) and the pattern of Coomassie Brilliant Blue-stained polypeptides (not shown) for VSV grown on all 4 cell lines.

Detailed analysis of [^{3}H]glucosamine-labeled virus glycopeptides was undertaken to determine possible differences in the carbohydrate moieties which could reflect abnormalities of glycosyltransferases of either "heterozygous carrier" or "homozygous" cystic fibrosis fibroblasts. Pronase-digested glycopeptides from VSV grown on each cell type were subjected to gel filtration on Bio Gel P-4 (Fig. 2). In all 4 cases, the glycopeptides were resolved into 3 major peaks [designated S_1, S_2, and S_3 (12)] and several minor peaks corresponding to undigested material in the void volume and "S_0" glycopeptides eluting just before the S_1 glycopeptides.

Treatment with neuraminidase to remove terminal sialic acid resulted in a single major peak comigrating with the S_3 glycopeptides of untreated glycopeptides with all 4 samples (Fig. 2). Therefore, heterogeneity in terminal sialic acid (see postulated structure in Introduction) was responsible for the multiple glycopeptide peaks, as previously observed for VSV grown in HeLa cells (12) and BHK cells (17). This elution pattern was due to a negative charge exclusion property of Bio Gel resins. These results suggested that the cystic fibrosis cells (both "heterozygous carrier" and "homozygous" cystic fibrosis) have the enzymatic capacity to add both the oligomannosyl core and the branch chains terminating in sialic acid. The differences in the relative amounts of [^{3}H]glucosamine label in the 3 major peaks between the 4 samples were not considered significant with respect to the cystic fibrosis genotype(s) because: i) the difference between the VSV glycopeptides from the 2 cystic fibrosis cell lines (GM1012 and GM770, Fig. 2) was as great as the differences between the "normal" cell line (GM123) and the cystic fibrosis "heterozygous carrier" or "homozygous" cell lines, and ii) these differences were minor compared to the major differences in terminal sialic acid content and glycopeptide distribution on Bio Gel columns previously reported between HeLa-grown VSV and BHK-grown VSV (12–14, 17).

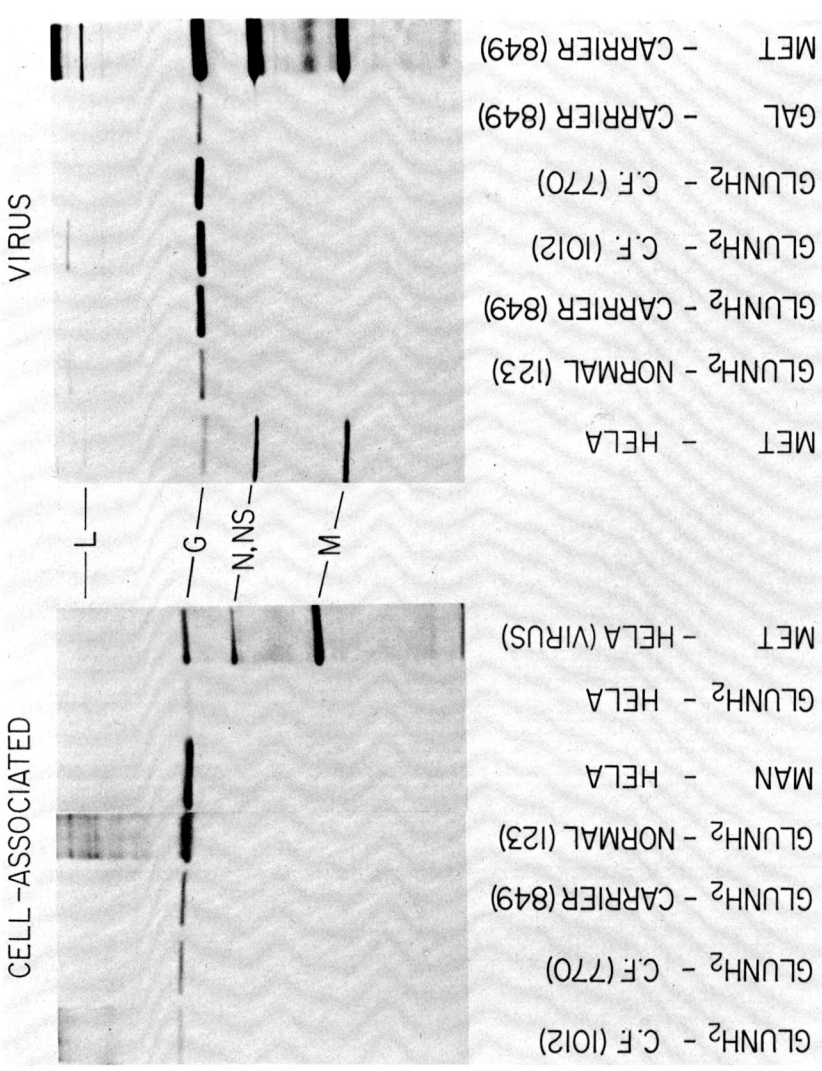

Fig. 1. SDS-polyacrylamide gel profile of residual cell-associated membrane protein and purified vesicular stomatitis virus grown in normal, "homozygous" cystic fibrosis, and "heterozygous carrier" human fibroblasts. The approximate molecular weights of the VSV (HeLa grown) marker proteins are: L, 200,000; G, 67,000; N, 50,000; M, 25,000–30,000 (8).

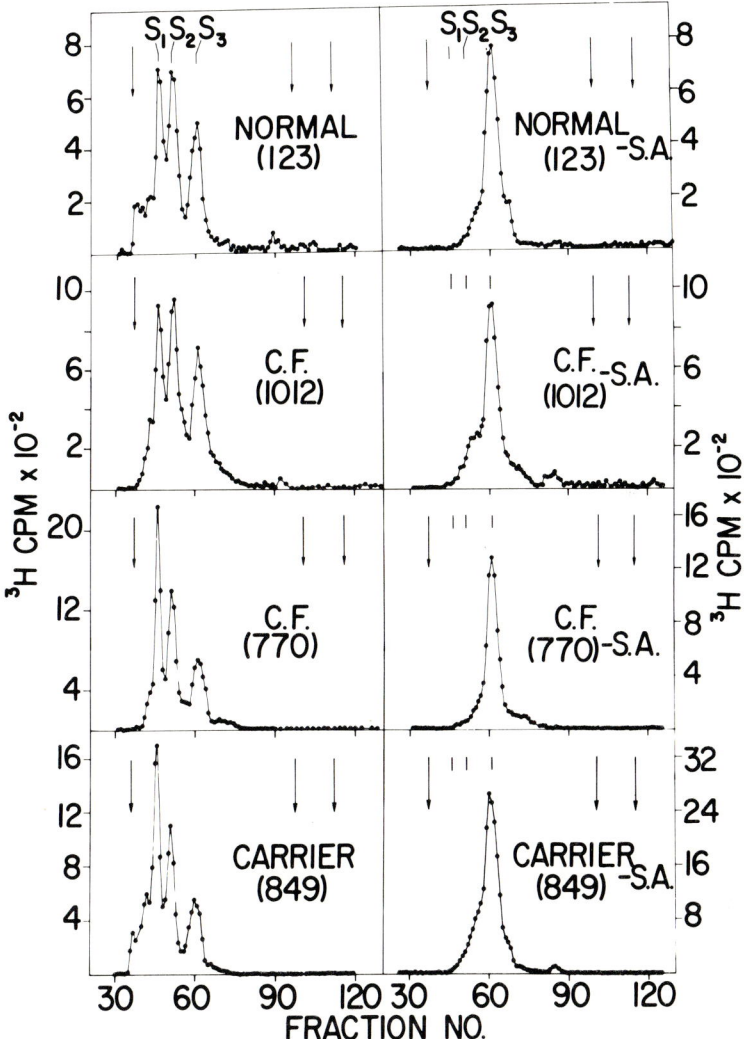

Fig. 2. BioGel P-4 gel filtration of pronase-digested glycopeptides from [^{3}H] glucosamine-labeled VSV grown in normal, "homozygous" cystic fibrosis, and "heterozygous carrier" human fibroblasts. The column markers are blue dextran (void volume, approximately 4,000 mol. wt.), stachyose (666 mol. wt.), and mannose (180 mol. wt.). S_0, S_1, S_2, and S_3 refer to glycopeptide peaks which differ in their content of terminal sialic acid (probably 3, 2, 1, and 0 residues), which can be enzymatically removed by neuraminidase.

The differences in the relative amounts of S_1-S_3 glycopeptides from VSV grown in GM1012, GM770, and GM849 correlated with the slight electrophoretic mobility differences observed for the undigested glycoproteins in Fig. 1.

In these human fibroblast cell lines, [^{3}H] glucosamine did not seem to be a significant precursor of sialic acid, since a major radioactive peak eluting in the position of sialic acid (approximately fraction No. 85) was not detected after treatment of glycopeptides with neuramidinidase. This was in contrast to studies with uninfected Chinese hamster cells in tissue culture (21) and with VSV grown in other mammalian tissue culture cells (12, 16, 17).

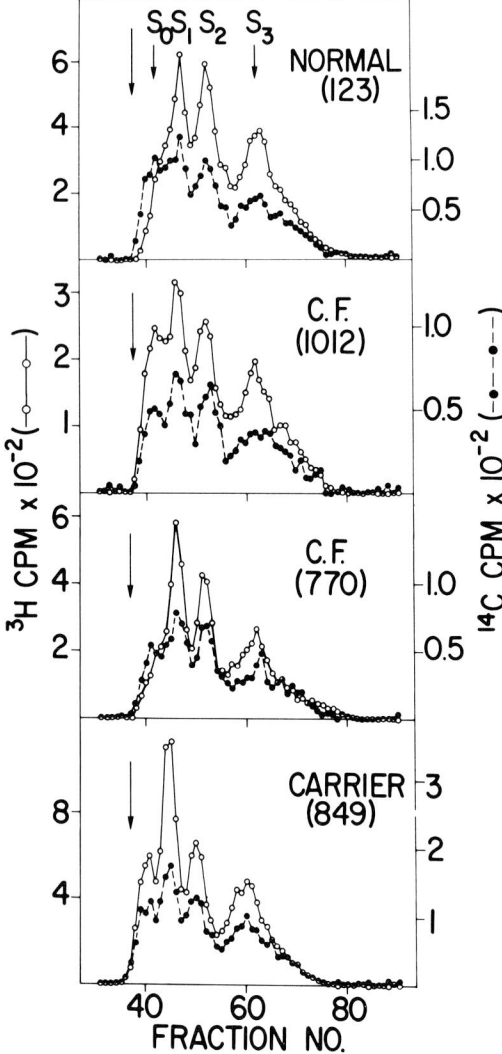

Fig. 3. BioGel P-4 gel filtration of pronase-digested glycopeptides from [^{3}H]glucosamine-labeled cell-associated glycoprotein from VSV-infected normal, cystic fibrosis, and carrier human fibroblasts. The [^{14}C]glucosamine glycopeptides are from VSV-infected normal cells (GM849).

Pronase-digested glycopeptides from residual cell-associated glycoprotein were also examined by gel filtration to rule out the possibility that a major portion of the total glycoprotein was not assembled into virions because it exhibited some deficiency or alternation in glycosylation, whereas the glycopeptides from released VSV were a specific subset of the total which were processed normally. The pattern of glycopeptides of cell-associated G protein was found to be similar to that of virion glycopeptides, with all 4 cell types exhibiting the major S_1-S_3 peaks in addition to variable amounts of S_0 material (Fig. 3). Neuraminidase treatment of the residual cell-associated glycopeptides converted the multiple peaks to the single major S_3 peak (not shown). The decreased resolution of the sialic acid containing peaks (S_0-S_2) and the broader S_3 peak in Fig. 3 compared to

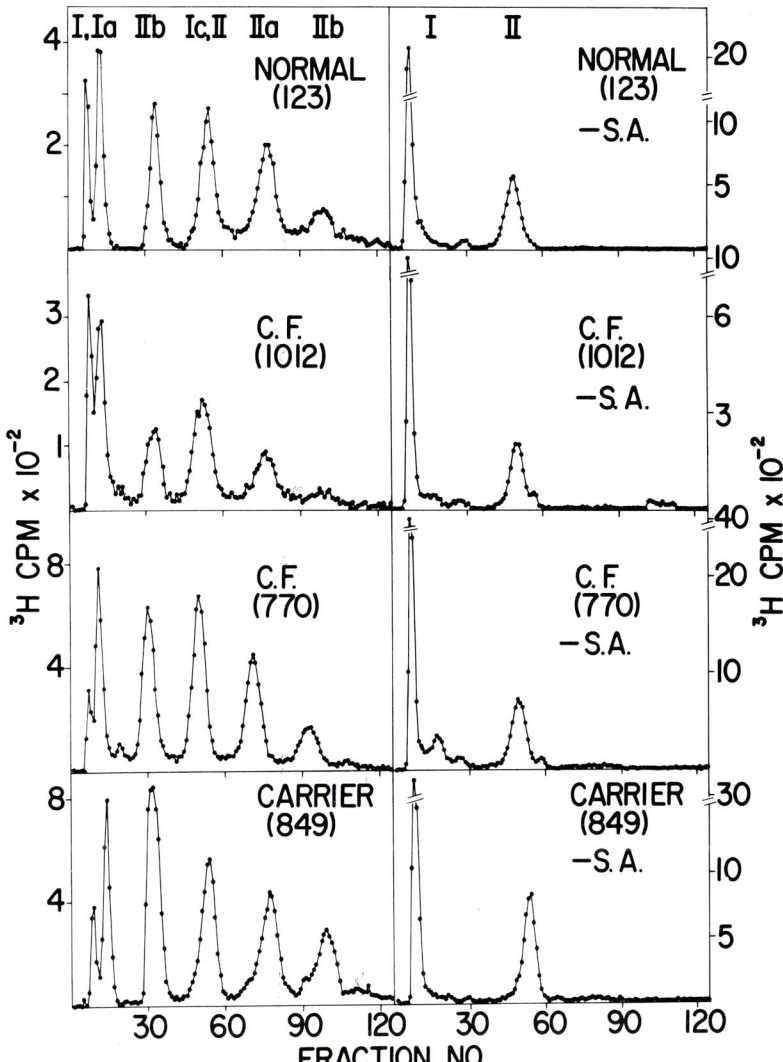

Fig. 4. DEAE-cellulose chromatography of trypsin-digested glycopeptides from [^{3}H]glucosamine-labeled VSV grown in normal, cystic fibrosis, and carrier human fibroblasts. Peptides were eluted with a 0.0–0.1 M NaCl gradient (in 10 mM Tris, pH 8.5) starting at fraction No 21. Neuraminidase treatment eliminates the charge heterogeneity due to terminal sialic acid [converting Ia, Ib, and Ic to I; and IIa, IIb, and (IIc) to II (16)].

virion glycopeptides in Fig. 2 probably could be attributed to the presence of oligosaccharide chains which lack all or part of the branch sugars, as previously described for VSV-infected HeLa cells (12). Other contributing factors could have been the presence of small amounts of heterogeneous, non-VSV glycopeptides and incomplete pronase digestion because of the large amount of unlabeled cellular protein in these samples.

The possibility of a defect in the recognition of the proper glycosylation sites on the VSV G polypeptide was examined by anion exchange chromatography of trypsin-digested glycopeptides before and after neuraminidase treatment (Fig. 4). There were 2 major glycosylated tryptic peptides (designated I and II) regardless of the cell line on which the VSV was grown, as reported earlier for VSV grown in HeLa suspension culture cells (16). Multiple peaks were obtained before neuraminidase treatment because the terminal sialic acid contributed to the charge heterogeneity of the glycopeptide mixture. The amounts of glucosamine label in peak I and peak II were not equal after neuraminidase treatment, and the reason for this inequality is not known. The ratio of radioactivity in peak I versus peak II was approximately 1.5 (ranging from 1.34 to 1.64 for the different cell lines), which was similar to the published value for VSV grown in HeLa suspension cells (16).

DISCUSSION

No significant differences in the oligosaccharide side chains of the VSV envelope glycoprotein have been detected with virus grown in normal human skin fibroblasts or fibroblasts of "heterozygous carrier" and "homozygous" cystic fibrosis patients. Since the VSV glycoprotein is processed by cellular enzymes and has oligosaccharide side chains with structures similar to complex, N-glycosidically linked side chains on cellular membrane and exported glycoproteins, the uninfected cystic fibrosis fibroblasts can be presumed to be normal with respect to the complex oligosaccharides on their own complement of glycoprotein.

If fibroblasts in tissue culture actually exhibit the metabolic defect(s) of cystic fibrosis, then the enzymes involved in the addition of these complex oligosaccharides are apparently not the site of the biochemical defect(s). These studies did not examine the biosynthesis of oligosaccharides which are O-glycosidically linked to threonine or serine, because they are not found in significant amounts in the membrane glycoproteins of enveloped animal viruses. This class of oligosaccharides is found in a large fraction of the mucous glycoproteins, and may be more important in cystic fibrosis than asparagine-linked, complex oligosaccharides (1, 2).

Although these studies were unable to detect any differences which might be correlated with the biochemical defect(s) in cystic fibrosis, the membrane glycoproteins of the lipid-enveloped animal viruses are powerful experimental tools in the investigation of altered glycoprotein metabolism in a wide variety of tissue culture cell lines. The advantages of the virus-infected cell system are twofold: elimination of the problems of i) a heterogeneous mixture of cellular glycoproteins and oligosaccharide side chains, and ii) the nonphysiological nature of cell-free glycosyltransferase assays. For example, the phenotype of tissue culture cells which have been selected for resistance to various plant lectins were reflected in the structure of the oligosacharide side chains which were smaller and lacked part or all of the branch chain sugars (–GlcNAc–gal±sialic acid) (22, 23; M. Robertson et al., manuscript in preparation). Sindbis virus was also used to examine possible defects in fibroblasts from patients with I-cell disease (mucolipidosis II), but no significant differences in virus glycopeptides were observed between Sindbis virus grown on normal or I-cell disease fibroblasts (24).

ACKNOWLEDGMENTS

The authors gratefully acknowledge the expert technical assistance of Ms. Margaret Robertson. This research was supported by Public Health Service grant R01 AI12316-01 from the National Institute of Allergy and Infectious Diseases and by National Science

Foundation Grant BMS 74-21128 A01 to D. F. Summers. L. A. Hunt was a postdoctoral Research Fellow of the Cystic Fibrosis Foundation. D. F. Summers is a recipient of an American Cancer Society Faculty Award.

REFERENCES

1. Sant'Agnese PA, Davis PB: N Engl J Med 295:481, 1976.
2. Wood RE, Boat TF, Doershuk CF: Am Rev Respir Dis 111:833, 1976.
3. Louisot P, Levrat C, Gilly R: Clin Chim Acta 48:373, 1973.
4. Singer L, Crozier D, Moscarello MA: Clin Biochem 7:146, 1974.
5. Butterworth J: Clin Chim Acta 56:159, 1974.
6. Howatson AF, Whitmore GF: Virology 16:466, 1962.
7. Zee YC, Hackett AJ, Talens L: J Gen Virol 7:95, 1970.
8. Wagner RR, Prevec L, Brown S, Summers DF, Sokol F, Macleod R: J Virol 10:1228, 1972.
9. Klenk HD, Choppin PW: J Virol 7:416, 1971.
10. Atkinson PH, Moyer SA, Summers DF: J Mol Biol 102:613, 1976.
11. Hunt LA, Summers DF: J Virol 20:637, 1976.
12. Hunt LA, Summers DF: J Virol 20:646, 1976.
13. Etchison JR, Holland JJ: Virology 60:217, 1974.
14. Etchison JR, Holland JJ: Proc Natl Acad Sci USA 71:4011, 1974.
15. Moyer SA, Tsang JM, Atkinson PH, Summers DF: J Virol 18:167, 1976.
16. Roberston JR, Etchison JR, Summers DF: J Virol 19:871, 1976.
17. Etchison JR, Robertson JR, Summers DF: Virology 78:375, 1977.
18. Mudd JA, Summers DF: Virology 42:328, 1970.
19. Grubman MJ, Summers DF: J Virol 12:265, 1973.
20. Bonner WM, Laskey RA: Eur J Biochem 46:83, 1974.
21. Draemer PM: J Cell Physiol 69:199, 1967.
22. Schlesinger S, Gottleib C, Feil P, Gleb N, Kornfeld S: J Virol 17:239, 1976.
23. Gottlieb C, Kornfeld S: J Biol Chem 251:7761, 1976.
24. Sly WS, Lagwinska E, Schlesinger S: Proc Natl Acad Sci USA 73:2443, 1976.

Cell Surface Carbohydrates of Preimplantation Embryos as Assessed by Lectin Binding

Anna G. Brownell

Laboratory for Development Biology, Ethel Percy Andrus Gerontology Center, University of Southern California, Los Angeles, California 90007

Preimplantation embryos were obtained from the uteri and oviducts of 2 strains of mice, Swiss CD-1 and B_6CBA. After removal of the zona pellucida by treatment with pronase, FITC-lectins were bound to the embryonic cell surfaces at either 4°C or 37°C. Both morula and blastocyst stage embryos bound the following lectins, FITC-ConA, FITC-WGA, FITC-RCA$_{II}$ and FITC-RCA$_I$. No difference in binding was observed between the morula stage and the blastocyst stage within each mouse strain for each specific lectin. However B_6CBA embryos bound less FITC-ConA and FITC-WGA than the corresponding Swiss CD-1 embryos. The topographical arrangement of the lectin receptors was observed to differ between 4°C and 37°C for FITC-Con A, FITC-RCA$_{II}$, and FITC-RCA$_I$. While lectins bound at 4°C showed a pattern of continuous labeling, the same lectin at 37°C showed aggregation of lectin receptors into patches indicating lateral mobility of these receptors within the embryonic cell membranes. In contrast FITC-WGA bound at 4°C and 37°C demonstrated continuous labeling of embryos at both temperatures. FITC-fucose binding protein did not bind to Swiss CD-1 embryos.

The invasiveness of trophoblastic cells of mouse blastocysts was studied by culturing isolated embryos without prior enzyme treatment on reconstituted collagen gels. After 4 days in BME containing only glutamine and bovine serum albumin as supplements, the embryos shed their zona pellucida and implanted into the collagen gel as indicated by zones of lysis in proximity to the embryonic cells when analyzed by scanning electron microscopy.

Key words: cell surfaces, carbohydrates, implantation, lectin binding

Implantation of the mammalian embryo into the maternal uterus occurs at the blastocyst stage of development. Prior to implantation, development of the embryo is characterized by cleavage of the fertilized ovum into successively smaller cells without concomitant cellular growth. Cleavage of the embryo occurs within the zona pellucida, a mucoprotein covering of the embryo which is shed just prior to implantation. At the blastocyst stage, the first morphological evidence of cell differentiation can be recognized

---

Abbreviations: BME – Basal Medium of Eagle; FITC – fluorescein isothiocyanate; ConA – concanavalin A; WGA – wheat germ agglutinin; RCA$_I$ – Ricinus communis agglutinin of mol. wt. 120,000; RCA$_{II}$ – Ricinus communis agglutinin of mol. wt. 60,000.

Received February 21, 1977; accepted June 6, 1977.

© 1977 Alan R. Liss, Inc., 150 Fifth Avenue, New York, NY 10011

by the segregation of 2 cell populations, the inner cell mass which gives rise to the embryo proper and the trophoblast cells which eventually form the embryonic part of the placenta (1). Although the trophoblastic cells are primarily involved in the early stages of implantation and invasion of the uterine wall, this process is necessary for the initiation of further differentiation of the inner cell mass into the various germinal layers, ultimately giving rise to tissues and organ systems typical of the mammalian embryo.

Although implantation has been studied morphologically in a number of animal species (2–6), biochemical events which mediate the process are virtually unknown. The discovery of implantation-associated proteases in uterine secretions of both the mouse (7, 8) and the rabbit (9) only at the time of implantation may be the first biochemical insight into this process. Present available evidence (7, 9) indicates these proteases are synthesized and secreted by the maternal tissues and may be involved in mediating attachment of the blastocyst to the uterus (7). Moreover, implantation probably involves changes in both the embryonic blastocyst as well as the maternal uterus. The first event in implantation, i.e., the attachment of embryonic cells to uterine cells, must involve a cell-cell interaction between these 2 cell types. The nature of this interaction must be specified by the external surface components of the 2 heterotypic cellular membranes. Later events in implantation such as those associated with trophoblastic invasion of the endometrium may be regulated by proteolytic activities either on the surface of, or secreted by the trophoblast cells.

Of the cell surface components most often thought to be involved with cell recognition phenomena, the carbohydrate-protein complexes have been the most extensively studied (10), albeit primarily in nonembryonic cells. One notable exception is the extensive work on embryonic avian cells done in the laboratory of Moscona (11). Studies on the cell surface carbohydrates of mammalian preimplantation embryos are even more limited. Pinsker and Mintz (12) showed that radioactive glucosamine could be incorporated into preimplantation mouse embryo cells in vitro, and that at least one of these components changed with preimplantation development between cleavage and blastocyst stages.

The use of lectin-binding properties of membranes of embryonic cells (11) as well as nonembryonic cells has yielded considerable amounts of information concerning the molecular architecture of animal cell surfaces (see review by Nicolson, Ref. 13). The biological significance of changes in lectin-binding capacities at different functional states such as those associated with transformation are still controversial (14). Nevertheless, recent studies on mammalian ova (15–18) and preimplantation embryos (18) have focused on the use of lectins to compare the carbohydrate components of the zona pellucida with those of the embryo proper as well as how these change during preimplantation development. These studies have demonstrated that most of the monosaccharides common to glycoproteins, i.e., α-D-mannose, N-acetyl-D-glucosamine, β-D-galactose, and N-acetyl-D-galactosamine, are also found on both the zona pellucida and the unfertilized, as well as the fertilized, hamster ovum. These monosaccharides are presumed to exist as part of the oligosaccharide structures of glycoproteins and/or glycolipids. Changes in the amounts of these sugar components, as measured by a difference in quantity of lectin bound and assessed microscopically are difficult to quantitate but seemed to decrease as the embryo reached the blastocyst stage (18).

The present study was undertaken to extend the previous findings (18) and especially to focus on the embryo just prior to implantation. This was done by comparing the fluorescent lectin binding properties of the blastocyst and the morula, the embryonic stage just prior to blastocyst formation. In addition, implantation in vitro was assessed using reconstituted collagen gel substrata and analyzed by scanning electron microscopy.

MATERIALS AND METHODS

Swiss CD-1 mice and the F_1 hybrid of C57BL/6 and CBA strains (hereafter called B_6CBA for brevity) were housed in vivarium facilities and kept on a 12 h dark-12 h light cycle. The females were superovulated by injection with 5 international units (IU) of follicle stimulating hormone (Gestyl^R, Organon, Oss, Holland) followed 48 h later by 4 IU of lutinizing hormone (Antuitrin "S"^{R}, Parke Davis). After the final injection, the females were mated with fertile males. The presence of a vaginal plug the following morning (day 0) was indicative of successful mating. In some experiments animals were mated without prior superovulation. No difference in experimental results was obtained between the 2 classes of embryos. On day 3.5 the animals were sacrificed and the embryos were flushed from the uterus and Fallopian tubes by insertion of a 30 gauge needle into the ovarian end of the tube and flushing isolation medium through the tube and uterus (19). Isolation medium consisted of basal medium of Eagle (BME, Gibco, Grand Island, New York) containing 1% Ficoll (Sigma, St. Louis, Missouri), and 0.3% bovine serum albumin and was also used for holding the embryos until collection was completed. Embryos from a number of females were combined and only morula and blastocyst stages were retained. The zona pellucida was dissolved by incubating the embryos in 0.2% RNAse-free Pronase (Calbiochem, LaJolla, California) in BME containing 0.5% polyvinylpyrollidone (K & K Laboratories, Inc., Plainview, New York) for 10–20 min at 37°C. The embryos were then washed with isolation medium.

While no dye exclusion tests for viability of the embryos were done, all embryos were assumed to be viable at the time of lectin binding. Two morphological observations supported this assumption: first, nonviable embryos were readily discerned and discarded at the time of isolation by arrested developmental stage and/or disintergrating blastomeres; second, patching phenomena, observed at 37°C with some lectins (see Results), requires a living cell.

Labeling of embryos with FITC-lectins (fluorescein isothiocyanate conjugated lectins) was done according to the method of Roberson et al. (20). All FITC-lectins were obtained from Miles-Yeda, Elkhart, Indiana. Control embryos were treated with specific monosaccharides to compete off the FITC-lectin in order to demonstrate specificity of the binding. α-Methylglucoside and D-(+)-galactose were obtained from Aldrich Chemical Co., Milwaukee, Wisconsin. N-Acetylglucosamine was from Nutritional Biochemicals, Cleveland, Ohio. After these incubations the embryos were washed 3 times with isolation medium, fixed in phosphate buffered formalin (4%) containing 1% Ficoll, and mounted on microscope slides. They were examined and photographed using a Zeiss Fluorescence Photomicroscope (light source HB0200; excitation filter KP500 and UG-2, barrier filter 47). Experiments were done at both 37°C and 4°C. Embryos were labeled for various time periods from 10 to 30 min. Within each experimental group all handling of embryos, i.e., lectin binding, washing, and fixing, was done at the indicated temperature.

Embryos were also incubated in a humid atmosphere of 5% CO_2–95% air for 4 days on reconstituted rabbit skin collagen fibrils (21) in BME to which glutamine (20 mM) and 0.3% bovine serum albumen had been added. These embryos were fixed and dehydrated through a graded alcohol series, critical point dried, coated with Au/Pd alloy (60:40) and examined in a Cambridge S4-10 scanning electron microscope.

RESULTS

Swiss CD-1 Embryos Labeled With FITC-Lectins

Figure 1 shows a blastocyst stage Swiss CD-1 embryo which was labeled with FITC-concanavalin A (ConA) at 37°C. Two points are illustrated by this photomicrograph. First the zona pellucida, the acellular mucoprotein covering of the embryo, was heavily labeled throughout. In contrast the embryo proper displayed discrete patches of fluorescent label suggesting that the ConA receptors within the embryonic cell membranes had aggregated at this temperature. To test this hypothesis, 2 sets of embryos were labeled with FITC-ConA, one at 4°C and the other at 37°C. Figure 2a shows that embryos labeled and processed at 4°C displayed an overall pattern of FITC-ConA labeling with no evidence of clusters or patches. When labeling was done at 37°C (Fig. 2b) patches of fluorescent lectin were again seen. Experiments in which lectin was bound at ambient room temperature gave similar results to those done at 37°C. Figure 2c is a photomicrograph of a control embryo which had been labeled with FITC-ConA at 37°C; the labeled lectin was then competed off of the embryo by incubation in 0.4 M α-methyl-D-glucoside, a monosaccharide known to bind tightly to the lectin. While not all of the lectin was displaced by this procedure, the fluorescence decreased drastically indicating that a great proportion of the lectin binding was specific. For all FITC-lectins studied competition experiments, conducted with monosaccharides known to specifically bind the lectin, gave similar results. In all instances the fluorescence due to FITC-lectin bound to the embryos decreased markedly but was never abolished.

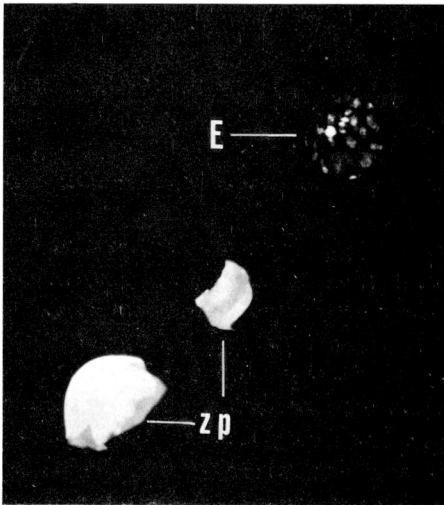

Fig. 1. Photomicrograph of Swiss CD-1 embryo and zona pellucida fragments labeled at 37°C with FITC-ConA. Note that the embryo (E) shows patching of lectin receptors while the zona pellucida (zp) does not. Magnification: 120 ×.

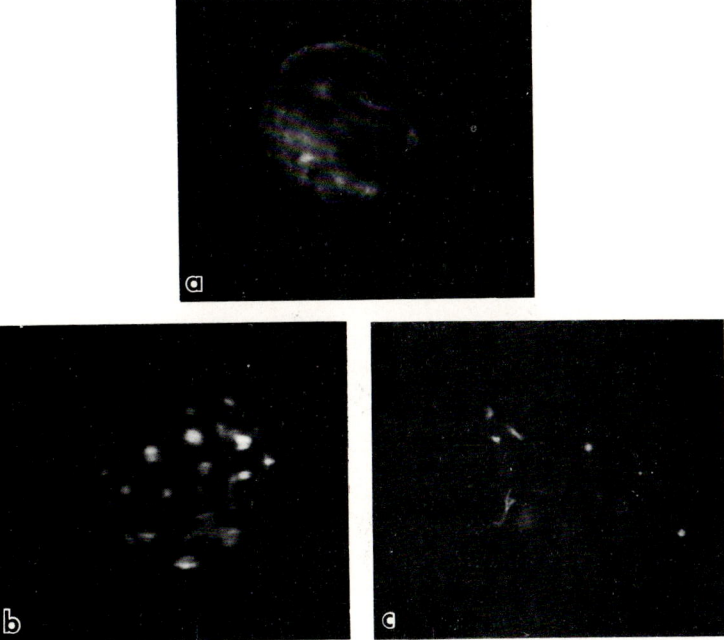

Fig. 2. Composite photomicrographs of Swiss CD-1 embryos treated at 4°C and 37°C with FITC-ConA and control embryo. a) Embryo labeled at 4°C. Note continuous distribution of label. The dark area is due to depth of focus problems in dealing with whole embryos. b) Embryo labeled at 37°C. Note patching of label on surface of embryo. c) Control embryo labeled with FITC-ConA, then incubated in 0.4 M α-methylglucoside to compete off the lectin. Note that only a small amount of fluorescence remains. Magnification: a, b) 270 ×; c) 300 ×.

Experiments in which FITC-wheat germ agglutinin (WGA) was used to label early mouse embryos are represented by the photomicrographs in Fig. 3. Embryos labeled at 4°C such as that shown in Fig. 3a displayed fluorescence on the entire surface of the embryo. In contrast to the results obtained with FITC-ConA labeling, embryos labeled with FITC-WGA at 37°C (Fig. 3b) did not show "patching" or "clustering" of fluorescent lectin-bound receptors. The reason for this inability of WGA to induce capping, patching, or clustering of receptors is unknown but has been noted also in the case of mouse fibroblasts (22).

Other lectins whose sugar specificities are well documented were used in experiments analogous to those just described. With each lectin examined, the results were unambiguous. If the lectin bound to the embryos, all embryos were labeled. In no case was it found that fluorescence was exhibited by only some of the embryos. In addition when patching was observed to occur, all embryos displayed this phenomena. Results of this series of experiments are summarized in Table I. Embryos at both the morula stage and the blastocyst stage were examined but no qualitative differences in the lectin binding patterns were ascertained and the results for both stages are grouped together. The 2 lectins from Ricinus communis beans have been shown to be differentially inhibited (23). The lower molecular weight lectin, RCA_{II} (formerly RCA_{60}) can be inhibited by β-D-galactose- and N-acetyl-D-galactosamine-like residues while the 120,000 molecular weight species, RCA_I is specifically inhibited by β-D-galactose. Both of these

TABLE I. FITC-Lectin Binding to Preimplantation Embryos

Lectin	Concentration (μg/ml)	Number of experiments	Number of embryos	Inhibitory monosaccharides*	Monosaccharide used and concentration	Binding to embryos	Patching
FITC-Concanavalin A	780	4	32	α-D-glucose α-D-mannose	α-methyl-glucoside (0.4 M)	+	+
FITC-Wheat Germ Agglutinin	800	2	23	(N-acetyl-D-glucosamine)$_2$	N-acetylglucosamine (0.4 M)	+	–
FITC-Fucose-binding protein from Lotus tetragonolobus	100 800	2	3	L-fucose		–	
FITC-Ricinus communis Agglutinin-I	575	3	26	β-D-galactose	D(+)galactose (0.05 M)	+	+
FITC-Ricinus communis Agglutinin-II	570	1	5	D-galactose N-acetyl-D-galactosamine	D(+)galactose (0.05 M)	+	+

*See excellent review on this subject (13).

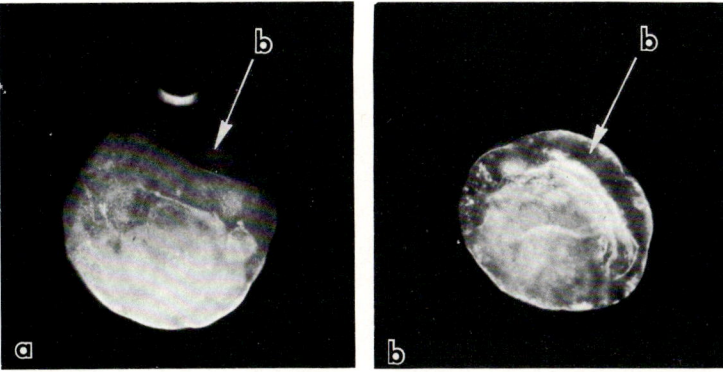

Fig. 3. Photomicrographs of 2 Swiss CD-1 embryos labeled with FITC-WGA at 4°C (a) and 37°C (b). Note that with this lectin the embryo labeled at 37°C does not show patching on cell surfaces in contrast to results obtained with FITC-ConA (see Fig. 1 and 2). The blastocoel, labeled b, a fluid-filled cavity characteristic of the blastocyst stage embryo is partially collapsed causing invagination of an area of the embryo. Magnification: 270 ×.

lectins bound to cell surfaces of the embryos and both lectins showed patching behavior when the labeling experiments were conducted at 37°C.

B_6CBA Embryos Labeled With FITC-Lectins

Additional experiments were done with B_6CBA embryos. The lectins used for binding to these embryos were FITC-ConA and FITC-WGA. Both of these lectins bound to B_6CBA embryos and embryos labeled at 37°C with FITC-ConA displayed patching phenomena. Shown in Fig. 4 is an embryo labeled with FITC-ConA. When compared to Swiss CD-1 embryos, fluorescence of B_6CBA embryos due to the FITC-lectin bound was diminished on the surfaces of these embryos (compare Fig. 4 with Fig. 1). Although these results are qualitative because of the method of assessment of lectin binding, i.e., microscopic examination, the implication of this finding is that strain specific differences in lectin binding capacity may exist. Quantitation of the amount of lectin bound to these embryos would aid in exploring this possibility.

Fig. 4. Photomicrograph of a B_6CBA embryo labeled at 37°C with FITC-ConA. Note the decreased fluorescence of this embryo compared to Swiss CD-1 embryos labeled with the same lectin. (Compare with Fig.1 or Fig. 2a, b). Magnification: 120 ×.

Culture of Swiss CD-1 Embryos on Reconstituted Collagen Gels

Embryos were flushed from the uterus and placed onto collagen gels without prior enzyme treatment. The zona pellucida was still intact after 14 h of incubation. After 4 days of incubation all of the embryos had shed their zona pellucida and implanted into the collagen gel. Figure 5 shows a scanning electron micrograph of 2 embryos and the zone of lysis around the cells which attached to the collagen gel. Zones of lysis such as this were not observed in the control collagen gel. i.e., gel incubated under exactly the same conditions except without embryos. In addition no lysis of the collagen gel was observed in the same dish as the embryos in any area not containing an embryo. Lysis was only observed in areas confluent with embryos. This indicates that the activity which lead to gel lysis was closely associated with the embryo and not disseminated throughout the culture medium.

Also demonstrated in Fig. 5 is the fact that cells of the embryo at this stage of culture are not uniform with regard to either size or surface topography. Very small cells as well as much larger ones were seen in all embryos examined. Smooth-surfaced and rough-surfaced cells were seen on all embryos probably reflecting various stages of the cell cycle (24). The significance and origin of the translucent bridge connecting the 2 embryos

Fig. 5. Low magnification scanning electron micrograph of 2 embryos "implanted" into a gel of reconstituted collagen fibrils. Note the variety of cells found in these embryos at this stage. A zone of lysis (indicated by arrowheads) can be seen around each embryo. C) collagen fibrils; E) embryo. Magnification: 1,000 ×.

is unknown. Figure 6 is a higher magnification scanning electron micrograph which illustrates the great profusion of microvilli on blastocyst cell surfaces. Of particular interest is the area in which knob-like projections rather than slender microvilli are seen. The significance of this surface specialization remains to be determined.

DISCUSSION

Binding of lectins to surfaces of mammalian embryos is a powerful tool for elucidating the chemical nature of carbohydrate moieties. This information would be virtually impossible to obtain by other means due to scarity of biological material. The use of lectins for which specificity of binding has been documented is essential for interpretation of results. While early studies on binding sites indicated that lectins bound primarily to terminal sugars of an oligosaccharide chain, more recent studies have demonstrated that some lectins bind to sugars within the oligosaccharide moiety (13). From the lectin binding experiments reported here, it can be concluded that α-D-glycosides (such as α-D-glucose and α-D-mannose), N-acetyl-D-glucosamine (singly or as the dimer), β-D-galactose and/or N-acetyl-D-galactosamine are all present, presumably as constituents of glycoproteins and

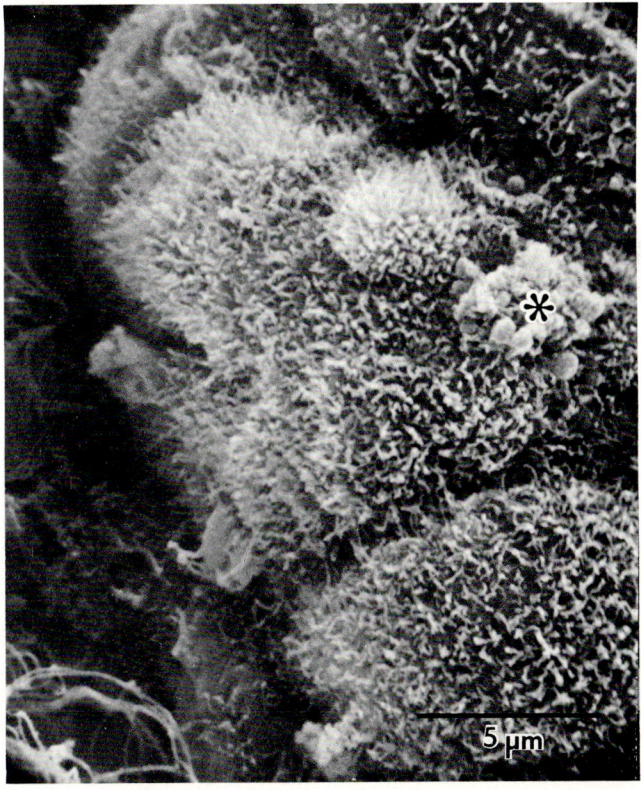

Fig. 6. Higher magnification scanning electron micrograph of a few cells of an embryo after 4 days in culture on reconstituted collagen fibrils. Note the profusion of microvilli, some of which are knob-like (*). Magnification: 5,000 ×.

glycolipids, on cellular surfaces of preimplantation embryos. These results agree with previously reported lectin studies from the laboratory of Nicolson (16–18) who along with his colleagues demonstrated that these lectins bound to both the zona pelucida (16, 17) and unfertilized egg plasma membranes (16) of hamster, mouse, and rat as well as to hamster preimplantation embryos (18). The present experiments further extend these data by focusing in depth on 2 later stages of preimplantation development, the morula stage and the blastocyst stage which contains about 64 cells (25).

In all stages of preimplantation embryos studied, no evidence for the existence of L-fucose-like residues could be obtained. Neither hamster zona pellucida, unfertilized and fertilized eggs (18), or mouse morula and blastocyst embryos bound lectins specific for this sugar. These results indicate that L-fucose may be absent or present in low amounts that cannot be detected by the present procedure. In this regard early avian embryos (blastoderm stage) showed no evidence of L-fucose-like residues (26).

Lateral mobility of membrane receptors has been postulated to account for the patching and capping phenomena observed when cells are incubated at physiological temperatures with molecules which can bind to the cell surface and at the same time form oligomeric structures (13). Tetrameric lectins such as ConA bind to sugar residues on cell surfaces; each subunit of the lectin can bind a sugar residue. Reaction of the multivalent lectin with sugar moieties on different molecules leads to aggregation of receptors yielding patches or caps (see review, Ref. 13). In the present study patching was observed to occur with ConA, RCA_{II}, and RCA_{I}. Thus preimplantation embryonic cells are similar to mature animal cells in this regard. Of importance is the fact that an energy dependent system, i.e., a living cell, is necessary for this phenomena to occur. This was strikingly demonstrated in Fig. 1 in which the zona pellucida, a nonvital extracellular coat, while labeling intensely did not show patching. In contrast to results obtained with ConA and RCA lectins, embryos labeled with WGA did not display lateral mobility of receptors for this lectin: embryos labeled at both $4°C$ (Fig. 2a) and $37°C$ (Fig. 2b) showed a continuous pattern of label. These results are in conflict with reports which indicated that this lectin induced patching of WGA-receptors on mammalian eggs (16, 18). However, mouse fibroblasts labeled with WGA did not show clustering of receptors (22). One reason for these discrepancies could be developmental — the unfertilized egg may have WGA-receptors whose mobilities are greater than those found within blastocyst stage embryos or mature cells.

Qualitative differences in the lectins bound or the labeling patterns at $4°C$ and $37°C$ between embryos from different mouse strains were not apparent. Species-specific differences between hamster, rat, and mouse zona pellucida have been reported (17). While all zona pellucida bound all the lectins examined, the quantity of lectin necessary to induce agglutination differed between the 3 species. This result may imply a quantitative difference in the number of lectin receptors present on zona pellucida from different animals. While microscopic examination is not a quantitative method for evaluating amounts of lectin bound, preliminary results from the present experiments suggest that different strains of mouse embryos may have different lectin-binding capacities. Whether these differences reflect genetic variance remains to be demonstrated.

While earlier studies found no developmental differences in lectin-binding properties between the fertilized egg, the 2-cell stage, and the 8-cell stage a decrease in lectin binding was noted at the blastocyst stage (18). As many biochemical events occur during development of the embryo from cleavage stages to the blastocyst stage, it was of interest to explore the question of whether this decrease in lectin binding occurred precipitously at the blastocyst stage or if it was a more gradual change. The experiments reported here demonstrated that both the blastocyst embryo and the morula stage embryo bound the same amounts of lectin as assessed microscopically. These results taken together with the previous ones indicate that decrease in lectin binding is a gradual change that occurs between the 8-cell stage and the blastocyst stage. This change in the number of binding sites for lectins could reflect either a change in total number of sites synthesized or masking of specific sites by the addition of other sugars to the oligosaccharide chain. Alternatively, degradation of binding sites by developmentally regulated glycosidases could account for the observed decrease in lectin receptors.

The biological significance of changes in the lectin-binding properties during preimplantation development of mammalian embryos remains to be determined, although it has been speculated that these changes are related to alterations in embryonic cell surfaces necessary for implantation (18). Perhaps unmasking of certain sugar residues on the embryo is necessary for recognition by cells of the endometrium. The in vitro implantation system described in this report should aid in elucidating the biochemistry of implantation and what role, if any, lectin receptors play in this important biological event.

The scanning electron micrographs demonstrate that collagen, one of the components of the endometrium, can serve as suitable substrata for culture of early mouse embryos (27) and for studying implantation in vitro. The finding of discrete areas of lysis surrounding the "implanted" embryos suggests that proteolytic activities arising from the trophoblast cells may be involved in this biological phenomena. The fact that no other areas of lysis were observed in this culture suggests that this activity is severely restricted perhaps by being bound to the cell membranes of the embryo.

ACKNOWLEDGMENTS

I wish to express my gratitude to Dr. Harold Slavkin for his encouragement of these studies and for the use of his research facilities. In addition I thank Drs. Lajos Pikó and Joseph Bonner for aid in the early phases of this work. The expert technical assistance of Ms. Peggy Wilson and Ms. Carol Ling was deeply appreciated during these studies. For photographic assistance, thanks to Mr. Pablo Bringas and for the typing of the manuscript to Ms. Joanne Leynnwood. This research was supported by a postdoctoral fellowship, 5 F22 DE01631, and research grants DE-03569 and DE-02848 from the National Institutes of Health.

REFERENCES

1. Davies J, Hesseldahl H: In Blandau RJ (ed): "The Biology of the Blastocyst." Chicago: University of Chicago Press, 1971, pp 27–48.
2. Denker H-W: J Embryol Exp Morphol 32:739, 1974.
3. Ljungkvist I, Nilsson O: J Endocrinol 60:149, 1974.
4. Parkening TA: Anat Embryol 147:293, 1975.
5. Parkening TA: J Anat 121:161, 1976.
6. Sinha AA, Mead RA: Am J Anat 145:331, 1976.
7. Pinsker MC, Sacco AG, Mintz B: Dev Biol 38:285, 1974.
8. Sacco AG, Mintz B: Biol Reprod 12:498, 1975.
9. van Hoorn G, Denker H-W: J Reprod Fertil 45:359, 1975.
10. Roseman S: In Moscona AA (ed): "The Cell Surface in Development." New York: John Wiley & Sons Inc., 1974, pp 255–271.
11. Moscona AA: In Moscona AA (ed): "The Cell Surface in Development." New York: John Wiley & Sons Inc., 1974, pp 67–99.
12. Pinsker MC Mintz B: Proc Natl Acad Sci USA 70:1645, 1973.

13. Nicolson GL: Int Rev Cytol 39:89, 1974.
14. Düzgünes N: Biosystems 6:209, 1975.
15. Johnson MH, Eager D, Muggleton-Harris A, Grave HM: Nature (London) 257:321, 1975.
16. Nicolson GL, Yanagimachi R, Yanagimachi H: J Cell Biol 66:263, 1975.
17. Oikawa T, Yanagimachi R, Nicolson GL: J Reprod Fertil 43:137, 1975.
18. Yanagimachi R, Nicolson GL: Exp Cell Res 100:249, 1976.
19. Rafferty KA Jr: "Methods in Experimental Embryology of the Mouse." Baltimore: Johns Hopkins Press, 1970, pp 29–32.
20. Roberson M, Neri A, Oppenheimer SB: Science 189:639, 1975.
21. Chandrakasan G, Torchia DA, Piez KA: J Biol Chem 251:6062, 1976.
22. Roos E, Temmink JHM: Exp Cell Res 94:140, 1975.
23. Nicolson GL, Blaustein J, Etzler ME: Biochemistry 13:196, 1974.
24. Porter K, Prescott D, Frye J: J Cell Biol 57:815, 1973.
25. Webb FTG, Surani MAH: In Talwar GP (ed): "Regulation of Growth and Differentiated Function in Eukaryote Cells." New York: Raven Press, 1975, pp 519–522.
26. Zalik SE, Cook GMW: Biochim Biophys Acta 419:119, 1976.
27. Hsu Y-C, Baskar J, Stevens LC, Rash JE: J Embryol Exp Morphol 31:235, 1974.

Biochemical Characteristics, Metabolism, and Antitumor Activity of Several Acetylated Hexosamines

R. J. Bernacki, M. Sharma, N. K. Porter, Y. Rustum, B. Paul, and W. Korytnyk

Department of Experimental Therapeutics, Roswell Park Memorial Institute, New York State Department of Health, Buffalo, New York 14263

We have synthesized several potential inhibitors and/or modifiers of the carbohydrate portion of plasma membrane glycoconjugates. These include fluorinated and actylated analogs of D-glucosamine, D-galactosamine, and D-mannosamine. These compounds have been tested to determine their effects on both [^{14}C]glucosamine and [^{3}H]leucine incorporation into glycoconjugate and on cell growth and viability using P-288 murine lymphoma cells maintained in tissue culture. The most cytotoxic agent tested was 2-acetamido-2-deoxy-1,3,4,6-tetra-O-acetyl-β-D-glucopyranose or simply β-pentaacetylglucosamine which prevented cell growth at $10^{-4} - 10^{-3}$ M. β-Pentaacetylglucosamine cytotoxicity was correlated with its high lipid solubility, having an octanol/water partition coefficient of 0.424 as compared with 0.278 for the α-anomer and 0.017 for N-acetylglucosamine. In vitro metabolism studies with [^{14}C]- and/or [^{3}H]-labeled pentaacetylglucosamine have indicated intracellular de-O-acetylation leading to the biosynthesis of UDP-N-acetylglucosamine, followed by the incorporation of this sugar into cellular glycoprotein. Concomitant with the formation of increased amounts of this nucleotide sugar, intracellular UTP and CTP pools fell to one third normal within 3 h after the administration of 1 mM pentaacetylglucosamine. At present it is unclear whether the cytotoxicity of β-pentaacetylglucosamine or other similar agents is due to alterations in nucleotide and nucleotide-sugar pools causing a decrease in energy charge and polynucleotide biosynthesis or is due to a direct effect on membrane glycoconjugate biosynthesis.

Key words: glucosamine, glycoproteins, chemotherapy, nucleotide sugars, ribonucleotide pools, lymphoma

A large body of evidence exists describing differences in cellular plasma membrane composition that develop following oncogenic transformation by viruses or carcinogens (1, 2). One of the major constituents of the membrane that is altered after transformation is the membrane glycoconjugate (3, 4). These membrane glycoconjugates have been implicated in the control of cell division, intercellular associations, and metastasis (5–9). These implications are based on the observations that changes in the biochemical and organizational structures of membrane-bound carbohydrates or related membrane-bound glycosyltransferases (10) occur during cell aggregation, mitotic division, cell surface capping, malignant transformation, and in response to stimuli such as contact inhibition of cell movement. Therefore it may be therapeutically advantageous to alter or inhibit the biosynthesis

Received March 7, 1977; accepted for publication June 14, 1977.

© 1977 Alan R. Liss, Inc., 150 Fifth Avenue, New York, NY 10011

of these tumor cell surface constituents. This might result in 1) tumor cell death caused by the inhibition of the biosynthesis of vital membrane components, 2) a change in tumor cell surface architecture leading to a reversion in cellular behavior to a more normal state, or 3) an increase in tumor cell immunogenicity due to the incorporation of sugar analogs into cell glycoconjugates.

Initial studies aimed at accomplishing these goals have utilized high concentrations of naturally occurring membrane sugars in hopes of creating a feedback inhibition on the biosynthesis of membrane glycoconjugates (11). The daily in vivo administration of large amounts of D-glucosamine to mice bearing experimental tumors resulted in an inhibition of tumor growth (12) while a continuous intravenous infusion of L-fucose resulted in a decrease in mammary tumor size in rats (13). The tumor toxicity of glucosamine has been attributed to a build-up of high concentrations of UDP-N-acetylhexosamines with a concomitant drop in intracellular UTP and ATP pool sizes (14). The effects of fucose on tumor growth have been attributed to an increase in tumor macromolecular-bound fucose. This increase is thought to effect a change in cell to cell communication resulting in a type of contact inhibition of tumor cell growth (15).

In addition to these studies using naturally occurring membrane sugars as antitumor agents, several reports have appeared describing the antitumor activity of various fluoro sugar analogs. 6-Deoxy-6-fluoro-D-glucose and 2-deoxy-2-fluoro-D-glucose affected leukemic cell growth in vitro but had little effect on tumor cell growth in vivo. The cytotoxic effects of these compounds may have been due to their inhibition of hexokinase since both were found to inhibit yeast hexokinase (16). Another fluoro sugar analog, 2-deoxy-2-(2-fluoroacetamido)-2-D-glucopyranose, has been shown to affect glycoprotein biosynthesis (17, 18) and increase the median life span of mice inoculated with L1210 leukemia by 25% (19). The only other sugar analog reported to have some antitumor activity was 2-deoxy-D-glucose (DOG), a glucose, mannose or, glucosamine analog.

DOG has been shown to lower intracellular adenine nucleotide pools (20) and to inhibit glycosylation in viruses (21) and mammalian cells (22). It has also been shown to be cytotoxic to tumor cells at low concentrations (23) and to alter cell surface architecture as detected by differences in plant-lectin induced agglutination (24). Along with these findings there have been reports that DOG is metabolized to a nucleotide-sugar and incorporated into membrane glycoconjugate (25).

We have extended these studies by synthesizing several fluorinated or acetylated sugar analogs such as 2-acetamido-2-deoxy-1,3,4,6-tetra-O-acetyl-β-D-glucopyranose or simply pentaacetylglucosamine (PAG), and assessing their antitumor activity, metabolism, and effects on ribonucleoside phosphate pools and on macromolecular biosynthesis in P288 murine lymphoma cells, in vitro.

METHODS

Syntheses of the Hexosamine Analogs

2-Trifluoroacetamido-2-deoxy-D-glucopyranose (F_3-GlcNac) was prepared according to Wolfrom and Conigliaro (26).

2-Acetamido-2-deoxy-1,3,4,6-tetra-O-acetyl-β-D-glucopyranose (β-PAG) was prepared by the method of Horton (27).

2-Acetamido-2-deoxy-1,3,4,6-tetra-O-acetyl-α-D-galactopyranose (PAGAL) was prepared by the method of Stacey (28).

2-Acetamido-2-deoxy-1,3,4,6-tetra-O-acetyl-β-D-mannopyranose (PAMAN) was prepared by the method of Levene (29).

Radioactive Syntheses

[U-^{14}C]-2-Acetamido-2-deoxy-1,3,4,6-tetra-O-acetyl-α-D-glucopyranose [^{14}C-uniformly labeled on the sugar carbons]. A solution of D-[U-^{14}C]glucosamine hydrochloride, (0.049 mg, specific activity 232 mCi/mmole) in ethanol-water, 0.5 ml, (supplied by New England Nuclear Corporation, Boston, Massachusetts) was transferred to a round bottom flask and evaported to dryness on a rotary evaporator and then dried in vacuo over phosphorus pentoxide. The residue was dissolved in dry pyridine (0.7 ml) and mixed with anhydrous sodium acetate (2 mg) to convert the hydrochloride to the free base. After 15 min acetic anhydride (0.2 ml) was added, mixed thoroughly, and left overnight. Ice (0.5 g) was added and after 10 min the solution was evaporated to dryness in vacuo until free from pyridine. The residue was extracted several times with chloroform solution and evaporated to dryness. The product was further purified by thin layer chromatography over silica gel (HF-254, E. Merck, Darmstadt, Germany) by developing the plate with methanol:ethyl acetate (1:9). The product (R_f 0.65) was extracted from the scrapings of the silica gel zone between R_f of 0.5 and 0.8 with methanol:chloroform (1:9). After removal of the solvent the residue was taken up in chloroform, filtered and evaporated to dryness to afford pure [U-^{14}C]-2-acetamido-2-deoxy-1,3,4,6-tetra-O-acetyl-D-glucopyranose. Finally it was dissolved in ethanol (95%, 0.5 ml) as a stock solution.

In order to determine optimal synthetic conditions and the anomeric nature of the product, the synthesis was carried out with cold D-glucosamine hydrochloride and its NMR spectrum determined. The NMR spectrum indicated the presence of only the α-anomer in the reaction mixture as determined by the position and coupling constants of the anomeric proton (27). Further confirmation was obtained by the crystallization of the oil and comparison of its mp with an authentic sample of "α-PAG."

2-Acetamido[N-acetyl-1-^{14}C]-2-deoxy-1,3,4,6-tetra-O-acetyl-[O-acetyl-^{3}H]-α-D-glucopyranose. A solution of D-[1-^{14}C]-N-acetylglucosamine (0.049 mg, specific activity 31.6 mCi/mmole) in ethanol-water, 0.5 ml (supplied by New England Nuclear Corporation), was evaporated to dryness in a round bottom flask on a rotary evaporator and then dried in vacuo over phosphorus pentoxide. The residue was dissolved in dry pyridine (0.5 ml) and [^{3}H] acetic anhydride (2.56 mg, specific activity 400 mCi/mmole, as an 80% benzene solution supplied by New England Nuclear Corporation) and cold acetic anhydride (0.1 mg) was added. The thoroughly mixed solution was left at room temperature overnight. Ice (0.5 g) was added and after 15 min the solution was evaporated to dryness in vacuo. A trap of solid KOH was inserted between the flask and the pump. The residue, freed from acetic acid and pyridine, was extracted several times with chloroform. The chloroform solution was evaporated to dryness and purified by thin layer chromatography as described in the previous experiment. A stock solution was made by dissolving the compound in ethanol (95%, 0.5 ml). The ratio of ^{3}H:^{14}C was found to be 3:1.

Octanol/Water Partition Coefficients

One hundred milligrams of the compound were added to octanol (20 ml) and water (20 ml). The mixture was shaken in a separatory funnel and allowed to stand for 48 h at

21°C to reach equilibrium. The resulting water solution (19.0 ml) was evaporated to dryness and the solid dried over P_2O_5 to a constant weight. The concentration of the solute in octanol was determined by the difference and the ratio of octanol to water concentrations was calculated.

Cell Cultures

Murine P288 lymphoma cells were maintained as an ascites tumor in DBA/2J female mice. Periodically cells were aseptically removed from mice, washed twice in RPMI 1640 medium, and cultured in RPMI 1640 containing 10% fetal calf serum. These cultures were grown in a 90% air/10% CO_2 incubator. These cells maintained their viability in culture and in log phase had a generation time of approximately 13 h. Cells were maintained in culture for periods of 1–3 months whereupon new in vitro cultures were initiated from the in vivo line. In this manner some constant selective genetic pressure was maintained on the cell line.

Generally, cell cultures were initiated at concentrations of $5 \times 10^4 - 1 \times 10^5$ cells/ml. These cultures grew logarithmically to about 2×10^6 cells/ml without the addition of fresh medium. Routinely these cell cultures were fed fresh medium containing 100 µg/ml neomycin every 2–3 days. Mycoplasma contamination was monitored by using the fluorescent dye, 4′,6-diamidino-2-phenylindole, (DAPI), provided by Professor Otto Dann of the Institut für Angewandte Chemie der Friedrich-Alexander-Universität, Schuhstrasse 19, 852 Erlangen, West Germany, as an indicator of plasma membrane associated nucleic acids found in mycoplasma (30). No indication of mycoplasma contamination was evident. Mycoplasma contaminated TA3 mammary cells were used as a positive control. They were provided by Drs. J Laskin and M. T. Hakala.

Biological Testing Systems

A. Cell growth and viability. For routine drug testing P288 murine leukemic cells were suspended at approximately 10^5 cells/ml in fresh RPMI 1640 minus glucose containing 10% fetal calf serum. One milliliter aliquots were then transferred to disposable polyethylene test tubes and placed in the CO_2 incubator. One hour later sugar analogs were added to a final concentration of 1 mM (or as otherwise stated). Cell growth and viability were monitored at later times by use of a Coulter cell counter and Trypan blue dye exclusion, respectively.

B. Macromolecular biosynthesis. [^{14}C] Glucosamine, 2 µM (1.1×10^6 DPM, Amersham Corporation, Arlington Heights, Illinois, and [^{3}H] leucine, 0.37 mM, (2.6×10^6 DPM, New England Nuclear Corporation) were added to cell cultures to assess the effects of sugar analogs on glycoprotein and protein biosynthesis. Two, five, or twenty-four hours later incubations were terminated by the addition of 2 ml of 10% trichloroacetic acid (TCA). The acid insoluble radioactivity was washed twice with 10% TCA and the resulting pellet was dissolved in NaOH and its radioactivity quantitated by scintillation counting methods (31).

Metabolism of 2-acetamido-2-deoxy-1,3,4,6-tetra-O-acetyl-α-D-glucopyranose (PAG)

The metabolism of PAG in P288 cells was studied utilizing freshly prepared [U-^{14}C] PAG prepared by acetylation of D-[U-^{14}C] glucosamine hydrochloride (232 mCi/mmole, NEN) or by use of penta[^{3}H]acetyl[^{14}C]glucosamine, prepared by acetylation of D-[1-^{14}C]-N-acetylglucosamine (31.6 mCi/mmole, NEN) with [^{3}H]acetic anhydride (400 mCi/mmole, NEN) as previously described.

The incorporation of these radiolabeled PAG analogs into glycoconjugate was monitored as previously described. Chromatographic analysis of intracellular soluble radioactivity was performed on both chloroform and ethanol extracts of cells incubated in the presence of both these radiolabeled analogs. S + S orange ribbon paper was spotted with aliquots of these extracts and developed in pyridine:HAc:EtAc:H_2O (5:1:5:3). Radioactivity was detected by scintillation counting methods (31).

Ribonucleotide Pool Size Analysis

Ribonucleotide pool size analysis was performed on P288 cells treated with 1 mM glucosamine, α-PAG, β-PAG, N-acetylglucosamine, 2-trifluoroacetamido-2-deoxy-D-glucopyranose or 2-deoxyglucose for various periods of time. Approximately $5 \times 10^6 - 10^7$ treated cells were washed twice with 2 ml of RPMI 1640 and extracted with 6% perchloric acid (100 μl/10^7 cells) and centrifuged at 800 × g for 2 min. The supernatant (acid soluble fraction) was neutralized with 2 N KOH to a pH of 7.0. Formed precipitate was removed by centrifugation and an aliquot of the resulting supernatant was analyzed for its ribonucleotide content using a Dupont 830 high pressure liquid chromatographic system equipped with a 254 and 280 nm detector. Ten microliters of the supernatant was absorbed onto an ABX column (1 m × 2 mm, U shaped) and eluted at room temperature using a phosphate buffer gradient (0.0025 M, pH 3.0 to 0.5 M) at a flow rate of about 1 ml/min (Y. Rustum, unpublished results). The retention time of each cellular peak was compared to the retention times of a mixture of authentic ribonucleotide and nucleotide-sugar standards. Each peak was identified and automatically integrated on a Spectra Physics Model 23,000-010 Autolab Minigrator. The sum total of all peak areas was also recorded.

RESULTS

Effects of Glucosamine Analogs on Tumor Growth and Synthesis

P288 cells incorporated 4–5 times more [^{14}C]glucosamine in medium depleted of glucose (Table I). Incorporation increased more than fivefold from 864 to 4,815 dpm in a 5-h incubation and more than fourfold in a 24-h incubation from 4,815 to 21,030 dpm. The viability, growth, and incorporation of leucine by cells maintained in RPMI 1640 minus glucose plus 10% fetal calf serum remained at approximately the same level as cells grown in RPMI 1640 containing 2 mg/ml glucose plus 10% fetal calf serum. Cells maintained in the complete absence of glucose (none provided by serum) stopped dividing within 24 h and their viability dropped to 79%. Therefore, all further testing was performed in RPMI 1640 minus glucose plus 10% fetal calf serum.

The effects of the addition of various hexosamine analogs (1 mM) on cell growth, viability, and the incorporation of glucosamine and leucine was investigated (Table II). It was found that most of the compounds were relatively noncytotoxic with the exception of β-PAG (III) which reduced cell number to one half and cellular viability to 83% within 24 h. Most of the other compounds eventually reduced cell number within 48 h as compared with the control (Fig. 1) with the exception of N-acetylglucosamine (II) which had no effect on cell growth or viability (not shown). All the pentaacetylhexosamines [α-and β-pentaacetylglucosamine (III and IV), pentaacetylgalactosamine (V) and pentaacetylmannosamine (VI)] and 2-trifluoro-N-acetylglucosamine (VII) specifically depressed

TABLE I. P288 Cell Growth and Incorporation of Glucosamine and Leucine in Glucose Free Medium

Medium[a]	Incubation time (h)	Cell number (cells/ml × 10^{-5})	Incorporation[c] (dpm)	
			[^{14}C]glucosamine	[^{14}C]leucine
RPMI 1640 + 10% FCS	5	1.18	864	5,313
RPMI 1640 + 10% FCS	24	2.87	4,815	37,582
RPMI 1640 + 10% HI dialyzed FCS	24	2.57	5,242	38,742
RPMI 1640 minus glucose + 10% FCS	5	1.13	4,573	4,914
RPMI 1640 minus glucose + 10% FCS	24	2.25	21,030	31,475
RPMI 1640 minus glucose + 10% HI dialyzed FCS	24	2.04	28,420	32,472

[a]Approximately 1 × 10^5 P288 cells were suspended in 1 ml of RPMI 1640 medium containing 10% fetal calf serum (FCS), or RPMI 1640 medium free of glucose plus 10% FCS or 10% heat-inactivated (HI, 56°C, 1 h), dialyzed (glucose free) FCS at time zero. Five or twenty-four hours later the incubation was terminated and the cell number determined with a Coulter counter.
[b]Cells were clumped when dialyzed FCS was used.
[c][U-^{14}C]glucosamine, 2 μM (1.1 × 10^6 dpm) or [1-^{14}C]leucine, 0.37 mM (1.1 × 10^6 dpm), was added to separate cultures at time zero. Incubations were terminated by the addition of 10% TCA. Acid insoluble radioactivity was quantitated.
All data are the average of 3–4 experiments performed with duplicate 1-ml cultures.

[^{14}C]glucosamine incorporation into glycoprotein while, with the exception of β-PAG, all had very little effect within 5 h on [^{3}H]leucine incorporation into protein. Depression of [^{14}C]glucosamine incorporation was as follows: β-PAG > α-PAG > F_3-GlcNAc > α-PAGAL > β-PAMAN ≫ GlcNAc and 2-deoxyglucose which had little effect on [^{14}C]-glucosamine incorporation.

Any of the sugar analogs manifesting effects in these test systems at 1 mM were then tested at lower concentration of 10^{-4} and 10^{-5} M. In all cases no effects were noted at 10^{-5} M after 24 h and slight effects were noted at 10^{-4} M. As an example, the effects of β-PAG at 10^{-3}, 10^{-4}, and 10^{-5} M on cell growth and macromolecular synthesis are shown in Table III. The effect of β-PAG on cell growth was not influenced by the concentration of glucose in the medium while its effect on glucosamine incorporation may have been amplified in low glucose containing medium. Little effect on cell growth was noted at 10^{-4} M after 24 h but became more apparent at 48 h.

Increasing incubation times to 48 h in the presence of 1 mM glucosamine or α- or β-PAG resulted in both a further decrease in cell growth and viability for cells treated with either α- or β-PAG but also with D-glucosamine (Fig. 1). At 24 h, cell viabilities for the control were 98%, glucosamine 98%, α-PAG 98%, and β-PAG 83%; at 48 h viabilities were control 95%, glucosamine 67%, α-PAG 92%, and β-PAG 19%. Forty-eight-hour exposures to 1 mM 2-deoxy-D-glucose also resulted in a decrease in cellular growth and viability intermediate between the effects of glucosamine and α-PAG (Fig. 1).

TABLE II. Effects of Various Sugar Analogs on P288 Mouse Lymphoma Cells

Compound[a]	Incorporation (% control)[b]				Growth[c] (% control) 24 h
	[^{14}C]glucosamine		[^{3}H]leucine		
	2 h	5 h	2 h	5 h	
I. 2-Amino-2-deoxy-D-glucopyranose (Glc NH$_2$)	–	–	97	95	99
II. 2-Acetamido-2-deoxy-D-glucopyranose (GlcNAc)	95	97	93	97	103
III. 2-Acetamido-2-deoxy-1,3,4,6-tetra-O-acetyl-β-D-glucopyranose (β-PAG)	33	10	77	62	53
IV. 2-Acetamido-2-deoxy-1,3,4,6-tetra-O-acetyl-α-D-glucopyranose (α-PAG)	30	23	90	91	88
V. 2-Acetamido-2-deoxy-1,3,4,6-tetra-O-acetyl-α-D-galactopyranose (α-PAGAL)	64	59	103	98	85
VI. 2-Acetamido-2-deoxy-1,3,4,6-tetra-O-acetyl-β-D mannopyranose (β-PAMAN)	76	77	96	91	83
VII. 2-Trifluoroacetamido-2-deoxy-D-glucopyranose (F$_3$-GlcNAc)	–	31	97	88	87
VIII. 2-Deoxy-D-glucopyranose (DOG)	90	98	89	96	92

[a]Approximately 1 × 10^5 P288 cells were suspended in 1 ml of glucose free RPMI 1640 medium containing 10% fetal calf serum and the indicated compound to a final concentration of 1 mM.
[b][^{14}C]glucosamine, 2 μM (1.1 × 10^6 dpm) and [^{3}H]leucine, 0.37 mM (2.6 × 10^6 dpm) were added at time zero. At the indicated times cells were extracted twice with 2 ml of 10% trichloroacetic acid and the acid insoluble radioactivity was quantitated and expressed as a percent of control incorporation.
[c]Growth was monitored by counting aliquots of cells with a Coulter counter.

Partition Coefficients

The octanol/water partition coefficient for β-PAG was found to be higher [0.424] than that of α-PAG [0.278] and much higher than that of GlcNAc [0.017].

Metabolism of α-Pentaacetylglucosamine (PAG)

The incorporation and metabolism of [U-^{14}C]-2-acetamido-2-deoxy-1,3,4,6-tetra-O-acetyl-α-D-glucopyranose, specific activity 232 mCi/mmole, and 2-acetamido[N-acetyl-1-^{14}C]-2-deoxy-1,3,4,6-tetra-O-acetyl [O-acetyl-^{3}H]-α-D-glucopyranose, specific activity ^{14}C (31.6 mCi/mmole) and ^{3}H (95 mCi/mmole), was studied with P288 lymphoma cells.

Cells incubated with trace amounts of [^{14}C]PAG, 0.1 μCi/ml, for various time periods were analyzed for incorporation of the label into putative cellular glycoprotein (Fig. 2). Incorporation into lipid-free macromolecular material was nearly linear for 20 h. Examination of the soluble pool of radioactivity following a 48 h exposure to [^{14}C]PAG indicated that most of the ethanol soluble label cochromatographed with authentic UDP-N-acetylglucosamine (UDP-GlcNAc), Fig. 3, while the cellular chloroform extract contained authentic [^{14}C]PAG. Chromatography of the extracellular tissue culture medium revealed only authentic [^{14}C]PAG.

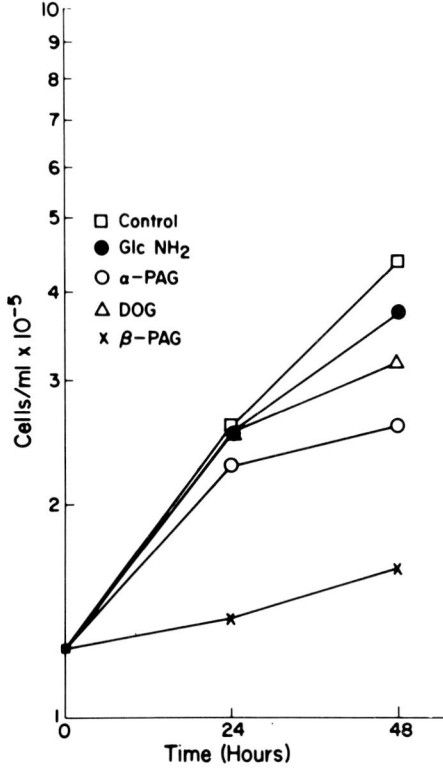

Fig. 1. The effects of various sugar analogs on P288 lymphoma cell growth. Glucosamine (GlcNH$_2$) and the other sugar analogs α- or β-pentaacetylglucosamine (PAG) and 2-deoxyglucose (DOG) were added at a final concentration of 1 mM to duplicate 1 ml P288 leukemic cell cultures. Cell growth and viability was monitored at 24 and 48 h as indicated in the text.

Similar experiments were performed with the double labeled [^{14}C,^{3}H] PAG. Cellular medium contained 2 μCi ^{3}H and 0.65 μCi ^{14}C per ml. Incorporation into acid insoluble material and chloroform or ethanol soluble pools were examined. A very rapid incorporation or exchange of some ^{3}H-labeled material into the acid insoluble pool occurred within the first few minutes of drug exposure. Overlooking this the ratio of incorporated ^{14}C:^{3}H increased with time (2–24 h) suggesting intracellular de-O-acetylation occurred before its incorporation into cellular glycoconjugate. This was later confirmed by chromatograms of soluble ethanol extracts (Fig. 4A). Only ^{14}C was detected in areas corresponding to UDP-GlcNAc while ^{14}C and ^{3}H were detectable in areas corresponding to authentic PAG. ^{14}C and ^{3}H were detectable in chloroform extracts (Fig. 4B) in areas corresponding to authentic [^{14}C,^{3}H] PAG. The ratio of ^{14}C:^{3}H in the partially de-O-acetylated PAG or front region of the chromatogram of the cellular chloroform extract was diminished as compared with the authentic PAG. This suggests some rapid exchange of ^{3}H in the cell.

Ribonucleotide Pool Size Analysis

Ribonucleotide pool size analysis was performed in P288 murine leukemic cells treated with 1 mM glucosamine (GlcNH$_2$), α- or β-pentaacetylglucosamine (α- or β-PAG), N-acetylglucosamine (GlcNAc), 2-trifluoro-N-acetylglucosamine (F$_3$-GlcNAc) and 2-deoxyglucose (DOG) for periods of 1, 3, 5, and 24 h.

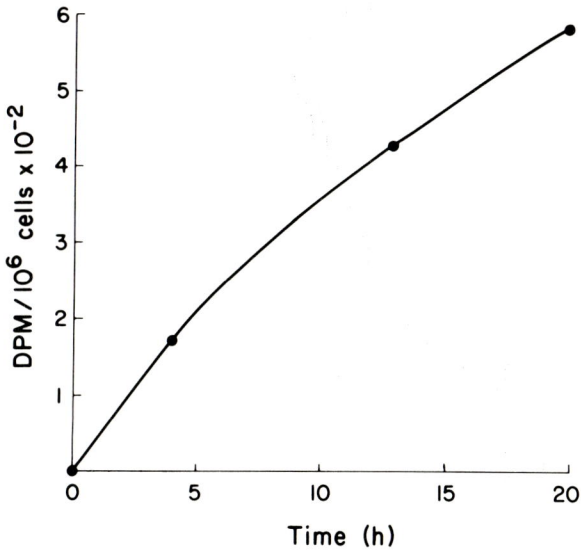

Fig. 2. Incorporation of [^{14}C] PAG in P288 murine lymphoma cells. Cells were seeded at 1×10^5/ml in RPMI 1640 medium plus 10% fetal calf serum containing 0.1 μCi [^{14}C] PAG/ml (specific activity, 232 Ci/mole). After 4, 12, and 20 h 25-ml cell aliquots were removed. The cell suspensions were centrifuged and extracted for 20 min with 2 ml of 1% phosphotungstic acid dissolved in 0.5 N HCl at 4°C. The mixtures were centrifuged and the resulting pellets washed with 2 ml of 5% TCA, centrifuged, and extracted with 2 ml of chloroform:methanol:ether (2:1:1). The resultant lipid free insoluble material was dissolved in 1.0 N NaOH and the radioactivity in this material was determined by scintillation counting methods.

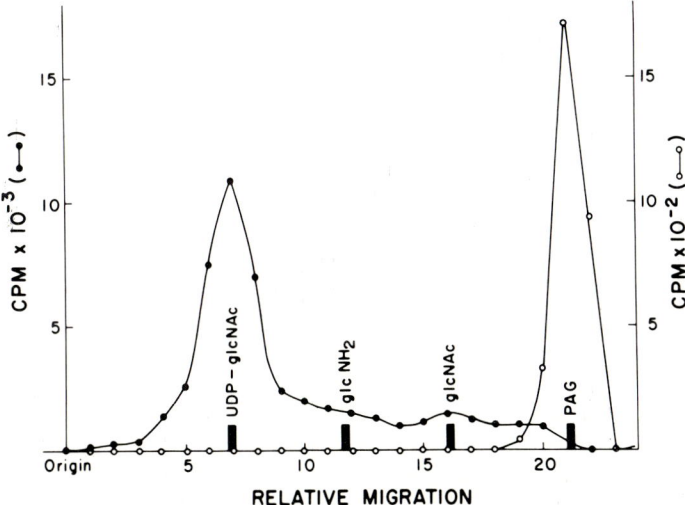

Fig. 3. Chromatography of P288 cell extracts. P288 murine lymphoma cells were incubated in the presence of [^{14}C] PAG (0.1 μCi/ml) for 48 h. Cells were harvested, washed twice with PBS, and extracted with 1 ml chloroform, centrifuged, and the pellet reextracted with 1 ml of 70% ethanol. Aliquots of these extracts were analyzed with descending paper chromatography using pyridine:HAc:EtAc:H$_2$O (5:1:5:3). Radioactivity was detected by scintillation counting methods. (●—●) ethanol extract; (○—○) chloroform extract.

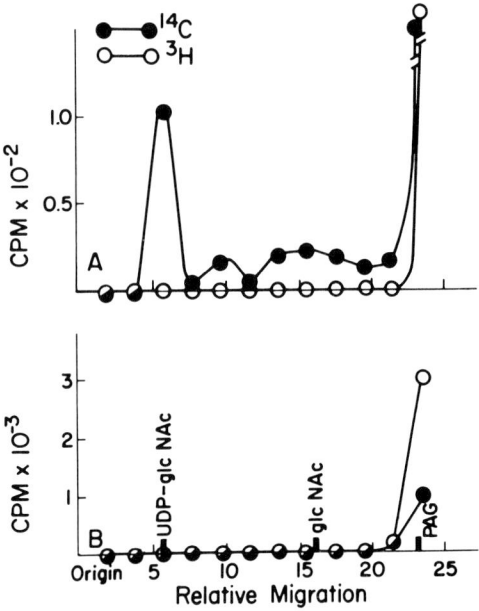

Fig. 4. Chromatography of P288 cell extracts. P288 murine lymphoma cells (10^5/ml) were incubated in the presence of [^{14}C,^{3}H]PAG, (^{14}C, 0.65 μCi/ml; ^{3}H, 2 μCi/ml) for 6 h. The ^{14}C is located in the N-acetyl group while the ^{3}H is located in the O-acetyl groups. Cells were harvested, washed twice with PBS, and extracted with 1 ml of chloroform, centrifuged, and the organic layer removed. The cellular pellet was then extracted with 1 ml 70% ethanol/H$_2$O. Chromatography of these extracts were performed as described in Fig. 3. A: Chromatogram of ethanol extract, (●—●) ^{14}C; (○—○) ^{3}H. B: Chloroform extract, (●—●) ^{14}C; (○—○) ^{3}H.

Treatment of P288 leukemic cells with 1 mM glucosamine for 3 h resulted in a large increase in the nucleotide-sugar pools (Fig. 5B and Table IV). The twin peaks probably consist mainly of UDP-N-acetylglucosamine and UDP-N-acetylgalactosamine and some UDP-glucose. After 3 h this pool was increased fourfold over control and after 24 h an eightfold increase was evident. Other early changes in pool sizes were a large decrease in the UTP pool (to 36% control) and a concomitant large decrease in the CTP pool size (to 23% control). Smaller decreases in the purine triphosphate pools were seen (GTP, 52% and ATP, 57%) but these decreases were more variable and were not discernable within 1 h of incubation. Both β-PAG (Fig. 5C) and α-PAG (Fig. 5D) treatment resulted in increases in the nucleotide-sugar pools. After 3 h the nucleotide-sugar pool in the cells treated with β-PAG was 167% of control while α-PAG treated cells had 202% of control nucleotide sugars (Table IV). Decreases in the UTP and CTP pools to one third of the control were evident in cells treated with β-PAG while little change in these pools was evident at 3 h in cells treat with α-PAG. However specific decreases in UTP and CTP pools were evident at 5 h and very prominent within 24 h in cells treated with α-PAG.

N-Acetylglucosamine (GlcNAc) treatment of P288 leukemic cells resulted in little change in pool sizes while F$_3$-N-acetylglucosamine (F$_3$-GlcNAc) administration resulted in more than doubling of the intracellular UDP-sugar content (Fig. 6D and Table IV). The only apparent pool size affected by a 3-h exposure to 1 mM GlcNAc was an increase in the UDP-ADP pool size (Fig. 6B and Table IV) to twice control. Again this change was variable

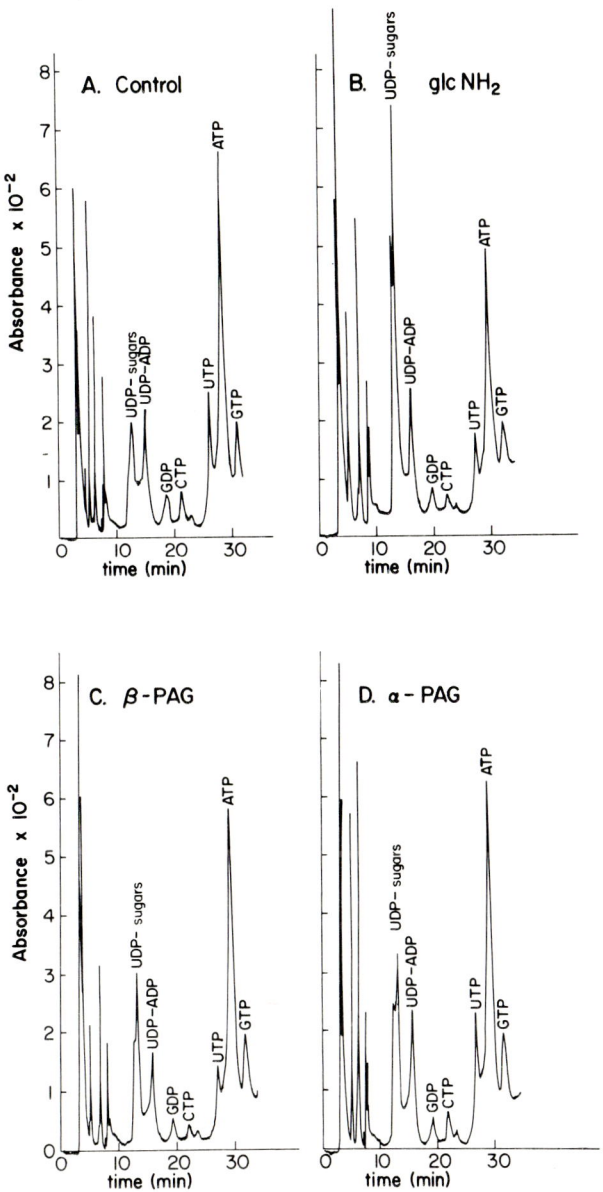

Fig. 5. High pressure liquid chromatography separation of intracellular nucleotide pools. Acid soluble extracts were prepared from 5×10^6 P288 leukemic cells. A) Control; B) treated with 1 mM glucosamine ($GlcNH_2$); C) β pentaacetylglucosamine (β-PAG); D) α-PAG for 3 h. Absorbance at 254 nm (0.0–0.08 units full scale) was monitored versus column retention time (min). Each peak was automatically integrated, normalized, and listed as the % control nucleotide pool (Table IV).

and was not evident in the other times studies. 2-Deoxy-D-glucose (Fig. 6C) treatment did affect pool sizes dramatically. Generally the ribonucleoside triphosphates were reduced (ATP and CTP to half of control) and the ribonucleoside diphosphates increased (GDP to more than 11 times control and UDP-ADP to more than eightfold higher). No increase in

TABLE III. Effects of β-PAG on P288 Lymphoma Cell Growth and Glucosamine Incorporation

Time	Incorporation of glucosamine (% control)[a]						Growth, % cell no.[c] β-PAG/control	
	2 h		5 h		24 h		24 h	
Glucose	+	−	+	−	+	−	+	−
β-PAG concentration (M)								
0	100[b]	100	100	100	100	100	100	100
10^{-3}	63	33	30	10	27	7	61	69
10^{-4}	82	99	74	73	85	50	95	86
10^{-5}	100	100	100	100	94	100	98	98

[a]Approximately 1×10^5 cells were suspended in 1 ml RPMI 1640 medium containing 11.1 mM glucose (+) or glucose free (−) plus 10% FCS. One hour later PAG was added to the desired final concentration and an additional hour later [^{14}C]glucosamine, 2 μM, (1.1×10^6 dpm), was added. Incubations were terminated with 2 ml 10% TCA and acid insoluble radioactivity was determined. Data are expressed as % control incorporation.
[b]Incorporation was calculated for each culture tube in duplicate and expressed as % dpm (drug treated/control). About a five-fold increase in glucosamine incorporation was apparent when glucose-free medium was employed. Growth was slightly lower in glucose-free medium as compared with control.
[c]Cells were seeded at 1×10^5 cells/ml at time zero. Twenty-four hours later cell number was determined with a Coulter counter. Data are expressed as % cell number (drug treated/control). Control cultures increased about 2.5 times in 24 h. All data are the average of 2–3 experiments.

TABLE IV. Effect of Various Sugars or Sugar Analogs on Nucleotide Pool Sizes

Nucleotide	Nucleotide pool size (% control)[a]					
	Sugar or sugar analog					
	GlcNH$_2$	GlcNAc	β-PAG	α-PAG	DOG	F$_3$-GlcNAc
GTP	52	73	80	88	83	88
ATP	57	103	91	86	53	98
UTP	36	117	33	82	80	103
CTP	23	110	32	86	47	110
GDP	84	122	73	95	1,122	83
UDP and ADP	100	200	100	165	845	118
UDP-sugars	434	108	167	202	83	252

[a]The % control nucleotide pool sizes were calculated from the integrated peak areas obtained from the high pressure liquid chromatograms scans, Fig. 5 and 6. To normalize the data each integrated peak was divided by the total integrated area and this value set to 100% for control cultures.

nucleotide-sugar pools was noted but this may be due to an inherent instability of UDP-2-DOG in acid extracts. Breakdown of large amounts of UDP-2-DOG may account for the high level of UDP and ADP found (Fig. 6C and Table IV). The high amount of GDP (11 times above normal) may also be accounted for by the instability of GDP-2-deoxyglucose in perchloric acid (21). At 0°C all the GDP-2-deoxyglucose is broken down to free 2-deoxyglucose within 5 min. Schmidt et al. (21) also concluded that after the administration of 2-deoxyglucose to virus-infected mammalian cells it is the GDP-2-deoxyglucose that is the agent responsible for the inhibition of glycosylation of viral glycoproteins.

DISCUSSION

It is evident that glucosamine rapidly enters tumor cells and is metabolized to UDP-N-acetylglucosamine and UDP-N-acetylgalactosamine resulting in a drop in both UTP and CTP pools (Fig. 5). The dramatic decrease in the CTP pool is not surprising since UTP

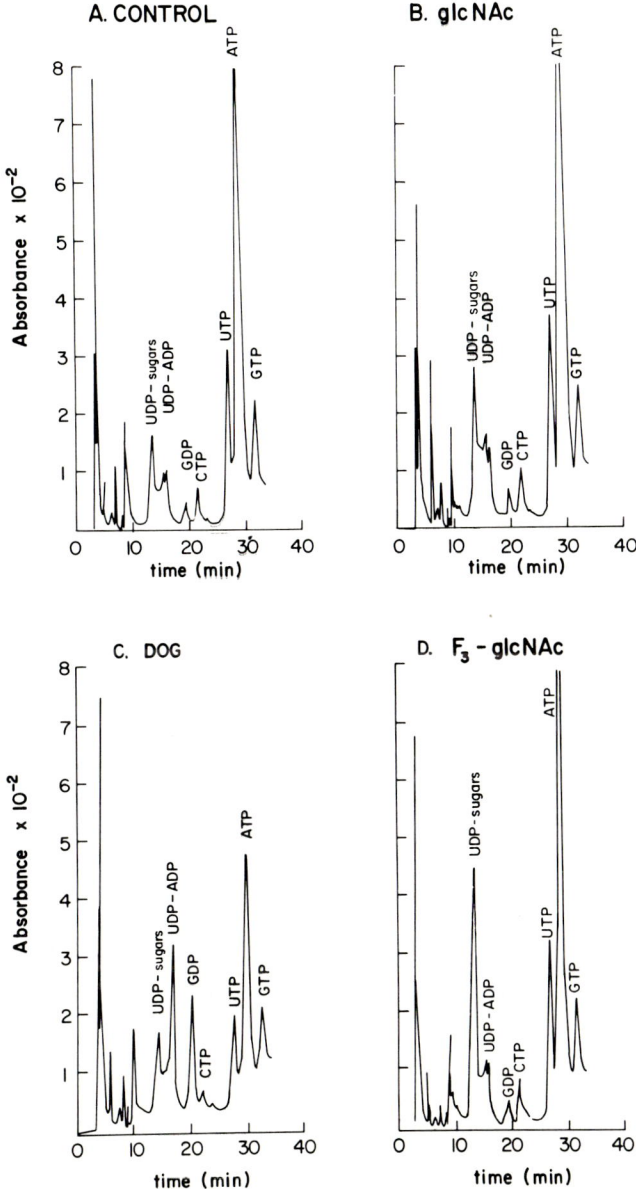

Fig. 6. High pressure liquid chromatography separation of intracellular nucleotide pools. Acid soluble extracts were prepared from 5×10^6 P288 leukemic cells, A) Control; B) treated with 1 mM N-acetylglucosamine (GlcNAc); C) 2-deoxy-D-glucose (DOG); D) 2-trifluoro-N-acetylglucosamine (F_3-GlcNAc). Absorbance at 254 nm was monitored versus column retention time (min). Each peak was automatically integrated, normalized, and listed as % control nucleotide pool (Table IV).

provides the de novo source of cytidine nucleotides through the CTP synthetase reaction (UTP + ATP + NH_3 = ADP + orthophosphate + CTP, E.C. 6.3.4.2). It is surprising that P288 murine leukemic cells are still able to grow at near normal levels (Fig. 1) for up to 48 h after the addition of 1 mM glucosamine even though the UTP and CTP pool sizes are diminished to one third normal within 3 h (Table IV). Recently this type of hexosamine induced nucleotide pool size reduction has been exploited in the chemotherapy of hepatic tumor cells in vitro (32). The addition of galactosamine and 3-deazauridine resulted in pharmacological synergism. This was a result of the enhancement of the toxicity of 3-deazauridine caused by the depletion of the normal intracellular UTP and CTP pools. These reductions were caused by the formation of large amounts of UDP-sugars (33). Therapeutic effects such as these may be further enhanced by agents such as β-PAG, a fully acetylated amino sugar analog which is more lipid soluble and has a greater inherent toxicity as compared to glucosamine or galactosamine.

It was found that β-PAG is de-O-acetylated and further metabolized to UDP-N-acetylglucosamine (Fig. 7) resulting in a concomitant lowering of the UTP and CTP pools. Its enhanced toxicity as compared to glucosamine or N-acetylglucosamine may be due to greater cellular uptake via passive diffusion since this fully acetylated sugar is more lipid soluble. The octanol/water partition coefficient for β-PAG was higher than α-PAG and much higher than that of GlcNAc. It is also possible that the enhanced toxicity of β-PAG may be due to its action as a potent inhibitor of hexokinase (34) although the addition of high concentrations of glucose to the cell medium did not reverse its toxicity (Table III). Whatever the case may be there is a difference in tumor toxicity between α- and β-anomers of PAG with the β form being much more toxic than the α form (Fig. 1). This may be due to the higher octanol/water partition coefficient for β-PAG compared to α-PAG resulting in the higher uptake of the former due to passive diffusion.

A comparison of the biological effects of F_3-GlcNAc (VII, Table II) and GlcNAc (II, Table II) is noteworthy. The addition of fluorine to the N-acetyl group permitted the compound to interfere with [^{14}C] glucosamine incorporation, decreasing it to 31% of control within 5 h (Table II). GlcNAc at a similar concentration of 1 mM did not interfere with [^{14}C] glucosamine incorporation and both compounds did not inhibit [^{3}H]-leucine incorporation. Ribonucleotide pool size analysis of P288 cells treated with 1 mM GlcNAc (Fig. 6B) or 1 mM F_3-GlcNAc demonstrated that F_3-GlcNAc was converted to nucleotide sugar whereas GlcNAc conversion was not detectable (Table IV), although it was probably occurring at a low rate. In 1 mM concentration GlcNAc was not

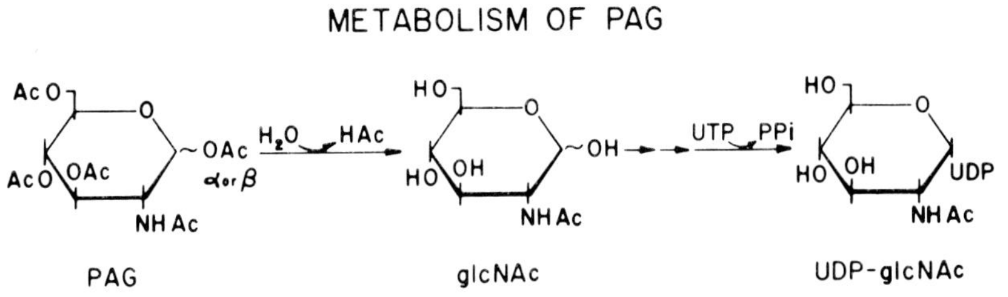

Fig. 7. Metabolism of PAG. The results of our studies have indicated that PAG (pentaacetylglucosamine) rapidly enters cells, and is de-O-acetylated to form naturally occurring N-acetylglucosamine (GlcNAc). The GlcNAc is then phosphorylated and later conjugated with UTP to form large amounts of UDP-GlcNAc.

cytoxic to P288 cells while F_3-GlcNAc (1 mM) did decrease cell number to 87% of control within 24 h and to 78% by 48 h. The differences in biological effects and metabolism of these compounds may be due to differences in cellular permeability or uptake as indicated by Bekesi et al. (35) or in rate of enzymatic phosphorylation reactions necessary in the formation of nucleotide sugar. Cellular uptake of glucosamine is carrier mediated as indicated by decreased macromolecular incorporation in the presence of glucose (Table I) while PAG and GlcNAc seem to rely on passive diffusion for entry into the cells. F_3-GlcNAc may be able to take advantage of the carrier mediated transport since it did compete with and was able to lower cellular glucosamine incorporation (Table II).

2-Deoxy-D-glucose, at 1 mM concentrations, was not a specific inhibitor of glycoprotein biosynthesis in P288 murine leukemic cells. In fact up to 24 h it had very little effect on any parameter studied except intracellular ribonucleotide pools which showed a dramatic decrease in ribonucleoside triphosphates along with a sharp increase in ribonucleoside diphosphates (Fig. 6, Table IV). There was no indication of further metabolism of 2-deoxy-D-glucose phosphate to UDP-2-deoxy-D-glucose as compared with the increases observed in UDP-sugars after glucosamine or α- or β-PAG administration (Figs. 5, 6). Since UDP-2-deoxy sugars and GDP-2-deoxyglucose are not stable in acid extracts, any nucleotide sugar formed might have broken down by the conditions used for the quantitation of intracellular ribonucleotide pools (21, 33).

Chemotherapy based on the inhibition or modification of plasma membrane glycoconjugate may indeed be feasible. Alterations in membrane glycoconjugate may stimulate immune responsiveness or inhibition of glycoconjugate biosynthesis may alter cellular behavior. Our first approach to this goal has been the synthesis of several simple acetylhexosamines. We were initially very excited when we observed that PAG caused a specific inhibition of macromolecular glucosamine incorporation with little or no inhibition of leucine. Further studies with acetylated sugars such as PAG indicated that once these moieties enter cells they are de-O-acetylated and further metabolized to nucleotide sugars (Fig. 7). Once deacetylated they can easily compete with and essentially dilute the radiolabeled GlcNAc pool synthesized from [^{14}C]glucosamine explaining the specific inhibition of [^{14}C]glucosamine incorporation. Once O-deacetylated, the GlcNAc can be converted to UDP-GlcNAc resulting in a reduction of the UTP and CTP pools. This specific ribonucleotide pool size reduction caused by PAG may be further exploited in combination chemotherapy with pyrimidine antagonists.

ACKNOWLEDGMENTS

We would like to acknowledge the valuable contributions of Mr. M. Hanchak, Mr. E. Kelly, and Ms. M. Hillman to this study. This work was supported by Public Health Service Grants CA-15757, CA-08793, CA-19814, CA-18420, and CA-13038 from the National Cancer Institute, National Institutes of Health, Department of Health, Education and Welfare.

REFERENCES

1. Nicolson GL: Biochim Biophys Acta 458:1, 1976.
2. Nicolson GL, Poste G: New Engl J Med 295:253, 1976.
3. Gahmberg CG, Hakomori S: Proc Natl Acad Sci USA 70:3329, 1973.
4. Warren L, Fuhrer JP, Buck CA: Proc Natl Acad Sci USA 69:1838, 1972.

5. Kraemer PM: In Manson LA (ed): "Biomembranes." New York: Plenum Press, 1964, vol 1, p 67.
6. Roseman S: In Weissman G, Claiborne R: "Cell Membranes, Biochemistry, Cell Biology and Pathology." New York: Hospital Practice, 1975, p 55.
7. Curtis ASG: In Butler JAV, Noble D (eds): "Progress in Biophysics and Molecular Biology." Oxford: Pergamon Press, 1973, vol 27, p 317.
8. Bosmann HB, Bieber GF, Brown AE, Case KR, Gersten DM, Kimmerer TW, Lione A: Nature (London) 246:487, 1973.
9. Bernacki RJ, Kim U: Science 195:577, 1977.
10. Porter CW, Bernacki RJ: Nature (London) 256:648, 1975.
11. Kornfeld S, Kornfeld R, Neufeld EF, O'Brien PJ: Proc Natl Acad Sci USA 52:371, 1964.
12. Quastel JH, Cantero A: Nature (London) 171:252, 1953.
13. Seltzer MH, Roseman JM, Wolfe DE, Tsou KC, Miller EE, Rosato FE: Growth 33:353, 1969.
14. Bekesi JG, Winzler RJ: J Biol Chem 244:5663, 1969.
15. Cox RP, Gessner BM: Cancer Res 28:1162, 1968.
16. Bessel EM, Courtenay VC, Foster AB, Jones M, Westwood JH: Eur J Cancer 9:463, 1973.
17. Schultz AM, Mora PT: Carbohydr Res 40:119, 1975.
18. Kent PW, Mora PT: In Kent PW (ed): "Membrane Mediated Information." New York: American Elsevier Publishing Company, 1973, vol 1, p 129.
19. Pero RW: Babiarz P, Rittman L, Simon P, Fondy TP: 170th National Meeting, American Chemical Society, Chicago, Illinois, August, 1975.
20. McComb RB, Yushok WD: Cancer Res 24:193, 1964.
21. Schmidt MFG, Schwartz RT, Scholtissek C: Eur J Biochem 70:55, 1976.
22. Havell EA, Vilcek J, Falcoff E, Berman B: Virology 63:475, 1975.
23. Myers MW, Sartorelli AC: Biochem Biophys Res Commun 63:164, 1975.
24. Hwang KM, Sartorelli AC: Biochem Pharmacol 24:1149, 1975.
25. Steiner M, Somers K Steiner S: Biochem Biophys Res Commun 61:795, 1974.
26. Wolfrom ML, Conigliaro PJ: Carbohydr Res 11:63, 1969.
27. Horton DJ: J Org Chem 29:1776, 1964.
28. Stacey M: J Chem Soc 272, 1944.
29. Levene PA: J Biol Chem 57:323, 1923.
30. Russell WC, Newman C, Williamson DH: Nature (London) 253:461 (1975).
31. Bernacki RJ: Eur J Biochem 58:477, 1975.
32. Jackson RC, Williams JC, Weber G: Cancer Chemother Rep 60:835, 1976.
33. Keppler DO, Rudigier JFM, Bischoff E, Decker KFA: Eur J Biochem 17:246, 1970.
34. Korytnyk W, Bernacki RJ, Danhauser L, Hanchak M, Paul B, Rustum Y, Sharma M, Sufrin J: Fed Proc Fed Am Soc Exp Biol 35:1639, 1976.
35. Bekesi JG, Molnar Z, Winzler RJ: Cancer Res 29:353, 1969.

Role of Mannosyl Phosphoryl Polyisoprenol in Biosynthesis of Mammary Glycoproteins

Inder K. Vijay and Steven R. Fram

Department of Dairy Science, University of Maryland, College Park, Maryland 20742

When a membrane preparation from the lactating bovine mammary gland is incubated with GDP-[^{14}C]mannose, mannose is incorporated into a [^{14}C]mannolipid, a [Man-^{14}C]oligosaccharide-lipid, and metabolically stable endogenous acceptor(s). The rate of mannosyl incorporation is the fastest into [^{14}C]mannolipid, intermediate in [Man-^{14}C]oligosaccharide-lipid, and least into]Man-^{14}C] endogenous acceptor(s).

The [^{14}C]mannolipid has been partially purified and characterized. Mild acid hydrolysis of this compound gives [^{14}C]mannose, whereas alkaline hydrolysis yielded [^{14}C]mannose phosphate as the labeled product. The $t_{½}$ of hydrolysis of the mannolipid under the acidic and basic conditions are comparable to values obtained for mannosyl phosphoryl dolichol in other systems. The mannolipid is chromatographically indistinguishable from calf brain mannosyl phosphoryl polyisoprenol and chemically synthesized β-mannosyl phosphoryl dolichol. Exogenous dolichol phosphate stimulates the synthesis of mannolipid in mammary particulate preparations 8.5-fold. Synthesis of mannolipid is freely reversible; in the presence of GDP, the transfer of mannosyl moiety from endogenously labeled mannolipid to GDP-mannose is obtained. All of these results indicate that the structure of mannolipid is mannosyl phosphoryl polyisoprenol. Even though the precise chain length of the polyisoprenol portion has not been established, it is tentatively suggested to be dolichol.

Partially purified [^{14}C]mannolipid can directly serve as a mannosyl donor in the synthesis of [Man-^{14}C]oligosaccharide-lipid and [Man-^{14}C] endogenous acceptor(s). Pulse and chase kinetics utilizing GDP-mannose to chase the mannosyl transfer from GDP-[^{14}C]mannose in the mammary membrane incubations caused an immediate and rapid turnover of [^{14}C]mannose from [^{14}C]mannolipid while the incorporation of label in [Man-^{14}C]oligosaccharide-lipid and radioactive endogenous acceptor(s) continued for a short period before coming to a halt.

Both gel filtration and electrophoresis indicate that the endogenous acceptor(s) are a mixture of 2 or more glycoproteins since incubation with proteases releases all of the radioactivity into water soluble low-molecular-weight components, perhaps glycopeptides.

All of the above evidence is consistent with the following precursor-product relationship:

GDP-mannose ⇌ mannosyl phosphoryl polyisoprenol → mannosyl-oligosaccharide-lipid → mannosyl-proteins.

The exact structure of the oligosaccharide-lipid and the endogenous glycoproteins is unknown.

Key words: mammary, glycoprotein, biosynthesis, mannosyl phosphoryl polyisoprenol

Abbreviations: C/M (2:1) – CHCl$_3$-CH$_3$OH, 2:1; C/M/W (10:10:3) – CHCl$_3$-CH$_3$OH-H$_2$O 10:10:3; NeuNAc – N-acetylneuraminic acid; Gal – galactose; GlcNAc – N-acetylglucosamine; Man, mannose.

Received March 18, 1977; accepted June 16, 1977.

© 1977 Alan R. Liss, Inc., 150 Fifth Avenue, New York, NY 10011

Presently, there are 2 types of mechanisms known for the assembly of carbohydrate moieties of glycoproteins. The process whereby single carbohydrate residues are transferred directly from their respective nucleotide donors to the distal end of the incomplete growing portions of glycoproteins containing complex heterooligosaccharide chains has been thoroughly documented (1, 2). The terminal carbohydrate sequence of many glycoproteins is the trisaccharide residue NeuNAc→Gal→GlcNAc. By sequential treatment of a number of glycoproteins with neuraminidase, β-galactosidase, and hexosaminidase, the stepwise and orderly transfer of terminal sugars from nucleotide donor substrates to the modified acceptor proteins, catalyzed by specific glycosyltransferases, has been established (1, 2).

A number of glycoproteins have the sequence . . . . . $(Man)_n$ -GlcNAc-GlcNAc-Asn, $n = 2–5$, for a portion of their prosthetic groups. This oligosaccharide region has been termed the "core" region to distinguish it from the commonly found NeuNAc→Gal→GlcNAc "terminal" trisaccharide. The attachment of core region of some glycoproteins has been shown to involve the transfer of sugars from sugar-nucleotides to a "carrier" lipid, dolichol phosphate, preassembly or oligomerization on this lipid, and finally a transfer en bloc to the acceptor protein (3). The identify of proteins glycosylated by this mechanism has not been established for most systems and the evidence indicates that endogenous glycosyl acceptor proteins are membrane bound. Recently, evidence has come forth from the laboratories of Heath and Lennarz to support that an oligosaccharide-lipid can indeed serve as an intermediate in the transfer of core region sugar residues to the nonglycosylated soluble proteins. With the mouse myeloma tumor system, 2-deoxy-D-glucose was used to block the first step in the glycosylation of K-46 immunoglobulin light chain (4). The secreted nonglycosylated protein was then shown to serve as an exogenous acceptor of core region sugars from the oligosaccharide-lipid intermediate. In the hen oviduct system, it was shown that the antibiotic tunicamycin inhibits the synthesis of N-acetylglucosaminyl pyrophosphoryl dolichol and oviduct slices incubated in the presence of this antibiotic synthesized nonglycosylated ovalbumin (5). Further, the enzymatic activity in the oviduct membrane preparation is capable of transferring the oligosaccharide moiety from the oligosaccharide-lipid intermediate to several exogenously added proteins after the latter have been denatured to expose potential sites for glycosylation (6).

We have recently initiated studies for the biosynthesis of glycoproteins in the lactating bovine mammary gland. The lactating mammary was chosen as an interesting tissue for the following reasons: i) this tissue elaborates large amounts of proteins including several glycoproteins into its secretion, i.e., milk; ii) it provides an excellent system for investigations into membrane glycoprotein biosynthesis. The milk fat globule membrane which is known to arise from the apical plasma membrane of the mammary secretory cell (7, 8), contains a number of glycoproteins (7–11). This report presents evidence that the lactating bovine mammary possesses all the basic parameters for a study of glycoprotein synthesis via lipid intermediates. A membrane preparation of this tissue catalyzes the transfer of $[^{14}C]$ mannose from GDP-$[^{14}C]$ mannose through the following sequence of steps:

$$\text{GDP-}[^{14}C]\text{ mannose} \underset{}{\overset{Mn^{2+}}{\rightleftarrows}} [^{14}C]\text{ mannosyl phosphoryl polyisoprenol}$$

$$\rightarrow [\text{Man-}^{14}C]\text{ oligosaccharide-lipid} \rightarrow [\text{Man-}^{14}C]\text{ protein(s)}.$$

MATERIALS AND METHODS

The bovine mammary tissue from a freshly slaughtered cow was routinely obtained from Frederick County Products, Frederick, Maryland. When sliced, the lactating gland oozes milk rather profusely. The tissue was placed on ice and transported to the laboratory immediately. The secretory tissue was substantially freed from the collagenous connective material and diced into pea sized cubes in the cold room (4°C). Unless stated otherwise, all subsequent operations were performed at ice temperature. After several washes in buffer consisting of 50 mM Tris-HCl, pH 7.2, 0.25 M sucrose, 1 mM EDTA, and 5 mM mercaptoethanol (homogenizing buffer) to remove milk and blood, the tissue was suspended in an equal volume of homogenizing buffer and homogenized in a Polytron homogenizer (Brinkmann) equipped with PT 10 ST head and operating at setting 8 for 60 sec. Since homogenization in Polytron results in excessive foaming and possible protein denaturation, some experiments were conducted using Lourdes MM-1A homogenizer, operating at maximum speed for 90 sec. The latter instrument gives a preparation with a slightly higher activity and is also more thoroughly homogenized.

The crude homogenate was filtered through 4 layers of cheesecloth and centrifuged at 750 × g for 10 min in a Sorvall RC-2B centrifuge. The supernatant was filtered through cheesecloth wetted with the homogenizing buffer to remove the thick layer of floating lipid and centrifuged at 9,000 × g for 15 min. The supernatant from the second spin was filtered through cheesecloth and centrifuged at 48,000 × for 30 min. The 48,000 × g pellet was resuspended in homogenizing buffer to a protein concentration of 20–30 mg/ml and designated P-3. The particulate preparation loses ~50% of the mannosyltransferase activity overnight upon storage at −20°C, but is stable with only a slight loss in activity thereafter for 7 weeks. All the experiments reported below were conducted with fresh P-3 preparations.

The transfer of mannose from GDP-[^{14}C] mannose into endogenous acceptors was assayed in standard incubations that contained 2.5 μl of 50 mM Tris-HCl, pH 7.2, 2.5 μl of 400 mM Mn^{2+}, 1 μl of 50 mM EDTA, 5 μl of 200 mM DTT, 20 μl of GDP-[^{14}C] mannose (417 cpm/pmole) and 65 μl of P-3 in a total volume of 100 μl. The reaction was initiated with the addition of GDP-[^{14}C] mannose and carried out at 37°C. At appropriate times, incubations were stopped by the addition of 20 volumes of C/M (2:1) and vortexed. Further processing to obtain [^{14}C] mannolipid, [Man-^{14}C] oligosaccharide-lipid and [Man-^{14}C] endogenous acceptors was conducted according to Waechter et al. (12).

For preparation and purification of [^{14}C] mannolipid, the standard incubation mixture was scaled up 100-fold. After incubation for 2 min at 37°C, the reaction was stopped with the addition of 50 volumes of C/M (2:1). Further processing was carried out through the completion of the C/M (2:1) extractions and washings as in the standard assay. After concentrating in a rotary evaporator, the organic phase containing the crude [^{14}C]-mannolipid (9.0 × 10^5 cpm) was chromatographed on a column of DEAE-cellulose [DE-52, Whatman, converted to the acetate form according to Rouser et al. (13)] as described by Forsee and Elbein (14). The radioactive fractions were pooled, desalted, and finally the partially purified [^{14}C] mannolipid (7.7 × 10^5 cpm) was taken up in C/M (2:1) and used in all subsequent experiments.

To study the transfer of mannose from [^{14}C] mannolipid into [Man-^{14}C] oligosaccharide-lipid and [Man-^{14}C] endogenous acceptors, incubations were performed exactly as described by Waechter et al. (15) except that 10 mM DTT was also included. Subsequent processing to separate the unreacted substrate, [Man-^{14}C] oligosaccharide-lipid and [Man-^{14}C] endogenous acceptors was performed as in the standard assay.

The partially purified [^{14}C] mannolipid was characterized for its susceptibility to acid and base hydrolysis under the conditions given in the figure legends. It was also chromatographed on both silica gel G TLC plates in: A) $CHCl_3$-CH_3OH-H_2O, 60:25:4; and B) $CHCl_3$-CH_3OH-50% NH_4OH, 60:35:8, and EDTA impregnated Whatman SG-81 paper (16) in the developing solvent systems: C) 2,6-dimethyl-4-heptanone-glacial acetic acid-H_2O, 60:45:6; D) $CHCl_3$-CH_3OH-concentrated NH_4OH, 36:13:3; and E) $CHCl_3$-CH_3OH-H_2O, 65:25:4. Chemically synthesized β-mannosyl phosphoryl dolichol (kindly provided by Dr. C. D. Warren, Harvard Medical School) and calf brain [^{14}C] mannosyl phosphoryl polyisoprenol (generous gift of Dr. C. J. Waechter, University of Maryland, School of Medicine) were used as reference standards. After the chromatographic run, 0.5-cm segments of the gel were scraped along the chromatogram and counted in Scinti-Verse (Fisher Scientific Company, Silver Spring, Maryland). Similarly, 0.5-cm segments of the SG-81 paper chromatogram were cut out and counted. Unlabeled β-mannosyl phosphoryl dolichol was detected by the phosphate (17) and polyisoprenol (18) specific spray reagents.

The labeled sugar in the oligosaccharide-lipid and endogenous acceptors was examined by paper chromatography on the acid hydrolysates of these products (14). After development, the chromatograms were dried and cut into 1-cm segments for counting in ScintiVerse. The unlabeled reference sugars were identified by an alkaline $AgNO_3$ reagent (19).

An aliquot (24,000 cpm, 50 mg protein) of the labeled endogenous acceptors prepared from GDP-[^{14}C] mannose was subjected to proteolytic digestion as described by Baynes et al. (20). Gel filtration on 50 mg (24,000 cpm) of the labeled endogenous acceptors and the proteolytic digest (given above), was carried out according to Gold and Hahn (21).

Aliquots of untreated and protease-treated endogenous acceptor(s) were applied to Whatman No. 3MM paper and subjected to high voltage electrophoresis in a Savant FP 30B unit using 1.5 M formic acid (pH 1.8) as the electrolyte in the electrode vessels. After electrophoresis at 30 volts/cm for 2 h, the paper was dried, cut up into 0.5-cm segments, and counted.

GDP-[^{14}C] mannose (specific activity 210 $\mu Ci/\mu mole$) was purchased from New England Nuclear Corporation (Boston, Massachusetts) and unlabeled GDP-mannose from Sigma Chemical Company (St. Louis, Missouri). All other reagents were from commercial sources and of the highest purity available.

Radioactivity was measured in a Packard Tri Carb 3375 liquid scintillation counter. Protein was determined by the Lowry procedure (22) with bovine serum as the standard.

RESULTS

The kinetics of mannosyl transfer into 3 products when crude mammary membranes are incubated with GDP-[^{14}C] mannose is shown in Fig. 1. On the basis of incorporation and turnover of radioactivity into products soluble in C/M (2:1), C/M/W (10:10:3), and the insoluble residue fraction, mannosyl transfer from GDP-[^{14}C] mannose into the 3 products is consistent with the following precursor-product sequence of reactions:

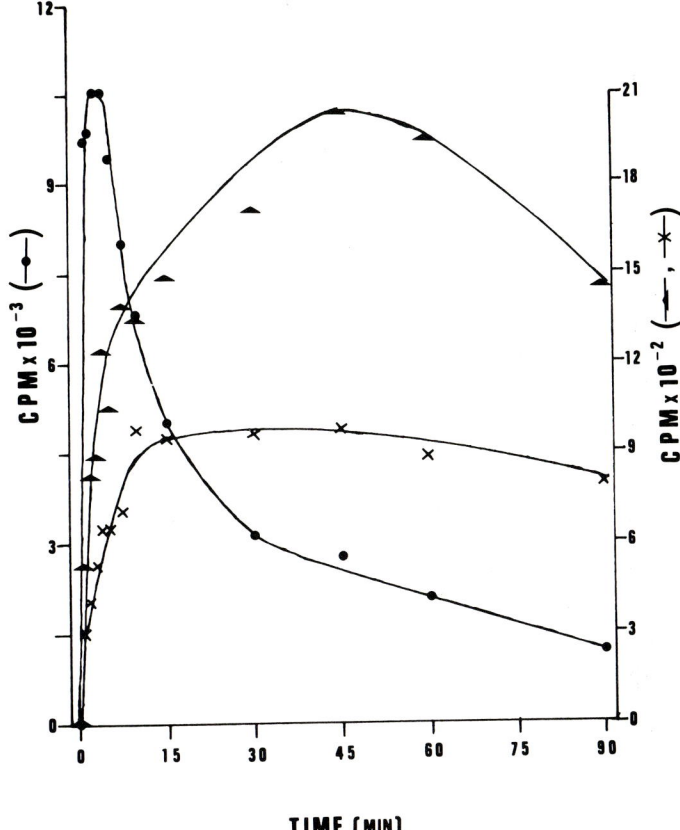

Fig. 1. Incorporation of radioactivity into C/M (2:1) soluble (–•–), C/M/W (10:10:3) soluble (–▲–), and into endogenous acceptors(s) fraction (–x–) upon incubation of GDP-[^{14}C]mannose with bovine mammary membrane preparation. Standard incubations were run in individual tubes and processed as given under Materials and Methods.

GDP-[^{14}C]mannose → [^{14}C]Mannolipid → [Man-^{14}C]Oligosaccharide-lipid
→ [Man-^{14}C] Endogenous "Acceptors" (Protein).

Analogous studies with other systems have made similar observations (14, 15, 20, 23–26). These results do not, however, rule out the possibility that mannose may be transferred directly from GDP-[^{14}C]mannose to endogenous acceptors in the mammary membrane preparation.

REVERSIBILITY OF MANNOLIPID SYNTHESIS

An experiment was conducted to examine the reversibility of the mannolipid synthesis, starting with GDP-[^{14}C]mannose (details are given in the legend to Fig. 2). There is a significant transfer of mannose from mannolipid to give GDP-[^{14}C]mannose indicating that mannose is bound to the lipid by an activated phosphodiester linkage and GDP, rather than GMP is the other product of this reaction.

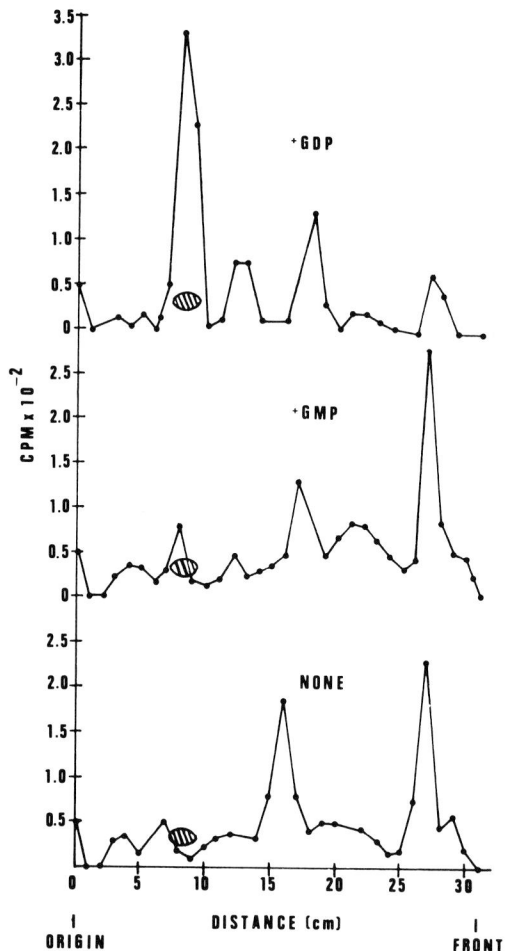

Fig. 2. Reversibility of mannolipid synthesis. Standard incubations containing GDP-[^{14}C]mannose and mammary membrane preparation were incubated for 2 min at 37°C, chilled on ice, and centrifuged in a cold room. Subsequent operations to wash and resuspend the membranes with guanine nucleotides were conducted according to Waechter et al. (15). After 10-min incubation at 37°C, 1 μmole of GDP-mannose was added to the incubation mixtures and the entire contents applied to Whatman 3MM paper as 1¼-inch-wide bands. In a parallel lane, 5 μmoles of GDP-mannose was also applied. After chromatography in solvent system F the GDP-mannose (hatched area) was located in the experimental lanes by examination under UV light and in the control lane by both UV absorption and alkaline AgNO$_3$ reagent (19). The experimental lanes were then cut up into 1-cm segments and counted in ScintiVerse.

In Fig. 2, minor peaks corresponding to mannose and mannose-1-phosphate are also observed. These are probably breakdown products of the hydrolysis of GDP-[^{14}C]-mannose and [^{14}C]mannolipid that have also been observed in other membrane systems (12, 15). These products may have resulted from enzymatic hydrolysis. Recently, it has been reported that divalent metal ions can also cause a nonenzymatic hydrolysis of sugar-nucleotides (27).

CONDITIONS FOR THE BIOSYNTHESIS OF [^{14}C] MANNOLIPID

The enzymatic transfer of mannose from GDP-[^{14}C] mannose to give [^{14}C] mannolipid was stimulated by a divalent metal ion and was inhibited by EDTA (data not shown). Mn^{2+} was the most efficient ion among the 7 ions tested (approximately twofold stimulation). Mammary membranes appear to have a reasonably high endogenous level of metal ions since there was a significant mannosyltransferase activity even in the absence of a metal^{2+}. In this regard, a considerable difficulty was encountered in keeping some of the mammary membrane preparations in a homogeneous suspension at 37°C and it was observed that precipitation could be avoided by the exclusion of metal^{2+} from these incubations. This might parallel the change in ionic composition and balance of milk depending upon feeding of the animal and the stage of lactation (28). Also since milk is a rich source of Ca^{2+}, it is possible that mammary membranes have significant amounts of this ion. Mannolipid synthesis was linearly dependent on the amount of membrane protein up to 65 mg/ml (Fig. 3, inset) and the pH optimum was 7.2. The dependence of mannolipid formation on the concentration of GDP-mannose gave an apparent K_m of 4.8×10^{-6} M as calculated from a reciprocal plot.

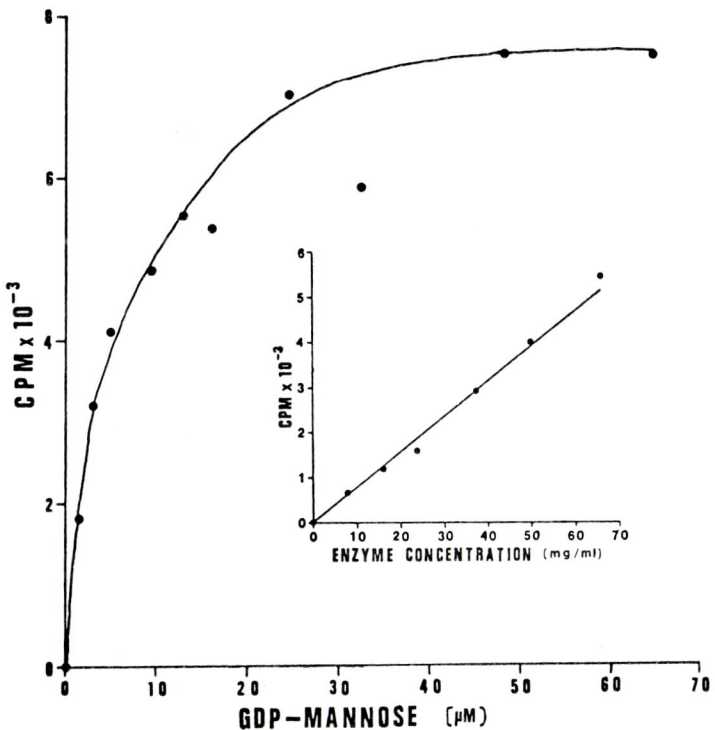

Fig. 3. Effect of GDP-[^{14}C] mannose concentration on the synthesis of [^{14}C] mannolipid. Standard incubations were run at 37°C for 2 min and processed to obtain [^{14}C] mannolipid as given under Materials and Methods. Inset: Formation of mannolipid as a function of mammary membrane concentration. Standard incubation mixtures, but with varying amounts of membrane preparation were incubated for 2 min at 37°C and processsed to obtain [^{14}C] mannolipid as given under Materials and Methods.

CHARACTERIZATION OF MANNOLIPID AS MANNOSYL PHOSPHORYL POLYISOPRENOL

When the mannolipid isolated from a scaled-up incubation mixture by extraction with C/M (2:1) was chromatographed on a column of DEAE-cellulose, more than 85% of the radioactivity was eluted as a sharp and fairly symmetrical single peak with a maximum at ~ 50 mM ammonium acetate.

The partially purified [^{14}C] mannolipid was chromatographed on both TLC plates of silica gel G and EDTA impregnated Whatman SG-81 paper in a number of solvent systems along with authentic β-mannosyl phosphoryl dolichol and calf brain mannosyl phosphoryl polyisoprenol. The results presented in Table I strongly indicate that the mammary mannolipid is a mannosyl phosphoryl polyisoprenol.

The half life of mannolipid under acidic and basic conditions of hydrolysis as given in Fig. 4 is compatible with similar results obtained for the hydrolysis of mannosyl phosphoryl polyisoprenol in several other systems (12, 15, 21).

Paper chromatography on the hydrolytic products revealed mannose and mannose phosphate (a product that cochromatographed with mannose-1-phosphate), respectively, as the products of acid and base hydrolysis. All of these data strongly imply that the mannolipid formed by mammary membranes is a mannosyl phosphoryl polyisoprenol.

STIMULATION OF MANNOLIPID BIOSYNTHESIS BY EXOGENOUS DOLICHOL PHOSPHATE

Since the [^{14}C] mannolipid synthesized in the mammary membrane system was chromatographically identical to synthetic β-mannosyl phosphoryl dolichol, it was of interest to see if exogenously added dolichol phosphate could serve as an acceptor of mannose from GDP-[^{14}C] mannose. Dolichol phosphate was found to stimulate the synthesis of [^{14}C] mannolipid at a linear rate up to 500 μM concentration (Fig. 5). As much as 8.5-fold stimulation in the synthesis of mannolipid was obtained at the highest concentration (500 μM) of exogenous dolichol phosphate. The radioactive mannolipid formed in the presence of exogenous dolichol phosphate had the mobility of mannosyl phosphoryl dolichol when chromatographed on thin layer plates of silica gel G in solvent systems A and B.

TABLE I. Mobility of [^{14}C] Mannolipid on TLC and Silica Gel Loaded Whatman SG 81 Paper Pretreated With EDTA (14)*

Glycolipids	Solvent system				
	A	B	C	D	E
Mammary endogenous lipid	0.31	0.60	0.59	0.31	0.52
Synthetic β-mannosyl phosphoryl dolichol	0.33	0.60	–	–	–
Calf brain mannosyl phosphoryl polyisoprenol	0.29	0.62	0.58	0.31	0.56

*Homemade TLC plates of silica gel G, approximately 0.5 mm gel thickness, were used. The plates and the SG 81 paper were activated at 110°C for 60 min immediately before use. Other details are given in the text.

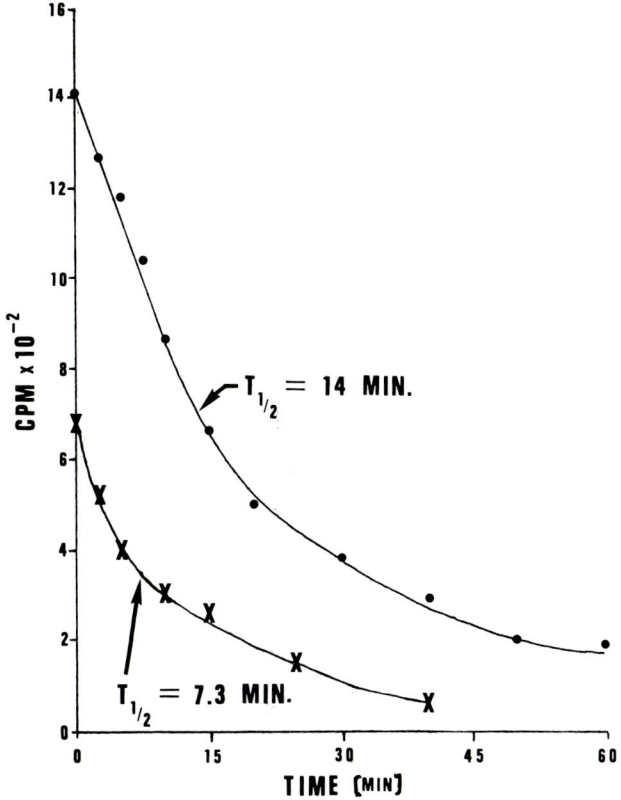

Fig. 4. Hydrolytic studies on [^{14}C] mannolipid. The radioactive glycolipid was dissolved in 1.0 ml of 0.1 N HCl in 50% 1-propanol and the reaction was carried out at 50°C (–●–). At indicated times, 0.1-ml aliquots were taken out and added to a mixture of 2 ml of C/M (2:1) and 0.5 ml of 20 mM NaOH. After vortexing, the 2 phases were separated by centrifugation and the radioactivity remaining in the organic phase was counted. For studying alkaline hydrolysis (–x–) reaction was carried out in 0.5 ml of 0.1 N NaOH in 90% ethanol at 85°C in screw cap tubes. At indicated times a 50-µl aliquot was transferred to a mixture of 4.0 ml of C/M (2:1) and 1.0 ml of 5 mM acetic acid. After vortex mixing, the 2 phases were separated by centrifugation and the radioactivity remaining in the organic phase was counted.

DIRECT TRANSFER OF MANNOSE FROM MANNOLIPID TO [MAN-^{14}C]-OLIGOSACCHARIDE-LIPID AND [MAN-^{14}C] ENDOGENOUS ACCEPTOR(S)

When a micellar suspension of [^{14}C] mannolipid was provided as a substrate in the mammary particulate system, there was a transfer of mannosyl moiety into a product soluble in C/M/W (10:10:3) and a product insoluble in C/M (2:1), C/M/W (10:10:3), water, or 10% TCA. The results in Fig. 6 clearly indicate that this transfer is virtually quantitative at short incubation times; with prolonged incubation there appears to be some loss of radioactivity, perhaps due to hydrolysis of either the [^{14}C] mannolipid or the [Man-^{14}C] oligosaccharide lipid. The amount of incorporation into [Man-^{14}C] endogenous acceptor(s) fraction is smaller than that obtained when GDP-mannose is the substrate and levels off after 20 min of incubation. These results do not rule out the possibility that initially there might have been transfer of [^{14}C] mannose to an endogenous pool of GDP

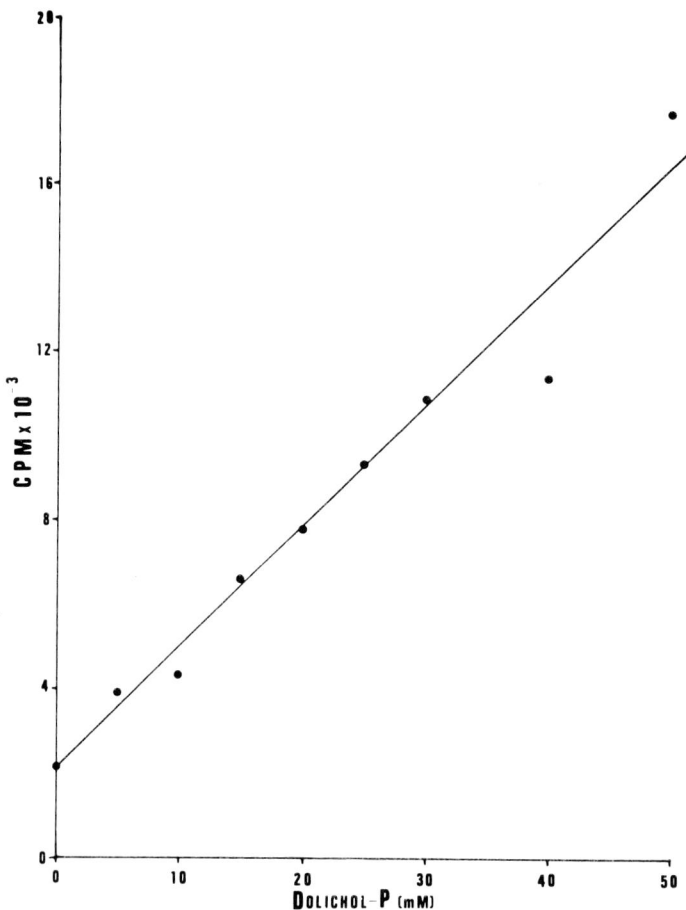

Fig. 5. Effect of dolichol phosphate on the incorporation of radioactivity from GDP-[^{14}C]mannose into the product soluble in C/M (2:1). Standard incubation mixtures that contained only 1 mg protein of the mammary membrane preparation were incubated for 2 min at 37°C with varying amounts of dolichol phosphate sonically dispersed in Ammonyx (at a final detergent concentration of 0.05%). [^{14}C]Mannolipid was isolated as given in the text.

to give rise to GDP-[^{14}C]mannose which might have then served as mannosyl donor for the glycosylation of the protein fraction. There is also the possibility that a part of mannose is incorporated into endogenous protein acceptor(s) directly from GDP-[^{14}C] mannose. When the [^{14}C]oligosaccharide-lipid was subjected to strong acidic hydrolysis and the products examined by paper chromatography, mannose was found to be the only labeled product.

PULSE-CHASE KINETICS

When the GDP-[^{14}C]mannose was chased with an excess of unlabeled substrate prior to the maximum incorporation into the mannolipid fraction, there was a rapid loss of radioactivity from the C/M (2:1) soluble fraction. This loss was much faster than a

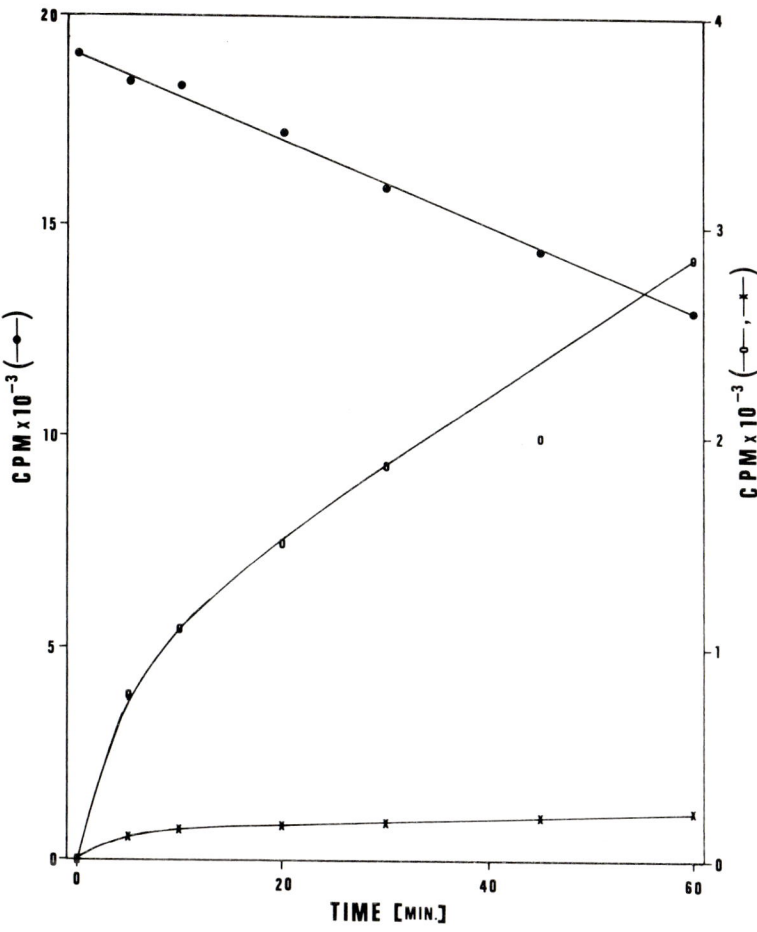

Fig. 6. Direct transfer of [^{14}C]mannose from [^{14}C]mannolipid into product soluble in C/M/W (10:10:3) and the endogenous acceptor protein(s). Details of experimental procedure are given in the text. (–●–) Product soluble in C/M (2:1); (–○–) product soluble in C/M/W (10:10:3); and (–x–) insoluble delipidated membrane pellet [endogenous acceptor(s)].

corresponding disappearance of the label from the control incubation (Fig. 7A). While the chase affected the incorporation of label into the oligosaccharide-lipid and the glycoprotein fractions as well (Figs. 7B and 7C), its effect on these reactions was not immediate and was much less pronounced. This might be expected from a pulse and chase in a precursor-product sequence of reactions. The unlabeled GDP-mannose would prevent any further appearance of label in the mannolipid fraction, but the radioactive mannolipid that would have been formed prior to the chase would continue to transfer [^{14}C]mannose into the oligosaccharide-lipid and endogenous acceptor(s) fractions for a time until all of the label from the mannolipid has been transferred to these products. The reduced incorporation of radioactivity into the oligosaccharide-lipid and endogenous acceptor(s) in the experimental fractions as compared to control incubation is also consistent with the above interpretation.

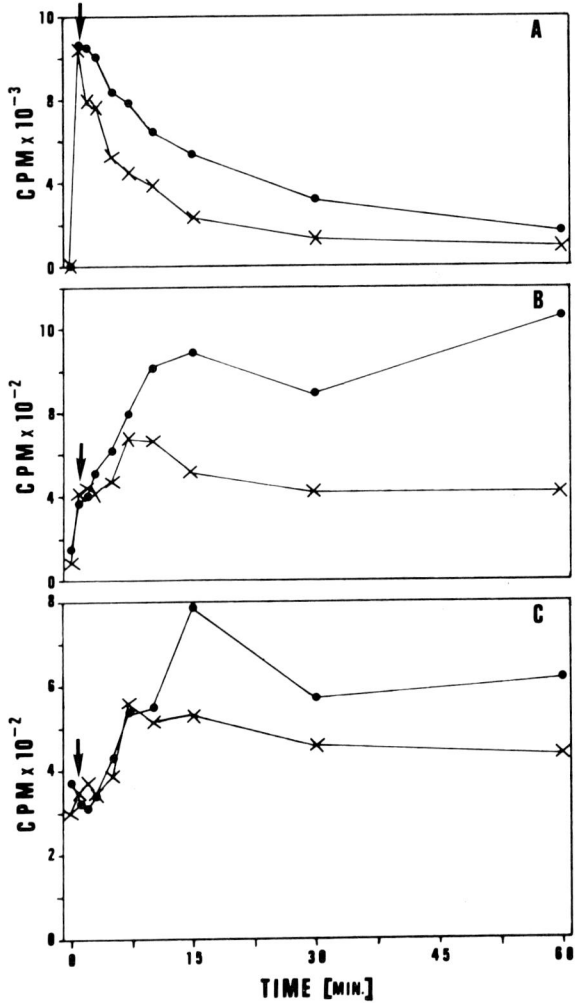

Fig. 7. Pulse-chase kinetics of mannosyl transfer into products soluble in C/M (2:1), C/M/W (10:10:3), and endogenous acceptor(s). Two tubes containing large scale incubation mixtures were prepared. Aliquots from each tube were withdrawn at different times and processed as for the standard assay. One min after the start of incubation (↓), a 100-fold excess of unlabeled GDP-mannose dissolved in 10^{-3} M Tris, pH 7.2, was added to the experimental tube. At the same time, an equal amount of 10^{-3} M Tris, pH 7.2, was added to the control tube (-●-).

NATURE OF METABOLICALLY STABLE ENDOGENOUS ACCEPTOR(S)

The radioactivity incorporated into the metabolically stable endogenous acceptor(s) was released as mannose after strong acid hydrolysis followed by paper chromatography in solvents F and G.

Of the several ionic and nonionic detergents tested for their ability to solubilize the radioactivity in the endogenous acceptor(s), only sodium dodecyl sulfate was effective. When the radioactivity in these products was solublized by sodium dodecyl sulfate and chromatographed on a column of Sephadex G-150 (Fig. 8), ~ 90% of the counts were recovered in 2 peaks eluting immediately after V_0 and a small amount was eluted in the V_i

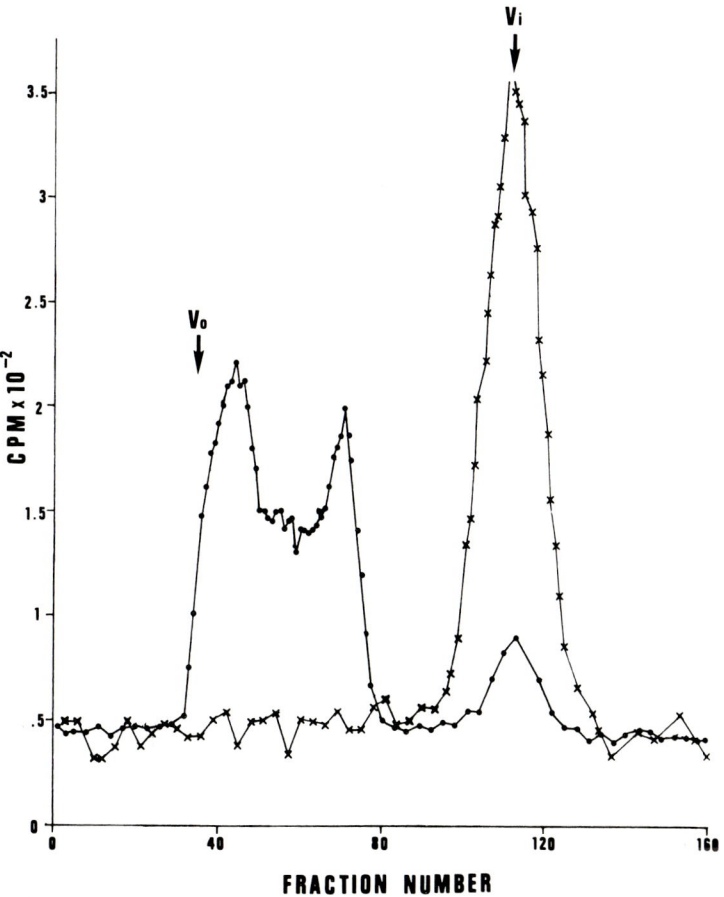

Fig. 8. Sephadex G-150 profile of protease treated (–x–) and untreated (–•–) radioactively labeled endogenous acceptor(s). After incubation with proteases as given in the text, solid sodium dodecyl sulfate was added to give a concentration of 1% and the tubes were heated at 50°C for 90 min. Conditions for gel filtration are given in the text.

region. After extensive digestion of these acceptors with proteolytic enzymes, all of the radioactivity was recovered as a single peak eluting near V_i. These results clearly imply that [^{14}C] mannose is linked to protein.

Another aliquot of the proteolytic digest was subjected to high voltage electrophoresis at pH 1.8. Two major and several minor peaks of radioactivity migrating towards the cathode, as expected of glycopeptides under these conditions of electrophoresis, were obtained. Since it was not possible to solubilize the labeled protein in a nonionic detergent, the sodium dodecyl sulfate-solubilized protein was therefore examined for its electrophoretic behavior in the control run. The solubilized labeled membrane products showed a slight anodic migration. Perhaps, in line with the Sephadex G-150 profile, the endogenous acceptors labeled in the mammary system are relatively low-molecular-weight proteins. These products might then exhibit an anionic behavior in the presence of sodium dodecyl sulfate and show some migration under these conditions of electrophoresis.

DISCUSSION

The studies presented in this report provide evidence that a 48,000 × g particulate fraction from the secreting mammary catalyzes the incorporation of mannose from GDP-[^{14}C]mannose into lipid soluble products and an insoluble glycoprotein fraction. The kinetics of mannosyl transfer are consistent with the following sequence of reactions:

$$\text{GDP-[}^{14}\text{C] mannose} \underset{}{\overset{Mn^{2+}}{\rightleftarrows}} \text{[}^{14}\text{C] mannosyl phosphoryl polyisoprenol}$$
$$\rightarrow \text{[Man-}^{14}\text{C] oligosaccharide-lipid} \rightarrow \text{[Man-}^{14}\text{C] endogenous protein(s)}.$$

The first reaction is freely reversible since in the presence of endogenously synthesized [^{14}C]mannosyl phosphoryl polyisoprenol and GDP, but not GMP, the particulate enzyme fraction catalyzes the formation of GDP-[^{14}C] mannose. Different divalent metal ions, especially Mn^{2+}, stimulate this reaction though there is a significant activity even in the absence of an added metal^{2+}. A number of different lines of evidence, as presented under Results, indicate that the product soluble in C/M (2:1) is mannosyl phosphoryl polyisoprenol. The observation that this compound cochromatographs with authentic β-mannosyl phosphoryl dolichol in a solvent system capable of separating such glycolipids on the basis of lipid moiety (29) indicates that the lipid component is probably dolichol.

The second intermediate in this pathway is assumed to be an oligosaccharide-lipid on the basis of its solubility in C/M/W (10:10:3). Glycolipid intermediates with similar solubility properties have been identified to be oligosaccharide-lipids in a number of different systems (3, 30). The rate of synthesis of this intermediate indicated that it might arise from the transfer of mannose from mannosyl phosphoryl polyisoprenol. Direct evidence for mannosyl transfer from [^{14}C]mannosyl phosphoryl polyisoprenol to an oligosaccharide-lipid and the glycoprotein fraction was obtained when mammary membranes were incubated with partially purified [^{14}C]mannosyl phosphoryl polyisoprenol in the presence of nonionic detergent Ammonyx. One might argue that mannosyl moiety could first be transferred to endogenous GDP to synthesize GDP-[^{14}C]mannose since this reaction is freely reversible; later, the GDP-[^{14}C]mannose might directly transfer mannose into the oligosaccharide-lipid and the glycoprotein. Our results have not ruled out this mode of mannosyl transfer; however, in light of the results of similar studies with other systems, it appears to be an unlikely possibility.

At present, we have no detailed structural information for the oligosaccharide-lipid soluble in C/M/W (10:10:3) apart from the fact that all of the water soluble radioactivity recovered after strong acid hydrolysis of this fraction cochromatographs with authentic mannose in systems F and G. The oligosaccharide-lipid migrates as a broad peak (R_f = 0.45) on Whatman 3MM in system F (S.R. Fram and I.K. Vijay, unpublished observations). Whether or not this is due to heterogeneity arising from several oligosaccharide-lipids due to oligosaccharide moieties of different sizes, is presently unknown. Such a heterogeneity has been reported for the oligosaccharide-lipid intermediates in a number of other systems (14, 24–26). Although analogous studies in other systems and the kinetic experiments as outlined here do implicate the oligosaccharide-lipid as an intermediate, more direct evidence is needed to definitively establish this glycolipid as an intermediate in the biosynthesis of mammary glycoproteins.

The endogenous acceptors for mannose incorporation have been shown to be glycoproteins on the basis of susceptibility to digestion with proteases followed by the appearance of mannosyl radioactivity in low-molecular-weight fragments as indicated by gel filtration and by electrophoresis.

The solubility properties of the endogenous acceptor(s) glycoproteins are indicative that these are tightly bound components of the membranes. Other than this, no information can be given on the nature of these products. In this regard, the lactating bovine mammary gland appears to be particularly attractive as a tissue of choice. It has been calculated that the entire apical membrane of the secretory cell must be replaced within 8–10 hours to replenish the material lost as fat globule membrane during secretion of fat droplets into milk (31). Further, a number of glycoproteins have been identified as components of the milk fat globule membrane (7–11).

ACKNOWLEDGMENTS

This work was supported by Biomedical Research Support Grant RRO-7042-11 given to the University of Maryland by the United States Public Health Service, and NSF Grant PCM76-16958; Scientific Article No. A2297 and Contribution No. 5296 of the Maryland Agricultural Experiment Station, Department of Dairy Sciences.

REFERENCES

1. Roseman S: In Lee EYC, Smith EE (eds): "Biology and Chemistry of Eucaryotic Cell Surfaces: Miami Winter Symposia." New York: Academic Press, 1974, vol 7, pp 317–354.
2. Spiro RG, Spiro MJ: In Scow RO (ed): "Endocrinology, Proceeding of the Fourth International Congress of Endocrinology." Amsterdam: Excerpta Medica, 1973, pp 554–560.
3. Waechter CJ, Lennarz WJ: Annu Rev Biochem 45:95, 1976.
4. Eagon PK, Hsu A-F, Heath EC: Fed Proc Fed Am Soc Exp Biol 24:678, 1975.
5. Struck DK, Lennarz WJ: J Biol Chem 252:1007, 1977.
6. Pless DD, Lennarz WJ: Proc Natl Acad Sci USA 74:134, 1977.
7. Patton S, Keenan TW: Biochim Biophys Acta 415:273, 1975.
8. Anderson M, Cawston TE: J Dairy Res 42:459, 1975.
9. Ebner KE, Schanbacher KE: In Larson BL, Smith VR (eds): "Lactation. A Comprehensive Treatise." New York: Academic Press, 1974, vol 2, pp 77–114.
10. Kanno C, Shimizu M, Yamauchi K: Agric Biol Chem 39:1835, 1975.
11. Newman RA, Harrison R, Uhlenbruck G: Biochim Biophys Acta 433:344, 1976.
12. Waechter CJ, Kennedy J, Harford J: Arch Biochem Biophys 176:724, 1976.
13. Rouser G, Kritchevsky G, Yamamoto A, Simon G, Galli C, Bauman A: Methods Enzymol 14:272, 1969.
14. Chambers J, Elbein AD: J Biol Chem 250:6904, 1975.
15. Waechter CJ, Lucas JJ, Lennarz WJ: J Biol Chem 248:7570, 1973.
16. Steiner SM, Lester RL: J Bacteriol 109:81, 1972.
17. Dittmer JC, Lester RL: J Lipid Res 5:126, 1964.
18. Dunphy PJ, Kerr JD, Pennock JF, Whittle KJ, Feeney J: Biochim Biophys Acta 136:136, 1967.
19. Trevelyan WE, Procter DP, Harrison JS: Nature (London) 166:444, 1959.
20. Baynes JW, Hsu A-F, Heath EC: J Biol Chem 248:5693, 1973.
21. Gold MH, Hahn HJ: Biochemistry 15:1808, 1976.
22. Lowry OH, Rosebrough NJ, Farr AL, Randall RJ: J Biol Chem 193:265, 1951.
23. Hsu A-F, Baynes JW, Heath EC: Proc Natl Acad Sci USA 71:2391, 1974.
24. Forsee WJ, Elbein AD: J Biol Chem 250:9283, 1975.
25. Forsee WT, Valkovich G, Elbein AD: Arch Biochem Biophys 174:469, 1976.
26. Herscovicz A, Golovtchenko AM, Warren CD, Rugge B, Jeanloz RW: J Biol Chem 252:224, 1977.
27. Hunez HA, Barker R: Fed Proc Fed Am Soc Exp Biol 25:1441, 1976.
28. Gordon WG, Kalan EB: In Webb BH, Johnson AH, Alford JA (eds): "Fundamentals of Dairy Chemistry." 2nd Ed. Westport, Connecticut: The AVI Publishing Company, 1974, pp 87–124.
29. Tkacz JS, Herscovicz A, Warren CD, Jeanloz RW: J Biol Chem 249:6372, 1974.
30. Behrens NH, Parodi AJ, Leloir LF: Proc Natl Acad Sci USA 68:2857, 1971.
31. Franke WW, Linder MR, Kartenbeck J, Zerban H, Keenan TW: J Cell Biol 69:173, 1976.

Plant Lectins Detect Age and Region Specific Differences in Cell Surface Carbohydrates and Cell Reassociation Behavior of Embryonic Mouse Cerebellar Cells

Mary E. Hatten and Richard L. Sidman

Department of Neuropathology, Harvard Medical School, and Department of Neuroscience, Children's Hospital Medical Center, Boston, Massachusetts 02115

When plated at high cell density in a microwell culture system, freshly dissociated embryonic mouse cerebellar cells assemble into reproducible, 3-dimensional patterns. The addition of the dimeric lectin Succinyl Concanavalin A blocks reversibly the formation of the microwell pattern, suggesting that cell surface carbohydrates affect the reassociation behavior of embryonic mouse cerebellar cells.

Agglutination studies of dissociated cell populations harvested from different regions of the embryonic brain reveal that different lectins agglutinate cell populations from different embryonic brain regions. Cells from E13 cerebellum are agglutinated with Concanavalin A, wheat germ agglutinin, Ricinus communis agglutinin, mol wt 60,000, Ricinus communis agglutinin, mol wt 120,000, and Lens culinaris, but not by soybean agglutinin or a fucose-binding protein. Cells from the midbrain are agglutinated only with Concanavalin A, Ricinus communis agglutinin, mol wt 60,000 and Ricinus communis agglutinin, mol wt 120,000; those from the cerebral cortex are agglutinated only with Lens culinaris; and those from the medulla are agglutinated only with Ricinus communis agglutinin, mol wt 60,000, and Ricinus communis agglutinin, mol wt 120,000. In addition, agglutination of cerebellar cells with Concanavalin A, wheat germ agglutinin, and Ricinus communis agglutinin is diminished over the course of development from embryonic day 13 to postnatal day 7. These studies suggest regional differences in the cell surfaces of the developing brain that are further modulated during the differentiation of the tissues.

On a poly(D-lysine) treated substrate in microwell cultures, cell migration is unique to the cerebellum of the 4 brain regions studied. Surfaces treated with carbohydrate-derivatized poly(D-lysine) are currently being tested for their efficacy as substrates for differential cell migration.

Key words: plant lectins, microwell cultures, cell migration

Abbreviations: CMF – calcium- and magnesium-free-balanced salt solution; ConA – Concanavalin A; E13 – embryonic day 13; LCA – Lens culinaris hemagglutinin A; LCB – Lens culinaris hemagglutinin B; PBS – phosphate-buffered saline; RCAI – Ricinus communis agglutinin, mol wt 60,000; RCAII – Ricinus communis agglutinin, mol wt 120,000; SBA – soybean agglutinin; Succ-ConA – Succinyl Concanavalin A; WGA – wheat germ agglutinin.

Received March 18, 1977; accepted June 6, 1977.

© 1977 Alan R. Liss, Inc., 150 Fifth Avenue, New York, NY 10011

During embryonic mouse cerebellar development, immature neurons migrate from the sites where they are generated to the positions they will occupy in the mature neuronal network (1, 2). When embryonic cerebellar tissue is removed and plated as a single-cell suspension of high cell density in microwell cultures, a particular pattern of cell reassociation is evident (3). Microwell cultures will be used as a model system to monitor the reassociation behavior of suspensions of single cells from embryonic mouse cerebellum, as well as to measure in vitro cell migration.

Plant lectins are useful probes of micromolecules with exposed carbohydrates at cell surfaces (4–6). With 8 lectins the agglutination of cells harvested from various brain regions at different developmental stages was assayed to survey cell surface carbohydrates. In addition, nontoxic dimeric lecitns were used in microwell cultures to assess the influence of cell surface carbohydrates on cell reassociation behavior (7).

MATERIALS AND METHODS

Preparation of Single Cell Suspensions From Embryonic Mouse Brain

The embryos from litters at the 13th day of gestation were removed by laparotomy from freshly killed C57B1/6J female mice. The cerebellum was removed by dissection (3), and transferred to 2.0 ml of CMF (NaCl, 8.00 g; KCl, 0.30 g; Glc, 2.00 g; $NaH_2PO_4 \cdot 2H_2O$, 0.5 g; KH_2PO_4, 0.25 g; 2.00 ml of 5% $NaHCO_3$; 0.5 ml of a 0.5% solution of phenol red in 480 ml distilled H_2O. The tissue was washed 3 times by sedimentation at $1 \times g$, and a single-cell suspension was prepared by gentle trituration with 2 or 3 fire-polished Pasteur pipets of decreasing bore size. This cell suspension was 85–95% single cells, and any remaining clumps of cells were removed by passing the cells through a Swinnex filter fitted with a double thickness of lens paper. The cells were sedimented (600 rpm in a Sorvall, model 6CC-1, Rotor HL-4 [Sorvall, Newton Conn.], 5 min, $4°C$), and the pellet was resuspended in Eagle's basal medium (Hank's salts) supplemented with horse serum (10%), glucose (6 mM), glutamine (4 mM), and penicillin-streptomycin (25 units/ml each) for microwell cultures, or in PBS (NaCl, 8 g; KCl, 0.200 g; $Na_2HPO_4 \cdot 2H_2O$, 1.440 g; KH_2PO_4, 0.200 g; $CaCl_2 \cdot 2H_2O$, 0.132 g; $MgCl_2 \cdot 6H_2O$, 0.100 g per liter distilled H_2O) for agglutination assays. For culture assays, the cells were resuspended at $3–4 \times 10^6$ cells/ml. For agglutination assays, the cells were resuspended at $1–2 \times 10^6$ cells/ml. The viability of this preparation was 85–95% by Trypan blue exclusion (3).

Primary Microwell Cultures

Falcon microtest plates were pretreated with poly(D-lysine) (mol wt 150,000; 1.0 μg/ml in H_2O) for 2 h at $37°C$ as described (3). The plates were washed with distilled H_2O and air dried. Cells (10 μl) from the cell suspension prepared as above were introduced into the microwells with a Finnpipette fitted with a sterile tip. The viability of the cells 24 h after plating (3) was 60–80%. Incubation of the cultures was at $35.5°C$ with 5% CO_2 and 100% humidity.

One major advantage of microtest plates is the number of duplicate cultures possible (60 per plate). From each single-cell preparation from a litter of 6–8 embryos, 80–100 individual microwell cultures were established. The results reported are based upon cultures from 40 E13 litters (2,400–4,000 individual microwell cultures).

In 4 microwell experiments (240 microwell cultures), Succinyl-ConA (Succ-ConA) (1–500 μg/ml in Eagle's basal medium supplemented as above) was added to microwell cultures 4 h after plating by exchanging 5 μl of the growth medium for 5 μl of the Succ-ConA solution.

Quantitation of the incidence of reaggregates, monolayer, and cables of processes were assayed on growing cultures by phase contrast on an inverted microscope.

Lectin-induced Agglutination Assays

These were performed according to Burger (8) with the single-cell preparation described above. The cells were tested for Trypan blue exclusion at the beginning and conclusion of the agglutination assays, and all results reported were from cell suspensions with 85–95% viability at both time points.

For tissues from postnatal animals, single-cell suspensions were prepared as described (3) with gentle trypsinization [crystalline 10 μg/ml, 10 min, 20°C, reaction stopped by ovomucoid (1 mg/ml), and trypsin removed by washing 3 times at 600 rpm, 5 min, 4°C]. Prior to the agglutination assay, the cells were preincubated in Eagle's basal medium supplemented with horse serum (1%) for 6 h at 35.5°C. The cells were washed 3 times with PBS to remove the serum. In some experiments, agglutination was assayed directly after the trypsinization step. All of the results reported in Tables I and II derive from 6 or more separate cell preparations. For each cell preparation, agglutination assays were performed in quadruplicate.

Materials

Microtest plates were obtained from Falcon Plastics (Brooklyn, New York) (No. 3034). Basal Medium Eagle's (Hank's salts), penicillin-streptomycin (25,000 units per ml, Lot 91035) and L-glutamine (200 mM in 0.85% NaCl) were purchased from Microbiological Associates (Bethesda, Maryland). Sera were purchased from GIBCO (horse serum, lot numbers 8350120 and 165322) and from Microbiological Associates (horse serum lot numbers 81428, 90111, 90467; fetal calf serum lot numbers 85260 and 86141; calf serum lot number 88985). Poly(D-lysine) (mol wt 150,000, type 1-B, hydrobromide, No. P-7761), methyl-α-D-mannoside, d-biotin, and ovomucoid (trypsin inhibitor, type 11-0, T9253) were

TABLE I. Region-specific Lectin-induced Agglutination of E13 Cells Harvested From Several Brain Regions*

Lectin	Hapten	Cerebellum	Midbrain	Cerebral cortex	Medulla	Liver
ConA	α-M-Man	200	200	1,000	1,000	115
WGA	D-Glc-NAc	50	500	500	500	25
RCAI	D-Gal	150	200	1,000	125	250
RCAII	D-Gal	75	125	1,000	80	175
SBA	D-Gal-NAc	1,000	1,000	1,000	1,000	1,000
Lotus	D-Fuc	1,000	1,000	1,000	1,000	1,000
LCA	D-Glc	50	250	75	1,000	250
LCB	D-Glc	100	500	75	1,000	500

*Cell suspensions were prepared as described in Materials and Methods without the use of enzymes. Agglutination was assayed according to Burger (8). Values are given as lectin required for half maximal agglutination (μg/ml). For WGA, values less than 400 represent good agglutination. For the other lectins tested, values less than 500 indicate agglutinable cells.

TABLE II. Developmental-stage-specific Agglutination of Cerebellar Cells*

Lectin	Hapten	E13	P0	P7
ConA	α-M-Man	200[a]	800 (200)	1,000 (200)
WGA	D-Glc-NAc	50[a]	450 (50)	500 (50)
RCAI	D-Gal	75[a]	1,000 (150)	1,000 (150)
SBA	D-Gal-NAc	1,000[a]	1,000[a]	1,000[a]
Lotus	D-Fuc	1,000[a]	1,000[a]	1,000[a]
LCA	D-Glc	50[a]	50[a]	60[a]
LCB	D-Glc	1,000[a]	150[a]	120[a]

*Cells were harvested from animals at postnatal days zero (P0) or 7 (P7) as described (3). Agglutinations were performed as described by Burger (8). Values are given as lectin required for half maximal agglutination (μg/ml). In some experiments, cells were trypsinized (0.01%, 10 min), pelleted, resuspended, and assayed for agglutination. Values in parentheses represent half maximal agglutination after trypsinization. For WGA, values less than 400 indicate good agglutination. For the other lectins tested, values less than 500 indicate good agglutination.
[a]Agglutinability not increased by trypsinization

purchased from Sigma Chemical Company (St. Louis, Missouri). Trypsin (crystalline, TRL 36C897, 194 μ/mg) was obtained from Worthington Biochemicals.

Highly purified ConA and WGA were the generous gifts of Dr. Max M. Burger, Biozentrum, Basel, Switzerland. All other lectins were purchased from Miles and further purified by affinity chromatography (3). Succ-ConA was the gift of Dr. R. J. Mannino, Biozentrum, Basel, Switzerland and its purity was as described (7).

In vitro cell migration was measured by a modification of the method of Bürk (9). A wound was made in the culture by gently vacuuming off a lane of cells with a micropipet mounted on a micromanipulator. The growth medium was removed and replaced with fresh medium without serum. The number of cells migrating into the wound was assayed by phase contrast with an inverted microscope.

RESULTS AND DISCUSSION

The studies reported here with a microwell culture assay of cell interactions in vitro and with plant lectins as probes of the molecular architecture of embryonic and postnatal brain cells reveal differences in cell reassociation behavior and cell surface features among tissues from several brain regions at different developmental ages.

A reproducible 3-dimensional pattern of reaggregates and interconnecting structures (Fig. 1) was formed by dissociated embryonic mouse cerebellar cells in microwell cultures. Within 4 h after the addition of the single-cell suspension to the microwells, reaggregates of 2,000–10,000 cells each had formed. After 6–24 h, the reaggregates were larger and straight interconnecting structures ("cables") were observed between them. These cables were invariably populated with a large number of neurons as well as cell processes. After 24–48 h in culture, a monolayer of phase-bright, highly refractile cells with processes was observed as a skirt around individual reaggregates. No further change in the organization of the pattern was observed for periods up to 6 weeks in culture.

These results suggest that embryonic mouse cerebellar cells have, under the described conditions, a particular reassociation behavior in microwell cultures. As described elsewhere (3), the details of this behavior depend on the serum supplement and the growth

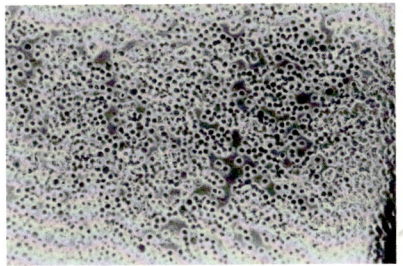

0 hr

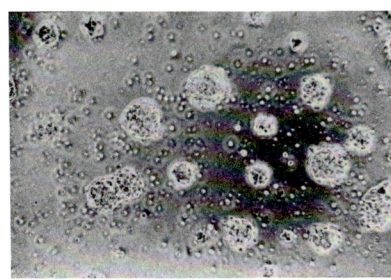
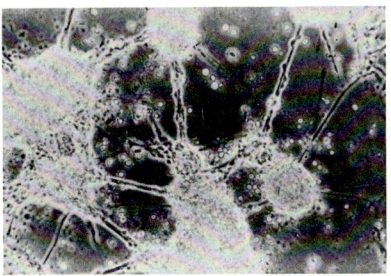

4 hr   24 hr

Fig. 1. The course of pattern formation by E13 cerebellar cells in microwell culture. Cerebellum was removed by dissection, washed twice in CMF, and triturated into a single-cell suspension with a series of fire-polished Pasteur pipets of decreasing bore. The cells were pelleted and resuspended at 3×10^6 cells/ml in Eagle's medium supplemented with glucose (6 mM), penicillin-streptomycin (25 units/ml), and horse serum (10%). The substratum was pretreated with poly(D-lysine). The pattern of reaggregates (R) and interconnecting cables (C) was formed after 24 h in culture. For details, see Ref. 3. $\times$ 150.

substratum, as well as the brain region and developmental stage of the tissue from which the cells were harvested. The question as to which cell types or interactions in the microwell cultures are the primary organizers of the pattern remains to be clarified. Available electron microscopic and electrophysiologic evidence (Fischbach and Hatten, unpublished) do, however, indicate that the E13 cerebellar cell population is predominantly neuronal.

The reassociation events required to form the described microwell pattern were influenced strongly by addition of the nontoxic, dimeric lectin Succ-Con A. In a growth medium supplemented with 6 mM glucose, low concentrations of Succ-ConA inhibited cable extension in the cultures by a mechanism that was reversed within 4 h by the removal of Succ-ConA and was blocked by preincubation of the cultures with methyl-α-D-mannoside (Fig. 2). The dose dependence of this effect is given in Fig. 3. These observations suggest that cable extension in microwell cultures relates to cell surface carbohydrates.

The binding of Succ-ConA could interrupt cable extension by several general mechanisms. Since Succ-ConA is added 4 h after plating, it is unlikely that it interferes with cell attachment to the substratum or the initial aggregation of the cells. However, Succ-ConA could influence cell interactions (or movements) within the reaggregates subsequent to the initial aggregation that signals cable extension. Alternatively, the binding of

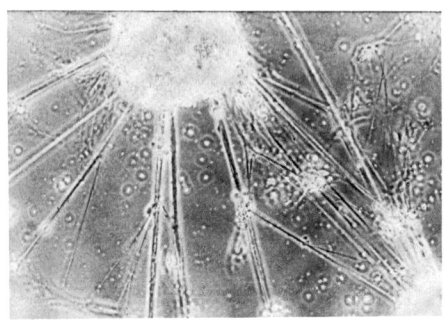

Control

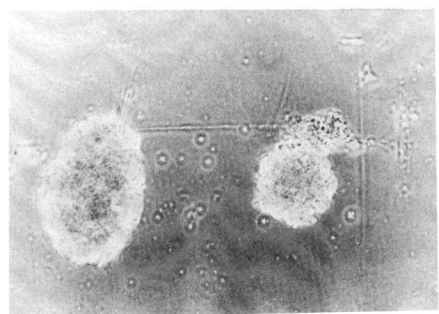

100 µg/ml S-ConA

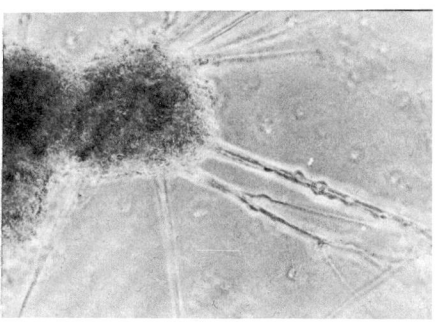

5 µg/ml S-ConA

Fig. 2. Effects of Succ-ConA on pattern formation by E13 cerebellar cells in microwell culture. Succ-ConA was added at the indicated dose 4 h after plating. × 250.

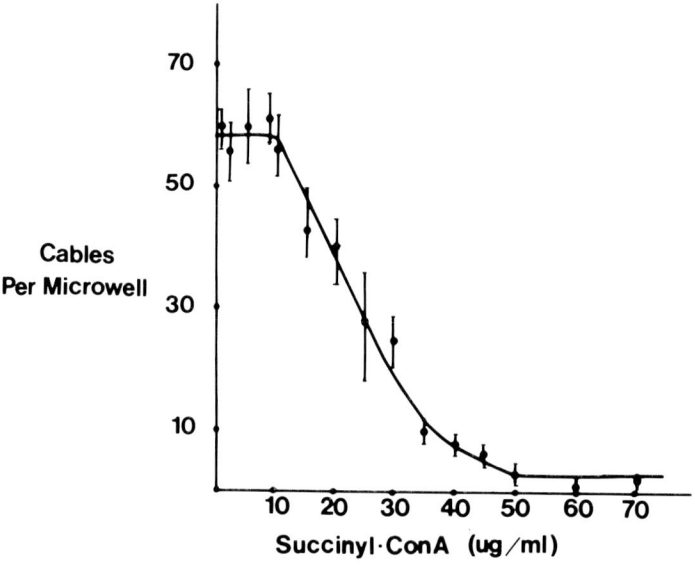

Fig. 3. Effect of Succ-ConA on cable formation in E13 cerebellar cultures. Succ-ConA was added at the indicated concentration 4 h after plating. Counts of the number of cables per microwell were made by phase contrast with an inverted microscope 96 h after plating.

Succ-ConA to the surface of fiber producing cells could inhibit the lengthening of cables per se. At any rate, these results suggests that conditions for cable extension relate, at least in part, to the cell surface.

Eight lectins were used to survey cell surface carbohydrate-containing macromolecules with cells harvested from several brain regions at different developmental stages.

Lectin-induced agglutination assays with single cell suspensions from various brain regions at several developmental stages revealed differences in agglutinating activity. E13 cells from the cerebellum, midbrain, and liver, but not cerebral cortex or medulla, were agglutinated with ConA (Table I). The reaction was temperature-dependent as previously reported for non-neuronal cells (10) and was inhibited by preincubation of the cells with methyl-α-D-mannoside. E13 cells from medulla and cerebral cortex were rendered agglutinable with ConA by gentle trypsinization (10 μg/ml, 10 min, 37°C).

E13 cells from cerebellum and liver, but not other brain regions, were agglutinated by WGA. The reaction was temperature-independent as reported for nonneuronal cells (10), and was inhibited by N-acetyl-D-glucosamine. E13 cells from midbrain, medulla, and cerebral cortex were rendered agglutinable with WGA, however, by gentle trypsin treatment (10 μg/ml, 10 min, 37°C). Cells from cerebellum, midbrain, medulla, and liver, but not cerebral cortex, were agglutinated by both Ricinus communis lectins I and II (RCAI and RCAII), a reaction that was inhibited by preincubation of the cells with D-galactose. Nonagglutinable E13 cells from cerebral cortex were rendered agglutinable with RCAI or RCAII by gentle trypsin treatment (10 μg/ml, 10 min, 37°C).

E13 cells from cerebellum, midbrain, and cerebral cortex were agglutinated with Lens culinaris A (LCA) lectin, a reaction that was inhibited by D-glucose. E13 cells from cerebellum and cerebral cortex, but not midbrain, medulla, or liver, were agglutinated by Lens culinaris B (LCB) lectin, a reaction inhibited by D-glucose. Trypsinization of cells that did not agglutinate with LCA or LCB lectins did not alter these results.

None of the cells tested were agglutinated with either the soybean agglutinin (SBA) or a fucose-binding protein from lotus seeds. Trypsinization (10 μg/ml, 10 min, 37°C) of the cells did not alter these results.

In addition to regional differences in the lectin-induced agglutination of developing neurons, developmental stage-specific alterations in lectin-induced agglutination were observed (Table II). The concentration of lectin required for the half maximal agglutination of E13 cerebellar cells was increased markedly for cerebellar cells harvested after embryonic day 16. As shown in Table II, postnatal cerebellar cells had little agglutinating activity with any of the 7 lectins tested except LCA. Unlike the other lectins tested in these studies, LCA agglutinates most nontransformed rather than transformed cell lines. (4, 5). Postnatal cerebellar cells were agglutinable with ConA, RCA, or WGA following gentle trypsinization (10 μg/ml, 10 min, 37°C).

The agglutination of embryonic neuronal populations appears to occur by a mechanism similar to that reported for nonneuronal cells, since the same hapten inhibition and temperature dependence is observed for both cell types (10, 12, 13). The differences reported here for lectin-induced agglutination probably do not reflect simple differences in intercellular adhesion, since recent studies (14, 15) have demonstrated that lectin-induced agglutination and spontaneous aggregation occur by separate mechanisms.

Agglutination with plant lectins is a complex process that can reflect, in addition to the carbohydrate composition of the receptor(s), more general cell surface features such as receptor availability, mobility, valence, and anchorage to a cytoskeletal network of submembranous elements (4–6). Although these results cannot be interpreted specifically as differences in carbohydrate-containing surface macromolecules among several brain

TABLE III. In Vitro Cell Migration of E13 Embryonic Cerebellar Cells*

Tissue origin	Migration (%)	Time (h)
Cerebellum	100	2
Midbrain	0	24
Cerebral cortex	0	24
Medulla	50	24

*In vitro cell migration was measured by a modification of the method of Bürk (9). A wound was made in the culture by gently vacuuming off a lane of cells with a micropipet mounted on a micromanipulator. The growth medium was removed, and replaced with fresh medium without serum. The number of cells migrating into the wound was assayed by phase contrast with an inverted microscope. Values are given as percent of wound area occupied by cells at the given time.

regions at different stages of development, they do suggest region-specific alterations in the cell surfaces of developing neurons.

Changes in lectin-induced agglutination over the course of chick retinal development have been suggested previously (11). However, one critical difference between those early studies and the present results is that these embryonic brain cells were not treated with proteases or with EGTA prior to the agglutination assays.

Studies with fluorescein-conjugated lectins are currently in progress to determine the amount of lectin bound by different cell types in particular brain regions. Preliminary evidence (Hatten, Lekić, and Schachner, unpublished) indicates that the E13 cerebellar population is heterogeneous with respect to lectin binding. It is possible that kinetic studies with lectins and carbohydrate haptens will detail further differences in carbohydrate-containing macromolecules during cerebellar development.

Cell migration is a major force in the modeling of the cerebellar neuronal network. We are therefore in the process of developing a model system to assess in vitro embryonic cerebellar cell migration in microwell cultures, and to ask whether migration events relate to the presence of particular carbohydrates in the substratum.

On a poly(D-lysine)-treated substratum, in vitro cells of the embryonic cerebellum migrated much more than cells of the other regions tested (Table III). In addition, considerable movement of whole reaggregates was observed. We are currently examining whether culture substrate treated with poly(D-lysine) derivatized with one or another of the cell surface carbohydrate moieties that had been detected in the agglutination assays will have different efficacies for cell migration.

REFERENCES

1. Ramón y Cajal S: Int Monatsch Anat Physiol (Leipzig), 7:12, 1890.
2. Sidman RL, Rakic P: Brain Res 62:1, 1973.
3. Hatten ME, Sidman RL: Manuscript submitted.
4. Lis H, Sharon N: Annu Rev Biochem 43:541, 1973.
5. Rapin AMC, Burger MM: Adv Cancer Res 20:1, 1974.
6. Chowdhury TK, Weiss AK (eds): "Concanavalin A." New York: Plenum Press, 1974.
7. Mannino RJ, Burger MM: Nature (London) 256:19, 1975.
8. Burger MM: In Colowick SP, Kaplan NO (eds): "Methods in Enzymology." New York: Academic Press, 1974, vol 32, pp 615–621.
9. Bürk RR: Proc Natl Acad Sci USA 70:369, 1973.

10. Horwitz AF, Hatten ME, Burger MM: Proc Natl Acad Sci USA 71:3115, 1975.
11. Kleinschuster SJ, Moscona AA: Exp Cell Res 70:397, 1972.
12. Inbar M, Ben-Bassat H, Sachs L: Proc Natl Acad Sci USA 68:2748, 1971.
13. Noonan KD, Burger MM: J Cell Biol 59:134, 1973.
14. Wright TC, Ukena TE, Campbell R, Karnovsky MJ: Proc Natl Acad Sci USA 74:258, 1972.
15. Steinberg MS, Gepner IA: Nature (London) New Biol 241:249, 1973.

Cell Surface Changes Accompanying Myoblast Differentiation

L. T. Furcht, G. Wendelschafer-Crabb, and P. A. Woodbridge

Department of Laboratory Medicine and Pathology, University of Minnesota, Minneapolis, Minnesota 55455 (L.T.F., G.W-C., and P.A.W.), and St. Paul Regional Red Cross Blood Center, St. Paul, Minnesota (L.T.F.)

Myoblasts are mononucleated cells and associated with differentiation undergo cell fusion and become multinucleated. The current studies have examined cell surface dynamic changes of Concanavalin A lectin receptor mobility and the role of hormones in modulating myoblast differentiation. A uniform distribution of Con-A receptors is observed in undifferentiated cells when reacted with Con-A at 37°C. Cells from differentiating cultures or fully differentiated myotubes reacted similarly at 37°C show a significant redistribution of Con-A into patches, "caps," and endocytic vesicles containing Con-A. If undifferentiated and differentiated cells are first prefixed with glutaraldehyde then reacted with Con-A continuous distribution of Con-A is seen across the cell surface. This suggests redistribution of Con-A and its receptors occurs in differentiated cells reacted with lectin at 37°C. It is further shown that insulin (10 μg/ml) significantly enhances myoblast differentiation but that this occurs after an apparent stimulation of proliferation. In contrast to insulin, dexamethasone (10 μM and 100 μM) profoundly inhibits myoblast differentiation while having different effects on proliferation; 10 μM dex stimulates cell growth while 100 μM dex suppresses cell proliferation. Lastly, an extracellular filamentous matrix which binds Con-A is observed at the ultrastructural level in high density cultures. No significant redistribution of Con-A is observed on this matrix in distinction to the redistribution observed on the cell membrane in differentiated cells.

Key words: myoblast differentiation, Con-A, extracellular filamentous matrix, insulin, dexamethasone

Recent work on cell membrane organization dynamics and composition has provided useful insights into studies of cell proliferation of various cell types. Early studies by Aub et al. (1) and Burger (2) showed differences in the agglutination of normal and transformed cells with lectins. This was followed by studies of Edidin (3) and dePetris and Raff (4) which showed the "fluid" nature of membrane receptors subsequent to binding of antibodies to histocompatibility antigens and xenoantibodies to surface immunoglobulin on lymphocytes, respectively.

Abbreviations: Con-A — Concanavalin A; CPK — creatine phosphokinase; dex — dexamethasone

Received April 5, 1977; accepted June 16, 1977.

© 1977 Alan R. Liss, Inc., 150 Fifth Avenue, New York, NY 10011

We have been interested in cell surface changes accompanying differentiation and have used L6 rat myoblast cell line originally isolated by Yaffe (5, 6). This cell line was isolated from embryonic rat striated muscle, and in its undifferentiated state consists of mononucleated cells. Differentiation is associated with cell fusion and myotube formation, increased synthesis of myofibrillar proteins and myofilaments, increased levels of creatine phosphokinase (CPK) and other enzymes, and the development of acetylcholine receptors (5–10). The current studies have examined cell surface topography and the ability for Concanavalin A (Con-A) to undergo redistribution at various stages of myoblast differentiation. In addition, we have examined the role of insulin and dexamethasone (dex) in affecting cell proliferation and differentiation in this system.

METHODS

Clonal isolates of L6 myoblasts were kindly supplied by Dr. David Schubert, Salk Institute. Cells are typically plated at $2,500/cm^2$ in 100-mm sterile Falcon plates, in Dulbecco's minimal essential media supplemented with 10% fetal calf serum (Flow, Rockville, Maryland) under conditions described for other cell types (11).

In addition to the formation of myotubes, differentiation is assessed by an increase in CPK activity (6). To assess the proliferation of the cells we have made sequential measurements of total DNA per culture dish. We have found the DNA assay more reliable than cell counts as once the cells start fusing it becomes rather difficult to get reliable data. In addition to sequential determinations of DNA to assess cell proliferation, $[^3H]$dThd incorporation was assayed. Cells are pulsed with 10 μCi of $[^3H]$dThd (New England Nuclear Corporation, Boston, Massachusetts, specific activity 40–60 μCi/M) for 2 h, harvested, and counted as described (11).

For biochemical assays cells are scraped from plates in 50 mM imidazole-phosphate buffer, pH 6.75, and homogenized at 4°C for 90–120 sec using a motorized Dounce homogenizer with a Teflon pestle at 600 rpm. Each sample is then split for determination of CPK and DNA. CPK activity is measured using a modification of the method of Oliver using a GEMSAC fast analyzer (12). DNA is extracted with 0.5 N perchloric acid and quantitated according to the method of Burton (13) by measuring the optical density at 600 nm in a Beckman 25 spectrophotometer using calf thymus DNA (Sigma Chemical Company, St. Louis, Missouri) as a standard.

Studies are also performed to examine the effects of insulin or dexamethasone on L6 proliferation and differentiation. Cells are allowed to grow up to a density of 20,000–40,000 cells/cm^2 and the medium is changed to one of the following: 1) medium + 10 μg/ml insulin (Sigma); 2) medium + 100 μM dex (Sigma); 3) medium + 10 μM dex; or 4) routine medium. Stock dex solutions are made 5×10^{-2} M in absolute ethanol with final concentrations of ethanol less than 0.2% in the dex and control medium samples.

Cells for Con-A localization are taken at 3 stages: 1) low density undifferentiated (approximately 20,000/cm^2); 2) predominantly mononucleated cells showing early signs of differentiation (cell density approximately 100,000/cm^2) and 3) advanced differentiation with preponderantly myotubes. At these various stages cells are washed 3 times in phosphate-buffered saline (PBS) (pH 7.4) and incubated with either 30 or 50 μg/ml Con-A (Sigma) for times ranging from 5 to 20 min. Samples are then washed 3 times with PBS and fixed in this buffer with 1.6% glutaraldehyde (Electron Microscopy Sciences, Fort Washington, Pennsylvania) for 20 min at 37°C. All above incubations and washings were performed at 37°C. To assess the native distribution of cell receptors for Con-A, parallel samples are prefixed with 2.5% glutaraldehyde-PBS for 20 min at 37°C, and then reacted with Con-A and fixed again as above. Con-A is then localized by the peroxidase reaction with diaminobenzidine after the method of Bernhard and Avaremas (14). Samples are processed for ultrastructural cytochemistry as described for routine electron microscopy (see below) with the omission of tannic acid and staining with lead citrate.

For routine electron microscopy samples are washed in 0.1 M cacodylate buffer, pH 7.2, fixed with 2.5% glutaraldehyde in 0.1 M cacodylate buffer, pH 7.2, at 37°C and subsequently postfixed for 1 h with 1% OsO_4 (Electron Microscopy Sciences) 0.1 M cacodylate (15). Cells are then treated with 1% tannic acid (Mallinckrodt, St. Louis, Missouri) (16), dehydrated in graded series of ethanol (30, 50, 70, 95, and 100%) and embedded in Epon 812 (17). Silver to gray sections (40–60 nm thick) are cut with a diamond knife on a LKB Ultratome III and mounted unsupported on 200-mesh copper grids. Thin sections are counterstained with Reynold's lead citrate (Baker) (18) and examined in a Philips 300 electron microscope.

To quantitate the surface distribution of Con-A, 1-μm thick epon sections are cut and examined with a Zeiss light microscope at 400 or 1,000 ×. The distribution of Con-A is denoted as either uniform showing a continuous distribution or showing redistribution, i.e., patching and/or capping evidenced by a segregation from a uniform distribution.

RESULTS

L6 myoblasts in their undifferentiated state at low density (approximately 10,000–20,000/cm^2) are mononucleated. These are very active cells having a very dilated rough endoplasmic reticulum, numerous ribosomes, and a paucity of myofilaments, though occasional microfilament bundles are observed (Fig. 1). Differentiation is dependent on medium conditioning, cell contacts, and on calcium ions among other things (5–10). At the early stages of differentiation the free ribosomes appear to be actively synthesizing the contractile proteins and most are associated with numerous thin and thick filaments (Fig. 2). Dense bodies which may correspond eventually to Z-line material are seen within these filament bundles and numerous microtubules, generally not associated with ribosomes, are observed (Fig. 2). At more advanced stages of differentiation the density of myofilaments increases (Fig. 3). At more advanced stages myofilaments organize and the relative number of microtubules appears to decrease compared to undifferentiated cells (Fig. 4).

Using this background information and biochemical estimations of differentiation (see below) we have examined Con-A receptor topography in cells at various stages of differentiation and the ability of cells to undergo receptor redistribution upon interacting with this ligand. Undifferentiated L6 myoblasts, interacted with 50 μg/ml Con-A at 37°C for 10 min, show a predominantly uniform distribution of Con-A (Fig. 5). Undifferentiated cells which are pre-fixed prior to reacting with Con-A also have a uniform distribution of Con-A at the cell surface (Fig. 6).

As cells begin to differentiate, we see an alteration of lectin distribution when compared to the uniform distribution seen in undifferentiated myoblasts. When these cells are reacted with Con-A at 37°C and then fixed, a predominantly uniform pattern of Con-A is seen at the cell surface; however, microvilli show markedly decreased staining for Con-A (Fig. 7). Interestingly, if samples from these cultures showing early differentiation are pre-fixed and then reacted with Con-A a uniform distribution is seen over the plasma membrane and the microvilli (Fig. 8). The uniform distribution of lectin receptors on pre-fixed cells suggests the native topographical array and that changes from this seen in samples reacted with Con-A prior to fixation represent redistribution (19).

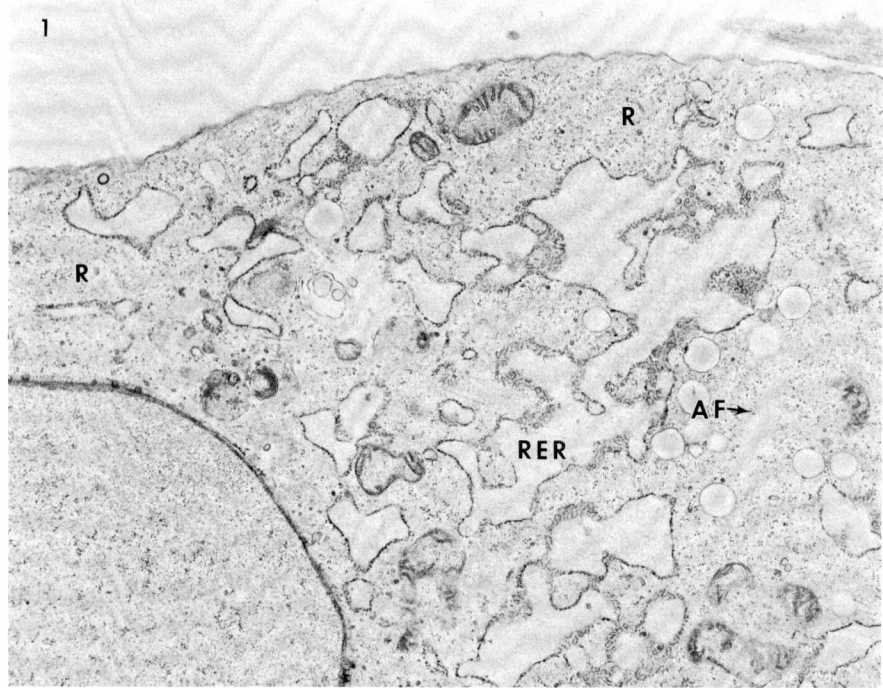

Fig. 1. Electron micrograph of undifferentiated L6 myoblast. Distended rough endoplasmic reticulum (RER) predominate in thin section of myoblast. Note great number of free ribosomes (R) and actin filaments (AF). Heavy metal counterstain (9,500 ×).

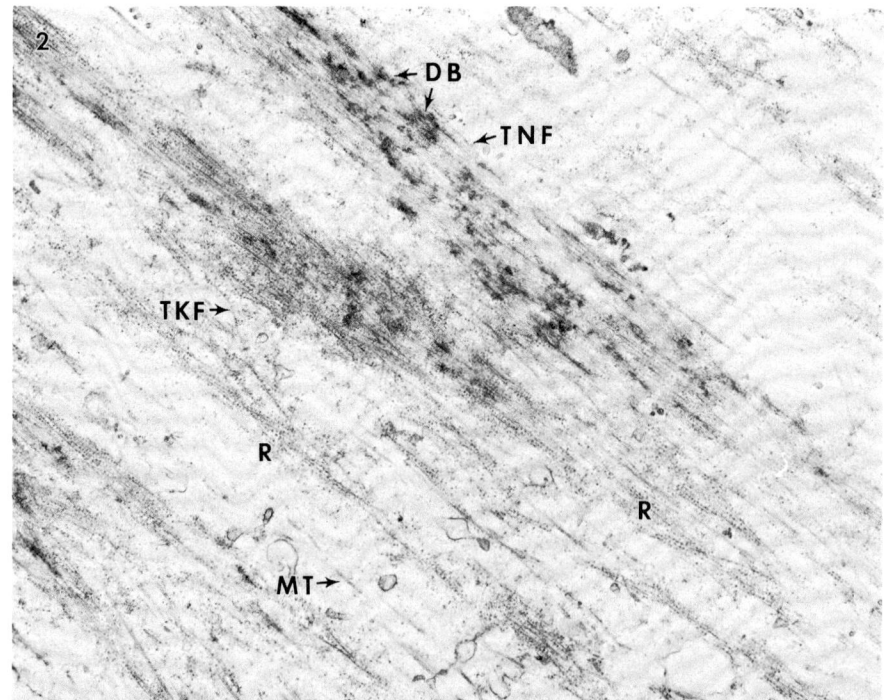

Fig. 2. Electron micrograph of differentiating L6 myoblast. Numerous ribosomes (R) are visible in close proximity to thin filaments (TNF) and thick filaments (TKF). Dense bodies (DB) are present in association with filament bundles while microtubules (MT) commonly course between filament bundles. Heavy metal counterstain (12,300 ×).

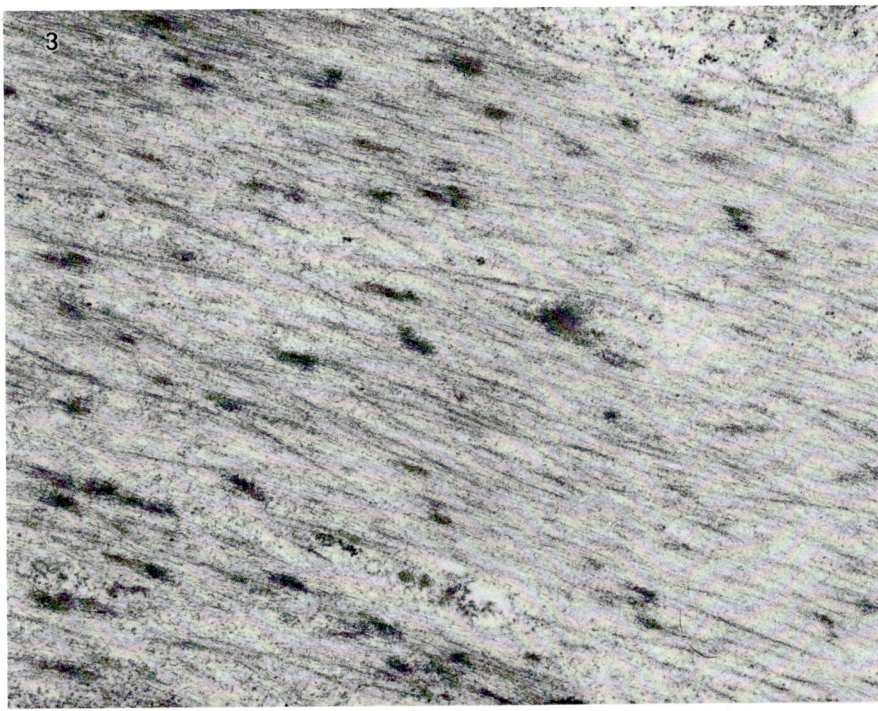

Fig. 3. Thin section micrograph of more highly differentiated L6 myotube. Increased numbers of myofilaments course in parallel through myotube. Heavy metal counterstain (17,900 ×).

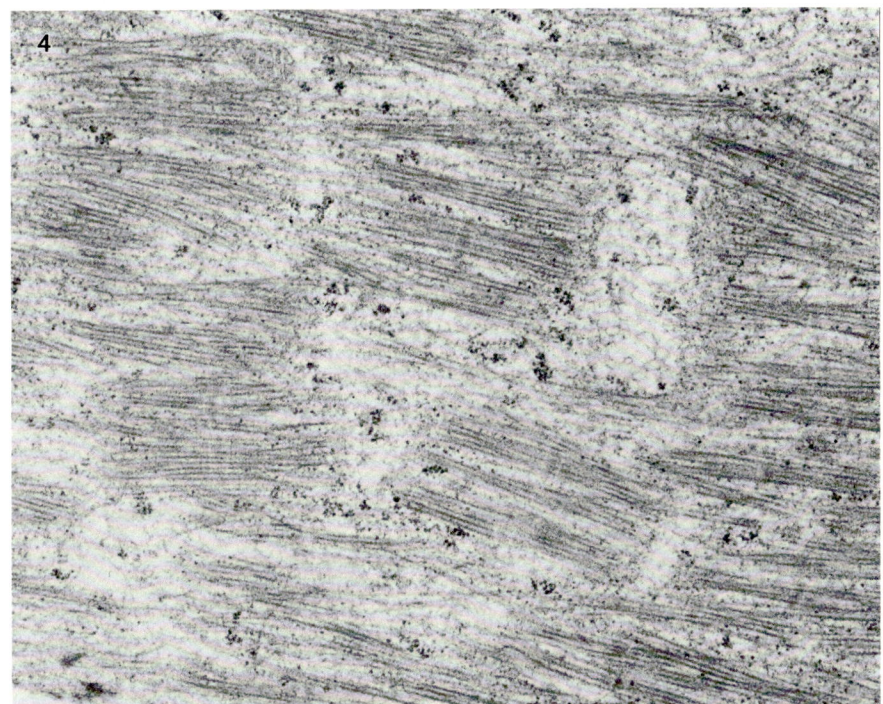

Fig. 4. Thin section of advanced stage of myotube differentiation. A more classic organization of myofilaments characterizes well-differentiated myotubes. Heavy metal counterstain (20,000 ×).

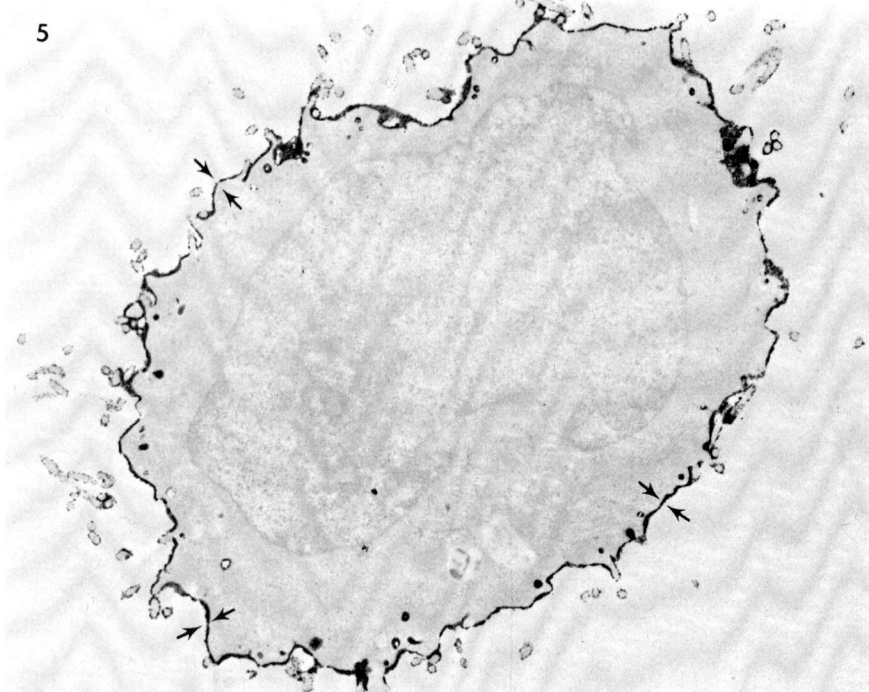

Fig. 5. Ultrastructural Con-A localization on low density L6 myoblast. Cells reacted with 50 μg/ml Con-A at 37°C for 10 min before fixation. Con-A was localized with peroxidase and diaminobenzidine as described in Methods. Con-A is uniformly distributed at the cell membrane (arrows). No heavy metal counterstain (9,600 ×).

As these cells differentiate into multinucleated myotubes they show greater redistribution of Con-A receptors. When myotubes are interacted with lectin at 37°C without prior fixation a "patching" of Con-A receptors occurs (Fig. 9), or a more pronounced redistribution occurs forming what others have termed a "cap" (Fig. 10). In addition to this redistribution the internal membrane surface of endocytic vesicles is labeled with Con-A (Fig. 10). When myotubes are first pre-fixed then reacted with Con-A none of these rearrangements (patching, capping, or endocytosis of Con-A receptors seen in unfixed samples), is observed (Figs. 11 and 12).

The binding of lectin and the reaction product show the expected specificity with the virtual complete absence of reaction in the presence of 50 mM α-methyl-D-mannoside (Fig. 13). Interestingly, in addition to the cell membrane binding of Con-A there is significant binding to an extracellular filamentous matrix in high density cultures (Fig. 11), the precise nature of which is unknown. We have failed to observe redistribution of receptors on this extracellular filamentous matrix as is observed on the plasmalemma of differentiated cells. It should be noted that the presence of the matrix is so extensive in high density cells in situ that it precludes definitive resolution of membrane receptor dynamics using fluorescein-conjugated Con-A (unpublished observation). The redistribution of receptors in differentiated cells or myotubes has been quantitated in Table I. For samples reacted with 50 μg/ml Con-A at 37°C for 20 min we observed: a uniform distribution in 99% of undifferentiated cells, redistribution into "patches" and/or caps in 37.5% of cells in a culture showing early differentiation, and a redistribution of lectin in 95% of the cells in a culture showing a high degree of differentiation and myotube formation. If parallel

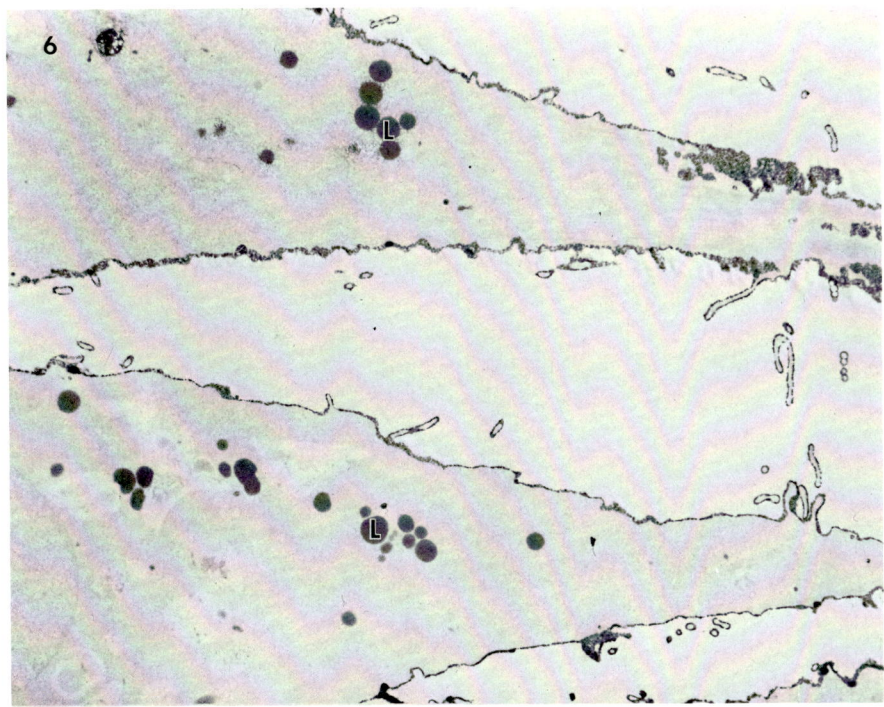

Fig. 6. Con-A localization on pre-fixed undifferentiated L6 myoblast. The uniform distribution of Con-A is interpreted as the native configuration of Con-A "receptors." Lysosomes (L) which appear dark are osmophilic. No heavy metal counterstain (10,200 ×).

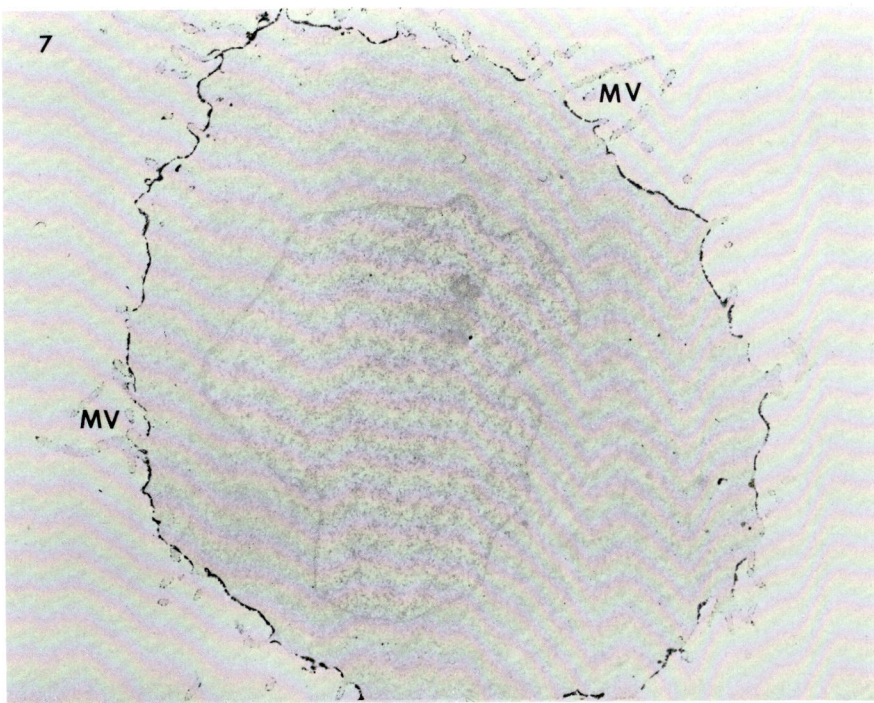

Fig. 7. Con-A localization on high density L6 culture showing early differentiation. Changes from the uniform distribution seen in Fig. 5 are observed as microvilli (MV) being less stained. No heavy metal counterstain (4,600 ×).

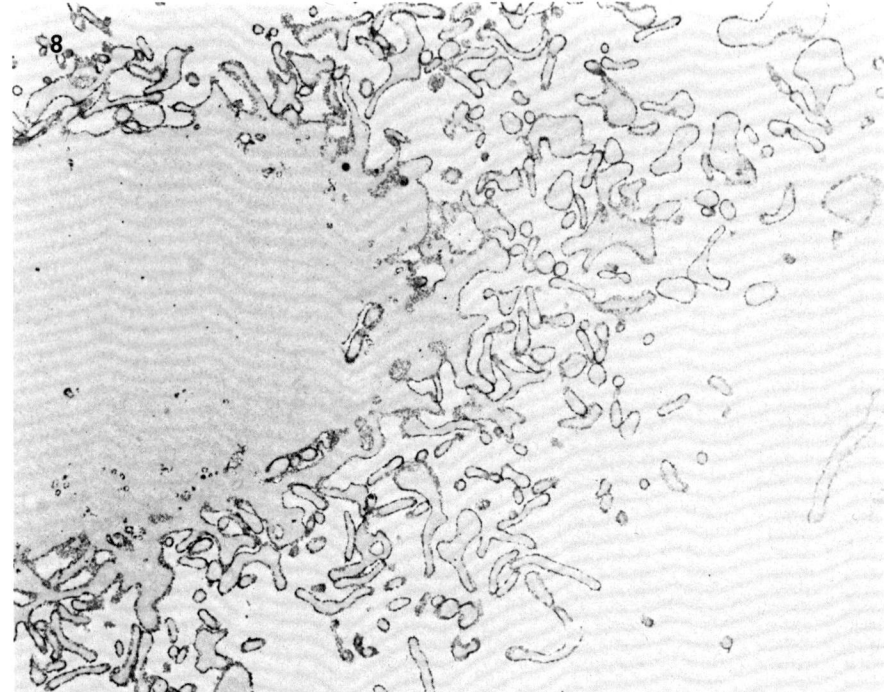

Fig. 8. Con-A localization on pre-fixed high-density L6 culture. Con-A is uniformly distributed over the entire surface of pre-fixed cells, including microvilli. No heavy metal counterstain (9,000 ×).

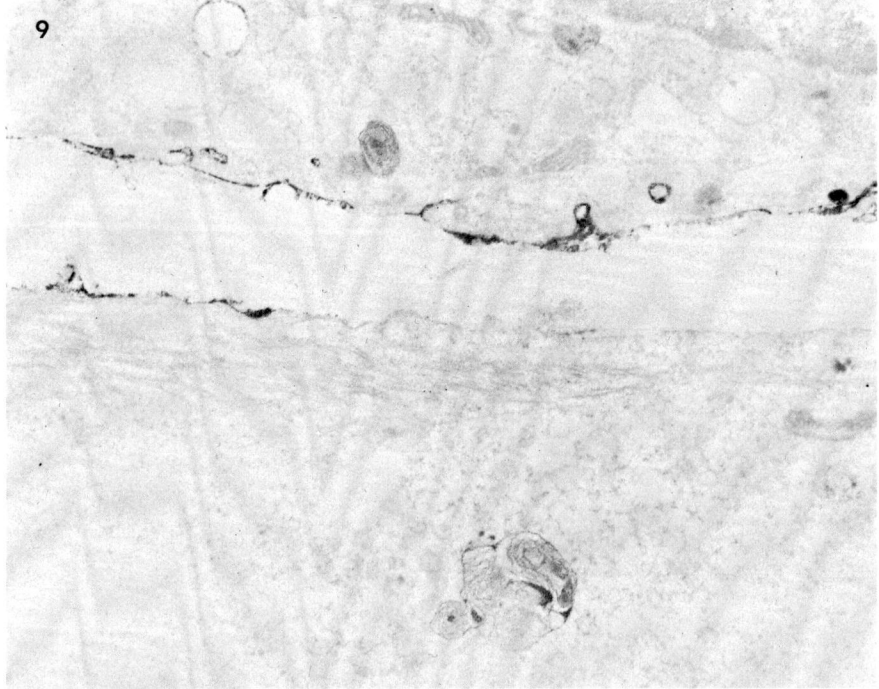

Fig. 9. Con-A localization on differentiated myotube. The Con-A reaction product is unevenly distributed into patches. No heavy metal counterstain (12,400 ×).

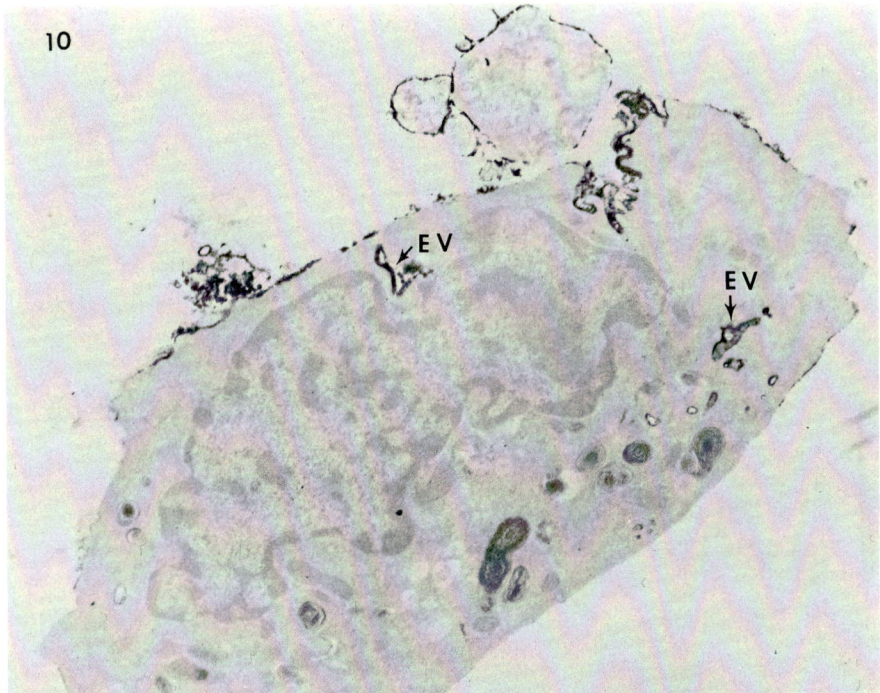

Fig. 10. Con-A localization on differentiated myotube. The Con-A reaction product is segregated (globally redistributed into a cap). Some Con-A appears in endocytic vacuoles (EV) within the cell. No heavy metal counterstain (10,300 ×).

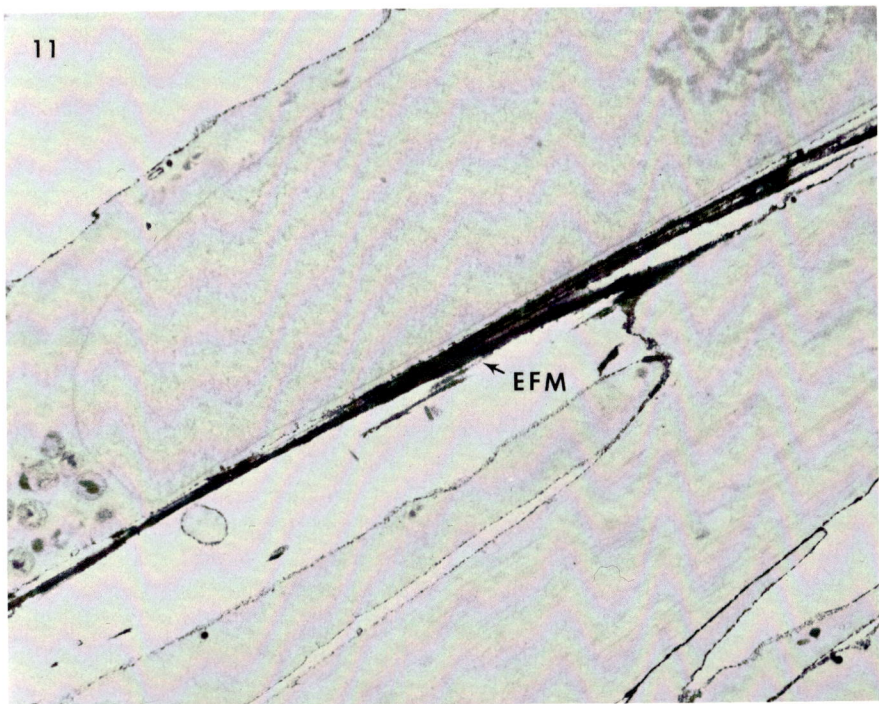

Fig. 11. Con-A localizalization on prefixed differentiated myotube. The Con-A reaction product is uniformly distrubuted. Note the extracellular filamentous matrix (EFM) which binds Con-A quite heavily. No heavy metal counterstain (8,000 ×).

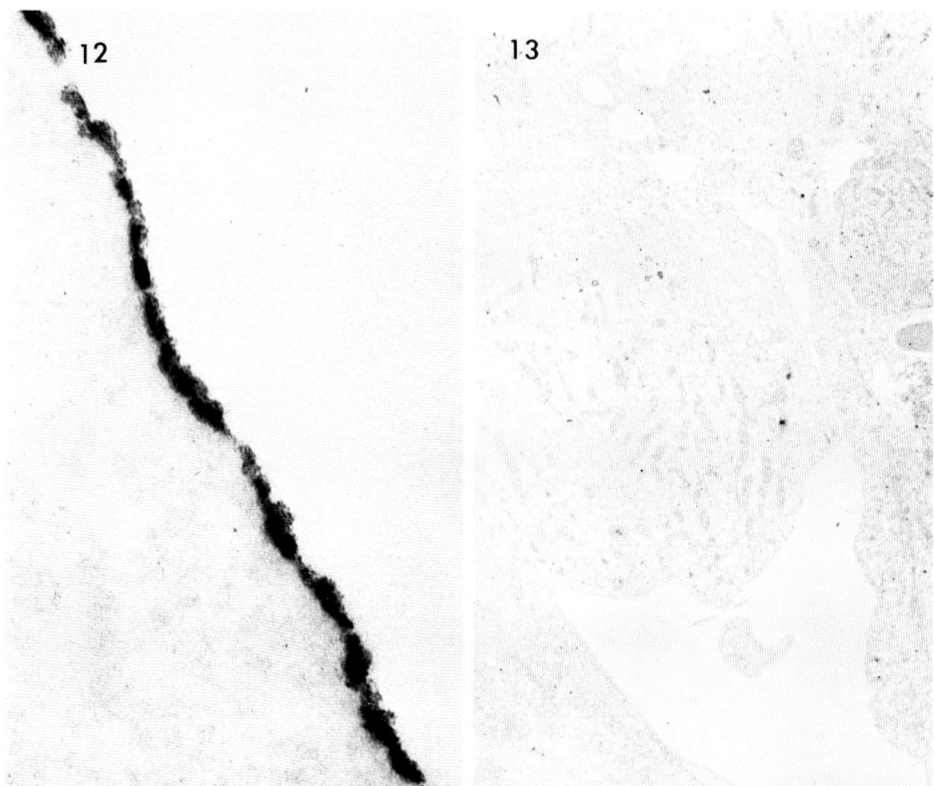

Fig. 12. High magnification micrograph of prefixed differentiated myotube. Con-A is distributed in a uniform pattern, continuously along the cell membrane. No heavy metal counterstain (93,000 ×).

Fig. 13. Micrograph of myotube reacted with Con-A in the presence of 50 mM α-methyl-D-mannoside. The mannoside competatively blocked all Con-A binding. No heavy metal counterstain (6,100 ×).

samples are first pre-fixed then reacted with Con-A a uniform distribution of membrane lectin receptor is seen in virtually all of the cells under all of the conditions (Table I).

Insulin (10 μg/ml) and dexamethasone (100 μM) have profound effects on proliferation of L6 myoblasts (Fig. 14). Even though addition of fresh medium and serum to cells moderately enhances [^{3}H] dThd incorporation, insulin produces a threefold elevation in incorporation. Following this initial burst in [^{3}H] dThd incorporation in insulin-treated samples incorporation falls to levels below 50 cpm/μg DNA at 5 days of treatment, while controls remained relatively high at approximately 200 cmp/μg DNA. This would be compatible with a larger percentage of insulin-treated cells arresting in the G_1 phase of the cell cycle. Dexamethasone (100 μM) profoundly inhibits proliferation and blocks the expected rise in [^{3}H] dThd incorporation from adding fresh medium. This blockade of proliferation by dexamethasone is completely reversible (unpublished observation).

Somewhat similar results are observed when sequential determinations of total DNA/plate are made (Fig. 15). Significant increases in DNA are observed in controls for the first 3–5 days. As would be expected from the enhancement of [^{3}H] dThd incorporation, insulin treatment produces a rapid rise in total DNA/plate which plateaus at the onset of differentiation. Dex (100 μM) produces a profound inhibition of proliferation

TABLE I. The Redistribution of Receptors in Differentiated Cells or Myotubes

Cells and conditions[a]	Con-A distribution[b]	
	% Uniform	% Nonuniform (Redistribution into patches and/or caps)
I. Low density undifferentiated L6 (N = 200)		
(A) Postfixed after reacting with Con-A	99%	1%
(B) Pre-fixed	99%	1%
II. High density confluent L6 showing (N = 110)		
(A) Postfixed	62.5%	37.5%
(B) Pre-fixed	99%	1%
III. Differentiated L6 myotubes (N = 200)		
(A) Postfixed	5%	95%
(B) Pre-fixed	99%	1%

[a]Postfixed: cells incubated with 50 µg/ml Con-A, 15 min, 37°C, then fixed with 1.6% glutaraldehyde. Pre-fixed: cells fixed with 2.5% glutaraldehyde, 15 min, 37°C, then reacted with Con-A as in postfixed specimens.
[b]These observations are made from 1-µm "thick sections" of samples used for electron microscopic localization and viewed at 1,000 ×.
N = number of cells.

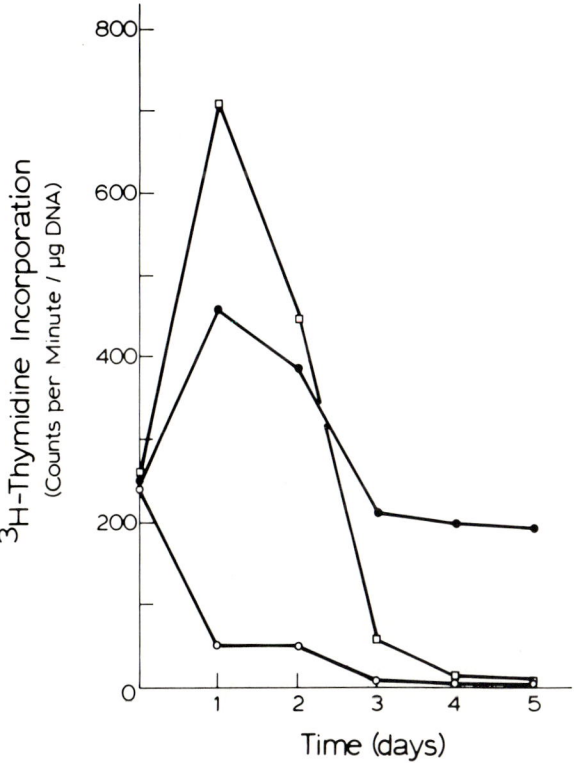

Fig. 14. [^{3}H] Thymidine incorporation (cpm/µg DNA) in L6 cells as a function of time in days. At time zero, the following are added: 1) medium (●—●); 2) medium + 10 µg/ml insulin (□—□); 3) medium + 100 µM Dexamethasone, (○—○)

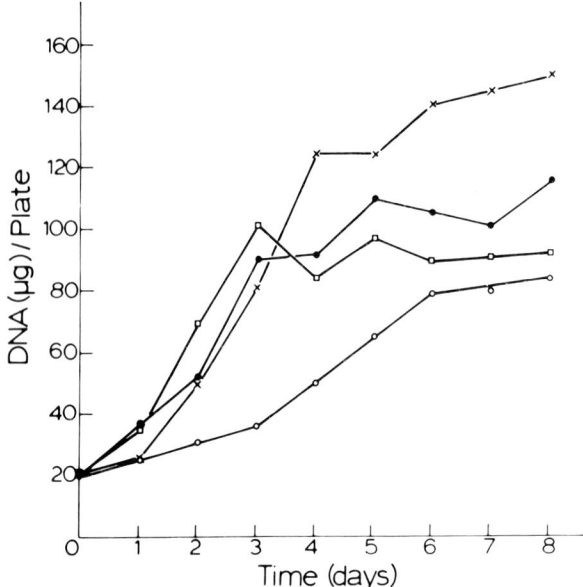

Fig. 15. Total DNA (μg) per 100 mM tissue culture plate as a function of time (in days) of culturing. Medium (●—●); or medium plus the following additions: 10 μg/ml insulin (□—□); 10 μM dex (x—x); 100 μM dex (○—○).

with only minimal increases in DNA with levels being consistently below controls. Treatment of cells with 10 μM dex has the opposite effect where an enhancement of proliferation occurs and elevated levels of DNA relative to controls are observed after 3–5 days of treatment.

L6 cells differentiate and increased levels of creatine phosphokinase are observed (Fig. 16). Significant increases of CPK/DNA are observed in controls from day 5 and later. As was shown above, insulin will stimulate DNA synthesis which then plateaus after day 3. Associated with the cessation of proliferation, insulin-treated cells show a greater rate of differentiation than controls as measured by CPK/DNA (Fig. 16). In contrast to the insulin effects both 100 μM and 10μM dex appear to inhibit differentiation with CPK levels remaining at baseline levels throughout the course of the experiment.

DISCUSSION

These studies have examined the cell surface topography for Con-A and its ability to redistribute at various stages of myoblast differentiation. A uniform distribution of Con-A receptors was observed in undifferentiated cells reacted with Con-A at 37°C, and in differentiated or undifferentiated cells pre-fixed with glutaraldehyde and then reacted with Con-A. Cells from differentiating cultures or fully differentiated myotubes reacted with Con-A at 37°C prior to fixation show a significant redistribution of Con-A into patches or caps and endocyte vesicles.

Studies of Singer and Nicholson (20) and Wallach (21) were the first to postulate the "fluid" or mobile nature of membrane components. Studies by Frye and Edidin (3) and dePetris and Raff (4) supported this hypothesis showing a redistribution of histo-

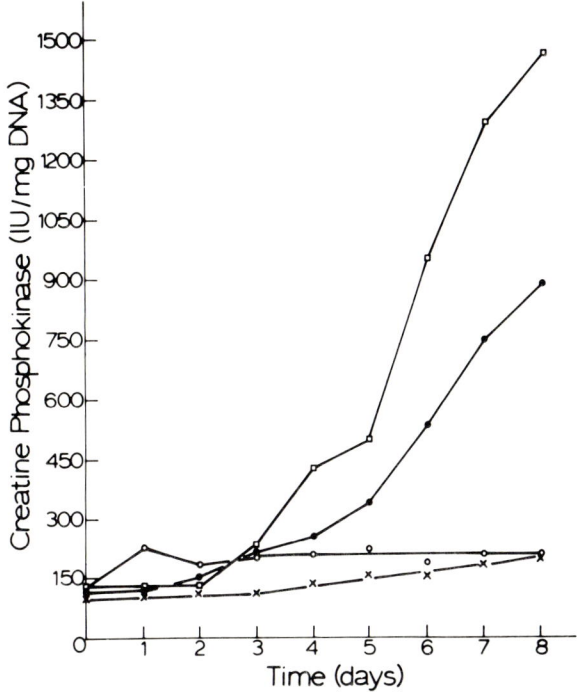

Fig. 16. Creatine phosphokinase activity in international units/mg DNA as a function of time in days of culturing. Medium (●—●); or medium plus the following additions: 10 μg/ml insulin (□—□); 10 μM dex (x—x); 100 μM dex (o—o).

compatibility antigens on heterokaryons and alloantibodies on lymphocytes. Other interesting studies have shown a difference in the topographical distribution or dynamic redistribution of lectin receptors in normal and transformed cells for Con-A and Ricinis communis agglutinin (19, 22). Both normal and transformed cells in their native state have a uniform distribution of Con-A receptors. This was determined by pre-fixing cells with glutaraldehyde then reacting with Con-A or reacting unfixed cells with lectin at 4°C without prior fixation (19, 22, 23). In both cases a uniform lectin distribution was observed in transformed cells. However, when cells are reacted with Con-A at 37°C a redistribution occurs on transformed cells but not on normal cells (19, 22). The interpretation of these, as well as our own studies, is that numerous receptors for multivalent ligands can undergo redistribution at the cell surface (23).

The nature of the Con-A receptor mobility in terms of the Singer and Nicholson membrane model could have 2 possible explanations: alterations of membrane lipid fluidity or interactions with cytoskeletal components. Redistribution as a result of enhanced membrane lipid fluidity would be a plausible explanation for a receptor site that is loosely associated with the cell membrane, or partially embedded in the lipid bylayer, or a glycolipid component of the membrane. Studies have shown that reduced temperature, thus presumably modulating membrane lipid environment, can effect the mobility of receptors or agglutination of cells by Con-A (24).

Redistribution of Con-A receptors as a result of interactions with cytoskeletal components would require that there be interactions between Con-A and the cytoskeletal structure via a transmembranous moiety (23, 25). However, there is no correlation between Con-A receptor sites and intramembranous particle distribution in myotubes as seen by freeze fracture (Furcht, in preparation). This does not rule out that the Con-A receptor may be a transmembranous component not represented within the intramembranous particle. Also, transformed cells which are thought to have more "fluid" or less "restricted" membranes will patch membrane receptors for Con-A and have few organized cytoskeletal components demonstrable by thin section electron microscopy or immunocytochemical localizing of contractile proteins (26–28). It is also interesting to note that Con-A receptors fail to patch or cap in lymphocytes unless microtubular disruptive drugs are utilized (29).

To account for the apparent inconsistencies we have postulated the existence of 2 operationally separable contractile systems in cells. The first is the classical cytoskeleton suggested by Porter (30) and the second is an undefined membrane-associated contractile network which may regulate cell surface dynamics (31, 32). This hypothesis is based on the ability of low concentrations of either cytochalasin B, vinblastine, or colchicine to independently modulate intrinsic membrane structure without affecting cell shape thought to be maintained by the classical cytoskeletal system (31). Until the true nature of Con-A receptors is known it would be speculation to suggest which of the 2 or perhaps other mechanisms may be occurring to produce the redistribution observed in these differentiated myoblasts or myotubes.

In this system, differentiation is associated with an increase in Con-A mobility. It could be argued that differentiation and fusion of cells leads to a more "active" or less restricted state of the cell membrane so that developing myotubes can fuse and incorporate new cells and cell membranes. A more restricted or rigid membrane in cells could inhibit cell fusion thereby inhibiting myotube formation. This hypothesis is supported by observations that agents which may enhance fusion have been suggested to have at least locally some membrane disorganizing or detergent-like effect (33). Whether this is the case or not awaits further experimentation. However, recent work using the laser bleaching of fluorescent-labeled acetylcholine receptors on myotubes suggests a fluid nature for this intrinsic membrane protein (34).

In view of the clearly different cell surface lectin receptor dynamics of the L6 myoblasts and myotubes, we are investigating the interactions of various agents that modulate differentiation and the distribution of Con-A receptors. As reported above, our preliminary studies show that insulin enhances differentiation while dexamethasone at 10 μM and 100 μM concentration inhibits differentiation.

The insulin promotion of differentiation has been reported by others (35). However, our studies suggest that this is somehow secondary to the initial enhancement of proliferation demonstrated by increased [^{3}H] thymidine uptake, DNA synthesis, and by the several-day delay in the appearance of differentiation following insulin treatment. This explanation would reconcile the observation that insulin and other mitogens are known to raise cGMP in other cell systems (36–38), whereas differentiation is normally associated with an elevation in cAMP (39–40). In contrast to insulin, dexamethasone at 10 μM and 100 μM concentrations inhibits differentiation. More interestingly, the 2 doses seem to act through mechanisms different from insulin as 10 μM dex stimulates cell growth (without differentiation) while 100 μM dex suppresses proliferation.

It is interesting to note that in undifferentiated cultures, there is an extensive network of an extracellular filamentous matrix which binds Con-A. The distinction between this network and membrane receptors for Con-A is readily seen using the peroxidase method of localization at the ultrastructural level. The nature of this matrix is unknown and the Con-A agglutination of certain cells cultured in vitro has been suggested to be a function of an extracellular matrix (41). Myoblasts are known to synthesize collagen; thus, the extracellular network could be collagen fibers. Possibly of greater interest, at least currently, is that this extracellular filamentous matrix may represent the glycoprotein fibronectin (LETS, Z, etc.). These proteins are a class of high-molecular-weight (250,000 dalton), externally disposed, loosely attached glycoproteins shown to be present in contact inhibited fibroblasts and myoblasts (42, 43). Studies are in progress to define the biochemical nature of this extracellular filamentous matrix which binds Con-A, and its relationship to fibronectin or LETS. Studies in progress are also examining the interrelationship of various hormone treatments, dynamic alterations in membrane structure and function, and the role of this extracellular matrix in myoblast differentiation.

NOTE ADDED IN PROOF

We have recently observed at the light and ultrastructural level using immunocytochemistry that antibodies made against purified fibronectin react with the extracellular filamentous matrix and the plasma membrane. The matrix is most pronounced in high density contacted cultures and appears to decrease with myotube formation (Furcht, Mosher, Wendelschafer-Crabb — submitted).

ACKNOWLEDGMENT

We wish to thank S. Gentry and R. Scott for technical assistance, C. Furcht for preparation of this manuscript, and Ellis Benson for his advice and support. This work is supported by a Basil O'Connor Starter Research Grant from the National Foundation-March of Dimes; and Grants 1R01 CA21463-01 and 5P01 CA16228-03 from the National Cancer Institute, DHEW, and the Minnesota Leukemia Task Force to L.T.F.

REFERENCES

1. Aub JC, Sanford BH, Cote MN: Proc Natl Acad Sci USA 54:396, 1965.
2. Burger MM, Goldberg AR: Proc Natl Acad Sci USA 57:359, 1967.
3. Frye CD, Edidin M: J Cell Sci 7:313, 1970.
4. dePetris S, Raff MC: Eur J Immunol 2:523, 1972.
5. Yaffe D: Proc Natl Acad Sci USA 61:477, 1968.
6. Yaffe D: Curr Top Dev Biol 4:37, 1973.
7. O'Neill MC, Stockdale FE: Dev Biol 29:410, 1972.
8. Shainberg A, Yagil G, Yaffe D: Dev Biol 25:1, 1971.
9. Yaffe D: Exp Cell Res 66:33, 1971.
10. Fluck RA, Strohman RC: Dev Biol 33:417, 1973.
11. Furcht LT, Scott RE: Exp Cell Res 96:271, 1975.
12. Oliver IT: Biochem J 61:116, 1955.
13. Burton K: Biochem J 62:315, 1956.
14. Bernhard W, Avarameas S: Exp Cell Res 64:232, 1974.
15. Terzakis JA: J Ultrastruct Res 22:168, 1968.
16. Simonescu N, Simonescu M: J Cell Biol 70:608, 1976.
17. Luft JH: J Biophys Biochem Cytol 9:409, 1961.
18. Reynolds ES: J Cell Biol 17:208, 1963.

19. Nicholson GL: Nature (London) New Biol 243:218, 1973.
20. Singer J, Nicholson G: Science 175:720, 1972
21. Wallach DFH, Zahler HP: Proc Natl Acad Sci USA 56:1552, 1966.
22. Nicholson GL, Lacorbiere M: Proc Natl Acad Sci USA 70:1672, 1973.
23. Nicholson GL: Biochim Biophys Acta 457:59, 1976.
24. Noonan K, Burger M: J Cell Biol 59:134, 1973.
25. Rittenhouse HG, Fox CF: Biochem Biophys Res Commun 57:323, 1974.
26. McNutt NS, Culp LA, Black PH: J Cell Biol 50:691, 1971.
27. Pollack RM, Osborn M, Weber K: Proc Natl Acad Sci USA 72:994, 1975.
28. Schollmeyer JE, Furcht LT, Goll DE, Robson RM, Stromer MH: In Goldman R, Pollard T, Rosenbaum J (eds): "Cell Motility." New York: Cold Spring Harbor Press, 1976, pp 361–388.
29. Yahara I, Edelman GM: Nature (London) 246:152, 1973.
30. Porter KR: In "Locomotion in Tissue Culture Cells." Porter R, Fitzsimons DW (eds). Ciba Foundation Symposium. Amsterdam: Elsevier Publishing Company, 1973, vol 14, pp 149–170.
31. Furcht LT, Scott RE, Maerklein P: Cancer Res 36:4584, 1976.
32. Furcht LT, Scott RE: Biochem Biophys Acta 401:213, 1975.
33. Ahkong QF, Fisher D, Tampion W, Lucy JA: Nature (London) 253:194, 1975.
34. Axelrod D, Ravidin P, Koppel DE, Schlessinger J, Webb W, Elson EL, Podleski TR: Proc Natl Acad Sci USA 73:4594, 1976.
35. Mandel JL, Pearson ML: Nature (London) 25:618 (1974).
36. Goldberg ND, Haddox M, Dunham E, Lopez C, Hadden JW: In Clarkson B, Baserga R (eds): "Control of Proliferation in Animal Cells." New York: Cold Spring Harbor Press, 1974, pp 627–634.
37. Hadden JW, Hadden EM, Haddox MK, Goldberg ND: Proc Natl Acad Sci USA 69:3024, 1972.
38. Haddox MK, Furcht LT, Gentry SR, Moser ME, Stephenson JH, Goldberg ND: Nature (London) 262:146, 1976.
39. Prasad K, Vernadakis A: Exp Cell Res 70:27, 1972.
40. McMahon D: Science 185:1012, 1974.
41. Skehan P, Friedman SJ: Exp Cell Res 92:350, 1975.
42. Hynes RO, Martin GS, Shearer M, Gritchley DR, Epstein CJ: Dev Biol 48:35, 1976.
43. Yamada KM, Weston JA: Proc Natl Acad Sci USA 71:3492, 1974.

Surface Antigens of the Embryonic Chick Myoblast: Expression on Freshly Trypsinized Cells

Martin Friedlander and Donald A. Fischman

Departments of Anatomy and Biology and the Committee on Developmental Biology, University of Chicago, Chicago, Illinois 60637

Using an antiserum raised in rabbits against embryonic chick skeletal myoblasts (Anti-M-24), we have examined the trypsin and neuraminidase sensitivity and physiological expression of myogenic cell surface antigens. It was found that trypsin-released muscle cells more effectively inhibited, on a cell to cell basis, the cytotoxicity of Anti-M-24 for 24-h-old myoblast monolayers than did identical cells that had received a 3–4 h suspension culture recovery period from trypsinization. There was no such difference in absorptive capacities observed for any other embryonic chick tissue tested (e.g. brain, retina, liver, heart, and red blood cells) when freshly trypsinized cells were compared to ones which were given a 3–4 h culture period. If freshly trypsinized muscle cells were treated with high concentrations (30,000 international units (IU)/0.1 ml packed cells) of trypsin or with neuraminidase (30,000 IU/ml packed cells), there was a selective loss of myoblast-specific surface antigens. When single cells that had been in suspension culture for 3.5 h were re-exposed to low concentrations (10,000 IU/0.1 ml packed cells) of trypsin, more antigenic sites were revealed on their surfaces as detected by an increased absorptive capacity in removing myoblast-binding antibodies from Anti-M-24. This increase in antigenic expression was time-dependent and inversely related to the length of culture time after trypsinization. Immunofluorescence studies revealed that tissue specific myoblast cell surface antigens are present on both muscle cells that were freshly dissociated and those that had been in suspension culture for 3–4 h. Furthermore, freshly trypsinized myoblasts possessed cell surface components that were highly antigenic; antiserum to such cells reacted extensively with both trypsinized and recovered muscle cells as detected by complement-dependent ^{51}Cr release cytotoxicity assays and immunofluorescence. We conclude that embryonic chick myoblasts possess surface antigens that may be selectively removed by neuraminidase or high concentrations of trypsin. These antigens may be progressively masked, with increasing time of culture after protease-dissociation, by molecules that are sensitive to low concentrations of trypsin. Such masking of tissue-specific cell surface antigens could result in the display of molecular mosaics which may play a role in facilitating intercellular recognition and subsequent differentiation and histogenesis.

Key words: embryonic muscle, cell surface antigens, myogenesis, cytotoxicity assays

Martin Friedlander is presently at Rockefeller University, New York, New York 10021.
Donald A. Fischman is presently at the Department of Anatomy and Cell Biology, SUNY, Downstate Medical Center, Brooklyn, New York 11203.

Received April 28, 1977; accepted June 28, 1977.

© 1977 Alan R. Liss, Inc., 150 Fifth Avenue, New York, NY 10011

Recently we reported the immunological detection of tissue and developmental stage specific surface antigens in embryonic chick skeletal muscle in vitro (1—3). These studies were initiated to detect subtle compositional changes at the cell surface which might accompany differentiation of a particular embryonic tissue. We chose to examine developing embryonic muscle because this tissue provides an excellent system for studying the relationship between cell surface modifications and the nuclear events which underlie the phenotypic modifications we term differentiation. Cell recognition prior to myotube formation is highly specific (4, 5), both prefusion myoblasts and multinucleated myotubes exhibit cellular migration and motility (6), and cell-cell and cell-substratum adhesions clearly influence the course of myodifferentiation (7, 8). The aggregation of myogenic cells and the formation of intercellular junctions at interfacing plasma membranes prior to cell fusion (9, 10) are poorly understood, but probably significant, steps in muscle differentiation. In addition, there are well established methods available for culturing muscle which produce a developing tissue that faithfully mimics in vivo myogenesis (11—13).

Antiserum to the prefusion myoblast (Anti-M-24) was produced in rabbits and extensively characterized by serological and immunohistochemical methods (3, 14). The results of these experiments are summarized in Table I. Such criteria have also been used to identify tissue specific surface antigens in other embryonic (15—18), transformed (19—22), or adult (23, 24) tissues. If the molecular basis for such operationally defined specificities is to be understood, we felt it would be necessary to examine the biochemical properties of the myoblast cell surface antigens. As a preliminary step towards immunochemically isolating and characterizing these antigens, we have examined their physiological expression and protease sensitivity. A number of proteases have been used in the isolation and characterization of histocompatibility antigens from mouse (25) and human (26) tissues. Chymotrypsin may be used to selectively expose tumor associated antigens on infected avian cell surfaces (27), while the major surface glycoprotein of fibroblasts is markedly trypsin sensitive (28, 29). Trypsin has also been used to explore the physiological expression (30, 31) and membrane fluidity (32—34) of lectin binding sites on embryonic cell surfaces. Others have used neuraminidase to examine the expression of a number of cell surface antigens (35, 36). While such enzymatic treatments of living cells is known to have marked effects on a number of cell properties (37, 38), it is still of interest to consider their use in the analysis of potentially cryptic, or masked, membrane antigens.

In this communication we wish to present new data we have obtained concerning the protease and neuraminidase sensitivity and the physiological expression of myoblast surface antigens. These data show that myoblast antigens may be selectively removed from the cell surface after treatment with neuraminidase or high concentrations of trypsin. Freshly trypsinized muscle cells apparently have greater quantities of these antigens on their surface than do cells permitted a period of culture after trypsinization; mild retrypsinization of such cells reveals the same amount of antigen as found on freshly dissociated cells. Parallel studies have involved the immunochemical analysis of detergent-extracted antigens by radioimmunoprecipitation and sodium dodecyl sulfate-polyacrylamide gel electrophoresis. The results of those investigations will be reported elsewhere (39).

MATERIALS AND METHODS

Cell Culture

Muscle was obtained from the hindlimbs of 12-day-old embryonic White Leghorn chicks by methods routinely used in our laboratory (40). All visible nervous, vascular, and

TABLE I. Summary of the Characteristics of Antiserum (Anti-M-24) Prepared in Rabbits Against Embryonic Chick Prefusion Myoblasts

1. Complement-mediated ^{51}Cr release cytotoxicity experiments indicate:
   A. Absorption of Anti-M-24 with live cell monolayers, live cell suspensions, acetone powders, or freshly homogenized tissue from embryonic red blood cells, liver, brain, retina, or heart or adult skeletal muscle will remove all cytotoxicity of this antiserum for these tissues while lowering the cytotoxicity for embryonic muscle by only 20–60%. We interpret these results to demonstrate TISSUE SPECIFICITY.
   B. Absorption of Anti-M-24 with myotubes or skeletal muscle fibroblasts removes all cytotoxicity of this antiserum for these cell types at a point where there is still significant cytotoxicity remaining for myoblasts. We interpret these results to demonstrate DEVELOPMENTAL STAGE SPECIFICITY.
   C. Light microscopical observations of cultures treated with appropriately absorbed antiserum and guinea pig complement confirm the serological data; in such cultures only the spindle-shaped myoblasts are lysed.
2. Immunohistochemical staining (indirect immunofluorescence or immunoperoxidase) of monolayers or single cell suspensions further demonstrates that, after appropriate absorption, Anti-M-24 will stain only prefusion myoblasts. Shared, muscle tissue-specific, antigens may also be detected on embryonic muscle at all stages of differentiation.
3. Myoblast-specific antigens are cell surface components: 1) intact, viable cells inhibit cytotoxicity of Anti-M-24 for myoblasts and 2) observations by indirect immunofluorescence demonstrate that, under appropriate conditions, these antigens are mobile within the plasma membrane.
4. Myoblast cell surface antigens detected with Anti-M-24 are not an artefact of tissue culture: 1) fresh, homogenized chick tissues will absorb reactivity of Anti-M-24 with cultured cells or cells freshly released from the embryo and 2) Anti-M-24 binds to frozen sections of fresh embryonic hindlimb muscle as detected by indirect immunofluorescence.

connective tissue was removed from the limbs by careful dissection and the remaining muscle finely minced with scissors. After washing the tissue with calcium- and magnesium-free Tyrodes solution (CMF), it was dissociated into single cells by a 50 min incubation period at 37°C in 0.3% trypsin (Armour Pharmaceuticals, St. Louis, Missouri, "Tryptar") in CMF. Proteolytic dispersion of the cells was stopped by washing the cell suspension 3 times with Eagle's basal medium containing 10% heat inactivated (30 min at 56°C) horse serum (Grand Island Biologicals, Grand Island, New York), 1% glutamine, 1% penicillin (E-HS) and soybean trypsin inhibitor (10 μg/ml). After passing the cell suspension through a metal mesh filter, single cells suspended in E-HS with 5% embryo extract (complete medium, CM) were plated at varying densities on gelatin coated plastic culture dishes (Falcon Plastics, Oxnard, California). Such primary muscle cell cultures contained approximately 90% myoblasts as assessed by morphologically scoring 24-h-old monolayers established from comparable cell suspensions (47, 48). Other embryonic chick tissues (e.g. heart, brain, liver, retina) were obtained in similar fashion. Red blood cells were collected from the chorioallantoic vessels of the embryo, washed with phosphate-buffered saline, and used immediately for absorptions or cytotoxic assays.

Antiserum

Freshly dissociated embryonic muscle cells were cultured after trypsin dissociation for 3.5 h by placing $5-10 \times 10^6$ cells/4.0 ml E-HS into 25-ml Ehrlenmeyer flasks and rotating the flasks at 110 rpm on a New Brunswick Gyratory Shaker. Such a culture period has been operationally defined as "preaggregation" to emphasize the observation that during the 3–4 h time period multicellular aggregates do not form and greater than 95%

of the cells remain as single, viable cells. To monitor the state of aggregation during such a time period, parallel suspension cultures were established and samples taken at 1-h intervals from 0 to 6 h of culture. For each time point, the cells were pelleted by centrifugation, resuspended with a known volume of E-HS, and observed in a light microscope. Representative fields were scored for the number of single cells and multicellular aggregates. By 2–3 h small clusters of cells were observed, but these were easily dissociated into single cells by gentle pipetting. Only after 4–5 h were larger multicellular aggregates present that could not be easily dispersed into single cells by pipetting. Such aggregates rapidly increased in number from 6–24 h of culture, even at gyratory shaker speeds of 120 rpm. Such a period of suspension culture presumably permitted the renewal of cell surface materials which may have been denuded by tryptic treatment during the dissociation procedure (15, 41).

After 3.5 h of suspension culture, cells were collected, washed several times with Tyrodes solution, and injected into rabbits. The immunization protocol and bleeding schedule has been reported elsewhere and will not be detailed here (3, 14). Antiserum raised in rabbits against prefusion embryonic chick myoblasts will be referred to as Anti-M-24 and represents serum pooled from 3 individual rabbits.

Cytotoxicity Assay

The complement-dependent ^{51}Cr release cytotoxicity assay is a modification of that described by Wigzell (42) and Sanderson (43). Target cells were plated at $1-5 \times 10^4$ cells/ 0.1 ml CM/well on Falcon microtiter plates. Six hours later, 1.0 µCi of ^{51}Cr (New England Nuclear Corporation, Boston, Massachusetts) in 0.1 ml E-HS was added to each well and after an additional 12 h incubation at 37°C, the excess ^{51}Cr was removed by several washes of the wells with E-HS. Antibody (0.05 ml) and guinea pig complement (0.05 ml) diluted to a final concentration of 1/10 (Beckman Diagnostics, Fullerton, California) was added and the cultures incubated for an additional 50 min. The reaction was stopped by adding 0.1 ml of cold Tyrodes solution to each well. Following a 10 min centrifugation at 2,800 rpm to remove cellular debris, 0.1 ml samples from each of triplicate wells were taken to be counted on a Searle Autogamma Counter. The results were calculated as a percent cytotoxicity:

$$\text{Percent Cytotoxicity} = \frac{\text{cpm }^{51}\text{Cr released by antibody} - \text{background}}{\text{cpm }^{51}\text{Cr released by TX-100} - \text{background}} \times 100$$

For each set of triplicate samples the mean and standard deviation were determined. The variability of replicates in a single experiment was never greater than ± 10% cytotoxicity and generally was between 5 and 8. All experiments were repeated in their entirety at least twice and most were repeated 3–5 times. For Figs. 2–7, the mean values from 2 to 5 experiments were averaged and presented as a single data point.

For inhibition of cytotoxicity assays, 0.5 ml of antiserum diluted 1/10 or 1/100 with E-HS was absorbed with increasing numbers of cells overnight on a wrist shaker at 4°C. Cells used for absorption were removed by centrifugation and the absorbed serum used immediately. For trypsinization of absorbing muscle cells, 10^8 freshly dissociated or pre-aggregated myoblasts were suspended in 4.0 ml of trypsin-CMF (0.05–0.3%), rotated at 110 rpm for 50 min in a gyratory shaker bath at 37°C, washed extensively with E-HS and soybean trypsin inhibitor (10 µg/ml), and then used to absorb antiserum as described above. Neuraminidase treatment was conducted under identical conditions as for trypsinization;

30,000 IU/ml of neuraminidase Type I from Cl. perfinigens (Sigma Chemical Company, St. Louis, Missouri) was used.

Immunofluorescence

Cells to be stained by immunofluorescence were either live or fixed with 2.0% paraformaldehyde for 10 min at 50°C. All subsequent procedures were conducted on ice (for live cells) or at room temperature (for fixed cells). Five million cells were washed with E-HS, treated for 15 min with Anti-M-24 or preimmune serum from the same rabbit diluted 1/10 with E-HS, washed 3 times with E-HS, incubated for an additional 15 min with goat antirabbit IgG conjugated to fluorescein isothiocyanate (GAR-FITC) (Cappel Laboratories, Downingtown, Pennsylvania), washed 2 times with E-HS, mounted in glycerol under a glass coverslip and observed under an Apo 40X oil immersion lens mounted to a Zeiss epifluorescent microscope. Photomicrographs were taken after a 30 sec exposure on Kodak Tri-X film and subsequently developed in Diafine developer. All prints were processed under identical conditions.

RESULTS

Anti-M-24 was prepared by immunizing rabbits with myoblasts that had been allowed a 3–4 h culture period after primary trypsin-dissociation under the assumption that such a period of time was necessary for the cells to renew surface components that may have been denuded by the protease dissociation procedure (15, 41). To further test this assumption, we decided to compare the ability of freshly dissociated and preaggregated myoblasts to inhibit cytotoxicity of Anti-M-24 for 24-h-old myoblast monolayers. At this point, 2 observations should be emphasized: 1) muscle cultures consist of 2 predominant cell types, myoblasts and fibroblasts, the former accounting for approximately 90% of the cells present at 24 h of monolayer culture (40, 44) and 2) on a cell to cell basis, fibroblasts take up 3–4 times the amount of ^{51}Cr as do myoblasts (14). These are important considerations when attempting to interpret data obtained from ^{51}Cr-release experiments. For example, while myoblasts account for nearly 90% of the total cell population in a 24-h-old muscle culture, they only contain 40–60% of the ^{51}Cr bound within the cells of such a culture.

Single cells that had been cultured in suspension for 3.5 h inhibited cytotoxicity of Anti-M-24 for myoblast target cells in a linear fashion; 12×10^6 cells reduced the cytotoxicity to 10% (Fig. 1). When freshly trypsinized cells were used, only 4×10^6 cells were necessary to obtain comparable inhibition (Fig. 1). Thus, antigenic sites detected by Anti-M-24 were present on freshly trypsinized cells and, on a cell to cell basis, there were fewer of them detected on muscle cells that had been cultured for 3.5 h. If freshly dissociated muscle possessed surface antigens also found on cells that had received a 3.5 h period of culture following trypsinization, then these cells should be immunogenic when injected into rabbits. If cells were injected into rabbits immediately following trypsinization, the resultant antiserum was highly cytotoxic for 24-h-old muscle monolayers (Fig. 2). Apparently, a 3–4 h culture period following trypsin dissociation was not necessary for the regeneration or expression of myoblast cell surface antigens as detected by Anti-M-24. We hypothesized that with time after trypsinization there may have been a progressive masking of myoblast surface antigens and to test this assumption the following experiments were conducted.

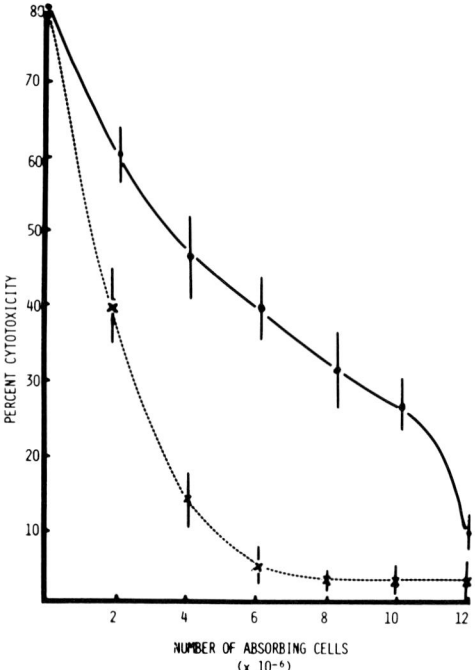

Fig. 1. Inhibition of cytotoxicity of Anti-M-24 for myoblast target cells after absorption with increasing numbers of suspension cultured (●—●) or freshly trypsinized (x- - -x) myoblasts. One-half milliliter of antiserum was diluted 1/100 with E-HS and absorbed overnight on a shaker at 4°C with either 2, 4, 6, 8, 10, or 12 million cells from 12-day-old embryonic chick hindlimb muscle that had been freshly dissociated with 0.3% Tryptar/ml of minced muscle or with identical cells that had been cultured for 3.5 h at 110 rpm on a gyratory shaker. The absorbed antisera were tested, in the presence of guinea pig complement, for cytotoxicity on 24-h-old muscle monolayers. Twice as many suspension cultured cells as freshly trypsinized ones were needed to remove all cytotoxicity of the antiserum for myoblast target cells. Vertical lines at each point represent the range of 1 standard deviation from the mean as calculated for 12 replicate samples obtained from 4 independent experiments.

Cells which had been cultured for 3.5 h following trypsinization ("preaggregated" cells) were treated with high (30,000 IU/0.1 ml packed cells) or low (10,000 IU/0.1 ml packed cells) concentrations of trypsin and then assessed for their ability to inhibit cytotoxicity of Anti-M-24 for 24-h-old muscle monolayers. Exposure of preaggregated cells to high concentrations of trypsin apparently removed most of the antigens while a comparable group of cells that had been exposed to low concentrations of this protease were as effective in the inhibition of cytotoxicity as freshly dissociated cells (Fig. 3). It should be emphasized that a single group of cells was used for the data obtained in Figs. 1 and 3; Anti-M-24 was absorbed with either the freshly dissociated cells or cells from the same dissociation which were then preaggregated or preaggregated and retrypsinized. Identical results were obtained when the experiment was repeated.

If freshly dissociated muscle was reexposed to higher concentrations of trypsin, the single cells were greatly reduced in the ability to inhibit cytotoxicity of Anti-M-24 for muscle monolayers (Fig. 4). However, if such absorbed antiserum was tested on pure fibroblast monolayers, cytotoxicity for these target cells was removed entirely after absorption with 8×10^6 cells. These results suggested that trypsin may have selectively removed the

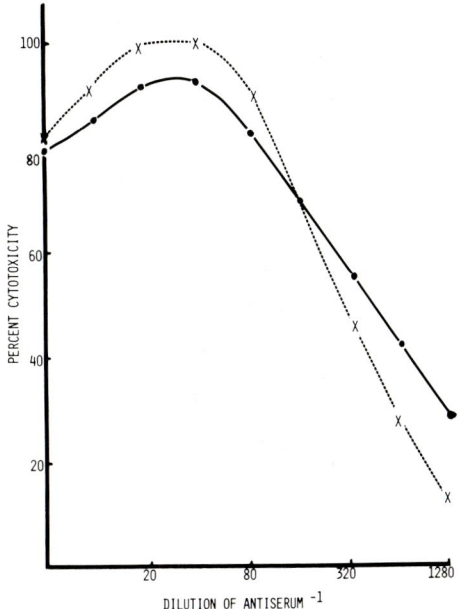

Fig. 2. Cytotoxicity for 24-h-old muscle monolayers of antiserum prepared in rabbits against freshly trypsin-dissociated embryonic chick hindlimb (Anti-M-24T, x- - -x) or identical cells that had been cultured in suspension for 3.5 h at 110 rpm before immunization (Anti-M-24, ●—●). Each dilution curve represents pooled serum obtained from 2–3 rabbits. Preimmune serum from identical rabbits used to obtain immune serum induced less than 10% cytotoxicity at any dilution tested (1–1/800). See Materials and Methods in the text for discussion of variability of data and experimental reproducibility for data presented in this and subsequent figures.

myoblast-specific surface antigens. To further test this hypothesis, several other embryonic chick tissues were examined for the effects of trypsin treatment on the inhibition of Anti-M-24 cytotoxicity. Two representative tissues are presented in Fig. 5. For both embryonic brain and liver there was no difference in the ability of freshly dissociated or preaggregated cells to inhibit the cytotoxicity of Anti-M-24 for 24-h-old muscle monolayers. When such absorbed antiserum was tested back on the homologous tissue with which it was absorbed, all cytotoxicity was removed after absorption with $2-4 \times 10^6$ cells. Similarly, there were no differences in the inhibition of cytotoxicity curves of Anti-M-24 for myoblasts after absorption with either preaggregated or freshly trypsinized embryonic chick retina, heart, or red blood cells.

If indeed low concentrations of trypsin reexposed myoblast surface antigens which were progressively masked with culture after trypsinization, one might predict a correlation between antigen masking and the culture time. Such a correlation was observed when Anti-M-24 was absorbed with myoblasts that were either freshly trypsinized or preaggregated for 1, 2, 3, or 3.5 h and tested for cytotoxicity on myoblast monolayers (Fig. 6). Cells preaggregated for 1 h exhibited an equivalent inhibition of cytotoxicity in Anti-M-24 for myoblasts as did freshly dissociated cells. With increasing time in suspension culture there was a progressive decrease in the ability to inhibit antimyoblast cytotoxicity. Thus, when compared to freshly trypsinized cells, the myoblasts after 3 h in suspension culture exhibited fewer antigenic sites per cell that were available for binding to antibodies directed

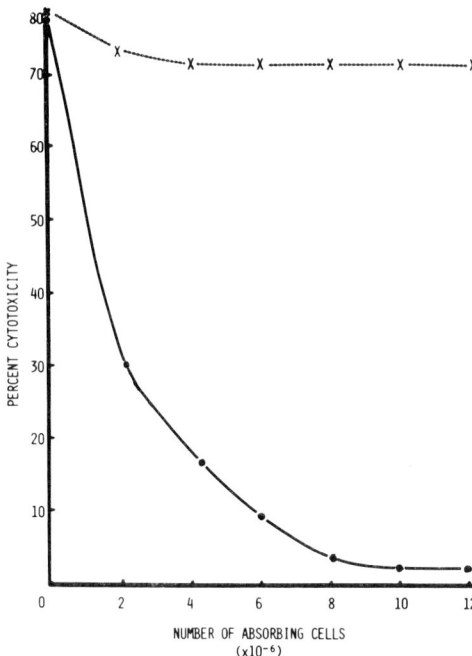

Fig. 3. Inhibition of cytotoxicity of Anti-M-24 for 24-h-old muscle cell monolayers after absorption with increasing numbers of myoblasts that had been suspension cultured for 3.5 h and then treated with high (x- - -x) or low (●——●) concentrations of trypsin as described in the text. Note 1) the lack of inhibition of cytotoxicity after treatment with high concentrations of trypsin and 2) the similarity in the inhibition curve of the low trypsin concentration treatment and the curve obtained after absorption with freshly dissociated cells as seen in Fig. 1.

against myoblast cell surface components. Periods of time longer than 3.5 h were not studied since these cells rapidly formed multicellular aggregates after approximately 4 h of rotary suspension culture. These varying sized aggregates would be very difficult to compare in absorption capacity to single cells since the exposed surface areas would not be equivalent.

Suspensions of freshly dissociated muscle cells were also incubated in neuraminidase and then assessed for the inhibition of cytotoxicity of Anti-M-24 for muscle or pure skeletal muscle fibroblast monolayers. After absorption with as many as 10^7 neuraminidase-treated cells, Anti-M-24 retained 75% cytotoxicity for 24-h-old muscle monolayers (Fig. 7). However, the cytotoxicity of this same antiserum for pure fibroblast monolayers began to fall precipitously after absorption with only 4×10^6 neuraminidase-treated cells (Fig. 7). Presumably, the myoblast-specific antigens are more susceptible to neuraminidase hydrolysis than those antigens shared with fibroblasts.

Immunofluorescence of Freshly Trypsinized and Suspension Cultured Muscle Cells

Both "preaggregated" and freshly trypsinized muscle cells stained intensely over the entire cell surface after incubation in unabsorbed Anti-M-24 and GAR-FITC (Fig. 8a–d). Absorption with preaggregated (Fig. 8e) or freshly trypsinized (Fig. 8f) embryonic liver cells diminished the intensity of the fluorescence and the pattern of staining became less regular. Unabsorbed Anti-M-24 intensely stained embryonic liver cells (Fig. 8g), but following absorption with liver cells no longer stained hepatocytes (Fig. 8h).

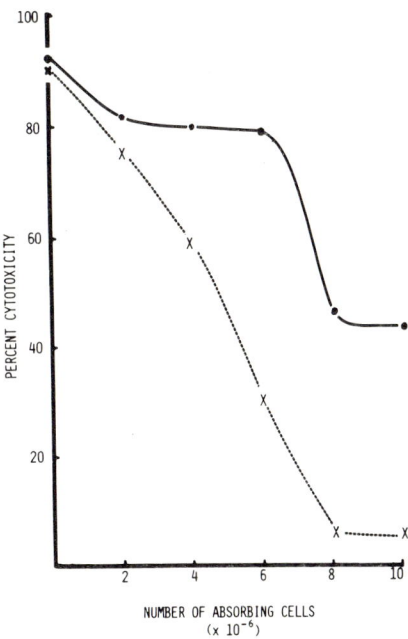

Fig. 4. Inhibition of cytotoxicity of Anti-M-24 for 24-h-old muscle (●——●) or pure skeletal muscle fibroblast (x- - -x) monolayers after absorption with increasing numbers of freshly dissociated myoblasts that had been retrypsinized. Muscle was dissociated into single cells by a 50 min incubation period at 37°C in 0.3% trypsin, washed 4 times with E-HS and soybean trypsin inhibitor, and then retrypsinized with a comparable amount of trypsin for an additional 50 min at 37°C in a gyratory shaker water bath. After 4 more washes with E-HS and soybean trypsin inhibitor, the cells were used to absorb Anti-M-24. Absorption with 8×10^6 treated cells removed all cytotoxicity of this antiserum for fibroblasts while as many as 10^7 only lowered the cytotoxicity for myoblasts to 50%.

Myoblasts incubated in neuraminidase or high concentrations of trypsin and subsequently stained with Anti-M-24 and GAR-FITC failed to demonstrate specific immunofluorescence (Fig. 8i). When freshly dissociated (Fig. 8j) or preggregated (Fig. 8k) myoblasts were stained with unabsorbed antiserum raised against freshly dissociated cells, a uniformly intense fluorescence pattern was observed around their periphery. Preimmune serum failed to stain any of the cell types used in this study (not illustrated). All immunofluorescence experiments were conducted on paraformaldehyde-fixed cells or on live cells at 4°C so the patterns of fluorescence could not be attributed to capping phenomena which occur as a result of mobile surface receptors (45). If cells were stained at 37°C, capping did occur, followed by the complete removal of surface antigen-antibody complexes (Friedlander, unpublished observations).

DISCUSSION

The role of surface glycoproteins as recognition molecules on developing embryonic cells has been explored in several systems and it is clear that such macromolecules are responsible for, or at least associated with, tissue specific aggregation phenomena of embryonic chick retina (46, 47) and the promotion of differentiation in cartilage (48), muscle (49), and glia (50). The manner in which such macromolecules exert their influence on cellular interactions and the promotion of differentiation is not clear; there have been

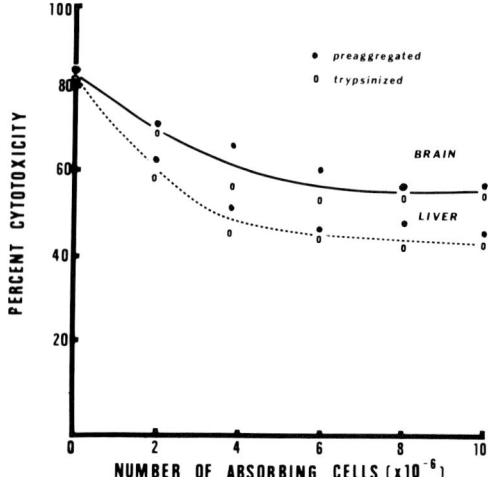

Fig. 5. Inhibition of cytotoxicity of Anti-M-24 for 24-h-old muscle monolayers after absorption with increasing numbers of 12-day-old embryonic brain or liver cells that had been freshly dissociated (○) or suspension cultured for 3.5 h (●). There was no significant difference in the rate at which freshly dissociated or preaggregated cells from liver or brain inhibited cytotoxicity.

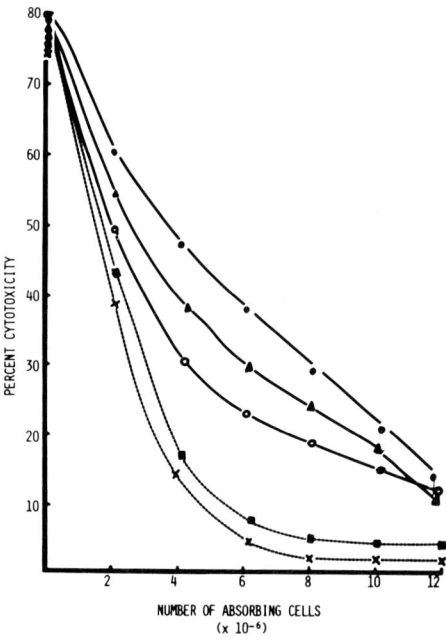

Fig. 6. Inhibition of cytotoxicity of Anti-M-24 for 24-h-old muscle cell monolayers after absorption with increasing numbers of myoblasts that were either freshly dissociated with trypsin (x- - -x) or cultured in suspension for 1 (■ - - - ■), 2 (○——○), 3 (▲——▲), or 3.5 (●——●) h. Cells to be used for absorption were divided into aliquots of 2, 4, 6, 8, 10, or 12 million and shaken gently overnight with 0.5 ml of antiserum diluted 1/100 with E-HS. After removal of the cells by centrifugation, the serum was assessed for cytotoxicity, in the presence of guinea pig complement, on myoblast target cells. Note that the cytotoxicity curves obtained after absorption of antiserum with cells preaggregated for 2 and 3 h were intermediate to those obtained with serum absorbed with cells preaggregated for 0 and 3.5 h.

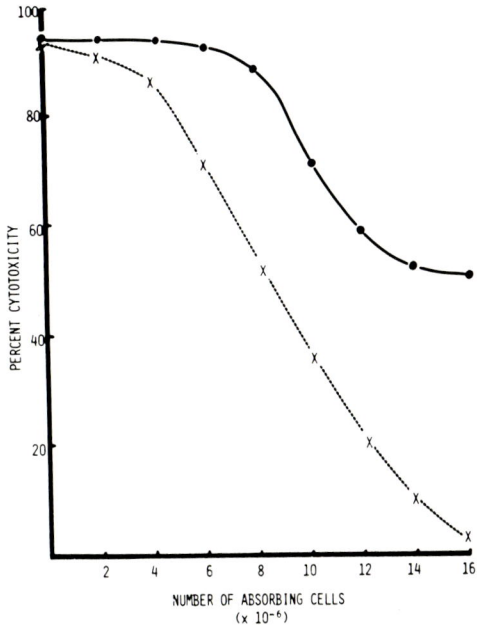

Fig. 7. Inhibition of cytotoxicity of Anti-M-24 for 24-h-old muscle (●——●) or pure skeletal muscle fibroblast (x - - - x) monolayers after absorption with increasing numbers of neuraminidase-treated myoblasts. Cells used for absorption were treated with enzyme as described in the text, washed with E-HS and then assessed for the ability to absorb out antibodies that normally bind to and, in the presence of complement, lyse myoblasts. Note that 16 million treated cells inhibited all cytotoxicity of this antiserum for pure fibroblast monolayers, but that this same number only reduced the cytotoxicity for myoblast monolayers to 50%. The fall in cytotoxicity for myoblast targets seen between absorption with 6 and 12 million cells reflects the loss of ^{51}Cr from fibroblasts that contaminate the 24-h-old muscle cultures and account for 40–60% of the ^{51}Cr in these cultures (see text for discussion).

suggestions that these glycoproteins may act as cell ligands (51), protease sensitive cell couplers (47), or enzymes that catalyze the transfer of sugars (52) and thus stabilize intercellular contacts. Our observations suggest that there is a class of trypsin- and neuraminidase-sensitive molecules which serve to uniquely identify the prefusion myoblast cell surface and that these antigens are revealed in greater quantity by mild trypsinization of pre-aggregated muscle cells. Whether these antigens are involved in facilitating intercellular recognition is uncertain at this time, but it is tempting to consider this possibility.

Mild retrypsinization of cultured cells revealed 3–5 times the number of antigenic sites found on cells which, after dissociation, were placed in suspension culture for 3–4 h. Prior to such a culture period, these cells do not exhibit aggregation (53, 54) or histiotypic cell sorting (55) or possess surface glycoproteins detected by ruthenium red staining (56). Based on these biological constrictions, we found it surprising that myoblast surface antigens should be present in greater quantities at a time when the surface of these cells exhibited lower biological specificity than cells in culture for 3–4 h. Of interest was the failure to detect such differences in antigenic expression, as measured by the capacity to absorb out cross-reacting, species-related antigens, on other tissues of the chick embryo. Perhaps only the muscle-related antigens are exposed upon mild trypsinization. Furthermore, after suspension culture of freshly dissociated myoblasts, the original quantity of

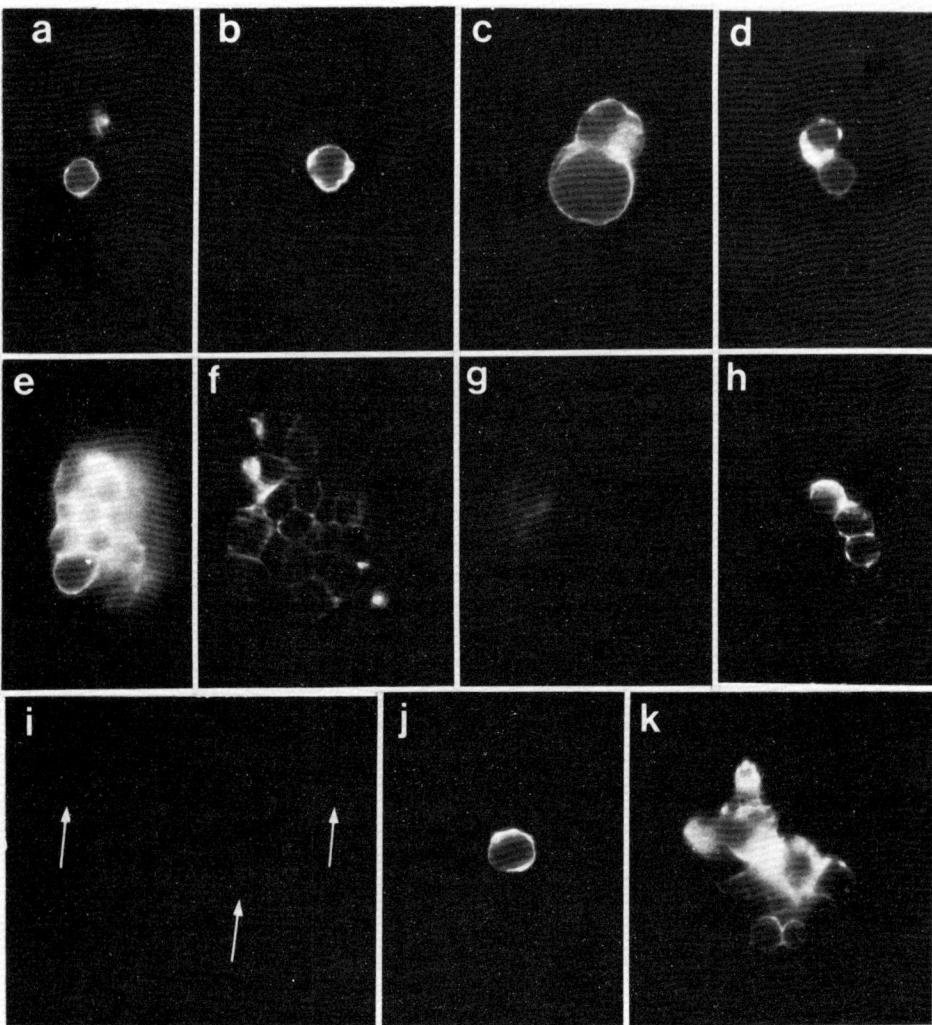

Fig. 8. Immunofluorescence staining of muscle or liver cell suspensions that had been treated with rabbit antiserum to freshly dissociated or trypsin-recovered (preaggregated) myoblasts. Unabsorbed Anti-M-24 diluted 1/10 with E-HS, followed by GAR-FITC, was used to stain either freshly dissociated (a, e) or preaggreated (b, f) muscle cells. This same antiserum intensely stained embryonic liver cells (c), but following absorption with embryonic liver homogenate or acetone powder no longer stained this cell type (g). If such absorbed antiserum was used to stain freshly dissociated (d) or preaggregated (h) muscle cells, a discontinuous pattern of surface staining was observed. When muscle cells were pretreated with high concentrations of trypsin or neuraminidase and then stained with Anti-M-24 and GAR-FITC, no fluorescence was observed (i). Antiserum prepared against freshly dissociated myoblasts also intensely stained the surfaces of both freshly dissociated (j) and preaggregated (k) muscle cells.

these antigens could be reexposed by mild trypsinization. This uncovering of antigen was time dependent and suggested a progressive masking of myoblast antigens upon repair of the cell surface and concommittant myodifferentiation. Cell surface antigens of normal and trypsinized myotubes must be compared before a statement can be made concerning the developmental significance of the trypsin effect. Such a study is in progress. Preliminary observations indicate that more muscle-specific antigens, as detected by inhibition of Anti-M-24 for myoblast monolayers, are exposed on mildly trypsinized myotubes than on untrypsinized cells (Friedlander, unpublished observations).

Our conclusion that only myoblast-specific surface antigens were enhanced by mild trypsinization, but eliminated by neuraminidase or extensive trypsinization, was based on experiments which warrant additional discussion. Myoblast-specific antiserum, as operationally defined, retained high cytotoxicity for 24-h-old muscle monolayers at levels of absorption where cytotoxicity was lost for other embryonic chick tissues. Thus, Anti-M-24 exhibited 40–60% cytotoxicity for myoblast monolayers after absorption, a 30–40% reduction from values obtained with unabsorbed antiserum. Although myoblasts account for 80–90% of the cells in 24-h-old monolayers, these cells only account for approximately 50% of the ^{51}Cr taken up in such cultures. Based on this observation and the morphological scoring of cell types that were specifically lysed in such heterologous cultures, we concluded that specifically absorbed Anti-M-24 binds predominantly to spindle-shaped myoblasts (3). Similarly, Fig. 4 and 7 are more readily interpreted when the heterogeneity of 24-h-old muscle cultures is considered. If myoblast-specific antigens were selectively removed with high levels of trypsin or neuraminidase, why was there an initial decline, then a plateau in the inhibition of cytotoxicity of Anti-M-24 for myoblast monolayers after absorption with protease- or neuraminidase-treated muscle cell suspensions? Since the inhibition of cytotoxicity plateaued between 40 and 60% cytotoxicity in both series of experiments, and since myoblasts only account for 40–60% of the ^{51}Cr in such cultures, it seemed reasonable to conclude that the fibroblasts in such cultures were lysed by the antiserum and complement. In other words, absorption of Anti-M-24 with trypsin- or neuraminidase-treated muscle cell suspensions failed to remove antibodies reactive with both myoblasts and fibroblasts. However, we cannot be certain if the myoblast-specific antigens were totally removed by enzymatic treatment. Similarly, the shared antigens might also be removed upon prolonged exposure of the cells to sialidase or trypsin. To better understand the relative labilities of these 2 classes of antigens, enzymatic treatment for longer periods of time and with higher concentrations of trypsin and neuraminidase will be required. Furthermore, absorption with higher cell numbers will be required to establish if the observed differences in absorptive capacity of enzymatically treated cells are qualitative or quantitative phenomena. If cells after 3–4 h of suspension culture were incubated in high concentrations of trypsin, all surface antigens were apparently removed since these cells failed to inhibit the cytotoxicity of Anti-M-24 for myoblast targets. It is conceivable that residual trypsin, present during the preaggregation periods that preceded retrypsinization, rendered the cell surface more labile to a subsequent trypsin digestion. Such residual activity might explain the difference in the cytotoxicity curves observed after absorption of Anti-M-24 with myoblasts treated with high concentrations of trypsin either immediately following tissue dissociation (Fig. 4) or after 3.5 h of suspension culture (Fig. 3).

It has been suggested that with progressive differentiation various antigenic sites on embryonic cells surfaces become preferentially masked or unmasked, thus creating cell surface mosaics which are tissue and developmental stage specific (57, 58). Our data are consistent with the existence of protease-sensitive, developmentally regulated, cell surface mosaics. At this time we are uncertain as to how protease treatment preferentially exposes myoblast cell surface antigens; whether trypsin unmasks "cryptic" antigens or somehow alters the organization of these molecules within the membrane, and thus affects their expression, remains unanswered. Our observation of differential staining patterns obtained with unabsorbed and absorbed Anti-M-24 suggest that there is a mosaicism to the distribution of the antigens detected by this antiserum. We previously reported that myoblasts possess surface antigens not detected on muscle of later stages of development (1, 2), but at that time could not rule out the possibility that these sites were simply masked or present in greatly diminished quantities on myotubes. We now present evidence that myoblast surface antigens were revealed in greater quantities by mild proteolysis and there are preliminary indications that similar treatment of myotubes makes these cells more effective inhibitors of cytotoxicity of Anti-M-24 for myoblast and myotube monolayers.

We conclude from these experiments that neuraminidase-sensitive antigens, which are also removed by high concentrations of trypsin, but revealed in greater quantities by mild proteolysis, may serve to distinguish the myoblast cell surface from other embryonic chick tissues and muscle at later stages of development. The potential biological roles for such developmental stage specific markers are many: intercellular recognition, the regulation of cellular proliferation, or the initiation of myofibrillogenesis by providing subplasmalemmal insertion sites for contractile filaments. All of these phenomena characterize the myogenic developmental program and probably involve surface associated events (59, 60). While the data presented here do not directly support any such role for the antigens we have detected, the information we now have about their neuraminidase and protease sensitivity should prove valuable in designing procedures for isolating and characterizing these molecules. We are currently exploring the immunochemical nature of these antigens, their fluidity within the cell membrane, and their possible relationship to cytoplasmic structures and physiological activities of the muscle cell.

ACKNOWLEDGMENTS

The authors gratefully acknowledge the invaluable technical assistance of Lovenia Williams, Marian Daniels, and Furman Davis. We would also like to thank Dr. Frank Fitch for the generous use of certain laboratory facilities and Sally Hoskins for her assistance in the preparation of the antisera used in this study. Our discussions with Eric Beyer and Sheila Fallon and Drs. Linda Marton, Theodore Steck, Michael Edidin, Frank Fitch, and Hewson Swift were most helpful and greatly appreciated. Martin Friedlander was a postdoctoral fellow of the Muscular Dystrophy Association of America and was also supported by an institutional allowance from American Cancer Society grant 1N-41-P. This research was also funded by grants from the University of Chicago Cancer Research Center and the Harry Levine Memorial Foundation, and United States Public Health Service grant NHLI-13505.

REFERENCES

1. Friedlander M, Fischman DA: J Cell Biol 63:105a, 1974.
2. Friedlander M, Fischman DA: J Cell Biol 67:124a, 1975.
3. Fischman DA, Doering J, Friedlander M: In Marois A (ed): "Tests of Teratogenicity in vitro." Amsterdam: North Holland Publishing Company, 1976, p 233.
4. Bischoff R, Holtzer H: J Cell Biol 41:188, 1969.
5. Yaffe D: Exp Cell Res 66:33, 1971.
6. Powell JA: Exp Cell Res 80:251, 1973.
7. Hauschka SD, Konigsberg IR: Proc Natl Acad Sci USA 55:119, 1966.

8. Yaffe D: Curr Top Dev Biol 4:37, 1969.
9. Rash JE, Staehelin LA: Dev Biol 36:455, 1974.
10. Kalderon N, Gilula NB: Neurosci Abstr 2:412, 1976.
11. Stockdale FE, Holtzer H: Exp Cell Res 24:508, 1961.
12. Konigsberg IR: Science 140:1273, 1963.
13. Shimada Y, Fischman DA, Moscona AA: J Cell Biol 35:445, 1967.
14. Friedlander M, Fischman DA: Manuscript submitted.
15. Goldschneider I, Moscona AA: J Cell Biol 53:435, 1972.
16. Wiley LD, Calarco PG: Dev Biol 47:407, 1975.
17. Edidin M, Gooding LR, Johnson M: In "Karolinska Symposia on Research Methods in Reproductive Endocrinology." 7th Symposium, Immunological Approaches to Fertility Control, 1974, p 336.
18. O'Rand MG, Romrell LJ: Dev Biol 55:347, 1977.
19. Kurth R, Bauer H: Virology 47:426, 1972.
20. Phillips ER, Perdue JF: J Supra mol Struct 4:27, 1976.
21. Artzt K, DuBois P, Bennett D, Condamine H, Babinet C, Jacob F: Proc Natl Acad Sci USA 70:2988, 1973.
22. Akeson R, Herschman HR: Proc Natl Acad Sci USA 71:187, 1974.
23. Lloyd KO, Darnule TV: J Immunol 112:311, 1974.
24. Croissile Y: In Marois A (ed): "Tests of Teratogenicity in vitro." Amsterdam: North Holland Publishing Company, 1976, p 149.
25. Nathenson SG, Cullen SE: Biochim Biophys Acta 344:1, 1974.
26. Springer TA, Strominger JL, Mann D: Proc Natl Acad Sci USA 71:1539, 1974.
27. Kurth R: Personal communication.
28. Robbins PW, Wickus GG, Branton PE, Gaffney BJ, Hirschberg CB, Fuchs P, Blumberg PM: Cold Spring Harbor Symp Quant Biol 39:1173, 1975.
29. Hynes RO, Wyck JA, Bye JM, Humphreys KC, Pearlstein ES: In Reich E, Rifkin DB, Shaw E (eds): "Proteases and Biological Control." New York: Cold Spring Harbor Laboratories, 1975, p 931.
30. Burger MM: Proc Natl Acad Sci USA 62:994, 1969.
31. Kleinschuster SJ, Moscona AA: Exp Cell Res 70:397, 1972.
32. Noonan KD, Burger MM: J Cell Biol 59:134, 1973.
33. Martinozzi M, Moscona AA: Exp Cell Res 94:253, 1975.
34. Singer SJ: In Meints RH, Davies E (eds): "Control Mechanisms in Development." New York: Plenum Press, 1976, p 181.
35. Parham P, Humphreys RE, Turner MJ, Strominger JL: Proc Natl Acad Sci USA 71:3998, 1974.
36. Ostrand-Rosenberg S, Edidin M, Jewett M: Manuscript submitted.
37. Weiss L: Exp Cell Res 14:80, 1958.
38. Maslow DE: In Poste G, Nicolson GL (eds): "The Cell Surface in Animal Embryogenesis and Development." Amsterdam: Elsevier/North Holland Publishing Company, 1976, p 697.
39. Marton L, van der Westhuyzen D, Friedlander M: Manuscript in preparation.
40. Friedlander M, Beyer E, Fischman DA: Manuscript submitted.
41. Shimada Y, Fischman DA: In Lieberman M, Sano T (eds): "Developmental and Physiological Correlates of Cardiac Muscle." New York: Raven Press, 1975, p 81.
42. Wigzell H: Transplantation 3:423, 1965.
43. Sanderson AR: Br J Exp Pathol 45:398, 1964.
44. Friedlander M: PhD thesis, University of Chicago, 1976.
45. Schreiner GF, Unanue ER: Adv Immunol 24:37, 1976.
46. Hausman RE, Moscona AA: Proc Natl Acad Sci USA 72:916, 1975.
47. Rutishauser U, Thiery J-P, Brackenbury R, Sela B, Edelman GM: Proc Natl Acad Sci USA 73:577, 1976.
48. Solursh M, Meier S: Dev Biol 30:279, 1973.
49. Doering J: PhD thesis, University of Chicago, 1975.
50. Lim R, Turrif DE, Troy SS, Kato T: In Federoff S (ed): "Cell, Tissue and Organ Cultures in Neurobiology." New York: Academic Press (In press).
51. Moscona AA: In Moscona AA (ed): "The Cell Surface in Development." New York: John Wiley and Sons, 1976, p 67.
52. Roth S, McGuire EJ, Roseman S: J Cell Biol 51:536, 1971.
53. Roth S, Dev Biol 18:602, 1968.
54. Moscona AA: In Bittar EE (ed): "Cell Biology in Medicine." New York: Wiley-Medical, 1973, p 571.

55. Steinberg MS, Granger RE: Am Zool 6:337a, 1966.
56. Caravita S, Zachei AM: J Embryol Exp Morphol 32:25, 1974.
57. Bennett D, Boyse EA, Old LJ: In Silvestri LG (ed): "Cell Interactions." Amsterdam: North Holland Publishing Company, 1972, p 247.
58. Singer SJ: In Bradshaw RA, Frazier WA, Merrell RC, Gottlieb DI, Hogue-Angeletti RA (eds): "Surface Membrane Receptors." New York: Plenum Press, 1976, p 1.
59. Edidin M: In Poste, G Nicolson GL (eds): "The Cell Surface in Animal Embryogenesis and Development." Amsterdam: Elsevier/North Holland Publishing Company, 1976, p 127.
60. Edelman GM: Science 192:218, 1976.

Organization and Polysaccharides of Sponge Aggregation Factor

Susie Humphreys, Tom Humphreys, and James Sano

University of Hawaii, Pacific Biomedical Research Center, Kewalo Marine Laboratory, Honolulu, Hawaii 96813

Aggregation factor, the macromolecular complex which mediates species-specific aggregation of dissociated sponge cells, was isolated from several species, partially characterized, and visualized by electron microscopy. All factors were large fibrous complexes with a backbone and side chains or arms. In some factors, the backbone is linear. In others it is circular and the complex appears as a sunburst with arms extending like rays from the circle. The size and location of the polysaccharide chains have been studied using purified preparations of Microciona prolifera. "Sunbursts" treated with ethylenediaminetraacetate (EDTA) for 4 weeks at 0°C dissociate into 3 protein- and polysaccharide-containing components. Sodium dodecyl sulfate does not cause the sunburst to dissociate nor does it inhibit dissociation in the presence of EDTA suggesting that dissociation is not due to hydrolytic enzymes. The dissociation products were fractionated on a 977-Å pore size micropore glass column. Fifteen percent of the material is excluded and appears in the electron microscope as the central circle of the sunburst. Digestion of the circles with 10^{-3} M dithiothreitol (DTT) and 0.5 mg/ml proteinase K for 72 h at 37°C produces 2 polysaccharide chains of 65,000 and 6,000 daltons as fractionated and sized on a 233-Å pore size micropore glass column using Pharmacia dextrans as standards. The included fractions of the EDTA-treated material are subunits of the arms which contain 70% of the polysaccharide. A single polysaccharide of 6,000 daltons as measured on 233-Å size glass beads and Sephadex G-75 is released from these subunits by proteinase digestion. Pharmacia dextrans are used as standard on both columns. We calculate that there would be four 65,000-dalton chains and one hundred 6,000-dalton chains per circle and fifty 6,000-dalton chains per arm. The third component of the EDTA-treated preparation is partially included on the column. It appears as linear fibrils in the electron microscope and contains polydisperse polysaccharides of several-hundred-thousand daltons. It may be an impurity since there is apparently less than 1 of the large polysaccharide chains per sunburst.

Key words: aggregation factor, proteoglycans, polysaccharides, aggregation factor, glycoconjugates, glycoproteins, sponges

Sponge aggregation factor is a large glycoprotein complex which functions in reaggregation of dissociated marine sponge cells. It is isolated from marine sponge tissue by soaking pieces of tissue in calcium- and magnesium-free sea water until the cells dissociate, thereby releasing the aggregation factor into the supernatant. When the cells are removed

Received for publication April 4, 1977; accepted July 8, 1977

© 1977 Alan R. Liss, Inc., 150 Fifth Avenue, New York, NY 10011

from the supernatant and returned to sea water they aggregate very slowly, but they aggregate rapidly if the material from the supernatant of dissociation is returned to the aggregating cells. This factor carries the recognition sites and the specificity for species-specific sorting that occurs when the cells from certain different species are mixed (1, 2).

Previously, we isolated and characterized the aggregation factor from Microciona parthena (2, 3). A large protein and polysaccharide complex of about 20,000,000 daltons, it consists of a central circle $\sim$ 800 Å in diameter and 15 or 16 arms radiating outward 1,100 Å in a unique sunburst organization. Our laboratory has now succeeded in isolating the aggregation factor from Terpios zeketi, Haliclona occulata, Halichondria bowerbankii, and purifying the factor complex from Microciona prolifera. The aggregation factor of each species appears large and fibrous with a backbone and side chains and is composed of protein and polysaccharide.

Aggregation factor from each species is basically similar, yet there is species specificity in aggregation of dissociated sponge cells. Where does this specificity lie? How are the sugars and proteins arranged in each part of the complex, how do they participate in the interactions of the factor complex, and what is the basis of the specificity? This report describes our initial efforts to fragment M. prolifera aggregation factor and characterize the macromolecular subunits.

MATERIALS AND METHODS

Buffers

Ca^{2+}- and Mg^{2+}-free sea water CMF) was prepared as previously described (1). We used 0.5 N NaCl buffer. CaCMF is CMF to which 10^{-3} M $CaCl_2$ is added.

Preparation and Purification of Sponge Aggregation Factor

The aggregation factor (AF) was prepared from Microciona prolifera, Haliclona occulata, and Halichondria bowerbankii at the Marine Biological Laboratory, Woods Hole, Massachusetts by published methods (1, 2, 4). Aggregation factor from Terpios zeketi was prepared by Charles Cauldwell at the Pacific Biomedical Research Center, Honolulu, Hawaii. Microciona prolifera aggregation factor was purified (4) until the preparation was essentially only sunbursts when viewed with the electron microscope. It was dissolved at about 1,000 units of activity per ml in CaCMF and had about 1.5–5 μg protein and 0.5–2 μg polysaccharide per unit activity depending on the preparation. The preparations ranged from 60 to 80% protein and from 20 to 40% polysaccharide.

Gel Filtration and Chromatography

The 977-Å and 233-Å pore glass columns (Electro-Nucleonics, Fairfield, New Jersey) and the Sephadex G-75 (Pharmacia, Uppsala) column employed in this study were poured and eluted as recommended by the manufacturers. All columns were 48 × 0.7 cm. The 977-Å pore glass bead column equilibrated with CMF had an excluded volume of 8.33 ml, an included volume of 16.90 ml, a flow rate of 16.2 ml/h, and was used in fractionating ethylenediaminetetracetate (EDTA)-treated AF. The 233-Å pore glass bead column, equilibrated with 0.5 N NaCl, had an excluded volume of 8.16 ml, an included volume of 16.45 ml, a flow rate of 12 ml/h, and was used in fractionating the proteinase K-dithiothreitol (DTT)-, and EDTA-treated AF. The Sephadex G-75 column equilibrated with 0.5 N NaCl had an excluded volume of 7.02 ml, an included volume of 18.36 ml, a flow rate of 12 ml/h, and was used to compare the included run of the proteinase K-, DTT-, and EDTA-treated AF from the 233-Å pore glass column. The 233-Å pore column and the Sephadex G-75 column were standardized with Pharmacia dextran standards of 10,000, 40,000 and 70,000 daltons.

EDTA Treatment

Two milliliters of purified AF was mixed with 0.5 M EDTA, pH 7, to a final concentration of 0.7 mM and left at 0°C for a 4-week period. Following this treatment the AF was run over the 977-Å pore glass column; thirty-five 0.54-ml fractions were collected and 25-μl aliquots were taken from fractions 11–35 for analysis of neutral hexoses by the phenol-sulfuric acid method (5).

Digestion With Proteinase K and DTT

The included, intermediate, and excluded peaks (1.51 ml each) of the 4-week EDTA-treated AF run over the 977-Å pore glass column were each treated with 13 mM DTT and 0.9 mg proteinase K (EM Labs, Elmsford, New York) by adding 41 μl of 0.5 M DTT, pH 7, and 91 μl of 10 mg/ml proteinase K in H_2O. These mixtures were left in a 37°C water bath for 24 h. An equal amount of proteinase K was added 2 more times at 24-h intervals. The digested fractions were then passed over a 233-Å pore glass column or over a Sephadex G-75 column. From each column, thirty-five 0.54-ml fractions were collected; 200-μl aliquots were taken from fractions 11–23 and analyzed for neutral hexoses by the phenol-sulfuric acid method (5).

Gel Electrophoresis of EDTA-Treated Aggregation Factor

One-half percent sodium dodecyl sulfate (SDS) and 10 mM DTT were added to included and excluded fractions from the aggregation factor treated with EDTA for 4 weeks and run over the 977-Å glass bead column. The eluates were treated at 100°C for 5 min and electrophoresed on 4.8% SDS-polyacrylamide gels (6).

Electron Microscopy

Positive staining. Five microliters of aggregation factor or factor fragment at 5 μg/ml in CaCMF or 10^{-3} M $CaCl_2$, with or without EDTA, was placed on a grid with a thin carbon film for about 15 sec, rinsed, and stained with freshly prepared 1% uranyl formate. Excess stain was withdrawn with filter paper.

Shadowing. Aggregation factor at 5 μg/ml in CaCMF was sprayed or dropped onto a grid or piece of mica coated with a carbon film, rinsed in distilled water or volatile buffer (0.5 M NH_4CO_3 with 0.1 M $NH_4C_2H_3O_2$) to eliminate salt, and dried. Aggregation factor in 10^{-3} M $CaCl_2$ was sprayed onto a thin carbon film on mica, not rinsed, and dried. Some preparations were spread by the aqueous basic protein film technique with the spreading solution consisting of 2.5 μg/ml aggregation factor, 0.05 mg/ml cytochrome C in 0.5 M $NH_4C_2H_3O_2$, and 1 mM EDTA, pH 7.5, with or without calcium excess to chelate the EDTA. The hypophase was 0.25 M $NH_4C_2H_3O_2$, pH 7.5. These samples were dipped into 100% ethanol and air dried. Preparations were then shadowed with C-Pt pellets at an angle of about 6–10°. Films of aggregation factor shadowed on mica were then floated onto grids.

RESULTS

Electron Microscopy

Figure 1 shows an electron micrograph of aggregation factor from Microciona prolifera applied to a grid and stained with uranyl formate without drying. The factor complex clearly consists of a circle with a number of radiating arms. This structure is very similar to the structure of the aggregation factor previously described from Microciona parthena (2). The circumference of the circle is ~ 0.4 μm depending on preparative conditions with about 16 arms or rays extending outward.

Stained preparations are of low contrast and show the aggregation factor with tangled arms; these characteristics make counting of the arms and precise measurements difficult. M. parthena was much better delineated in shadowed preparations. However, we have been unable to obtain shadowed preparations of the M. prolifera factor which are as well spread as our preparations of the M. parthena aggregation factor. Prolifera aggregation factor attached to the grids, dried, and shadowed invariably revealed collapsed structures with the circle collapsed or partially collapsed and the arms aggregating together (Fig. 2). The collapse seems to be greatly increased by drying the aggregation factor onto the substrate without support, such as that given by stain, as evidenced by the fact that aggregation factor dried on the grid before staining shows much more condensation than when stained without drying. Thus far the collapse of the circle and contraction of the arms upon drying has not been prevented by treatment of the carbon substrate by ionizing, by spreading the aggregation factor on the grids in various concentrations of volatile buffers or by chemical fixation of the aggregation factor.

Several treatments of the aggregation factor, such as incubation in ethylenediaminetetraacetate, guanidine hydrochloride, and sodium dodecyl sulfate with dithiothreitol, do

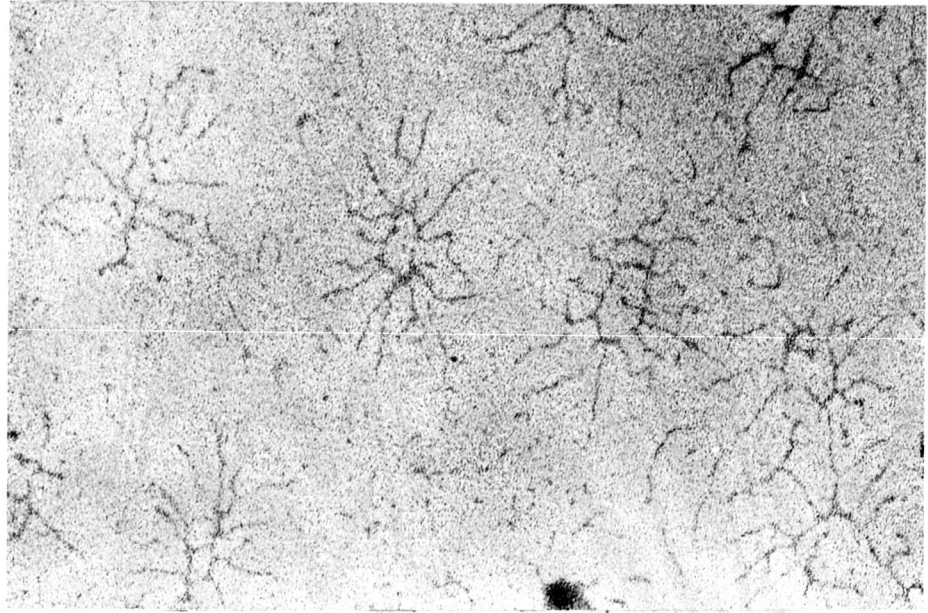

Fig. 1. Microciona prolifera aggregation factor in CaCMF positively stained with uranyl formate. The factor has a circular backbone about 0.4 μm in circumference with about 12–16 arms radiating outward giving a "sunburst" appearance. Magnification 81,000 ×.

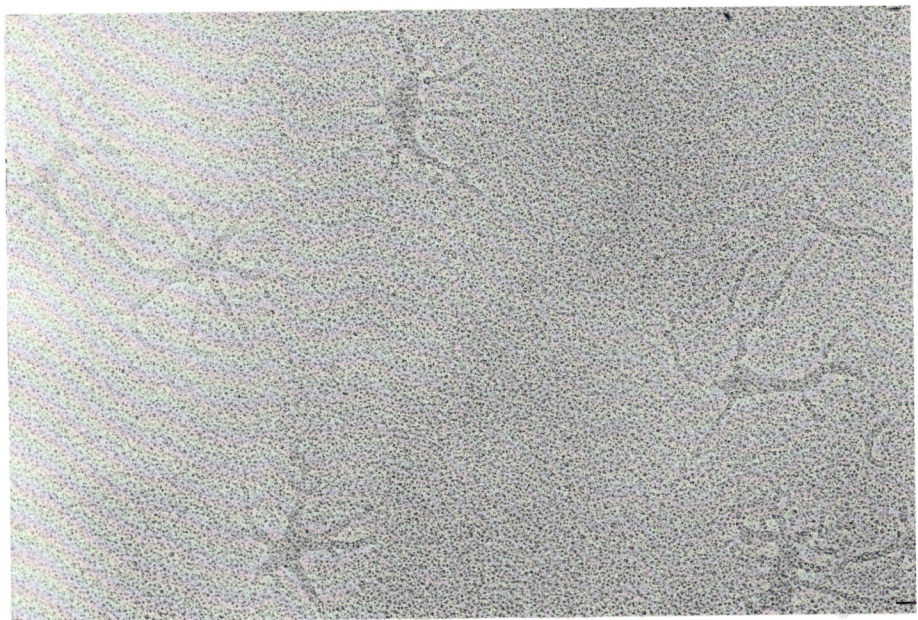

Fig. 2. Microciona prolifera aggregation factor in CaCMF rinsed in distilled water, dried, and shadowed with a C-Pt pellet. The circle and the arms collapse into solid stellate forms and more elongate forms. These condensed figures may polymerize. Magnification 81,000 ×.

seem to improve the spreading of the circular backbone in shadowed preparations. So far, however, this prevention of collapse has been associated with the removal of the arms. The nature of the reactions occurring when the sunburst "collapses" remains unknown. The collapse upon drying is real and demonstrates that the macromolecule is not rigid. Currently the method which best displays the arms of M. prolifera is a basic protein film technique. Molecules spread in this way (Fig. 3) show about 16 arms.

The circular configuration of the backbone has been observed in aggregation factors from M. prolifera (4), M. parthena (2), and Geodia cydonium (7). However, it does not seem to be a universal characteristic of aggregation factors. Halichondria bowerbankii, Terpios zeketi (Fig. 4), and Haliclona occulata (Fig. 5), extracted by methods parallel to those used for Microciona parthena (2), have a very similar size and structure of a backbone fiber with side arms. However, the backbone of the aggregation factor complexes is linear. Without further evidence, we presume that the linear backbone is the native configuration of these aggregation factors since it was extracted by rather gentle methods which did yield circular backbones from Microciona factors.

EDTA Dissociation

When calcium is removed from the active M. prolifera aggregation factor complex, activity is lost very rapidly. For example, aggregation factor treated for less than 1 min with 10^{-3} M EDTA has irreversibly lost all activity. If 2×10^{-3} M $CaCl^2$ is added, the EDTA has no effect on the aggregation factor activity. Thus the removal of calcium and not other polyvalent metal ions is the mechanism of inactivation of the aggregation factor. Although EDTA quickly inactivites aggregation factor, it has no initial obvious effect on the structure of the aggregation factor. Electron micrographs indicate that it is still a circle

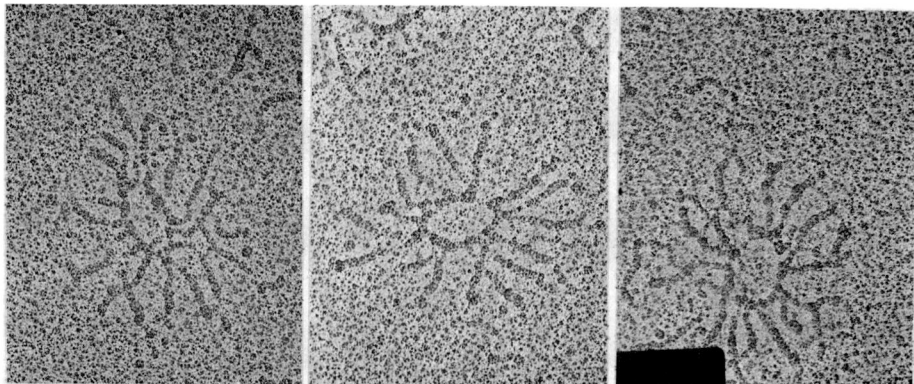

Fig. 3. Microciona prolifera aggregation factor spread with cytochrome C and shadowed display about 16 arms. Magnification 70,000 ×.

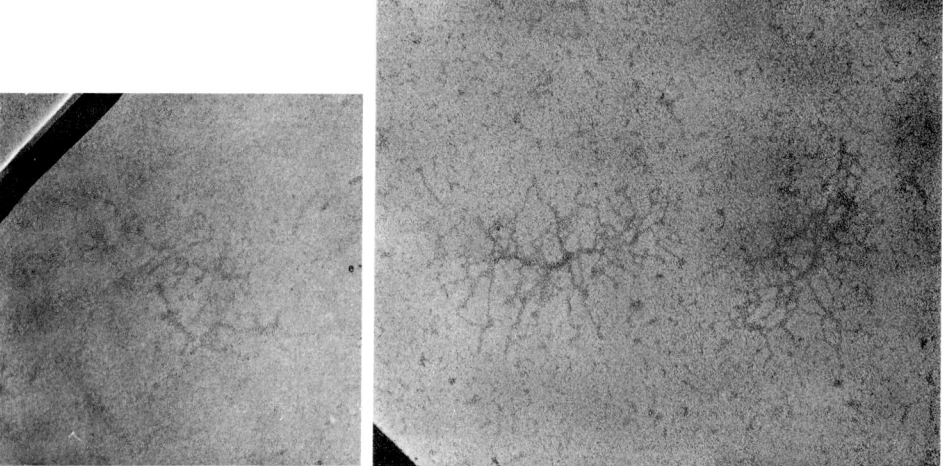

Fig. 4. Terpios zeketi aggregation factor in CaCMF positively stained with uranyl formate. The backbone is linear. Magnification 90,000 ×.

Fig. 5. Haliclona occulata aggregation factor in CaCMF positively stained with uranyl formate. The backbone is linear. Magnification 90,000 ×.

with arms and permeation chromatography shows that the aggregation factor is still excluded. However, in EDTA the aggregation factor does begin to dissociate with time. Figure 5 shows chromatography on 977-Å pore glass beads of aggregation factor treated for 2 weeks and 4 weeks at 4°C with EDTA. Active, untreated factor is completely excluded by this column, but by 2 days (not shown) about half of the protein and polysaccharide of the aggregation factor runs in the included fractions of the column. By 2 weeks (Fig. 6A) 70% of the material and by 4 weeks (Fig. 6B) 85% of the material runs in the included fractions. Most of the included material is totally included but there is always a distinct shoulder of partially included material which apparently represents a third component.

We asked how these fractions relate to the original sunburst? Protein and polysaccharide run equally distributed across the column so we are not separating protein- or polysaccharide-rich molecules. Excluded and included fractions were examined with the electron microscope. With time, the excluded fractions appeared as circles with fewer and fewer arms and by 4 weeks it was circles devoid of arms. Such a preparation is shown in Fig. 7. The completely included fractions were very small pieces without a distinct morphology. They were not whole arms. Examination of the fractions intermediate between included and excluded showed larger pieces of linear material. It appears the arms are falling off the circles and dissociating into small subunits.

The requirement of 4 weeks for the dissociation of aggregation factor seemed curious. We wondered if it was the results of slow hydrolytic degradation of the aggregation factor by enzymes that contaminated the preparation. We had previously observed that the factor was stable in 0.5% SDS and reasoned that if SDS inhibited the EDTA effect this might indicate that the dissociation of the factor was due to an enzyme whose activity was destroyed by SDS. Figure 8 shows permeation chromatography of aggregation factor treated with SDS for 5 days and aggregation factor treated with SDS plus EDTA for 5 days. It can be seen that the aggregation factor was completely stable in SDS alone but dissociated as usual when EDTA was added to the SDS-treated aggregation factor. These results do not lend support to the idea that the complex is degraded by a hydrolytic enzyme which is denatured by SDS.

We attempted to dissociate the aggregation factor further and to measure the size of the subunits using standard SDS-polyacrylamide gel chromatography. The included fraction and the excluded fraction were each taken, treated with SDS and dithiothreitol, heated 5 min at $100°C$, and electrophoresed on 4.8% polyacrylamide gels. The excluded fraction, the circle, apparently was not further degraded by treatment with SDS and dithiothreitol and failed to enter the gel. This was confirmed by passing such treated material over the 977-Å pore column. Again, the material ran as an excluded peak. The included fraction, the subunits of the arms, electrophoresed as a broad band (stained for protein and polysaccharide) ranging between 13,000 and 300,000 mol. wt. using phosphorylase A, equine heart cytochrome C, and myosin as standards. This heterodispersion is probably due to variation in length of polysaccharide chains although we have not established this point. The exact size of this subunit cannot be determined from these measurements since polysaccharide runs anomalously on acrylamide gels.

Proteinase K Digestion

Thirty percent of the material of the sunburst is polysaccharide. Polysaccharides have long been associated with cell surfaces and are believed to be intimately involved in cell surface reactions. Therefore, it seemed of special interest to analyze the polysaccharides of the factor to determine their number, size, and characteristics. To do this, we digested away the protein using proteinase K and analyzed the resultant polysaccharide chains on 233-Å pore glass bead permeation chromatography columns. When the excluded fraction of the 977-Å pore glass bead column, which contained the circles, was digested and passed over the 233-Å pore size glass bead columns 2 peaks were obtained as shown in Fig. 9A. Twenty-five percent of the material ran as an excluded peak, 50% of the material ran as an included peak, and 25% appeared as a large shoulder on the included peak. The apparent

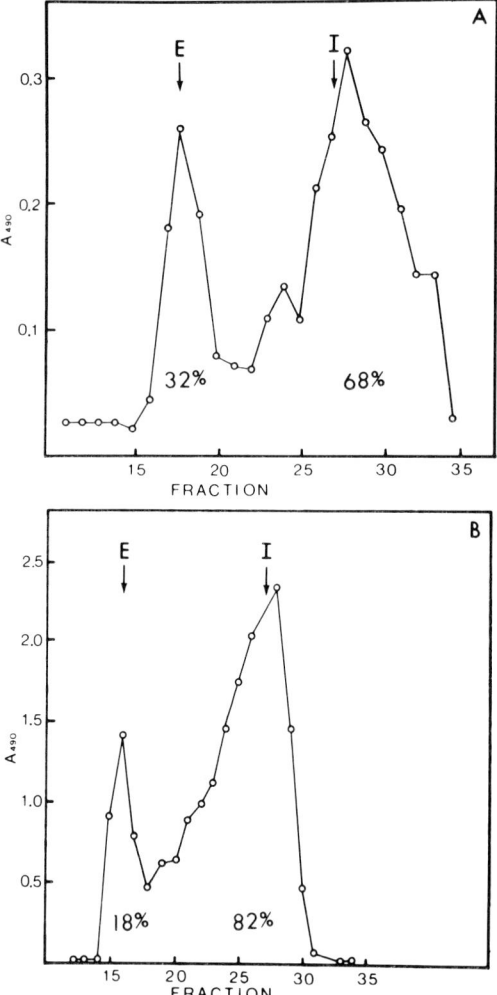

Fig. 6. Gel filtration chromatography of 2-week EDTA-treated Microciona prolifera AF (A) and 4-week EDTA-treated AF (B). A 977-Å pore glass bead column (48 × 0.7 cm) was eluted with CMF at 16.2 ml/h. Aliquots were removed for analysis for sugar (5). OD_{490}) I: included fraction. E: excluded fraction.

molecular weight of the shoulder is 65,000 and of the included peak is 6,000 when the column is standardized with Pharmacia dextrans.

Most of the mass of the sunburst appears as an included peak in the column run after EDTA dissociation and presumably represents fragments of the arms. When these are digested with proteinase K and the digest passed over a 233-Å pore size glass bead column, a single peak of polysaccharide is obtained (Fig. 9B). Its apparent molecular weight is about 6,000 when compared to dextran standards. To estimate the accuracy of this molecular weight determination, we prepared a Sephadex G-75 column and standardized it with the same dextran preparations. On this column the apparent molecular weight was also

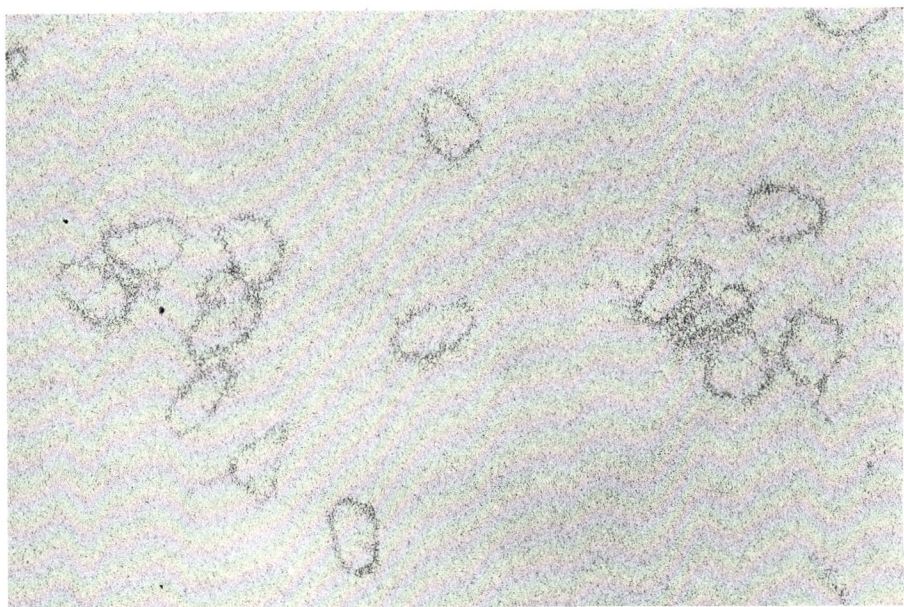

Fig. 7. The excluded peak of Microciona prolifera aggregation factor treated with EDTA for 4 weeks and run over a 977-Å glass bead column consists of circles. These circles are the backbone of the "sunbursts." The preparation was positively stained with uranyl formate. Magnification 81,000 ×.

6,000. It is difficult to judge the accuracy of either of these determinations; further analysis of the polysaccharide and comparison to other standards are required before a firm figure for the molecular weight can be achieved.

The shoulder of material that ran between the included and excluded fractions of the 977-Å pore size glass bead columns of EDTA-treated material was digested also with proteinase K and passed over the 233-Å pore size glass bead column (Fig. 9C). It ran as an excluded peak and a peak of polysaccharide equivalent to the small polysaccharide from the arms. When the excluded peak was run on larger pore size columns, it spread out and proved to be very heterogeneous and large, ranging from 100,000 to 1,000,000 daltons. The excluded peak of polysaccharide from the circles (Fig. 6A) appears to be similar to this heterodisperse polysaccharide from the intermediate shoulder. A calculation of the number of polysaccharide chains in this large heterogeneous class indicate less than 1 per 4 sunburst molecules. This led us to speculate that this polysaccharide derives from a contaminant of the aggregation factor preparation which is also released from the sunbursts in the presence of EDTA.

Other reagents were used in attempts to break the sunburst complex down into smaller subunits. When treated with 4 M guanidine hydrochloride, the reagent used to dissociate cartilage proteoglycan aggregates, the factor broke down into the same subunits as described above. Treatment of the aggregation factor with 0.1 M NaOH overnight at 0°C did not release any of the polysaccharide chains. These conditions are sufficient to hydrolyze sugar linkages to serine and thus indicate that the polysaccharide is not attached to the protein by serine linkages.

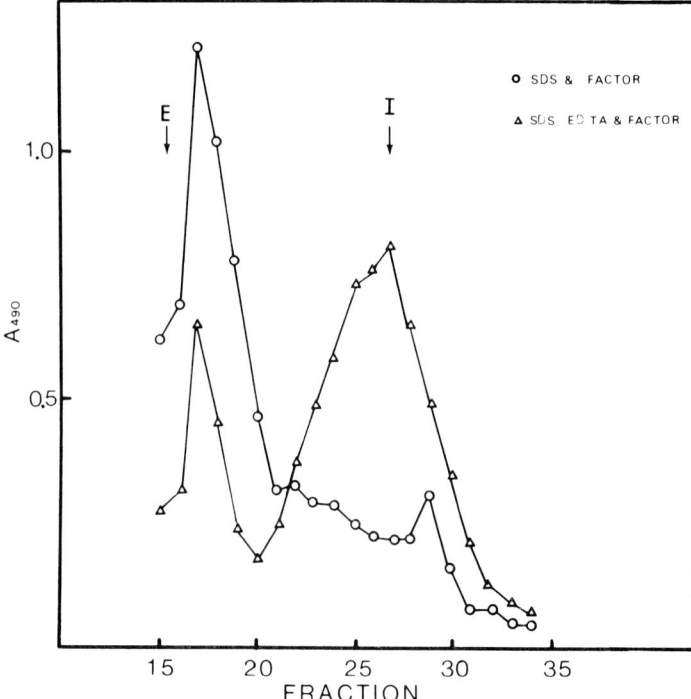

Fig. 8. Gel filtration chromatography of SDS-treated Microciona prolifera AF (o) and SDS plus EDTA-treated AF (△). A 977-Å pore glass bead column (48 × 0.7 cm) was eluted with 0.1% SDS in CMF. Fractions were analyzed for sugar (5). OD_{490}) I: included fraction. E: excluded fraction.

DISCUSSION

These results show that the aggregation factors isolated and assayed by the procedures we developed are large, fibrous macromolecular complexes with a backbone and a number of side chains. In Microciona parthena and prolifera, the backbone is a closed circle and the side chains radiate as arms from the circle producing a sunburst-like structure. However, in the other species the backbone appears linear, with arms extending from both "sides" of the backbone. All the factors we have analyzed have from 20 to 50% polysaccharide and all have been large enough to be excluded by Sepharose 2B or 3,000-Å pore size glass beads with exclusion limits of about 2×10^7 daltons.

The sunburst complex is very large, about 0.4 μm in diameter in M. prolifera when fully extended, and is much larger than the usual distance between the closely apposed membranes of adjacent cells. We have as yet been unable to localize it among aggregated cells by microscopy. The molecule could function as an essentially planar ligand between adjacent membranes, flat in the intercellular space, and binding either to receptor on apposing cell surfaces or other factor molecules. However, this planar configuration is not the only possibility. Aggregation factor molecules may be confined to specialized regions of the cell surface where membranes are not apposed or bind in a collapsed configuration.

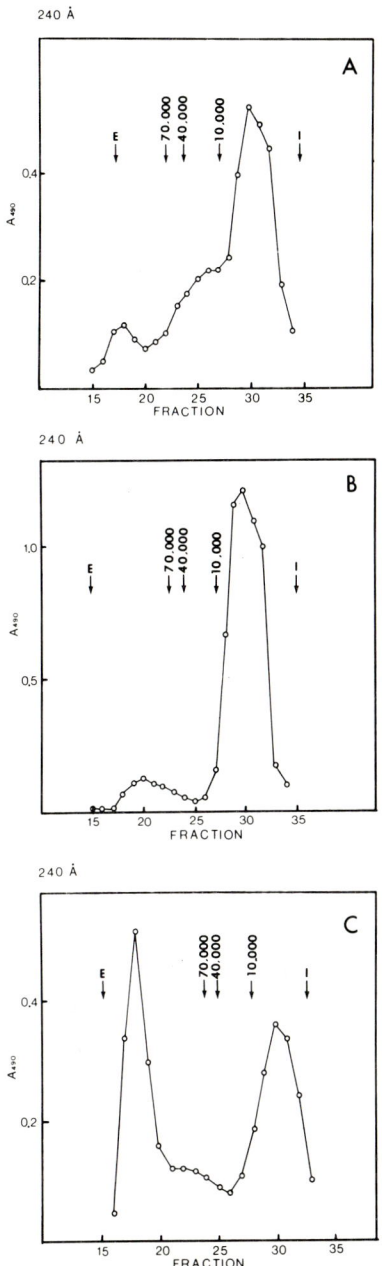

Fig. 9. Gel filtration chromatography of the excluded (A), included (B), and intermediate (C) peaks of the 4-week EDTA-treated Microciona prolifera AF from the 977-Å pore glass column, incubated in DTT and proteinase K (3 additions at 24-h intervals) at 37°C for 72 h, and run over a 233-Å pore glass bead column. The 233-Å column (48 × 0.7 cm) was eluted with 0.5 N NaCl at 12 ml/h. Fractions were analyzed for sugar (5). OD_{490}) I: included fraction. E: excluded fraction.

The development of adhesions due to apposed membranes could be secondary to the adhesion accelerated by the factor molecules.

The association of polysaccharide with cell surface reactions and the presence of polysaccharide in the factor has led to the idea that polysaccharides are intimately involved in the active sites of aggregation factors. For other such cases that are known, the specific polysaccharide of the active site inhibits the activity of the molecule. We looked for specific polysaccharide inhibition by digesting a partially purified aggregation factor preparation with pronase and testing the polysaccharides for inhibitory activity and found none (8). It will be interesting to test the individual polysaccharides of the aggregation factor for their effect on activity of intact aggregation factor.

The most analogous macromolecule and the nearest model for aggregation factor is the proteoglycan aggregate of cartilage (10). In this case, hyaluronic acid provides the backbone with proteoglycan monomer protein and link proteins attached. Each proteoglycan monomer produces a radiating side chain carrying polysaccharides of chondroitin sulfate and keratan sulfate. The sponge aggregation factor may be similar.

When treated with EDTA, the M. prolifera aggregation factor breaks down into a protein- and polysaccharide-containing circle and a series of fragments of the arms which also contain protein and polysaccharide. The organization of these fragments is far from established. However, the large polysaccharide isolated from the circle after proteinase K digestion could be equivalent to the hyaluronic acid backbone of the cartilage of the proteoglycan. We calculate that a backbone of 2,500 Å, the size we have normally seen in these aggregation factors, would have a molecular weight of 120,000–200,000 if it were a single, linear polysaccharide chain. This is larger than the apparent 65,000 daltons of the large polysaccharide on porous glass beads but it is difficult to evaluate the accuracy of this determination. If the polysaccharide were circular, this could affect the apparent molecular weight on the column.

The protein in the circular backbone may be linking proteins which attach the arms to the polysaccharide of the circular backbone. Subunits of the arms may bind end to end to produce the arms of the aggregation factor which have the smaller 6,000 dalton polysaccharide. However, the arms clearly differ in 2 respects from the cartilage proteoglycan monomers. The protein of the proteoglycan monomer of cartilage is a single polypeptide chain while the arms of the aggregation factor must have several subunits because EDTA dissociates the arms into smaller pieces. Since we do not have a molecular weight for these pieces, we cannot calculate how many subunits there might be per sunburst arm. Secondly, the polysaccharides of the aggregation factor are probably much smaller than the sulfated glycosaminoglycan side chains on the cartilage proteoglycan. The polysaccharide of the aggregation factor arms is about 6,000 while chondroitins of cartilage are about 30,000 daltons.

Sponge aggregation factor and cartilage proteoglycan may represent different evolutionary stages of the same basic type of primordial cell component. The cartilage proteoglycan is a very specialized complex developed to be a major volume occupying molecule of cartilage matrix. Sponge aggregation factor is isolated by purification procedures that give about a thousandfold purification and thus the aggregation factor is a minor component of the tissue. At present we cannot ascertain the importance of the similarities and differences between these macromolecules. It is interesting to speculate but only further analysis will tell.

ACKNOWLEDGMENTS

This work was supported by National Science Foundation grant PCM 76-09309. The expert help of Rick Lee Smith, Wes Yonemoto, and Anthony J. Zukowski in preparing the purified factor and conducting the initial phases of these experiments is gratefully acknowledged.

REFERENCES

1. Humphreys T: Dev Biol 8:27, 1963.
2. Henkart P, Humphreys S, Humphreys T: Biochemistry 12:3045, 1973.
3. Cauldwell C, Henkart P, Humphreys T: Biochemistry 12: 3051, 1973.
4. Humphreys T, Yonemoto W, Humphreys S, Anderson D: Biol Bull 149:430, 1975.
5. Dubois M, Gilles KA, Hamilton JK, Rebers PA, Smith F: Anal Chem 28:350, 1956.
6. Fairbanks G, Steck TL, Wallach DFH: Biochemistry 10:2606, 1971.
7. Muller W, Zahn R: Exp Cell Res 80:95, 1973.
8. Humphreys T, Humphreys S: Unpublished data.
9. Humphreys T: In "Cellular Membranes and Tumor Cell Behavior." Baltimore: Williams & Wilkins, 1975, 173–192.
10. Hascall V, Heinegard D: In Slavkin H (ed): "Extracellular Matrix Influences on Gene Expression." New York: Academic Press, 1975, pp 423–433.

Membrane Assembly: Synthesis and Intracellular Processing of the Vesicular Stomatitis Viral Glycoprotein

Flora N. Katz, James E. Rothman, David M. Knipe, and Harvey F. Lodish

Department of Biology, Massachusetts Institute of Technology, Cambridge, Massachusetts 02139

The glycoprotein (G) of vesicular stomatitis virus (VSV) is synthesized on membrane-bound polyribosomes. Approximately 30 min after its synthesis, it reaches the surface plasma membrane where it is incorporated into budding virus. The first part of this paper focuses on the 2 intracellular, membrane-bound, glycosylated forms of the glycoprotein which are intermediates in its biogenesis. All glycosylation and processing is completed in the smooth microsome fraction before the protein reaches the surface.

Next, we turn to the mechanism by which G is synthesized on membrane-bound polyribosomes. All of the G mRNA is bound to membranes, and studies with puromycin suggest that this attachment of G mRNA is mediated by the nascent glycoprotein chain. After its synthesis G is a transmembrane protein with about 30 amino acids at the carboxyl terminus remaining on the cytoplasmic side of the endoplasmic reticulum. Since 95% of the glycoprotein, containing the carbohydrate residues, is resistant to attack by external proteases, it appears to be within the lumen of the endoplasmic reticulum or embedded within the lipid bilayer. Finally, we show that synthesis, glycosylation, and proper asymmetric insertion of G into the ER can be achieved in cell-free extracts. Both glycosylation of G and proper insertion into the ER membrane in this cell-free system require concomitant protein synthesis.

Key words: VSV, glycoprotein, membranes, cell-free synthesis

The envelope of vesicular stomatitis virus (VSV) contains a lipid bilayer in which external spikes, made up of a single glycoprotein (G), are embedded (1). This structure surrounds the RNA-containing nucleocapsid of the virus, with the virus matrix protein (M) forming a layer between them (2–4) (Fig. 1). The virus matures by budding from the host cell plasma membrane and contains lipids of the host cell surface membrane in closely conserved proportions (5). The glycoprotein is synthesized on membrane-bound polyribosomes (6–9) and, as we shall describe, later migrates to the surface membrane of the host cell where it is incorporated into the budding virus (10–12). As the genome of VSV encodes only 5 proteins, all of which are structural proteins of the virion particle (3), it

David M. Knipe is presently with the Department of Biology, University of Chicago, Chicago, IL 60637

Received March 22, 1977; accepted June 6, 1977.

© 1977 Alan R. Liss, Inc., 150 Fifth Avenue, New York, NY 10011

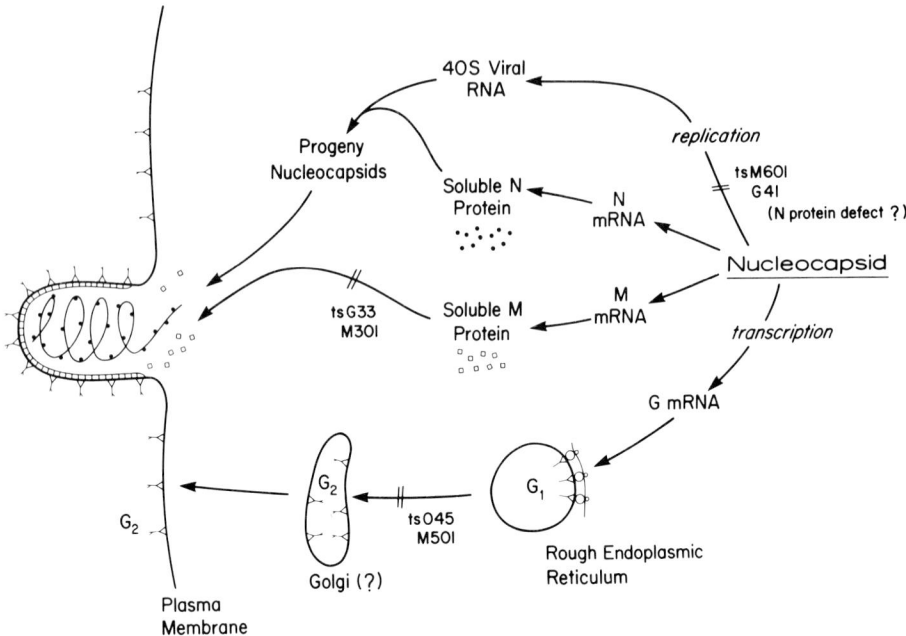

Fig. 1. Schematic diagram illustrating the pathways of maturation of the major structural proteins of VSV and the proposed site of block in virion assembly for certain temperature-sensitive mutants.

must make use of the host cell for the biosynthesis and assembly of viral components, including the viral membranes. Furthermore, the VSV G protein is, we shall show, an example of a major class of cellular membrane proteins which asymmetrically span the lipid bilayer. Thus, a study of the biogenesis of the G protein should provide insight into the mechanisms by which a large class of cell surface glycoproteins are manufactured and processed.

In this paper, we shall focus on the mechanism by which the G protein is inserted asymmetrically into the endoplasmic reticulum (ER) membrane. We shall show that G is synthesized as a transmembrane protein with about 30 amino residues at the carboxyl terminus remaining on the cytoplasmic face of the endoplasmic reticulum membrane. Most or all of the amino terminal 95% of the G protein, containing carbohydrate residues, appears to be within the lumen of the endoplasmic reticulum since it is resistant to attack by external protease. Synthesis, glycosylation, and proper asymmetrical insertion of G into the ER can be achieved in cell-free extracts. Both glycosylation of G and transmembrane insertion into the ER membrane in this cell-free system require concomitant protein synthesis. Attachment of G mRNA to membranes is apparently mediated by the nascent glycoprotein chain.

Multiple Forms of the VSV Glycoprotein: Intracellular Processing

Before discussing experiments on the mode of biosynthesis of G, it is first necessary to describe and characterize in some detail the several forms of G which can be isolated from infected cells or produced in a wheat germ cell-free system using exogenous VSV mRNA (13). Indeed, the existence of multiple forms of the G protein with distinct subcellular localizations has revealed a complex sequence of steps in which the G protein participates between its synthesis in the rough endoplasmic reticulum and its incorporation into budding viral particles. This intracellular processing mechanism is believed to be of general significance and is, therefore, interesting in its own right.

Infected cultures were labeled with [^{35}S] methionine for 5 min at 4 h postinfection, chased for various periods of time, and the total cellular protein analyzed by sodium dodecyl sulfate (SDS)-polyacrylamide gel electrophoresis (Fig. 2). Following the pulse label, the G protein was observed as a single species of apparent molecular weight 65,000 which we call G_1. During the period of chase, this species was progressively converted to a more slowly migrating species, called G_2, with an apparent molecular weight 67,000. The difference in mobility is related, in part, to differences in the extent of glycosylation (see below). No changes in mobility of the other VSV proteins (N, NS, M, L) were apparent.

Both G_1 and G_2 appeared to have molecular weights greater than that of the polypeptide synthesized in vitro under the direction of G mRNA (6, 14, 15). To investigate this

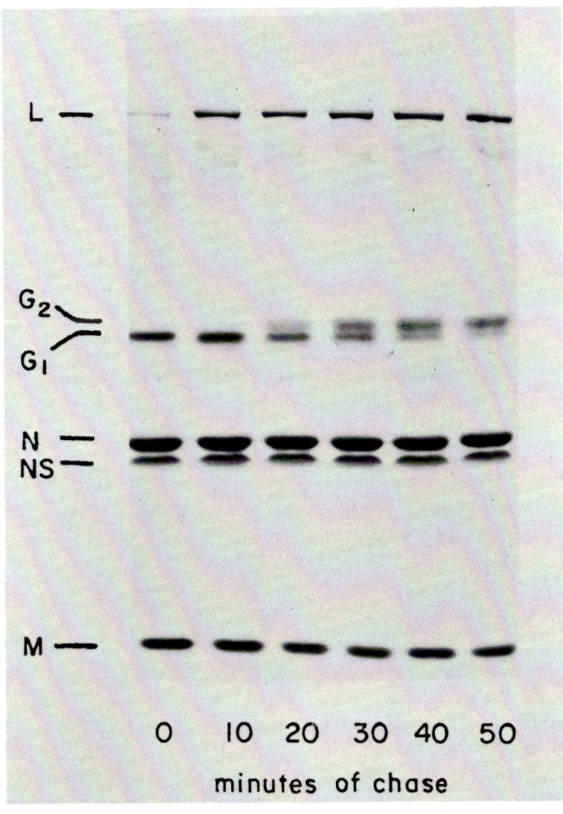

Fig. 2. Pulse-chase labeling of total cellular proteins of cells infected with VSV. Chinese hamster ovary (CHO) cells were infected at an MOI of 10 with VSV and incubated at 37°C. The culture was labeled for 5 min at 4 h postinfection with [^{35}S] methionine as described in Ref. 13, and then excess unlabeled methionine was added. Portions of the culture were removed at the times shown and transferred into cold Earle's saline. The cells were washed and directly solubilized in gel sample buffer, and the cellular proteins were resolved by gel electrophoresis on a 10% SDS-polyacrylamide slab gel. The autoradiogram shown was a 3 day exposure of the dried gel.

question directly, the polypeptide made in a wheat germ cell-free system programmed with total VSV mRNA (Fig. 3, lane b) was compared by slab gel electrophoresis with the proteins of infected cell membranes. The in vitro synthesized G protein, called G_0, migrated more rapidly than any G protein found in infected cells. Even after 2.5 min of labeling of cells, the shortest feasible time, no G protein migrating with G_0 could be observed (data not shown). G_0 has all of the methionine-containing tryptic peptides of G_1 or G_2 (6, 8, 14, 15); its rapid migration relative to G_1 and G_2 would be consistent with the absence of carbohydrate in G_0. As is shown below, G_0 will not adhere to lectin columns which will adsorb G_1 and G_2. If the extents of glycosylation of G_0 and G_1 are different, it would therefore appear that the G protein is glycosylated either during or very soon after its synthesis, so the G_0 form is never evident in infected cells.

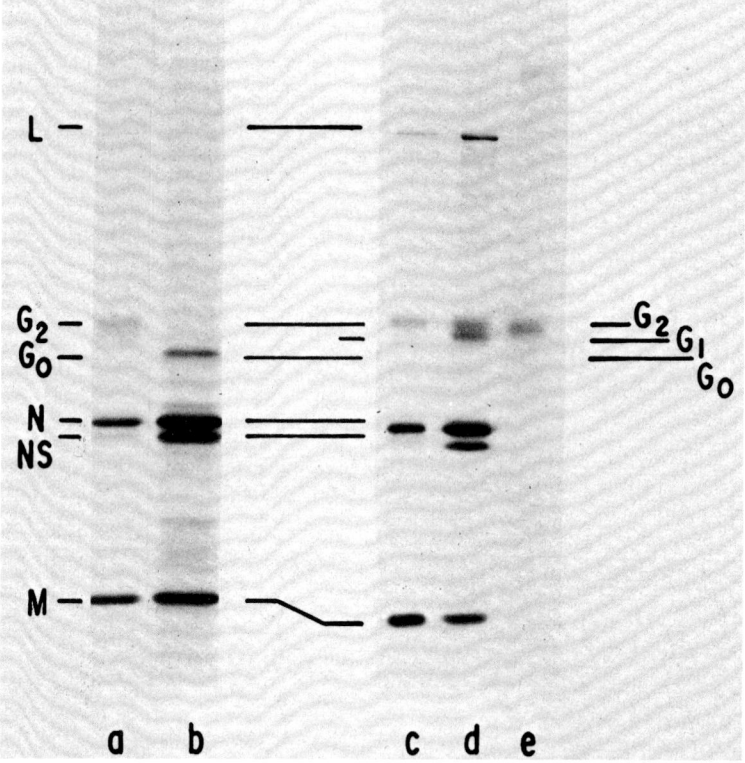

Fig. 3. Comparison of the electrophoretic mobilities of the various forms of the VSV glycoprotein on 10% SDS-polyacrylamide slab gels. Lanes a and b are from one gel, and c–e are from a second gel (run in parallel). Lane a. [^{35}S] methionine-labeled cytoplasmic VSV proteins from infected cells labeled for 30 min and chased for 30 min (10). 48-h exposure. Lane b. [^{35}S] Methionine-labeled proteins synthesized in a wheat germ cell-free extract in response to addition of total cytoplasmic VSV mRNA using the conditions of Morrison and Lodish (6). 16-h exposure. Lane c. [^{35}S] Methionine-labeled virion proteins. 48-h exposure. Lane d. [^{35}S] Methionine-labeled total cellular proteins from infected cells labeled from 4 to 4.5 hr post-infection. 48-h exposure. Lane e. ^{125}I-labeled surface proteins of infected CHO cells. Cells were labeled in a peroxide-lactoperoxidase-catalyzed reaction as detailed in Ref. 13. 48-h exposure. The few minor bands at the top of the gel are normal CHO cell surface proteins and are similar in intensity to iodinated proteins from uninfected cells.

The forms of the glycoprotein in virions and on the surface of infected cells (Fig. 3, lanes c—e) were also examined. Virions contained only the G_2 species (Fig. 3, lane c). In addition, lactoperoxidase-catalyzed iodination of the surface of infected cells revealed a virus-specific protein comigrating with G_2 (Fig. 3, lane e). Thus, the only form of the G protein present on the surface of the cells was G_2, a result consistent with the acquisition of only G_2 by virions as they bud.

In cells labeled with [^{3}H]glucosamine or [^{3}H]galactose, radioactivity is incorporated into only 2 species, comigrating with G_1 and G_2. Thus, both forms are at least partially glycosylated (13). Neuraminidase converts G_2 into a form comigrating with G_1; thus the difference in electrophoretic mobility between G_1 and G_2 is due largely, if not completely, to the presence of N-acetylneuraminic acid residues on the G_2 form (13).

Subcellular fractionation studies showed that all cellular forms of the glycoprotein are bound to membranes (10, 16). G protein labeled during a short pulse with radioactive amino acid was partially glycosylated (G_1) and was localized predominantly in fractions enriched in rough microsomes. Completion of glycosylation [conversion of G_1 to G_2] occurred only following movement to smooth membrane fractions. Addition of the terminal sialic acid residues and movement of G to smooth membranes appear to be coordinate events since some thermolabile mutant glycoproteins both remain localized in the rough microsomes and remain in the partially glycosylated G_1 form (17).

It is of considerable interest that glycosylation appears to be complete before the G protein reaches the surface membrane (13). To develop an assay for the time of appearance of the G protein on the cell surface, the ability of chymotrypsin to remove specifically proteins at the cell surface was exploited. Surface G protein, iodinated by lactoperoxidase catalysis (Fig. 3, lane 3), can be removed by chymotrypsin treatment (Fig. 4) without affecting the pattern of Coomassie blue-stained cell proteins or of N, NS, or L proteins in infected cells. Cells were labeled with [^{35}S]methionine for 5 min and then chased for various periods of time. Samples were taken and half of each was treated with chymotrypsin. The treated and control samples were then analyzed in parallel by gel electrophoresis (Fig. 4) to determine when G protein reaches the cell surface (i.e., when G protein becomes sensitive to treatment of intact cells with protease). As is most evident between 10 and 30 min of the chase, there was protein migrating as G_2 that was not sensitive to protease treatment. The conversion of G_1 to G_2 had a half-time of about 30 min, but there was a 10—20 min lag period before the G_2 protein was converted to a protease-sensitive form. By 90 min of chase, all of the G protein was in the G_2 form, and nearly 80% of it was removed by chymotrypsin and appeared to be on the surface of the cell. At all times, the precursor G_1 protein was resistant to treatment of intact cells with protease. Since sialic acid is the terminal sugar on each of the 2 carbohydrate chains bound to G, it appears that the G protein is completely glycosylated inside the infected cell approximately 10—20 min prior to its exposure on the surface of the cell. Glycosylation is evidently not the limiting factor in movement of G protein to the cell surface.

INTERACTION BETWEEN G mRNA AND THE ENDOPLASMIC RETICULUM

Newly made glycoprotein is associated with the rough microsome fraction, and all of the G mRNA is attached to the ER membrane (6—10, 16). This suggests a close coupling of G biosynthesis to insertion into the membrane. As we shall show in this section, puromycin, a drug which causes premature termination of growing peptide chains, also causes release of the G mRNA from the membrane fraction, presumably the result of dissociation

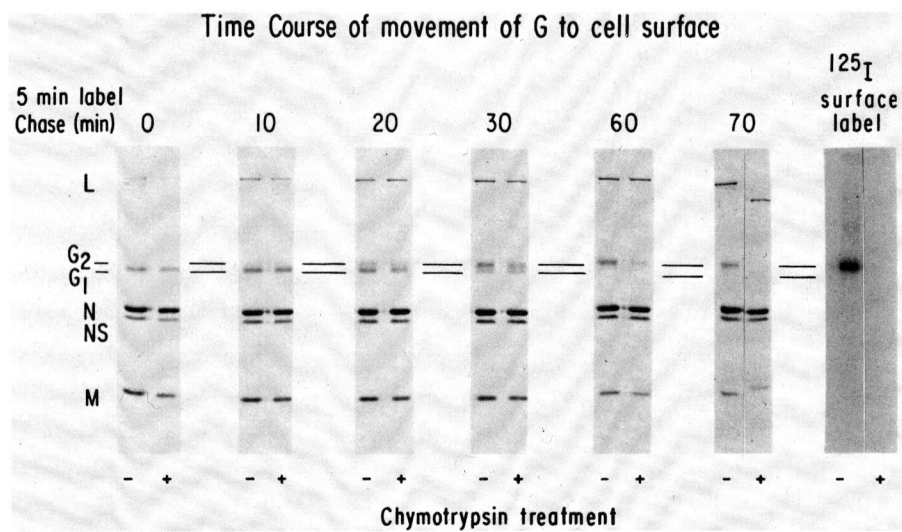

Fig. 4. The time course of movement of the G protein to the cell surface. A culture of infected cells was exposed to [^{35}S] methionine for 5 min at 4 h postinfection at 37°C. Following the labeling period, excess unlabeled methionine was added and the incubation continued at 37°C. At the times indicated, portions of the culture were removed and transferred into 4 volumes of ice-cold Earle's saline. Each sample of cells was washed with phosphate-buffered saline (PBS) 3 times. The cells were resuspended in PBS, and one half of each sample was incubated in PBS with 1 mg/ml chymotrypsin at 37°C for 10 min, while the remainder was incubated in PBS only for the same period. Following this incubation, the inhibitor phenylmethylsulfonylfluoride was added to 2 mM and incubation continued for 5 min at 37°C. The cells were washed with PBS and dissolved directly in gel sample buffer. The sample of iodinated cells was labeled as described in Ref. 13 using infected cells at 4 h postinfection. Following the labeling reaction, the cells were extensively washed and suspended at the same concentration as the [^{35}S] methionine-labeled cells in PBS for the protease treatment. The remainder of the protease reaction was conducted as above. The samples were subjected to SDS-polyacrylamide gel electrophoresis on 10% slab gels. The autoradiogram shown is from a 4 day exposure of the dried gel.

of ribosomes from the ER membrane. This suggests that the attachment of ribosome-G mRNA complexes to membranes is mediated primarily by the nascent glycoprotein chain, a conclusion which is consistent with the close coupling of G protein polypeptide synthesis and insertion into the membrane, as revealed by in vitro biosynthesis studies (see below).

To determine the distribution of VSV G mRNA in the free and membrane-bound compartments of infected CHO cells, cells were labeled with [^{32}P]PO$_4$ from 1 to 4 h postinfection. Postnuclear supernatants from control cultures and from cells treated with puromycin were prepared in a buffer containing a low concentration of salt (0.01 M KCl). They were then analyzed in a sucrose density gradient of the same ionic composition; these contained a cushion of 55% (wt/wt) sucrose overlaid with a linear 15–40% sucrose (wt/wt) gradient. The centrifugation conditions were chosen to separate membranes, which banded isopycnically at the 40–55% sucrose interface, from free polysomes and ribosomes which sedimented in the top half of the gradient. Analysis of ^{32}P radioactivity in each gradient fraction which was extractable into a solution of chloroform:methanol (2:1) provided a measure of the distribution of phospholipids; as expected, the great majority of the labeled phospholipids were found at the 40–55% interface, coincident with the visible band of membranes (Fig. 5, panel II).

In control cells, about 35% of the VSV-specific mRNA cosedimented with the membrane fraction (Fig. 5, panel I). Polyacrylamide gel electrophoresis resolves VSV RNA into 4 classes (c.f., Fig. 6). Band 2 (mol. wt. = 7.4×10^5) is the mRNA for VSV G protein. Band

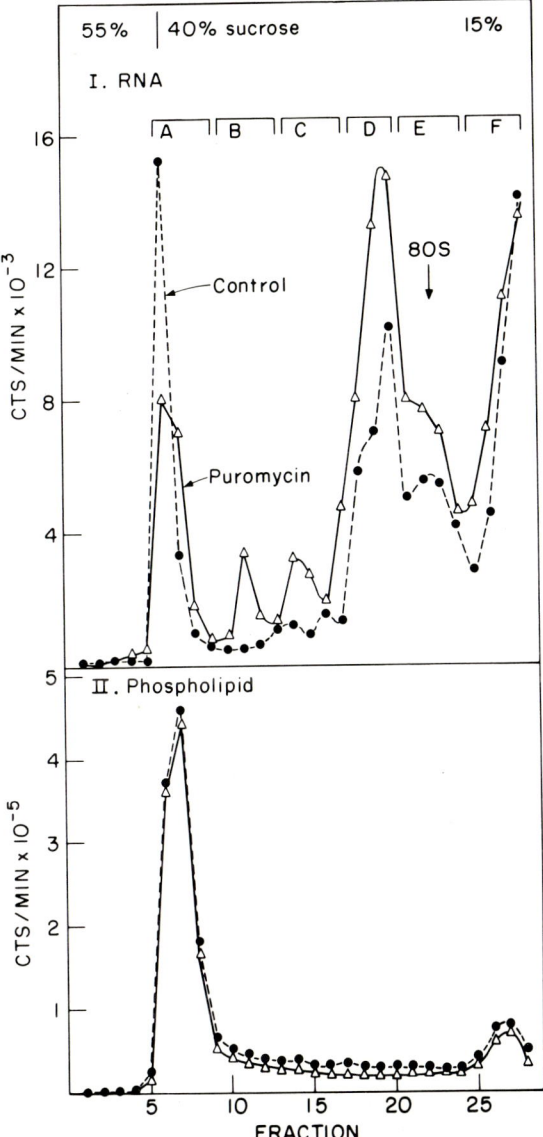

Fig. 5. Fractionation of ^{32}P-labeled RNA and phospholipid in VSV-infected cells. A postnuclear supernatant prepared from ^{32}P-labeled VSV-infected cells treated (triangles) or not (circles) with puromycin was fractionated on a sucrose gradient as detailed in Ref. 19 and in the text. Aliquots of each fraction were extracted with a mixture of chloroform and methanol. Shown in the top panel (I) is the distribution of radioactivity which was not extracted into the organic phase but which was precipitable with trichloroacetic acid. The bottom panel (II) shows the distribution of radioactivity which was extracted into the organic solvent. Sedimentation is from right to left. Brackets indicate fractions which were pooled for isolation of RNA.

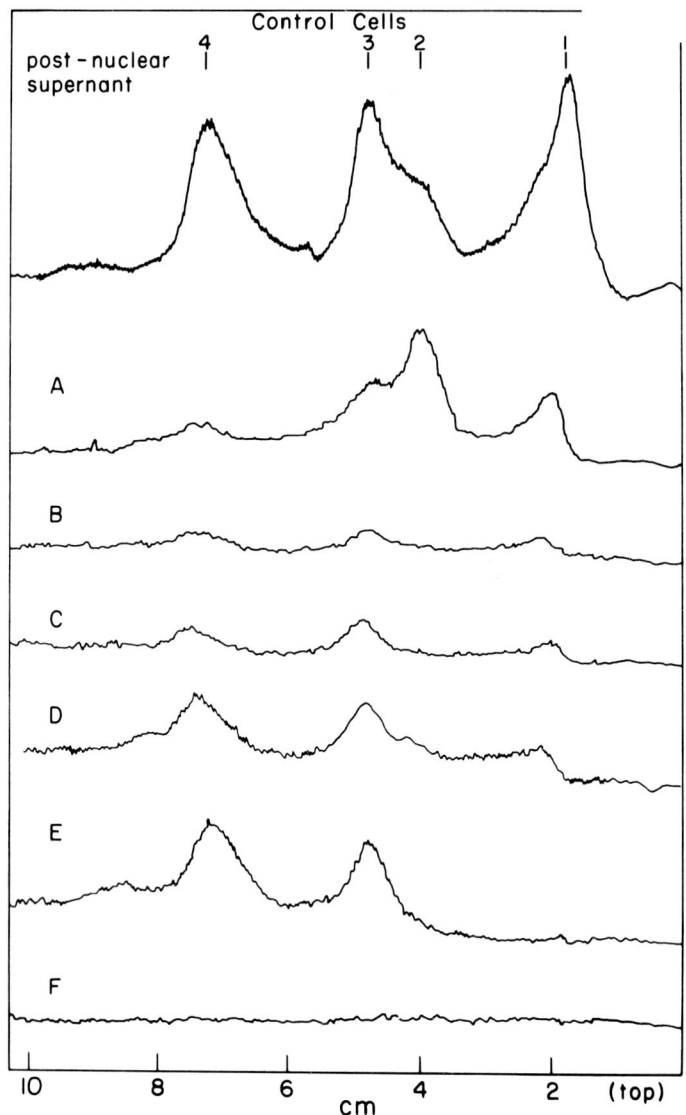

Fig. 6. Formamide-polyacrylamide gel analysis of ^{32}P-labeled RNA isolated from subcellular fractions from control VSV-infected cells. RNA was extracted from the postnuclear supernatant or from pooled gradient fractions of the experiment depicted in Fig. 5. Shown are scans of the radioautograms of the polyacrylamide gel.

1 is a mixture of 42S virion RNA and 26S mRNA for L protein. Band 3 (mol. wt. = 5.5 × 10^5) encodes the N protein, and band 4 (mol. wt. = 2.8 × 10^5) is a mixture of mRNAs for the M and NS proteins (15, 18). As is shown in Fig. 6, over 90% of the G mRNA (band 2) is associated with the membrane fraction. Over 80% of the ^{32}P radioactivity in bands 3 and 4, by contrast, sediments with free polyribosomes, in agreement with our earlier results (6). The interaction of G mRNA with membranes is maintained when the latter are floated to equilibrium in a sucrose density gradient (19).

Treatment of cells with puromycin 15 min before harvesting results in selective release of G mRNA from membranes (Figs. 5, 7). Only 20% of G mRNA remains associated with membranes; 80% sediments as free polysomes (fractions B–D) or ribosomes (panel E). In both puromycin-treated and control cells, about 20% of the VSV mRNA bands 3 and 4 — encoding the soluble proteins N, NS, and M — cosediments with the membrane

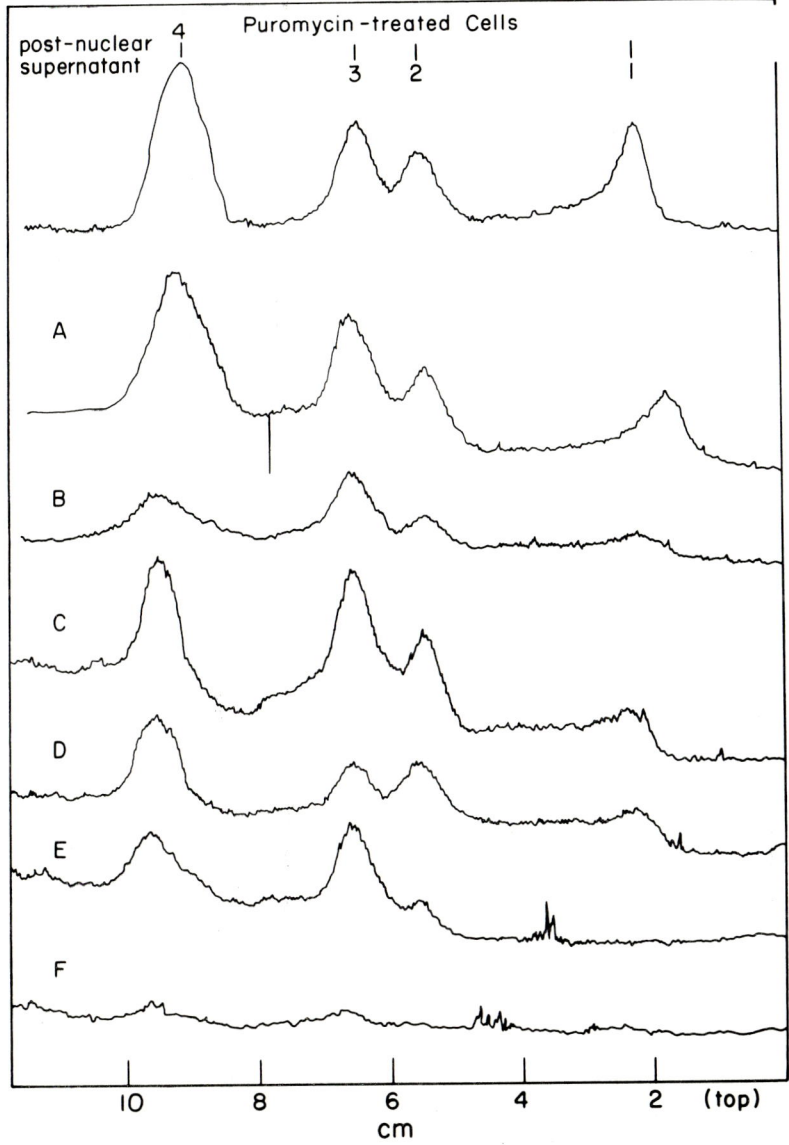

Fig. 7. Polyacrylamide gel analysis of ^{32}P-labeled RNA isolated from subcellular fractions from puromycin-treated VSV-infected cells. RNA was extracted from the postnuclear supernatant or from pooled gradient fractions of the experiment depicted in Fig. 5. One twentieth of each preparation of RNA was analyzed, except for fraction A, for which 0.1 of the preparation was used. Hence, to compare A with the other fractions, areas under the peaks in A should be divided by 2.

fraction. Presumably this is due to nonspecific adsorption since mRNAs for bands 3 and 4 are selectively lost during equilibrium centrifugation of the membranes. The 20% of the G mRNA remaining in the membrane fraction of puromycin-treated cells might also be due to nonspecific adsorption.

Treatment with puromycin in a buffer of high ionic strength (100–500 mM KCl) is required to dissociate from pancreatic, myeloma, and liver ER membranes most of the ribosomes and also mRNA active in directing synthesis of secretory proteins (20–22). A similar result was obtained with the Sindbis 26S mRNA, which directs the synthesis of 2 virus glycoproteins and 1 soluble protein: removal of ribosomes and of 26S mRNA from ER membranes required treatment of cells with puromycin and also addition of high-salt buffer to the isolated membranes (23). Presumably, ionic linkages between the ribosomes and/or nascent chains and the ER membrane, supposedly those disrupted by high-salt solutions, are of less importance for binding of VSV G mRNA than for Sindbis 26S RNA or mRNAs encoding secretory proteins.

These experiments suggest that there is a direct interaction between the nascent G chain and the ER membrane and also indicate an absence of a direct interaction between G mRNA and membranes.

THE TOPOLOGY OF NEWLY MADE G — SYNTHESIS AS A TRANSMEMBRANE PROTEIN

Our efforts were next directed toward elucidating the manner in which the newly completed G protein is bound to the ER membrane. To this end, infected cells were labeled for 10 min with [^{35}S]methionine, and a crude microsomal membrane fraction was prepared. By comparison with marker proteins, it was shown that this preparation (Fig. 8, lane 7) contained only the G_1 form of the G protein and also various amounts of other VSV structural proteins, L, N, NS, and M. To determine which polypeptides were accessible to added protease, a portion of the preparation was digested with trypsin. As can be seen (column 9), essentially all of the labeled L, N, M, and NS proteins are sensitive to proteolysis. [The amount of N which is resistant to trypsin remains so when the membranes are destroyed by sodium deoxycholate (column 8); it is probably in nucleocapsids.]

By contrast, G is largely resistant to proteolysis, and this protection is afforded by the permeability barrier of membrane vesicles. Trypsin does convert G to a form which migrates slightly faster on polyacrylamide gels; this change in mobility is consistent with a loss of 30–50 amino acids from the polypeptide chain. All of the methionine-containing tryptic peptides found in this fragment are also found in authentic G protein; this establishes that this fragment indeed is derived from G (Fig. 9). The resistance of newly made G to protease is dependent on the integrity of a membrane barrier; treatment of the preparation with the detergent sodium deoxycholate (Fig. 8, lane 8) before proteolysis results in essentially complete digestion of G.

Note that the proteolytic fragment of G lacks 3 methionine-containing tryptic peptides characteristic of authentic G (arrows in Fig. 9). By the use of a procedure devised by Dintzis (24), it has been possible to establish the relative order of the methionine-containing peptides of the G protein (25). The 3 peptides which are lost from G upon proteolysis are located closer to the carboxyl terminus of the protein than are any other

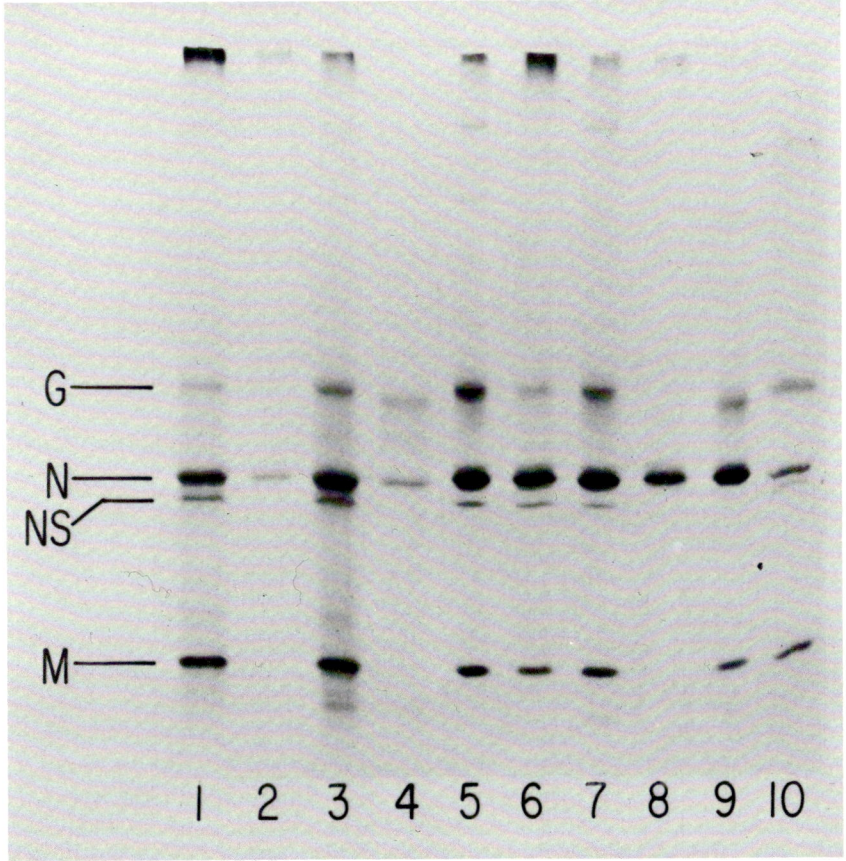

Fig. 8. Transmembrane biogenesis of VSV glycoprotein. Lanes 1–4: CHO cells infected with VSV (3 PFU/cell) in the presence of actinomycin D were harvested at 4 h postinfection. The cells were swelled in a hypotonic buffer (10) and broken by 40 strokes of a tight dounce homogenizer. Nuclei were removed by low speed centrifugation. The postnuclear supernatant was centrifuged at 20,000 × g for 30 min and the pellet resuspended in buffer S (50 mM KCl, 50 mM Tris-HCl, pH 7.6, 5 mM $MgCl_2$). This fraction was added to an S100 preparation from uninfected cells and a protein synthesis mixture containing [^{35}S] methionine, as previously described (23). After a 30 min incubation at 37°C, aliquots were treated for 30 min at 37°C with protease and/or detergent as indicated below. All reactions were subsequently treated with soybean trypsin inhibitor at 100 μg/ml for an additional 30 min at 37°C before being run on 10% SDS-polyacrylamide gels. Additions: 1) 1% DOC (sodium deoxycholate); 2) 1% DOC + trypsin (100 μg/ml); 3) trypsin (100 μg/ml) + soybean trypsin inhibitor (100 μg/ml) added simultaneously; 4) trypsin (100 μg/ml).

Lanes 5–10: At 4 h postinfection, cells were pelleted, resuspended in MEM supplemented with 1% dialyzed calf serum but lacking methionine, and labeled for 10 min with [^{35}S] methionine (100,000 mCi/mmole) at 24 μCi/ml. These cells were harvested, broken, and a 20,000 × g pellet prepared as above. The pellet was resuspended in buffer S and incubated for 30 min at 37°C in the presence of protease and/or detergent as indicated below. All reactions were incubated an additional 30 min in the presence of soybean trypsin inhibitor (100 μg/ml) and run on 10% SDS-polyacrylamide gels. Additions: 5) H_2O, 6) 1% DOC + soybean trypsin inhibitor (100 μg/ml); 7) trypsin (100 μg/ml) + soybean trypsin inhibitor (100 μg/ml); 8) trypsin (100 μg/ml) + 1% DOC; 9) trypsin (100 μg/ml); 10) marker from VSV-infected cells.

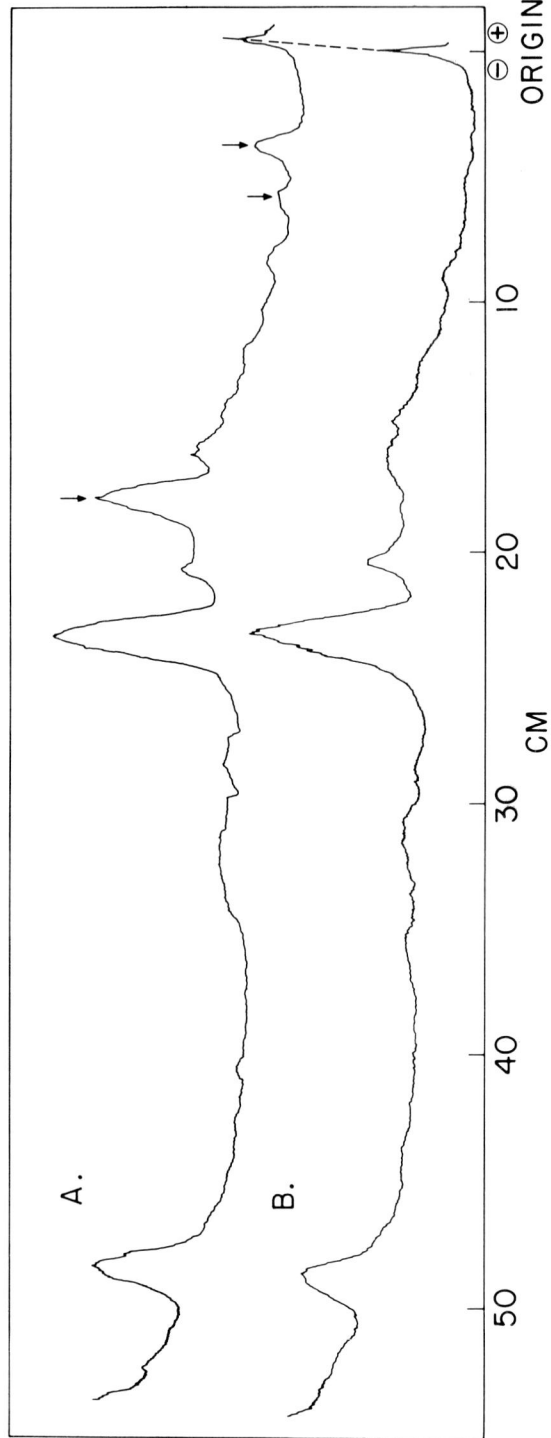

Fig. 9. Tryptic digests of [^{35}S]methionine-labeled G_1 and its protease-resistant fragment. Tryptic peptides of G protein (A) and it protease-resistant derivative (B) found associated with microsomal membranes derived from labeled infected CHO cells were prepared and paper ionophoresis at pH 3.5 performed as previously described (15). Microdensitometer tracings were made from autoradiograms of the electrophoretograms.

methionine-containing peptides. Indeed, these peptides must reside in the carboxy-terminal 5% of G_1 since about 95% of the polypeptide chain is protected from proteolysis.

Similar results were obtained if a microsome fraction was prepared from (unlabeled) cells and allowed to complete synthesis of nascent chains initiated in vivo in a cell-free system. Labeled G_1 protein is made in vitro (Fig. 8, lane 1); digestion of the reaction products with trypsin again results in conversion of G to a smaller fragment (Fig. 8, lane 4). Again, treatment with deoxycholate results in complete digestion of G (lane 2).

It is important to point out that all of the G protein completed in the cell-free system is in the G_1 form. Since it is unlikely that any movement between membrane fractions takes place in this extract, and since G is made on rough microsomal vesicles, this glycosylation of G takes place in the rough microsomes.

We conclude that G is a transmembrane protein immediately after its synthesis in the rough ER. The 30–50 amino acid residues, and the 2 methionine-containing peptides removed from the extreme carboxy-terminal region, are located on the outside of the vesicle, that side which faced the cytoplasm in the cell. This follows from the fact that the factors required for completion of G chain in vitro could only have interacted with ribosomes located on the outside of the membrane vesicles, the same side to which protease was added. From the requirement for disruption of the lipid bilayer by treatment with detergent for the extensive proteolysis of G_1, it is inferred that G must have protease-sensitive sites on the lumenal side of the ER. Thus, it appears to be transmembrane. While it may also be tentatively concluded that most or all of the protected amino-terminal 95% of the polypeptide is either facing the lumen of the ER or is imbedded in the lipid bilayer, these studies do not rigorously eliminate the possibility that more than 5% of the protein is exposed to the cytoplasmic face.

As was noted previously, G is partially glycosylated (G_1 form) concomitantly with or immediately after its synthesis. To show this (25), we took advantage of the finding that G_1 is specifically adsorbed to a column of Sepharose to which is covalently linked the lectin Concanavalin A (Con A). Presumably, this is due to the specific binding of α-mannosyl residues of G_1 to the Con A since the binding is prevented by α-methylmannoside but not by galactose, a sugar which does not bind to Con A. [Fig. 11 illustrates a similar experiment on G_1 made in the reconstituted cell-free system.]

IN VITRO GLYCOSYLATION AND ASYMMETRIC INSERTION OF THE GLYCOPROTEIN INTO THE ENDOPLASMIC RETICULUM MEMBRANE

To understand how specific membrane proteins are inserted only into the appropriate cytoplasmic membrane, to elucidate the molecular basis of asymmetry of membrane proteins such as G (26–28), and to understand how the extracytoplasmic portions of G cross the membrane, it is necessary to develop a cell-free system capable of proper synthesis and insertion of this protein. On the hypothesis that secreted proteins and the external portions of transmembrane glycoproteins such as G cross the endoplasmic reticulum membrane by similar mechanisms (28, 29), we have employed a modification of the cell-free system devised by Blobel and Dobberstein for secretion (30). In this system, secreted proteins such as immunoglobin are specifically transferred into the lumen of vesicles from pancreatic rough endoplasmic reticulum. This segregation process requires that the light chains be synthesized in the presence of the membrane vesicles. Free ribosomes, initially unbound to membranes, will suffice for proper segregation (30). In this section, we report experiments (31) done in collaboration with G. Blobel and V. Lingappa in which the VSV

G protein is translated by extracts of wheat germ in the presence of canine pancreatic rough ER vesicles. The G protein is incorporated into the membrane during or immediately after its synthesis, spans the ER membrane asymmetrically and properly, and is glycosylated.

It will be recalled that translation of VSV mRNA by wheat germ extracts results in production of a form of G, termed G_0, which is apparently unglycosylated (Fig. 3). Figure 10A, lane 3, shows that when VSV mRNA is translated by wheat germ ribosomes in the presence of "stripped" rough pancreatic microsomes (the ribosomes having been removed with 20 mM EDTA), a polypeptide appears which is absent when membranes are not included during protein synthesis (lane 1). This protein comigrates with G_1 found in the cytoplasm of infected cells and has thus far proven to be indistinguishable from G_1 in all respects. Increasing concentration of membranes results in increasing yields of G_1 relative to G_0 (Fig. 10A, lanes 3, 5, 7). Production of G_1 is not accomplished by ribosomes endogenous to the pancreatic membrane since these were removed by treatment with 20 mM EDTA before and since the total level of G made does not increase with increasing membrane concentration. Note that when membranes are added after synthesis, only G_0 is found (Fig. 10B, lanes 3, 6).

As is the case for the cytoplasmic G_1, the G_1 formed in this cell-free system is glycosylated. Figure 11, panels 1 and 2, shows that of the products made in the cell-free system, only G_1 but not G_0 or other VSV proteins will adhere specifically to Con A Sepharose. Binding of G_1 is specifically prevented by the ligand α-methylmannoside (lane 3) but not by the nonspecific sugar galactose (lane 4). Thus, G_1 formed in vitro contains at least the mannose-rich core region of the mature G_2 form of the glycoprotein. By inference, G_0 contains few, if any, mannose residues.

As is the case with the cytoplasmic form of G_1, G_1 formed in vitro asymmetrically spans the membrane bilayer. Figure 10A, lane 2, shows that the G_0 form of G is completely sensitive to exogenous protease, irrespective of whether membranes are present or not (Fig. 10B, lanes 5, 6). By contrast, the G_1 formed in vitro is resistant to the action of trypsin, apart from the removal of a discrete portion of G_1, resulting in a fragment which is shorter by 30–50 amino acids (Fig. 10A, lanes 4, 6, 8). This partial fragment of G comigrates with the analogous fragment of G_1 produced by protease digestion of ER vesicles isolated from infected cells (Fig. 8). Furthermore, tryptic maps of these 2 [^{35}S]-methionine-labeled G_1 fragments are similar (data not shown). The protection afforded most of the G_1 polypeptides formed in vitro results from the permeability barrier of the pancreatic membranes since if membranes are treated with 1% Triton X-100 (Fig. 10B, lane 8) or 1% deoxycholate (not shown) before proteolytic digestion, no G_1 is protected. Because Triton X100 is a nondenaturing detergent which disrupts membranes by displacing native lipid molecules from hydrophobic binding sites of native membrane proteins (32), it is extremely unlikely that the increased susceptibility of G_1 to protease induced by it is due to partial denaturation of the protein.

Figure 11 (lanes 5–8) shows that this proteolytic fragment of G contains at least some mannosyl residues. It binds specifically to columns of Con A, and its binding is inhibited by the ligand α-methylmannoside, but not by the nonspecific sugar galactose. Thus, some, if not all, of the mannose residues on G_1 are protected by the ER membrane from proteolytic digestion and are almost certainly located within the lumen of ER-derived vesicles.

We conclude that when the VSV glycoprotein is synthesized in vitro in the presence of pancreatic microsomes, a form of the glycoprotein is produced which is indistinguishable

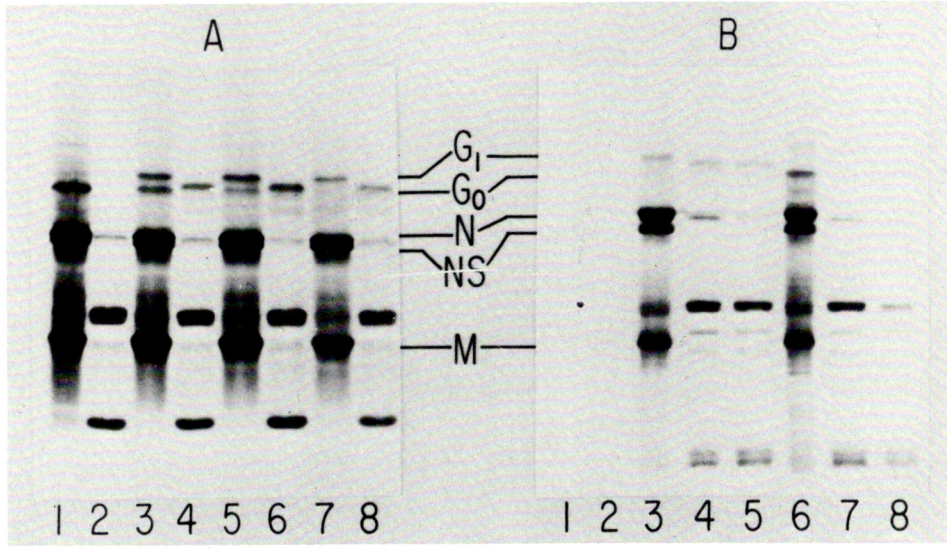

Fig. 10. Synthesis of the vesicular stomatitis viral glycoprotein in the presence of pancreatic endoplasmic reticulum. A) The effect of increasing concentrations of membranes and of trypsin treatment. Proteins were synthesized in 100 μl incubations for 1 h in the presence of pancreatic membranes. The latter were prepared (31) by treatment of pancreatic microsomes with EDTA (0.01 M), followed by centrifugation through a sucrose cushion, a procedure modified from Ref. 30. Reactions in lanes 1 and 2 contained no membranes; those in 3 and 4, 0.5 A_{280} units; 5 and 6, 1.0 A_{280} units; 7 and 8, 2.5 A_{280} units. Standard wheat germ reaction conditions (6, 36) containing VSV mRNA (37) were employed. After 1 h of incubation, pancreatic ribonuclease A (10 μg/ml) was added to inhibit further protein synthesis, and each incubation was divided in half. To one half of each incubation (lanes 2, 4, 6, and 8), trypsin (0.5 mg/ml final concentration) was added in 5-μl volume. After 20 min of subsequent incubation at 23°C, proteins were analyzed by SDS-polyacrylamide gel electrophoresis. Shown is a radioautogram of the dried gel. B) Features of the reconstitution reaction: Lane 1, no VSV mRNA or membranes; Lane 2, 0.025 A_{280} units of membranes per 50-μl reaction; Lanes 3–5, A 150-μl reaction containing 0.375 A_{280} units of membranes and VSV mRNA was incubated for 60 min at 23°C. Then pancreatic ribonuclease (10 μg/ml) was added. The incubation was divided into 3 portions which were incubated for 30 min more without trypsin (lane 3), 15 min with 0.5 mg/ml trypsin (lane 4), or 30 min with 0.5 mg/ml trypsin (lane 5).

Lanes 6–7: Proteins were synthesized for 30 min in an incubation (100 μl) to which no membranes were added. At 30 min, pancreatic ribonuclease (10 μg/ml) was added. Five minutes later, membranes (0.05 A_{280} units) were added, and incubation was continued for another 30 min. Then one half of the reaction was incubated with 0.5 mg/ml trypsin for 15 min (lane 7) while the other half was incubated for the same period without trypsin (lane 6).

Lane 8: Proteins were synthesized in the presence of 0.025 A_{280} units of membrane in a total volume of 50 μl for 60 min. The pancreatic ribonuclease, Triton X-100 (1% wt/vol) and trypsin were added, in that order, and incubation was continued for 15 min more.

from the G_1 form found in association with intracytoplasmic membranes from infected cells (Figs. 2–4, 8). Furthermore the product of the cell-free system has the same topographical relationship to the pancreatic microsomal vesicles as does the G_1 form of intracytoplasmic membranes (Figs. 8, 10). Specifically, both cell-free and in vivo G_1 forms span the membrane asymmetrically with their carboxy termini outside the vesicles, are glycosylated, and appear to have most, and possibly all, of their N-terminal portion inside the

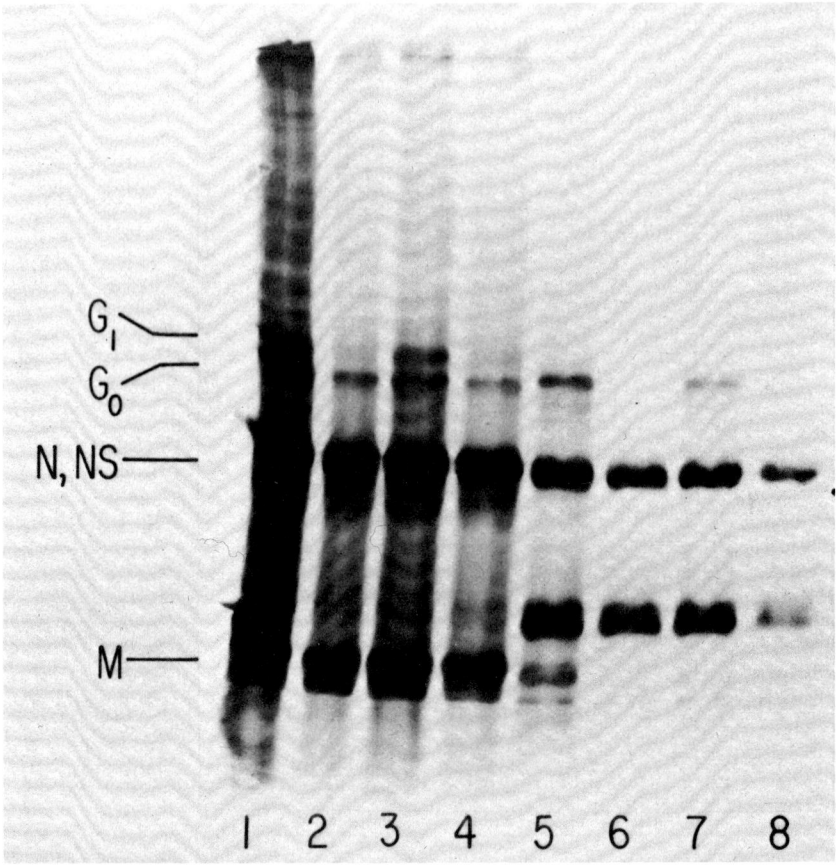

Fig. 11. Binding of vesicular stomatitis viral glycoprotein forms to columns of Concanavalin A Sepharose. Columns of Concanavalin A-derived Sepharose 4B (Pharmacia), 2 cm in length, were formed in Pasteur pipettes. All manipulations were at room temperature. Each column was washed with 5 ml of a buffer containing Hepes (25 mM, pH 7.5), $MnCl_2$ (1 mM), $CaCl_2$ (1 mM), $MgCl_2$ (1 mM), bovine serum albumin (0.1 mg/ml), and Triton X-100 (1% w/v). Then columns were washed with 5 ml of Hepes (25 mM, pH 7.5) containing Triton X-100 (1%) and, where appropriate, either galactose (0.2 M) or α-methylmannoside (0.2 M). Samples (25 μl) of radioactive polypeptides from protein synthesis incubations were loaded onto the columns and were washed with 2 ml of a buffer containing Hepes (25 mM, pH 7.5) and Triton X-100 (1%). Where appropriate, sugar (galactose or α-methylmannoside) was added to the sample to a final concentration of 0.2 M and was also included (0.2 M) in the buffer used to wash unbound proteins from the column. Proteins were precipitated from the 2-ml fraction containing unbound proteins by adding 8 ml acetone. After 15 min at $-20°C$, the precipitates were collected by low speed centrifugation, were lyophilized, and were subjected to gel electrophoresis. Gel represents proteins not bound to column.

Lanes 1–4: VSV proteins were synthesized in a total volume of 100 μl which contained 0.05 A_{280} units of membranes. After 60 min, pancreatic ribonuclease (10 μg/ml) was added, and incubation was continued for 5 min more. Then Triton X-100 (1%) was added. Lane 1, total reaction, before affinity chromatography on Concanavalin A Sepharose. Lane 2, chromatography in the absence of sugar. Lane 3, chromatography in the presence of α-methylmannoside. Lane 4, chromatography in the presence of galactose.

Lanes 5–8: VSV proteins were synthesized in a total volume of 100 μl which contained 0.25 A_{280} units of membranes. After 60 min, ribonuclease was added, as above, followed by trypsin (0.25 mg/ml) treatment for 15 min. Then soybean trypsin inhibitor (0.5 mg/ml) was added, and incubation was continued for 10 min more. Immediately before chromatography, Triton X-100 (1%) was added. Lane 5, proteins before chromatography. lane 6, chromatography in the absence of sugar. Lane 7, chromatography in the presence of α-methylmannoside. lane 8, chromatography in the presence of galactose.

vesicle permeability barrier. The glycosylation of an identified membrane protein in a cell-free system represents a novel observation which may potentially be of value in elucidating the temporal sequence, biochemical mechanisms, and the basis for specificity in protein glycosylations.

The observation that synthesis of the glycoprotein in the presence of membranes is required for protection of portions of the glycoprotein from external proteolysis shows that the intravesicular portions of the glycoprotein can only cross the membrane during or immediately following protein synthesis. Together with the fact that the C-terminus of the protein is exposed on the surface of vesicles following synthesis, these observations support a model (28, 29) in which the intravesicular portions of the glycoprotein pass N-terminus first across the membrane during polypeptide chain synthesis. The passage of the chain through the membrane must certainly depend on membrane proteins since when membranes are first treated with N-ethylmaleimide or are heated at 60°C for 10 min, only the G_0 form of the glycoprotein is found, and this is not protected from proteolysis (J.E. Rothman and H.F. Lodish, unpublished experiments).

All of the sugar in the G protein in virions is readily released with external proteases (5), yet sites for binding to Con A are retained in the large fragment of G_1 protected by membranes from trypsin in vitro. Taken together with the observation that synthesis of G in the presence of membranes is absolutely required to yield a glycosylated product, this suggests that glycosylation is restricted to the intravesicular surface (26) corresponding to the lumenal side of the endoplasmic reticulum in cells. In relating the apparently opposite overall orientations of the glycoprotein when synthesized in vitro to that found in virions, it is concluded that the plasma membrane glycoprotein is derived from glycoprotein in intracellular vesicles by a membrane fusion process (33, 34) (Fig. 1).

It cannot be excluded that other forms of the glycoprotein exist as transient intermediates in the process of incorporation into the membrane, or in glycosylation. Thus, a broad spectrum of proteins secreted by pancreas have been shown to possess short-lived hydrophobic N-terminal sequences which are cleaved off during or immediately after protein synthesis (35). These sequences have been proposed (30) to function as "signals" which direct the ribosomes which carry them to specific receptors in the rough endoplasmic reticulum and set up a ribosome-membrane junction which permits the nascent chain to cross the membrane during synthesis. It is entirely possible that a similar N-terminal cleavage may take place during the insertion of the VSV glycoprotein, but is masked by the anomalously high apparent molecular weight of the G_1 form due to its carbohydrate constituents. If so, intermediates which, for example, are cleaved at the N-terminus but not glycosylated (or vice versa) would have to exist, but probably would not be detected in the present experiments since these would be short-lived forms, comprising only a small fraction of the total in the steady state.

The cell-free system for membrane assembly described here appears to be capable of faithfully reproducing all known aspects of the topography and structure of an important class of membrane proteins, ectoproteins (28). These characteristics include an absolute asymmetry in transmembrane orientation, substantial mass located on the extracytoplasmic side of the lipid bilayer, and generally, the presence of sugar substituents on the extractyoplasmic portions of the polypeptide (28). The mechanisms used in this system in the synthesis and assembly of the VSV glycoprotein are almost certainly not unique to this one ectoprotein. This is indicated both by the vast evolutionary disparity of the sources of materials (plants and animals) which function together in the cell-free extracts, and by the fact that none of these components is obtained from viral-infected cells, so that these steps could not depend on the prior expression of viral genes. Therefore, a detailed biochemical and kinetic examination of the steps involved in the synthesis, processing, and incorporation into membranes of the VSV glycoprotein in the cell-free system described here can be expected to yield valuable insight into fundamental aspects of membrane assembly.

ACKNOWLEDGMENTS

This research was supported by Grant AI-08814 from the U.S. National Institutes of Health and Grant NP-180 from the American Cancer Society. J. E. R. is a postdoctoral fellow of the Damon Runyon-Walter Winchell Cancer Fund. We thank Ms. Susan Froshauer and Mr. Martin Brock for expert technical assistance.

REFERENCES

1. McSharry JJ, Compans R, Choppin P: J Virol 8:722, 1971.
2. Nakai T, Howatson AF: Virology 35:268, 1968.
3. Wagner RR, Prevec L, Brown F, Summers DF, Sokol F, MacLead R: J Virol 10:1228, 1972.
4. Eger R, Compans RW, Rifkin DB: Virology 66:616, 1975.
5. Lenard J, Compans RW: Biochim Biophys Acta 344:51, 1974.
6. Morrison T, Lodish HF: J Biol Chem 250:6955, 1975.
7. Grubman MJ, Moyer SA, Banerjee AK, Ehrenfeld E: Biochem Biophys Res Commun 62:531, 1975.
8. Both GW, Moyer SA, Banerjee AK: J Virol 15:1012, 1975.
9. Toneguzzo F, Ghosh HP: FEBS Lett 50:369, 1975.
10. Knipe DM, Baltimore D, Lodish HF: J Virol (In press).
11. Wagner RR, Kiley MP, Synder RM, Schnaitman C: J Virol 9:672, 1972.
12. Lafay F: J Virol 14:1220, 1974.
13. Knipe DM, Lodish HF, Baltimore D: J Virol (In press).
14. Both G, Moyer S, Banerjee A: Proc Natl Acad Sci USA 72:274, 1975.
15. Knipe DM, Rose JK, Lodish HF: J Virol 15:1004, 1975.
16. Hunt LA, Summers DF: J Virol 20:637, 1976.
17. Knipe D, Baltimore D, Lodish HF: J Virol (In press).
18. Rose JK, Knipe D: J Virol 15:994, 1975.
19. Lodish HF, Froshauer S: J Cell Biol (In press).
20. Blobel G, Sabatini D: In Manson LA (ed): "Biomembranes." New York: Plenum Press, 1971, vol 2, pp 193–195.
21. Borgese N, Mok W, Kreibich G, Sabatini D: J Mol Biol 88:559, 1974.
22. Harrison TM, Brownlee GG, Milstein C: Eur J Biochem 47:621, 1974.
23. Wirth DF, Katz FN, Small B, Lodish HF: Cell 10:253, 1977.
24. Dintzis HM: Proc Natl Acad Sci USA 47:247, 1961.
25. Katz FN, Lodish HF: Manuscript submitted.
26. Bretscher MS: Science 181:622, 1973.
27. Steck TL: J Cell Biol 62:1, 1974.
28. Rothman J, Lenard J: Science 195:743, 1977.
29. Blobel G, Dobberstein B: J Cell Biol 67:835, 1975.
30. Blobel G, Dobberstein B: J Cell Biol 67:852, 1975.
31. Katz FN, Rothman JE, Lingappa V, Blobel G, Lodish HF: Proc Natl Acad Sci USA (In press).
32. Helenius A, Simons K: Biochim Biophys Acta 415:29, 1975.
33. Palade G: Science 189:347, 1975.
34. Hirono H, Parkhouse B, Nicolson GC, Lennox ES, Singer SJ: Proc Natl Acad Sci USA 69:2945, 1972.
35. Devillers-Thiery A, Kindt T, Scheele G, Blobel G: Proc Natl Acad Sci USA 72:5016, 1975.
36. Lodish HF, Small B: Cell 7:59, 1976.
37. Rose JK, Lodish HF: Nature (London) 262:32, 1976.

Comparison of the Carbohydrate of Sindbis Virus Glycoproteins With the Carbohydrate of Host Glycoproteins

Kenneth Keegstra and David Burke

Department of Microbiology, State University of New York, Stony Brook, New York 11794

The carbohydrate portions of the Sindbis virus glycoproteins were compared with the carbohydrate portions of cell surface glycoproteins from uninfected host cells. Comparisons of the size of glycopeptides were made using gel filtrations. Comparisons of sugar linkages were made by methylation analysis. The conclusion was that the Sindbis carbohydrate is similar to a portion of the host carbohydrate. Thus, the Sindbis carbohydrate structures appear to be structures normally made in the uninfected host cell, but which are added to the Sindbis glycoproteins in virus-infected cells.

Key words: Sindbis, glycoproteins, cell surface

Considerable progress has been made in the past few years toward elucidating the structure of the carbohydrate portion of glycoproteins, principally soluble glycoproteins (1, 2). Of the membrane glycoproteins studied to date, most have carbohydrate structures similar to carbohydrate structures found on soluble glycoproteins (1, 2). The most common type of carbohydrate found on glycoproteins are oligosaccharides which are attached to the polypeptide chain by an N-glycosidic linkage to an asparagine residue (1). This type of oligosaccharide can be subdivided into 2 distinct groups (1). The first contains only glucosamine and mannose and is usually rich in mannose. The second is more complex, containing galactose, fucose, and sialic acid as well as glucosamine and mannose. Within each group most oligosaccharides have many structural features in common, with some differences observed between the oligosaccharides of various glycoproteins. One difficulty

---

David Burke is presently with the Department of Biochemistry, University of California, Berkeley, California 94720.

Received March 29, 1977; accepted June 30, 1977.

© 1977 Alan R. Liss, Inc., 150 Fifth Avenue, New York, NY 10011

in assessing the importance of the gross similarities or subtle differences in the oligosaccharides is that the glycoproteins which have been examined are not only from diverse sources, but also have different amino acid sequences. Thus, the determinants of the observed similarities and differences may be the alteration in the polypeptide sequences or in the enzymes which glycosylate them. This problem, when combined with the natural heterogeneity of the carbohydrate portion of glycoproteins (1), has made it difficult to draw conclusions about the importance of differences or similarities which have been observed.

To circumvent this problem, we have attempted 2 different approaches. The first is to examine the structure of the carbohydrate found on a particular polypeptide sequence when it is glycosylated in 2 different cells. The second, which is the focus of this paper, is to examine the structure of the carbohydrate found on different polypeptide sequences glycosylated within the same cell.

The experimental system we have chosen for the first approach is to grow Sindbis virus in cultured cells from 2 different organisms and then to examine the carbohydrate portion of the Sindbis virus glycoproteins. Sindbis virus is a small lipid-enveloped virus which contains only 3 polypeptides: 1 nonglycosylated capsid protein (designated C) and 2 glycosylated envelope proteins (designated E1 and E2). The genome of Sindbis is a single-stranded RNA of 4×10^6 daltons, with coding capacity for a maximum of 400,000 daltons of protein. At least 75–80% of the coding capacity is utilized in coding for the viral structural proteins and the polymerase enzymes necessary to make new viral RNA. This leaves little or no viral information to code for enzymes involved in adding carbohydrate to the viral glycoproteins. Based on this argument, it has been postulated that the envelope glycoproteins of Sindbis virus are glycosylated by host-specific enzymes (3–5).

Strauss et al. (4) and Burge and Strauss (5) have examined the glycoproteins of Sindbis virus grown in chicken embryo cells and baby hamster kidney (BHK) fibroblasts. They observed that Sindbis virus grown in BHK cells contains more sialic acid than virus grown in chick cells (4), but in other respects the carbohydrate from virus grown in the 2 hosts was basically similar (5). We have extended their original observations and have examined the carbohydrate of each of the glycoproteins separately (6). Both glycoproteins contain more galactose and sialic acid when virus is grown in BHK cells (6). In addition, glycoprotein E1 from virus grown in BHK cells is deficient in a mannose-rich oligosaccharide that is present when virus is grown in chick cells (6). In other respects, the carbohydrate added by the 2 hosts appears identical.

More detailed structural studies indicate that glycoprotein E2 contains 2 distinct oligosaccharides (7). One glycopeptide, a mannose-rich structure, is indistinguishable from virus grown in either host (7). A second glycopeptide, similar to the complex glycopeptide of IgG (1), is also identical for virus grown in both hosts (7). However, this latter glycopeptide demonstrates microheterogeneity in the extent of sialylation, having forms which contain 0, 1, or 2 sialic acid residues (6). When virus is grown in BHK cells a larger proportion of this glycopeptide consists of the structure containing 2 sialic acid residues (6). Other than this difference in the relative amounts of the structures differing in the amount of sialic acid, the Sindbis glycoprotein E2 has exactly the same carbohydrate structure regardless of the host in which the virus was grown (7).

This finding can be interpreted in at least 2 ways. The first interpretation is that glycosylation mechanisms have been sufficiently conserved during evolution so that these 2 widely divergent cells glycosylate the same polypeptide sequence in the same way. Alternatively, it is possible that Sindbis virus in some way alters the host glycosylation mechanisms so that Sindbis-specific oligosaccharides are added to the viral polypeptide. In an attempt to distinguish between these 2 possibilities we have compared Sindbis glycopeptides with glycopeptides from uninfected host cells. If the first assumption is correct, and Sindbis is simply picking up oligosaccharide structures normally made in the host, then uninfected cells should contain some structures identical to the Sindbis glycopeptides. The results of the comparisons are presented here.

MATERIALS AND METHODS

Growth of Cells and Virus

Primary chick cells were prepared from 10-day-old chicken embryos (8). The cells were transfered 1 time before using them for the growth of Sindbis. Chick cells and BHK 21/13 cells were maintained as described previously (6). Sindbis virus was grown and purified as described previously (6, 8).

Preparation of Trypsinate From Uninfected Chick Cells

A confluent monolayer of secondary chick cells was washed 5 times with phosphate-buffered saline. The cells were then treated with a trypsin solution (crystalline trypsin, obtained from Worthington Biochemicals Corporation, 0.1 mg/ml in phosphate-buffered saline) for 15 min at 37°C followed by the addition of an excess of soybean trypsin inhibitor. The cells, which were removed from the plate, were collected by centrifugation at 1,000 × g for 5 min. The supernatant solution containing the solubilized cell surface molecules was then subjected to a second centrifugation at 40,000 × g for 30 min to remove additional particulate material. The supernatant solution from this second centrifugation was designated as trypsinate.

Preparation of Glycopeptides and Analysis on BioGel P-6

Proteolytic digestion of glycoproteins was carried out using the protease mixture from Streptomyces griseus as described previously (8). The glycopeptides were fractionated using gel filtration on BioGel P-6 and fractions analyzed by liquid scintillation counting as described previously (3, 8).

Methylation Analysis of Glycopeptides

Complete methylation of glycopeptides was achieved using the procedure described by Hakomari (9). The methylation reactions were carried out and the methylated products were purified on an LH-20 column as described previously (10). The methylated glycopeptides were hydrolyzed under nitrogen atmosphere using 2 N trifluoroacetic acid (TFA) at 121°C for 2 h. The alditol acetate derivatives were prepared as described previously (6). The resulting partially methylated alditol acetates were separated on a Hewlett-Packard 5710A gas chromatograph using a glass column packed with either 1% OV-225 on Gas Chrom Q or 3% ECNSS-M on Gas Chrom Q (both from Applied Science Laboratories, State College, Pennsylvania). Quantitation was achieved by integration of peak areas with a Hewlett-Packard 3370-B integrator. The derivatives were identified using combined gas chromatography-mass spectrometry as described by Bjorndal et al. (11). The combined gas chromatography-mass spectrometry was carried out on a Hewlett-Packard system; a 5710A gas chromatograph was interfaced to a 5982 mass spectrometer. Spectra were recorded every 10 sec and stored in a 5933 A data system for later retrieval.

The sugar linkages were deduced from the observed derivatives assuming the pyranose ring form for all sugars. Thus the derivative 2,3,4,6-tetramethyl-1,5-diacetylmannitol is assumed to arise from terminal mannose.

Molar response factors were assumed to be equal for all of the neutral sugar derivatives (11). This was clearly not correct for the amino sugar derivative; therefore, no attempt was made to quantitate the derivatives of glucosaminitol (12).

RESULTS

Previous results had shown that the Sindbis glycopeptides could be separated into 4 distinct size classes using gel filtration on BioGel P-6 (2). In order to determine whether uninfected host cells had glycopeptides of the same size as Sindbis, gel filtration was used to compare the cell glycopeptides with the Sindbis glycopeptides. Subconfluent monolayers of secondary chick cells were labeled for 24 h with [^{14}C]glucosamine. Trypsin solubilized surface components were then prepared as described in Materials and Methods. An aliquot of the [^{14}C]glucosamine-labeled trypsinate was mixed with an aliquot of [^{3}H]glucosamine-labeled Sindbis virus grown in secondary chick cells. This mixture was digested exhaustively with pronase and one half of the digested mixture was analyzed by gel filtration as shown in Fig. 1. The remaining half of the mixture was heated in a boiling water bath for 5 min to inactivate the pronase, adjusted to pH 5.2 with citric acid, and then treated with neuraminidase for 2 h at 37°C as described previously (3). The neuraminidase-treated glycopeptides were then analyzed by gel filgration on BioGel P-6 as shown in Fig. 2. Experiments in which the isotopic labels were reversed gave results identical to those shown in Fig. 1 and 2.

The Sindbis glycopeptides give rise to 4 peaks designated S1 to S4 (8). The uninfected host trypsinate gives rise to a heterogeneous mixture of glycopeptides, including a large peak at the column void volume and a peak which elutes at the position of authentic glucosamine. It is important to note however, that included in this mixture of glycopeptides are species of the size of the Sindbis glycopeptides. When the mixture of glycopeptides is treated with neuraminidase, the pattern of the Sindbis glycopeptides simplifies to only 2 peaks, S3 and S4, as demonstrated previously (3). At the same time, the heterogeneity of trypsinate glycopeptides is greatly reduced. There is still a peak comigrating with glucosamine and a new peak of radioactivity which moves at the position of authentic sialic acid. About two thirds of the remaining radioactivity elutes as 3 poorly resolved peaks, one eluting considerably before S3, one approximately the same as S3, and one approximately with S4. Here the conclusion seems clear that at least a portion of the cell surface glycopeptides from unifected host cells are similar in size to the glycopeptides of Sindbis virus.

Further support for the contention that the Sindbis glycopeptides are structures found in unifected host cells was sought from chemical analysis of the glycopeptides from the 2 sources. Previous studies had shown that the Sindbis glycoproteins contained the sugars normally found in glycoproteins, i.e., glucosamine, mannose, galactose, fucose, and sialic acid (1, 2, 8). Since these sugars are so widely distributed in glycoproteins this information would be of little assistance in supporting the hypothesis that Sindbis was not altering host glycosylation. Therefore it was decided to examine the sugar linkages found in Sindbis and host cell glycopeptides. This procedure was used to ascertain if any new linkages were present in the Sindbis glycopeptides or whether all the linkages found in Sindbis were also present in glycopeptides of uninfected cells.

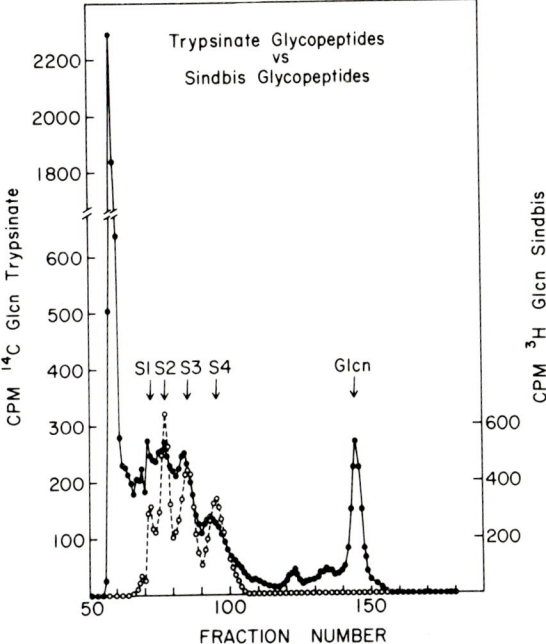

Fig. 1. Comparison of the glycopeptides from Sindbis virus with those from uninfected chick cell typsinate. [^{3}H] Glucosamine-labeled Sindbis virus (o - - - o) was mixed with [^{14}C] glucosamine-labeled trypsinate (•—•) from uninfected chick cells. This mixture was digested exhaustively with protease and the resultant glycopeptides applied to a BioGel P-6 column (1.0 × 110 cm). The column was eluted with 0.1 Tris, pH 8, collecting 0.5-ml fractions. The amount of radioactivity in each fraction was determined by liquid scintillation counting.

To obtain adequate quantities of viral glycopeptides, 25 mg of purified Sindbis virus was mixed with purified [^{3}H] glucosamine-labeled Sindbis and subjected to exhaustive proteolytic digestion. The radiolabel was used to locate the glycopeptide containing fractions in the subsequent steps described below. The glycopeptides were separated from degraded peptides and amino acids by gel filtration on BioGel P-2. The glycopeptides, which eluted in the column void volume, were pooled, concentrated, and resubjected to exhaustive proteolytic digestion. The glycopeptide fraction was again purified and desalted by gel filtration on BioGel P-2 using the column eluent. The resulting mixture of Sindbis glycopeptides was subjected to methylation analysis as described in Materials and Methods. Glycopeptides were prepared and analyzed from Sindbis grown either in chick cells or BHK cells. The results of these analyses are shown in Table I.

Glycopeptides were prepared from trypsinate of uninfected secondary chick cells in an analogous manner. Specifically, trypsinate was prepared from 16 roller bottles of secondary chick cells using the procedure described in Materials and Methods. This trypsinate was mixed with [^{3}H] glucosamine-labeled trypsinate prepared separately and a glycopeptide fraction was prepared exactly as described for the Sindbis glycoprotein. The results of a methylation analysis of the trypsinate glycopeptides are also shown in Table I.

Terminal fucose is present in all 3 preparations of glycopeptide, but becasue it is present in small quantities it was difficult to accurately quantitate the levels of this derivative. Mannose is present in only 3 different linkages in the Sindbis glycopeptides. In the

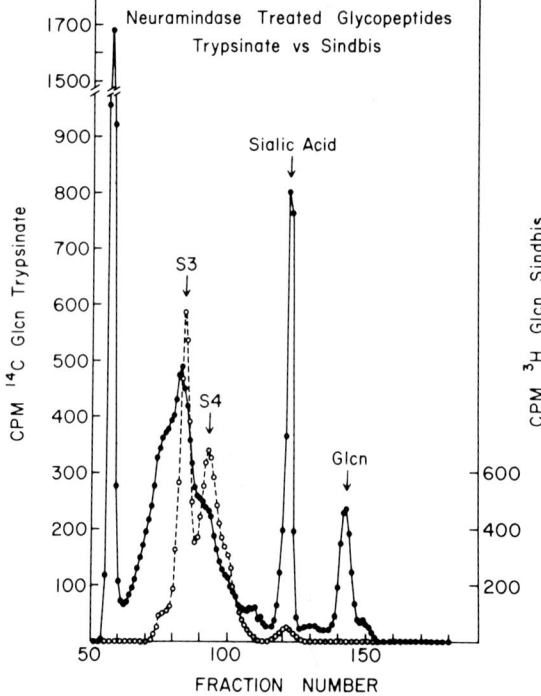

Fig. 2. Comparison of neuraminidase treated glycopeptides from Sindbis virus with those from uninfected chick cell trypsinate. [^{3}H]Glucosamine-labeled Sindbis virus (o - - - o) was mixed with [^{14}C]-glucosamine-labeled trypsinate (●—●) from uninfected chick cells. This mixture was digested exhaustively with protease, followed by boiling, then treatment with neuraminidase. The desialylated glycopeptides were analyzed on a BioGel P-6 column as described in Fig. 1.

cell surface glycopeptides, these same 3 linkages account for most of the mannose; however, there is, in addition, a small amount of 2,4-linked mannose. Other mannose derivatives, if present, represent considerably less than 10% of the total mannose in the glycopeptides.

Galactose is found in only 2 different linkages in the Sindbis glycopeptides. The cell surface glycopeptides contain these same 2 linkages in addition to 6-linked galactose. No branched galactose residues were found. Glucosamine was found only in the single linkage shown in Table I. No attempt was made to quantitate this derivative due to the lack of a molar response factor as discussed in Materials and Methods. The derivative corresponding to 4-linked glucosamine was a major peak in the chromatogram. Peaks one tenth as large could have been detected. Thus one can conclude that any other derivatives of glucosamine must have constituted a minor component of the total glucosamine present in the glycopeptides.

DISCUSSION

The data presented here suggest that the oligosaccharides added to the Sindbis glycoproteins by both chick cells and BHK are not unique, Sindbis-specific structures but rather are similar or identical to some of the structures found in uninfected host cells. The uninfected host cell has additional oligosaccharide structures which are not added to the Sindbis glycoprotein. Thus, the Sindbis glycoproteins act not only as an accurate probe of

TABLE I. Molar Ratios of Sugar Linkages

Deduced linkage	Sindbis-BHK	Sindbis-CEC	Uninfected chick trypsinate
Terminal Fucose	+	+	+
Terminal mannose	2.6	3.6	2.0
2-Linked mannose	6.7	7.0	7.6
3,6-Linked mannose	4.0	4.0	4.0
2,4-Linked mannose	0	0	+
Terminal galactose	2.1	1.4	1.8
3-Linked galactose	3.7	3.1	4.8
6-Linked galactose	0	0	1.7
4-Linked glucosamine	+	+	+

host glycosylation and reflect structures normally made in the host, but they also are selective probes and reflect only a portion of the structures made in the host.

These conclusions are supported by both types of evidence presented here. The comparison of glycopeptides by gel filtration shows that uninfected cells contain glycopeptides of the same size as the Sindbis glycopeptides. The host cell surface glycopeptides consists of a heterogeneous mixture of glycopeptides ranging from those as small as Sindbis glycopeptide S4 to structures considerably larger than glycopeptide S1. Experiments using galactose and fucose to label cell surface glycopeptides show that the peak at the void volume of the column labels poorly with these sugars. This suggests that the peak at the void volume includes some glycopeptides but consists mainly of glucosamine-rich materials such as glycosaminoglycans.

Sindbis glycopeptide S4 has previously been shown to be rich in mannose (6, 8), with a structure typical of the high mannose N-glycosidically-linked glycopeptides (1, 7). The uninfected host glycopeptides include structures of the same size as S4 (Fig. 1). Sindbis glycopeptides S1, S2, and S3 have been shown to have a more complex carbohydrate composition (6, 8), typical of the complex N-glycosidically-linked glycopeptides (1). Glycopeptides S1, S2, and S3 differ principally in the degree of sialylation, containing 2, 1, and 0 residues of sialic acid, respectively (6, 7). This is consistent with the observation that neuraminadase treatment converts glycopeptides S1 and S2 to species which coelute with glycopeptide S3 (Ref. 3; Fig. 2). The heterogeneity of the host surface glycopeptides is also greatly reduced by treatment with neuraminadase. The resulting broad peak of desialyted glycopeptides probably consists of a mixture of a number of different structures. Included in this mixture are structures the same size as Sindbis glycopeptides S3 and S4.

While it is necessary to show that the uninfected host cells have glycopeptides the same size as the Sindbis glycopeptides, further evidence is necessary to support the hypothesis that the structures are the same. The data in Table I demonstrate that all of the sugar linkages present in the Sindbis glycopeptides are also present in glycopeptides of uninfected cells. The similarity of the linkages found is quite striking. Uninfected chick cell glycopeptides contain all of the linkages found in the Sindbis glycopeptides. In addition, the chick cell glycopeptides contain at least 2 linkages not found in the Sindbis glycopeptides. Again these data support the hypothesis that Sindbis is acting as an accurate but selective probe of host glycosylation mechanisms. Final proof of this hypothesis will require the demonstration that a host glycoprotein contains glycopeptides identical to the Sindbis glycopeptides. We are currently examining this possibility.

The data presented here support, but do not prove, the original assumption that Sindbis is glycosylated by host specified enzymes. This conclusion, combined with the observations that Sindbis protein E2 is glycosylated nearly identically in 2 different hosts (7), suggests that the glycosylation mechanisms are highly conserved between these 2 hosts. While this conclusion is currently limited to only 2 cells, chick cells and BHK cells, preliminary results using other cells suggest that it may be more general. This apparent conservation of glycosylation mechanisms has broad implications for understanding the structure, biosynthesis, and function of the N-glycosidically-linked carbohydrates found on glycoproteins.

The observation described earlier that the host cell makes structures which are not utilized to glycosylate the Sindbis proteins raises the interesting question of how many different N-glycosidically-linked glycopeptide structures are made in a cell. One approach to answering this question is to examine different polypeptide sequences glycosylated in the same cell. This can be accomplished by growing different viruses in the same host cell and analyzing the carbohydrate added to the different viral glycoproteins by the host cell. This approach has been taken by Burge and Huang (13) and Sefton (14). The data presented here also give some indication of the diversity encountered within one cell. There seems to be a complex mixture of glycopeptides in chick cells as analyzed by gel filtration (Fig. 1). This pattern is greatly simplified if the glycopeptides are treated with neuraminidase. This suggests that a considerable portion of the diversity is caused by heterogeneity in the amount of sialic acid, as in the case with the Sindbis glycopeptides (Fig. 2). The function of this heterogeneity, if any, remains unclear; however, in some systems the absence of sialic acid has been shown to have important consequences (15, 16).

The diversity of sugar linkages found in the cell glycopeptides is surprisingly small, with most of the sugars existing in only 1 or 2 different linkages. This observation, combined with the recent observations of Muramatsu et al. (17, 18), provides evidence that there are a relatively limited number of N-glycosidically-linked glycopeptide structures. Muramatsu et al. (17, 18) demonstrated that most of the mannose-labeled (N-glycosidically-linked) glycopeptides from human fibroblasts were hydrolyzed by either endoglucosaminidase D or endoglucosamidase H. They interpret this to mean that most of the glycopeptides are similar in structure to thyroglobulin Unit A (Sindbis S4) or thryglobulin Unit B (Sindbis S1, S2, and S3) glycopeptides. Further experiments will be necessary to define exactly how many different types of N-glycosidically-linked oligosaccharides are made within a particular cell.

ACKNOWLEDGMENTS

This work was supported by Public Health Service grant CA 15630 and a National Science Foundation grant BMS 7505378. The authors wish to acknowledge the excellent technical assistance of John Bruno.

REFERENCES

1. Kornfeld R, Kornfeld S: Annu Rev Biochem 45:217, 1976.
2. Hughes RC: "Membrane Glycoproteins. A Review of Structure and Function." London: Butterworth Publishers, Inc., 1976.
3. Keegstra K, Sefton B, Burke D: J Virol 16:613, 1975.
4. Strauss JH, Burge BW, Darnell JE: J Mol Biol 47:437, 1970.

5. Burge BW, Strauss JH: J Mol Biol 47:449, 1970.
6. Burke DJ, Keegstra K: J Virol 20:676, 1976.
7. Burke D: PhD Thesis, State University of New York at Stony Brook, 1976.
8. Sefton B, Keegstra K: J Virol 14:522, 1974.
9. Hakomori S: J Biochem 55:205, 1964.
10. Talmadge KW, Keegstra K, Bauer WD, Albersheim P: Plant Physiol 51:158, 1973.
11. Bjorndal H, Hellerquist GG, Lindberg B, Svensson S: Angew Chem Int Ed 9:610, 1970.
12. Stellner K, Saito H, Hakomori S: Arch Biochem Biophys 155:464, 1973.
13. Burge BW, Huang AS: J Virol 6:176, 1970.
14. Sefton BM: J Virol 17:85, 1976.
15. Ashwell G, Morell AG: Adv Enzymol Relat Areas Mol Biol 41:99, 1974.
16. Schloemer RH, Wagner RR: J Virol 14:270, 1974.
17. Muramatsu T, Koide N, Ceccaroni C, Atkinson P: J Biol Chem 251:4673, 1976.
18. Muramatsu T, Ogata M, Koide N: Biochem Biophys Acta 444:53, 1976

The Biosynthesis of Mannolipids and Mannose-Containing Complex Glycans by the Retina

Edward L. Kean

The Lorand V. Johnson Laboratory for Research in Ophthalmology, Department of Surgery, Division of Ophthalmology and the Department of Biochemistry, Case Western Reserve University, School of Medicine, Cleveland, Ohio 44106

Large-scale incubations were carried out with homogenates of the retinas of the 15–16-day-old chick embryo in the presence of GDP[U-^{14}C]mannose, from which there were isolated a mannolipid (Lipid I), oligosaccharide-lipids (Lipid II), and glycoprotein (residue). These incubations were performed in the presence of endogenous acceptors as well as dolichyl phosphate. [^{14}C]Mannolipid I was subjected to chromatography on DEAE cellulose and silicic acid. The response to these, as well as TLC, enzymatic, and chemical treatments, were consistent with the product being dolichyl phosphomannose. [^{14}C]Lipid II was purified by DEAE cellulose chromatography and gel filtration on LH-20. Responses to these treatments, as well as TLC and paper chromatography, were consistent with this product being of the class of the oligosaccharide-pyrophosphate-lipids. The residue remaining after removal of the lipids was shown to contain glycoproteins by conversion of high-molecular-weight radioactive material to low-molecular-weight [^{14}C]mannose-containing glycopeptides by the action of pronase. These reactions and their products are consistent with there being in the retina, the pathway for glycoprotein synthesis involving the participation of the lipid-activated carbohydrates.

When the incubations were performed in the presence of ATP or ADP there was a decrease in the labeling of Lipid I, accompanied by an increase in the labeling of Lipid II and glycoprotein. When incubated in the presence of dolichyl phosphate and detergent, however, the stimulatory effect of ATP did not occur. The effect on these activities of a variety of other nucleotide phosphates was also examined.

Key words: dolichyl phosphomannose, glycoproteins, mannosyltransferases, polyprenyl phosphosugars, retina

It has been suggested (1) that the pathway involving the lipid activation of carbohydrates via their polyprenyl monosaccharides and oligosaccharides is involved in the biosynthesis of the core region of membrane-bound as well as secretory glycoproteins in animal tissues. The major protein in the membranes of the discs of the rod outer segments

Abbreviations: TES – N-tris[hydroxylmethyl]methyl-2-aminoethanesulfonic acid; TX-100 – Triton X-100; C/M – chloroform/methanol (volume to volume); TCA/PTA – trichloroacetic acid/phosphotungstic acid; SDS – sodium dodecyl sulfate; TLC – thin layer chromatography.

Received March 22, 1977; accepted July 10, 1977.

© 1977 Alan R. Liss, Inc., 150 Fifth Avenue, New York, NY 10011

of the retina is the unique glycoprotein, rhodopsin. This molecule has been shown to contain 9 moles of mannose and 5 moles of glucosamine per mole (2). Studies from this laboratory have shown the presence of, optimal conditions for, and many of the properties of lipid and glycan mannosyltransferases in preparations from the retinas of several species (3–7). Three major classes of products are formed: a mannolipid, oligosaccharide(-pyrophospho-)lipids, and mannose-containing glycoproteins. The present report is concerned with aspects of the characterization of these materials synthesized by large-scale preparations from the retina of the embryonic chick. These reactions and products are consistent with the presence in the retina of the pathway involving the participation of the lipid-activated carbohydrates in the biosynthesis of its glycoproteins. Further studies may reveal whether these types of reactions are involved in the biosynthesis of the carbohydrate chains of rhodopsin. This report also describes the differential responses by these mannosyltransferases to the presence of nucleotides, principally ATP, in the formation of these 3 products.

MATERIALS AND METHODS

Enzyme Preparation

Retinas were obtained from the eyes of 15–16-day-old chick embryos (White Leghorn chickens), after the eggs had been incubated at 37°C. From 120 eggs, about 4 g wet weight of retinas can be recovered. The tissue was homogenized in 0.25 M sucrose (2 volumes of sucrose to 1 g of tissue) in the cold using a hand operated Ten Broeck all-glass homogenizer. The homogenate was dialyzed in the cold for 4 h against at least a 100-fold volume excess of 0.01 M TES buffer, pH 7.0, with one change of dialysis medium. After dialysis, the retentate was rehomogenized and either used at once or stored at −20°C.

Assay for Mannosyl Transferase Activities

Incubations were carried out as described previously (3–7) in 12-ml glass centrifuge tubes for 5 min at 37°C. The mixture contained: GDP[$U^{14}C$] mannose, 3.3 µM (specific activity 160–221 µCi/µmole); $MnCl_2$, 3.3 mM; TES buffer, 0.2 M, pH 7.0; and enzyme (0.6–0.9 mg protein) in a total volume of 0.15 ml. When incubations were performed in the presence of exogenously supplied dolichyl phosphate, the lipid was dispersed in TX-100 [final concentration, 0.24% (wt/vol)] by vigorous agitation with the Vortex. The reaction was stopped by the addition of 3 ml of a mixture of cold 6% trichloracetic acid/ 0.5% phosphotungstic acid (TCA/PTA) (8). All of the assay steps were carried out in the cold, except where indicated. The mixture, after standing 15–30 min was centrifuged at 1,000 × g for 5 min, and the supernatant solution discarded. The pellet was rewashed twice by resuspending in 2 ml of TCA/PTA, followed by centrifugation. After the excess fluid had been drained for a few minutes, the assay could be interrupted at this point and the pellet stored at −20°C. Lipids were extracted from the washed pellet by vigorous mixing with 2 ml of C/M (2:1) on the Vortex. After centrifugation for 5 min at 1,000 × g the C/M extract was transferred to a ground-glass centrifuge tube. The C/M extraction of the pellet was carried out 3 times. The pooled C/M extract was then equilibrated with either water or 0.1 M KCl, and the lower phase washed twice with theoretical upper phase prepared with water or KCl, according to the procedures described by Folch, Lees, and Sloane-Stanley (9). The washed lower phase, containing Lipid I, was evaporated to dryness in counting vials and the amount of radioactivity present determined by liquid scintillation spectrometry.

The acid-washed, C/M (2:1) extracted pellet was then mixed vigorously with 2 ml of C/M/H$_2$O (10:10:3) and the supernatant solution recovered after centrifugation. This extraction was performed twice, and this material (Lipid II) was evaporated to dryness in counting vials on a steam bath, and the amount of radioactivity measured.

The residual pellet (referred to as the "residue") was then dissolved in 0.5 ml of Unisol overnight at room temperature. After the addition of 0.25 ml of methanol, the solution was transferred to a counting vial and the radioactivity determined in the presence of 5 ml of "Complement" (the Unisol-Complement scintillation counting system was obtained from Isolab, Inc., Akron, Ohio).

Every incubation was performed in duplicate, accompanied by a zero time control, i.e., a complete incubation to which enzyme was added after the addition of TCA/PTA. Values obtained with the latter type of incubation were similar to that obtained using boiled enzymed controls.

CHEMICALS

All solvents used in the chemical synthesis and in the preparative procedures for the mannolipids were redistilled before use. GDP[U^{14}C]Mannose was purchased from New England Nuclear Corporation, Boston, Massachusetts. Nucleotides were obtained from Cal Biochemicals, LaJolla, California and Sigma Chemical Company, St. Louis, Missouri. DEAE-cellulose Selectacel, No. 70, Standard type, was purchased from Schleicher and Schuell Company, Keen, New Hampshire. Sephadexes G-200, G-25, and LH-20, were purchased from Pharmacia, Piscataway, New Jersey. Silicic acid (Unisil) was obtained from Clarkson Chemical Company, Inc., Williamsport, Pennsylvania.

Dolichyl phosphate was prepared according to the procedure described by Wedgewood, Strominger, and Warren (3, 8) using O-phenylene phosphochloridate as phosphorylating reagent (purchased from Aldrich Chemical Company, Milwaukee, Wisconsin).

PREPARATION AND PURIFICATION OF MANNOLIPIDS

Large scale incubations were carried out using either the endogenous acceptor alone or in the presence of exogenously supplied dolichyl phosphate. When the latter compound was used, it was evaporated to dryness with nitrogen, then dispersed by vigorous vortex mixing in the presence of TX-100. The large scale incubations were either multiple incubations performed as indicated in Materials and Methods, or incubations increased proportionately in scale.

Purifications of the products of the enzymatic reactions were carried out by the chromatographic procedures described in general below, and in more detail in the legends to the pertinent figures.

Lipid I

Mild deacylation: DEAE-cellulose and silicic acid chromatography. The C/M (2:1) extract of the acid-washed pellet was partitioned according to the procedure of Folch et al (9), and the lower phase was then subjected to a mild deacylation according to the method of Lester and Steiner (11). After deacylation, the lipid phase was again partitioned according to the procedure of Folch et al (9), and then placed over a column of DEAE-cellulose acetate prepared as described by Rouser, Kritchevsky, and Yamamoto (12) or by Dankert, Wright, Kelley, and Robbins (13) (see Figs. 1 and 2).

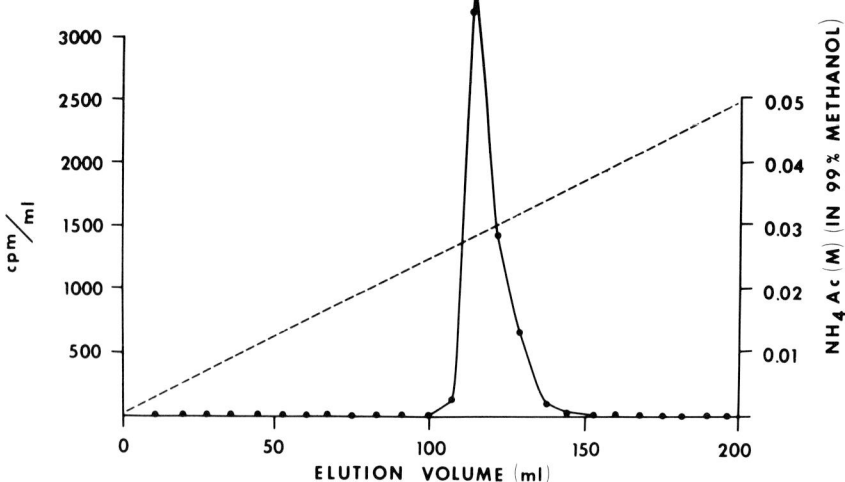

Fig. 1. Elution profile of Lipid I from DEAE-cellulose acetate. The "Folch washed," C/M (2:1) extract of the pellet obtained after treating large-scale incubations with TCA/PTA as described in Materials and Methods, was placed onto a DEAE-cellulose column prepared as described by Rouser et al (12). After washes with C/M (2:1), methanol, and 99% methanol, a linear gradient between 99% methanol and 99% methanol containing 0.1 M NH_4Ac was performed. The total volume of the system was 400 ml. Gradient (- - -).

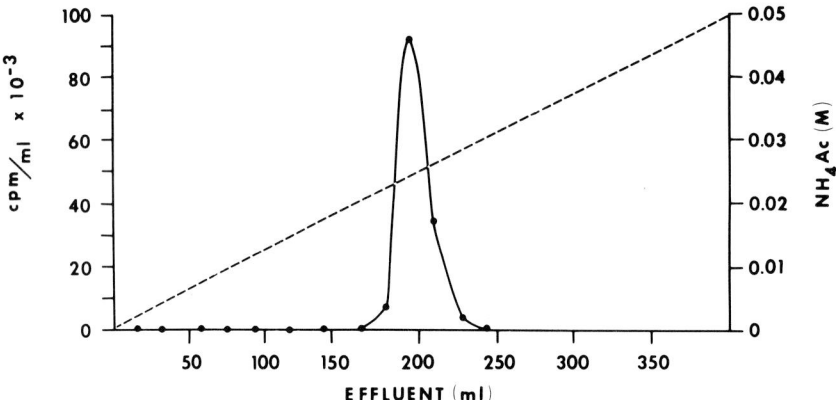

Fig. 2. Elution profile of dolichyl phosphomannose from DEAE-cellulose. The "Folch washed" C/M (2:1) extract of the TCA/PTA pellet obtained after large-scale incubations were performed in the presence of dolichyl phosphate, as described in Materials and Methods, was placed onto a column of DEAE-cellulose, After washes with C/M (2:1), methanol, and 99% methanol, a linear gradient was performed between 99% methanol (400 ml) and this solvent containing 0.1 M NH_4Ac (400 ml). Gradient (- - -).

The product obtained from DEAE-cellulose was evaporated to dryness, redissolved in C/M (2:1), and partitioned according to the procedure of Folch et al (9). The washed lower phase was evaporated to dryness, redissolved in C/M (95:5), and placed over a column of silicic acid (Fig. 3).

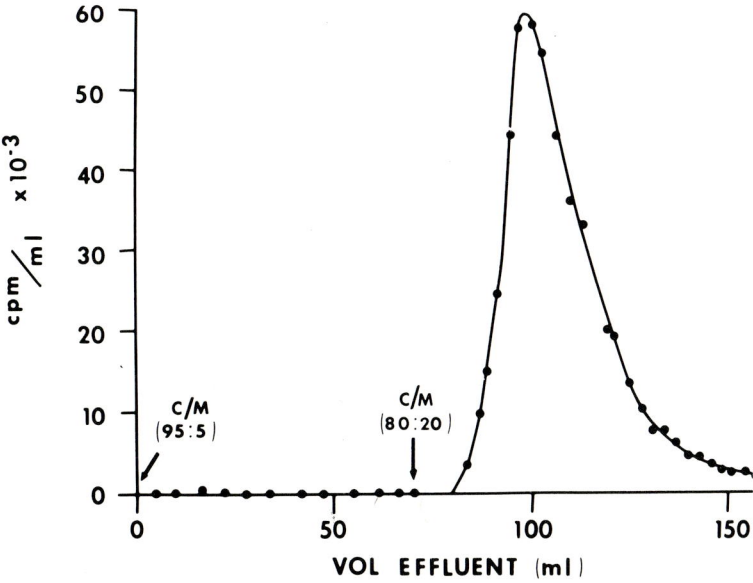

Fig. 3. Silicic acid chromatography of dolichyl phosphomannose synthesized by the retina. The radioactive material obtained after DEAE-cellulose chromatography, described in Fig. 2, was placed onto a column of silicic acid (Unisil) in C/M (95:5). Where indicated by the arrow, the solvent was changed to C/M (80:20). The flow rate was maintained at 0.6 ml/min, and 3-ml fractions were collected.

Lipid II

DEAE-cellulose chromatography and gel filtration on Sephadex LH-20. As described in the assay procedure (Materials and Methods), Lipid II is the material extracted with $C/M/H_2O$ (10:10:3) from the pellet remaining after the extensive extraction with TCA/PTA and C/M (2:1). The $C/M/H_2O$ (10:10:3) extract was then fractionated on DEAE-cellulose acetate (see Fig. 5). The product recovered from DEAE-cellulose was evaporated to dryness on a rotary evaporator, redissolved in 5ml of $C/M/H_2O$ (10:10:3), and placed over a column (2.5 × 41 cm) of Sephadex LH-20, which had been packed and equilibrated with $C/M/H_2O$ (10:10:3). This same solvent was used to elute the product from the gel, the flow rate being regulated at about 0.5 ml/min. Lipid II isolated by these procedures was stored at $-20°C$.

The elution patterns from these columns were monitored by measuring the amount of radioactivity in aliquots of the fractions, after evaporating to dryness in counting vials.

Residue: treatment with SDS and pronase. The residue remaining after TCA/PTA treatment followed by extraction with C/M (2:1), and $C/M/H_2O$ (10:10:3) was solubilized by incubating with 1% SDS at 50°C for 4 h. The solubilized material was examined by gel filtration on a column of Sephadex G-200 (Fig. 8).

Chromatography and electrophoresis. Thin layer chromatography (TLC) was carried out on precoated plates of silica gel 60 (without fluorescent indicator) 0.25 mm thick, (EM Reagents, Merck, Darmstadt, Germany), of dimensions 20 × 20 cm or 5 × 20 cm. TLC was also performed with 0.25 mm thick plates of microcrystalline cellulose (Avicel, Analtech, Inc., Newark, Delaware). The following solvents were used: A) C/M/acetic acid/H_2O (25:15:4:2), B) isobutyric acid/concentrated ammonium hydroxide/H_2O (57:4:39), C) 1-butanol/95%

ethanol/H_2O (10:1:2), D) 1-butanol/pyridine/0.1 N HCl (5:3:1). Paper electrophoresis was carried out with a Gilson high voltage electroforator using borate buffer as described previously (14). Lipids were visualized after TLC by exposure to iodine vapor, spraying with anisaldehyde reagent [anisaldehyde/sulfuric acid/95% ethanol (1:1:18)] followed by heating the plate to about 150°C, or spraying with the phosphate reagent described by Dittmer and Lester (15). Carbohydrates were visualized on paper by the benzidine-periodate (16) and the alkaline silver nitrate procedures (17). Radioactivity was detected after TLC by scraping zones into counting vials and adding 10 ml of a TX-100 toluene solution described previously (18), or 10 ml of a similar solution obtained commercially, Formula 950A (New England Nuclear Corporation, Boston, Massachusetts), and measuring the amount of radioactivity by scintillation spectrometry. Radioactivity was detected after paper chromatography and electrophoresis by scanning the strips.

OTHER PROCEDURES

Protein was determined by the method of Lowry, Rosebrough, Farr, and Randall (19). Phosphodiesterase was measured as described previously (18). Pyrophosphatase activity was determined by following the cleavage of β-DPN in the presence of L+ lactate (Hohorst, Ref. 20). α-Mannosidase was purified from jack bean meal and assayed as described by Li and Li (21).

RESULTS

Isolation of Lipid I and Dolichyl Phosphomannose Synthesized by the Retina

DEAE-cellulose and silicic acid chromatography. The product formed from GDP-[^{14}C]mannose when incubated in the presence of exogenously added dolichyl phosphate and in its absence (Lipid I) were both retained by DEAE-cellulose. No radioactivity was eluted from the columns with the washes of C/M (2:1), methanol, or 99% methanol. Quantitative recovery of radioactivity was obtained by the linear gradient elution, typical patterns of which are seen in Figs. 1 and 2. In different preparations both mannolipids were eluted between 0.025 and 0.04 M ammonium acetate in 99% methanol using DEAE-cellulose prepared by the procedure of Dankert et al (13) or by that of Rouser et al (12).

After extraction by the procedure of Folch et al (9) Lipid I and dolichyl phosphomannose from DEAE-cellulose were subjected to silicic acid (Unisil) chromatography, a typical elution pattern of which is seen in Fig. 3. A single peak of radioactive material was obtained in the C/M (80:20) fraction with dolichyl phosphomannose (Fig. 3) and with Lipid I (data not shown). A similar elution pattern was obtained when carried out by gradient elution. Of the radioactivity placed onto the column in C/M (95:5), quantitative recovery was obtained in the C/M (80:20) fraction. The components in the peak tubes were pooled, evaporated to dryness, redissolved in C/M (2:1) and partitioned according to the procedure of Folch et al (9).

TLC. The product synthesized by the retina using dolichyl phosphate as acceptor, and the product from the endogenous acceptors, Lipid I, extracted with C/M (2:1) from the TCA/PTA pellet, migrated on TLC in several solvent systems in a manner similar to one another and to standard dolichyl phosphomannose as seen in Fig. 4, using the acidic solvent system, A. While the migrations in neutral and basic solvents were greatly reduced compared to the acidic system, the 2 products showed identical R_f values (3, 4, 7).

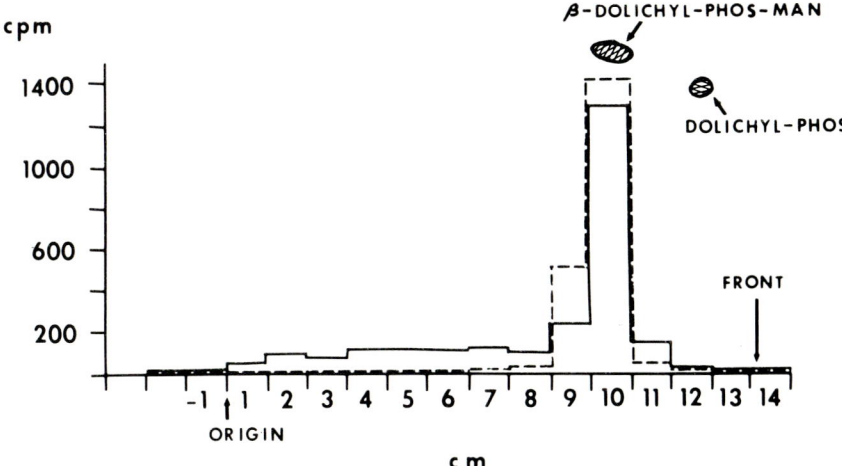

Fig. 4. TLC of mannolipid I and dolichyl phosphomannose. After purification by the column chromatographic procedures described in Materials and Methods, [^{14}C] mannolipid I and dolichyl phosphate-[^{14}C] mannose synthesized by the retina were chromatographed on silica gel plates using as solvent system, C/M/acetic acid/H$_2$O (25:15:4:2). Their migrations were detected by scintillation spectrometry of 1-cm zones of the gel scraped into counting vials. The standards were visualized by the anisaldehyde spray reagent. Lipid I (- - -); dolichyl phosphomannose (——).

Acid hydrolysis. After strong acid hydrolysis (2 N HCl, 100°C, 4 h, sealed tube), of Lipid I or dolichyl phosphomannose, the hydrolysate was placed over a mixed bed resin column (Ag-50- X 8(200–400 mesh)H$^+$ and Ag-1- X 8(200–400 mesh) formate). The radioactivity was recovered quantitatively in the water eluate of the column. When examined by borate electrophoresis and by paper chromatography in solvent systems B, C, and D, only one radioactive spot was observed which migrated with standard mannose. Lipid I and dolichyl phosphomannose synthesized by the retina were rapidly hydrolyzed by dilute acid (0.1 N HCl) at room temperature ($t_{1/2}$, about 12 min) (3).

Resistance to phosphodiesterases; α-mannosidase. The mannolipids synthesized by the retina were not cleaved by phosphodiesterase preparations from snake venom either in the presence or absence of detergents. In addition to phosphodiesterase these preparations contained alkaline phosphatase and inorganic and nucleotide pyrophosphatases. The action of these enzymes on the mannolipids synthesized by the retina was also examined by the procedure described by Ghalambor, Warren, and Jeanloz (22) in which a chaotropic agent, KCNS, was present to aid in the availability of the substrate to the enzyme. The retina mannolipid was also resistant to cleavage when treated in this manner. Neither Lipid I nor dolichyl phosphomannose synthesized by the retina were cleaved by treatment with α-mannosidase purified from jack bean meal incubated at either pH 6.0 or 4.5 in the presence or absence of detergents. Control studies with all of these enzymes showed that the mannolipids did not interfere with their activities.

In addition to the chromatographic properties described above, Lipid I and the product synthesized using dolichyl phosphate as exogenous acceptor showed identical responses to treatment with mild and strong acids and bases to those described for dolichyl-β-mannosyl phosphate (3, 7).

Lipid II — Chromatography

The C/M/H$_2$O (10:10:3) extract of the acid-washed, C/M (2:1) extracted pellet was chromatographed on DEAE-cellulose (Fig. 5). No radioactivity was obtained in the preliminary wash of the column with C/M/H$_2$O (10:10:3), but the produce was recovered by the linear gradient using 0.2 M ammonium acetate in this solvent. The radioactive material eluted as a single peak from ~ 0.08 to 0.11 M ammonium acetate, with recoveries of 80–95% in different preparations. Using a linear gradient of NH$_4$CHO$_2$ in C/M/H$_2$O (10:10:3) the radioactivity was eluted at 0.11 M salt, with a recovery of about 75% of the radioactivity applied to the column (data not shown).

Lipid II, recovered from DEAE-cellulose, was desalted by gel filtration on a column of LH-20 (2.5 × 20 cm) packed in C/M/H$_2$O (10:10:3) and eluted with this solvent. Over 90% recovery of radioactivity was obtained in a single sharp peak at about 0.36 of the calculated V_t.

The migration of purified Lipid II after paper chromatography in solvent system B was also similar to that of the oligosaccharide-lipid described by Lucas, Waechter, and Lennarz (23).

TLC of Lipid II (recovered from DEAE-cellulose) on microcrystalline cellulose (Avicel) as described by Chambers and Elbein (24) showed the possible presence of 2 components which migrated slower than dolichyl phosphomannose in this system (Fig. 6). This pattern was similar to, but simpler than that seen with, the oligosaccharide-lipid derived from aorta.

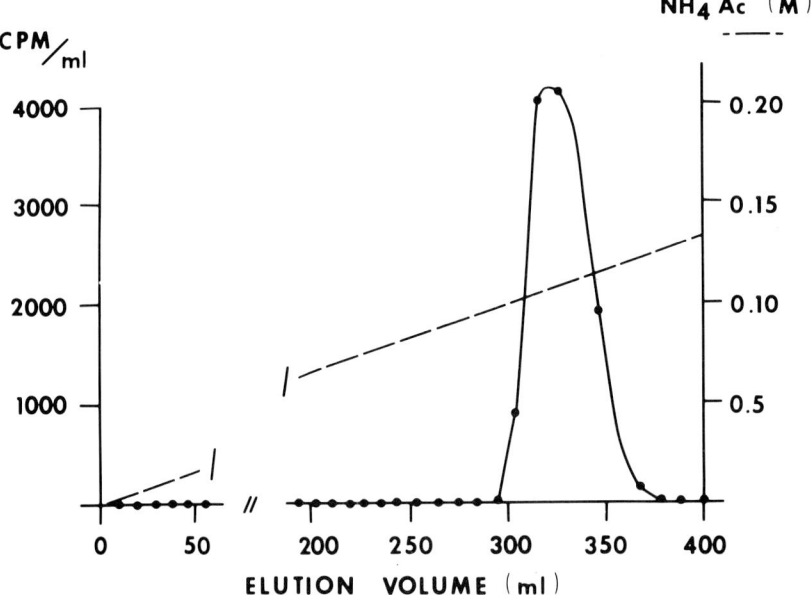

Fig. 5. DEAE-cellulose chromatography of Lipid II. The C/M/H$_2$O (10:10:3) extract of the TCA/PTA pellet obtained after a large-scale incubation using the endogenous acceptors, as described in Materials and Methods was placed over a column of DEAE-cellulose acetate prepared by the method of Rouser et al, (12), and packed in 10:10:3. After an initial wash with 200 ml of this solvent, a linear gradient was carried out using 200 ml of 10:10:3 in the mixing chamber and 300 ml of this solvent containing 0.2 M NH$_4$Ac in the reservoir. Fractions of 5 ml each were collected at a flow rate of 1 ml/min. Gradient (- - -).

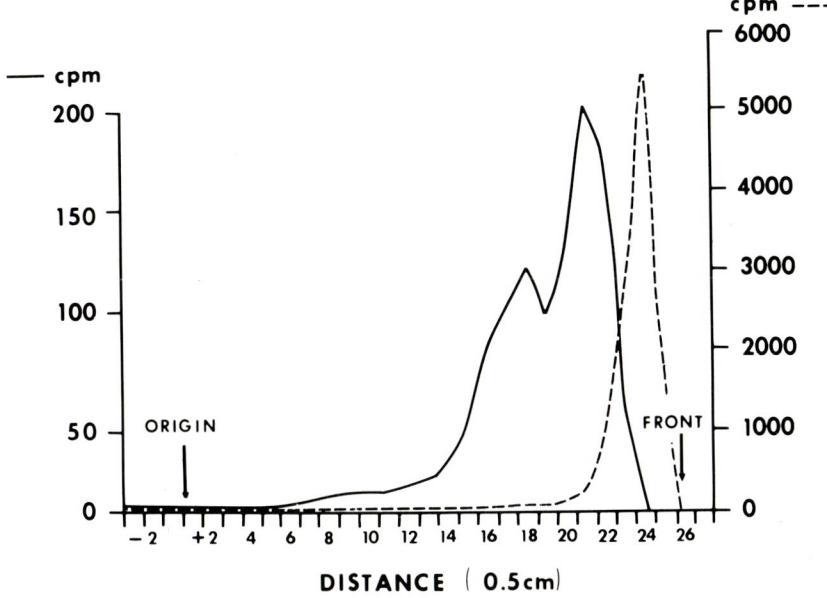

Fig. 6. TLC of Lipid II. Lipid II, purified by chromatography of DEAE-celluose and gel filtration of LH-20, was applied to a 250 micron thick layer of microcrystalline cellulose (Avicel) and the plate irrigated with solvent system B. Dolichyl phosphate-[^{14}C] mannose was also applied to the same plate. After air drying, 0.5-cm zones were scraped into counting vials and the radioactivity determined by liquid scintillation spectrometry. The ordinate on the left refers to Lipid II (——); that on the right, to dolichyl phosphomannose (- - -).

Strong acid hydrolysis of purified Lipid II followed by borate electrophoresis and paper chromatography (solvent system D), showed that the radioactivity was present in a single component which migrated with standard mannose.

These chromatographic characteristics of Lipid II are consistent with the suggestion that this product is an oligosaccharide(-pyrophospho-)lipid similar to those synthesized by other tissues (23, 24). Studies are in process to characterize this product further.

Residue

The residue remaining after the isolation of Lipid I and Lipid II, was solubilized in 1% SDS as described in Materials and Methods. Within 4 h at $50°C$, ~70% of the radioactivity was solubilized. The elution pattern of this material from Sephadex G-200 is seen in Fig. 7. Two large radioactive components can be seen: one eluting at the void volume and a second for which an apparent molecular weight of 110,000 was calculated. After treatment with pronase, the high-molecular-weight components were converted to those of lower molecular weight. When examined further by gel filtration on Sephadex G-75, a major radioactive component was obtained whose apparent molecular weight was calculated to be 7,000 (data not shown). Quantitative recovery of radioactive material was obtained from both columns.

These findings, of the conversion of the high-molecular-weight components to a low-molecular-weight component by the action of pronase, together with the observation that

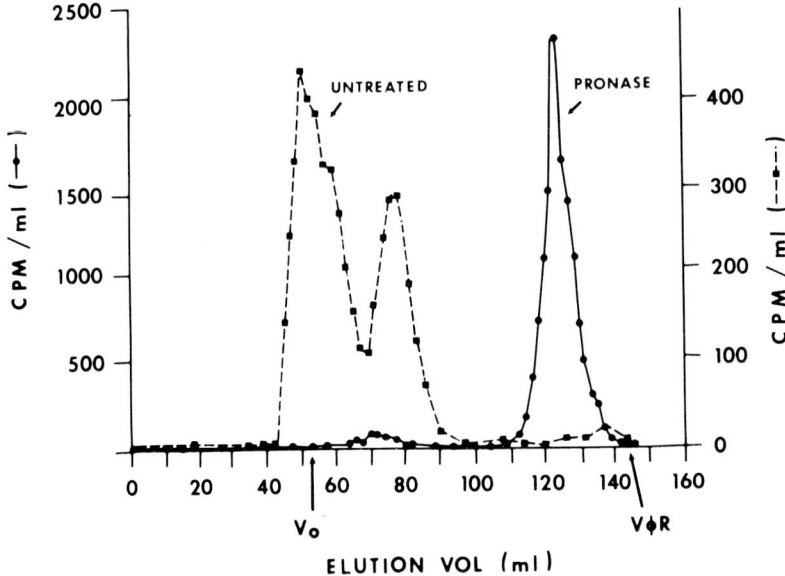

Fig. 7. Elution patterns from Sephadex G-200 of the "residue" in SDS before and after treatment with pronase. The residue (containing 15,400 cpm) remaining after the removal of Lipid I and Lipid II, was dissolved in 4 ml of 1% SDS by heating at 50°C for 4 h. The solution was fractionated on a Sephadex G-200 column packed in and eluted with 0.005 M Tris-HCl buffer, pH 7.0, containing 0.2% SDS. A separate sample of the residue (containing 65,000 cpm) was incubated at 37°C with 0.3 mg pronase in 0.2 M Tris-HCl, pH 7.9, containing 0.0015 M $CaCl_2$. After 4 h, 78% of the radioactivity was solubilized. The mixture was centrifuged and the supernatant solution adjusted to 1% SDS and then kept at 50°C for 4 h. The solution was then fractionated on Sephadex G-200 as above. The column dimensions were: 1.5 × 84 cm; V_0 = 52 ml; V(phenyl red) = 143 ml. The flow rate was maintained at 8 ml/h. Treated with pronase (●); not treated with pronase (■).

the only radioactive component was mannose (3), are consistent with the conclusion that the sugar was a component of glycoproteins in the fraction described in this report as the "residue."

Effect of Nucleotides

The effect of nucleotides on these enzymatic activities was examined. ATP inhibited the incorporation of radioactivity from GDP[^{14}C] mannose into Lipid I, while stimulating the incorporation into Lipid II and the residue (Fig. 8). ATP at greater than 1 mM stimulated the incorporation of radioactivity into Lipid II by twofold and into the residue by 30%, while the incorporation into Lipid I was inhibited by 50%. A similar inhibition of mannolipid formation by ATP has been observed previously by Richards and Hemming (25) although the effect on the other products was not noted. However, stimulation, instead of the inhibition observed here, of the formation of mannolipids by ATP (26) has been reported.

When the incubations were performed in the presence of 20 μM dolichyl phosphate, which stimulated the incorporation of radioactivity into the fraction extracted by C/M (2:1) (dolichyl phosphomannose) about 27-fold over the endogenous level, the same extent

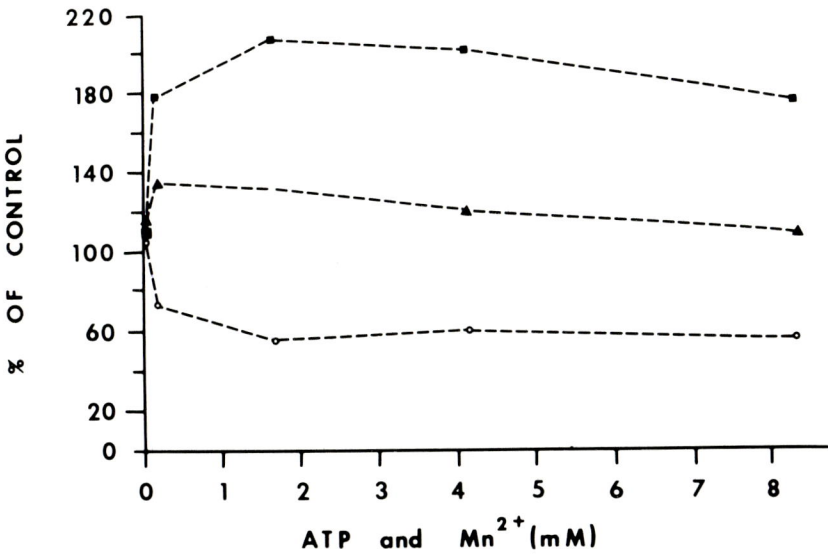

Fig. 8. Effect of the concentration of ATP on the endogenous incorporation of radioactivity into the 3 products. Incubations were performed for 5 min using GDP[U-^{14}C]mannose, TES buffer, pH 7.0, and enzyme (0.80 mg protein) and assays were carried out as described in Materials and Methods. In addition to the usual concentration of Mn^{2+} (3.3 mM), additional $MnCl_2$ was added equal to that indicated for ATP. The control values in the absence of added ATP were: Lipid I, 3,120 cpm; Lipid II, 1,380 cpm; Residue, 3,400 cpm. Lipid I, (○); Lipid II, (■); residue, (▲).

of inhibition in the presence of ATP was seen (Fig. 9). Unlike the response with the endogenous acceptors, however, the stimulation by ATP of the incorporation of radioactivity into the oligosaccharide-lipid fraction (Lipid II) and the residue did not occur.

The loss of the stimulatory effect of ATP was probably the result of the inhibition by TX-100 on the incorporation of radioactivity into Lipid II and the residue. Thus, using the endogenous acceptor system in the presence of 0.24% TX-100, there was a 32% decrease in the labeling of Lipid II and a 62% reduction in the radioactivity incorporated into the residue. These decreases in incorporation were not overcome by the presence of optimal concentrations of ATP (1.7 mM). These observations suggest that it may be necessary for the polyprenyl-monosaccharide as well as the oligosaccharide-lipid and glycoprotein acceptors to be present in a membrane-bound form in order for the stimulation by nucleotides to be exerted.

In the present studies, additional amounts of $MnCl_2$ equal to that of ATP were also present during the incubations. Similar results were observed, however, over this same range in ATP concentration without the addition of additional Mn^{2+} over that usually present (3.3 mM). The time course of these reactions in the presence of ATP (1.67 mM) was the same as in its absence (3, 7). These effects were not observed for $MnCl_2$ alone at these concentrations.

Studies were also performed in the presence of a constant concentration of ATP (1.67 mM) with varying amounts of Mn^{2+} (0–6.67 mM) (data not shown). In the absence of metal, all 3 activities were inhibited. While inhibition of the incorporation of radioactivity into Lipid I persisted as the ratio of Mn^{2+} to ATP was increased to 4:1, the stimulation

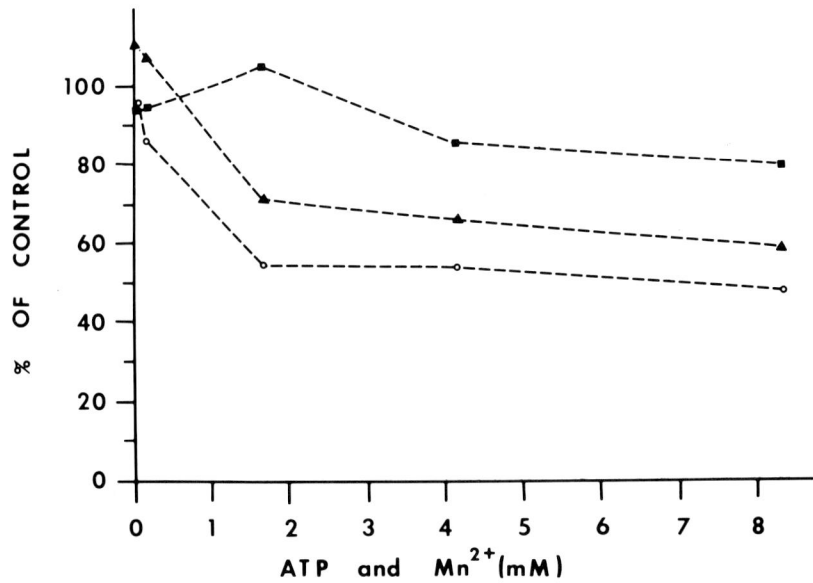

Fig. 9. Effect of ATP on the mannosyltransferases incubated in the presence of dolichyl phosphate. Incubations were carried out for 5 min as described in Fig. 8, (using a different enzyme preparation), but contained in addition, dolichyl phosphate (19 μM) and TX-100 (0.24%). The control values, in the absence of added ATP were: Lipid I, 80,600 cpm; Lipid II, 1,720 cpm; Residue, 3,292 cpm. Lipid I (○); Lipid II (■); residue (▲).

into Lipid II and into the residue occurred as previously. Thus, the inhibition of Lipid I activity was not due solely to chelation effects of ATP on the metal, but was due to an effect on the activity of the enzyme itself.

When ADP was added to 1.67 mM, the stimulation of incorporation of label into Lipid II and the residue, and the inhibition of incorporation into Lipid I were similar to those observed with ATP (Table I). AMP, cyclic-AMP, and inorganic phosphate had little effect on these activities.

The inhibition by ATP and ADP of the formation of Lipid I, accompanied by the stimulation of incorporation into Lipid II and the residue, suggested that ADP[^{14}C]-mannose might have been formed by the reversal of the reaction and that this compound might be more directly involved in the formation of Lipid II and the glycoproteins present in the residue. That this was probably not the case may be concluded from experiments in which up to 150-fold molar excess of nonradioactive ADP-mannose over GDP[^{14}C]-mannose was present during the incubations. There was no effect on the incorporation of radioactivity into Lipid I, II, or the residue from the presence of ADP-mannose.

In Table I also is shown the effects of a variety of other nucleotides examined at this same concentration. Although these experiments were carried out under conditions approximating the initial rates, their interpretation in terms of specificity is difficult due to the possible influence of phosphokinases and phosphatases in the enzyme preparation. Nonetheless, some differential response to these reagents on the formation of the 3 products was evident. Of the compounds tested GDP and GTP inhibited all of these activities to the greatest extent. Some of this inhibition may be due to reversal of the reaction. This has been observed by incubating dolichyl phosphate [^{14}C]mannose in the presence of GDP,

TABLE I. Influence of Various Nucleotides on the Enzyme Activities*

Addition (1.67 mM)	Percentage of radioactivity in products		
	Lipid I	Lipid II	Residue
ATP	57	207	126
ADP	71	229	125
AMP	99	98	98
Cyclic-AMP	85	97	91
CTP	96	125	89
dCTP	96	168	103
CDP	82	164	118
GTP	4	7	7
GDP	3	4	5
GMP	29	27	22
Cyclic-GMP	103	104	94
ITP	22	51	50
TTP	100	108	97
UTP	85	47	61
UDP	81	49	52
PEP	98	–	117[a]
P_i	89	78	98

*Incubations were carried out for 5 min using 0.84 mg of dialyzed homogenates of embryonic chick retina, and assayed for the incorporation of radioactivity into products as indicated in Materials and Methods. The added compounds were all present at 1.67 mM. The data is presented in terms of the percentage of the radioactivity incorporated compared to the controls incubated in absence of these compounds which in different enzyme preparations and different experiments varied from: Lipid I, 2,800–3,500 cpm; Lipid II, 1,300–1,800 cpm; residue, 1,960–3,300 cpm.
[a] Refers to Lipid II plus residue.

and following the appearance of a new uv quenching, radioactive area which migrated with GDP-mannose (3) (data not shown). However, in addition to reversal of the reaction which produced dolichyl phosphomannose, an inhibitory effect on the mannosyltransferases by the guanosine nucleotides may also be involved. The decrease in the labeling of Lipid II and the residue observed in the presence of UDP and UTP with relatively little effect on Lipid I also indicates influences by these compounds other than reversal of reaction. While cyclic-GMP was without effect, GMP also resulted in considerable inhibition.

DISCUSSION

The endogenous acceptor-lipid in the retina was shown to have the properties of dolichyl phosphate as indicated by the characteristics of the product of the mannosyltransferase reaction. This material and that synthesized using exogenously added dolichyl phosphate were identical to one another by a variety of criteria, and to authentic dolichyl phosphomannose. Studies from this laboratory with the retina from the embryonic chick have revealed that dolichyl phosphate was the only polyprenyl phosphate among a variety tested, including retinyl phosphate, which was capable of stimulating the incorporation of mannose from GDP[^{14}C] mannose into material extractable by C/M (2:1) (3, 7). These findings, in addition to the formation of material which had the properties of the oligosaccharide-pyrophosphate-lipids described in other tissues, are consistent with the presence in the retina of the pathway for the biosynthesis of its glycoproteins involving the participation of the lipid-activated carbohydrates as intermediates.

In a previous report from this laboratory [Kean and Plantner, (27)], the biosynthesis by preparations from bovine retina of a high-molecular-weight mannose, glucosamine-containing oligosaccharide-lipid was observed. It was suggested that this component may function as an intermediate in the biosynthesis of the carbohydrate groups of bovine rhodopsin.

Several nucleotide di- and triphosphates have been shown in this study to influence these mannosyltransferase reactions. Most extensively studied was the effect of ATP. The mechanism whereby its presence inhibited the incorporation of mannose into Lipid I, while at the same time stimulating over twofold the labeling of the oligosaccharide-lipid fraction, Lipid II, and also stimulating the labeling of the glycoprotein, is not known. The stoichiometry of these relationships was not simple. Much more (up to twofold) radioactivity was found incorporated into the complex glycans than could be accounted for by the decrease in labeling of Lipid I after a 5 min incubation carried out in the presence of ATP. The similarity in response to ADP as with ATP may reflect the presence of adenylate kinase in the preparation from the retina. The nucleotides may function as participants in the formation of as yet unknown intermediates for the biosynthesis of the oligosaccharide-(-pyrophosphate-)lipids (Lipid II) and glycoproteins, or act as modifiers of these enzymatic reactions.

Previous studies have shown (3, 7) that detergent is necessary in order to obtain stimulation from exogenously added dolichyl phosphate. The concentration of TX-100 which was optimal for this stimulation (0.24%) was, however, inhibitory for the endogenous labeling of Lipid II and the residue. This probably accounts for the failure of ATP to stimulate the labeling of Lipid II and the residue when the reactions were performed in the presence of dolichyl phosphate and TX-100. Nonetheless, the inhibition of the labeling of dolichyl phosphomannose when incubated in the presence of ATP was still observed, and to the same proportionate extent as with the endogenous product, Lipid I. These observations suggest that in order for the nucleotide stimulation of the formation of the complex glycan products to be effected, the activated lipids may be required to be present in the form of membrane-bound components rather than as detergent suspensions, a relationship which may reflect more accurately the situation within the cell. Further studies are in process to examine more fully the influence of the nucleotides on these reactions.

ACKNOWLEDGMENTS

The excellent technical assistance of Ms. Maxine Klein throughout this work is acknowledged. Appreciation is expressed to Dr. James J. Plantner for assistance in studies on the "residue." The author is very grateful to Dr. Christopher D. Warren for supplying dolichyl β-mannosyl phosphate used as a chromatographic standard.

This work was supported by Public Health Service Research Grant EY 00393 from National Eye Institute and the Ohio Lions Eye Research Foundation.

REFERENCES

1. Waechter CJ, Lennarz WJ: Annu Rev Biochem 45:95, 1976.
2. Plantner JJ, Kean EL: J Biol Chem 251:1548, 1976.
3. Kean EL: J Biol Chem 252:5622, 1977.

4. Kean EL: Exp Eye Res (In press).
5. Kean EL, Bruner WE, Sherwood PC: Biol Bull 143:466, 1972.
6. Kean EL, Plantner JJ, Bruner WE: Fed Proc Fed Am Soc Exp Biol 33:1368, 1974.
7. Kean EL: Fed Proc Fed Am Soc Exp Biol 35:1539, 1976.
8. Zatz M, Barondes SH: Biochem Biophys Res Commun 36:511, 1969.
9. Folch J, Lees M, Sloane-Stanley GH: J Biol Chem 226:497, 1957.
10. Wedgewood JF, Strominger JL, Warren CD: J Biol Chem 249: 6316, 1974.
11. Lester RL, Steiner MR: J Biol Chem 243:4889, 1968.
12. Rouser G, Kritchevsky G, Yamamoto A: In Marinetti GV (ed): "Lipid Chromatographic Analysis." New York: Marcel Dekker, 1967, p 99.
13. Dankert M, Wright A, Kelley WS, Robbins PW: Arch Biochem Biophys 116:425, 1966.
14. Kean EL: J Biol Chem 245:2301, 1970.
15. Dittmer JC, Lester RL: J Lipid Res 5:126, 1964.
16. Gordon HT, Thornburg W, Werum LN: Anal Chem 28:849, 1956.
17. Trevelyan WE, Procter DP, Harrison JS: Nature (London) 166:444, 1950.
18. Kean EL, Bighouse KJ: J Biol Chem 249:7813, 1974.
19. Lowry OH, Rosebrough NJ, Farr AL, Randall RJ: J Biol Chem 193:265, 1951.
20. Hohorst H-J: In Bergmeyer H-U (ed): "Methods of Enzymatic Analysis." New York: Academic Press, 1963, p 260.
21. Li Y-T, Li S-C: Methods Enzymol 28:702, 1972.
22. Ghalambor MA, Warren CD, Jeanloz RW: Biochem Biophys Res Commun 56:407, 1974.
23. Lucas JJ, Waechter CJ, Lennarz WJ: J Biol Chem 250:1992, 1975.
24. Chambers J, Elbein AD: J Biol Chem 250:6904, 1975.
25. Richards JB, Hemming FW: Biochem J 130:77, 1972.
26. DeLuca L, Maestri N, Rosso G, Wolf G: J Biol Chem 248:641, 1973.
27. Kean EL, Plantner JJ: Exp Eye Res 23:89, 1976.

Synthesis, Secretion, and Attachment of LETS Glycoprotein in Normal and Transformed Cells

Richard O. Hynes, Antonia T. Destree, Vivien M. Mautner, and Iqbal U. Ali

Center for Cancer Research and Department of Biology, Massachusetts Institute of Technology, Cambridge, Massachusetts 02139

LETS glycoprotein is a surface glycoprotein which is absent or greatly diminished on the surfaces of transformed cells. Normal cells secrete large amounts of this protein into the medium; transformed cell medium contains much less. The difference is not due to degradation of the soluble LETS protein. Biosynthesis of LETS protein can be studied by analysis of cell extracts by detergent extraction and immune precipitation and appears to proceed in transformed cells at a reduced rate compared with normal cells. Addition of inhibitors of protein synthesis to transformed cell cultures causes the small amount of LETS protein in the medium to attach to the cells. Addition of normal conditioned medium, which contains LETS protein, to transformed cells alters their morphology towards normal. Addition of purified LETS protein to transformed cells causes the cells to attach, spread, align with one another, and regain actin cables. The results indicate that LETS protein can exchange between cell surface and medium and that it affects cellular adhesion, morphology, and cytoskeleton.

Key words: LETS protein, biosynthesis, adhesion, transformation, cytoskeleton

Much recent research has focused on the question of surface changes in cells transformed by viruses or chemicals. In particular, changes in surface proteins have been reported (1) and the most prominent of these is the loss or reduction of a large external transformation-sensitive (LETS) glycoprotein from the surfaces of transformed cells (2). LETS protein is a major surface protein of fibroblasts and myoblasts and it varies in amount on normal cells depending on their growth state and position in the cell cycle (3).

The absence of LETS protein from the surfaces of transformed cells raises the question of whether this is as a result of reduced or altered synthesis or of other factors such as degradation. Since LETS protein is known to be very sensitive to proteases (1, 2), the latter possibility has received some attention but remains unproven (4). It is known that prelabeled LETS protein turns over into the medium if cells are returned to culture

Vivien M. Mautner is presently with the MRC Virology Unit, Institute of Virology, University of Glasgow, Glasgow G11 5JR, Scotland

Received February 23, 1977; accepted July 15, 1977.

© 1977 Alan R. Liss, Inc., 150 Fifth Avenue, New York, NY 10011

(4–6), so it is possible that some alteration in the rate of turnover occurs on transformation such that the balance between LETS protein on the cells and in the culture medium is altered. The question also arises as to whether LETS protein in the medium can bind to cells. Indeed it is not clear that this is not the normal route by which it reaches the cell surface, as is the case for collagen (7). If this were the case, then reduced attachment could provide an explanation for the reduced surface levels of LETS protein in transformed cells.

In this paper, we collect together several experiments bearing on the questions of synthesis, secretion, and attachment of LETS protein in normal and transformed cells and discuss their implications for the understanding of the transformation-induced reduction in surface levels of this protein.

METHODS

Cells and Culture Conditions

The cells used were a normal hamster fibroblast line, NIL8, and its derivative, NIL8-HSV, transformed by hamster sarcoma virus (8). Cells were cultured in Dulbecco's modified Eagle's medium plus 5% fetal calf serum. For labeling with [^{35}S] methionine (New England Nuclear Corporation, Boston, Massachusetts, 22 Ci/mmole) the methionine concentration was reduced to 10% of normal.

Lactoperoxidase Iodination and Gel Electrophoresis

Iodination was performed on monolayers as described (2). Sodium dodecyl sulfate-(SDS)-polyacrylamide slab gels were run in the buffers described by Laemmli (9). Gels were dried and autoradiographed on Kodak x-ray film (NS-2T or RP-R2). For detection of [^{35}S] methionine, gels were impregnated with PPO prior to drying (10).

Purification of LETS Protein

LETS protein was purified from conditioned medium by ammonium sulfate precipitation and Sephadex G200 chromatography or from cells by urea extraction as described elsewhere (11, 12). For preparation of antisera, LETS protein was further purified on preparative SDS-polyacrylamide gels before injection.

Immunofluorescence

Cells grown on coverslips were stained for LETS protein or actin using indirect immunofluorescence as described elsewhere (13, 14). For staining surface proteins, cells were fixed with formaldehyde alone; for staining internal proteins, they were also permeabilized with acetone. Slides were viewed and photographed on a Zeiss microscope equipped with epifluorescent illumination.

Immune Precipitation

Cells labeled with [^{35}S] methionine were lysed in 2% deoxycholate, 0.05 M NaCl, 0.02 M Tris-HCl, pH 8.3 at 4°C, vortexed, and centrifuged at 10,000 × g for 10 min. Aliquots of the supernatant were incubated with 5–10 μl antiserum to LETS protein or with preimmune serum for 1 h at 37°C. Fifty to one hundred milliliters of goat antirabbit immunoglobulin (Cappel Labs, Cochraneville, Pennsylvania) were added and the incubation continued for 1 h at 37°C and then overnight at 4°C. The precipitates were collected and washed 3 times by centrifugation and then dissolved for SDS-polyacrylamide gel electrophoresis.

RESULTS

LETS Protein Synthesized by Transformed Cells

In an earlier paper we reported that chicken cells transformed by temperature-sensitive Rous sarcoma virus at the permissive temperature could regain surface LETS protein on shift to restrictive temperature even in the absence of protein synthesis (15). This result suggested that the transformed cells synthesize LETS protein but for some reason it is not found on the cell surface. No temperature-sensitive mutants were available in the NIL8 hamster system but we observed that addition of cycloheximide or puromycin to transformed NIL8-HSV cells led to the appearance of LETS protein on the surfaces of these cells and to some flattening of the cells (4). Although the explanation for this effect of inhibitors of protein synthesis remains obscure, it suggests the idea that transformed cell cultures contain a pool of LETS protein. Since the transformed hamster line is an established cloned line we have pursued this result in this system.

It turns out that the pool of LETS protein in these cultures is in the medium. This conclusion arises from experiments such as that shown in Fig. 1. If cycloheximide (2 or 20 μg/ml which inhibit protein synthesis ≥ 98% and 95%, respectively) was added to cultures of NIL8-HSV without medium change then, 24 h later, small amounts of LETS

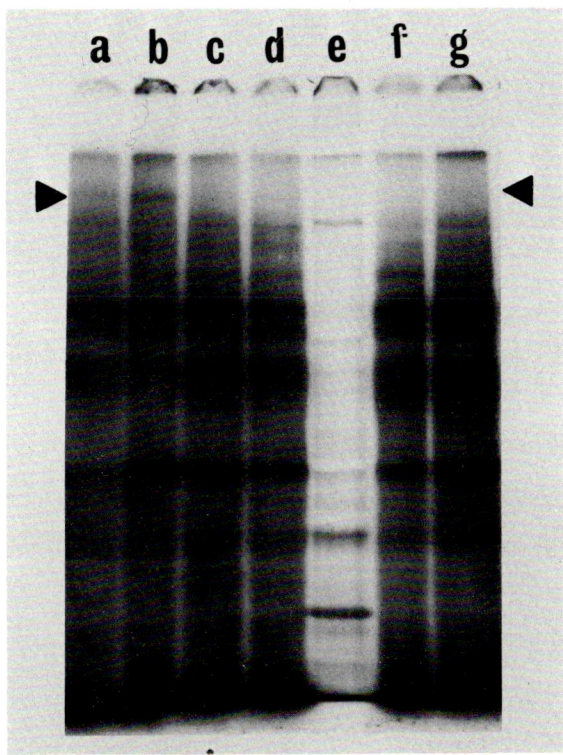

Fig. 1. Effect of cycloheximide on binding of LETS protein to transformed cells. NIL8-HSV cultures were iodinated 24 h after treatments as below and equal amounts of protein analyzed on an SDS-polyacrylamide slab. a, b) 2 and 20 μg/ml cycloheximide added to the medium. c,d) 2 and 20 μg/ml cycloheximide added to fresh medium. e) Metabolically labeled cell lysate run as molecular weight marker. Major bands at 200, 125, 65, 58, and 42 thousand daltons. f, g) Controls: medium changed or not, respectively. Arrow marks position of LETS protein.

protein could be detected on their surfaces by iodination (Fig. 1a, b). However, if the cells were changed to fresh medium at the time of addition of the cycloheximide, then no LETS protein appeared on the cells (Fig. 1c, d): their iodination patterns remained like those of controls (Fig. 1f, g).

This result suggested that the transformed cells were secreting LETS protein into the medium. A metabolic labeling study confirmed this conclusion. Figure 2a shows that NIL8-HSV medium after 24 h of labeling contains a labeled band comigrating with the LETS protein in conditioned medium of normal NIL8 cells (Fig. 2b), although much less was present in the transformed cell culture. Other differences were observed between the labeling patterns of the 2 cell types. NIL8 cells contained prominent bands at 185,000, 130,000, ~100,000, and around the position of the serum albumin. In contrast, the most prominent labeling in the transformed cell culture was of a doublet at about 60,000 daltons.

To test whether the reduced amount of LETS protein and other larger proteins in the NIL8-HSV medium might be due to degradation, prelabeled NIL8 conditioned medium was mixed with fresh unlabeled conditioned medium from normal or transformed cells and incubated. No degradation was observed (Fig. 2c, d, e). A similar result was obtained when prelabeled NIL8 conditioned medium was incubated with NIL8-HSV cells (data not shown) suggesting that degradation was not the explanation for the differing patterns.

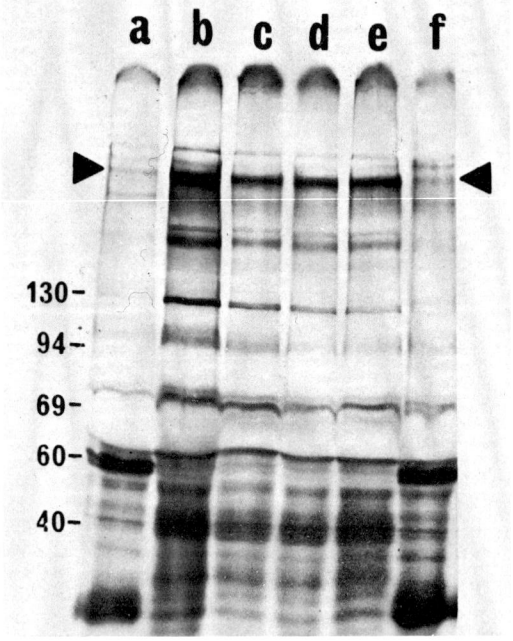

Fig. 2. Metabolically labeled conditioned medium. Analyzed on an SDS-polyacrylamide slab gel. a,b,f) Conditioned medium (65 μl) of NIL8-HSV (a, f) and NIL8 (b) cells after 24 h labeling with [^{35}S]methionine. c-e) Equal aliquots (50 μl) of ^{35}S-labeled NIL8 conditioned medium incubated for 18 h alone (c) or with equal volumes of unlabeled conditioned medium from NIL8 (d) or NIL8-HSV (e) before loading onto gel. Markers were β-galactosidase (130), phosphorylase A (94), bovine serum albumin (69), catalase (60), and creatine kinase (40) as shown on left. Arrow marks position of LETS protein.

Process of Synthesis of LETS Protein

Since it appeared that transformed cells released much less LETS protein into the medium we investigated the early stages of synthesis inside the cells. Figure 3 shows a pulse-chase experiment in NIL8 cells. After a 20 min pulse, LETS protein could be clearly seen in the profile (track 1) and the majority of it was resistant to trypsin (track 4). After a 60 or 120 min chase, LETS protein was still present (tracks 2, 3) but now much of it was sensitive to trypsin treatment of intact cells (tracks 5, 6). Other changes in profile were also observed: a band at 185,000 daltons increased in size during the chase and bands at 145,000 and 65,000 disappeared. One of these proteins is presumably related to the rapidly labeled protein reported by Kuusela et al (16) in chicken cells. None of these proteins was present at the surface after the pulse as judged by their insensitivity to trypsin (track 4). In other experiments of the same sort, it was found that LETS protein was largely insensitive to trypsin treatment up to about 50–60 min of labeling or chase, suggesting that during this period it was being processed to the cell surface. This long processing time is presumably related to the time taken to add the carbohydrate residues and to transport the protein from the rough endoplasmic reticulum (ER) to the surface. In this context, it is interesting to note that in chicken cells there is a minor band comigrating with LETS protein which labels with glucosamine but not with fucose and is trypsin-insensitive, whereas the surface LETS protein which is sensitive to trypsin labels with both sugars (5, 17).

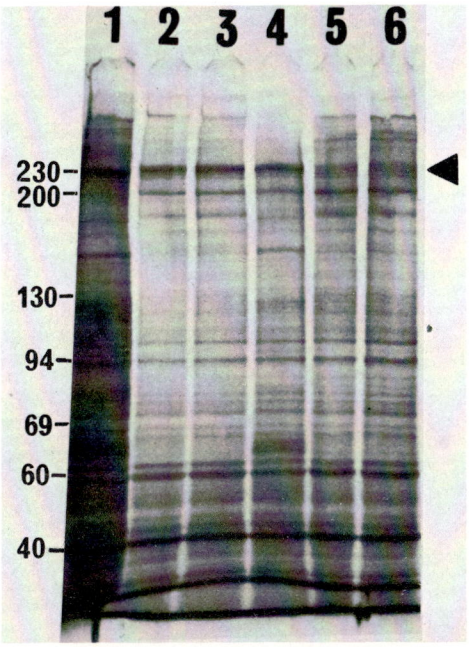

Fig. 3. Pulse-chase labeling of NIL8 cells. Analysis of total SDS lysates on SDS-polyacylamide slab gel. 1) 20 min pulse label. 2) 20 min pulse, 60 min chase. 3) 20 min pulse, 120 min chase. 4–6) As 1–3 respectively but treated at the end of the incubation with trypsin (10 µg/ml, 10 min, room temperature stopped with excess soyabean trypsin inhibitor). Markers on left as Fig. 2 plus myosin and LETS proteins at 200,000 and 230,000 daltons respectively.

To characterize further the course of synthesis and processing of LETS protein, cells were fractionated by detergent extraction. The majority of the iodinatable surface LETS protein is insoluble in deoxycholate (18). However, after a short pulse or a pulse followed by a chase, material comigrating with LETS protein and precipitable by specific antisera was detectable in the deoxycholate soluble material. Analysis of a pulse-chase experiment in this way is shown in Fig. 4. Both normal and transformed cells show synthesis

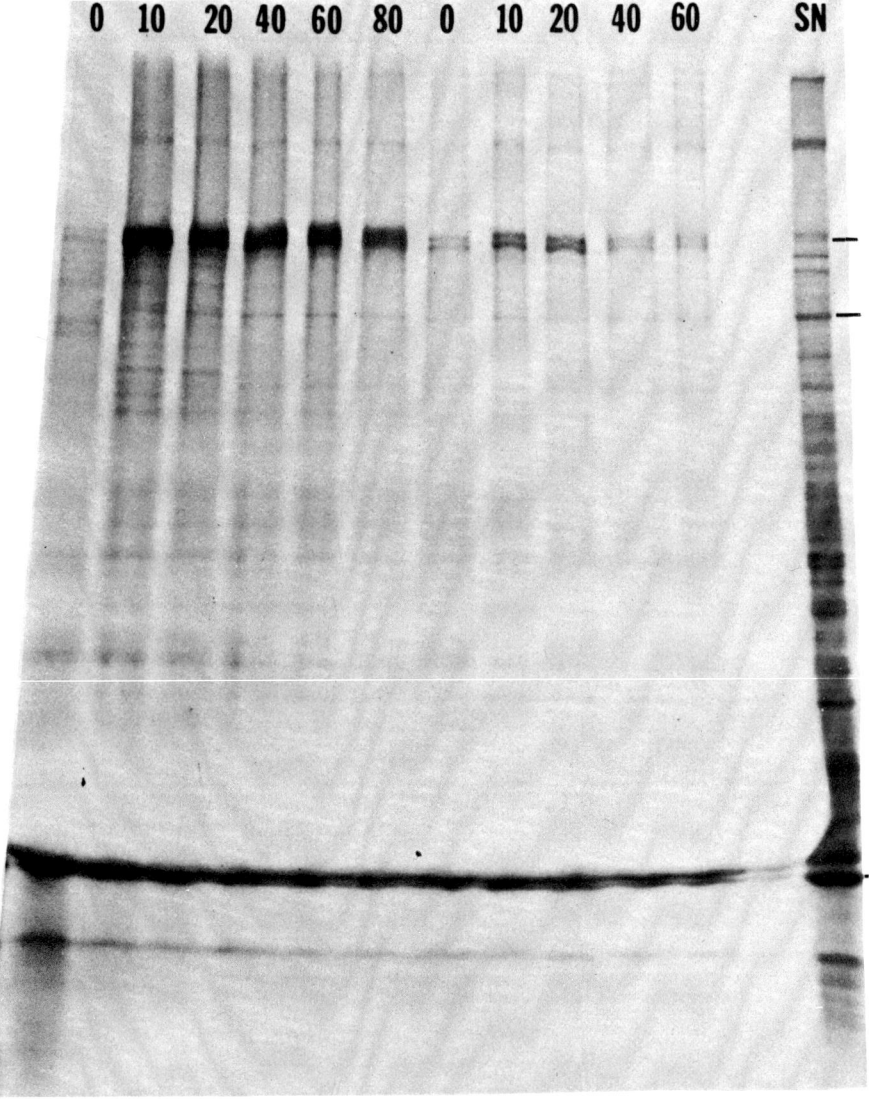

Fig. 4. Analysis of anti-LETS protein immune precipitates of deoxycholate-soluble cell extracts on a 5–10% gradient gel. SN) Total extract. All other samples are immune precipitates (see Materials and Methods). Comparison of the immune precipitates with the total extract shows the selectivity of the immune precipitation. Actin and other minor bands observed in the immune precipitates were also seen in precipitation with preimmune serum (not shown). Cells were pulse labeled for 10 min with [^{35}S] methionine and chased for periods of time shown (min) before harvesting. NIL8 cells on left. NIL8-HSV cells on right. Marks at right indicate positions of LETS protein, myosin, and actin.

of a doublet band comigrating with surface-labeled LETS protein and precipitable by antibody. In both cell types, more of this material was evident after 10 min of chase than immediately after the 10 min pulse. Thereafter, the amount of label precipitable in this band decreased with increasing times of chase, in both cell types. Gel analysis of the deoxycholate-insoluble material (without immune precipitation) showed a labeled doublet band comigrating with surface-labeled LETS protein which increased in amount with time of chase (Fig. 5). These results are consistent with synthesis of LETS protein in some precursor form not identified as yet, followed by a time-dependent processing through a deoxycholate-soluble form to a deoxycholate-insoluble form. This process appears to occur in both normal and transformed cells. Quantitation of autoradiograms such as those in Figs. 4 and 5, followed by correction for total incorporation of radioactivity in each cell sample gave estimates of the rate of LETS protein synthesis relative

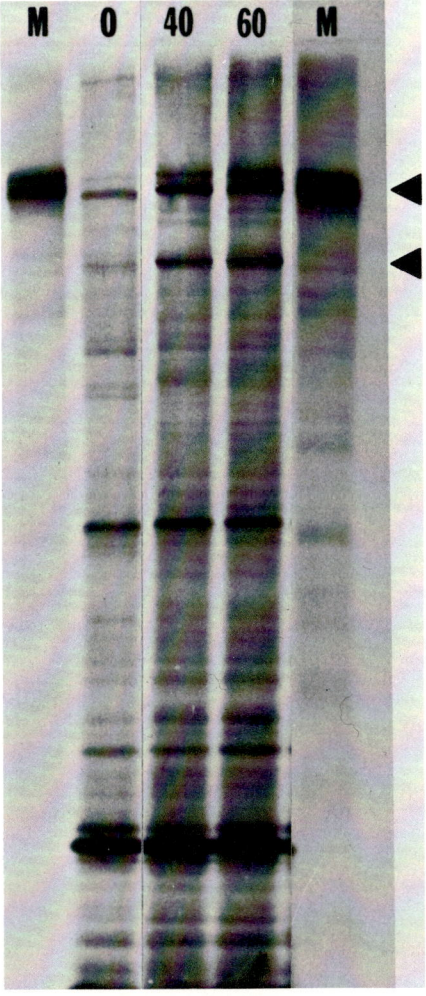

Fig. 5. Deoxycholate insoluble material from pulse-chase labeled NIL8 cells analyzed on 5–10% gradient gel. M) Iodinated NIL cell marker showing LETS protein doublet. Other samples are cells labeled for 10 min with [^{35}S] methionine and chased for periods of time shown (min) before harvesting. Arrows mark position of LETS protein and myosin.

to total protein synthesis. Different experiments gave relative synthesis rates of the putative deoxycholate soluble precursor form of LETS protein by transformed cells which were approximately 10% of those in normal cells. The rate of accumulation of label in the comigrating deoxycholate insoluble band was similarly reduced in transformed cells. This reduction must be contrasted with the very low levels of LETS protein found at the surface and in the medium of transformed cells, roughly estimated at $\leqslant 1\%$ and $\sim 5\%$, respectively.

The proviso must be made that it is not proven that the doublet band quantitated in these experiments is in fact LETS protein or a precursor to it. In support of this hypothesis is the comigration, in particular the fact that a comigrating doublet is observed on the gradient gels used here both for the iodinated surface protein (Fig. 5M) and for the various metabolically labeled samples. Furthermore, a specific antiserum prepared against LETS protein isolated from conditioned medium specifically precipitates these 2 labeled intracellular bands from the total deoxycholate supernatent (Fig. 4). Final proof of relatedness will depend upon peptide fingerprinting studies in progress.

Assuming for the present that the labeled bands do represent LETS protein in the course of intracellular synthesis and processing, these results suggest that reduced synthesis in the transformed cells is not by itself a sufficient explanation for the low levels of LETS protein in external positions (surface and medium) and suggest that some other alteration must exist, perhaps in the processing or transport to the cell surface or in retention at the surface.

Properties of LETS Protein in the Medium

Since in the cycloheximide experiment described earlier, it was found that LETS protein in the medium could bind to cells it is of interest to consider the effect of increasing the amount of LETS protein in the medium of NIL8-HSV cells. The simplest way to do this is to transfer these cells into medium conditioned by the growth of NIL8 cells which as shown earlier (Fig. 2) secrete intact LETS protein into the medium. Figure 6 shows the result of this experiment. NIL8-HSV cells normally grow as rather rounded, refractile cells and also grow detached in the culture medium (Fig. 6a). The floating cells are viable and will regenerate a similar mixed culture if passaged on their own. When NIL8 conditioned medium is added to NIL8-HSV cells, the number of floating cells is greatly reduced and the attached cells become more flattened and show a tendency to line up with each other more like normal cells (Fig. 6b, c). In contrast, cultures changed to fresh medium or medium conditioned by NIL8-HSV cells remain as controls (Fig. 6a, d). This result suggested that the LETS protein secreted by NIL8 cells could affect the properties of transformed cells. However, other components of the conditioned medium could equally well be responsible.

Accordingly, LETS protein was purified from NIL8 conditioned medium as described elsewhere (11) and added to cultures of NIL8-HSV cells. The result is shown in Fig. 7b, c. All the floaters attached to the dishes and the attached cells spread, elongated, and aligned with one another. A similar result was obtained with LETS protein extracted from cells with urea (Fig. 7d) as reported elsewhere by Yamada et al (12) and confirmed by us.[1] Thus, both LETS protein from the surface of NIL8 cells and in their conditioned medium is able to affect the adhesion and morphology of transformed cells.

---

[1] In these experiments much greater amounts of LETS protein were detected on the surface by iodination (11) than were observed in the cycloheximide experiments discussed earlier.

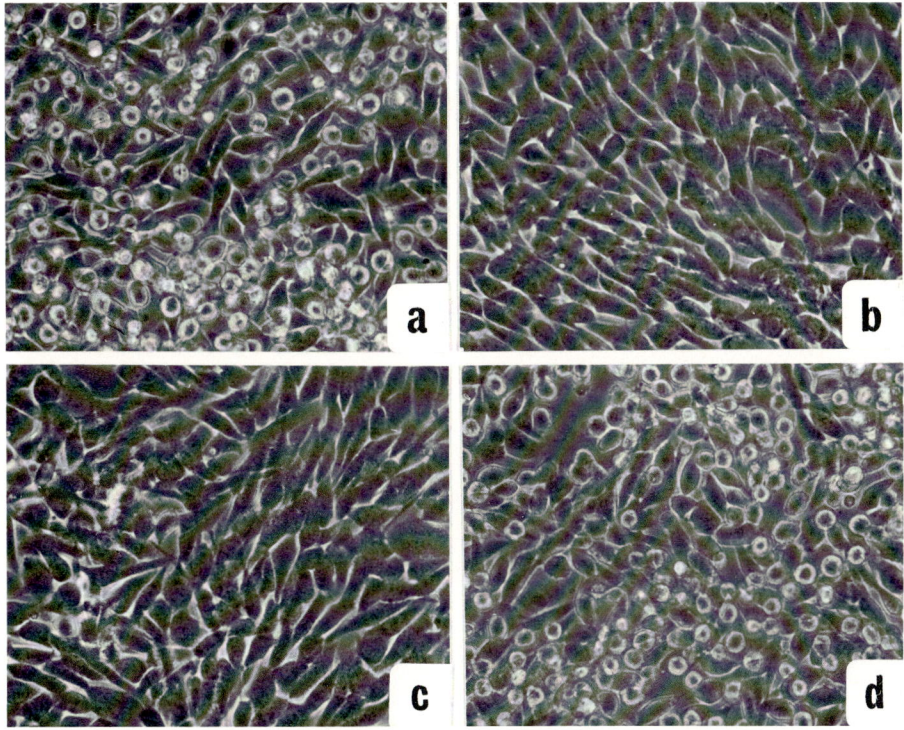

Fig. 6. Effects of conditioned medium on transformed cells. NIL8-HSV cells were cultured for 3 days and then changed for 2 more days into a) fresh medium or to medium conditioned for 5 days by b, c) NIL8 cells or d) NIL8-HSV cells. Note the attachment of floaters and flattening and alignment of cells in (b) and (c). Unchanged controls looked like (a) and (d). Phase contrast, approximately 200 ×.

Immunofluorescence studies show that added LETS protein binds to NIL8-HSV cells in a fibrillar network similar to that observed on normal cells (Fig. 8b and Refs. 11, 14, 19). Furthermore in addition to its effects on cell shape described above, addition of LETS protein to transformed cells leads to the appearance of actin cables within the cells (Fig. 8d) whereas these are absent from control NIL8-HSV cells (Fig. 8c).

All these effects of LETS protein on transformed cells are blocked or reversed by treatment with low levels of trypsin, such as cleave LETS protein, and are unaffected by chondroitinase ABC or hyaluronidase (11). Furthermore, preincubation of the LETS protein preparation with specific antisera to LETS protein blocks the effect (11).

These results suggest that several of the properties of transformed cells (reduced adhesion of cells, morphology and absence of actin cables) are a result of their low levels of LETS protein and that these properties can be altered towards normal by addition of LETS protein from normal cells or their conditioned medium.

DISCUSSION

The data presented indicate that both normal and transformed cells release LETS protein into the culture medium. Vaheri and Ruoslahti (20) have reported the detection of material immunologically cross-reactive with LETS protein in the medium of normal

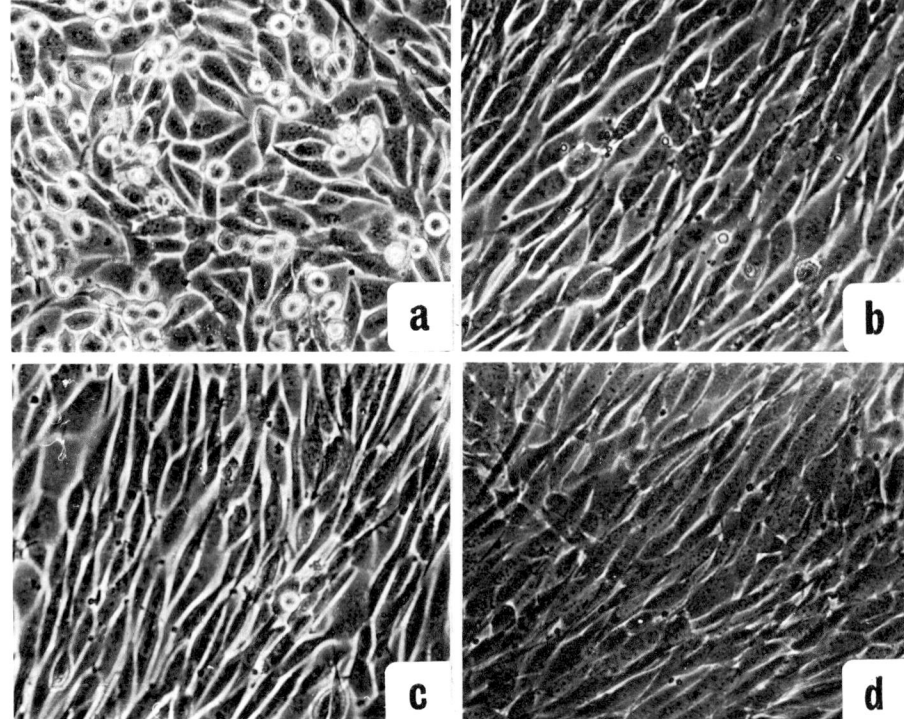

Fig. 7. Effects of purified LETS protein on transformed cells. Two-day-old cultures of NIL8-HSV cells received additions as below and were photographed 24 h later. a) Buffer alone. b) 30 μg/ml of LETS protein purified from conditioned medium of NIL8 cells as described (11). c) 40 μg/ml of LETS protein purified from conditioned medium and further fractionated on Sephadex G-200. d) 100 μg/ml of LETS protein extracted with urea from chick embryo fibroblasts (12). Panel (d) was from a separate experiment from (a)–(c). Phase contrast, approximate magnification, 200 ×.

and transformed cells. The present results show that both cell types secrete apparently intact LETS protein which is capable of binding to cells. The transformed cells studied have markedly reduced quantities of LETS protein in their conditioned medium as well as on their surfaces, when compared with normal cells. There is no evidence that this is due to degradation of material secreted into the medium, since added NIL8 conditioned medium is not degraded and purified LETS protein added to the cells binds to them and alters many of their properties towards those of normal cells.

Studies of the biosynthesis of LETS protein show a reduced rate of synthesis in the transformed cells. The reduction in synthesis rate observed, while considerable, was not as great as that in the levels of LETS protein on the surface or in the medium of transformed cells. This discrepancy suggests that while reduced synthesis of LETS protein contributes to the reduction in surface levels of LETS protein in the transformed cells, it is not the complete explanation. Increased turnover (5, 15) into the medium also appears to contribute, although the reason for this is as yet unclear. Possibilities include synthesis of an altered form of LETS protein by the transformed cells or alteration in other components at the transformed cell surface which affect retention of LETS.

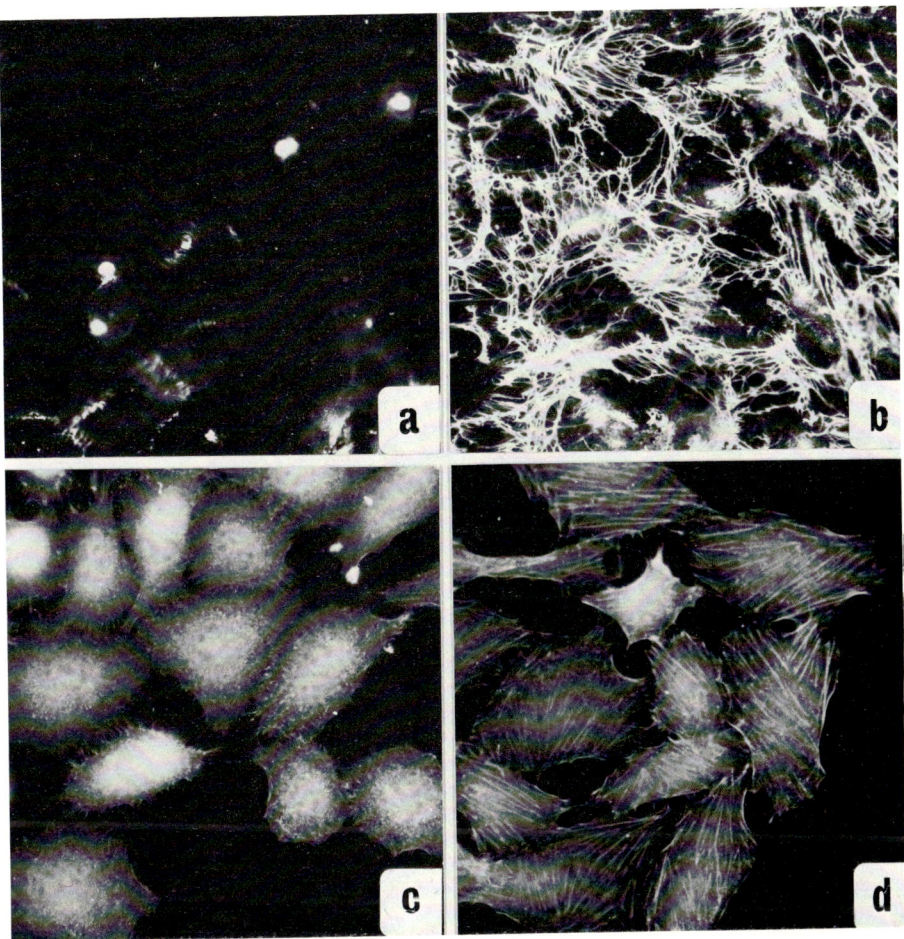

Fig. 8. Immunofluorescent staining of NIL8-HSV cells treated or not with 100 μg/ml LETS protein from chicken cells and fixed 24 h later. a) No LETS protein added, stained for LETS protein. b) LETS protein added, stained for LETS protein. c) No LETS protein added, stained for actin. d) LETS protein added, stained for actin. Bar represents 25 μm.

The results presented also raise the question of the normal route of processing of LETS protein. Is it internal → surface → medium or internal → medium → surface? Evidence is available for exchange in both directions between cell surface and medium. It was reported previously that prelabeled LETS protein falls off into the medium (4, 5, 15), and it is reported here and elsewhere (6, 11, 12) that LETS protein added to the culture medium binds to cells. It is possible that an equilibrium exists between the 2 compartments.

Another possibility raised by the experiments reported here concerns the effect of protein synthesis inhibitors on the transformed phenotype. Ash et al (21) have reported that cycloheximide or puromycin treatment of transformed NRK cells leads to reversion of their morphology and of their arrangement of myosin filaments towards normal. We have shown that these same inhibitors lead to the reattachment of LETS protein onto the surfaces of NIL8-HSV cells. Furthermore, the addition of exogenous LETS protein leads to reversion of morphology and of arrangement of actin towards normal. The parallels suggest that the 2 sets of observations may be related.

It is clear that the presence of LETS protein in the culture medium has profound effects on cellular properties and it seems likely that the absence of this protein from transformed cells and their medium is involved in their altered behavior in vitro.

ACKNOWLEDGMENTS

This research was supported by grants R01-CA-17007 and P01-CA-140-051, from the National Institutes of Health. V.M. was supported by a travelling fellowship from the Medical Research Council (United Kingdom).

NOTE ADDED IN PROOF

After submission of this manuscript, it was reported by Olden and Yamada (1977, Cell II, 957–969) that transformed chicken cells show reduced synthesis of LETS protein but that surface levels of this protein were reduced even further.

REFERENCES

1. Hynes RO: BBA Reviews on Cancer 458:73, 1976.
2. Hynes RO: Proc Natl Acad Sci USA 70:3170, 1973.
3. Hynes RO, Bye JM: Cell 3:113, 1974.
4. Hynes RO, Wyke JA, Bye JM, Humphryes KC, Pearlstein ES: In Reich E, Shaw E, Rifkin DB (eds): "Proteases and Biological Control." New York: Cold Spring Harbor Laboratory, 1975, pp 931–944.
5. Robbins PW, Wickus GG, Branton PE, Gaffney BJ, Hirschberg CB, Fuchs P, Blumberg PM: Cold Spring Harbor Symp Quant Biol 39:1173, 1974.
6. Yamada KM, Weston JA: Cell 5:75, 1975.
7. Goldberg B, Sherr CJ: Proc Natl Acad Sci USA 70:361, 1973.
8. Zavada J, Macpherson IA: Nature (London) 225:24, 1970.
9. Laemmli UK: Nature (London) 227:680, 1970.
10. Bonner WM, Laskey RA: Eur J Biochem 46:83, 1974.
11. Ali IU, Mautner VM, Lanza RP, Hynes RO: Cell 11:115, 1977.
12. Yamada KM, Yamada SS, Pastan I: Proc Natl Acad Sci USA 73:1217, 1976.
13. Lazarides E, Weber K: Proc Natl Acad Sci USA 71:2268, 1974.
14. Mautner VM, Hynes RO: J Cell Biol (in press).
15. Hynes RO, Wyke JA: Virology 64:492, 1975.
16. Kuusela P, Ruoslahti E, Vaheri A: Biochim Biophys Acta 379:295, 1975.
17. Critchley DR, Wyke JA, Hynes RO: Biochim Biophys Acta 436:335, 1976.
18. Hynes RO, Destree AT, Mautner VM: In Marchesi VT (ed): "Membranes and Neoplasia: New Approaches and Strategies." New York: Alan R Liss, Inc, 1976, pp 189–201.
19. Wartiovaara J, Linder E, Ruoslahti E, Vaheri A: J Exp Med 140:1522, 1975.
20. Vaheri A, Ruoslahti E: J Exp Med 142:530, 1975.
21. Ash JF, Vogt PK, Singer SJ: Proc Natl Acad Sci USA 73:3603, 1976.

The Turnover of a Tissue Specific Cell Surface Ligand Which Inhibits Lectin Induced Capping

James McDonough and Jack Lilien

Department of Zoology, University of Wisconsin, Madison, Wisconsin 53706

Ten-day-old embryonic chick neural retina release into the environment glycoprotein ligands which bind to homologous cells, inhibiting the lectin-induced redistribution of cell surface receptors. Material with identical activity is released from trypsin-dissociated neural retina cells that are allowed to repair in culture for 2 h and are then transferred to fresh medium. Release of ligand is inhibited by cytosine arabinoside, hydroxyurea, UDP, and EDTA, and is potentiated by $MnCl_2$. These data suggest that a glycosyltransferase reaction plays a critical role in the turnover of the cell surface ligand. Reactivation of enzymatically deglycosylated ligand solutions by intact cells provides further support for this hypothesis.

Release of ligand is also accompanied by a loss of the agglutinability of the cells by a tissue-specific component which accumulates in monolayer conditioned medium. Conditions which inhibit release maintain maximal agglutinability suggesting similar mechanisms mediate both processes.

Key words: capping of surface receptors, adhesive ligand, glycosyltransferase

Cell surface proteins accumulate in cell and tissue culture media, presumably due to the continual "shedding" or turnover of the cell surface (1). Our laboratory has been studying such components released from embryonic chick neural retina and cerebral lobe cell and tissue cultures. Our observations strongly suggest that among the components released are tissue specific ligands which can participate in the formation of intercellular adhesions (2, 3) and which inhibit the redistribution of a variety of cell surface receptors into "caps" on single cells freshly prepared by trypsinization (4).

While there has been a great deal of effort by our own laboratory and others (5, 6) at purification and characterization of the role of such "shed" components in intercellular adhesion, little or no information is available on the mechanism mediating release into the environment. To examine this process we have used trypsin-prepared single cells which were allowed to repair in culture; such cells have lost the ability to cap Concanavalin A

James McDonough is presently with the Department of Psychiatry, University of California, San Diego, California 92037.

Received March 23, 1977; accepted June 10, 1977.

© 1977 Alan R. Liss, Inc., 150 Fifth Avenue, New York, NY 10011

(Con A) receptors (7). Capping ability is regained following transfer of the cells to fresh medium containing cycloheximide; furthermore, activity which inhibits capping on freshly dissociated cells is concomitantly released into the medium. Quantitative evaluation of capping ability of cells and the amount of released ligand together with the effects of various chemicals suggest that release is accomplished by addition of a terminal sugar to the ligand via a glycosyltransferase reaction at the cell surface.

Repair of neural retina cells in culture is also accompanied by an increase in their agglutinability by a tissue-type specific component which accumulates in serum-free monolayer cultures (3). The same conditions which prevent release of ligand and the regaining of capping ability also prevent loss of agglutinability. Therefore, either the same component mediates both processes, or the various components are turned over by similar mechanisms.

MATERIALS AND METHODS

Cell Preparation and Repair

Trypsin dissociated 10-day-old chick neural retina cells were prepared as described previously (7). 2×10^7 cells were allowed to repair in still culture at $37°C$ for 2 h in 60-mm Falcon bacteriological dishes in 4 ml of Eagle's basal medium containing an additional 2 mg/ml glucose, 2% nonessential amino acids, 1 mM glutamine, and 50 µg/ml gentamycin (Schering Diagnostics, Union, New Jersey) under an atmosphere of 10% CO_2 in air.

Assay for Redistribution of Con A Receptors

Cells in suspension were assayed by methods previously described (4, 7) with the following modifications: 0.1 ml of cells (2.5×10^7 cells/ml) suspended in HBSG-CH (Hepes-buffered saline, pH 7.4, with 1 mg/ml glucose and 5 µg/ml cycloheximide) was added to 0.05 ml of FITC-Con A (fluorescent labeled Con A, Miles Laboratories, 100 µg/ml in HBSG) and 0.4 ml of HBSG containing various additives. Following 10 min at $0°C$, the samples were incubated for 60 min at $37°C$ with intermittent shaking, pelleted for 5 min at $200 \times g$ at $5°C$, and resuspended in 0.4 ml of 2% glutaraldehyde in 0.02 M Na_3PO_4 (pH 7.4) at $5°C$. The proportion of the cell populations exhibiting cap fluorescence was determined using a Zeiss Universal microscope modified for epifluorescence.

Assay for Activity of Released Material

To collect the released material (RM) 10^8 cells repaired as above were resuspended in 1 ml of HBSG-CH and incubated for 10 min at $37°C$ (release period). The cells were then pelleted at $200 \times g$ at $5°C$ for 10 min. The supernatant was collected and recentrifuged for 30 min at $10,000 \times g$. The high speed supernatant was then dialyzed overnight at $5°C$ against 0.15 M NaCl–0.01 M Na_3PO_4 (pH 7.4). Protein concentration in the dialyzed RM was determined by the method of Lowry et al. (8) using crystalline bovine serum albumin as a standard. Capping inhibition activity was determined by incubating various concentrations of RM in 0.5 ml of HBSG-CH containing freshly dissociated neural retina cells (5×10^6 cells/ml) and FITC-Con A (5 µg/ml). Incubation conditions were as described above. Released material was considered active if it inhibited capping of Con A receptors by greater than 50% at a concentration of less than 30 µg/ml. Retina tissue culture supernatants inhibit capping by greater than 50% at a concentration of 5 µg/ml and maximally inhibit capping (80% inhibition) at a concentration of 25 µg/ml (see McDonough and Lilien, Ref. 4).

Preparation of Ligand and Deglycosylated Ligand

Ligand which inhibits the induced redistribution of Con A receptors was collected from tissue culture medium as previously described (4). Ligand-containing solutions were terminally deglycosylated by incubating 7 ml of the culture supernatant solution with 3 ml of 0.15 M sodium acetate (pH 4.2) containing 0.033 Units/ml of purified β-N-acetyl-hexosaminidase (Miles Laboratories, Turbo cornutus) for 30 min at 30°C, boiling for 10 min, and dialyzing overnight at 5°C against 0.01 M Na_3PO_4–0.15 M NaCl (pH 7.4). Boiled control preparations lacking enzyme were carried through the same procedure and retained greater than 95% of initial activity.

Fixed Cell Agglutination Assay (see Ref. 3)

Cells incubated at reduced temperature or in the presence of drugs during the release period were fixed in 2% glutaraldehyde (Ladd, EM grade) in 10 mM Na_3PO_4-buffered saline (0.12 M) (pH 7.4) for 30 min at 4°C. The cell suspension was then pelleted at 1,000 × g for 5 min at 4°C, resuspended in 0.2 M glycine in 10 mM Na_3PO_4 (pH 7.4), and incubated for 10 min at 22°C. Following 4 washes with 0.15 M NaCl – 0.01 M Na_3PO_4 (pH 7.4), aliquots of 10^6 cells/3 ml/35-mm Falcon dish were incubated at 70 rpm for 24 h in serum-free medium conditioned by monolayers of 10-day-old neural retina cells. Agglutination was scored by Coulter Counter determination of the number of single cells remaining.

Monolayers were prepared by aliquoting 2×10^7 freshly dissociated cells to 35-mm Falcon plastic tissue culture dishes in 4.0 ml of Eagle's basal medium containing an additional 2 mg/ml glucose, 2% nonessential amino acids, 1 mM glutamine, and 50 μg/ml gentamycin. After 24 h the medium was collected, centrifuged at 10,000 rpm for 30 min and made 2 mM with phenyl methyl sulfonyl fluoride in 2-propanol. This additive prevents the rapid loss of agglutination mediating activity (3). It should be stressed that this activity is distinct from the tissue-specific ligand which accumulates in organ culture conditioned medium previously described by our laboratory (11).

RESULTS

I. Characterization of the Release Process

Embryonic chick neural retina cells prepared by trypsinization are able to redistribute Con A receptors into caps (7). With increased time of repair in culture such cells progressively lose capping ability and by 2 h are unable to cap Con A receptors (7). When such cells are transferred to fresh medium containing cycloheximide (5 μg/ml) for as little as 10 min they regain completely their initial capacity to cap Con A receptors. That this reacquisition of capping ability is due to the release of surface associated macromolecules is demonstrated in 2 ways: 1) culture medium from the 10-min incubation period is active in inhibiting capping of Con A receptors on freshly dissociated neural retina cells, while medium collected from freshly trypsinized cells subjected to a similar 10-min incubation is inactive as is medium from repaired cells subjected to a second 10-min incubation period. 2) Treatments which prevent the reacquisition of capping ability also prevent the accumulation of capping inhibitory activity in the supernatant medium (see below).

We have established that there exists an intercellular pool of the capping inhibitory ligand which is mobilized to the cell surface in the presence of cycloheximide (8). In addition, release of surface-bound ligand does not occur until the entire pool has been mobilized to the surface (8). Thus the amount of surface ligand is not affected by the presence of cycloheximide during the 2-h repair period. The presence of cycloheximide during the 10-min release period prevents further synthesis of surface ligands. Such synthesis reduces the capacity of the cells to cap Con A receptors but allows release of capping inhibitory material, altering the one to one relationship between capping ability and activity of the released material.

To further characterize the release process, various drugs, cations, and altered temperature were assayed for their effect during the 10-min release period following a 2-h repair period in the presence of cycloheximide. Table I summarizes these results. The data are normalized relative to the extent of capping recorded following a release period of 10 min at 37°C in medium containing only HBSG-CH. Low temperatures, cytosine arabinoside, and hydroxyurea all inhibit release while treatment with either cytochalasin B or puromycin has no effect. The effect seen with cholchicine is not due to inhibition of release, since colchicine inhibits capping and its effect is partially irreversible (7). Cells incubated with colchicine either during the repair period only or during the release period have a similar degree of inhibition of capping following the release period incubation. In all cases where release is inhibited no capping inhibitory activity is found in the supernatant medium.

These data suggest that a process which is inhibited by low temperatures, cytosine arabinoside, and hydroxyurea mediates release of the cell surface ligand. It has been shown that both cytosine arabinoside and hydroxyurea can inhibit the activity of glycosyltransferases (10). Experiments were therefore undertaken to test the hypothesis that glycosyltransferase activity is required for release of the cell surface ligand.

As above, cells repaired for 2 h in medium containing cycloheximide were harvested, washed, and incubated for 5 or 10 min at 37°C in HBSG containing various additives. After washing, the cells were assayed for the extent of capping. Since release appears to be maximal in 10 min it was necessary to reduce the release period to 5 min in order to test the effects of cations, as certain of them might be expected to accelerate the rate of a glycosyltransferase reaction. As shown in Fig. 1, neither calcium nor magnesium when present during release has any effect on the extent of capping after the release period. However, manganese dramatically increases the extent of capping, its effect being maximal at a concentration of approximately 5 mM.

Other compounds which might be expected to inhibit glycosyl transferase activity were tested during 10-min release periods. Ethylenediaminetetraacetic acid (EDTA, 0.005%) completely prevents release. Manganese ion at concentrations above 10 mM can overcome the inhibition caused by EDTA and stimulate activity while calcium and magnesium ions restore activity to the basal level.

Since nucleotides have been reported to affect both soluble (11) and membrane-bound (12) glycosyltransferases they were also tested for their effect on release. GDP, GMP, UTP, UDP, UMP, ADP, CMP, cyclic AMP, and dibutyryl cyclic AMP all inhibit release (Table II). However, UDP is approximately 1,000 times more effective in inhibiting the release process than GDP and ADP and approximately 10,000 times more effective than the other nucleotides tested.

These data support the contention that glycosyltransferase activity catalyzes the release of cell surface ligands into the surrounding medium. If release is mediated through glycosylation at the cell surface, repaired cells that have been stripped of ligand should retain the transferase at their surface, where it should be directly demonstrable. While ligand released from repaired cells inhibits capping of freshly prepared cells, it has little

TABLE I. Effects of Additives on Release of the Cell Surface Ligand

Treatment during release period[a]	% Caps following release period
Cycloheximide (5 μg/ml, CH)	100
CH + colchicine (20 μg/ml)	65 ± 10
CH + cytochalasin B (40 μg/ml)	107 ± 9
CH + NaN$_3$ (2.5 mM)	80 ± 6
CH, 22°C	75
CH, 4°C	0
CH + cytosine arabinoside (1 mM)	4 ± 8
CH + cytosine arabinoside (0.25 mM)	79
CH + hydroxyurea (1.3 mM)	7 ± 5
CH + hydroxyurea (0.65 mM)	32
Puromycin (10 μg/ml)	93

[a]The release period incubation was carried out for 10 min at 37°C except as indicated. Following the release period the cells were washed with ice-cold HBSG-CH and assayed for their extent of capping at 37°C as described in Methods.

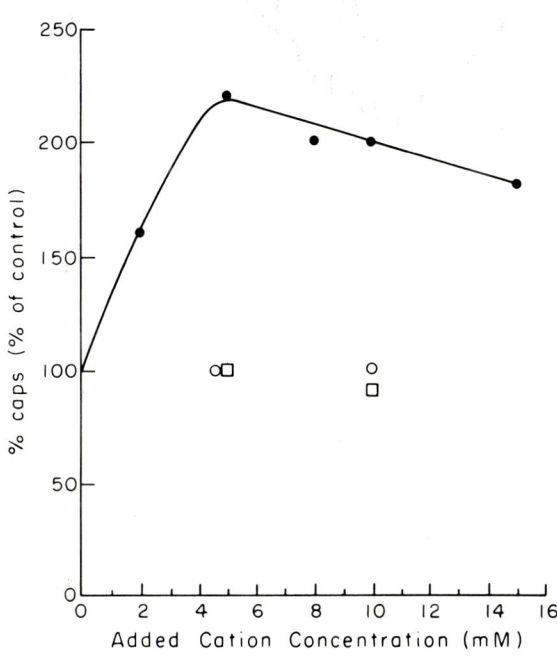

Fig. 1. Effects of different cations on release of cell surface material from 10-day-old chick neural retina cells. Cells were cultured for 2 h at 37°C in the presence of cycloheximide, harvested, washed with ice-cold HBSG/CH, then incubated for 5 min in HBSG-CH at 37°C containing various additions (10^8 cells/ml). The cells were then pelleted, washed with HBSG-CH, and assayed for their extent of capping of Con A receptors. Values shown are means from duplicates from 2 experiments. The control level of capping was that for a sample incubated with no additional ions. ●) Mn^{2+}; ○) Ca^{2+}, □) Mg^{2+}.

CSCBR: 385

TABLE II. Effects of Exogenously Added Nucleotides on Release of the Cell Surface Ligand

Nucleotide added	Effective concentration[a] (μM)
UDP	10^{-1}
UTP	10^3
UMP	10^3
GDP	10^2
GMP	10^3
CMP	$> 10^3$
ADP	10^2
Cyclic AMP	10^3
Dibutyryl cyclic AMP	10^3

[a]Effective concentration is considered to be the lowest concentration of nucleotide which inhibits by greater than 90% the release of cell surface material that restricts capping of Con A receptors. Nucleotides were added only to the release period incubation medium. Following the release period the cells were washed with ice-cold HBSG-CH and assayed for their extent of capping at 37°C. Results are expressed as the means of duplicate samples from 2–3 separate experiments.

effect on the repaired stripped cells. Conversely, ligand solutions treated with β-N-acetylhexosaminidase inhibit capping on repaired, stripped cells but not on freshly prepared cells (McDonough and Lilien, unpublished). Therefore, if there is a glycosyltransferase active during release of ligand, then repaired, stripped cells should catalyze reglycosylation of hexosaminidase-treated ligand, reactivating it with respect to freshly prepared cells and inactivating its ability to inhibit capping of repaired, stripped cells.

The results of these experiments are shown in Table III. In the absence of any added deglycosylated ligand the cells release into the medium material which can only inhibit capping minimally. Similarly when cells are incubated with deglycosylated ligand, MnCl$_2$, and 1 μM UDP (which inhibits release and therefore should inhibit the transferase) a small amount of activity is seen. However, when the cells are incubated with MnCl$_2$ and deglycosylated ligand alone the resultant medium contains capping inhibition activity for freshly trypsinized cells while that for 2-h cultured cells is much reduced. The presence of capping inhibition activity in the medium requires an interaction between deglycosylated ligand and some cell surface component(s). The supernatant medium from cells was unable to reactivate deglycosylated ligand in a 10-min incubation at 37°C. Moreover, treatment of the reactivated ligand with purified β-N-acetylhexosaminidase reduces capping inhibition activity to the initial minimal level (Table III). In addition, reactivated deglycosylated ligand solutions do not inhibit capping on freshly dissociated 10-day-old chick cerebral lobe cells. Thus the specificity of the reactivated ligand mimics that previously reported for tissue culture conditioned medium (4).

The observation that 2-h repaired and stripped cells will reactivate deglycosylated ligand prompted us to determine whether prior to the repair period such activity was also present. The ability of fresh, trypsin-prepared single cells to reactivate deglycosylated ligand was compared to 2-h repaired, stripped cells. The results are illustrated in Fig. 2; freshly prepared single cells lack the ability to reactivate deglycosylated ligand. Similarly, 2-h repaired, stripped cells subjected to a brief trypsinization also lack activity. Furthermore, such trypsin-treated repaired cells are no longer subject to inhibition of cap formation by deglycosylated ligand.

TABLE III. Reactivation of Deglycosylated Ligand*

	% Capping inhibition	
Activity of ligand preparations prior to incubation with cells	Freshly trypsinized cells	2 h cultured cells
Glycosylated ligand	68 ± 4	10
Deglycosylated ligand	30 ± 6	59
Reactivation of deglycosylated ligand		
Cells alone	26 ± 5	14
Cells + deglycosylated ligand (75% vol/vol) + 1 μM UDP	25 ± 15	–
Cells + deglycosylated ligand (75% vol/vol)	62 ± 7	23
Supernatant from cells following 10 min at 37°C + deglycosylated ligand (75% vol/vol) incubated 10 min at 37°C	30	–
Cells + deglycosylated ligand (75% vol/vol) 10 min at 37°C followed by treatment of supernatant with β-N-acetylhexosaminidase[a]	30 ± 8	55 ± 5

*Cells were incubated for 2 h at 37°C in medium containing cycloheximide (5 μg/ml), harvested, incubated for 10 min at 37°C in HBSG-CH and washed with cold HBSG. Cells were then incubated in HBSG-CH, 5 mM $MnCl_2$ containing the indicated additives in a total volume of 1 ml (10^8 cells/ml). After 10 min at 37°C the cells were pelleted at 5°C for 10 min at 200 × g and the supernate collected and assayed (at 80% vol/vol) for its Con A induced capping inhibition activity on 10-day-old neural retina cells either immediately following dissociation or after incubation for 2 h in medium containing cycloheximide followed by a release period. Values shown are mean ± standard deviation from 3–5 separate experiments.

[a] The release period medium was treated for 30 min at 30°C with purified β-N-acetylhexosaminidase (Miles Laboratories, Turbo cornutus), boiled for 10 min, and dialyzed overnight against 0.01 M Na_3PO_4-0.5 M NaCl (pH 7.4).

II. Agglutinability of Fixed Cells

During the 2-h repair period, cells coordinately lose the ability to cap lectin receptors and gain agglutinability (3). To characterize this relationship trypsin dispensed single cells were allowed to repair for 2 h in the presence of cycloheximide. The cells were then transferred to HBSG-CH containing the various affectors of release. Following the 10-min release period cells were fixed and tested for agglutinability by the tissue specific component of monolayer conditioned medium. Table IV shows that agglutinability is lost following a 10-min release period. Agglutinability is maintained under conditions which inhibit release and, therefore, prevent acquisition of capping ability.

DISCUSSION

Embryonic chick neural retina cells release into the environment a cell surface ligand which restricts the capping of lectin receptors. The data are consistent with the notion that release is mediated by the enzymatic addition of a terminal sugar to the ligand.

The ion specificity of release and inhibition of release by cytosine arabinoside, hydroxyurea, and UDP, are consistent with known properties of glycosyltransferases. However, it is the restoration by intact cells of biological activity of the enzymatically deglycosylated ligand which is most convincing. This interpretation of the mechanism of release is further strengthened by the result that after digestion with purified β-N-acetyl-

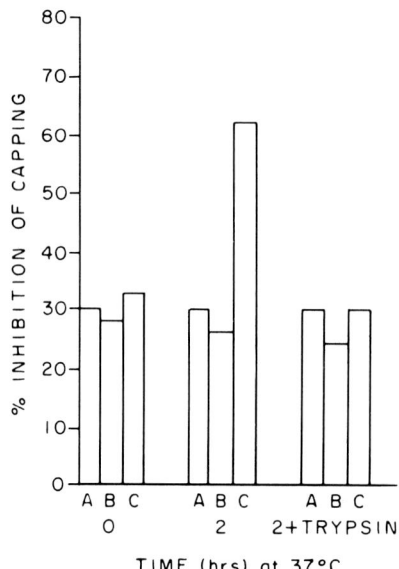

Fig. 2. Reactivation of deglycosylated ligand on interaction with cells. Freshly dissociated cells (0 time), and cells incubated for 2 h in the presence of cycloheximide (5 μg/ml) with or without subsequent trypsin treatment (4,000 National Formulatory Units/10^8 cells/ml, 3x crystallized trypsin, Miles Laboratories) were harvested, incubated in a total volume of 1 ml (10^8 cells/ml) of HBSG containing 5 mM $MnCl_2$ with or without deglycosylated ligand (75% vol/vol). After 10 min at 37°C the cells were pelleted at 5°C for 10 min at 200 × g and the supernate was collected and assayed (80% vol/vol) for its ability to inhibit Con A induced capping on freshly dissociated 10-day-old chick neural retina cells [1]. Results shown are the mean of duplicates from 2 experiments. A) DL solution alone; B) supernate from cells + 5 mM $MnCl_2$; C) supernate from cells + 5 mM $MnCl_2$ + DL.

TABLE IV. Agglutinability of Fixed Cells Following Release

Treatment during the 10-min release period	% of maximal[a] agglutination
No release	100
Cycloheximide (5 μg/ml, CH)	15
CH + NaN_3 (2.5 mM)	27
CH, 22°C	65
CH, 4°C	82
CH + cytosine arabinoside (2 mM)	70
CH + hydroxyurea (1.3 mM)	79
CH + UDP (1 μM)	71
CH + GDP (1 μM)	26

[a]Experimental variation is within 10% of the expressed values.

hexosaminidase the biologically reactivated ligand loses capping inhibition activity. Thus, with respect to its sensitivity to β-N-acetylhexosaminidase and its tissue specificity, the reactivated ligand is identical to the ligand originally released from the cell surface.

That UDP is the most effective nucleotide in inhibiting release of the retina ligand is consistent with the donor requirements for transferases catalyzing the addition of N-acetylgalactosamine, the previously identified terminal sugar of the retina ligand (4, 14). It is interesting that UDP is effective in inhibiting the glycosylation reaction at 0.1 μM, a concentration notably less than that observed for partially isolated or solubilized transferases (13). This extreme sensitivity may be due to the fact that the enzyme is in its native configuration at the cell surface. As far as we are aware the effect of nucleotides in other intact cell systems has not been examined so that no comparisons can be made.

In similar studies being conducted on cerebral lobe cells (Hermolin and Lilien, unpublished data), we have found that GDP is the most effective nucleotide in inhibiting release of the cerebral lobe ligand. The nucleotide specificity in this system is consistent with the donor requirements for mannosyltransferases and thus is also in agreement with our previous report that binding and inhibition of capping by the cerebral lobe ligand are α-mannosidase labile (14).

An obvious implication of these studies is that glycosyltransferase acts as a cell surface receptor for deglycosylated ligand, and that upon binding of deglycosylated ligand the ability of cells to cap lectin receptors is lost. This is consistent with release of ligand only under conditions which support catalysis and with reactivation of deglycosylated ligand by repaired, stripped cells. Further support for this hypothesis derives from the experiments showing that trypsinization of repaired, stripped cells destroys not only the capacity to reglycosylate but also sensitivity to inhibition of capping by deglycosylated ligand. However, fully glycosylated ligand both binds to fresh, trypsin-dispersed cells and inhibits capping among them. This observation suggests that there is a separate receptor for this form of ligand or that the receptor for deglycosylated ligand (transferase) has a trypsin-stable portion which recognizes fully glycosylated ligand.

In addition to ligand which inhibits lectin-induced cap formation, tissue culture conditioned media contains a component essential for the formation of intercellular adhesions (2, 3). That the 2 activities reside in the same molecule is suggested by the sensitivity of both capping inhibitory activity and agglutinability enhancing activity to β-N-acetylhexosaminidase. Furthermore, single cells acquire agglutination-mediating activity and lose capping ability concomitantly during repair (3). In addition our preliminary data indicate that both activities copurify. The data presented here are also indicative of an identity between the macromolecular species which mediate both processes. Agglutinability is lost following release and is retained under conditions which inhibit release.

By considering the data presented in the context of cell adhesion it is possible to integrate the notion that glycosyltransferases are involved in intercellular adhesion (15, 16) with the model previously proposed by our laboratory (2, 3). The model suggests that intercellular adhesions are mediated by 3 components: a cell surface receptor which interacts with the oligosaccharide moiety of the ligand and a third component which interacts with the polypeptide moiety of ligands on opposing cells to establish an adhesion. The data presented here suggest that the ligand receptor is a glycosyltransferase. While this model is substantially different from that proposed by Roseman (15) and Roth et al. (16) it does incorporate the virtue of known, highly specific enzyme-substrate interactions as partial determinants of adhesive specificity.

ACKNOWLEDGMENTS

This work was supported by National Science Foundation grant BMS 75-02824. J. M. was supported by Training Grant HD00409 from the NICHD.

REFERENCES

1. Doljanski F, Kapeller M: J Theor Biol 62:253, 1976.
2. Balsamo J, Lilien J: Nature (London) 251:522, 1974.
3. Lilien J, Rutz R: In Lash J, Burger M (eds): "Cell and Tissue Interactions." Soc of General Physiologists Symposium. New York: Raven Press, p 187, 1977.
4. McDonough J, Lilien J: Nature (London) 256:216, 1975.
5. Hausman RE, Moscona AA: Proc Natl Acad Sci USA 72:916, 1975.
6. Rutishauser N, Thiery J-P, Brackenbury R, Sela B-A, Edleman G: Proc Natl Acad Sci USA 73:577, 1976.
7. McDonough J, Lilien J: J Cell Sci 19:347, 1975.
8. McDonough J, Lilien J: J Cell Sci (In press).
9. Lowry OH, Rosebrough NJ, Farr AL, Randall RJ: J Biol Chem 193:265, 1951.
10. Hawtrey AO, Scott-Burden T, Robertson G: Nature (London) 252:58, 1974.
11. Ku GKW, Raghupathy E: Biochim Biophys Acta 313:277, 1973.
12. Ishibashi T, Atsuta T, Makita A: Biochim Biophys Acta 429:759, 1976.
13. Baker AP, Sawyer JL, Munro JR, Weiner GP, Hillegass LM: J Biol Chem 247:5173, 1972.
14. Balsamo J, Lilien J: Biochemistry 14:167, 1975
15. Roseman S: Chem Phys Lipids 5:270, 1970.
16. Roth S, McGuire EJ, Roseman S: J Cell Biol 51:536, 1971.

Glycoprotein Synthesis as a Function of Epithelial Cell Arrangement: Biosynthesis and Release of Glycoproteins by Human Breast and Prostate Cells in Organ Culture

Zoltán A. Tökés

Cell Membrane Laboratory, Los Angeles County/University of Southern California Cancer Center and Department of Biochemistry, University of Southern California School of Medicine, Los Angeles, California 90033

Gerald B. Dermer

Department of Pathology, Hospital of the Good Samaritan and University of Southern California School of Medicine, Los Angeles, California 90033

We demonstrate that a technique is available to investigate glycoprotein synthesis in organ cultures of human breast and prostate surgical specimens where the 3-dimensional epithelial cell arrangement remains intact. Malignant breast and prostate epithelium maintained their capacity to synthesize glycoproteins for at least 3 days as followed by the incorporation of [^{3}H]glucosamine into macromolecules. Over 70% of incorporation was by malignant cells as judged by autoradiography. Labeled glycoproteins were released into glandular lumina and consequently into the culture fluid. Sodium dodecyl sulfate-polyacrylamide gel electrophoresis revealed predominantly one group of macromolecules released with an apparent molecular weight of 48,000 ± 6,000 daltons. This glycoprotein was found in all of the breast specimens studied, which included 1 medullary, 1 infiltrating lobular, and 8 infiltrating duct carcinomas. The pattern was independent of the availability of estrogen receptors. A similar glycoprotein was also observed in the culture media from a Grade I and a Grade II well-differentiated infiltrating prostate carcinoma. Incorporation was below the level of detection in 4 of 6 cases of benign prostatic hyperplasia. A more complex pattern of labeled glycoproteins was found in the media of a Grade II and a Grade III poorly-differentiated prostate carcinoma. The established human mammary carcinoma cell line MCF-7 synthesized and released a similar 48,000 molecular weight glycoprotein but additional components with larger molecular weights were also released. An intriguing interpretation that 3-dimensional tissue integrity restricts some glycoprotein synthesis is discussed. Cells grown in 2-dimensional monolayers could escape from such a topographic restriction and express additional families of glycoproteins.

Key words: breast, prostate, carcinoma, glycoproteins, organ culture

Received for publication April 14, 1977; accepted July 29, 1977

© 1977 Alan R. Liss, Inc., 150 Fifth Avenue, New York, NY 10011

Human breast and prostate epithelial cells are arranged in vivo in 3-dimensional glandular structures. Glycoproteins are important for organ-specific cell aggregation, for cell-cell recognition (1), for cell surface antigenicity (2), and may play an important role in the arrangement of 3-dimensional cell structures (3). Therefore conditions must be established for studying their synthesis and turnover where the 3-dimensional arrangement of cells is maintained. In organ cultures tissue integrity is largely unaltered, therefore biosynthetic events can be followed in vitro, under conditions which resemble the in vivo environment.

Cancer cells derived from secretory cell types appear to retain secretory capacity, but to date no complete study is available on the molecular nature of glycoproteins secreted by benign hyperplastic prostatic epithelium and by various grades of malignant prostatic or breast epithelium. These glycoproteins are particularly important since they may be markers responsible for immune recognition (4) and in their released form or in combination with their corresponding antibodies (5) could block immune cytotoxicity. Mammary glands from pregnant mice maintained in organ culture have already been shown to be a good model system for studying the hormone-dependent biosynthesis and secretion of proteins (6, 7). Two recent studies, one autoradiographic (8) and the other biochemical (9), used organ cultures of human breast carcinoma to show that the malignant cells retained biosynthetic and secretory capacity. Glycoprotein synthesis by epithelial cells is under hormonal regulation (10) and it is probably influenced by the degree of cell differentiation and the 3-dimensional arrangement of cells as well. Although prostate and breast epithelial cells can be maintained in cell culture using fetal calf serum and various hormones (11, 12), extrapolation to an in vivo glycoprotein-synthesizing activity must be undertaken with caution.

We have investigated the feasibility of using human prostate and breast surgical specimens in short-term organ cultures. The present report describes the biochemical nature of [^{3}H]glucosamine-labeled material released in these organ cultures and from the established human breast carcinoma cell line MCF-7 maintained in a 2-dimensional monolayer culture.

MATERIALS AND METHODS

Organ Cultures

Tumor samples were obtained from 10 mastectomy and 11 prostatectomy specimens under sterile conditions within 15 min after completion of surgery. Adjacent tissue was fixed in formalin for diagnostic purposes. Light microscopy revealed that 8 of the breast tumors were infiltrating duct carcinomas, exhibiting varying degrees of differentiation. The other 2 were a medullary and an infiltrating lobular carcinoma. In 6 prostate specimens, cut surfaces revealed pale hyperplastic nodular areas, portions of which were selected for this study. Microscopic examination revealed nodular and fibroglandular hyperplasia without any evidence of carcinoma. In the other 5 prostate specimens, cut surfaces revealed firm, yellow carcinomatous areas, portions of which were selected. The microscopic diagnosis was Grade I, well-differentiated infiltrating adenocarcinoma in 2 of these cases; Grade II, moderately well-differentiated infiltration adenocarcinoma in 2 cases; and Grade III, poorly-differentiated infiltrating adenocarcinoma in 1 case.

Tissue samples were cut into approximately 1-mm cubes and placed on steel grids in 60 × 15-mm petri dishes (Falcon Plastics, Oxnard, California) containing 5 ml of media which reached to the upper surface of the explants. Usually 0.5 g of tumor was sliced into more than 50 cubes per incubation. The cultures were carried out at 37°C in a water-saturated atmosphere of 95% air:5% CO_2. Medium 199 (Flow Laboratories, Rockville, Maryland), supplemented with 125 mg/100 ml of glucose, 2 mM glutamine, 100 units/ml of penicillin and 100 μg/ml of streptomycin, was the basic medium. Twenty-five microcuries of D-[6-^{3}H]glucosamine hydrochloride (specific activity 10.1 Ci/mM, New England Nuclear Corporation, Boston, Massachusetts) was added to the media within 1 h of culture. During the first 24-h labeling period, the cultures were supplemented either with fetal bovine serum (Flow Laboratories) or with patient serum, obtained a day before surgery, to a final concentration of 10%. After 24 h, the labeled explants were removed from isotope-containing media, rinsed with basic media, and added back to petri dishes containing 5 ml of basic media supplemented with serum or bovine serum albumin (BSA) to a final concentration of 1%. The culture fluid containing labeled material was collected at various time intervals and centrifuged to remove cellular debris. Solid urea was added to the supernatant to 10 M final concentration and after 30 min at room temperature the urea-treated media was extensively dialyzed at 4°C for 2 days against 0.01% ammonium bicarbonate and lyophilized.

Estrogen Receptors

The presence of estrogen receptors in the breast tumors was determined commercially by Bioscience Laboratories (Van Nuys, California) using the dextran-coated charcoal assay.

Cell Culture

The established human mammary carcinoma cell line MCF-7 (12), mycoplasma-free, was kindly provided by Dr. John A. Sykes, California Hospital and Medical Center, Los Angeles. These cells were maintained in RPMI-1640 medium (Flow Laboratories) supplemented with 10% fetal bovine serum in monolayer cultures. Six million cells, from passages 189 to 193, in confluent monolayer cultures, were maintained for 24 and 48 h in the presence of 25 μCi of D-[6-^{3}H]glucosamine hydrochloride. Postlabel incubation was carried out for various lengths of time in isotope-free media supplemented with 160 μunit/ml of bovine pancreatic insulin (Sigma Chemical Company, St. Louis, Missouri). The cell-free supernatant was processed in the same manner as the organ culture samples.

Autoradiography

At the end of organ culture, several explants from 3 infiltrating duct carcinomas of the breast, the Grade III prostate carcinoma, and explants from 1 case of benign prostatic hyperplasia were taken for autoradiographic analysis to determine the distribution of radioactivity within the explants. The explants were fixed for 1 h in 2% glutaralydehyde in 0.2 M sodium cacodylate buffer (pH 7.3) and postfixed for 1 h in 1.0% osmium tetroxide in cacodylate buffer. They were then rapidly dehydrated in acetone and embedded in a polyester resin. One-micron-thick unstained sections of the embedded explants were coated with Kodak NTB 2 emulsion (Eastman Kodak Company, Rochester, New York) by dipping the slides on which the sections had been placed into melted emulsion diluted 1:1 with distilled water. The slides were stored at 4°C in lightproof boxes for 1 week and then developed. After development, the sections were stained through the emulsion with toluidine blue. A measure of the amount of label associated with each compartment within the sections was calculated by first estimating the percentage area over random sections of explants occupied by tumor cells and connective tissue. This was done essentially by the method of Whur (13, 14) and involved placing at random over photographic enlargements of entire sections a stencil containing a large number of randomly-placed small circles of the same diameter. The contents of 50 circles were then analyzed for each section by a point system. The average number of grains over each compartment was then determined on enlargements of micrographs taken at 1,000 × magnification. A measure of the amount of label associated with each compartment was given by multiplying the percentage area of the section occupied by each compartment by the average number of grains over that compartment.

Gel Electrophoresis

Portions of the urea-treated, dialyzed, and lyophilized media samples were subjected to 3 M urea-1% sodium dodecyl sulfate (SDS) electrophoresis, using 7.5% precast polyacrylamide gels (Bio-Rad, Richmond, California) (15). Molecular weight markers were heavy- and light-polypeptide chains from human immunoglobulins. After electrophoresis the gels were cut into 1- or 2-mm slices and the radioactivity within each slice was determined, after treatment with Beckman Tissue Solubilizer, using scintillation counting. The samples were counted for 10 min; the efficiency of counting was 33%.

Enzymatic Treatment

A portion of the cell-free culture medium, 2 mg dry protein equivalent, was incubated in 600 μl of 0.05% NH_4HCO_3 with 0.2 mg of trypsin, TPCK, or with 0.1 mg of chymotrypsin (Sigma Chemical Company) overnight at 37°C. A separate 2-mg sample was treated with 100 μg of pronase (Merck & Co., Inc., Rahway, New Jersey) in 0.1 M phosphate buffer, pH 7.4, and incubated for 1 h at 37°C. The digested samples were dialyzed against water and lyophilized prior to SDS-polyacrylamide gel electrophoresis.

RESULTS

Breast Epithelial Cells

In Fig. 1, the hematoxylin- and eosin-stained paraffin sections of 4 different breast epithelial cell arrangements are illustrated. The normal duct from a nonlactating breast has a single layer of epithelial cells, resting on a basement membrane (Fig. 1A). The epithelium consists of 2 or 3 rows of cells resting on a basement membrane (Fig. 1B) in a hyperplastic duct; however, normal cell morphology and glandular architecture is preserved. In an intraductal carcinoma (Fig. 1C), the duct is filled with many layers of malignant epithelial cells. In this condition the basement membrane remains intact and cells have not invaded the connective tissue stroma. A representative field of an infiltrating duct carcinoma is illustrated in Fig. 1D. The connective tissue stroma is filled with rows of single malignant cells which form small gland-like structures.

Fig. 1. Hematoxylin- and eosin-stained paraffin sections of 4 different breast epithelial cell arrangements (magnification 800 ×). A) Normal duct. A single layer of cells rests on a basement membrane.
B) Hyperplastic duct. Epithelium consists of 2 or 3 rows of cells resting on a basement membrane.
C) Intraductal carcinoma of the breast. Duct is filled with many layers of malignant epithelial cells. Thin line in center of the picture is intact basement membrane; cells have not invaded the connective tissue stroma. D) Infiltrating duct carcinoma. Stroma is filled with malignant cells. They are found in rows of single cells or form small gland-like structures.

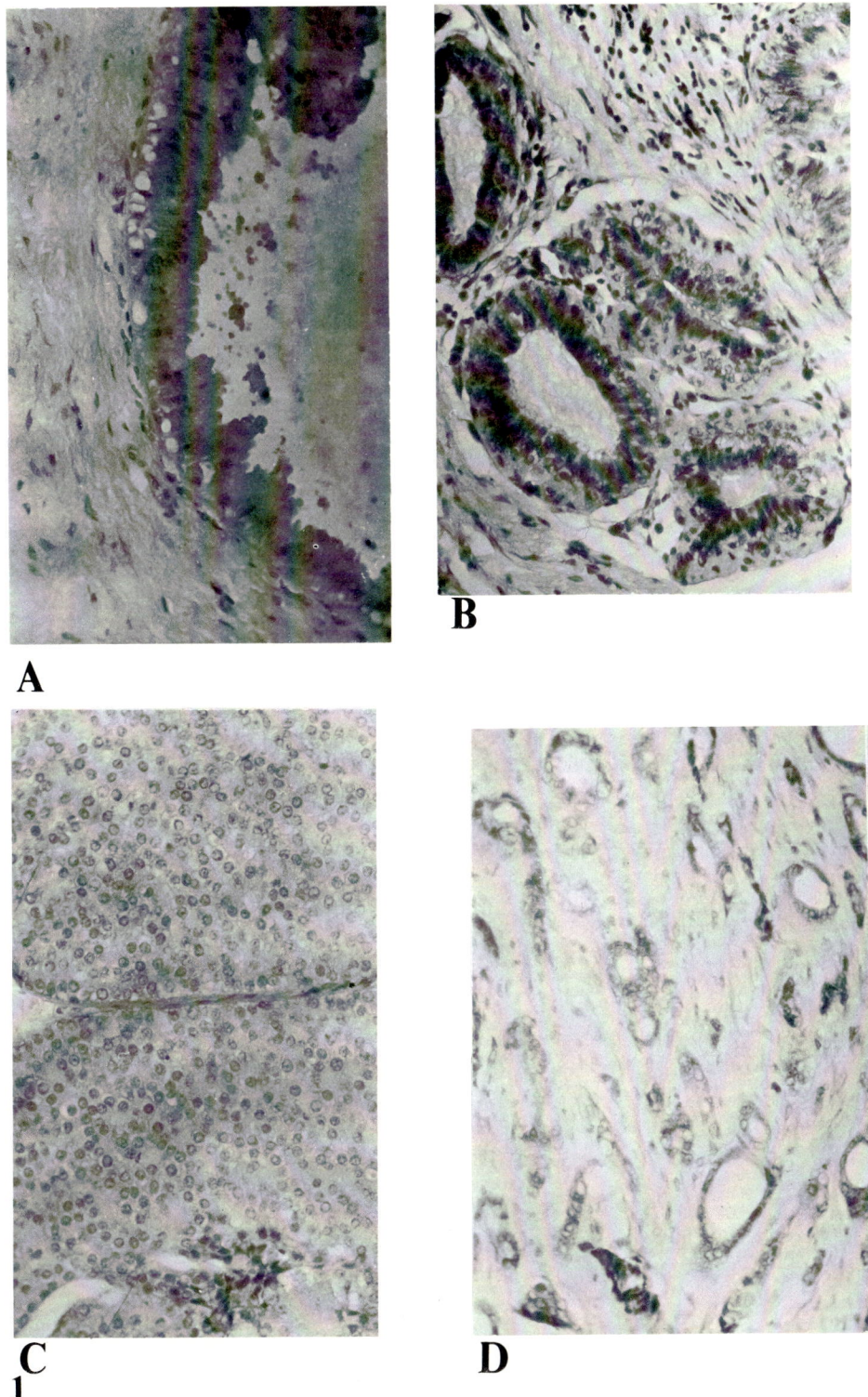

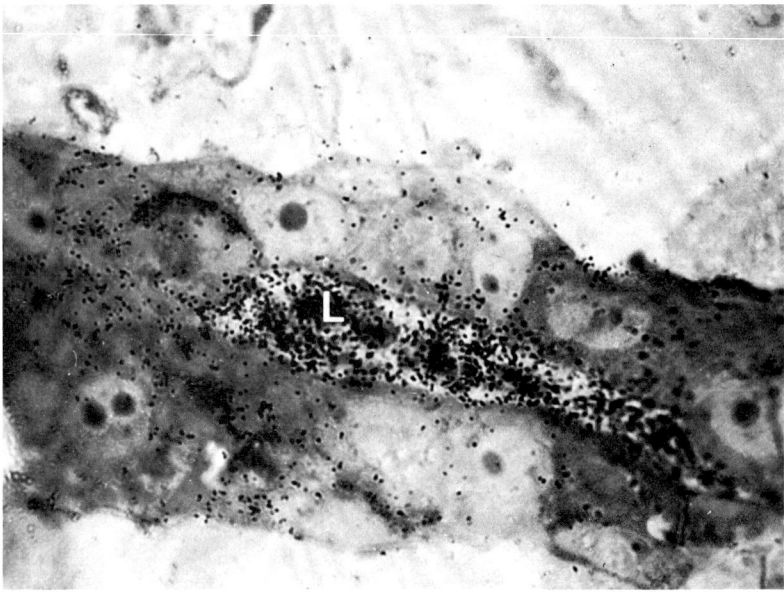

Fig. 2. Autoradiography of an infiltrating duct carcinoma of the breast. Most of the silver grains are found over the cellular elements and lumen (L) of a small malignant gland after 24-h incubation with [^{3}H]glucosamine followed by 48 h in nonisotopic media. Connective tissue stroma around gland contains few grains (enlarged from 1,000 × original magnification).

Explants from the 3 breast tumors chosen for autoradiography were found to contain well-maintained epithelial cells after organ culture for 3 days (24-h incubation with [^{3}H]glucosamine followed by 48 h in nonisotopic media). Most cells contained large nuclei with prominent nucleoli and were often arranged in gland-like structures. Isolated islands of neoplastic cells were also found within the connective tissue stroma. Autoradiography revealed significant amounts of radioactivity within the cellular and luminal components of malignant glands (Fig. 2). In the 3 tumors studied, an average of 25% of the area of sections was occupied by malignant cells and 70% of the total number of silver grains were associated with these cells and their glandular lumina. Thirty percent of the grains were found over noncellular connective tissue. Few fibroblasts and leukocytes were found within the explants of the 3 tumors and less than 3% of the total grains were found over them. These observations indicate that at least 70% of the label was incorporated by neoplastic breast epithelium.

Urea-treated and dialyzed cell-free organ culture media were subjected to SDS-polyacrylamide gel electrophoresis. In Fig. 3, we illustrate representative results from different incubations of surgical specimens from 5 breast cancer patients. These gels were run on standard 7.5% precast acrylamide gels and calibrated with immunoglobulin light and heavy chains. Each sample represents 5% of a culture supernatant. A striking similarity of macromolecular patterns was observed with all the samples studied to date. A predominant component with an apparent molecular weight of 48,000 ± 6,000 was released from all organ cultures. The exact resolution of the major peak is technically not feasible using the 1- or 2-mm slicing techniques. Minor variations in slice thickness may result in artifacts which resemble 2 or more apparent peaks (Graph E in Fig. 3).

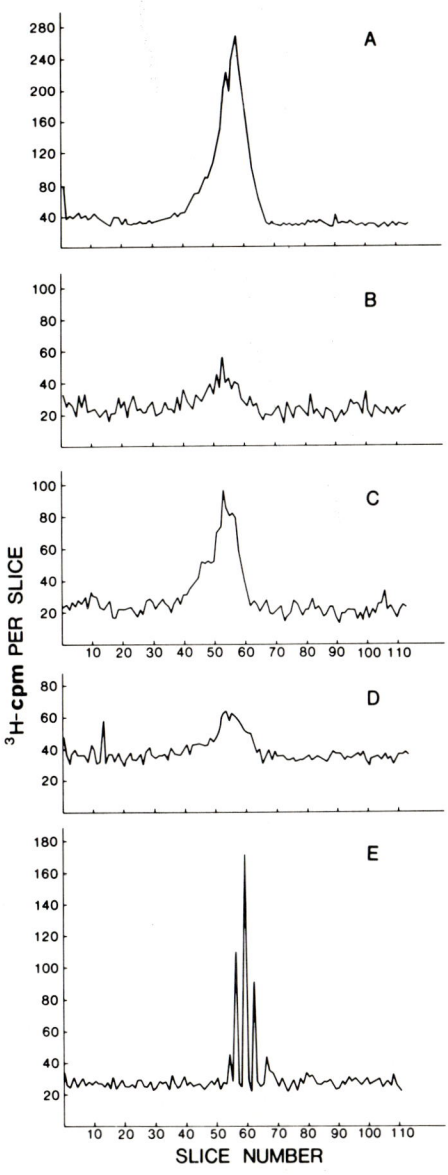

Fig. 3. SDS-polyacrylamide gel electrophoresis of dialyzed breast cancer organ culture media using standard precast 7.5% gels. Slices were 1 mm thick. The dye marker position was between slices 95 and 108. Immunoglobulin light chains migrated to slice positions 68 to 72, and heavy chains to positions 43 to 47. Graph A: Medullary carcinoma. 24 hour post-label incubation. Graph B: Infiltrating duct carcinoma. Second 24-hour post-lable incubation. Estrogen receptor negative. Graph C: Infiltrating duct carcinoma. 48-hour post-label incubation. Graph D: Infiltrating lobular carcinoma. Third 24-hour post-label incubation. Estrogen receptor-positive. Graph E: Infiltrating duct carcinoma. 48-hour post-label incubation. Estrogen receptor-positive.

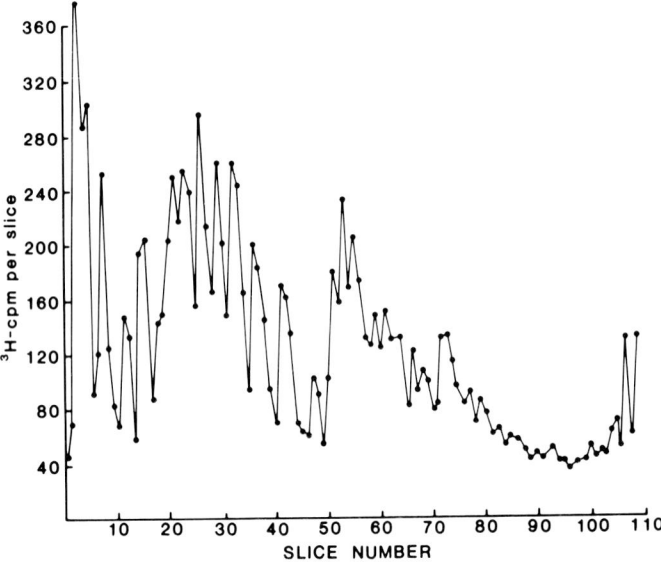

Fig. 4. SDS-polyacrylamide gel electrophoresis of MCF-7 cell culture media from 48-h postincubation. Five to six million cells were grown in a 2-dimensional monolayer culture to confluency. Five percent equivalent of urea-treated, dialyzed, and lyophilized supernatant was applied to the gel. Slices were 1 mm thick. Calibration was the same as in Fig. 3.

An established human mammary carcinoma cell line, MCF-7, was grown in culture to a confluent monolayer in order to have maximum cell-cell contact in this 2-dimensional array of cells. [^{3}H] Glucosamine was introduced for 24 h, and postlabel incubation was carried out in serum-free media. The cell-free culture media from the 48-h postlabel period was treated with urea, dialyzed, and lyophilized. A 5% equivalent was analyzed by SDS-polyacrylamide gel electrophoresis. The results illustrated in Fig. 4 indicate a greater complexity of labeled glycoproteins than observed in any of our organ cultures of human breast carcinoma, even though a family of glycoproteins with apparent molecular weights of 48,000 daltons was also released from this cell line.

Prostate Epithelial Cells

Figure 5 illustrates 3 different prostatic cell arrangements as seen in hematoxylin- and eosin-stained paraffin sections. A single layer of columnar epithelial cells rests on a basement membrane (Fig. 5A) in a normal prostate gland. Normal glandular architecture is preserved in benign hyperplasia (Fig. 5B), but the columnar epithelium is 2 or 3 layers thick. In a Grade III prostatic carcinoma, the stroma is filled with malignant cells which occasionally form small gland-like structures (Fig. 5C).

Fig. 5. Hematoxylin- and eosin-stained paraffin sections of 3 different prostatic cell arrangements (magnification 800 ×). A) Normal prostatic gland. A single layer of columnar epithelial cells rests on a basement membrane. B) Benigh prostatic hyperplasia. Columnar epithelium is 2 or 3 layers thick. Normal architecture of gland is preserved. C) Grade III prostatic carcinoma. Stroma is filled with malignant cells which occasionally form small glands.

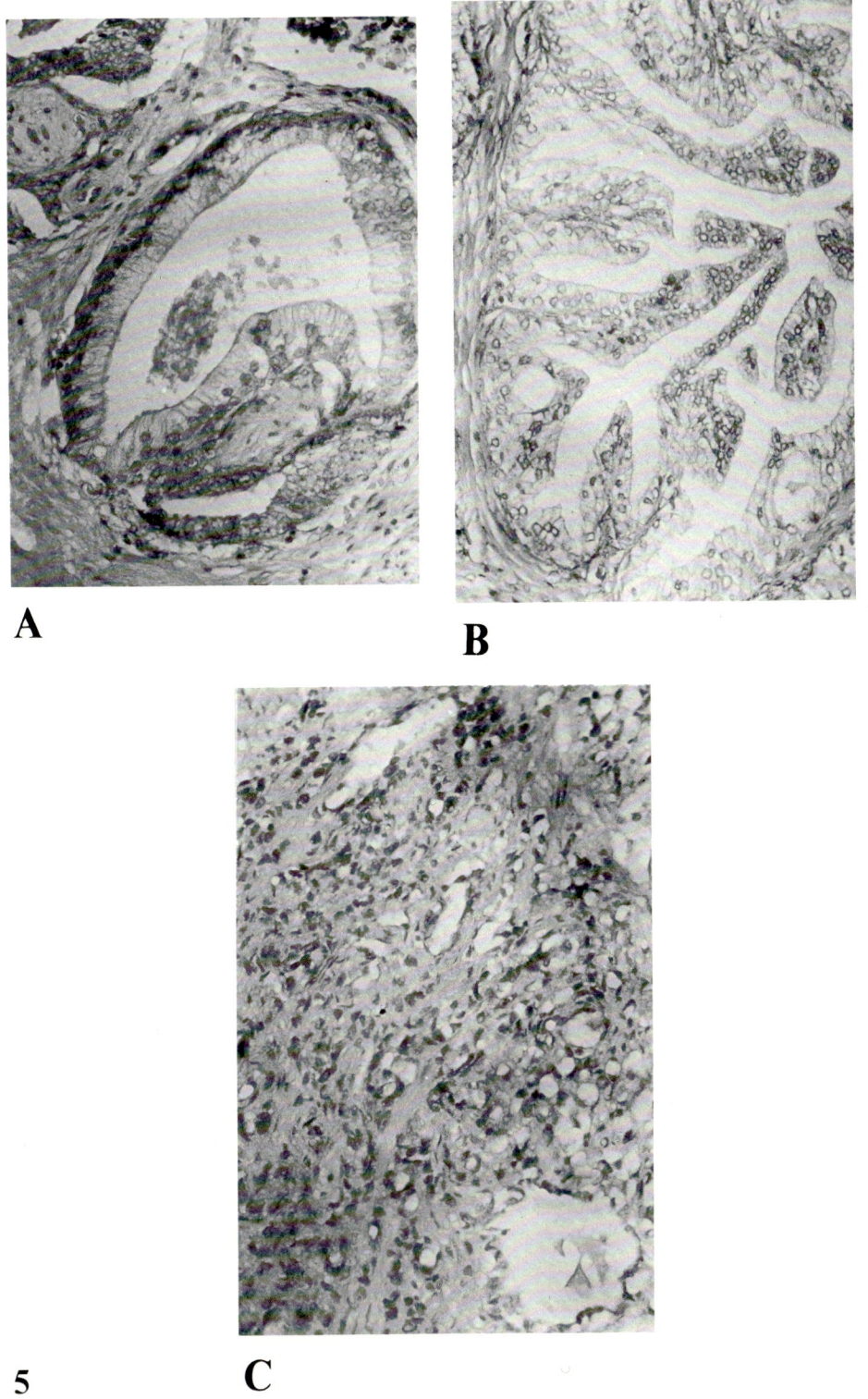

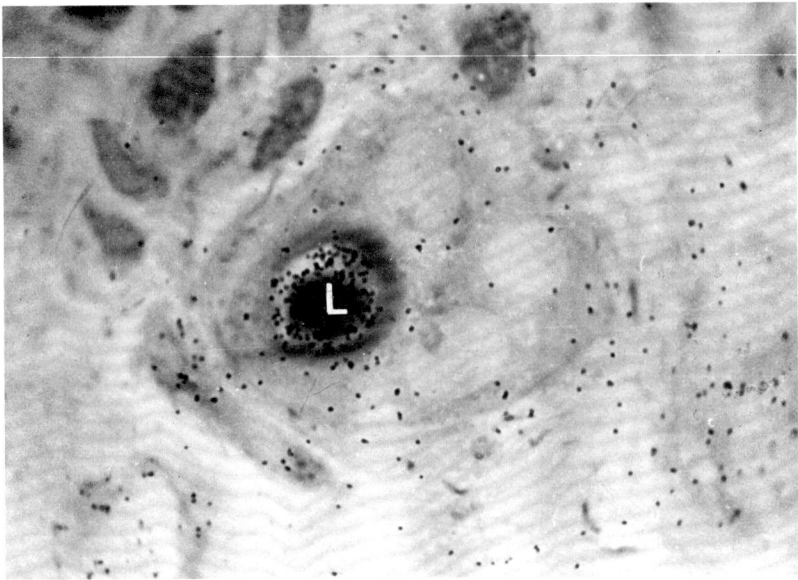

Fig. 6. Autoradiography of a poorly-differentiated Grade III prostate carcinoma. Most of the silver grains are found over the lumen (L) of a small malignant gland. Incubation as in Fig. 2 (enlarged from 1,000 × original magnification).

Well-maintained glandular and stromal structures were found in the hyperplastic explants after organ culture for 3 days, which included 24-h incubation with [^{3}H]glucosamine followed by 48 h in nonisotopic media. Autoradiography revealed significant amounts of radioactivity within the cellular and luminal components of the glands. After organ culture for 3 days, explants from the Grade III, poorly-differentiated carcinoma also contained well-maintained epithelial and stromal components. These explants were characterized by the presence of minute neoplastic acini haphazardly arranged, and neoplastic cells without any distinct glandular formation within the connective tissue stroma. Autoradiography revealed particularly intense reactions over the lumens of the small neoplastic glands (Fig. 6). Quantitation of grains revealed that at least 62% of the incorporation took place by neoplastic epithelium.

Organ culture media from prostate specimens were analyzed by SDS-polyacrylamide gel electrophoresis. No significant release of [^{3}H]glucosamine-labeled macromolecules was found with explants from 4 cases of benign hyperplasia and with one Grade I well-differentiated infiltrating carcinoma. One major family of glycoproteins with an apparent molecular weight of 48,000 ± 6,000 daltons was released by a Grade I well-differentiated and a Grade II moderately well-differentiated infiltrating carcinoma. One fibroglandular and one nodular hyperplasia sample also released similar glycoproteins. Characteristic patterns are illustrated in Fig. 7. A more complicated pattern of labeled glycoproteins was observed in the organ culture media of a Grade II moderately-differentiated and a Grade III poorly-differentiated carcinoma (Fig. 8). In addition to the 48,000 molecular weight glycoprotein, a high-molecular-weight component was observed near the top of the gels and there were additional peaks with an apparent molecular weight range of 30,000–120,000.

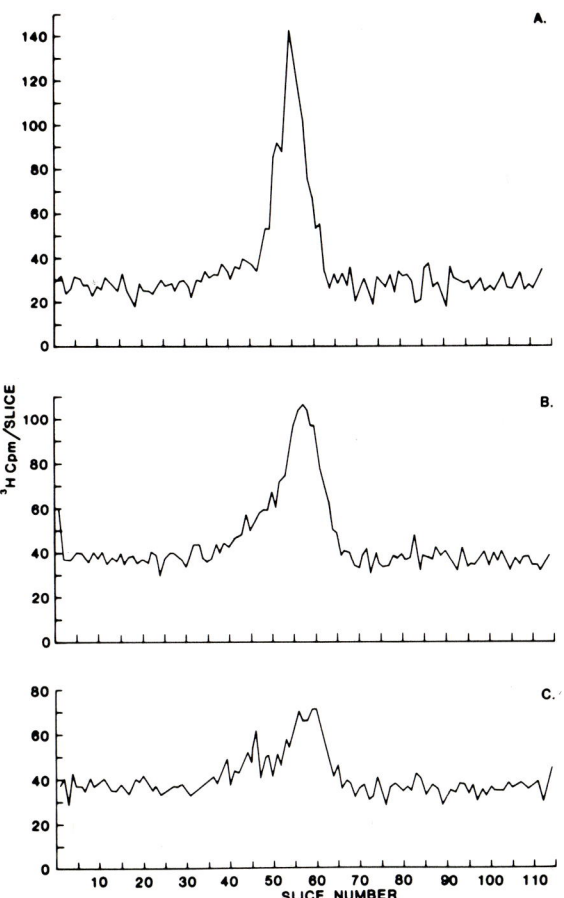

Fig. 7. Distribution of incorporated [^{3}H]glucosamine from prostate organ-culture media in SDS-polyacrylamide gel slices. Slices were 1 mm thick. Conditions and calibrations were the same as in Fig. 3. Graph A: Grade II, moderately well differentiated infiltrating adenocarcinoma. Graph B: Nodular hyperplasia. Graph C: Grade I, well differentiated infiltrating adenocarcinoma.

The dialyzed and lyophilized medium from explants of one case of benign prostatic hyperplasia was treated with trypsin, chymotrypsin, or pronase (Fig. 9). The [^{3}H]glucosamine-labeled glycoproteins were completely digested to fragments smaller than 5,000–10,000 molecular weight, since all of the detectable radioactivity was abolished in SDS gels. Trypsin digestion decreased the apparent molecular weight by approximately 15–20%. All of the SDS-polyacrylamide gel patterns of untreated glycoproteins remained the same when reducing agents were omitted from the sample buffer, indicating that disulfide-linked aggregation was not responsible for the pattern.

DISCUSSION

Breast Epithelial Cells

Active synthesis and release of glycoproteins was observed for at least 48 h in organ cultures of human breast carcinomas. In some samples, for which data is not presented

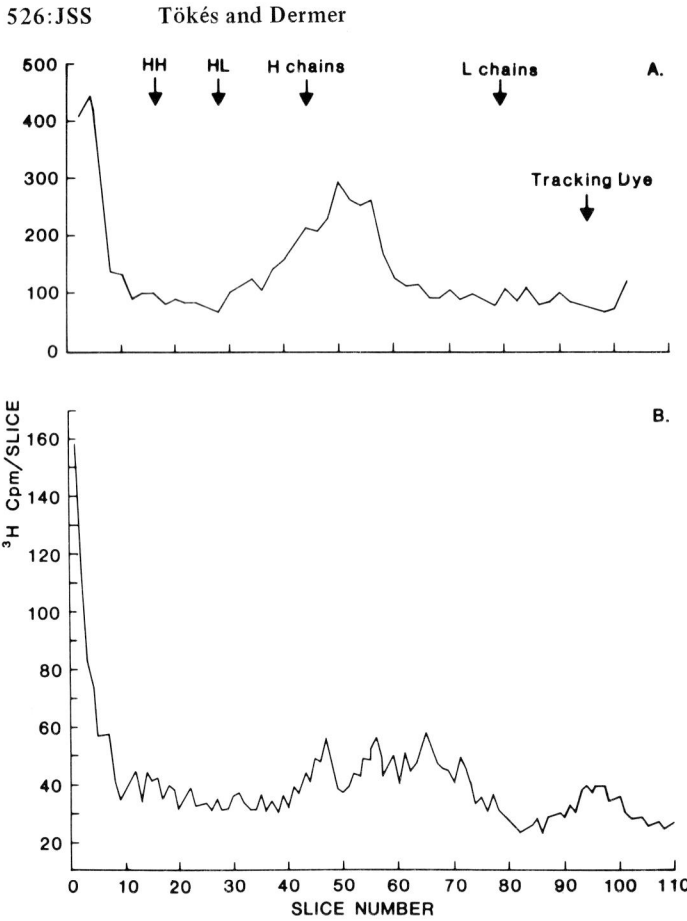

Fig. 8. Distribution of incorporated [^{3}H]glucosamine from prostate organ-culture media in SDS-polyacrylamide gel slices. Slices were 1 mm thick. Graph A) Grade III, poorly-differentiated adenocarcinoma. Graph B) Grade II, moderately-differentiated adenocarcinoma.

here, this activity remained for 5–7 days. Autoradiography performed on explants of 3 infiltrating duct carcinomas of the breast revealed that an average of 70% of the total number of silver grains over random sections were associated with the malignant cells. When glandular structures were present, more intense reactions were found over lumina than epithelial cytoplasm, indicating that the glycoproteins synthesized by malignant epithelium were released into glandular lumina. Similar experiments with early rat mammary carcinomas induced by a carcinogen also indicated the production of an unidentified glycoproteins which has been interpreted as an early detectable sign of malignant transformation (16).

SDS-polyacrylamide gel electrophoresis patterns of the glycoproteins released into the media by 10 different breast tumors, 5 of which are illustrated in Fig. 3, revealed a striking similarity of labeled and released macromolecules. Medullary carcinoma (Graph A), infiltrating lobular carcinoma (Graph D), and 8 infiltrating duct carcinomas, of which only 3 representative samples are illustrated (Graphs B, C, and E), all produced a group of macromolecules with an apparent molecular weight of 48,000 ± 6,000 daltons. Since at this time the percent of carbohydrate is not known, their exact molecular weight cannot

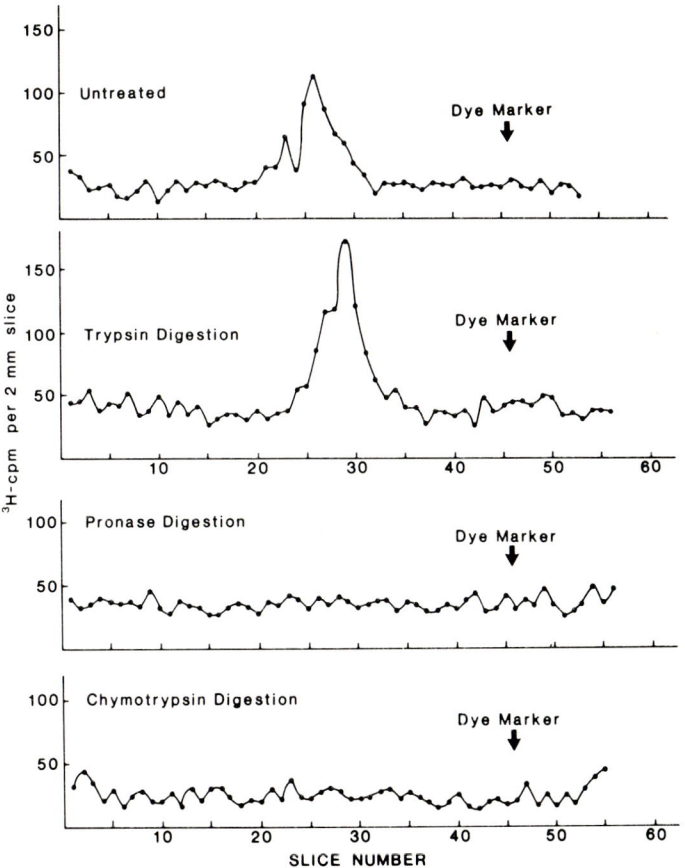

Fig. 9. The effect of proteolytic enzyme treatment on the organ culture supernatant. A media sample from benign prostatic hyperplasia was treated with trypsin, pronase, or chymotrypsin and analyzed by SDS polyacrylamide gel electrophoresis. Slices were 2 mm thick. Conditions and calibrations were the same as in Fig. 3.

be determined. Neuraminidase treatment of these glycoproteins decreased the apparent molecular weight by 3–5% (17) suggesting the presence of sialic acid on these molecules. Since these molecules were also susceptible to pronase digestion, we conclude that the isotope label is incorporated into a glycoprotein (17). The labeled glycoprotein patterns (Fig. 3) appear to be independent of the availability of estrogen receptors since only 50% of the tumors were estrogen receptor positive.

A complex pattern of released glycoproteins, labeled with [^{3}H]glucosamine, was observed in the supernatant of the established human mammary carcinoma cell line MCF-7, grown in 2-dimensional monolayer cultures (Fig. 4). These cells also produced a group of glycoproteins with an apparent molecular weight of 48,000 ± 6,000 daltons. Whether these molecules are antigenically related to the organ culture products is currently under investigation. It is interesting to note, however, that in addition to this group of glycoproteins, the majority of incorporation was into larger macromolecules. This is not due to a less extensive digestion of glycoproteins after their release in monolayer cultures since these components are not further degraded when introduced in organ cultures for 24 h (Tökés and Lam, unpublished observation).

Prostate Epithelial Cells

Autoradiography performed on explants of a representative case of benign prostatic hyperplasia that had been cultured for 1 day with [^{3}H]glucosamine followed by a chase of 2 days in nonisotopic media revealed that 70% of the total number of silver grains over random sections were associated with glandular-epithelial structures. Usually more intense reactions were found over the lumina of glands than epithelial cytoplasm. Autoradiographic data obtained from explants of the Grade III, poorly-differentiated prostate carcinoma were similar to those obtained from hyperplastic tissue. The major contribution of this data is that the glycoproteins recovered from the media of organ cultures of hyperplastic and carcinomatous human prostate and breast tissue are predominantly products of epithelium.

Labeled glycoproteins were below the level of detection in culture supernatant fractions in 4 of 6 cases of benign prostatic hyperplasia and from 1 well-differentiated Grade I prostate carcinoma. In these samples the number of epithelial cells was similar to or higher than in other biosynthetically-active samples. Therefore the low levels of incorporation are not due to decreased cell numbers but may reflect the generally poor metabolic state of the cells. These cells may also have their biosynthetic activity under more stringent regulation which is inactivated by organ culture conditions. Two of the six hyperplastic prostate cultures synthesized and released a major 48,000 molecular weight glycoprotein fraction, similar to that found in cultures of breast cancer specimens. These glycoproteins were also observed in a Grade I well-differentiated infiltrating prostate carcinoma and in a Grade II, moderately well-differentiated infiltrating prostate carcinoma (Fig. 7). In 2 of the samples illustrated in Fig. 7, we examined the molecular weight patterns released after 24 and 48 h of postlabel incubation. Even though quantitative differences were observed as expected, both patterns were identical signifying the lack of extensive degradation.

Trypsin treatment of the released glycoproteins from a benign prostatic hyperplasia sample slightly decreased their molecular weight, but both chymotrypsin and pronase treatment resulted in complete digestion (Fig. 9). These observations are identical to the previously reported sensitivity of glycoproteins produced by breast carcinoma specimens to various proteases (17).

The most complex glycoprotein patterns were obtained from one Grade II, moderately-differentiated and a Grade II, poorly-differentiated prostate carcinoma (Fig. 8). The large molecular weight component did not migrate faster in a 5% polyacrylamide gel when introduced with dithiothreitol and urea. Mucins and glycoproteins larger than 500,000 molecular weight behave in this manner. In both of these samples, 48,000 molecular weight components also were released. More samples will be necessary to establish whether the complexity of glycoprotein patterns is related to the undifferentiated state of Grades II and III carcinomas of the prostate.

GENERAL COMMENTS

In organ culture samples of benign hyperplastic prostate where the glandular structure and the basal membrane appeared normal, glycoprotein synthesis and release was usually below the level of detection. Moderately- or well-differentiated malignant epithelial cells from both prostate and breast released a strikingly similar pattern of glycoproteins. A 48,000 molecular weight glycoprotein was the predominant component in all of the samples where glandular structures and the basal membrane were still recognizable. The most complex pattern was observed with poorly-differentiated prostatic adenocarcinomas or with breast epithelial cells grown in the complete absence of any 3-dimensional glandular structures. These observations suggested to us a hypothesis that the 3-dimensional integrity of glandular structures may restrict synthesis and release of several glycoproteins. With progressive alteration in glandular structure the synthesis and the release of a 48,000 molecular weight glycoprotein would increase. Epithelial cells grown in 2-dimensional monolayer cultures or in vivo without basement-membrane-supported glandular structures would escape from such a topographic restriction and express additional families of glycoproteins. The data presented here is in agreement with this hypothesis; further quantitation of glycoprotein turnover by epithelial cells as a function of progressive disattachment from glandular structures is in progress in our laboratory.

Although the molecular nature of the major 48,000 molecular weight glycoprotein is not yet elucidated, a few observations reveal that it may not be the product of a differentiated cell function, such as the production of a specific milk or prostatic fluid protein. Supplementing organ culture media with hormones, and exchanging the patient's own serum with fetal bovine serum failed to alter the released glycoprotein pattern. Similarly, the availability of estrogen receptors did not alter the pattern. Therefore it is reasonable to hypothesize that these glycoproteins represent an expression of an "epithelial state" of differentiation.

A number of molecular markers have been reported to be associated with prostate and breast cancer (18–23). In addition to these markers, 2 membrane-associated glycoproteins need to be evaluated for their possible relation to the labeled components we have found in organ culture. Human histocompatibility antigens, normal components of most if not all cell types, have approximate molecular weights of 45,000 (24) and cell membrane turnover results in the shedding of these entities. Therefore it is possible that at least a fraction of the labeled glycoproteins belong to this category. A second membrane-associated component, the major envelope glycoprotein of Type C viruses, should also be considered. A polymorphism of these virion-associated and differentiation antigens, encoded by a multigene family, was reported (25) with the murine Type C viruses. This study demonstrated that the tryptic peptides of glycoproteins with molecular weights of 69,000 and 70,000, containing 32% carbohydrate, and the peptides of the 45,000 molecular weight glycoprotein with 6–9% carbohydrate were identical. Therefore gp45 can be considered an incompletely-glycosylated glycoprotein (25). Recent reports (26) also claim that the presence of the gp52 viral glycoprotein associated with mouse mammary tumor can serve in the plasma as a diagnostic indicator of the presence of a solid tumor. Since similar molecular weight glycoproteins, labeled in our experiments, are released, it is reasonable to postulate that they accumulate in the plasma and could therefore act as a diagnostic indicator of neoplasia.

ACKNOWLEDGMENTS

We wish to thank Dr. F. Pincus for her histopathological characterizations, Ms. E. Gilbert for the preparation of this manuscript, and Mr. C. Csipke and Ms. J. Lam for their valuable participation in the execution of these experiments. This work was supported by National Cancer Institute grant CA-14089.

REFERENCES

1. Cook GMW, Stoddart RW: Surface Carbohydrates of the Eukaryotic Cell. London: Academic Press, 257, 1973.
2. Cook GMW, Stoddart RW: Surface Carbohydrates of the Eukaryotic Cell. London: Academic Press, 257:140, 1973.
3. Beug H, Gerisch G, Kempff S, Riedel V, Cremer G: Exp Cell Res 63:147, 1970.
4. Baldwin RW, Harris JR, Price MR: Int J Cancer 11:385, 1973.
5. Baldwin RW, Price MR, Robins RA: Nature (London) 238:185, 1972.
6. Juergens WG, Stockdale FE, Topper YJ, Elias JJ: Proc Natl Acad Sci USA 54:629, 1965.
7. Turkington RW, Lockwood DH, Topper YJ: Biochim Biophys Acta 148:475, 1975.
8. Dermer GB, Sherwin RP: Cancer Res 35:63, 1975.
9. Hurlimann J, Lichaa M, Ozzello L: Cancer Res 36:1284, 1976.
10. Baulieu EE, LeGoascogne C, Groyer A, Feyel-Cabanes T, Robel P: Vitam Horm (NY) 33:1, 1975.
11. Kaighn ME, Babcock MS: Cancer Chemother Rep 59:59, 1975.
12. Soule HD, Vazquez J, Long A, Albert S, Brennan M: J Natl Cancer Inst 51: 1409, 1973.
13. Whur P, Herscovics A, LeBlond CP: J Cell Biol 43:289, 1969.
14. Dermer GB: J Ultrastruct Res 22:312, 1968.
15. Fairbanks G, Steck TL, Wallach DFH: Biochemistry 10:2606, 1971.
16. Russo I, Saby J, Isenberg W: Proc Am Assoc Cancer Res 16:116, 1976.
17. Dermer GB, Tökés ZA: J Natl Cancer Inst (Manuscript submitted).
18. Chu TM, Bhargava AK, Barnard EA, Ostrowski W, Varkaris MJ, Morrin C, Murphy GP: Cancer Chemother Rep 59:97, 1975.
19. Chu TM, Nemoto T: J Natl Cancer Inst 51:119, 1973.
20. Steward AM, Nixon D, Zamcheck N, Aisenberg A: Cancer 33:1246, 1974.
21. Rosato FE, Seltzer M, Mullen J, Rosato EF: Cancer 28:1575, 1971.
22. Amaral L, Werthamer S: Nature (London) 262:589, 1976.
23. Muller M, Grossman H: Nature (London) New Biol 237:116, 1972.
24. Terhorst C, Parham P, Mann DL, Strominger JL: Proc Natl Acad Sci USA 73:910, 1976.
25. Elder JH, Jensen FC, Bryant ML, Lerner RA: Nature (London) (In press).
26. Ritzi E, Martin DS, Stolfi RL, Spiegelman S: Proc Natl Acad Sci USA 73:4190, 1976.

The Area-Code Hypothesis: The Immune System Provides Clues to Understanding the Genetic and Molecular Basis of Cell Recognition During Development

L. Hood, H. V. Huang, and W. J. Dreyer
Division of Biology, California Institute of Technology, Pasadena, California 91125

Numerous studies of embryogenesis have provided evidence for highly specific cell-surface recognition phenomena. These include both the interactions of neighboring cells and the specific cellular migrations which occur as the developmental program of the embryo progresses. The area-code hypothesis elaborated here is an attempt to provide a framework for understanding cell-recognition phenomena in development.

This hypothesis is based on extensive genetic, molecular, and cellular studies of the immune system. These studies suggest that the following events occur during the differentiation of antibody-producing cells. 1) Somatic cell lines of antibody-producing cells undergo a modification of their DNA as they become committed to synthesize a particular type of antibody molecule. This chromosomal modification event is probably a DNA translocation which leads to a somatic rearrangement of certain antibody genes. 2) In each of the specific cell lineages the new arrangement of DNA is inherited by all subsequent generations of cells. 3) The developmental programs which control these genetic alterations may be employed in a programmed and reproducible fashion. This programming of antibody development is suggested because different embryos appear to become committed to the production of identical antibody molecules in the same developmental sequence. 4) Antibody molecules are initially displayed on the cell surface where they serve as highly specific receptors to trigger the cell to proliferate and differentiate upon interacting with appropriate external molecular signals. 5) Antibody-producing cells display combinations of different molecules on their surfaces which cause each of a very large number of different cells to interact differently with their environment. 6) The genes which code for many of these cell-surface molecules are organized into multigene families.

These observations as well as information from other developmental systems have led us to propose the area-code hypothesis. This hypothesis is concerned with the structure, function, and regulation of cell-surface molecules that mediate recognition phenomena during embryogenesis. Area-code molecules are cell-surface molecules which are involved in the specific recognition phenomena during growth and development. These molecules provide cells with distinct cell-surface addresses or phenotypes, and provide the basis for the specificity in cell-cell recognition during cell migrations and cell-cell interactions, as well as serving as receptors for diffusible

Received for publication June 22, 1977; accepted August 15, 1977

© 1977 Alan R. Liss, Inc., 150 Fifth Avenue, New York, NY 10011

differentiation signals. The area-code hypothesis has 3 main postulates. i) There is a progressive display of specific combinations of area-code molecules on the surfaces of cells during development. ii) The genetic programs which determine the specific expression of area-code molecules are in part controlled by DNA modifications. These chromosomal modifications are believed to channel cells into specific lineages with progressively restricted developmental options. iii) Many of the area-code systems are organized into multigene families. Rapid evolutionary increases in complexity may proceed by the duplication and subsequent independent evolution of multigene families. In short, many of the remarkable events which occur during the development of the immune system may form a basis for understanding other developmental systems. Some experimental approaches toward testing this hypothesis are discussed.

Key words: area-code hypothesis, combinations of cell-surface recognition molecules, chromosomal modifications, DNA translocation, multigene families, immune system as developmental model

Development is the orderly process whereby a single cell, the zygote, generates a large diversity of cell types (Fig. 1). These cells migrate to appropriate locations and interact with one another to give rise to the supracellular organization of the adult organism. The adult human has about 10^{14} cells. If all cells were to divide at an equal rate, the average adult cell would be separated from the zygote by at least 48 cell divisions ($2^{47} \simeq 10^{14}$). All adult cells have a cell lineage whose origin can be traced back to the zygote. As embryogenesis proceeds, cell lineages develop which become increasingly limited in their future development options (Fig. 1). Individual cells acquire specific developmental programs which limit the developmental fate of their progeny cells. Individual cells may become committed to a particular developmental program long before they differentiate to acquire the phenotypic characteristics of that cell lineage. These committed cells may later be induced to differentiate by hormones or other external signals. Little is known about the nature of the developmental programs or the mechanisms of cellular commitment. However, information has accumulated on the nature of cell-surface changes which occur as a cell lineage differentiates.

The unfolding of the developmental programs of individual cells leads to the expression of new gene products including molecules on the cell surface. Some of these cell-surface molecules are involved in cell recognition processes and may encompass a variety of functions in the developing embryo. Combinations of these molecules displayed on the surface may play a vital role in providing an address system for the massive cellular migrations that are characteristic of the developing embryo (Fig. 2). They are also vital to the myriad of specific cell-cell interactions occurring during growth and development. In the next section we propose a hypothesis that describes the general features of area-code molecules and the genes which encode them.

THE AREA-CODE HYPOTHESIS

The area-code hypothesis was formulated from an analysis of the vertebrate immune system. This hypothesis deals both with the role played by cell-surface recognition or area-code molecules in development and with the genetic events which place this address system on cell surfaces. The area-code hypothesis has several interrelated postulates: i) During development combinations of area-code molecules are displayed on the surface of cells of specific lineages. These cell-surface displays provide the specificity for cell-cell interactions. They also provide cell-surface receptors essential for sensing diffusible differentiation

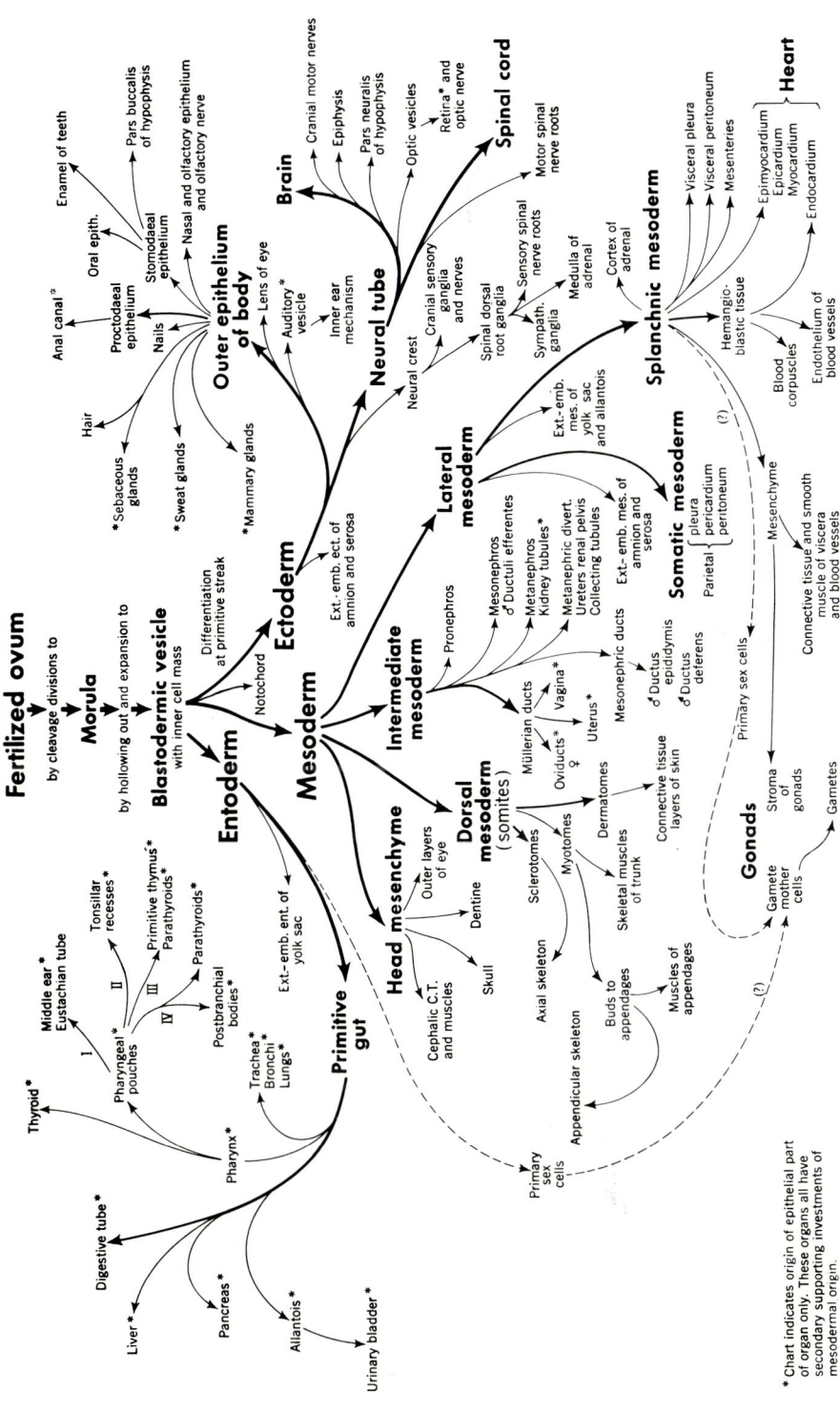

Fig. 1. Cell lineages of a developing organism. [Reprinted from Patten BM, Carlson BM: "Foundations of Embryology." 3rd Ed. San Francisco: McGraw-Hill, 1974, p 141, with permission.]

*Chart indicates origin of epithelial part of organ only. These organs all have secondary supporting investments of mesodermal origin.

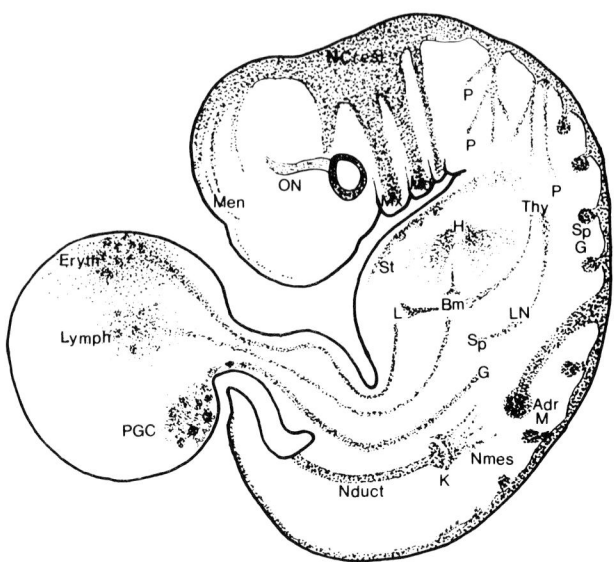

Fig. 2. Cellular migrations in the embryo. This diagram displays events taking place at different stages of development.

	Origin	Route	Target(s)
Erythrocyte precursors (Eryth)	Yolk sac	Blood stream	Liver (L), Bone marrow (Bm)
Lymphoid cell precursors (Lymph)	Yolk sac	Blood stream	Bone marrow, Thymus (Thy), Lymph nodes (LN), Spleen (Spl)
Primordial germ cells (PGC)	Yolk sac		Gonadal ridges, forming gonads (G)
Mesenchyme cells	Dorsal embryo	Migrate ventrally	Form sternum (St)
Retinal ganglion cell axons	Eye	Optic nerve (ON)	Brain visual centers
Neural crest cells (NCrest)	Neural crest	Migrate ventrally	Form meninges (Men), embryonic skull cartilages (Mx,Md), Pigment cells (P), Spinal ganglia (SpG), Adrenal medulla (AdrM)
Heart cells		"Heart-forming territory" Aggregate to	Form heart (H)
Nephric ducts		Elongate and meet aggregations of nephric mesenchyme cells (Nmes) to form kidney rudiments (K)	
Not included in figure:		The intricate cell translocations in histogenesis of the nervous system, and other examples of cell translocations and aggregations (e.g., those resulting in the formation of hair follicles, teeth, limb rudiments).	

[Reprinted from Ref. 99, with permission.]

signals such as those provided by hormones. Thus differentiating cells acquire distinct combinations of area-code molecules which serve as specific cellular addresses, not unlike those of the telephone or postal systems. ii) The genetic programs which govern the expression of area-code molecules are controlled in part by alterations in somatic cell chromosomes such as the translocation of DNA sequences. These chromosomal modifications occur at branch points of differentiation which define cell lineages and limit the

TABLE I. Features of the Immune System Facilitating Its Study

1. Lymphocytes are freely wandering cells easily separated from other cell types and fractionated into functional subclasses.
2. Developmental decisions made in the immune system generate specific cell lineages in which the future options are genetically programmed.
3. Lymphocyte tumors represent clones derived from single progenitor cells. Many different stages of lymphocyte differentiation are represented in the available tumor lines.
4. Serologic reagents are available to identify different lymphocyte lineages. Thus specific tags are available for studying the differentiation, migration, interactions, and triggering of lymphocytes.
5. The cells of the immune system undergo programmed migrations during development and interact in highly specific ways within various tissues such as the liver, spleen, bone marrow, thymus, and lymph nodes.
6. Lymphocytes form a regulatory network with one another and with other cells. This network can be studied in vitro as well as in vivo. Thus it is perhaps easier to study cell cooperation in this system than in any other.
7. Triggering of lymphocytes to differentiate and divide by mitogens, antigens, and hormones can readily be studied in vitro and in vivo.
8. Antibody molecules and mRNAs have been isolated from myeloma tumors and studied by chemical, serological, genetic, and functional techniques.

future options open to particular cell lineages. iii) In many cases the genes encoding area-code systems are organized into closely linked multigene families. During evolution, multigene families can be duplicated to produce new multigene families. These new families can assume new functions in programming development. Presumably these reprogrammed multigene families can produce major evolutionary changes in complex systems such as the brain. Thus the area-code gene families from diverse developmental systems may share common evolutionary originals.

It is our belief that area-code molecules play an important role in the recognition phenomena of embryogenesis for a variety of simple and complex systems including the neuroectoderm (1), the nervous system (2), the optic nerve (3), and the immune system.

THE IMMUNE SYSTEM IS A MODEL FOR STUDYING DEVELOPMENT IN COMPLEX EUKARYOTIC SYSTEMS

A. General Comments

The immune system is the first complex eukaryotic system that has been investigated in depth at the genetic, molecular, and cellular levels. Its study has revealed some remarkable mechanisms and strategies for development that are embodied in the area-code hypothesis. Although some of these mechanisms appear unique to the immune system, we feel that this is so only because no other complex eukaryotic system has been studied in similar detail. We suggest that many of the developmental strategies of the immune system will be employed by other area-code systems.

The immune system has several features which facilitate its experimental study (4). These are summarized in Table I. The major cells of the immune system, lymphocytes, are freely wandering cells that can be readily separated from other cell types and even fractionated into functional subclasses. Migration pathways of embryonic lymphocytes can be traced as development proceeds in vivo. The differentiation of lymphocytes also can be followed in vitro. The availability of large quantities of antibody molecules and antibody mRNA from myeloma tumors has permitted a detailed analysis of the antibody molecules and their genes. The antibody molecule appears to play an important role in mediating cellular interactions among lymphocytes and thus serves as a model for the prototype area-code molecule. Cell-cell interactions and receptor-mediated triggering of lymphocytes to differentiate can be studied in vivo or in vitro. We will discuss in some detail the developmental, cellular, molecular, genetic, and evolutionary strategies of the immune system as they have formed the basis for our thinking about the area-code hypothesis.

B. The Immune System Employs Antibody Molecules to Recognize Foreign Molecular Patterns

The vertebrate immune system shows 2 cardinal features of area-code systems, namely it utilizes recognition molecules exhibiting specificity and diversity.

Specificity. The immune system recognizes and destroys foreign molecules or antigens (5). The fundamental unit of recognition in this process is the antibody molecule. This molecule can be affixed to a lymphocyte as a specific cell-surface receptor or it can be secreted into the blood or lymphatic circulations. The antibody binds antigen through a molecular complementarity similar to that which an enzyme exhibits for its specific substrate. This interaction leads by a variety of mechanisms to the specific destruction or elimination of the antigen.

Diversity. Lymphocytes and their antibody molecules recognize a virtually limitless number of different antigenic determinants because almost any macromolecule that is foreign to a particular vertebrate organism can evoke an immune response. The average man has approximately 10^{12} lymphocytes circulating throughout his body and 10^{20} antibody molecules in his circulation. Estimates as to the number of different molecular species of antibody molecules a vertebrate can synthesize range between 10^5 and 10^8. Hence the immune system is capable of generating an enormous array of different types of specific cell-surface recognition molecules. Let us consider how distinct cell lineages develop in the immune system.

C. Lymphocytes Differentiate Along One of Two Discrete Developmental Pathways to Produce B Cells and T Cells

Differentiation in lymphocytes requires specific cellular migration and hormonal induction, two features shared by other developmental systems.

B- and T-cell lineages. The development of the immune system begins with stem cells arising in the yolk sac and later migrating to the fetal liver and finally to the bone marrow (Fig. 2) (6). In the adult, stem cells for lymphocytes divide in the bone marrow and there become committed to the B- or T-cell pathway. These are termed pre-B or pre-T cells. Later these committed lymphocytes migrate to an appropriate microenvironment where hormonal inducers trigger the subsequent expression of the precommitted B- or T-cell developmental programs.

The pre-T cells migrate to the thymus. Under the influence of thymic hormones, they differentiate and later enter the circulation as mature T cells (7). T cells undergo additional differentiation steps on interaction of antigen with the antibody-like receptors on their cell surface.

In birds, the pre-B cell migrates to the bursa of Fabricius and presumably under hormonal induction differentiates to a mature B cell (8). In mammals, the pre-B cell probably differentiates in the bone marrow (9). B cells migrate to the blood and lymphatic circulation. There the interaction of antigen with antibody receptors induces terminal differentiation to the plasma cell, a highly efficient factory for the synthesis of antibody molecules.

B and T cells both employ antibody or antibody-like molecules as specific cell-surface receptors (10). The library of T-cell antibody-like receptors is believed to be comparable in diversity to those of its B-cell counterpart.

Functions of B and T cells. B cells synthesize antibody molecules. These molecules are employed as cell-surface receptors and they are also secreted into the serum. B cells constitute the basis of the humoral immune response which depends on the secreted antibody molecules to fight acute viral and bacterial infections. T cells synthesize antibody-like molecules which are employed as cell-surface receptors for the diverse reactions of the cellular immune response. These include the surveillance for and destruction of cells altered by neoplastic transformation or viral infection. T and B cells display characteristic cell-surface molecules, some of which mediate cell-cell interactions.

D. As Lymphocytes Differentiate, Distinct Combinations of Cell-Surface Molecules Are Displayed

The area-code hypothesis suggests that cell-surface recognition molecules play a fundamental role in growth and development. The successive acquisition of cell-surface molecules during differentiation has been clearly demonstrated in lymphocytes.

Cell-surface molecules on T and B cells. The display of a variety of cell-surface molecules on mouse lymphocytes at various stages of development has been studied by detailed genetic and serological analyses (11). In the bone marrow all precursor cells of lymphocytes express the transplantation (H-2) antigens (Fig. 3). The pre-T cell migrates to the thymus and there is induced by thymic hormones to express at least 4 cell-surface molecules: TL, Ly 1, Ly 2, and Thy 1 (12). The T-cell receptors, designated IgT, probably appear at this stage. As the T cell migrates to the periphery, TL disappears, and the cell-surface concentration of Thy 1 decreases while that of H-2 increases (13). There are 3 Ly phenotypes of T cells in the periphery circulation, Ly 1, Ly 2, and Ly 1, 2, and it appears likely that the Ly 1 and Ly 2 cells are derived from the Ly 1,2 cells (4). The pre-B cell migrates to the bursa or its equivalent and there presumably acquires the antibody receptor, designated IgB (Fig. 3). Antigen induces the expression of the PC-1 antigen in plasma cells (14).

These cell-surface molecules are designated differentiation antigens (11) because they are molecules that have been expressed during successive developmental stages and have been studied by serological techniques.

Expression of differentiation antigens. Differentiation antigens show various modes of expression. i) The differentiation antigens induced in pre-T cells by thymic hormones can be expressed within 5 h of induction. Moreover, this expression can occur without cell division (12). Accordingly, inducers trigger the expression of previously committed developmental programs. ii) Certain differentiation antigens are lost during the course of subsequent differentiation (e.g., TL), and others change markedly in their cell-surface concentrations (e.g., Thy 1 and H-2). Hence genes coding for differentiation antigens can be turned off or altered in their rate of expression.

Sets of genes determine phenotype. The genes encoding the differentiation antigens of lymphocytes are present on at least 8 different chromosomes (Fig. 3b). Thus, an array of cell-surface molecules encoded by unlinked genes determines the cell-surface state of differentiation in B and T cells. Different subsets of these genes must be expressed in a coordinated fashion by the developmental programs for each distinct cellular phenotype.

Differentiation antigens and lymphocyte-lineage relationships. Differentiation antigens provide clues as to the lineage relationships of various lymphocyte clones. T cells can be readily distinguished from B cells based on their cell surface molecules (Fig. 3a). Indeed, 3 distinct subclasses of T cells can be distinguished in the peripheral circulation

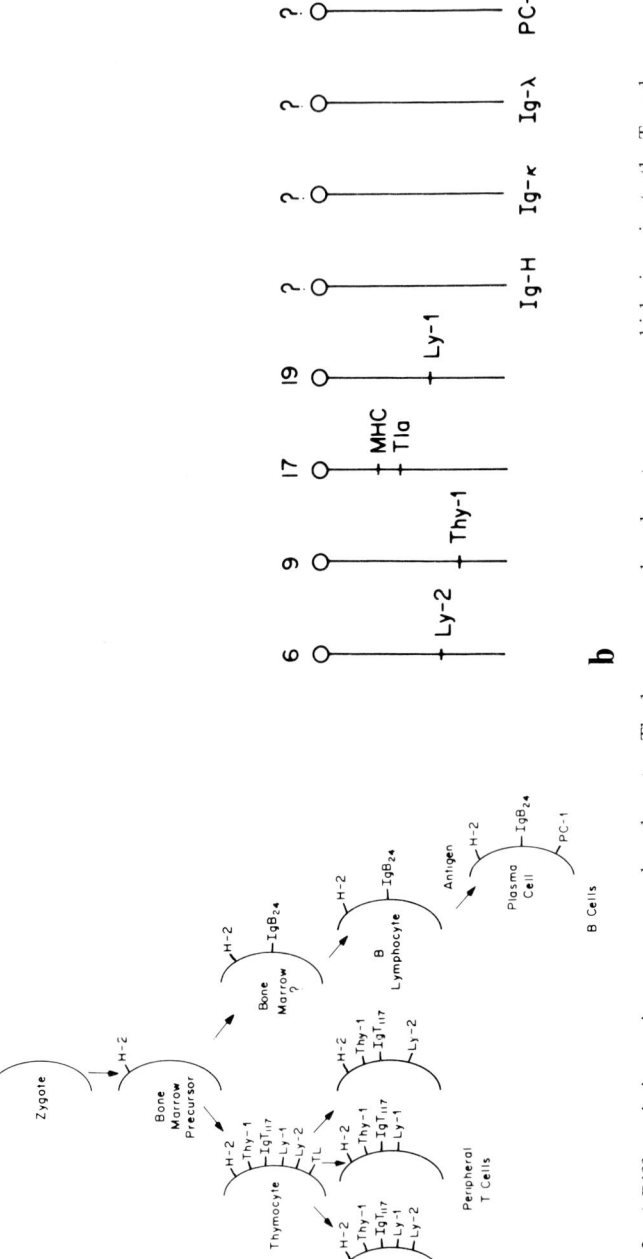

Fig. 3. a) Differentiation antigens on mouse lymphocytes. The bone marrow lymphocyte precursor, which gives rise to the T- and B-lymphocyte lineages, expresses H-2 antigens on its surface. In the respective inductive microenvironments, lymphocytes display new sets of differentiation antigens. On seeding to the periphery some differentiation antigens may be selectively lost (e.g., TL, Ly1, or Ly2). Terminal differentiation may be correlated with new differentiation antigens (e.g., PC-1). H-2 designates the H-2 K and D transplantation antigens. Thy-1 denotes thymus-derived lymphocyte antigen. IgT and IgB designate the antigen receptors of T and B cells, respectively, with subscripts of arbitrary numbers used to emphasize the great diversity of these molecules, and to stress the one cell-one antigen-binding site feature of these cells. Ly 1 and Ly2 represent lymphocyte antigens 1 and 2 which are present exclusively on T cells. TL denotes thymus leukemia antigen. PC-1 represents plasma cell antigen. Other known lymphocyte differentiation antigens (e.g., Ia, Qa-2, Aka-1) are not shown. b) Chromosomal map of genes coding for differentiation antigens on mouse lymphocytes. The positions on the chromosomes are approximate for loci whose position is known. Chromosomal assignment for the immunoglobulin (Ig) families (H, λ, κ) and PC-1 are not known.

based on the Ly phenotypes (Fig. 3a). Finally, clones of lymphocytes acquire their unique functional specificities based on their expression of individual antibody or antibody-like cell-surface receptors. Thus, each individual lymphocyte clone expresses a unique set of cell-surface molecules beginning, for example, with those shared by all lymphocytes (e.g., H-2), to those shared by T cells (e.g., Thy 1), to those shared by T-cell subclasses (e.g., Ly1, Ly 2, or Ly 1,2), and finally to those conferring clonal individuality (e.g., IgT_{117}) (Fig. 3). These combinations of cell-surface molecules serve as molecular addresses to distinguish individual clones of lymphocytes.

Area-code molecules and differentiation antigens. Area-code molecules are defined as those involved in cell-surface recognition processes. On the other hand, the differentiation antigens are any cell-surface molecule that appears during the course of differentiation in a particular cell lineage. Some differentiation antigens may be area-code molecules, though presumably not all. Many differentiation antigens may carry out cell-surface roles unrelated to cell recognition such as enzymatic reactions, structural support, and transport functions. The functions of the differentiation antigens of lymphocytes, apart from those of the IgT or IgB receptors, are unknown. Since the B-cell system serves as a model for an area-code system, let us consider how it is stimulated to undergo the final stages of differentiation by antigen.

E. Antigen Triggers Clones of Lymphocytes With Complementary Antibody Receptors

One of the unresolved questions about complex eukaryotic systems employing cell-surface recognition molecules is how are they triggered to differentiate. While the molecular details of this triggering process are not understood for lymphocytes, a reasonable phenomenological description of this process is available.

In a system that encompasses 10^{12} lymphocytes expressing 10^5 to 10^8 different antibody molecules, how are appropriate antibody molecules expressed in response to individual antigens? There are several aspects to the triggering of a specific immune response to antigen (15) (Fig. 4). i) Individual lymphocytes can synthesize only one molecular species of antibody molecule. This commitment of each lymphocyte to the synthesis of one type of antibody molecule is an antigen-independent process. ii) Antigen triggers the clonal expansion of individual lymphocytes through interaction with complementary antibody receptors at the cell surface. iii) The clonal descendants of a particular lymphocyte are all committed to the expression of antibody molecules of precisely the same specificity as those of the parent lymphocyte. iv) Antigenic triggering of a lymphocyte results in 2 general classes of clonal descendants. Effector cells are terminally differentiated and mediate the immediate response to antigen. Memory cells constitute a greatly expanded specific lymphocyte compartment for enhanced secondary immune responses on reencounter with antigen. Thus antigen is one of the final inductive triggers in lymphocyte differentiation. Moreover, antibody molecules, the prototype area-code molecules, play a fundamental role in this differentiation process.

Once antigen triggers clones of specific lymphocytes, antibody production from the individual lymphocyte clones must be regulated. Specific antibody molecules and antigen play an important role in this process. The effects of lymphocytes interacting with one another in a regulatory network are also very important in regulating the immune response.

F. Lymphocyte Interactions Regulate the Immune Response

One might argue that the immune system fails as a model for many potential area-code systems because lymphocytes are mobile and do not exhibit the fixed cellular

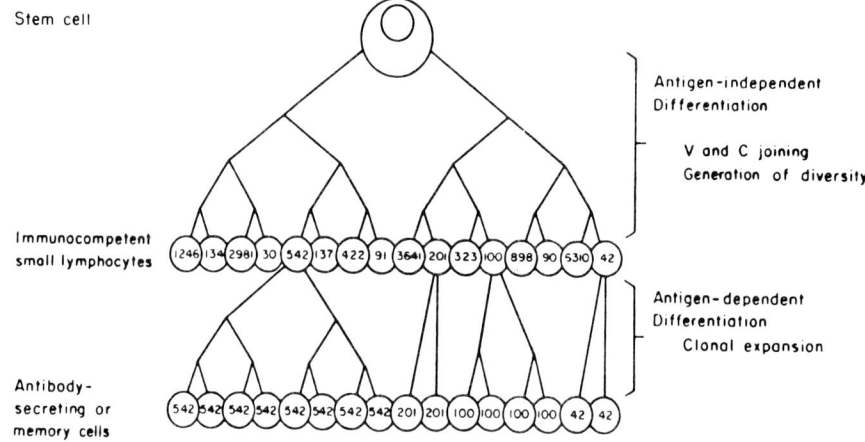

Fig. 4. A model for clonal selection in the immune system. Adapted from Ref. 34.

TABLE II. Cellular Interactions Among Sets of Lymphocytes*

	Cooperating sets	
Effector	Inducer	Function
B cell	Ly 1:H[a]	↑ antibody
Ly 2:KS[a]	Ly 1:H	↑ cytotoxicity
Ly 1,2:ARC[a]	Ly 2:KS	↑ suppressing
Macrophage	Ly 1:H	Delayed hypersensitivity
Macrophage	Ly 2:KS	Macrophage killers

*Adapted from Ref. 31.
[a]H indicates helper; KS denotes killer-suppressor; ARC designates antigen receptor cell.

interactions characteristic of most differentiated tissues. However, lymphocyte interactions with one another and with other tissues play a fundamental role in lymphocyte differentiation.

Lymphocyte interactions in the immune response. Clones of lymphocytes interact with one another to regulate the immune response and produce a finely balanced lymphocyte network (16–18). The cellular basis for these interactions rests in part on the presence of 3 functionally distinct subclasses of T cells with differing Ly phenotypes (19). The Ly 1 helper-T cells cooperate with B cells or other T cells to produce an immune response. The Ly 2 cells fall into 2 distinct categories. The Ly 2 suppressor-T cells inhibit immune responses of B cells or other T cells. The Ly 2 killer-T cells destroy foreign cells by lysing them through unknown mechanisms. The Ly 1,2 cell also plays a role in the suppression process. These cellular interactions are summarized in Table II. Indeed, a third type of cell, the macrophage, also interacts with T cells to mediate certain aspects of the immune response (20). Macrophages share a common progenitor lineage with lymphocytes and, accordingly, are closely related in an embryological sense (21). Thus lymphocytes interact with one another and with macrophages to facilitate or suppress immune responses in a delicately balanced network of cellular regulation. Accordingly, lymphocyte networks may be an ideal model system for studying various aspects of cell-cell interaction — a cardinal feature of development during embryogenesis and of the area-code hypothesis.

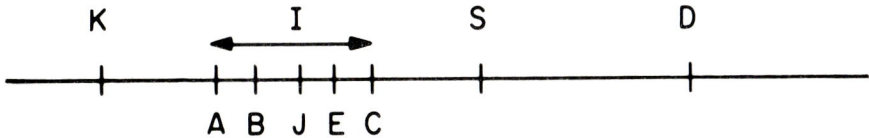

Fig. 5. A genetic map of the major histocompatibility complex or H-2 complex of the mouse.

The lymphocyte network and cell-surface molecules. Lymphocyte interactions are mediated by several classes of cell-surface molecules including those coded by the major histocompatibility complex (22) and the antibody genes (17).

i) In the mouse, the major histocompatibility complex, or H-2 complex, was defined on the basis of its ability to mediate strong graft rejections (13). This chromosomal complex encodes at least 2 different categories of cell-surface molecules (Fig. 5). First, the K and D regions encode the classical transplantation antigens, cell-surface glycoproteins which appear to play a fundamental role in T-cell surveillance of infected or transformed cells (23). Second, the I region encodes a number of cell surface glycoproteins collectively designated the Ia antigens (see Ref. 24). These molecules are present on B cells and, at lower concentrations, on at least some T cells. The I region regulates a bewildering array of immune-related traits. They include the control of the ability of mice to respond to a wide variety of different antigens (immune responsiveness) and some of the cell-surface interactions of T cell-B cell cooperation. The role of the Ia antigens in these functions is uncertain.

Both the transplantation antigens and the I region gene products mediate specific lymphocyte interactions. The key to understanding these interactions is the realization that the genes of the K, D, and I regions are extremely polymorphic in mice. For example, inbred mice have at least 11 different K alleles and 10 different D alleles (25). Thus cells with one combination of K, D, and I phenotype can be tested against those with other combinations. Some experiments demonstrate that in order for specific lymphocyte interactions to occur, these gene products must be identical on the 2 interacting cell types (23). For example, killer-T cells and target cells must share the same K or D gene products in order for the destruction of infected or transformed cells to occur. Also, certain helper-T cells must share I region identity with the cells that stimulate their participation in the immune response (26). The molecular basis for these cell-cell interactions is uncertain; however, it does not appear to be a simple like-like interaction of identical molecules coded by the H-2 complex. Thus, gene products from the K, D, and I regions are involved in cellular recognition in the immune system and, accordingly, are area-code molecules.

ii) The antibody molecules displayed on B cells and the antibody-like molecules on T cells also mediate cellular interactions among lymphocyte clones. The collection of antigenic determinants on the receptor portion (V domain) of an individual antibody molecule is known as its idiotype (27). An idiotype defines serologically each distinct molecular species of antibody molecule. Three interesting observations related to cellular interactions in the immune system have been made about idiotypes. First, animals can make antibodies to their own idiotypes (antiidiotype antibodies) (28). Second, antiidiotype antibodies may enhance or suppress immune response (29). Third, the network of B cells, helper-T cells, and suppressor-T cells involved in a particular immune response exhibit antibody receptors with related idiotype and antiidiotype specificities (30). Thus cellular interactions in the immune response also are mediated by idiotype-antiidiotype interactions. Since the antibody molecule participates in cellular recognition phenomena, it also is an area-code molecule.

iii) Specific cellular interactions occur between lymphocytes and the stromal and endothelial cells of various lymphoid organs (31). For example, the lymph nodes and the spleen appear to have specific regions to which B cells or T cells migrate. Indeed, lymph nodes appear to have 3 separate regions, respectively, for B, Ly 1, and Ly 2 cells (32). Presumably specific sets of lymphocytes migrate to these areas because of specific cell-surface reactions with the underlying endothelial or stromal cells.

In summary, cellular interactions as well as precise cellular migrations are fundamental features of the immune system. Lymphocytes form a network of interacting cells that presumably play a fundamental role in regulating the immune response. Moreover, lymphocytes can specifically interact with other cell types such as macrophages and the stromal or endothelial cells of specific lymphocyte regions. Many of these interactions are mediated by cell-surface molecules. The programs for differentiation that lead to these specific cellular interactions are contained in T cells, in B cells, and in the other cell types with which lymphocytes specifically interact. Clearly these programs and their expression of appropriate area-code molecules must be coordinated with one another to produce these functionally interacting cellular networks.

Since the antibody molecule serves as a model for our thinking about area-code molecules, it is important to understand the molecular strategies it employs for its functions.

G. The Antibody Molecule Is Composed of Discrete Globular Domains Which Carry Out Distinct Functions

Light and heavy chains (33). Individual antibody molecules are composed of 2 identical light and 2 identical heavy polypeptide chains associated by noncovalent and disulfide interactions (Fig. 6). Each antigen-binding site requires a single pair of light and heavy chains.

Variable and constant regions (34). All immunoglobulin polypeptides can be divided into an amino-terminal portion, the variable (V) region, and a carboxy-terminal portion, the constant (C) region (Fig. 6). The V regions exhibit extensive amino acid sequence diversity and the C regions far more limited diversity. The variable regions encode the antigen-binding or recognition function, whereas the C regions encode a more limited number of effector functions such as complement fixation.

Homology units (35). Antibody polypeptides can be divided into homology units about 110 residues in length on the basis of amino acid sequence similarities. The heavy chain in Fig. 6 is comprised of 4 homology units (V_H, C_H1, C_H2, C_H3) and the light chain has 2 homology units (V_L and C_L). The V_H and V_L homology units exhibit extensive sequence homology with one another as do the C region homology units, These homology relationships indicate that the homology units of antibody genes share a common evolutionary ancestry.

Domains (35). Each pair of homology units (e.g., V_H-V_L, C_H1-C_L, C_H2-C_H2 and C_H3-C_H3) folds into a compact globular domain (Fig. 6). The V_H and V_L homology units fold together to form a large crevice for antigen binding, whereas individual C domains carry out various effector functions. For example, the carboxy-terminal domain affixes the antibody molecule to the lymphocyte cell surface and is involved in the antigen-stimulated triggering of differentiation. Likewise, the C_H2 domain mediates complement fixation (36). Thus the antibody molecule is a sophisticated molecular machine that folds into discrete globular domains each of which may carry out different functions.

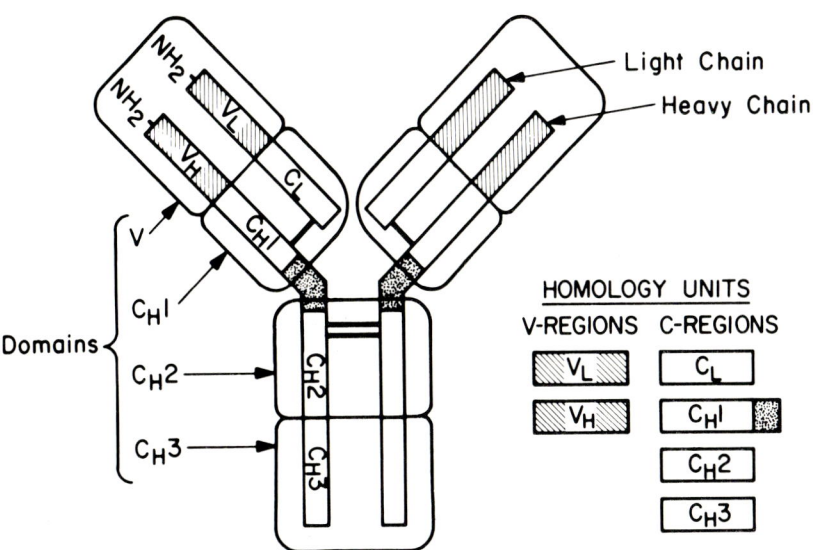

Fig. 6. Structure of an antibody molecule. [Reprinted from Ref. 74, with permission.]

The antibody molecule has fused together 2 discrete types of functions — recognition and effector. This functional dichotomy of the antibody molecule is reflected in the structure and organization of the antibody genes.

H. Antibody Genes Are Encoded as Three multigene Families With Two Distinct Types of Genes — Variable and Constant

The organization of the antibody gene families provides important insights into the possible organization of gene families for other area-code systems.

Three gene families. Classical genetic studies have demonstrated that 3 clusters or families of antibody genes, λ, κ, and H, for B cells are present in all mammals studied to date (Fig. 7). These gene families are genetically unlinked to one another. The λ and κ gene families code for light chains whereas the heavy gene family codes for heavy chains. Thus multiple gene families encode the antibody-receptor molecules.

Antibody gene families are multigenic (38). The number of antibody genes present in the germ line (or zygote) of a vertebrate organism is still a matter of controversy. However, most immunologists agree that the average antibody gene family has somewhere between 20 and thousands of genes. The antibody gene families are multigenic in nature.

Separate V and C genes. The variable and constant regions of antibody polypeptides appear to be encoded by separate germ line genes (Fig. 7). The evidence for this surprising gene organization is compelling. Initially this supposition was based on genetic, serological, and amino acid sequence analyses (40). Subsequently this problem has been approached directly by the use of restriction endonucleases which cleave DNA at specific recognition sites (41, 42). When genomic DNA is digested by such an enzyme and the resulting fragments are separated by size, a particular gene present as a single copy in the genome will be found only in one or a few fractions. Accordingly, linkage relationships between 2 or more genes (or the lack thereof) can be determined by examining these fractions with appropriate nucleic acid probes (radio-labeled mRNA or cDNA). In this manner, it has

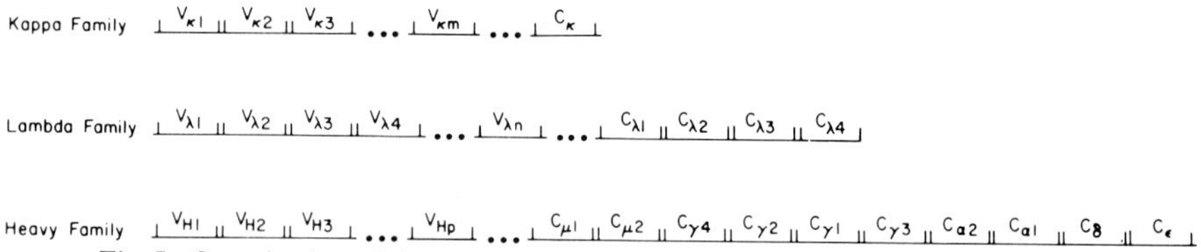

Kappa Family $\quad\mid V_{\kappa 1} \mid\mid V_{\kappa 2} \mid\mid V_{\kappa 3} \mid \ldots \mid V_{\kappa m} \mid \ldots \mid C_{\kappa} \mid$

Lambda Family $\quad\mid V_{\lambda 1} \mid\mid V_{\lambda 2} \mid\mid V_{\lambda 3} \mid\mid V_{\lambda 4} \mid \ldots \mid V_{\lambda n} \mid \ldots \mid C_{\lambda 1} \mid\mid C_{\lambda 2} \mid\mid C_{\lambda 3} \mid\mid C_{\lambda 4} \mid$

Heavy Family $\quad\mid V_{H1} \mid\mid V_{H2} \mid\mid V_{H3} \mid \ldots \mid V_{Hp} \mid \ldots \mid C_{\mu 1} \mid\mid C_{\mu 2} \mid\mid C_{\gamma 4} \mid\mid C_{\gamma 2} \mid\mid C_{\gamma 1} \mid\mid C_{\gamma 3} \mid\mid C_{\alpha 2} \mid\mid C_{\alpha 1} \mid\mid C_{\delta} \mid\mid C_{\epsilon} \mid$

Fig. 7. Organization of the antibody gene families in man. Adapted from Ref. 34.

been shown that the V and C genes for light chains from the mouse are in separate restriction fragments in undifferentiated mouse embryo DNA, but on the same restriction fragment (and presumably joined) in differentiated myeloma tumor DNA. More recently, DNA sequence analysis of a V gene isolated from mouse embryonic DNA has confirmed that the C gene is not adjacent to the V gene in undifferentiated DNA (43). Obviously, the separate organization of V and C genes has important implications for mechanisms of lymphocyte differentiation.

I. The Translocation of V and C Genes in Lymphocytes During Their Differentiation Appears to be a Fundamental Mechanism of Commitment

V-C translocation and antibody polypeptides. Since the V and C genes are separted in the embryo, these genes or mRNAs must undergo a rearrangement during somatic differentiation to form a contiguous VC gene or mRNA that is translated into a single polypeptide chain. This probably occurs by a DNA translocation event that joins the V and C genes (Fig. 8).[1]

V-C translocation and cellular commitment. Each mature lymphocyte expresses one type of antibody molecule. We feel that DNA translocation is a fundamental component of the molecular mechanism for committing a single lymphocyte to the expression of one type of antibody molecule (44) (Fig. 8). The implications of this hypothesis are extremely interesting with regard to development. The new arrangement of genetic material, including the V and C genes which have been joined as a part of the developmental process, becomes a heritable property of the daughter cell lines. The process of joining a specific V and a specific C gene is a definitive example of a commitment event. It provides a simple mechanism for limiting the future options open to a given cell lineage. It also provides a mechanism to explain how the memory of developmental decisions can be maintained throughout cell division and passed on to subsequent generations of somatic cell lines of the same lineage. The new arrangement of DNA in differentiated lymphocytes is simply replicated and passed to the daughter cells along with the newly activated genetic programs which control future events in the lineage.

---

[1] The nucleic acid studies on antibody genes provide unequivocal evidence for 2 suppositions. First, the DNA sequence data on the embryonic V gene (43) provide compelling evidence that the V and C genes are distinct in the germ line. Second, the restriction enzyme studies on embryonic and differentiated DNA (44, 45) argue that a DNA modification event has occurred during lymphocyte differentiation. This DNA modification could be anything which alters the restriction enzyme sites in the adult DNA with respect to embryonic DNA. The obvious DNA modification to explain these restriction enzyme patterns would be a translocation of the V and C gene sequences (Fig. 7). Other DNA modifications that do not rearrange V and C gene sequences require 2 assumptions. i) DNA modification such as base methylation changes multiple restriction enzyme sites. ii) The separate V and C genes are transcribed as a single mRNA. In this regard, recent observations suggest that a single mRNA can be obtained from separate DNA segments on the adenovirus chromosome (105, 106). These alternative possibilities for DNA modification can be tested by examining the organization of V and C genes in myeloma DNA. However, we tend to favor the simpler hypothesis of DNA translocation and assume throughout this paper that the DNA modification event in lymphocyte differentiation is a joining of the V and C genes.

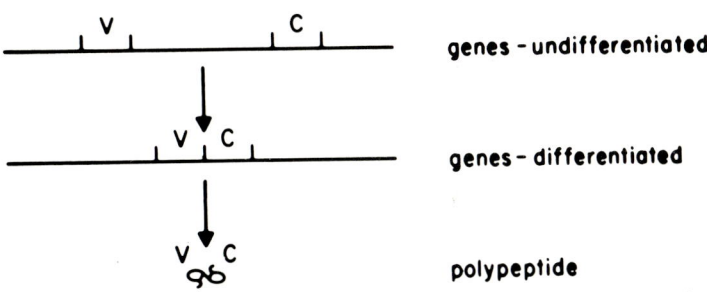

Fig. 8. Illustration of the differentiation of a lymphocyte through V-C joining. [Reprinted from Ref. 38, with permission.]

V-C translocation may alter developmental programs. The V-C translocation during somatic differentiation in lymphocytes has 2 distinct functions. First, it joins together separate V and C genes into a single, contiguous V-C sequence which codes for a complete antibody chain.[1] Second, it changes the developmental program of the lymphocyte so that it is hereafter committed to the expression of a single antibody polypeptide. Thus V-C translocation may alter the organization of control as well as structural elements and thereby alter the future developmental options of that particular cell lineage. We view this DNA modification event as a fundamental feature of differentiation in the antibody system and as we shall discuss subsequently in the differentiation of other area-code systems. Moreover, DNA translocation may be involved in the orderly read out of antibody genes during development.

J. Antibody Genes May Be Expressed During Differentiation in an Orderly and Programmed Manner

Development requires the orderly expression of phenotypic information. For example, during the differentiation of the T-cell lineage, the differentiation antigens are expressed at precise developmental stages (Fig. 3a). Accordingly, the set of unlinked genes that codes for these molecules (Fig. 3b) must be expressed in a coordinated and orderly manner.

Antibody V genes may also be expressed during differentiation in a coordinated and orderly manner (Fig. 9). Although the issue is controversial (see Ref. 45), several lines of evidence suggest that animals acquire the ability to respond to different antigens at different stages of development of the immune system. i) The development of individual clones of specific antibody-producing cells can be followed using x-irradiated mice as an in vivo tissue culture system (46). When lymphocytes are taken from neonatal mice, the ability to respond to the dinitrophenyl and trinitrophenyl groups appears by the first day after birth, to fluorescein by 3 days after birth, and to phosphorylcholine at 6–7 days after birth (47). All mice in an inbred line seem to follow this same developmental progression. Moreover, isoelectric-focusing analyses of these antibody molecules suggest that the same major molecular species are expressed in each mouse at each developmental stage. ii) In the bursa of Fabricius of individual chicken embryos, specific antigen-binding lymphocytes for keyhole limpet hemocyanin and poly-L(Tyr,Glu)-poly-D,L-Ala-poly-L-Lys appeared earlier than those binding sheep erythrocytes (48). Once again this implies that certain antibody molecules are expressed during differentiation before others. iii) Since the bursa appears to be the central organ for B-cell differentiation in chickens, an attempt has been made to remove this organ at various stages of chicken development and determine whether there is a corresponding loss of the ability to express certain antibody polypeptides (Huang and Dreyer, in preparation). Early ablation should delete most of the antibody repertoire, whereas later ablations should permit the reproducible expression of an ever increasing fraction of the total repertoire. Preliminary studies of this type employing the very sensitive technique of two-dimension gel electrophoresis bear out these predictions and further suggest that there may be an ordered and reproducible expression of light chains.

Each of these individual studies can be given alternative interpretations (45), but taken together they raise the intriguing possibility that antibody genes may be read out during development in a programmed and orderly fashion (Fig. 9). If so, several interesting questions are raised. During the differentiation of lymphocytes, are the V genes in a given multigene family sequentially translocated to their corresponding C genes? Is the V-C translocation mechanism an integral part of the development program for reading out V genes? Is there some type of mechanism for coordinating the readout of 2 multigene families (e.g., light and heavy chain gene families) so that particular molecular combinations of these 2 distinct polypeptide chains may be expressed in a programmed and reproducible fashion? How might genetic translocation help set up future developmental programs? Clearly those area-code systems mediating cell-cell recognition during embryogenesis must be capable of expressing their information in an orderly and programmed manner consistent with the orderly nature of growth and development.

K. Antibody Genes Appear to Evolve From a Common Precursor Gene

The evolution of antibody genes gives important insights into the possible evolutionary pathways of other area-code systems and raises the intriguing possibility that some distinct area-code systems may share common gene ancestors.

Antibody polypeptides are linearly differentiated into homology units of about 110 amino acid residues (Fig. 6). The constant homology units demonstrate significant amino acid sequence homology to one another as do the variable homology units (35). The existence of variable and constant region homology units and the observation by x-ray crystallographic analysis that the tertiary structures for the V and C homology units are very similar (49, 50) indicate that antibody genes probably evolved from a precursor gene coding for a single ancestral homology unit. One hypothetical evolutionary scheme is depicted in Fig. 10. The hypothetical precursor gene duplicated at a very early time to produce ancestral V and C genes. These gene products presumably assumed primitive area-code functions on membranes. Once a V-C translocation mechanism evolved, the V gene library could be expanded to generate a primordial multigene family.

This original family may have coded for primitive membrane area-code molecules (Fig. 10). This multigene family was in turn duplicated either by polyploidization or duplication and translocation of a chromosomal fragment to produce primitive antibody gene families. Subsequent duplication of this primitive multigene family produced the 3 families that evolved to become contemporary λ, κ, and H families. Contiguous or fused gene duplication in the heavy chain family led to the different C_H genes, each comprised of 3 or 4 homology units. Homology units, folding to comprise domains with different functions, can be added to (or deleted from) C_H genes in the course of evolution. Accordingly, the evolution of antibody genes employed all the major mechanisms of gene evolution — point mutation, discrete duplication, contiguous duplication, polyploidization,

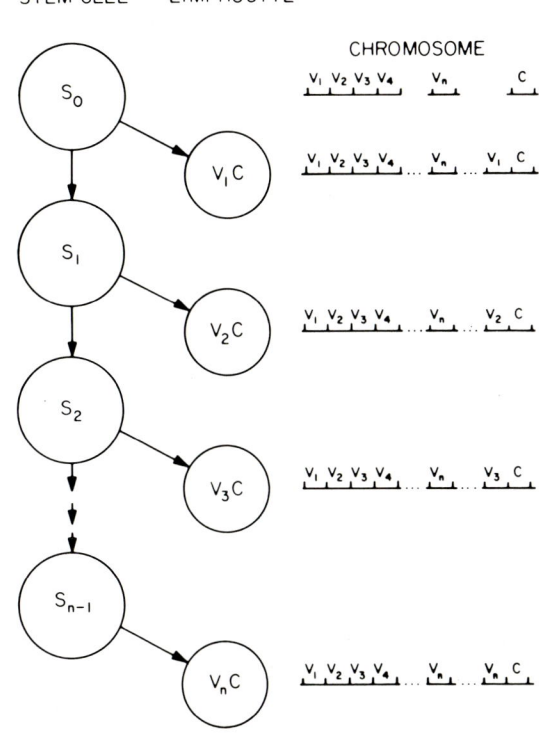

Fig. 9. A model of the linear and programmed read out of V genes in a multigene family.

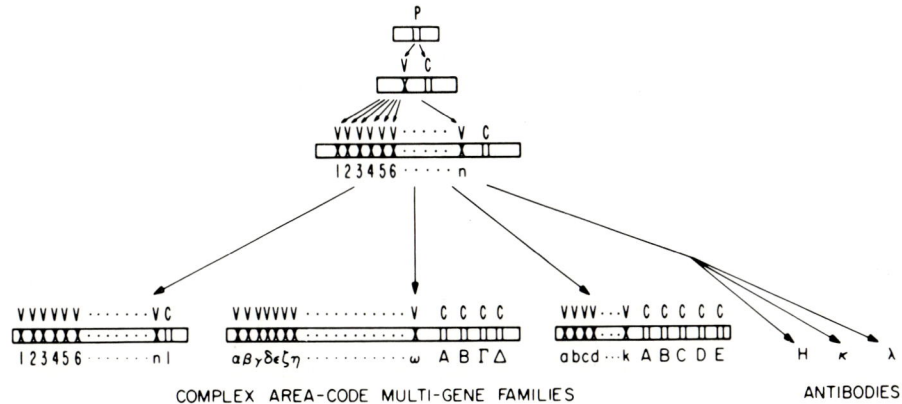

Fig. 10. A hypothetical model for the evolution of the antibody and other area-code gene families.

and/or translocation. As we shall discuss subsequently, the antibody gene families may share a common evolutionary origin with other cell-surface recognition systems (Fig.10).

L. The Immune System Employs a Variety of Mechanisms for the Amplification of Information

An enormous amount of information is required to develop a eukaryotic organism. Thus it is of interest to analyze those strategies the immune system employs for the amplification of information. Once again, many of these strategies will almost certainly be employed by other area-code systems.

The vertebrate immune system can respond specifically to a universe of different antigenic determinants by virtue of molecular interactions with complementary antibody molecules. How then can the gene products from a finite number of antibody genes react with untold numbers of different antigenic determinants? The basic strategies for the amplification of antibody information fall into 2 broad categories — genetic and molecular (Table III). Genetic strategies amplify information by producing multiple antibody genes, whereas molecular strategies amplify information by employing certain fundamental characteristics of antibody molecules themselves. i) The antibody gene families are multigenic and thus encode directly multiple receptor molecules (38). ii) Somatic mutation may occur in germ line antibody genes during the differentiation of individual lymphocytes to produce additional antibody genes (52, 53). iii) At the genetic level different V genes may associate with the same C gene (e.g., $V_1 C_\mu, V_2 C_\mu, \ldots V_n C_\mu$). In addition, the same V gene may associate with different C genes (e.g., $V_1 C_\mu, V_1 C_{\gamma 1}, V_1 C_{\gamma 2}, V_1 C_{\gamma 3}, \ldots V_1 C_\epsilon$). Thus the combinatorial-joining mechanism of DNA translocation allows one library of recognition sites to be combinatorially associated with a second library of effector functions. Moreover, during evolution crossing-over can add (or delete) homology units to C_H genes. Thus various combinations of domains can be associated in a single molecule by evolutionary mechanisms. iv) The diversity of antigen-binding sites can be increased by the combinatorial association of light and heavy chains. For example, if 10^3 different L chains could associate with 10^3 different heavy chains, 10^6 different antibody molecules will be produced ($10^3 \times 10^3 = 10^6$). Thus unrestricted light and heavy chain interactions will generate an amplification factor of $p \times q$, where p equals the number of light chains and q the number of heavy chains. v) Multispecificity is defined as the ability of a single antibody molecule to interact with a variety of different antigens, some presumably related in tertiary structure and others possibly unrelated. Thus the inherent degeneracy of the antigen-binding site is an important mechanism for amplifying the information contained in a discrete number of antibody V genes. vi) At the supramolecular level, cell-cell recognition phenomena may involve combinations of 2 or more distinct species of cell-surface molecules. For example, T-cell surveillance of virus-transformed cells requires the simultaneous recognition of both viral and transplantation antigens on the transformed cells (23). One explanation for this simultaneous dual recognition of 2 distinct molecules is that the viral antigen and the transplantation antigen associate at the cell surface to form a supramolecular complex. Obviously, the combinatorial association of cell-surface molecules can lead to significant amplification of the number of distinct cell-surface recognition units.

Thus the immune system displays a variety of strategies at the gentic, protein, and surface-display levels for the amplification of information (Table III). Other membrane recognition systems will certainly employ similar strategies.

TABLE III. Levels at Which Information Amplification Occurs

Genetic level
1. Multiple germ line genes
2. Somatic mutation
3. Combinatorial joining of V_H and C_H genes
4. Association of different homology units by crossing over during evolution.

Protein level
1. Combinatorial association of subunits
2. Multispecificity

Cell-surface displays
1. Combinations of 2 or more distinct species of cell-surface molecules

M. Summary of Features of the Immune System Which Relate to the Area-Code Hypothesis

Area-code molecules. The immune system is an intriguing microcosm of the differentiating organism. Lymphocytes develop along 2 separate cell lineages — B cells and T cells. Cell-surface molecules are expressed at various developmental stages of these lineages. Area-code molecules confer upon lymphocytes specific cell-surface addresses that direct 2 important recognition phenomena — cell-cell interactions and migration to specific tissues. The antibody molecule, a prototype area-code molecule, is divided into discrete molecular domains which carry out distinct recognition and effector functions. The immune system employs combinatorial mechanisms at the genetic, evolutionary, molecular, and supramolecular levels to amplify information.

DNA modification. The commitment of a lymphocyte to express a particular antibody polypeptide appears to require the DNA translocation of distinct V and C genes. This somatic modification of chromosomes can be passed on to progeny in a stable and heritable fashion. The DNA translocation even may create new structural genes and also alter the organization of control elements, thus modifying the future developmental options of a particular cell lineage.

Multigene families. Antibody molecules are coded for by 3 multigene families. Multigene families are a fundamental unit of chromosomal organization and evolution in the immune system and presumably in other complex eukaryotic systems.

Let us now consider the area-code hypothesis in more general terms.

IMPLICATIONS OF THE AREA-CODE HYPOTHESIS

A. The Area-Code Hypothesis Suggests That Cell-Surface Recognition Molecules Form a Recognition Code on the Cell Surface

General. As a cell undergoes successive stages of differentiation, changing patterns of area-code molecules are displayed on the cell surface (Fig. 11). These molecules, singly or in groups, constitute a cell-surface display system that gives a cell or a group of cells an individual address much as the collective digits in a phone number or postal zip code identify individual locations. Area-code molecules expressed early in the differentiation of cell lineages if not lost, will constitute a portion of the cell-surface address that identifies those cell lineages with an earlier embryonic origin than other area-code molecules ex-

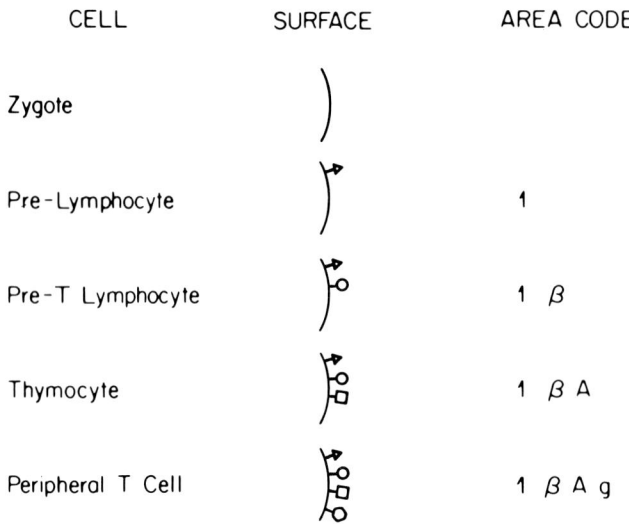

Fig. 11. The model of the display of area-code molecules. For the sake of simplicity the possibilities of loss of area-code molecules, simultaneous appearance of 2 or more area-code molecules, sharing of area-code molecules, etc., are not indicated. The use of the T lymphocyte lineage is for illustration purposes only.

pressed late in differentiation. For example, the α molecules expressed early in the differentiation of the hypothetical cells in Fig. 12 identify earlier lineage relationships than do the A molecules expressed somewhat later. Thus one can understand how the early digits in the area-code address specify general tissue relationships, whereas the later digits confer individual specificity (Fig. 12).

The area-code molecules which constitute these cell-surface displays may serve as molecular addresses to direct the migrating embryonic cells to appropriate locations or to permit specific cell-cell interactions. From Fig. 12 it can be seen how a cell which specifically recognizes, for example, the α, 1, B cell, could search and find that cell. The searching process may be a random one, or it may take advantage of the adjacent cells with slightly different area codes. For example, the cell could first find α, 10, B, then α, 9, B, and so on up the "cell-surface gradient" until it contacts α, 1, B. Thus a cell's search might be towards an increasingly better match between its area-code molecules and those of the target cell. If it strays off the path, to, e.g., α, 8, A, then the matching would decrease, and the cell could act accordingly by returning to the appropriate cell-surface gradient. Once at the target site, with its area-code maximally matched, the cell would lose its motility, perhaps through a process analogous to contact inhibition. Thus cell-surface as well as diffusible molecule gradients may play an important role in development.

Diverse cell-surface addresses may be generated by the various mechanisms employed by the immune system (Table III). These include genetic, protein, and cell-surface display strategies. These mechanisms are capable of generating an enormous array of distinct cell-surface addresses from relatively few germ-line genes. Cells may interact with one another via area-code molecules by 1 of 2 general mechanisms — self-self recognition or lock-and key recognition (Fig. 13). Thus area-code molecules play a fundamental role in the highly precise cell-cell interactions of embryogenesis.

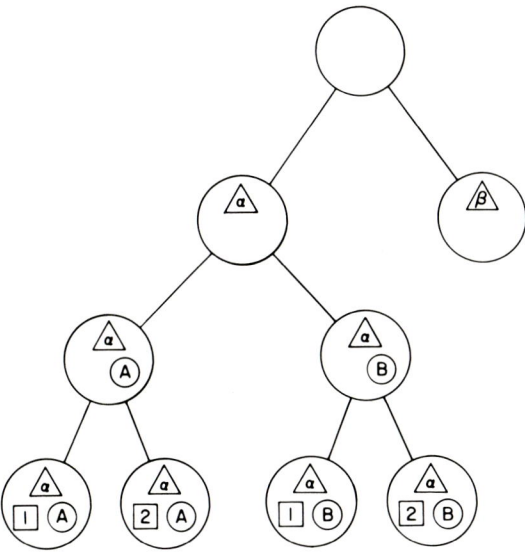

Fig. 12. A model of the lineage of a differentiation cell line. α, β, A, B, 1, and 2 denote area code molecules.

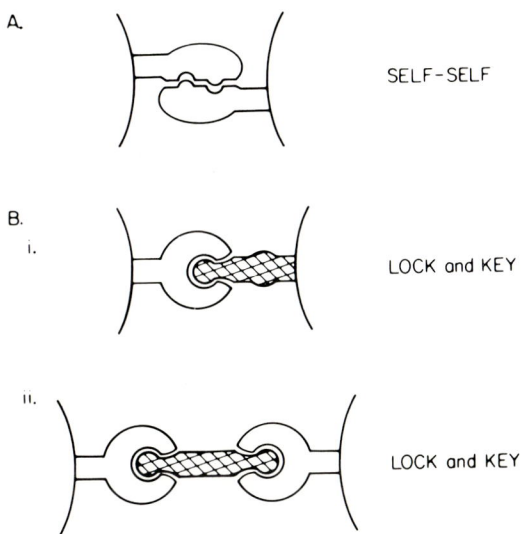

Fig. 13. Models for specific cell-cell interactions. A) Self-self. B) Lock-and-key.

B. DNA Modification May Be a Fundamental Mechanism for Cellular Differentiation

The generality of DNA modification in somatic cell lines. In addition to the translocation of V and C genes, there is evidence for DNA modification during differentiation in a variety of developmental systems. For example, chromosomal diminution or loss occurs in organisms as diverse as dipteran insets (43), ascarid nematodes (55), and copepodes

(56). In these cases, somatic cells discard much of the DNA carried by germ-line cells. This loss is due to the fragmentation or reduction in size of chromosomes when somatic cell lineages diverge from the germ line. In addition, various types of chromosomal modifications are seen in the somatic cells of a variety of organisms such as maize (57, 58), snapdragons (59), and Drosophila (60). These genetic events, apparently due to the insertion and excision of DNA elements, include the turning on and off of genes, and the transposition, deletion, and inversion of chromosomal segments. Some of these DNA modifications appear to occur in a more or less programmed fashion while others are viewed as genetic instabilities leading to random genetic changes during development. Some of these phenomena appear to be very similar to those exhibited by the insertion sequences of prokaryotes (61). Insertion sequences are specific DNA sequences which insert into and excise from bacterial chromosomes presumably by crossing over (62). Indeed, the alternative expression of 2 genes in Salmonella for flagellar proteins is mediated by DNA inversions (63). Accordingly, it appears that DNA modifications may play a role in gene expression in a wide variety of systems. Admittedly few examples have been described for eukaryotes, but the search for DNA modification during somatic differentiation is an extremely active area of research (64). Our feeling is that DNA modification will be a fundamental mechanism of gene expression in simple as well as complex gene systems (65, 66).

DNA modification and regulatory elements. B cells never express a V gene or a C gene unless they have been joined. Thus the translocation process creates a complete structural gene as well as activating (and rearranging) the regulatory elements necessary for its expression. So far the modification or rearrangement of structural genes has been found only in the immune system. However, the modification or rearrangement of regulatory elements may be more common. Single genes can be turned on or off by appropriate modification of its regulatory element, as seen with the Salmonella flagellar proteins (63). In addition, a whole gene complex may be activated by the modification of a regulatory element. For example, a reversible inversion of a promoter element has been postulated for the alternate and mutually exclusive expression of 2 closely linked gene complexes which determine alternate mating types in the fission yeast (67). If one of the activated genes has a regulatory product (e.g., an inducer), then genes controlled by that product would be affected, even if they are on another part of the genome. This appears to be the case with the *Spm* system in maize (58). In general, whole batteries of genes, linked or unlinked, could be controlled by the DNA modification of control elements. Thus areacode systems may regulate gene expression through the DNA modification of control elements.

Do the nuclear transplantation experiments of Gurdon and others argue against DNA modification as a general mechanism of differentiation? It is possible to transplant a somatic nucleus into an enucleated egg and then to stimulate this chimeric cell to begin embryogenesis (68, 69). When nuclei from blastula cells are transplanted to enucleated eggs, a high fraction of the eggs develop normally to form fertile, adult frogs. When nuclei from the intestinal cells of young embryos are used, most become defective embryos but a very small number produced fertile, adult frogs (70). These experiments suggest to some that differentiation is a reversible process and therefore the DNA in all somatic cells is identical to that of the germ cells. However, an important reservation about these experiments should be noted. There are striking differences in the results obtained with blastula and somatic cell nuclei. Clearly these nuclei have quite different potentials for generating adult frogs. Perhaps germ cells (or early stem cells), which are known to migrate through the intestinal mucosa, are the source of the nuclei which produce the few fertile adults in the nuclear transplantation experiments carried out with putative intestinal nuclei. To rule out this possibility, nuclei from lymphocytes and differentiated epidermal cells of adults have been transplanted into enucleated eggs (71, 72). In these cases, no fertile adult frogs are obtained. All embryos died at or before the early tapole stage. Indeed, the nuclei from an aneuploid liver cell line from a frog supports development almost as well as the adult somatic cell nuclei (73). These results are not inconsistent with the proposition that DNA modifications occur as somatic cell lineages undergo development. Perhaps somatic nuclei support development in nuclear transplantation experiments up until the stage at which the genes which have been modified are needed.

We feel that the nuclear transplantation experiments do not rule out the possibility that DNA modifications occur as a general phenomena in the differentiation of somatic cells. Certainly the experiments already discussed in relation to the development of the immune system provide compelling evidence for DNA modification in lymphocytes. The evidence for DNA modifications is also compelling in a variety of other systems as discussed above.

C. Multigene Families Appear to be a Fundamental Unit of Eukaryotic Gene Organization and Evolution

General. Multigene families display 4 fundamental characteristics: multiplicity, close linkage, sequence homology, and similar or overlapping functions (see Ref. 74 for review). A variety of eukaryotic genes exhibit these properties including the ribosomal RNA genes (75), tRNA genes (76), histone genes (77), the β-like globin genes (78), and the DNA satellites (79). Multigene families may range in size from a few gene members to thousands of gene members. Indeed, simple multigene families appear to code for several of the differentiation antigens found on lymphocytes including TL (80), Qa (81), and Ia (82). Accordingly, area-code families may range from few to many gene members.

Evolution of primordial multigene families. Multigene families are found in eukaryotes but not in prokaryotes (74). Multigene families may have evolved in response to the informational requirements of differentiation in multicelluar organisms (see Fig. 1). This supposition suggests that multigene families will be employed in cell-recognition systems and accordingly, will code for a variety of area-code molecules.

New evolutionary unit. New multigene families can arise in 2 ways: 1) gene duplication from a single gene, or 2) duplication of all or part of a preexisting multigene family. The duplication of a preexisting multigene family may occur in 2 ways: tetraploidization (see Ref. 83) or duplications and translocation of a portion of a chromosome. It obviously requires a long period of evolutionary time to produce a large multigene family starting with a single gene. In contrast, the duplication of an entire multigene family presumably generates new gene families by a single genetic event. Thus evolution can proceed at a more rapid pace by the instantaneous generation of entire families of new genes. The duplication of multigene families is illustrated by the homology relationships of the antibody gene families which suggest that all 3 gene families were derived from a common ancestral multigene family. Thus the multigene family is a basic unit of eukaryotic evolution. New and complex area-code gene families can obviously be created rapidly.

Homologies among multigene families. A new multigene family arising by duplication from a primordial gene family may interact with its sister-gene family as occurs between the light and heavy chain gene families of antibodies. Alternatively, it may evolve to assume new functions. Evolutionary relationships among multigene families might be discerned from 2 aspects of homology: i) amino acid sequence homology among gene products and ii) shared complex control mechanisms such as V-C translocation. Once more data are available on a variety of area-code systems, these homology analyses should allow us to construct a genealogical tree of relationships among the multigene families encoding these area-code systems.

POTENTIAL SYSTEMS EMPLOYING THE AREA-CODE STRATEGY

A. The Immune System

Antibodies. Antibody molecules and genes provide a model system for posing questions about other complex area-code systems (Table IV). Other area-code systems may employ some but not necessarily all of the strategies of the antibody system. Indeed, some will probably present novel strategies and characteristics. However, the questions posed in Table IV help us to approach experimentally other area-code systems, both simple and complex, at the protein, genetic, regulatory, and evolutionary levels.

The study of antibody molecules and genes is currently under intensive investigation in many different laboratories (see Ref. 84). Important unresolved experimental questions include the following: i) What are the mechanisms responsible for antibody diversity? ii) What is the molecular mechanism of V-C translocation or joining? iii) How are antibody genes and their control elements organized on vertebrate chromosomes? iv) How many V genes are present in various antibody gene families? v) Are V genes translocted to C genes according to an orderly developmental program? vi) What developmental strategy is used for programming specific combinations of light chain genes and heavy chain genes? Certainly the investigation of these and many other questions will provide important insights into this model area-code system.

T-cell receptors. The T-cell receptor has been an elusive entity. Recently, however, serological and genetic studies suggest that the T-cell receptor for antigen binding employs a V_H gene from the B-cell H chain family presumably associated with a new C_H gene (see Ref. 85). The presence of a light chain in the T-cell receptor is still a matter of uncertainty (86). Thus T and B cells both employ the same V_H gene family. However, the expression of V_H genes on T or B cells appears to be controlled by the nature of the C_H gene to which they are translocated. The various C_H genes given in Fig. 7 are expressed on B cells, whereas the putative C_H gene(s) for T cells is expressed only on that cell lineage. Thus the same area-code gene library can be employed for 2 distinct receptor systems by virtue of the DNA translocation mechanism and the differential expression of C_H genes in the T- and B-cell lineages.

The study of T-cell receptors is in its infancy. Very little is known about the structure of these molecules. All of the questions raised for antibody molecules and genes of B cells can be asked of the T-cell receptor (see above and Table IV).

H-2 complex. Chromosome 17 of the mouse appears to code for a series of multi-gene families (87) (Fig. 14). For example, by simplistic genetic calculations the region between K and Tla has sufficient DNA to code for about 16,000 polypeptides 100 residues in length (e.g., V region size). The T/t family with at least 6 complementation groups appears to encode cell-surface proteins regulating neuroectodermal development during embryogenesis (1). The S family codes for the structural and/or regulatory elements of several complement components (88). The Qa region appears to have 2 or more genes coding for differentiation antigens on lymphocytes (89). The D (90), I (87), and Tla (91)

TABLE IV. Questions About Potential Area-Code Molecules and Genes Based on an Analysis of the Immune System

A. Protein level
   1. Are the molecules composed of multiple subunits?
   2. If so, do they exhibit combinatorial association?
   3. Are homology units present? How large are they?
   4. Do they show homology to immunoglobulins? To other known gene systems?
   5. Are V and C regions present?
   6. What type of molecular recognition functions do they perform?
B. Genetic level
   1. Are the polypeptides encoded by multigene families?
   2. Are V and C genes present?
   3. How large is the gene family?
   4. How are the various control and structural genes organized?
C. Regulatory level
   1. Is there V-C translocation?
   2. What is the pattern of clonal expression of individual area-code molecules?
   3. What is the nature of the programs which govern development?
D. Evolutionary level
   1. Are the multigene families within the system homologous to one another?
   2. Are they related to other known multigene systems?

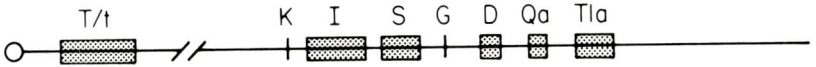

Chromosome 17

Fig. 14. Chromosome 17 of the mouse. Shaded bars represent 2 or more closely linked homologous genes.

gene families also appear to have 2 or more homologous gene members. Moreover, indirect evidence suggests that the genes coding for the transplantation antigens may be present in multiple copies (92). Accordingly, chromosome 17 of the mouse appears to be a library of gene families coding for cell-surface molecules.

The gene products of the K, D, Qa, and Tla gene families show 2 interesting relationships to one another. i) All are cell-surface glycoproteins of about 45,000 molecular weight (89, 93). ii) Each is associated with β_2-microglobulin, a polypeptide of about 100 residues that demonstrates significant homology with homology units of antibody constant regions (94, 95). These relationships suggest that these differentiation antigens may be homologous and thereby share a common evolutionary origin. This supposition is supported by preliminary amino acid sequence analyses of the N-terminal regions of the K and D molecules that demonstrate sequence homology. A similar analysis has not yet been carried out on the Qa and Tla gene products. The association between these differentiation antigens and β_2-microglobulin provides a fascinating relationship between the antibody system and these cell-surface molecules. Indeed, amino acid sequence data have demonstrated a striking homology between an internal 17-residue portion of a human transplantation antigen and a portion of the antibody V region (C. Terhorst and J. Strominger, personal communication). If further sequence analyses confirm this observation, it appears that antibodies and transplantation antigens may be evolutionarily i̇̇ ed. It will be interesting to determine whether the same is true of the Qa and Tla gene products.

The functions of these cell-surface molecules are uncertain, but the Qa, Tla, and transplantation antigens are expressed in different patterns on different populations of lymphocytes. Thus these cell-surface molecules may be encoded by small multigene families that diverged from a common ancestor and which have diverged to carry out distinct cell-surface functions.

One gene product from the T/t system has been characterized and it also appears to be 45,000 in molecular weight and associated with a β_2-microglobulin-like molecule (96). The I region molecules differ in molecular weight from the transplantation antigens and do not show obvious sequence hemology at their N termini (97). Preliminary amino acid sequence studies indicate that the non-membrane-associated S region gene products also fail to show homology to immunoglobulins (98). Clearly additional data are necessary before any convincing relationships can be established among these gene families.

Certain of these differentiation antigens carry out cell-surface recognition functions (e.g., K, Ia, and D) and, accordingly, are area-code molecules. The others may be area-code molecules, although their functions are generally unknown. We are in a position to ask of these gene families many of the questions listed in Table IV.

B. Other Developmental Systems

General. There are a variety of developmental systems where area-code molecules appear to play an important role. For example, embryonic tissues appear to have cell-surface recognition molecules that mediate tissue-specific cellular interactions (99). The imaginal disks of Drosophila larva can be disassociated and reaggregated in a manner that suggests area-code molecules may be involved (100). The retinal nerves of the goldfish upon sectioning appear capable of reattaching to their specific tectal counterparts (101). One possibility is that these specific interactions are mediated through the cell-surface addresses provided by area-code molecules. The difficulty with each of these systems has been that large quantities of homogeneous area-code molecules are difficult to obtain. A promising approach is to create homogeneous cell lines from individual cells in these systems. Homogeneous cell lines could be produced from cells transformed by viral, chemical, or physical agents. In addition, permanent cell lines may be generated by fusing primary cells with established cell lines. This latter approach has been successful in creating cell lines secreting homogeneous anti-H2 antibodies by fusing normal B cells synthesizing anti-H-2 antibodies with B-cell tumors (102). The production of large quantities of cells from homogeneous cell lines appears to be one of the primary prerequisites for the detailed biochemical analysis of many interesting developmental systems.

Tumor-specific antigens. Certain tumor-specific antigen systems appear to be potential area-code systems which can be studied currently. One of the key experimental approaches to the study of the immune system was the use of tumors cultured in vivo and in vitro. These tumors were homogeneous clones of lymphoid cells and provided large amounts of experimental material. Since we believe that different embryonic cells have different cell-surface arrays of area-code molecules, we might ask if there are presently available tumor cell lines which display a large diversity of line-specific surface molecules or antigens. In fact, such individually specific cell-surface antigens have been found on a wide variety of tumors induced by physical or chemical agents. In certain inbred strains of mice both methylcholanthrene (103) and ultraviolet light (104) induce tumors with individually unique tumor antigens. The repertoire of these unique tumor antigens appears very large (103). Moreover, large quantities of tumor cells can be raised for analysis of the tumor antigens and their genes. It will be important to determine whether these antigens are abnormal molecules related only to tumor transformation or whether they might indeed be area-code molecules playing a fundamental role in some stage of development. For these systems it will be interesting to answer many of the questions raised in Table IV.

CODA

An analysis of the immune system has led to the area-code hypothesis. The strategies employed by the antibody system for gene organization, regulation and evolution provide a model for asking experimentally testable questions about other potential area-code systems (Table IV). Investigations of area-code systems have 2 basic requirements. First, homogeneous clones of cells expressing the potential area-code molecules must be available. Second, microchemical techniques must be employed to characterize these molecules which are generally available in very small quantities. Our laboratories have been engaged in developing microsequencing techniques for the last several years. We are now applying these techniques to the analysis of several potential area-code systems (93, 97) including the ultraviolet light-induced tumors of C3H mice. This tumor system may provide a second detailed view of a complex area-code system.

ACKNOWLEDGMENTS

We thank the Caltech secretaries for their patience. This work was supported by grants NSF PCM71-00770, USPHS A1-10781, USPHS CA15926, and GM06965. H.V.H. is a Danforth Fellow.

REFERENCES

1. Bennett D: Cell 6:441, 1975.
2. Changeux J-P, Danchin A: Nature (London) 264:705, 1976.
3. Sperry RW: Proc Natl Acad Sci USA 50:703, 1963.
4. Cantor H, Boyse EA: Immunol Rev 33:105, 1977.
5. Jerne NK: Sci Am 229:52, 1973.
6. Metcalf D, Moore MAS: In Neuberger A, Tatum EL (eds): "Haemopoietic Cells." Amsterdam: North Holland Research Monographs, 1971, Frontiers of Biology Series, vol 24, p 172.
7. Cantor H, Weisman I: Prog Allergy 20:1, 1976.
8. Houssaint E, Belo M, Le Douarin NM: Dev Biol 53:250, 1976.
9. Good RA: Harvey Lect 67:1, 1973.
10. Cazenave P-A, Cavaillon JM, Bona C: Immunol Rev 34:34, 1977.
11. Boyse EA, Old LJ: Annu Rev Genet 3:269, 1969.
12. Storrie B, Goldstein C, Boyse EA, Hammerling U: J Immunol 116:1358, 1976.
13. Klein J: "Biology of the Mouse Histocompatibility-2 Complex." New York: Springer-Verlag, 1975, p 135.
14. Takahashi T, Olds J, Boyse EA: J Exp Med 131:1325, 1970.
15. Raff M: Nature (London) 242:19, 1973.
16. Jerne NK: Ann Immunol (Inst Pasteur) 125C:373, 1974.
17. Jerne NK: Harvey Lect 70:93, 1976.
18. Jerne NK: "27th Mosbacher Colloquium." Berlin: Springer-Verlag, 1976, p 259.
19. Cantor H, Shen FW, Boyse EA: J Exp Med 143:1391, 1976.
20. Shevach EM: Fed Proc Fed Am Soc Exp Biol 35:2048, 1976.

21. Metcalf D, Moore MAS: In Neuberger A, Tatum EL (eds): "Hemopoietic Cells." Amsterdam: North Holland Research Monographs, 1971, Frontiers of Biology Series, vol 24, p 70.
22. Katz DH, Benacerraf B (eds): "The Role of Products of the Histocompatibility Complex in Immune Responses." New York: Academic Press, 1976.
23. Doherty PC Götze D, Trinchieri G, Zinkernagel RM: Immunogenetics 3:517, 1976.
24. Möller G (ed): "Biochemistry and Biology of Ia Antigens." Immunol Rev 30, 1976.
25. Klein J: Annu Rev Genet 8:63, 1974.
26. Rosenthal AS, Shevach EM: J Exp Med 138:1194, 1973.
27. Cosenza H, Julius MH, Augustin AA: Immunol Rev 34:3, 1977.
28. Rodkey LS: J Exp Med 139:712, 1974.
29. Eichmann K, Rajewsky K: Eur J Immunol 5:661, 1975.
30. Cantor H: In Sercarz EE, Herzenberg LA, Fox CF (eds): "The Immune System. II. Regulatory Genetics." New York: Academic Press (In press).
31. Boyse EA, Cantor H: In "First International Conference on the Molecular Basis of Cell-Cell Interaction" (San Diego). New York: Academic Press (In press).
32. Huber B, Cantor H, Shen FW, Boyse EA: J Exp Med 144:1128, 1976.
33. Gally J: In Sela M (ed): "The Antigens." New York: Academic Press, 1973, vol 1, p 162.
34. Gally J, Edelman GM: Annu Rev Genet 6:1, 1972.
35. Edelman GM, Cunningham BA, Gall W, Gottlieb P, Rutishauser U, Waxdal M: Proc Natl Acad Sci USA 63:78, 1969.
36. Kehoe JM, Fougereau M: Nature (London) 224:1212, 1969
37. Mage R, Lieberman R, Potter M, Terry WD: In Sela M (ed): "The Antigens." New York: Academic Press 1973, vol 1, p 300.
38. Hood L, Kronenberg M, Early P, Johnson N: In Sercarz EE, Herzenberg LA, Fox CF (eds): "The Immune System. II. Regulatory Genetics." New York: Academic Press (In press).
39. Hood L, Loh E, Hubert J, Barstad P, Eaton B, Early P, Fuhrman J, Johnson N, Kronenberg M, Schilling J: Cold Spring Harbor Symp Quant Biol 41:817, 1976.
40. Hood L: Fed Proc Fed Am Soc Exp Biol 31:177, 1972.
41. Hozumi N, Tonegawa S: Proc Natl Acad Sci USA 73:3632, 1976.
42. Tonegawa S, Hozumi N, Matthyssens G, Schuller R: Cold Spring Harbor Symp Quant Biol 41:877, 1976.
43. Tonegawa S, Brack C, Hirama M, Hozumi N, Matthyssens G: Cold Spring Harbor Symp Quant Biol (In press).
44. Dreyer WJ, Gray W, Hood L: Cold Spring Harbor Symp Quant Biol 32:353, 1967.
45. Edelman GM: Cold Spring Harbor Symp Quant Biol 41:891, 1976.
46. Klinman NR, Press JL: J Exp Med 141:1133, 1975.
47. Sigal NH, Gearhart PJ, Press JL, Klinman NR: Nature (London) 259:51, 1976.
48. Lydyard PN, Grossi CE, Cooper MD: J Exp Med 144:79, 1976.
49. Amzel L, Poljak R, Saul F, Varga J, Richards F: Proc Natl Acad Sci USA 71:1427, 1974.
50. Padlan EA, Segal DM, Cohen GH, Danes DR: In Sercarz E, Williamson A, Fox CF (eds): "The Immune System: Genes, Receptors, Signals." New York: Academic Press, 1974, p 7.
51. Edmunson AB, Ely KR, Girling RL, Abola EF, Schiffer M, Westholm FA, Fausch ND, Deutsch HF: Biochemistry 13:3816, 1974.
52. Cohn M, Blomberg B, Geckeler W, Raschke W, Riblet R, Weigert M: In Sercarz E, Williamson A, Fox CF (eds): "The Immune System: Genes, Receptors, Signals." New York: Academic Press, 1974, p 89.
53. Tonegawa S: Proc Natl Acad Sci USA 73:203, 1976.
54. Bantock CR: J Embryol Exp Morph 24:257, 1970.
55. Goldstein P: J Morphol 152:141, 1977.
56. Beermann S: Chromosoma (Berlin) 60:297, 1977.
57. McClintock B: Brookhaven Symp Biol 18:162, 1965.
58. Nevers P, Saedler H: Nature (London) 268:109, 1977.
59. Fincham JRS, Harrison BJ: Heredity 22:211, 1967.
60. Green NM: Genetics (Suppl) 73:187, 1974.
61. Starlinger P: Crit Rev Microbiol (In press).
62. Cohen SN: Nature (London) 263:731, 1976.
63. Zieg J, Silverman M, Hilmen M, Siman M: Science 196:170, 1977.
64. Bukhari AJ (ed): "DNA insertion elements, plasmids and episomes." Cold Spring Harbor, New York: Cold Spring Harbor Laboratories (In press).

65. Holliday R, Pugh JE: Science 187:226, 1975.
66. Sager R, Kitchin R: Science 189:426, 1975.
67. Egel R: Nature (London) 266:172, 1977.
68. Gurdon JB, Laskey RA: J Embryol Exp Morphol 24:227, 1970.
69. Briggs R, King TF: Proc Natl Acad Sci USA 38:455, 1972.
70. Gurdon JB, Uehlinger V: Nature 210:1240, 1966.
71. Gurdon JB, Laskey RA, Reeves OR: J Embryol Exp Morphol 34:93, 1975.
72. Wabl MR, Brun RB, DuPasquier L: Science 190:1310, 1975.
73. Kobel HR, Brun RB, Rischberg M: J Embryol Exp Morphol 29:539, 1973.
74. Hood L, Campbell JH, Elgin SCR: Annu Rev Genet 9:305, 1975.
75. Birnstiel ML, Chipchase M, Spiers J: Prog Nucleic Acid Res Mol Biol 11:351, 1971.
76. Clarkson SG, Birnstiel ML: Cold Spring Harbor Symp Quant Biol 38:451, 1973.
77. Schaffner W, Gross K, Telford J, Birnstiel M: Cell 8:471, 1976.
78. Kitchen H, Bayer S (eds), "Hemoglobin: Comparative Molecular Biology Models for the Study of Disease." Ann NY Acad Sci 24, 1974.
79. Gall JG, Cohen EH, Atherton DD: Cold Spring Harbor Symp Quant Biol 38:417, 1973.
80. Boyse EA, Stockert E, Olds LJ: In Rose NR, Milgrom F (eds): "International Convocation on Immunology." Basel, Switzerland: S Karger, 1968, p 353.
81. Flaherty L: Immunogenetics 3:533, 1976.
82. McDevitt HO, Delovitch TL, Press JL, Murphy DB: Transplant Rev 30:197, 1976.
83. Ohno S: "Evolution by Gene Duplication," New York: Springer-Verlag, 1970.
84. "Origins of Lymphocyte Diversity." Cold Spring Harbor Symp Quant Biol 41:627, 1977.
85. Janeway CA, Wigzell H, Binz H: Scand J Immunol 5:993, 1976.
86. Rajewsky K: In Sercarz EE, Herzenberg LA, Fox CF (eds): "The Immune System. II. Regulatory Genetics." New York: Academic Press (In press).
87. McDevitt HO: Fed Proc Fed Am Soc Exp Biol 35:2168, 1976.
88. Schreffler DC: Transplant Rev 32:140, 1976.
89. Michaelson J, Flaherty L, Vitetta E, Poulik MD: J Exp Med 145:1066, 1977.
90. Hansen TH, Cullen SE, Sachs DH: J Exp Med 145:438, 1977.
91. Klein J: "Biology of the Mouse Histocompatibility-2 Complex." New York: Springer-Verlag, 1975, p 241.
92. Garrido F, Festenstein H, Schirrmacher V: Nature (London) 261:705, 1976.
93. Silver J, Hood L: In Eisen HN, Reisfeld RA (eds): "Contemporary Topics in Immunobiology." New York: Plenum Press, 1976, vol 5, p 35.
94. Smithies O, Poulik MD: Proc Natl Acad Sci USA 69:2914, 1972.
95. Cunningham BA, Wang JL, Berggård I, Peterson PA: Biochemistry 12:4811, 1973.
96. Jacob F: Immunol Rev 33:3, 1977.
97. Silver J, Cecka JM, McMillan M, Hood L: Cold Spring Harbor Symp Quant Biol 41:369, 1976.
98. Bolotin C, Morris S, Tack B, Prahl J: Biochemistry 16:2008, 1977.
99. Moscona AA, Housman RE: In Lash J, Burger M (eds): New York: Raven Press (In press).
100. Garcia-Bellido A: In Ursprung H, Nöthiger R (eds): Berlin: Springer-Verlag, 1972, vol 5, p 59.
101. Sperry RW: In DeHaan, Urspring H (eds): "Organogenesis." New York: Holt, Reinhart, and Winston, 1965, p 161.
102. Galfre A, Howe SC, Milstein C, Butcher AW, Howard JC: Nature (London) 266:550, 1976.
103. Basombrio MA: Cancer Res 30:2458, 1970.
104. Kripke ML: J Natl Cancer Inst 53:1333, 1974.
105. Berget SM, Berk AJ, Harrison T, Sharp P: Cold Spring Harbor Symp Quant Biol (In press).
106. Broker TR, Chew LT, Dunn A, Gelinas RE, Hassell J, Kleosig DF, Lewis J, Roberts RJ: Cold Spring Harbor Symp Quant Biol (In press).

Ganglioside Structures and Distribution: Are They Localized at the Nerve Ending?

R. W. Ledeen

Departments of Neurology and Biochemistry, Albert Einstein College of Medicine, Bronx, New York 10461

Gangliosides generally provide a small portion of the complex carbohydrate content of cell surfaces. An exception is the central nervous system where they comprise up to 5–10% of the total lipid of some membranes. This tissue is unique in that the quantity of lipid-bound sialic acid exceeds that of the protein-bound fraction. Over 30 different molecular species have been characterized to date. These range in complexity from sialosylgalactosyl ceramide with 2 sugars to the pentasialoganglioside of fish brain with 9 carbohydrate units. Virtually all cellular and subcellular fractions of brain that have been carefully examined contain gangliosides to one degree or another, but the majority of brain ganglioside is located in the neurons. Their mode of distribution within the neuron has not been entirely clarified by subcellular studies. Calculations based on reported values for axon terminal density and synaptosomal ganglioside concentration in the rat reveal that nerve endings contribute less than 12% of total cerebral cortical ganglioside. It is concluded that the plasma membranes of neuronal processes contain most of the neuronal ganglioside. These and other considerations suggest the possibility that gangliosides may be distributed over the entire neuronal surface.

Key words: gangliosides; glycosphingolipids; oligosaccharide structures; nervous system; neurons; subcellular distribution

Gangliosides and other glycolipids usually comprise a small proportion of the total glycoconjugates that exist on the surface of cells. Major exceptions to this general rule are found in the central nervous system of mammals where the quantity of lipid-bound sialic acid exceeds that of protein-bound sialic acid, and where the myelin membrane contains substantial glycolipid but only a minor amount of glycoprotein. Similarities in the oligosaccharide structures of gangliosides and sialoglycoproteins have led to speculation that

Received April 14, 1977; accepted June 23, 1977.

0091-7419/78/0801-0001$03.20 © 1978 Alan R. Liss, Inc

both groups may subserve similar membrane functions. In containing a single oligosaccharide chain per molecule and a ceramide unit in place of a peptide, glycosphingolipids such as the gangliosides comprise simpler and hence more easily characterized structures than the glycoproteins.

The discovery that gangliosides are receptors for a variety of bacterial toxins and possibly some viruses (reviewed in Refs. 1 and 2) has renewed interest in their possible role as receptors for natural agonists. Demonstration of regions of peptide homology between the B chain of cholera toxin and the β subunits of such glycoprotein hormones as thyrotropin, luteinizing hormone, human chorionic gonadotrophin, and follicle-stimulating hormone has led to the proposal that these proteins might utilize a common mechanism based on ganglioside receptors for transporting subunits across and possibly within the plane of the membrane (2–4). More generally, gangliosides might have as one of their functions the transfer of information from the exterior to the interior of the cell.

If this is indeed a correct view of ganglioside function the large variety of oligosaccharide structures present in both neural and extraneural tissues could provide the specificity required for recognition and binding of an equally large variety of biologically active peptides. At the same time it may be pointed out that several other functions have been proposed, particularly in the central nervous system (CNS) where the high concentration of gangliosides in neuronal elements has led to much speculation regarding a possible role in nerve conduction and/or synaptic transmission. Aside from intriguing hints that gangliosides help to retain the excitability of isolated cerebral tissues (5), very little evidence has come forth to support this or any other hypothesis concerning their role in the CNS. It seems likely that a variety of functions could be involved, consistent with the widespread distribution of these substances throughout the myriad components of the nervous tissue and the impressive diversity of structure that is now apparent. Following a review of ganglioside structures some aspects of their CNS distribution will be critically examined, particularly in regard to the question of localization within the neuron.

STRUCTURES

The distinguishing feature of gangliosides, as opposed to the large group of neutral glycosphingolipids, is the presence of one or more sialic acid units in the oligosaccharide chain. These generally occupy a terminal position, being linked to either galactose or another sialic acid. Some 17 or 18 different sialic acids have been found in nature (6) but only a few of these have been detected thus far in gangliosides. The most commonly occurring forms are N-acetylneuraminic acid (NAN) and N-glycolylneuraminic acid (NGN). The former is virtually the only type found in brain gangliosides of most mammals, although in some cases (e.g., bovine) a small percentage of NGN has been detected (7, 8). Both NAN and NGN, and often acylated forms of each, are found in extraneural tissues, the types and amounts showing considerable species specificity. NAN is the only sialic acid that has been reliably identified in man.

Glycosphingolipids may be divided into 2 general categories based on the carbohydrate immediately linked to ceramide. The first category is derived from galactosylceramide (Fig. 1). It is a relatively small family with a single ganglioside ($G_7 = G_{M4}$). The large majority of glycosphingolipids, including virtually all gangliosides except G_7, are derived from the second family originating with glucosylceramide (Fig. 2). This family diverges into 4 major branches from lactosyl ceramide, based on the nature of the third

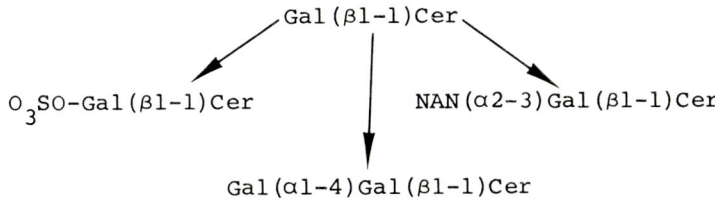

Fig. 1. Family of glycosphingolipids derived from galactosylceramide.

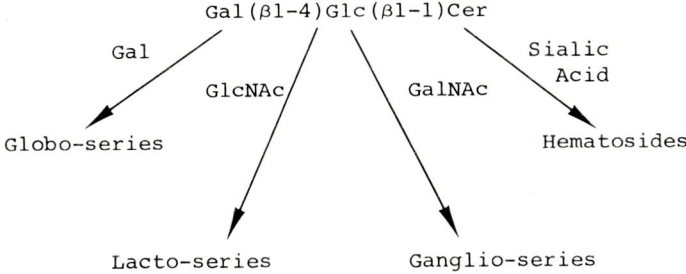

Fig. 2. Family of glycosphingolipids derived from glucosylceramide.

attached sugar. The globo series (following the currently adopted nomenclature, see Ref. 9) has galactose as the third sugar and includes such substances as globoside I (10, 11) and Forssman glycolipid (12):

GalNAc (β1-3) Gal (α1-4) Gal (β1-4) Glc (β1-1) Cer
Globoside I

GalNAc (α1-3) GalNAc (β1-3) Gal (α1-4) Gal (β1-4) Glc (β1-1) Cer
Forssman glycolipid

The added galactose in most cases has an α-glycosidic linkage of the type shown, but it has also been found to have a different linkage (e.g., β1-4) in certain of the blood group A-active fucolipids of hog gastric mucosa (13–15). The globo series is the only one of the 4 groups depicted in Fig. 2 which has not yet been found to include gangliosides.

The lacto, ganglio, and hematoside series (Fig. 2) contain most of the 30-odd gangliosides characterized to date. Their structures are summarized in Tables I–III along with some of the sources from which they were obtained. The tissues listed are representative rather than comprehensive. Animal species are indicated in some instances where this appears to be a factor affecting structural specificity.

Hematosides (Table III) are defined in the broad sense as those gangliosides which lack hexosamine. This widely used term arose from the fact that they were originally isolated from erythrocytes (16), but they have since been detected in virtually every

TABLE I. Structures of Vertebrate Gangliosides. Ganglio-N-Glycose Series

Structure	Symbol[a]	Source	Reference
Monosialo			
1. GalNAc (β1-4) Gal (β1-4) Glc (β1-1) Cer 　　　　　　　　3 　　　　　　　　↑ α 　　　　　　　　2 　　　　　　　NAN	G_{M2}	normal brain Tay-Sachs brain	83 84
2. GalNAc (β1-4) Gal (β1-4) Glc (β1-1) Cer 　　　　　　　　3 　　　　　　　　↑ α 　　　　　　　　2 　　　　　　　NGN	G_{M2}(NGN)	spleen, kidney, (bovine)	17
3. Gal (β1-3) GalNAc (β1-4) Gal (β1-4) Glc (β1-1) Cer 　　　　　　　　　　　　3 　　　　　　　　　　　↑ α 　　　　　　　　　　　2 　　　　　　　　　　NAN	G_{M1}	normal brain G_{M1} gangliosidosis spleen, kidney adrenal medulla	85 86 17 87
4. Gal (β1-3) GalNAc (β1-4) Gal (β1-4) Glc (β1-1) Cer 　　　　　　　　　　　　3 　　　　　　　　　　　↑ α 　　　　　　　　　　　2 　　　　　　　　　　NGN	G_{M1}(NGN)	spleen, kidney (bovine)	17
5. Gal (β1-3) GalNAc (β1-4) Gal (β1-4) Glc (β1-1) Cer 　　2　　　　　　　　　3 　↑ α　　　　　　　↑ α 　　1　　　　　　　　　2 　Fuc　　　　　　　　NAN		brain (bovine) testis (boar)	88 89
6. Gal (β1-3) GalNAc (β1-4) Gal (β1-4) Glc (β1-1) Cer 　　2　　　　　　　　　3 　↑ α　　　　　　　↑ α 　　1　　　　　　　　　2 　Fuc　　　　　　　　NGN		liver (bovine)	17
Disialo			
7. GalNAc (β1-4) Gal (β1-4) Glc (β1-1) Cer 　　　　　　　　3 　　　　　　　　↑ α 　　　　　　　　2 　　　　NAN (8←2α) NAN	G_{D2}	brain	90, 91
8. Gal (β1-3) GalNAc (β1-4) Gal (β1-4) Glc (β1-1) Cer 　　3　　　　　　　　　3 　↑ α　　　　　　　↑ α 　　2　　　　　　　　　2 　NAN　　　　　　　　NAN	G_{D1a}	brain adrenal medulla muscle retina	92–95 87 112 119
9. Gal (β1-3) GalNAc (β1-4) Gal (β1-4) Glc (β1-1) Cer 　　3　　　　　　　　　3 　↑ α　　　　　　　↑ α 　　2　　　　　　　　　2 　NAN　　　　　　　　NGN	G_{D1a} (NAN/NGN)	brain (bovine) adrenal medulla (bovine)	88 87
10. Gal (β1-3) GalNAc (β1-4) Gal (β1-4) Glc (β1-1) Cer 　　3　　　　　　　　　3 　↑ α　　　　　　　↑ α 　　2　　　　　　　　　2 　NGN　　　　　　　　NAN	G_{D1a} (NGN/NAN)	brain (bovine)	88

TABLE I. Structures of Vertebrate Gangliosides. Ganglio-N-Glycose Series (continued)

Structure	Symbol[a]	Source	Reference
Disialo (continued)			

11. Gal (β1-3) GalNAc (β1-4) Gal (β1-4) Glc (β1-1) Cer
      3   3
      ↑ α ↑ α
      2   2
    NGN   NGN
	$G_{D1a}(NGN)_2$	liver, spleen, kidney (bovine)	17

12. Gal (β1-3) GalNAc (β1-4) Gal (β1-4) Glc (β1-1) Cer
                           3
                         ↑ α
                         2
                 NAN (8←2α) NAN
	G_{D1b}	brain
		retina
			92, 96–98
			119

13. GalNAc (β1-4) Gal (β1-3) GalNAc (β1-4) Gal (β1-4) Glc (β1-1) Cer
           3   3
          ↑ α ↑ α
          2   2
         NAN   NAN
		brain	99

Trisialo			

14. Gal (β1-3) GalNAc (β1-4) Gal (β1-4) Glc (β1-1) Cer
      3   3
      ↑ α ↑ α
      2   2
 NAN (8←2α)NAN   NAN
	G_{T1a}	brain (human)	100

15. Gal (β1-3) GalNAc (β1-4) Gal (β1-4) Glc (β1-1) Cer
      3   3
      ↑ α ↑ α
      2   2
    NAN   NAN (8←2α)NAN
	G_{T1b}	brain	92, 96

16. Gal (β1-3) GalNAc (β1-4) Gal (β1-4) Glc (β1-1) Cer
                         3
                         ↑ α
                         2
        NAN(8←2α)NAN(8←2α)NAN
	G_{T1c}	brain (fish)	29

Tetrasialo			

17. Gal (β1-3) GalNAc (β1-4) Gal (β1-4) Glc (β1-1)Cer
      3   3
      ↑ α ↑ α
      2   2
 NAN (8←2α) NAN   NAN (8←2α) NAN
	G_{Q1b}	brain (human)	31, 101

18. Gal (β1-3) GalNAc (β1-4) Gal (β1-4) Glc (β1-1) Cer
      3   3
      ↑ α ↑ α
      2   2
    NAN   NAN (8←2α) NAN (8←2α) NAN
	G_{Q1c}	brain (fish)	29

Pentasialo			

19. Gal (β1-3) GalNAc (β1-4) Gal (β1-4) Glc (β1-1) Cer
      3   3
      ↑ α ↑ α
      2   2
 NAN(8←2α) NAN   NAN (8←2α) NAN (8←2α) NAN
	G_{P1}	brain (fish)	29

[a]The symbols are those of the Svennerholm system (Ref. 102). Additional symbols beyond those originally proposed have been added in a manner thought to be consistent with the system as a whole. Where more than one type of sialic acid is present in the same molecule the first designated within ( ) is that most distal from ceramide.

TABLE II. Structures of Vertebrate Gangliosides. Lacto-N-Glycose Series

Structure	Source	Reference
1. Gal (β1-4) GlcNAc (β1-3) Gal (β1-4) Glc (β1-1) Cer 3 $\uparrow \alpha$ 2 NAN	erythrocytes (human) peripheral nerve brain spleen, kidney, liver muscle plasma	79, 20 19 19 17 19 19
2. Gal (β1-4) GlcNAc (β1-3) Gal (β1-4) Glc (β1-1) Cer 6 $\uparrow \alpha$ 2 NAN	spleen, kidney, liver	17
3. Gal (β1-4) GlcNAc (β1-3) Gal (β1-4) Glc (β1-1) Cer 3 $\uparrow \alpha$ 2 NGN	spleen, kidney (bovine)	17, 103, 104
4. Gal (β1-4) GlcNAc (β1-3) Gal (β1-4) Glc (β1-1) Cer 3 3 $\uparrow \alpha$ $\uparrow \alpha$ 2 1 NAN Fuc	kidney (human)	105
5. Gal (β1-4) GlcNAc (β1-3) Gal (β1-4) Glc (β1-1) Cer 3 $\uparrow \beta$ 1 GlcNAc (4←1β) Gal (3←2α) NAN	spleen (human)	106

TABLE III. Structures of Vertebrate Gangliosides. Hematoside Series

Structure	Symbol[a]	Source	Reference
1. NAN (α2-3) Gal (β1-1) Cer	G$_{M4}$	human brain chimpanzee brain avian brain	34, 83, 107, 108 109 110
2. NAN (α2-3) Gal (β1-4) Glc (β1-1) Cer	G$_{M3}$	brain erythrocytes spleen, kidney, liver adrenal medulla muscle plasma platelets	90, 108 111 17, 121 87 112 113, 120 118
3. NGN (α2-3) Gal (β1-4) Glc (β1-1) Cer	G$_{M3}$(NGN)	erythrocytes (equine) spleen (bovine) adrenal medulla (bovine)	16, 114, 115 17 87
4. OAc-NGN (α2-3) Gal (β1-4) Glc (β1-1) Cer		erythrocytes (equine)	116
5. NAN (α2-8) NAN (α2-3) Gal (β1-4) Glc (β1-1) Cer	G$_{D3}$	brain, normal brain, SSPE spleen, kidney, liver platelets retina placenta	83 117 17, 121 118 119, 123 122
6. NGN (α2-8) NAN (α2-3) Gal (β1-4) Glc (β1-1) Cer	G$_{D3}$ (NGN/NAN)	spleen, kidney, liver (bovine)	17
7. NAN (α2-8) NGN (α2-3) Gal (β1-4) Glc (β1-1) Cer	G$_{D3}$ (NAN/NGN)	spleen, kidney, liver (bovine)	17
8. NGN (α2-8) NGN (α2-3) Gal (β1-4) Glc (β1-1) Cer	G$_{D3}$ (NGN)$_2$	erythrocytes (cat) spleen, kidney, liver (bovine)	124, 125 17

[a]See footnote to Table I.

vertebrate tissue and often occur as the major ganglioside(s). Brain contains only minor amounts of the mono- (G_{M3}) and disialosyl (G_{D3}) types and variable amounts of sialosylgalactosyl ceramide ($G_7 = G_{M4}$) depending on species (see below).

Several gangliosides have been detected in the lacto series (Table II), most of these occurring in visceral tissues such as liver, kidney, and spleen (17). Two basic tetraglycosyl structures occur in this series:

$$\text{Gal } (\beta1\text{-}3) \text{ GlcNAc } (\beta1\text{-}3) \text{ Gal } (\beta1\text{-}4) \text{ Glc } (\beta1\text{-}1) \text{ Cer}$$
lacto-N-tetraosylceramide
Type I

$$\text{Gal } (\beta1\text{-}4) \text{ GlcNAc } (\beta1\text{-}3) \text{ Gal } (\beta1\text{-}4) \text{ Glc } (\beta1\text{-}1) \text{ Cer}$$
lacto-N-neotetraosylceramide
Type II

Virtually all the well-characterized gangliosides of this kind are based on lacto-N-neotetraosylceramide (Type II). A ganglioside containing the Type I structure was initially described in bovine spleen and kidney by Wiegandt (18), but his later report on these substances (17) failed to include this particular structure. Wiegandt also pointed out in his second study (17) that one monosialosyl derivative of lacto-N-neotetraosylceramide ($G_{Lntet}1b$ by his nomenclature) had been incorrectly identified as a disialosyl derivative in the earlier study (18). The structure with NAN linked ($\alpha2$-3) to terminal galactose of lacto-N-neotetraosylceramide was shown to be the major ganglioside in human peripheral nerve (19) and erythrocytes (20) but only a minor type in human brain (19). To date it is the only glucosamine-containing ganglioside detected in brain.

Most of the known gangliosides, including the large majority in brain, belong to the ganglio series (Table I) with N-acetylgalactosamine as the third sugar. Since sialic acid is also substituted on the same galactose, these all contain branched structures. Biosynthetically, sialic acid is attached first to lactosylceramide followed by N-acetylgalactosamine. These and subsequent sugars were shown to be added sequentially by a series of glycosyltransferases, believed to function as a coordinated complex (21).

Mammals generally contain 4 major brain gangliosides, of which G_{M1} forms the basic structural unit (Fig. 3). The 3-dimensional projection shown is based on manipulation of molecular models, which indicated reduced steric crowding when the sialic acid group is perpendicular to the linear carbohydrate chain. When attached by the α-ketosidic linkage demonstrated for gangliosides (22), the sialic acid of G_{M1} would thus tend to be shielded to some extent by the adjacent sugars, while the additional sialic acids of G_{D1a}, G_{D1b}, and G_{T1b} would experience less shielding. A small difference in the acidity of these groups has been correlated with differences in their chromatographic behavior on ion-exchange resins (23).

Treatment of the di- and trisialogangliosides of Fig. 3 with neuraminidase converts these to G_{M1}, the sialic acid of which is resistant to this enzyme except in the presence of detergent (24). Despite such resistance the sialic acid of G_{M1} is linked through an α-ketosidic bond of the same configuration as the reactive sialic acids, and its resistance has been ascribed to steric hindrance by the adjacent hexosamine (25, 26). When such interference is absent, as in G_{M3}, sialic acid becomes quite reactive to the enzyme. It may be noted that the disialosyl grouping, NAN(2-8)NAN, such as is found in G_{D1b} and G_{T1b}, reacts very sluggishly with neuraminidase in comparison to the NAN(2-3)Gal unit present in G_{D1a}, G_{M3}, etc. This disialosyl unit, known for a long time to be present in gangliosides, was only recently detected in glycoproteins (27).

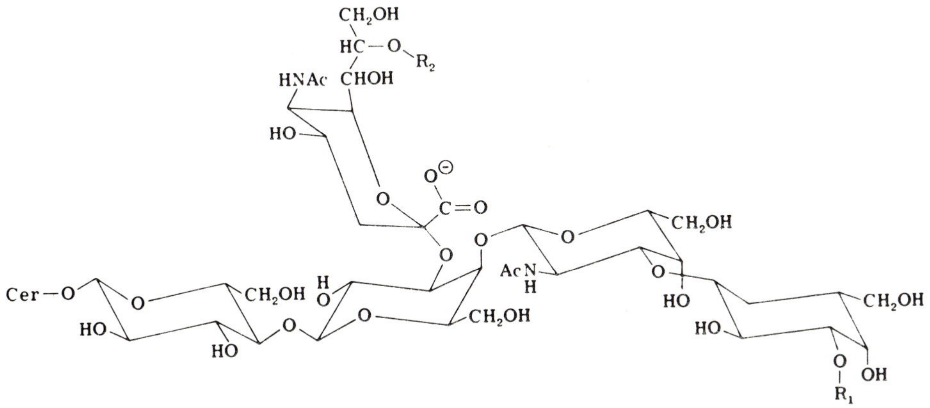

Fig. 3. Structures of the 4 major gangliosides of mammalian brain. G_{M1}: $R_1 = R_2 = H$; G_{D1a}: $R_1 = NAN$, $R_2 = H$; G_{D1b}: $R_1 = H$, $R_2 = NAN$; G_{T1b}: $R_1 = R_2 = NAN$.

While the 4 gangliosides depicted in Fig. 3 comprise the major brain types of most mammals, lower vertebrates have a different pattern in which more complex polysialogangliosides predominate and monosialo species such as G_{M1} are barely detectable. Fish brain was thus found to have a tetra- and a pentasialoganglioside as the 2 major forms (28, 29) while a similar pattern was suggested for carp, perch, and frog (30). The structure proposed (29) for the tetrasialoganglioside of fish brain (entry 18, Table I) differs from that of the tetrasialoganglioside of human brain (entry 17, Table I). The latter was recently shown (31) to have a disialosyl NAN(2-3)NAN unit attached to each galactose, corresponding to a structure proposed earlier by Klenk (32) on the basis of incomplete evidence. Attachment of the disialosyl unit to terminal galactose was also detected in a newly characterized trisialoganglioside G_{T1a} of human brain (33).

It should be pointed out that certain other vertebrates depart from the typical mammalian brain pattern in having considerable amounts of sialosylgalactosyl ceramide ($G_7 = G_{M4}$). In human white matter this was shown to be the third most abundant ganglioside on a molar basis (34), while recent work has revealed an unusually high concentration of this ganglioside in chicken and pigeon brains (35). In all cases it appears to be localized primarily in myelin.

Space does not permit detailed discussion of the lipophilic constituents of these glycosphingolipids. Some general trends may be noted, however, a principal one being the presence of substantial amounts of C_{20} long-chain bases in CNS gangliosides and little if any in extraneural species. Fatty acid differences have also been reported. Gangliosides from certain peripheral nerve sources provided an interesting hybrid of CNS and extraneural properties (36). Ganglioside G_7 from human myelin had unique lipophilic components that closely resembled those of myelin cerebrosides (34). Lipophilic composition has been considered in recent reviews of ganglioside biochemistry (37–40).

Although gangliosides are generally considered to be characteristic lipids of the vertebrates, a few types have been isolated from various species of Echinodermata. They were first detected in the gametes of the sea urchin Pseudocentrotus depressus (41), and most subsequent studies have employed either gametes or gonads of various echinoderms. This was the only one of many marine invertebrate phyla examined which was found to contain gangliosides (41, 42). The simplest of these had the structure NAN (2-6) Glc (1-1)

Cer (43). Virtually all the analyzed fractions contained glucose as the only carbohydrate besides sialic acid. One such substance from the gonads of Echinocardium cordatum contained an unusual sialic acid with a sulfate on the 8-hydroxyl of NAN (44). Disialogangliosides containing one (43) and 2 (45) glucose units have also been detected. The lipophilic constituents were shown to include phytosphingosine and α-hydroxy fatty acids, thus differing from most vertebrate gangliosides.

DISTRIBUTION

Early distribution studies (46, 47) recognized the substantially higher concentration of gangliosides in gray matter compared to white, and this finding has since been confirmed many times. The inference from this that gangliosides are uniquely neuronal constituents has been refuted by more recent studies in several laboratories which have revealed their presence in virtually every cell type and most subcellular fractions of the CNS. Like glycosphingolipids in general, they appear to be ubiquitous in vertebrate tissues and a few were detected in certain invertebrates as well (see above). However, the fact that the concentration in brain so greatly exceeds that of most other organs — approximately 15-fold when comparing gray matter to liver, for example — has served to focus special interest on their distribution and behavior in this organ.

While it now appears highly probable that the majority of brain ganglioside is localized in the neurons, their precise distribution within these complex cells is far from clear. When gangliosides were first quantified in isolated neuronal perikarya their concentration turned out to be surprisingly low: 1.24 μg NAN/mg protein for the rat in the study of Norton and Poduslo (48), and 2.88 μg NAN/mg protein for the rabbit in the study of Hamberger and Svennerholm (49). Both groups reported the astrocyte level to be higher than that of neurons from the same source (3.45 and 5.38 μg NAN/mg protein from rat and rabbit, respectively). It was suggested (48) that since astrocytes have a higher ratio of surface area to volume than neuronal perikarya their higher ganglioside content could reflect the greater quantity of plasma membrane. An alternative explanation proposed that these astrocyte values are artifically high due to contamination by nerve ending membranes (49, 50). In any case it is apparent that the cell bodies of these 2 cell types account for only a small part of total gray matter ganglioside.

This conclusion is supported by quantitative estimations utilizing the above concentrations and reported values for neuronal and glial numbers. The rat somatosensory cortex has been estimated (51, 52) to contain 1.32×10^8 neurons/g[1], and with a protein weight of 246 pg/neuronal cell body (135) the perikaryal protein amounts to 32.5 mg/g cortex. Using a ganglioside concentration of 0.8 μg NAN/mg protein obtained for rat neurons (136) the total contribution from this source is calculated as 26 μg NAN/g cortex. Similar calculations for glia, based on a density of 53×10^6 cells/g cortex (51, 52), a protein content of 307 pg/cell (48), and a ganglioside concentration of 2.1 μg NAN/mg protein (136), yield values of 16.3 mg glial protein/g rat cortex and 34.2 μg NAN/g rat cortex. These values pertain to glial cell bodies and the processes associated during their isolation. The content of the processes shorn off during isolation cannot be estimated at present. The result here based on the ganglioside value for astroglia should be considered an upper limit estimate since the cell count included oligodendroglia and microglia as well. Oligodendroglia, which are less numerous than astrocytes in the cortex, contain less ganglioside (53) and hence would tend to lower the above estimate. If the claim of synaptic membrane contamination is true this would further contribute to an overestimation of the glial contribution.

[1] All references to a gram of cortex denote fresh weight values.

Since ganglioside content of rat cerebral cortex is approximately 880 μg NAN/g (54) — slightly higher than the value (849 μg NAN/g) for whole rat brain (8) — it is evident that the contribution from neuronal plus glial cell bodies is probably less than 10% of the total. By inference, therefore, a large portion of cortical ganglioside resides in the neuronal (and to a lesser extent glial) processes. This was suggested by an earlier study (55) in which wet microdissection was employed to demonstrate that neuropil trimmed from an area adjacent to neurons (and therefore rich in neuronal processes) contained the highest concentration while untrimmed neuropil containing mainly glial cells had the lowest. Cleaned neurons were intermediate. More recently, microchemical analysis of sectioned layers of human and rat cortex demonstrated maximal ganglioside in regions of high concentrations of dendritic and axonal plexuses and their synaptic articulations (52).

An important question yet to be addressed, however, is the mode of ganglioside distribution within the processes. The idea has gained currency in recent years that the axon terminals are the major loci of ganglioside concentration. Support for this idea came from studies reporting appreciable ganglioside levels in isolated synaptosomes and, most particularly, synaptic plasma membranes. Critical examination of the data, however, reveals that these structures do not account for a major portion of neuronal ganglioside.

Published values for ganglioside content of synaptosomes and synaptic plasma membranes are summarized in Tables IV and V, respectively. The wide variations, particularly in membrane values, probably reflect to some extent differences in purity of the various preparations stemming from different isolation techniques. In a critical review Morgan (56) has estimated that synaptosomes are generally isolated in 40–50% purity while synaptic plasma membranes prepared by his method (57) are estimated as approximately 50–80% pure. Population selection is another possible source of variation since some laboratories (58, 59) have reported higher ganglioside content in "cholinergic" as opposed to "noncholinergic" synaptic membranes. Species differences might be an additional factor. Reports dealing with ganglioside levels in several components of the nervous system have been summarized in a recent review (60).

The approximate contribution of nerve endings to total cortical ganglioside may be calculated as follows. The ganglioside concentration in whole synaptosomes is taken as 9.0 μg NAN/mg protein, the average of the 5 values reported for rat (Table IV). The range (7.0–10.3) for this species was not excessive. The reliability of this value depends of course on the nature of the contaminants and their ganglioside content. Microsomes are usually considered a major source of contamination (56, 61) and since this heterogeneous fraction has a high ganglioside content (62–65) their presence would tend to elevate synaptosomal values. Mitochondria or mitochondrial membranes would have the opposite effect because of their low ganglioside content. Glial contamination is another possible factor whose magnitude has not been established.

In regard to nerve-ending density, values of 12.6×10^{11} (66) and 14×10^{11} (67, 68) were reported as the number of axon terminals per cubic centimeter of mature rat cortex. Clementi et al. (69), using a polystyrene bead tagging procedure, reported a value of 4×10^{11} for the number of synaptosomes produced per gram of guinea pig cortex on homogenizing under defined conditions. Since a certain percentage is destroyed by homogenization, the true number of axon terminals should be greater. Cragg's more recent figure of 21×10^{11} will be used here; this value, an average for rat visual and frontal cortex, was obtained by electron microscopy (70). Cragg also obtained a value of 0.452 μm for the average diameter of axon terminals in the same areas. The calculated volume (0.048 μm^3) in conjunction with a density (from gradient centrifugation) of 1.15, yields a weight of 0.055 pg/nerve ending. The 21×10^{11} axon terminals in a gram of rat cortex would thus

TABLE IV. Ganglioside Content of Synaptosomes

Species	µg NAN per mg protein	Reference
ox	16.3	126
rabbit	13.8	49
rabbit	11.9	127
rat	10.3	128
rat	9.7	129
rat	9.2	59
rat	9.2	64
rat	7.0	130
guinea pig	8–9	65
human (infant)	7.0	131
human (adult)	7–10	131

Several of the above values have been recalculated from the original data.

TABLE V. Ganglioside Content of Synaptic Plasma Membranes

Species	µg NAN per mg protein	Reference
guinea pig	16–18	132
rat ("cholinergic")	16.7	58
rat ("noncholinergic")	7.3	58
rat ("cholinergic")	45.2	59
rat ("noncholinergic")	18.5	59
rat	44.6	133
rat	19.3	134

weigh 116 mg and of this amount it may be assumed that approximately 10% is protein. Utilizing the above ganglioside concentration one obtains a calculated value of 104 µg NAN as the contribution of all nerve endings in 1 g of rat cortex.[2] This is roughly 12% of the total (880 µg NAN/g cortex). Since the calculation was based on whole synaptosomes it does not depend on the concentration in synaptic plasma membranes, about which there is still considerable uncertainty (Table V).

These results are summarized in Table VI. Such elements as epithelial cells, ependymal cells, myelin, and capillaries, which would probably contribute little, and a portion of astroglial processes are not included. The calculations indicate that neuronal and glial cell bodies plus nerve endings account for only about 50% of the protein and 20% of the

---

[2]This result is considered an upper limit estimate because of the relatively high values employed for ganglioside concentration and axon terminal density. A recent study in our laboratory (unpublished) indicated the ganglioside concentration of carefully washed rat brain synaptosomes to be significantly below the value of 9.0 µg NAN per mg protein, taken as an average from Table IV. Several of the synaptosome preparations represented in Table IV were not washed in a manner required to free the particles of microsomal contamination (57, 61), and this may have resulted in artificially high ganglioside concentrations.

TABLE VI. Contributions of Subfractions to Total Ganglioside of Rat Cerebral Cortex

	Protein mg/g cortex	Ganglioside concentration μg NAN/mg protein	Ganglioside quantity μg NAN/g cortex
Neuronal cell bodies	32.5	0.8	26.0
Glial cell bodies	16.3	2.1	34.2
Nerve endings	11.6	9.0	104.4
	60.1		164.6
Total/g cortex	123		880
Remainder	63		715

ganglioside in a gram of rat cerebral cortex. The large majority of cortical ganglioside would therefore reside in neuronal processes exclusive of axon terminals. The calculations are, of course, dependent on the reliability of the reported measurements of cell number, nerve-ending density, ganglioside concentrations, etc., but it may be noted that considerable variation in 1 or more of these parameters would be possible without altering the basic conclusion.

It is not possible at this stage to estimate the relative ganglioside content of dendrites and axons, the major components of the neuropil. Assuming that these processes contain in equivalent proportions the 715 μg ganglioside NAN and 63 mg protein unaccounted for ("remainder") in Table VI, the average concentration would be 715/63 = 11.3 μg NAN/mg protein, a value higher than that of whole synaptosomes. From these considerations one might speculate that the ganglioside content in plasma membranes of processes may be comparable to or greater than that of synaptosomal plasma membranes. Extending this concept to include the cell body (see below) the possibility exists that gangliosides may be distributed over a substantial part of the neuronal surface. This plasma membrane pool would likely comprise half or more of total neuronal ganglioside, additional though smaller pools being present in endoplasmic reticulum, cytoplasm, mitochondria, and possibly other intracellular compartments. As noted above, the microsomal fraction which is thought to originate from a diversity of plasma and reticular membranes has a rather high ganglioside content (62–65), approaching in some studies the value for synaptic plasma membrane. Our own studies (137) have verified this similarity between microsomes and synaptic plasma membranes, although the ganglioside concentrations in these preparations were well below the maximum value appearing in Table V. Owing to the uncertainties cited above as well as other potential methodological pitfalls, the precise concentrations of gangliosides in such membranes remain in some doubt.

To calculate the approximate ganglioside content of the perikaryal plasma membrane, one can utilize data which correlates surface area with protein content. This was found in the case of erythrocytes (71, 72) to be 0.37 μg protein per square centimeter (10^8 μm^2) of surface area, and it is assumed here that the neuronal membrane has roughly the same percentage of protein and hence the same protein to area ratio. A spherical neuron with a diameter of 18 μm has a surface area of 1,017 μm^2, while 1.32×10^8 neurons (the amount in a gram of rat cortex, Ref. 52) would have a total area of 1.34×10^{11} μm^2 corresponding to 495 μg protein. If it is assumed that 50–80% of cell body ganglioside (26 μg NAN/g rat cortex, Table VI) is localized in the plasma membrane, the ganglioside concentration of the latter is calculated to be 26–42 μg NAN/mg protein. These calculated values, while admittedly dependent on a number of assumptions, do appear to fall within the range of values reported for synaptic plasma membranes (Table V) and suggest the possibility that ganglioside concentration may approach uniformity over certain portions of the neuronal membrane. It is conceivable, in light of considerations discussed previously, that dendritic membrane gangliosides could also fall in the same concentration range. Axonal membranes, owing to the presence of myelin and altered surface characteristics, may or may not prove unique in this regard.

The hypothesis that gangliosides are distributed over a large part of the neuronal surface is consistent with current concepts of a fluid mosaic model of membranes (73, 74). If the neuronal plasma membrane can be viewed as a continuum, individual lipids inserted into the perikaryal membrane, for example, would be expected to diffuse into the adjoining plasma membranes of the processes, assuming the absence of diffusional barriers. The rate of lateral diffusion can be quite rapid, e.g., the diffusion constant for phospholipid molecules in sarcoplasmic reticulum has been estimated at 6×10^{-8} cm^2/sec (75). If the constant for neuronal membranes is comparable a molecule would diffuse 1 mm in about a day, a rate considerably more rapid than the turnover time of brain gangliosides (76, 77). Such a mechanism would therefore allow equilibration of surface molecules between the neuron and a number of its processes, while the more remote axonal and synaptic regions would be expected to receive ganglioside through the demonstrated mechanism of rapid axonal transport (78, 79). Lateral diffusion might then function to equalize ganglioside concentrations in these distal membranes as well. The shorter axons and associated terminals could receive and equilibrate their membrane gangliosides by a combination of lateral diffusion and axonal flow.

If such equilibration does in fact occur over large portions of the neuronal surface this should be reflected in similarity of ganglioside composition as well as concentration. It may therefore be significant that the ganglioside patterns of neurons and synaptosomes were similar when analyzed in the same laboratory (49), although somewhat different patterns were found for the synaptosomes themselves in other laboratories (59, 133). Comparison of microsomes, synaptosomes, and synaptic plasma membranes revealed similar ganglioside compositions, again when the analyses were carried out in the same laboratory (59). Whether the synaptic junction itself would participate in such equilibration, or present local barriers to diffusion, is open to question. It was previously shown (80) that the extrajunctional (axon terminal) synaptic membrane conforms to the fluid mosiac model in that at least some of its membrane-bound components exhibit lateral mobility, but certain other synaptic components were later found (81) to have greatly restricted mobility. Inhomogeneity of ganglioside content was suggested by the results of one study (79) showing the synaptic junctional complex to have significantly less of this lipid than the adjoining membrane, although the finding of no difference in another study (82) leaves the question open. In any case, the area encompassed by synaptic thickenings, where restricted mobility would be most likely to occur, comprises a minority of the total neuronal membrane. The hypothesis of ganglioside distribution over much if not all the neuronal surface will hopefully be amenable to testing with the aid of improved methods for isolating membranes of the perikaryon and processes.

ACKNOWLEDGMENTS

This work was supported by grants NS 04834, NS 03356, and NS 10931 from the National Institutes of Health, United States Public Health Service.

REFERENCES

1. Fishman PH, Brady RO: Science 194:906, 1976.
2. Ledeen RW, Mellanby J: In Bernheimer A (ed): "Perspectives in Toxicology." New York: John Wiley and Sons, 1977, p 15.
3. Olsnes S, Pappenheimer AM Jr, Meren R: J Immunol 113:842, 1974.
4. Mullin BR, Fishman PH, Lee G, Aloj SM, Ledley FD, Winand RJ, Kohn LD, Brady RO: Proc Natl Acad Sci USA 73:842, 1976.
5. McIlwain H: Biochem J 78:24, 1961.
6. Ledeen RW, Yu RK: In Rosenberg A, Schengrund C-L (eds): "Biological Roles of Sialic Acid." New York: Plenum Press, 1976, p 1.
7. Tettamanti G, Bertona L, Berra B, Zambotti V: Ital J Biochem 13:315, 1964.
8. Yu RK, Ledeen RW: J Lipid Res 11:506, 1970.
9. Wiegandt H: Adv Exp Med Biol 25:127, 1972.
10. Yamakawa T, Suzuki S: J Biochem (Tokyo) 39:373, 1952.
11. Yamakawa T, Nishimura S, Kamimura M: Jpn J Exp Med 35:201, 1965.
12. Siddiqui B, Hakomori S: J Biol Chem 246:5766, 1971.
13. Slomiany A, Slomiany BL, Horowitz MI: J Biol Chem 249:1225, 1974.
14. Slomiany BL, Slomiany A, Horowitz MI: Biochim Biophys Acta 326:224, 1973.
15. Slomiany BL, Slomiany A, Horowitz MI: Fed Proc Fed Am Soc Exp Biol 35:1443, 1976.
16. Yamakawa T, Suzuki S: J Biochem (Tokyo) 38:199, 1951.
17. Wiegandt H: Hoppe-Seyler's Z Physiol Chem 354:1049, 1973.
18. Wiegandt H, Bücking HW: Eur J Biochem 15:287, 1970.
19. Li Y-T, Månsson J-E, Vanier M-T, Svennerholm L: J Biol Chem 248:2634, 1973.
20. Wherrett JR: Biochim Biophys Acta 326:63, 1973.
21. Roseman S: Chem Phys Lipids 5:270, 1970.
22. Yu RK, Ledeen RW: J Biol Chem 244:1306, 1969.
23. Yu RK: Personal communication.
24. Wenger DA, Wardell S: J Neurochem 20:607, 1973.
25. Ledeen R: Chem Phys Lipids 5:205, 1970.
26. Huang RTC, Klenk E: Hoppe-Seyler's Z Physiol Chem 353:679, 1972.
27. Finne J, Krusius T, Rauvala H: Biochem Biophys Res Commun 74:405, 1977.
28. Ishizuka I, Kloppenburg M, Wiegandt H: Biochim Biophys Acta 210:299, 1970.
29. Ishizuka I, Wiegandt H: Biochim Biophys Acta 260:279, 1972.
30. Avrova NF: J Neurochem 18:667, 1971.
31. Ando S, Yu RK: Proc Internat Soc Neurochem (Abstract) 6:535, 1977.
32. Klenk E: Prog Chem Fats Other Lipids 10(4):411, 1969.
33. Ando S, Yu RK: Trans Am Soc Neurochem (Abstract) 8:183, 1977.
34. Ledeen RW, Yu RK, Eng LF: J Neurochem 21:829, 1973.
35. Cochran FB, Yu RK, Ledeen RW: Proc Internat Soc Neurochem (Abstract) 6:540, 1977.
36. Fong JW, Ledeen RW, Kundu SK, Brostoff S: J Neurochem 26:157, 1976.
37. Svennerholm L: Compr Biochem 18:201, 1970.
38. Wiegandt H: Adv Lipid Res 9:249, 1971.
39. Stoffel W: Annu Rev Biochem 40:56, 1971.
40. Ledeen RW, Yu RK: In Hers HG, van Hoof F (eds): "Lysosomes and Storage Diseases." New York: Academic Press, 1973, p 105.
41. Isono Y, Nagai Y: Jpn J Exp Med 36:461, 1966.
42. Vaskovsky VE, Kostetsky EY, Svetashev VI, Zhukova IG, Smirnova GP: Comp Biochem Physiol 34:163, 1970.
43. Hoshi M, Nagai Y: Biochim Biophys Acta 388:152, 1975.
44. Kochetkov NK, Smirnova GP, Chekareva NV: Biochim Biophys Acta 424:274, 1976.
45. Kochetkov NK, Zhukova IG, Smirnova GP, Glukhoded IS: Biochim Biophys Acta 326:74, 1973.
46. Klenk E, Langerbeins H: Hoppe-Seyler's Z Physiol Chem 270:185, 1941.
47. Svennerholm L: Acta Soc Med Ups 62:1, 1957.
48. Norton WT, Poduslo SE: J Lipid Res 12:84, 1971.
49. Hamberger A, Svennerholm L: J Neurochem 18:1821, 1971.
50. Morgan IG, Gombos G: In Barondes SH (ed): "Neuronal Recognition." New York: Plenum Press, 1976, p 179.
51. Bass NH, Hess HH, Pope A, Thalheimer C: J Comp Neurol 143:481, 1971.
52. Hess HH, Bass NH, Thalheimer C, Devarakonda R: J Neurochem 26:1115, 1976.

53. Poduslo SE, Norton WT: J Neurochem 19:727, 1972.
54. Yu RK, Chang NC: (In preparation).
55. Derry DM, Wolfe LS: Science 158:1450, 1967.
56. Morgan IG: Neuroscience 1:159, 1976.
57. Morgan IG, Wolfe LS, Mandel P, Gombos G: Biochim Biophys Acta 241:737, 1971.
58. Lapetina EG, Soto EF, DeRobertis E: J Neurochem 15:437, 1968.
59. Avrova NF, Chenykaeva E Yu, Obukhova EL: J Neurochem 20:997, 1973.
60. Ledeen RW, Yu RK: In Witting LA (ed): "Glycolipid Methodology." Champaign, Illinois: American Oil Chemists Society, 1976. p 187.
61. Gurd JW, Jones LR, Mahler HR, Moore WJ: J Neurochem 22:281, 1974.
62. Wolfe LS: Biochem J 79:348, 1961.
63. Wherrett JR, McIlwain H: Biochem J 84:232, 1962.
64. Seminario LM, Hren N, Gomez CJ: J Neurochem 11:197, 1964.
65. Eichberg J, Whittaker VP, Dawson RMC: Biochem J 92:91, 1964.
66. Armstrong-James MA, Johnson FR: J Anat 104:590, 1969.
67. Aghajanian GK, Bloom FE: Brain Res 6:716, 1967.
68. Cragg BG: Proc R Soc London Ser B 171:319, 1968.
69. Clementi F, Whittaker VP, Sheridan MN: Z Zellforsch Mikrosk Anat 72:126, 1966.
70. Cragg BG: Brain 95:143, 1972.
71. Jacobson BS, Branton D: Science 195:302, 1977.
72. Steck TL: J Cell Biol 62:1, 1974.
73. Singer SJ, Nicolson GL: Science 175:720, 1972.
74. Singer SJ: In Bradshaw RA, Frazier WA, Merrell RC, Gottlieb DI, Hogue-Angeletti RA (eds): "Surface Membrane Receptors." New York: Plenum Press, 1976, p 1.
75. Scandella CJ, Devaux P, McConnell HM: Proc Natl Acad Sci USA 69:2056, 1972.
76. Burton RM, Balfour YM, Gibbons JM: Fed Proc Fed Am Soc Exp Biol 23:230, 1964.
77. Suzuki K: J Neurochem 14:917, 1967.
78. Forman DS, Ledeen RW: Science 177:630, 1972.
79. Ledeen RW, Skrivanek JA, Tirri LJ, Margolis RK, Margolis RU: In Porcellati G, Ceccarelli B, Tettamanti G (eds): "Advances in Experimental Medicine and Biology." New York: Plenum Press, 1976, vol 17, p 83.
80. Matus AI, DePetris S, Raff MC: Nature (London) 244:278, 1973.
81. Matus AI, Jones DH, Mughal S: Brain Res 103:171, 1976.
82. Lapetina EG, DeRobertis E: Life Sci 7:203, 1968.
83. Kuhn R, Wiegandt H: Z Naturforsch Mikrosk Anat 19b:256, 1964.
84. Ledeen R, Salsman K: Biochemistry 4:2225, 1965.
85. Kuhn R, Wiegandt H: Chem Ber 96:866, 1963.
86. Ledeen R, Salsman K, Gonatas J, Taghavy A: J Neuropathol Exp Neurol 24:341, 1965.
87. Price H, Kundu S, Ledeen R: Biochemistry 14:1512, 1975.
88. Ghidoni R, Sonnino S, Tettamanti G, Wiegandt H, Zambotti V: J Neurochem 27:511, 1976.
89. Suzuki A, Ishizuka I, Yamakawa T: J Biochem (Tokyo) 78:947, 1975.
90. Kuhn R, Wiegandt H: Z Naturforsch Mikrosk Anat 19b:256, 1964.
91. Klenk E, Naoi M: Hoppe-Seyler's Z Physiol Chem 349:288, 1968.
92. Kuhn R, Wiegandt H: Z Naturforschg Mikrosk Anat 18b:541, 1963.
93. Kuhn R, Egge H: Chem Ber 96:3338, 1963.
94. Klenk E, Gielen W: Hoppe-Seyler's Z Physiol Chem 330:218, 1963.
95. Klenk E, Kunau W: Hoppe-Seyler's Z Physiol Chem 335:275, 1964.
96. Klenk E, Hof L, Georgias L: Hoppe-Seyler's Z Physiol Chem 348:149, 1967.
97. Klenk E: Prog Chem Fats Other Lipids 10(4):409, 1969.
98. Johnson GA, McCluer RH: Biochim Biophys Acta 84:587, 1964.
99. Svennerholm L, Månsson J-E, Li Y-T: J Biol Chem 248:740, 1973.
100. Ando S, Yu RK: Trans Am Soc Neurochem 8(2):183, 1977.
101. Ando S, Yu RK: (In press).
102. Svennerholm L: J Neurochem 10:613, 1963.
103. Kuhn R, Wiegandt H: Z Naturforsch Mikrosk Anat 19b:80, 1964.
104. Wiegandt H, Schulze B: Z Naturforsch Mikrosk Anat 24b:945, 1969.
105. Rauvala H: FEBS Lett 62:161, 1976.
106. Wiegandt H: Eur J Biochem 45:367, 1974.

107. Siddiqui B, McCluer RH: J Lipid Res 9:366, 1968.
108. Klenk E, Georgias L: Hoppe-Seyler's Z Physiol Chem 348:1261, 1967.
109. Yu RK, Ledeen RW, Gajdusek DL, Gibbs CJ: Brain Res 70:103, 1974.
110. Cochran FB, Yu RK, Ledeen RW: (In press).
111. Ando S, Yamakawa T: J Biochem (Tokyo) 73:387, 1973.
112. Svennerholm L, Åke B, Månsson J-E, Rynmark B-M, Vanier M-T: Biochim Biophys Acta 280:626, 1972.
113. Yu RK, Ledeen RW: J Lipid Res 13:680, 1972.
114. Yamakawa T: J Biochem (Tokyo) 43:867, 1956.
115. Klenk E, Lauenstein I: Hoppe-Seyler's Z Physiol Chem 295:164, 1953.
116. Hakomori S, Saito T: Biochemistry 8:5082, 1969.
117. Ledeen R, Salsman K, Cabrera M: J Lipid Res 9:129, 1968.
118. Marcus AJ, Ullman HL, Saifer LB: J Clin Invest 51:2602, 1972.
119. Holm M, Månsson J-E, Vanier M-T, Svennerholm L: Biochim Biophys Acta 280:356, 1972.
120. Tao RVP, Sweeley CC: Biochim Biophys Acta 218:372, 1970.
121. Puro K: Biochim Biophys Acta 189:401, 1969.
122. Svennerholm L: Acta Chem Scand 19:1506, 1965.
123. Handa S, Burton RM: Lipids 4:205, 1969.
124. Handa S, Yamakawa T: Jpn J Exp Med 34:293, 1964.
125. Handa N, Handa S: Jpn J Exp Med 35:331, 1965.
126. Wiegandt H: J Neurochem 14:671, 1967.
127. Tettamanti G, Preti A, Lombardo A, Bonali F, Zambotti V: Biochim Biophys Acta 306:466, 1973.
128. Caputto R, Maccioni HJ, Arce A: Mol Cell Biochem 4:97, 1974.
129. Dekirmenjian H, Brunngraber EG: Biochim Biophys Acta 177:1, 1969.
130. Yohe HC, Chang NC, Glaser GH, Yu RK: Trans Am Soc Neurochem 8(2):185, 1977.
131. Kornguth S, Wannamaker B, Kolodny E, Geison R, Scott G, O'Brien JF: J Neurol Sci 22:383, 1974.
132. Whittaker VP: In Lajtha (ed): "Handbook of Neurochemistry." New York: Plenum Press, 1969, p 327.
133. Breckenridge WL, Gombos G, Morgan IG: Biochim Biophys Acta 266:695, 1972.
134. Brunngraber EG, Dekirmenjian H, Brown BD: Biochem J 103:73, 1967.
135. Norton WT: Private communication (In preparation).
136. Skrivanek J, Ledeen R, Norton W, Farooq M: (In press).
137. Skrivanek J, Ledeen R: (In preparation).

Cell Surface Glycosyltransferases—Do They Exist?

Wolfgang Deppert and Gernot Walter

Tumor Virology Laboratory, The Salk Institute, San Diego, California 92112

The presence of glycosyltransferases on surfaces of mammalian cells has been reported by many investigators and a biological role for these enzymes in cell adhesion and cell recognition has been postulated. Critical analysis, however, showed 2 major complications regarding the assay for cell surface glycosyltransferases: 1) hydrolysis of the nucleotide sugar by cell surface enzymes and subsequent intracellular use of the free sugar and 2) loss of cell integrity if trypsinized or EDTA-treated cells were used in suspension asays. We have assayed intact, viable cells in monolayer for cell surface glycosyltransferases using conditions under which intracellular utilization of free sugars generated by hydrolysis of the nucleotide sugar was prevented. Our data demonstrate that the presence of galactosyltransferases on the surface of a variety of cells, including established (normal and virally transformed) as well as nonestablished cells, is unlikely. No evidence for the existence of cell surface fucosyl- and sialyltransferases could be obtained, but our data do not exclude the possibility that low levels of these enzymes are present.

Key words: **cell viability, nucleotide sugar hydrolysis, intracellular glycosylation**

A large number of publications have reported glycosyltransferase activities on the surface of whole cells (cf. Ref. 1 for a review). These studies were initated by a hypothesis put forward by Roseman (2) suggesting that glycosyltransferases on the cell surface are

---

Abbreviations: EDTA – (ethylenedinitrilo) tetraacetic acid; UDP-galactose – uridine diphosphate galactose; DME – Dulbecco and Vogt's modified Eagle's medium; BHK – baby hamster kidney cells; Py BHK – polyoma virus-transformed BHK cells; ME – mouse embryo cells; SV 3T3 – SV40-transformed BALB/c 3T3 cells; CHE – Chinese hamster embryo cells; AGMK – African green monkey kidney cells; Hepes – N-2-hydroxyethylpiperazine-N'-2-ethanesulfonic acid; SDS – sodium dodecyl sulfate; AFP – antifreeze proteins; Fetuin-desial. – desialized fetuin; AFP-degal. – degalactosized antifreeze proteins; Fetuin-desial.-degal. – desialized and degalactosized fetuin; Gal-1-P – galactose-1-phosphate; GDP-fucose – guanosine diphosphate fucose; CMP sialic acid – cytidine monophosphate sialic acid; 5'-AMP – 5'-adenosine monophosphate; ND – experiment not done; BSA – bovine serum albumin.

Wolfgang Deppert is presently with the Max-Planck-Institut für Biophysikalische Chemie, Postfach 968, D-3400 Göttingen, Federal Republic of Germany

Received for publication March 23, 1977; accepted August 17, 1977.

involved in intercellular adhesion and cell recognition. According to this hypothesis, glycosyltransferases on the cell surface interact with their substrate carbohydrate chains of glycoproteins or glycolipids on the surfaces of neighboring cells.

Evidence consistent with this idea has come from experiments in which single cell suspensions, prepared by trypsin or EDTA treatment of cells grown in monolayer cultures, were incubated with nucleotide sugars and the transfer of sugar residues to either endogenous or exogenous glycoproteins or glycolipids was observed. The conclusion drawn from these experiments that the transfer of the sugar residue from the added nucleotide sugar was mediated by cell surface glycosyltransferases is ambiguous for the following reasons: 1) The cell surface of many cells (see Results) contain 2 enzymatic activities, a nucleotide pyrophosphatase and a monophosphoesterhydrolase, which together catalyze the hydrolysis of nucleotide sugars to free sugar. The generation of free sugar, its transport into the cells, and its incorporation into carbohydrate-containing macromolecules would simulate the activity of cell surface glycosyltransferases. 2) The choice of assay systems using cells in suspension after trypsin or EDTA treatment further complicates the evaluation of the data reported in the literature. Cells which are normally cultured in monolayers are viable in suspension only for a limited period of time (see Results). Therefore, many of the experiments reported may have been performed with dead or dying cells. A small fraction of dead cells and the accompanying availability of intracellular membranes could account for the incorporation data described in many studies. 3) The possibility that trypsin or EDTA treatment might alter the permeability of cells for small molecules, and macromolecules as well, has to be considered.

In a previous report (3), we have shown that the incubation of BHK cells in monolayer with exogenous UDP-galactose does not result in a measurable transfer of galactose residues from UDP-galactose to cellular acceptor molecules if the utilization of free galactose as a hydrolysis product of UDP-galactose was blocked. Since we also could not find any detectable transfer to exogenous acceptor molecules, we concluded that the cell surface of BHK cells does not contain detectable cell surface galactosyltransferase activity. In this report, we demonstrate that the ability to hydrolyze exogenous nucleotide sugars is not confined to BHK cells and that cell surface glycosyltransferase assays using trypsinized cells in suspension may lead to artefacts due to the altered permeability and decreased viability of these cell preparations. Furthermore, various cells grown and assayed in monolayer do not show conclusive evidence for nucleotide sugar-dependent cell surface glycosylation if criteria for the differentiation between intracellular and cell surface glycosyltransferases (3) are applied.

MATERIALS AND METHODS

Cell Culture

BHK 21/13 were cultured in Dulbecco and Vogt's modified Eagle's medium (DME) (4) supplemented with 10% calf serum and 0.3% tryptose phosphate.

Polyoma virus-transformed BHK cells (Py BHK) were obtained from Dr. W. Eckhart (The Salk Institute), as were primary mouse embryo cells (ME). Mouse BALB/c 3T3 and Swiss 3T3 4A cells, SV40-transformed BALB/c 3T3 cells (SV 3T3), and Chinese hamster embryo cells (CHE) at passage 16 after establishing the primary culture were obtained from Dr. M. Vogt (The Salk Institute). Primary African green monkey kidney cells (AGMK) were obtained from Microbiological Associates, Inc., Los Angeles, California.

Py BHK, 3T3 (BALB/c and Swiss 4A), CHE, and AGMK cells were cultured in DME supplemented with 10% calf serum. BHK, Py BHK, 3T3 (BALB/c and Swiss 4A), SV 3T3, and CHE cells were free of mycoplasma as tested by labeling with [^{3}H] thymidine followed by autoradiography.

Incubation Assay

If not stated otherwise, subconfluent to confluent cells in monolayer were used for the experiments. Cells were seeded 2 days before the start of the experiments. After removal of the medium, the cells were washed 3 times with prewarmed Mg^{2+} and Ca^{2+} free isotonic salt solution, buffered at pH 7.4 with 20 mM Hepes (isotonic Hepes buffer). Cells were incubated in the same buffer, containing 1–5 mM $MnCl_2$ as stated. Uridine diphosphate D-[1-^{3}H] galactose (1.23 Ci/mmole, New England Nuclear Corporation, Boston, Massachusetts), D-[6-^{3}H] galactose (168 mCi/mmole, New England Nuclear Corporation), guanosine diphospho-2-[U-^{14}C] fucose (170 mCi/mmole, Amersham/Searle), 2-[U-^{14}C]-fucose (133 mCi/mmole, Amersham/Searle), and cytidine-5′-monophospho-[G-^{3}H] sialic acid (2.33 Ci/mmole, New England Nuclear Corporation) were added in the concentrations indicated. The plates were incubated at 37°C in a moist chamber mounted on a rocking shaker.

Paper Chromatography

The hydrolysis products of the nucleotide sugars were determined by analyzing aliquots of the incubation mixture by descending paper chromatography on Whatman No. 3 MM paper. The following solvent systems were used: ethanol:1 M ammonium acetate, pH 3.8, (5:2) for analysis of UDP-galactose; ethanol: 1 M ammonium acetate (7:3) for analysis of CMP-sialic acid; and ethanol:1 M ammonium acetate:acetic acid (50:20:33) for analysis of GDP-fucose. The running time was 17 h. The paper was then dried, cut into 1-cm strips, eluted with 1 ml of distilled water, and counted in Aquasol (New England Nuclear Corporation) using a Beckman liquid scintillation counter.

Determination of Incorporated Label

After the incubation times indicated, the medium was removed, the cells washed twice with isotonic Hepes buffer, dissolved in 0.5 ml of 1% SDS solution per 3-cm dish, and homogenized by sonication (10 sec at full power in a Branson sonifier equipped with a mictrotip). To 200-μl aliquots, diluted with 2 ml of ice cold water, 3 ml of 1% phosphotungstic acid in 0.5 N HCl were added. After 45 min at 0°C, the precipitates were collected on Whatman GF/C filters and washed with 10% trichloroacetic acid. The filters were then dried and their radioactivity was measured.

High-Molecular-Weight Acceptors

The freezing point depressing proteins AFP (antifreeze proteins), a mixture of 3 components with molecular weights of 10,500, 17,000, and 21,000 isolated from the serum of the Antarctic fish Trematomus borchgrevinki (5), and bovine fetuin were used as high-molecular-weight acceptors. Terminal galactose residues of AFP were removed by Smith degradation as described (6); terminal sialic acid and penultimate galactose residues were removed from fetuin by the method of Kim et al. (7). AFP and desialized fetuin (fetuin-desial.) were used as acceptors in sialyltransferase assays, degalactosized AFP (AFP-degal.), and desialized and degalactosized fetuin (fetuin-desial.-degal.) as acceptors for galactosyltransferase assays. AFP was a gift from Dr. T. Shier, The Salk Institute.

TABLE I. Hydrolysis of UDP[^{3}H]galactose by Various Cells*

Cells	Percent of radioactivity				Protein (μg)/incubation
	UDP-galactose	Gal-1-P	X	Galactose	
A. Established cell lines					
BHK 21/13	0	16.6	8.2	75.1	630
BALB/c 3T3	0	90.8	2.1	7.1	430
Swiss 3T3 4A	0	94.5	1.1	4.4	440
Py BHK	0	5.7	7.4	86.9	830
SV3T3	26.2	61.3	11.9	0.6	800
B. Nonestablished cells					
AGMK primary	28.4	32.4	3.0	36.2	560
ME secondary	0.6	14.9	2.6	81.9	720
CHE passage 18	2.1	24.7	, 4.2	69.0	480
C. BHK cells in suspension					
trypsin treated	4.7	73.0	0	22.3	430
EDTA treated	43.0	50.9	2.1	4.0	430
control (on plate)	0	55.0	3.8	41.2	430
No cells (control)	99.6	< 1.0	< 1.0	< 1.0	–

*2×10^5 cells were seeded on 3-cm dishes and tested for hydrolysis of UDP[^{3}H]galactose 2 days after seeding. Cells were incubated for 3 h with 0.5 ml of isotonic Hepes buffer containing 1.0 μCi UDP[^{3}H]-galactose and 1 mM MnCl$_2$, and hydrolysis products were determined as described in Materials and Methods. "X" is an unidentified compound, whose chemical nature was not further characterized. For testing BHK cells in suspension, the cells were removed from plates either by treatment with 0.1% trypsin or with 10 mM EDTA in isotonic Hepes buffer. The cell suspension was incubated for 30 min in 5 ml of growth medium and then for 3 h in 0.5 ml of isotonic Hepes buffer, containing 1.0 μCi UDP[^{3}H]galactose and 1 mM MnCl$_2$. Protein was determined according to Lowry et al. (29).

RESULTS

Hydrolysis of Nucleotide Sugars by Intact Cells

The cell surface of BHK 21/13 cells contains 2 enzymatic activities, a nucleotide pyrophosphatase and a monophosphoester hydrolase, which together hydrolyze exogenous UDP-galactose to free galactose in a 2-step mechanism with galactose-1-phosphate (gal-1-P) being the intermediate product (3). The ability of intact cells to hydrolyze UDP-galactose is not unique to BHK cells. Table I, A and B, shows that all cells exhibited considerable hydrolytic activity. Differences, however, were observed between the different types of cells in their ability hydrolyze gal-1-P to free galactose. Whereas with BHK, Py BHK, AGMK, CHE, and ME cells, the major hydrolysis product after a 3 h incubation period was galactose, with BALB/c 3T3, Swiss 3T3 4A, and SV 3T3 cells, the major hydrolysis product was gal-1-P. Single cell suspensions of BHK cells, obtained either by trypsin or EDTA treatment, also hydrolyzed UDP-galactose to free galactose, although to a lesser extent (Table IC).

We have also tested the hydrolysis of GDP-fucose, the substrate for fucosyltransferases, and CMP-sialic acid, the substrate for sialyltransferases by BHK cells in monolayer (Table II). GDP-fucose was hydrolyzed at about the same rate as UDP-galactose with fucose-1-phosphate being the intermediate, and free fucose being the end product of

TABLE II. Hydrolysis of UDP[^{3}H]Galactose, GDP[^{14}C]Fucose and CMP[^{3}H]Sialic Acid by BHK Cells*

	Percent of radioactivity		
Nucleotide sugar	Nucleotide sugar	Sugar-1-phosphate	Free sugar
UDP-galactose	0	51.1	46.8 (galactose)
GDP- fucose	0	48.4	50.7 (fucose)
CMP-sialic-acid	65.1	nonexistent	34.9 (sialic acid)

*Monolayers of BHK cells (1×10^6 per 3-cm dish) were incubated for 3 h with 0.5 ml of isotonic Hepes buffer, containing 1.0 μCi UDP[^{3}H]galactose, GDP[^{14}C]fucose, or CMP[^{3}H]sialic acid, respectively, and 1 mM MnCl$_2$. Hydrolysis products were determined as described in Materials and Methods.

hydrolysis. The hydrolysis of CMP-sialic acid, on the other hand, occurs at a much slower rate, but still yields a considerable amount of free sialic acid during a 3 h incubation period. Most of this hydrolysis reaction is not mediated by cell surface enzymes but is due to the chemical instability of CMP-sialic acid (3, 8).

The hydrolysis data shown in Table I and Table II were obtained by analysis of an aliquot of the incubation medium after 3 h of incubation. During that time period, a considerable amount of the free sugar, generated by hydrolysis, was taken up by the cells, whereas neither the nucleotide sugar nor the sugar-1-phosphate were able to permeate (3). The total amount of free galactose generated by hydrolysis, therefore, is higher than the amount of free galactose found in the incubation medium at the end of the assay. This to be considered in cases where the hydrolysis of UDP-galactose to free galactose is not very effective.

Comparison of the Incorporation of Galactose by BHK Cells in Monolayer and by Trypsinized BHK Cells in Suspension

In any assay for cell surface glycosyltransferases, the hydrolysis of nucleotide sugars to the free sugar by cell surface enzymes will simulate or obscure cell surface glycosyltransferase activity if the intracellular utilization of the free sugar is not prevented. Inhibition of the intracellular utilization of the radioactive free sugar can be achieved 1) by the addition of transport inhibitors, such as phloridzin (9), 2) by competition with high concentrations, e.g., at or above the K_m for the sugar transport system of the nonradioactive free sugar, or 3) by poisoning the cells with azide. We have shown in a previous study (3) that neither phloridzin, nor high concentrations of galactose, nor azide have any effect on the galactosyltransferase reaction in cell homogenate.

The effect of these inhibitors on the galactose incorporation of BHK cells in monolayer and in suspension assays with trypsinized cells was compared. Table III shows that high concentrations of nonradioactive galactose as well as the addition of 15 mM sodium azide inhibited galactose incorporation into BHK cells in monolayer and in suspension to a similar degree. In contrast, phloridzin inhibited galactose incorporation during a 3 h incubation period only by about 60% with BHK cells in a suspension assay, but inhibited incorporation by 95% with BHK cells assayed in monolayer. Phloridzin acts as an inhibitor of galactose incorporation by blocking the transport sites for galactose (9). The failure of phloridzin to inhibit galactose incorporation with trypsinized cells to the same extent as with BHK cells in monolayer, therefore, may indicate either altered binding properties of

TABLE III. Inhibition of [^{3}H]Galactose Incorporation With BHK Cells in Monolayer and With Trypsinized BHK Cells in Suspension*

	BHK cells in monolayer		BHK cells in suspension	
Addition	cpm/10^6 cells	% Incorporation	cpm/10^6 cells	% Incorporation
None	2,535	100	1,487	100
1.5 mM nonradioactive galactose	335	13	223	15
15 mM NaN$_3$	483	19	252	17
1.5 mM nonradioactive galactose, 10 mM NaN$_3$	228	9	45	3
5 mM phloridzin	127	5	565	38

*BHK cells (3 × 10^6 per assay) either in monolayer or in suspension after trypsin treatment (see Legend to Table I) were incubated for 1 h with 0.5 ml of isotonic Hepes buffer, containing 1.0 μCi [^{3}H]-galactose. Additions to the incubation medium were as indicated. The reaction was stopped by the addition of 50 μl of 10% SDS and incorporation of radioactive galactose was determined in 200-μl aliquots as described in Materials and Methods.

this inhibitor at the galactose transport sites or altered permeability properties of trypsinized cells. That the latter may indeed be the case is suggested by the different kinetics of galactose incorporation with BHK cells in monolayer and BHK cells in suspension. As shown in Fig. 1, the incorporation of [^{3}H]galactose by BHK cells in monolayer into phosphotungstic acid precipitable material showed a lag phase of 10–15 min which can be explained by the time required for the equilibration of the radioactive galactose with the intracellular sugar pool. With trypsinized cells, on the other hand, the incorporation occurred without an apparent lag. A possible explanation for this result is that the intracellular sugar pool had decreased due to leakiness of the trypsinized cells and, therefore, equilibrium with the transported radioactive sugar was reached much earlier. If sugar molecules can leak out of trypsinized cells as shown by Hirschberg et al (8), one can expect also that sugar molecules may permeate into the cells unspecifically, even if the specific transport sites are blocked.

Viability of BHK Cells in Monolayer and in Suspension Assays

An absolute requirement for an assay of cell surface glycosyltransferases is that the cells remain viable during the incubation period. Dead cells could either leak out intracellular glycosyltransferases or exogenous nucleotide sugar might be made available to glycosyltransferases inside the cell. We have previously found that trypan blue exclusion is not a very sensitive measure for cell viability (3) and therefore we decided to investigate the viability of BHK cells by their ability to grow after addition of growth medium at the end of the assay.

Two types of cell preparations were tested: BHK cells grown and assayed in monolayer cultures and BHK cells which were removed from the plate with trypsin, resuspended in complete growth medium for a "recovery period" of 30 min, and then assayed in suspension. BHK cells incubated in monolayer with isotonic Hepes buffer (see Materials and Methods) did not significantly increase in number during a 3 h incubation period, but doubled in number within 16 h following the readdition of growth medium. Trypsinized BHK cells which were reseeded immediately after the recovery period (control cells)

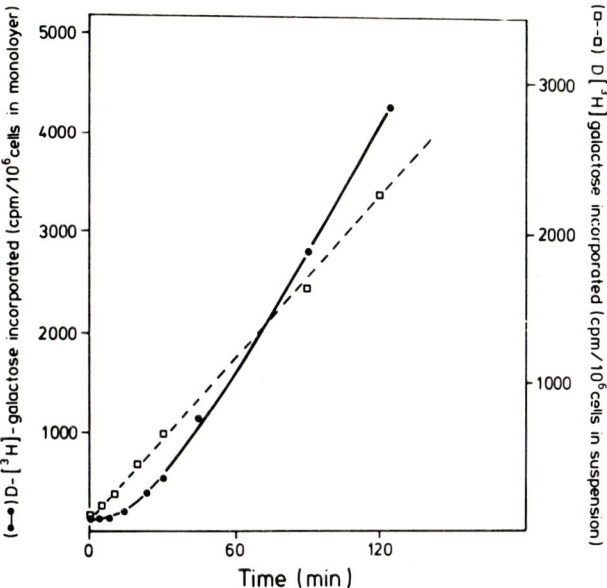

Fig. 1. Kinetics of [³H]galactose incorporation into BHK cells in monolayer and into BHK cells in suspensions. BHK cells (1×10^6) in monolayer on 3-cm dishes (●) or in suspension after trypsin treatment (see legend to Table I) were incubated with 1.0 µCi of [³H]galactose in isotonic Hepes buffer. At the times indicated, galactose incorporation was determined as described in Materials and Methods.

attached almost completely to the plate within 4 h and doubled within 20 h. However, these same cells if incubated for 3 h with isotonic Hepes buffer following the recovery period attached much more poorly (about 65%), and only the cells attached to the plate were able to resume normal growth. This indicated that after suspension of trypsinized cells in buffer for a 3-h incubation period about one third of the cells were no longer viable according to our standards.

The decreased viability of trypsinized cells maintained in suspension during an assay is a particular problem when highly unphysiological conditions, such as the presence of high concentrations of $MnCl_2$ and sodium azide, are used. When trypsinized BHK cells were incubated with isotonic Hepes buffer containing 10 mM $MnCl_2$ and 10 mM sodium azide (conditions routinely used by Roth and White, Ref. 10), all cells were killed during a 3 h incubation period as judged by their inability to reattach to the plate within 24 h after reseeding (Fig. 2B). BHK cells in monolayer, on the other hand, survived under the same assay conditions. Although these monolayer cells started to round up during the 3 h incubation period, they recovered completely after readdition of growth medium (Fig. 3). The altered permeability and decreased viability of trypsinized cells in suspension assays, in our opinion, renders such a cell preparation unsuitable for the assay of cell surface glycosyltransferases.

Assay for Cell Surface Glycosyltransferase Activity Towards Endogenous Acceptors

In a previous paper we have provided evidence that the incorporation of galactose from exogenous UDP-galactose with BHK cells can be inhibited if the intracellular utilization of free galactose is blocked by various inhibitors (3). We, therefore, concluded that

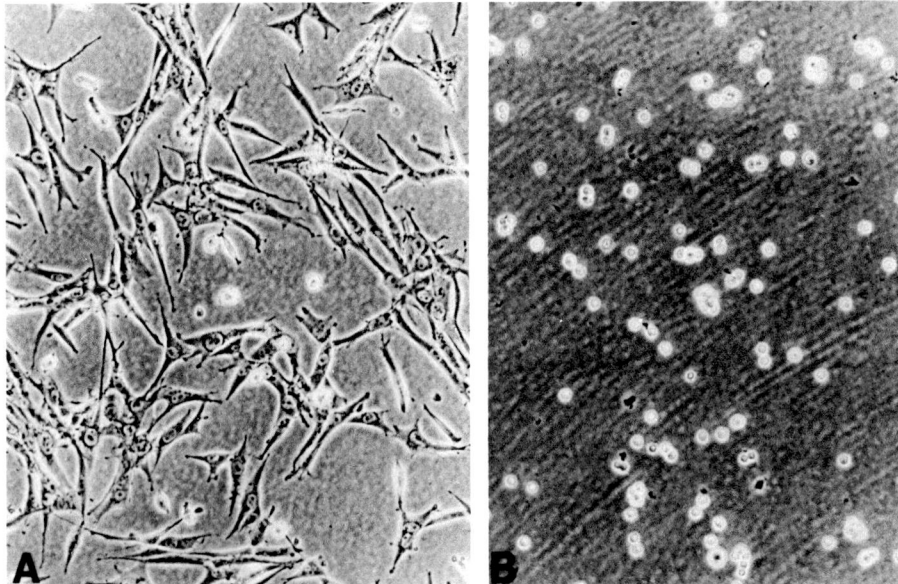

Fig. 2. Survival of BHK cells in suspension following incubation with isotonic Hepes buffer containing manganese chloride and sodium azide. BHK cells (1.3×10^6 per 3-cm dish) were removed from petri dishes by treatment with 0.1% trypsin in isotonic Hepes buffer. The cell preparations were preincubated with 1 ml of normal growth medium for 30 min, washed twice with isotonic Hepes buffer and then incubated for 3 h with isotonic Hepes buffer containing 10 mM $MnCl_2$ and 10 mM sodium azide. A) Control cells were incubated for approximately 5 min, washed twice with growth medium, and then reseeded onto 10-cm dishes. B) Cells were incubated for 3 h with occasional shaking, washed twice with growth medium, and then reseeded onto 10-cm dishes. Phase contrast microscopy pictures (magnification 165 ×) were taken 24 h after reseeding the cells.

the incorporation observed in the absence of inhibitors is due to hydrolysis of the nucleotide sugar and to the subsequent intracellular use of free galactose. This conclsuion is further substantiated by the experiment shown in Table IV. By cloning BHK cells, it is possible to obtain BHK subclones, whose ability to hydrolyze UDP-galactose is greatly reduced (11). If these BHK subclones are assayed for cell surface galactosyltransferase activity in the absence of inhibitors of galactose incorporation, a close relationship between the hydrolytic activity of these cells and the amount of galactose incorporated is found. BHK cells which exhibit strong hydrolytic activity, show high incorporation values. This incorporation can be inhibited by the addition of phloridzin. On the other hand, BHK cells which do not measurably hydrolyze UDP-galactose, also do not incorporate radioactive galactose to a significant amount, even in the absence of phloridzin.

The experiment shown in Tables IV further demonstrates that upon incubation of BHK cells with UDP-galactose a basal level of incorporation of radioactive galactose cannot be inhibited by the addition of phloridzin. Since this basal level of incorporation also is observed upon incubation of cells with radioactive galactose (Table III) it most likely does not represent specific incorporation by cell surface glycosyltransferases.

An argument against the validity of our finding that BHK cells do not exhibit endogenous cell surface galactosyltransferase activity (3) was put forward by Shur and Roth (1).

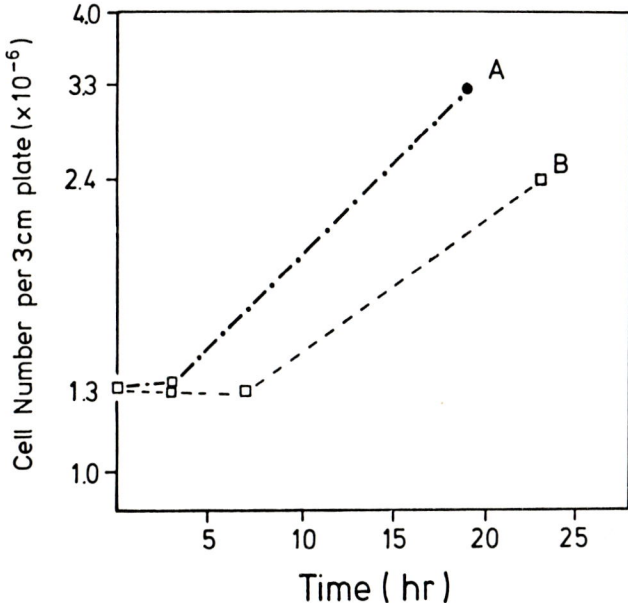

Fig. 3. Growth of BHK cells in monolayer following incubation with isotonic Hepes buffer containing manganese chloride and sodium azide. BHK cells in monolayer (1.3×10^6 per 3-cm dish) were incubated for 3 h with isotonic Hepes buffer (A) or with isotonic Hepes buffer containing 10 mM sodium azide and 10 mM $MnCl_2$ (B). The cells were washed twice with growth medium and 3 ml of growth medium was added. Cell numbers were determined from parallel plates at the times indicated.

They feel that monolayer cultures are not suitable for any assay of endogenous cell surface glycosyltransferase activity, because "cells from subconfluent cultures may have a low endogenous activity, because they were separate, while cells from confluent cultures may have already glycosylated their acceptors" (cited from Ref. 1). For the reasons given earlier in this paper, we believe that any assay system using trypsinized cells will possibly lead to artefacts due to the altered permeability and decreased viability of these cells. In order to test whether contact dependent cell surface galactosyltransferase activity could be detected with BHK cells in monolayer, we have seeded BHK cells at different concentrations, ranging from sparse (2.9×10^4 cells/cm^2) to dense (2.9×10^5 cells/cm^2). Approximately 6–8 h after seeding, when the cells were attached to the plate and no longer rounded up, e.g., at a time when cell contact was most likely to be (re)established in the more dense cultures, these cells were assayed for endogenous cell surface galactosyltransferase activity in monolayer. Table V shows that the incorporation of radioactive galactose from UDP-galactose by the BHK cell monolayers was independent of cell density and could be inhibited with phloridzin, suggesting the absence of contact-dependent cell surface galactosyltransferase activity on BHK cells.

We then examined whether the absence of cell surface galactosyltransferase activity on BHK cells in monolayer might be due to the fact that all endogenous acceptors were already glycosylated. BHK cells in confluent monolayers were pretreated with a) 0.5 µg/ml trypsin, b) 0.5 µg/ml pronase, or c) a glycosidase mixture containing 50 µg/ml β-galactosidase, 50 µg/ml β-glucosidase, 50 µg/ml hyaluronidase, and 10 µg/ml sialidase, for 45 min. This treatment did not affect the viability of these cells, although a few of the

TABLE IV. Incorporation of [^{3}H] Galactose From UDP[^{3}H] Galactose by BHK-Subclones With Different Hydrolytic Activities*

BHK subclone	Percent of free galactose generated by hydrolysis	[^{3}H] Galactose incorporated (cpm/10^6 cells)		Percent residual incorporation in the presence of phloridzin
		No inhibitor	8 mM phloridzin	
scl 3	58.8	10,671	127	1.2
scl 16	7.1	1,410	50	3.5
scl 15	2.6	227	46	20.3
scl 23	0.8	107	69	64.5

*Monolayers of BHK cells (1 × 10^6 per 3-cm dish) were incubated for 3 h with 0.3 ml of isotonic Hepes buffer, containing 1.0 µCi UDP[^{3}H]galactose and 5 mM of MnCl$_2$. Phloridzin was added as indicated. Hydrolysis of UDP[^{3}H]galactose and incorporation of [^{3}H]galactose was determined as described in Materials and Methods. Designation of BHK subclones is according to (11).

TABLE V. [^{3}H] Galactose Incorporation From UDP[^{3}H] Galactose by BHK Cells at Different Cell Densities*

Cell density (cells/cm^2)	[^{3}H] Galactose incorporated (total cpm/assay)		Percent residual activity in the presence of phloridzin
	No inhibitor	5 mM Phloridzin	
2.9 × 10^4	1,407	84	6.0
7.1 × 10^4	3,545	137	3.9
1.0 × 10^5	6,113	122	2.0
1.4 × 10^5	7,697	159	2.1
2.9 × 10^5	16,349	346	2.2

*BHK cells were seeded on 3-cm dishes at the cell densities indicated approximately 6–8 h before the experiment. Monolayers were incubated for 3 h with 0.3 ml of isotonic Hepes buffer, containing 2.0 µCi UDP[^{3}H]galactose and 5 mM MnCl$_2$. Phloridzin was added as indicated. Incorporation of [^{3}H]-galactose was determined as described in Materials and Methods.

cells treated with trypsin rounded up while remaining on the plate. The cells were then assayed for cell surface galactosyltransferase activity in the absence or presence of 5 mM phloridzin either directly after the pretreatment or after a recovery period of 45 min in normal growth medium. The result of this experiment is summarized in Table VI. None of the pretreatments caused an enhancement of galactose incorporation. Furthermore, in all samples the incorporation observed could be inhibited with phloridzin.

Cells of Various Origins

The absence of detectable surface galactosyltransferase activity of BHK cells is not an unique property of these cells. We have tested Py BHK, BALB/c 3T3, and SV 3T3 cells as representatives of established cell lines and primary AGMK cells, secondary ME cells, and CHE cells in passage 18 as representatives of nonestablished cells. Table VII shows that with all cells tested the incorporation of radioactive galactose from exogenous UDP-

TABLE VI. Assay for Endogenous Cell Surface Galactosyltransferase Activity With Protease (or Glycosidase) Pretreated BHK Cells on Plate*

Pretreatment	Recovery period	Additions to the incubation medium	[^{3}H] Galactose incorporation	
			cpm/10^6 cells	% incorporation
A. Trypsin	none	none	1,168	100
	none	5 mM Phloridzin	164	14.0
	45 min	none	1,365	100
	45 min	5 mM Phloridzin	130	9.5
B. Pronase	none	none	1,300	100
	none	5 mM Phloridzin	111	8.5
	45 min	none	1,289	100
	45 min	5 mM Phloridzin	97	7.5
C. Glycosidase mixture	none	none	1,205	100
	none	5 mM Phloridzin	94	7.8
	45 min	none	1,292	100
	45 min	5 mM Phloridzin	69	5.3
D. None (Control)	none	none	1,292	100
	none	5 mM Phloridzin	98	7.6

*Monolayers of BHK cells (3 × 10^6 per 3-cm dish) were pretreated for 45 min with A) 0.5 μg/ml trypsin in isotonic Hepes buffer, B) 0.5 μg/ml pronase in isotonic Hepes buffer, or C) a glycosidase mixture containing 50 μg/ml β-galactosidase, 50 μg/ml hyaluronidase, and 10 μg/ml sialidase in isotonic Hepes buffer. Cells incubated with isotonic Hepes buffer for 45 min were used as control (D). Following the pretreatment, the monolayers were washed twice with isotonic Hepes buffer and the cells assayed for cell surface glycosyltransferase activity either directly or after a recovery period of 45 min in normal growth medium. The incubation medium was 0.3 ml of isotonic Hepes buffer, containing 1.0 μCi UDP[^{3}H]galactose and 5 mM $MnCl_2$. Phloridzin was added as indicated. Incubation was for 3 h. Incorporation of [^{3}H]galactose was determined as described in Materials and Methods.

galactose could be inhibited by either phloridzin or azide to a similar extent as the incorporation resulting from the incubation with radioactive-free galactose. This result strongly suggests that none of the cells tested exhibits cell surface galactosyltransferase activity.

The criteria developed in our earlier study (3) to differentiate between intracellular and cell surface galactosyltransferase activity can also be applied for testing whether glycosyltransferases other than galactosyltransferases may be located on the cell surface.

Assay for Fucosyl- and Sialyltransferases

We decided to test for cell surface fucosyl- and sialyltransferases because both fucose and sialic acid are terminal sugars on carbohydrate chains of glycoproteins and glycolipids (2) and neither sugar is rapidly metabolized in cells.

Table VIII shows that the incorporation of radioactive L-fucose into BHK cells is not inhibited by either high concentrations (1.5 mM) of glucose or by the addition of 8 mM phloridzin. This indicates that L-fucose is transported into the cells by a different transport system than glucose or galactose. However, inhibition of radioactive fucose incorporation can be obtained by competing with nonradioactive fucose, by inhibition of the cellular metabolism with sodium azide (10), and most effectively by a combination of both (Table VIII). Similarly, the incorporation of radioactive fucose upon incubating BHK cells with GDP-fucose can be inhibited by the addition of either cold fucose or azide, or

TABLE VII. Assay for Endogenous Cell Surface Galactosyltransferase Activity With Various Cells*

Cells	Additions to the incubation medium	Incubation with [^{3}H]galactose		Incubation with UDP[^{3}H]galactose		μg Cell protein per assay
		cpm/assay	% Incorporation	cpm/assay	% Incorporation	
A. Established cells						
BALB/c3T3	none	1,781	100	203	100	170
	4 mM Phloridzin	358	20	32	16	170
	10 mM Sodium azide	610	34	58	28	170
	20 mM Sodium azide	432	24	38	19	170
SV 3T3	none	ND		696	100	270
	4 mM Phloridzin	ND		108	15	270
	8 mM Phloridzin	ND		78	11	270
Py BHK	none	21,347	100	9,204	100	350
	4 mM Phloridzin	2,989	14	1,473	16	350
	8 mM Phloridzin	1,281	6	828	9	350
B. Nonestablished Cells						
AGMK, primary	none	ND		2,132	100	220
	8 mM Phloridzin	ND		175	8	220
	20 mM Azide	ND		327	15	220
	1.6 mM nonradioactive galactose	ND		220	10	220
ME, secondary	none	4,163	100	3,202	100	290
	8 mM Phloridzin	281	7	95	3	290
CHE, passage 18	none	ND		8,497	100	190
	4 mM Phloridzin	ND		202	2.4	190
	8 mM Phloridzin	ND		119	1.4	190

*2×10^5 cells were seeded on 3-cm dishes and assayed for cell surface glycosyltransferases 2 days after seeding. The incubation medium consisted of 0.3 ml of isotonic Hepes buffer containing 5 mM MnCl$_2$ and 1.0 μCi [^{3}H]galactose or 1.0 μCi UDP[^{3}H]galactose, respectively. Additions to the incubation medium were as indicated. Incubation was for 3 h and incorporation of [^{3}H]galactose was determined as described in Materials and Methods. Protein was determined according to Lowry et al. (29).

TABLE VIII. Assay for Endogenous Cell Surface Fucosyltransferase Activity With BHK Cells*

Additions to the incubation medium	[^{14}C] Fucose (1.0 μCi)		GDP[^{14}C] Fucose (1.0 μCi)	
	cpm/10^6 cells	% Incorporation	cpm/10^6 cells	% Incorporation
None	375	100	153	100
8 mM Phloridzin	382			
1.5 mM D-glucose	376			
15 μM L-fucose	352	94		
75 μM L-fucose	217	58		
150 μM L-fucose	152	40		
750 μM L-fucose	95	25		
1,500 μM L-fucose	24	6	27	17
5 mM Sodium azide	86	22	23	15
10 mM Sodium azide	36	10	16	10
15 mM Sodium azide	34	9	10	6
5 mM Sodium azide +1,500 μM L-fucose	20	5	3	2

*BHK cells in monolayer (1 × 10^6 per 3-cm dish) were incubated for 3 h with 0.3 ml of isotonic Hepes buffer, containing 1 mM Mn Cl$_2$ and μCi 2-[^{14}C] fucose or 1.0 μCi GDP-2-[^{14}C] fucose, respectively. Additions to the incubation medium were as indicated. Incorporation of 2-[^{14}C] fucose was determined as described in Materials and Methods.

a combination of both, suggesting the absence of surface fucosyltransferases on BHK cells (Table VIII).

Datta (12) has reported that BHK cells in monolayer catalyze the transfer of sialic acid from exogenous CMP-sialic acid onto acceptor molecules on the surface of these cells. He found that neither a 1,000-fold molar excess of nonradioactive sialic acid (1 mM) in the labeling medium nor the addition of transport inhibitors such as phloridzin or a mixture of sodium cyanide and iodoacetate (13) reduced the incorporation of radioactivity into the acid-insoluble fraction. The addition of 20 mM sodium azide reduced the incorporation only by about 20%. Datta, therefore, concluded that the incorporation of sialic acid from CMP-sialic acid was mediated by cell surface sialytransferase.

Hirschberg et al. (8) recently have demonstrated that the uptake of sialic acid by BHK cells and other fibroblasts is a saturable process with an apparent K_m of 10 mM. The incorporation of radioactive sialic acid as a breakdown product of CMP-sialic acid, therefore, should not be affected by the addition of unlabeled sialic acid in concentrations lower than the K_m for sialic acid uptake. Table IX shows that radioactive sialic acid incorporation from radioactive CMP-sialic acid and from radioactive free sialic acid can be effectively inhibited by the addition of 10 mM or 20 mM unlabeled sialic acid. This result may confirm the conclusion of Hirschberg et al. (8) that the incorporation of sialic acid observed after incubation of monolayer cells with CMP-sialic acid is more likely due to the uptake and intracellular utilization of free sialic acid as a breakdown product of CMP-sialic acid than to cell surface transferase activity. On the other hand, our finding that the addition of 5 mM unlabeled sialic acid to the incubation medium hardly inhibited radioactive sialic acid incorporation, whereas the addition of 10 mM unlabeled sialic acid inhibited almost completely is difficult to understand. A possible explanation is that high concentrations of unlabeled sialic acid in the incubation medium may inhibit the incor-

TABLE IX. Assay for Endogenous Cell Surface Sialyltransferase Activity With BHK Cells*

Additions to the incubation medium	[^{3}H] sialic acid (1.0 μCi)		CMP[^{3}H] sialic acid (1.0 μCi)	
	cmp/10^6 cells	% Incorporation	cmp/10^6 cells	% Incorporation
None	256	100	224	100
5 mM sialic acid	213	83	205	92
10 mM sialic acid	36	14	16	7
20 mM sialic acid	8	3	3	1

*BHK cells in monolayer (1 × 10^6 per 3-cm dish) were incubated for 3 h with 0.3 ml of isotonic Hepes buffer containing 1.0 μCi [^{3}H] sialic acid or 1.0 μCi CMP[^{3}H] sialic acid, respectively. Nonradioactive sialic acid was added to the incubation medium as indicated. Incorporation of [^{3}H] sialic acid was determined as described in Materials and Methods.

poration of sialic acid unspecifically either because of the presence of a nonspecific inhibitor in the sialic acid preparation or because of toxic effects of the sialic acid itself.

Assay for Cell Surface Glycosyltransferase Activity Towards High-Molecular-Weight Exogenous Acceptors

A demonstration that viable, intact cells were able to catalyze the transfer of the sugar moiety from a nucleotide sugar directly onto an exogenous acceptor molecule present in the incubation medium would be strong support for the existence of cell surface glycosyltransferases. Experimental data suggesting this transfer reaction have been reported (see review by Shur and Roth, Ref. 1) but, unfortunately, in most of these studies trypsinized or EDTA-treated cells in suspension assays were used. In these studies, therefore, the transfer reaction may have been catalyzed by dead cells, where the acceptor molecules as well as the nucleotide sugars were available to intracellular glycosyltransferases. We have tested various cells in monolayer for cell surface galactosyltransferase and cell surface sialyltransferase activity using exogenous high-molecular-weight acceptors (see Materials and Methods). Table X shows that the degalactosized antifreeze protein as well as the desialized and degalactosized fetuin were active as acceptors in homogenate of BHK cells. Intact cells, however, were not able to transfer galactose onto these acceptor proteins (Table XI). Similar results were obtained, when acceptor proteins for the sialic acid transfer were tested in BHK cell homogenate and on intact cells. Whereas the antifreeze protein (AFP) and the desialized fetuin acted as acceptor proteins for sialic acid transfer in BHK cell homogenate (Table XII), no transfer reaction could be detected when these acceptor proteins were added to intact cells (Table XIII).

DISCUSSION

Prevention of intracellular utilization of the radioactive sugar as a breakdown product of nucleotide sugars and the use of intact viable cells are prerequisites for studies on the existence of cell surface glycosyltransferases. We have shown in this and in previous studies (3, 11) that the cell surface of various cells contains hydrolytic enzymes which catalyze the conversion of nucleotide sugars to the free sugar. Intracellular utilization of the free radioactive sugar generated by hydrolysis, however, can be prevented by com-

TABLE X. Galactosyltransferase Activity Towards Exogenous Acceptors in BHK Cell Homogenate*

Addition	[^{3}H] Galactose incorporation (cpm/100 μg cell homogenate)
None	3,870
100 μg AFP	3,756
100 μg AFP-degal.	18,212
100 μg fetuin	3,945
100 μg fetuin-desial.-degal.	23,109

*Cells were removed from plates with 0.1% trypsin in isotonic Hepes buffer, suspended in isotonic Hepes buffer, and homogenized by sonication (60 sec in 5-sec intervals at full power in a Branson sonifier equipped with a microtip). Cell homogenate at a concentration of 100 μg in a final volume of 50 μl was incubated with 1.0 μCi UDP[^{3}H] galactose in isotonic Hepes buffer containing 5 mM MnCl$_2$ and 0.1% Triton X-100. Acceptor protein (see Materials and Methods) was added as indicated. After 3 h incubation at 37°C the reaction was stopped by adding 10 μl of 10% SDS. Samples were precipitated and their radioactivities measured as described in Materials and Methods. Background incorporation with 1.0 μCi of [^{3}H] galactose was 200 cpm. Protein was determined according to Lowry et al. (29).

petition with the nonradioactive free sugar in concentrations higher than the K_m for the transport system of the free sugar or by inhibiting the transport of the radioactive sugar with appropriate inhibitors or by inhibiting cellular metabolism with sodium azide.

We have compared the permeability and the viability of trypsinized BHK cells in suspension and of BHK cells in monolayer. Our finding that phloridzin inhibits galactose incorporation to 95% with BHK cells in monolayer but only to about 60% with trypsinized BHK cells, as well as the different kinetics of galactose incorporation with these 2 cell preparations strongly suggests that trypsinization alters the permeability of BHK cells for galactose. In a more extensive study, Hirschberg et al. (8) have shown that trypsin or EDTA treatment of fibroblasts grown on plates leads to leakage of small molecules, as measured by the release of 2-[^{3}H]-deoxyglucose, as well as to leakage of macromolecules, as measured by the release of ^{51}Cr. In agreement with similar studies by the same authors, we found that trypsin treatment of BHK cells grown on plates significantly reduced the viability of these cells if they were kept suspended in buffer during the assay period. BHK cells on plates, on the other hand, stayed viable when incubated with buffer for the same period of time. The finding that trypsinized or EDTA-treated cells are leaky and lose their viability during the assay, severely questions the validity of all studies in which the assay for cell surface glycosyltransferases has been performed with such cell preparations.

In a previous paper (3), we have provided evidence that BHK cells in monolayer do not exhibit detectable cell surface galactosyltransferase activity. This conclusion is further substantiated in the present study. Under all experimental conditions applied, cell surface galactosyltransferase activity on BHK cells either endogenous or towards 2 different high-molecular-weight exogenous acceptor molecules could not be found. Similarly, all other cells tested, including transformed cells and primary cells, did not give any indication for the presence of galactosyltransferases on their surface. Our studies on the presence of cell surface fucosyl- and sialyltransferases were less extensive but point into the same direction. Inhibition of fucose incorporation from GDP-fucose into BHK cells could be obtained by adding high concentrations of unlabeled fucose, by metabolic inhibition with sodium azide, and by a combination of both, suggesting the absence of endogenous cell surface fucosyltransferases. On the other hand, inhibition with unlabeled fucose was not

TABLE XI. Galactosyltransferase Activity Towards Exogenous Acceptors With Cells in Monolayer*

Cells	Additions to the incubation medium	Precipitation I (cells + incubation medium) cpm/10^6 cells	Precipitation II (cells + incubation medium separately) cpm/10^6 cells	Precipitation II cpm total incubation medium
BHK	10.0 µCi UDP[^{3}H]galactose 50 µg AFP 5 mM Phloridzin	188	–	–
	10.0 µCi UDP-gal 50 µg AFP-degal. 5 mM Phloridzin	212	–	–
ME, secondary	1.0 µCi UDP[^{3}H]galactose no acceptor	–	2,202	1,300
	1.0 µCi UDP[^{3}H]galactose 50 µg AFP	–	2,136	1,363
	1.0 µCi UDP[^{3}H]galactose 50 µg AFP-degal.	–	1,971	1,623
AGMK, primary	1.0 µCi UDP[^{3}H]galactose	2,132	–	–
	1.0 µCi UDP[^{3}H]galactose 200 µg AFP-degal.	2,037	–	–
	1.0 µCi UDP[^{3}H]galactose 200 µg fetuin-desial-degal.	2,209	–	–

*2×10^5 cells were seeded on 3-cm dishes and assayed for cell surface galactosyltransferase activity 2 days after seeding. The cells were incubated for 3 h with 0.3 ml of isotonic Hepes buffer, containing 5 mM $MnCl_2$. UDP[^{3}H]galactose, acceptor proteins (see Materials and Methods), and phloridzin were added as indicated. Incorporation of [^{3}H]galactose was determined by the following procedures: Precipitation I: The reaction was stopped by the addition of 50 µl of 10% SDS, the cell lysate diluted with 2 ml of ice cold water, precipitated with 3 ml of 1% phosphotungstic acid in 0.5 N HCl, and incorporation of radioactive sugar into cellular and acceptor proteins determined as described in Materials and Methods. Precipitation II: The incubation medium was collected with a Pasteur pipette, as were 2 subsequent washes of the cells with isotonic Hepes buffer. Incubation medium and the washes were pooled, 100 µg of BSA was added as carrier, and proteins precipitated with 1% phosphotungstic acid in 0.5 N HCl. Radioactivity of the samples was measured as described in Materials and Methods. Incorporation of radioactive sugars into cellular proteins was determined as described in Materials and Methods.

complete, and due to lack of an appropriate high-molecular-weight fucosyl acceptor, cell surface fucosyltransferase activity toward exogenous acceptors could not be tested. We therefore cannot exclude low levels of cell surface fucosyltransferase activity on BHK cells. In testing for cell surface sialyltransferase activity we were unable to obtain significant sialic acid transfer onto 2 different high-molecular-weight sialic acid acceptors using BHK and Balb/c 3T3 cells. We, therefore, conclude that these cells do not express cell surface sialytransferase activity towards exogenous acceptors. Analysis of BHK cells for cell surface sialyltransferase activity towards endogenous acceptors however, failed to give a conclusive result. Inhibition of radioactive sialic acid incorporation could only be achieved by using high concentrations of unlabeled sialic acid. Although this is in accordance with data obtained by Hirschberg et al. (8), and could mean that concentrations of unlabeled sialic acid above the apparent K_M of sialic acid transport are nec-

TABLE XII. Sialyltransferase Activity Towards Exogenous Acceptors in BHK Cell Homogenate*

Addition	[^{3}H] Sialic acid incorporation (cpm/100 μg cell homogenate)
None	1,142
100 μg AFP	38,737
100 μg AFP-degal.	1,203
100 μg fetuin	1,089
100 μg fetuin-desial.	21,045

*BHK cell homogenate was prepared as described in the legend to Table X and incubated at a concentration of 100 μg in a final volume of 50 μl with 1.0 μCi CMP[^{3}H] sialic acid in isotonic Hepes buffer containing 0.1% Triton X-100. Acceptor protein (see Materials and Methods) was added as indicated. After 3 h incubation at 37°C the reaction was stopped by adding 10 μl of 10% SDS. Samples were precipitated and their radioactivities determined as described in Materials and Methods. Protein was determined according to Lowry et al. (29).

TABLE XIII. Sialyltransferase Activity Towards Exogenous Acceptors With Cells in Monolayer*

Cells	Addition	Precipitation I (cells + incubation medium) cpm/10^6 cells	Precipitation II (cells + incubation medium separately) cpm/10^6 cells	cpm total incubation medium
BHK	1.0 μCi CMP[^{3}H] sialic acid	205		
	2.0 μCi CMP [^{3}H] sialic acid	389		
	1.0 μCi CMP 200 μg AFP [^{3}H] sialic acid	263		
	1.0 μCi CMP [^{3}H] sialic acid 200 μg AFP degal.	228		
	2.0 μCi CMP [^{3}H] sialic acid 200 μg AFP	441		
BALB c/3T3	1.0 μCi CMP[^{3}H] sialic acid		152	120
	1.0 μCi CMP[^{3}H] sialic acid 100 μg AFP		157	132
	1.0 μCi CMP [^{3}H] sialic acid 200 μg AFP		131	143

*BHK cells and BALB/c/3T3 cells in monolayers (1 × 10^6 per 3-cm dish) were incubated with 0.3 ml of isotonic Hepes buffer. CMP[^{3}H] sialic acid and acceptor proteins (see Materials and Methods) were added as indicated. Precipitation I and Precipitation II were described in the legend to Table XI.

essary to suppress the incorporation of radioactive sialic acid, another possible explanation for the suppression of radioactive sialic acid incorporation only with very high concentrations of unlabeled sialic acid is a toxic effect of this compound on the cells.

Obviously, different results are obtained when the assay for cell surface glycosyltransferases is performed with cells attached to plates and when performed with EDTA- or trypsin-treated cells. The finding that cells in monolayer do not contain detectable cell surface glycosyltransferase activity (3, 14), whereas EDTA- or trypsin-treated cells appear to possess such an activity, is as easily explained by the increased permeability of EDTA- and trypsin-treated cells as by the speculation that the contact dependency of cell surface glycosylation might render it impossible to detect such activities with monolayer cells. Nevertheless, we have tried in this study to find conditions which might have allowed the detection of contact dependent endogenous cell surface galactosyltransferase activity on BHK cells but obtained a negative result.

Patt and Grimes (14) have tested BHK cells for cell surface glycosyltransferase activity in monolayer and in suspension after EDTA treatment. While no endogenous glycosyltransferase activity was found with BHK cell monolayers, if hydrolysis of the nucleotide sugars was inhibited by the addition of 5'-AMP, an inhibitor of nucleotide pyrophosphatase activity (15), the authors report a transfer reaction onto high-molecular-weight exogenous acceptors with BHK monolayer cells. Although Patt and Grimes do not offer direct proof for the transfer of the sugar specifically onto their acceptor molecules, they interpret their finding as evidence for the presence of cell surface glycosyltransferases on monolayer cells. In this study, we have assayed various cells for surface galactosyl-sialyltransferase activity using 2 different acceptors for each transferase reaction. In all cases we were unable to detect any significant transfer. In some experiments a slight stimulation of total incorporation was observed, but no correlation was found between the amount of acceptor or radioactivity added and the stimulation of incorporation.

Since the sialyltransferase reaction is very effective in cell homogenate (Table XII), less than 1% of broken cells would be sufficient to account for the incorporation data reported by Patt and Grimes (14) and would explain the small stimulation observed in some of our experiments.

Further support for our view that the surface of mammalian cells does not contain nucleotide sugar dependent glycosyltransferases comes from studies investigating the subcellular distribution of glycosyltransferases. If cell fractions of defined purity were analyzed, it was found that in rat liver cells the Golgi apparatus was the sole location of galactosyltransferase, N-acetylglucosaminetransferase, and transferases of ganglioside synthesis (16–23). A similar subcellular distribution of galactosyltransferase and cerebroside sulfotransferase was found in rat kidney cells (22, 23). In all of these studies, the plasma membrane fraction was either devoid of glycosyltransferases or the minute amounts detected could be assigned to contamination by Golgi membrane fragments. Glycosyltransferases were also not detected in bovine milk fat globule membrane (24, 25), a membrane known to be derived directly from apical plasma membrane of mammary secretory cells (26).

In sum, the experimental data available cause us to view with caution the concept of cell surface glycosyltransferases. We feel that the evidence presented in previous studies supporting the existence of cell surface glycosyltransferases is insufficient; recent studies (3, 27, 8, this paper) have provided evidence that either hydrolysis of the nucleotide sugars or altered permeability due to cell damage or a combination of both will account for the data presented in most reports of cell surface glycosyltransferases. On the other hand, failure to detect nucleotide sugar dependent cell surface glycosyltransferases does not necessarily rule out the possibility of other mechanisms of cell surface glycosylation. So, for instance, Yogeeswaran et al. (28) have suggested the existence of a hitherto unknown class of cell surface glycosyltransferases, which do not use nucleotide sugars as substrate but instead use another, yet unknown, sugar donor. Clearly, more experimental support for this kind of surface glycosylation would be desirable.

ACKNOWLEDGMENTS

We thank Drs. C. Hirschberg and B. Sefton for many helpful discussions during the preparation of this manuscript. The excellent technical assistance of Eric Sasso is gratefully acknowledged.

This work was supported by grants 5 R01 CA 15365-01A1 and GB-41879 from the National Science Foundation and NIH grant 14195. W.D. was a recipient of a postdoctoral fellowship from the Deutsche Forschungsgemeinschaft (DFG).

REFERENCES

1. Shur BD, Roth S: Biochim Biophys Acta 415:473, 1975.
2. Roseman S: Chem Phys Lipids 5:270, 1970.
3. Deppert W, Werchau H, Walter G: Proc Natl Acad Sci USA 71:3068, 1974.
4. Vogt M, Dulbecco R: Proc Natl Acad Sci USA 49:171, 1963.
5. DeVries AL, Komatsu SK, Feeney RE: J Biol Chem 245:2901, 1970.
6. Komatsu SK, DeVries AL, Feeney RE: J Biol Chem 245:2909, 1970.
7. Kim YS, Perdomo J, Nordberg J: J Biol Chem 246:5466, 1971.
8. Hirschberg CG, Goodman SR, Green C: Biochemistry 15:3591, 1976.
9. Alvarado F: Biochim Biophys Acta 135:483, 1967.
10. Roth S, White D: Proc Natl Acad Sci USA 69:485, 1972.
11. Deppert W, Walter G: J Cell Physiol 90:41, 1977.
12. Datta P: Biochemistry 13:3987, 1974.
13. Cunningham DD, Pardee AB: Proc Natl Acad Sci USA 64:3501, 1969.
14. Patt LM, Grimes WJ: Biochem Biophys Res Commun 67:483, 1975.
15. Bischoff E, Liersch M, Keppler D, Decker K: Hoppe-Seyler's Z Physiol Chem 351:729, 1970.
16. Schachter H, Jabbal I, Hudgin RL, Pinteric L, McGuire EJ, Roseman S: J Biol Chem 245:1090, 1970.
17. Wagner RR, Cynkin MA: J Biol Chem 246:143, 1971.
18. Morre DJ, Merlin LM, Keenan TW: Biochem Biophys Res Commun 37:387, 1969.
19. Fleischer B, Fleischer S: Biochim Biophys Acta 219:301, 1970.
20. Bergeron JJM, Ehrenreich JH, Siekevitz P, Palade G: J Cell Biol 59:73, 1973.
21. Bizzi A, Marsh JB: Proc Soc Exp Biol Med 144:762, 1973.
22. Fleischer B, Zambrano F: Biochem Biophys Res Commun 52:951, 1973.
23. Fleischer B, Zambrano F, Fleischer S: J Supramol Struct 2:737, 1974.
24. Keenan TW, Huang CM: J Diary Sci 55:1013, 1972.
25. Keenan TW: J Diary Sci 57:187, 1974.
26. Keenan TW, Morré DJ, Olson DE, Yunghans WN, Patton S: J Cell Biol 44:80, 1970.
27. Keenan TW, Morré DJ: FEBS Lett 55:8, 1975.
28. Yogeeswaran G, Laine RA, Hakomori S: Biochem Biophys Res Commun 59:591, 1974.
29. Lowry OH, Rosebrough NJ, Farr AL, Randall RJ: J Biol Chem 193:265, 1951.

Immunochemical Purification of Probe-Labeled Plasma Membrane Proteins: An Approach to the Molecular Anatomy of the Cell Surface

Paolo M. Comoglio, Guido Tarone, and Marilena Bertini

Department of Human Anatomy, University of Torino School of Medicine, Turin, Italy

The probe 2,4,6-trinitrobenzene sodium sulfonate may be used under appropriate conditions for selective labeling of plasma membrane proteins exposed at the outer cell surface. Labeled proteins, solubilized by detergents, can be purified by reverse immunoadsorption using antiprobe antibodies covalently linked to Sepharose 4B. This method has been applied to an investigation of the outer cell surface structure of chicken embryo and hamster fibroblasts. Coelectrophoresis in sodium dodecyl sulfate-polyacrylamide gels of probe-labeled membrane proteins purified from baby hamster kidney fibroblasts have shown that 7 major protein groups of different molecular weight are exposed on both control and Rous sarcoma or polyoma virus-transformed cells. Moreover, the transformed cells display a nonvirion component of 80–100 k daltons that is not labeled by the probe in normal cells. In fibroblasts transformed by a temperature sensitive Rous sarcoma virus mutant, that transforms at 37°C but not at 41°C, the expression of this component is related to the expression of the transformed phenotype.

Key words: affinity chromatography, plasma membrane, neoplastic transformation

The selective labeling of molecules exposed at the outer surface of intact cells has proved to be a suitable method for investigating the molecular structure of plasma membranes of eukaryotic cells (for review see Ref. 1). In the past few years this laboratory has developed a method that is both analytical and preparative. It involves the binding of the chemical hapten-2,4,6-trinitrophenol (TNP) to lysine ε-amino groups exposed at the outer cell surface on plasma membrane proteins, and the purification of labeled molecules by affinity chromatography with insolubilized antihapten antibodies.

We have previously shown that selective binding of TNP groups to proteins exposed at the cell surface may be achieved by labeling, under appropriate conditions, living cells with 2,4,6-trinitrobenzene sulfonic acid (TNBS). This has been confirmed by a variety of techniques, including immunofluorescence and immunoelectronmicroscopy with fluorescein or horseradish peroxidase-coupled anti-TNP antibodies, subcellular fractionation, digestion

Received May 3, 1977; accepted July 26, 1977.

P. M. Comoglio holds the chair of Histology and Embryology, University of Trieste, School of Medicine.

of cell surface by proteolytic enzyme, and comparative analysis of proteins labeled by TNP or radioiodinated with lactoperoxidase (1–7).

This paper reviews briefly the method devised and its application to the study of the structure of plasma membranes of normal fibroblasts grown in vitro as primary explants or as continuous cell lines. The structural alterations found by this method in membranes of virus-transformed cells will also be discussed.

MATERIALS AND METHODS

All the fibroblast lines used in this experiment were grown to subconfluency in Eagle's minimal essestial medium (MEM) supplemented with 10% calf serum (EuroBio, Paris, France), penicillin (50 IU/ml) and streptomycin (100 μg/ml). Cultures were incubated at 37°C in a 5% CO_2 atmosphere and periodically checked for pleuropneumonia-like organism (PPLO) infection (8). C13 cells were a subline of baby hamster kidney (BHK) fibroblasts derived from the BHK 21/C13 line established by Macpherson and Stoker (9). PY and B4 cells were sublines of BHK 21/C13 cells transformed in vitro by polyoma virus and Rous sarcoma virus (RSV) of the Bryan strain, respectively. Neither subline produced virus particles. Each had lost topoinhibition and acquired a "malignant" phenotype (10). 14B was a BHK 21/C13 fibroblast line transformed by the FU19 temperature-sensitive mutant of RSV (11); the full virus genome persisted in FU19-transformed fibroblasts since the virus could be rescued by fusing them with chicken cells (12). The FU19 mutant does not replicate in hamster cells but transforms them at 37°C.

Chicken embryo fibroblasts were primary cultures obtained by repeated passages of glass-adherent cells dissociated by trypsin digestion from 11-day-old embryo skeletal muscles. Selection and enrichment of myoblasts was performed as described in detail elsewhere (6).

Anti-DNP sera were prepared in rabbits immunized by 3 monthly injections of 1 mg of DNP_{47}-human serum albumin, in Freund's complete adjuvant. Anti-DNP antibodies were purified using an immunoadsorbent made with Sepharose 4B coupled to DNP_{60}-bovine gamma globulin by the cyanogen bromide technique (13). Fifteen milligrams of antigen were coupled to each gram of activated Sepharose. Purified antibodies were eluted by 100 mM dinitrophenol buffered at ph 8.0 chromatographed on Dowex 1-X8 ion-exchange resin (Cl^- form; 20–50 mesh), dialyzed against 0.1 M sodium bicarbonate, pH 9.0, and used to prepare reverted immunoadsorbents. Twenty-five milligrams of purified antibodies were reacted overnight at 4°C with each gram of activated Sepharose. Under these conditions, 80% of the antibodies were covalently linked to the resin. The same procedure was used to link to Sepharose immunoglobulins purified by DEAE-cellulose chromatography (2) from anticalf serum rabbit antisera.

Acrylamide gel electrophoresis was performed by the SDS-disk gel method of Maizel et al. (14), using 9% or 7.5% acrylamide gels with a 3% upper spacer. Purified TNP-labeled surface proteins solubilized from control and transformed cells and radiolabeled with ^{131}I or ^{125}I were coelectrophoresed in the same gel. Each gel was sectioned in 2-mm slices and the radioactivity of both isotopes counted in each slice. The cpm values of ^{125}I were corrected for interference of ^{131}I. Molecular weight calibration of the system was achieved by simultaneously running the following standards: egg albumin (43 k daltons); rabbit immunoglobulin μ chains (73.5 k daltons), γ chains (50 k daltons), and light chains (23.5 k daltons); bovine serum albumin (67 k daltons); lactoperoxidase (92 k daltons).

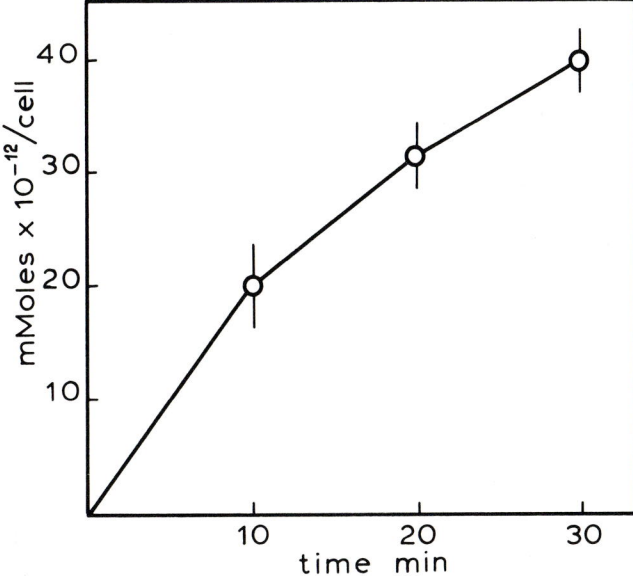

Fig. 1. Kinetic curve of the reaction between TNBS and its binding groups exposed on the outer surface of RSV-transformed BHK fibroblasts (B4). The number of TNP groups linked to the cells was measured by light adsorption at 348 nm of labeled cell preparations (5–10 × 10^6 cells) dissolved in 10% SDS, 2 M mercaptoethanol at 100°C for 3 min. Blanks were prepared by dissolving an identical number of unlabeled cells in the same way. The molar extinction coefficient of TNP-lysine (E_M^{1cm} = 15,400) was used in calculations.

Slab gel electrophoresis was performed according to Studier (15) with 3-mm thick 9% acrylamide plates. Autoradiography of dried slabs was performed with ^{131}I-labeled samples and Gevaert Curix RP-2FW x-ray film.

RESULTS AND DISCUSSION

Selective Labeling of Proteins Exposed at the Plasma Membrane Outer Surface by TNBS

Cell monolayers were carefully washed and incubated at 37°C with 5 mM trinitrobenzene sulphonic acid (Pierce) in Earle's solution buffered at pH 7.4. The reaction was stopped with iced 750 mM glycine diluted 1:1 with Earle's solution. Figure 1 shows a representative experiment in which the kinetics of the reaction between TNBS and its binding groups exposed at the cell surface of B4 hamster fibroblasts was measured. Saturation was not reached within 30 min of incubation; however, labeling was not further carried on because prolonged incubation in the absence of serum was found to result in loss of cell viability and changes in membrane impermeability to TNBS. Trypan blue was at any event excluded from more than 97% of the cells up to 45 min. Detachment of cells from tissue culture flasks by treatment with ethylenediaminetetraacetic acid (EDTA) or trypsin or by gentle scraping with a rubber policeman before reaction with TNBS always resulted in significant loss of cell viability and alteration in membrane permeability. Ten minutes of incubation of monolayers under the conditions described above were sufficient to bind a suitable number of TNP groups to surface proteins to allow their purification by

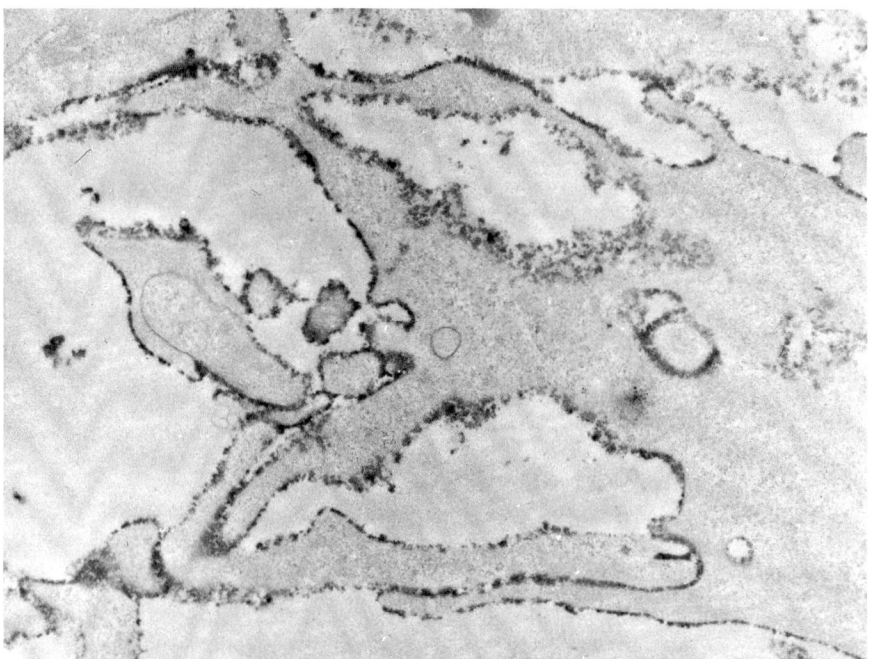

Fig. 2. Visualization of TNP groups bound to chicken embryo myoblasts by immunoelectronmicroscopy using anti-TNP antibodies conjugated with horseradish peroxidase. Spread cells in monolayers were fixed and sectioned parallel to the plane of cell growth. Peroxidase activity was developed by 15 min incubation with 3,3'-diaminobenzidine (0.5 mg/ml) in 0.1 M Tris-HCl, pH 7.4, in the presence of 0.01% hydrogen peroxide. (Courtesy of Dr. S. Sartore and Dr. S. Schiaffino).

immunoadsorption. This labeling protocol was followed in all experiments reported here. Under these conditions binding for TNP groups was restricted to the outer cell surface. This has been shown by a variety of techniques reported in previous papers (1–7). Moreover, direct visualization of cell bound TNP groups has been also recently achieved by immunoelectronmicroscopy, using goat anti-TNP antibodies coupled with horseradish peroxidase. The pattern obtained with chicken embryo fibroblasts and myoblasts (Fig. 2) confirmed the results obtained previously with hamster fibroblast and fluorescein-isothiocyanate labeled anti-TNP antibodies. All detectable TNP groups were found on the outer cell surface.

Purification of TNP-Labeled Surface Proteins

TNP-labeled cells were solubilized (in ice) with 1% of sodium dodecyl sulfate (SDS) in 0.25 M Tris-HCl buffer, pH 8.0. Tosylphenylalanylchloromethylketone, tosyllysyl-chloromethyl ketone, tosyl-L-arginine methyl ester and diisopropyl fluorophosphate were added, at final concentrations of 30, 200, 100, and 1,000 μM respectively, to prevent proteolytic degradation. After reduction with 1% mercaptoethanol and 50 mM dithiothreitol (DTT) for 3 h at 0°C in a nitrogen atmosphere, and alkylation with twofold molar excess of iodoacetic acid, samples were centrifuged at 150,000 $\times$ g for 1h. The supernatant was dialyzed against 50 mM Tris-HCl, pH 8.0, 250 mM NaCl, and 0.05% sodium deoxycholate and proteins were labeled with radioactive iodine by the chloramine-T method (16).

TABLE I. Binding of Solubilized Membrane Proteins (SMP) to Rabbit Antibody Immunoadsorbents

Sample	Adsorbed on:	% binding (total)	% binding (specific)[a]
TNP-labeled SMP	IgG[b]-Sepharose	0.7 ± 0.1	0
Unlabeled SMP	anti-DNP[c]-Sepharose	0.9 ± 0.1	0
TNP-labeled SMP	anti-DNP-Sepharose	5.5 ± 0.5	4.6
TNP-labeled SMP preadsorbed[d]	anti-DNP-Sepharose	3.6 ± 0.3	2.7

[a]cpm bound by the antibody-Sepharose column after subtraction of the cpm bound by the IgG-Sepharose column.
[b]Nonimmune rabbit immunoglobulins bound to Sepharose.
[c]Purified rabbit anti-DNP antibodies bound to Sepharose.
[d]Chromatographed through an immunoadsorbent made of Sepharose coupled to IgG purified from anti-calf serum proteins antisera.

Under these conditions 80% of the membrane proteins were solubilized from all cell types, as measured in experiments performed with plasma membranes isolated by sucrose gradients described elsewhere (17).

TNP-labeled surface proteins were purified by affinity chromatography on a column of rabbit anti-dinitrophenyl (DNP) antibodies covalently linked to Sepharose 4B, or by double immunoprecipitation using rabbit anti-DNP and goat antirabbit immunoglobulin antiserum. Anti-DNP antibodies cross-react with TNP and were used instead of anti-TNP antibodies to allow specific elution with DNP-glycine as discussed in previous papers (3, 4).

For the column procedure, 1 ml of packed immunoadsorbent was washed with 1 M acetic acid and equilibrated with 50 mM Tris-HCl, pH 8.0, 250 mM NaCl, and 0.05% sodium deoxycholate. It was found that membrane proteins solubilized by detergents have a certain affinity for Sepharose bound immunoglobulins (18). For this reason the immunoadsorbent was presaturated with excess nonradiolabeled proteins solubilized from the membrane of unrelated cells. Solubilized membrane proteins were slowly passed through the column; after extensive washing at 4°C to prevent proteolytic degradation, TNP-labeled molecules were eluted either by 100 mM DNP-glycine or by boiling Sepharose beads in 5% SDS plus 10% mercaptoethanol.

When double immunoprecipitation was used, 10 μl of rabbit anti-DNP antiserum were added to radiolabeled membrane samples in the presence of excess unlabeled proteins solubilized from mouse liver membranes and incubated for 1 h in ice. Goat antirabbit immunoglobulin G antiserum was then added at the equivalence and incubated for 15 min; the precipitate was washed twice with saline and eluted with DNP-glycine, or dissolved by boiling in SDS-mercaptoethanol as above.

About 5% of the proteins solubilized from cells were TNP labeled and specifically adsorbed on the anti-DNP immunoadsorbent. Controls were performed both with membranes not labeled with TNBS and with nonimmune rabbit immunoglobulins bound to Sepharose (Table I).

Removal of Exogenous Serum Proteins Adsorbed Onto the Cell Surface

It was found that about 40% of the TNP-labeled surface proteins reacted with antibodies prepared against the calf serum of the tissue culture medium (Table I). This showed that serum components were tightly adsorbed to the cell surface of fibroblasts grown in vitro. Immunochemical purification and analysis by gel electrophoresis showed that these were predominantly low-molecular-weight polypeptides, undetectable when native calf serum was coelectrophoresed for reference. As discussed elsewhere (7, 18) these polypeptides probably originate from the proteolytic degradation of higher-molecular-weight serum proteins. Since these polypeptides were still bound by specific antisera they were removed from solubilized membrane samples by chromatography on rabbit anticalf serum antibodies covalently linked to Sepharose, as described in detail previously (18). After removal of exogenous proteins adsorbed onto the cell surface, 2–3% of the membrane proteins solubilized from the TNP-labeled cells were specifically bound by the anti-DNP immunoadsorbent (Table I). The amount of proteins bound was virtually identical in all the cell lines studied, either control or virus-transformed. Similar results were obtained after purification of TNP-labeled surface proteins with double immunoprecipitation.

Exposed proteins are thus no more than 3% of the total cell proteins. Similar values have been obtained in other laboratories by cell surface iodination with lactoperoxidase under conditions where binding of exogenous serum proteins was cut down to the minimum (19).

Analysis of TNP-Labeled Membrane Proteins Purified From Control and Virus-Transformed Fibroblasts

TNP-labeled surface proteins eluted from the anti-DNP immunoadsorbent or from specific immunoprecipitates were analyzed in SDS-electrophoresis after removal of contaminating serum proteins. Determination of the radioactivity in the gel slices revealed a pattern which, though complex, was sufficiently reproducible to enable the detection of several major component classes of molecular weight 20 to over 150 k daltons. Figure 3 shows the TNP-labeled cell surface proteins purified from BHK and chicken embryo fibroblasts.

Peak radioactivity distribution varied from one cell preparation to another, whereas peak mobility was constant and reproducible. Further resolution in the electrophoretic separation of TNP-labeled surface proteins was achieved with slab gels. These showed that most peaks obtained in tube gels were composed of more than one electrophoretic band (Fig. 4).

The electrophoretic pattern of TNP-labeled proteins showed a small amount of radioactivity in components weighing more than 150 k daltons in either cells (Fig. 3). Several laboratories have used lactoperoxidase catalyzed surface radioiodination to identify a membrane glycoprotein, known as cell surface protein (CSP) or as large external transformation-sensitive (LETS) protein, of molecular weight 250 k daltons (20–23). CSP is present in large quantities in nongrowing cells and its expression is reduced in transformed cells; it is probably involved in intercellular adhesion (24, 25). The possibility that TNBS for some reason fails to label this protein has been ruled out by experiments on chicken embryo fibroblasts and myoblasts (Fig. 3 and Ref. 6). Moreover it is known that CSP is labeled by amino group-specific reagents (26). We concluded that the expression of CSP is low in our BHK-C13 subline.

Comparison between surface proteins purified from nontransformed C13 fibroblasts and those transformed by RSV or polyoma was carried out by running ^{131}I- and ^{125}I-labeled proteins on the same acrylamide gel. Perfect coincidence was noted in the case of 6 out of the 7 peaks, with slight differences in the radioactivity distribution. However,

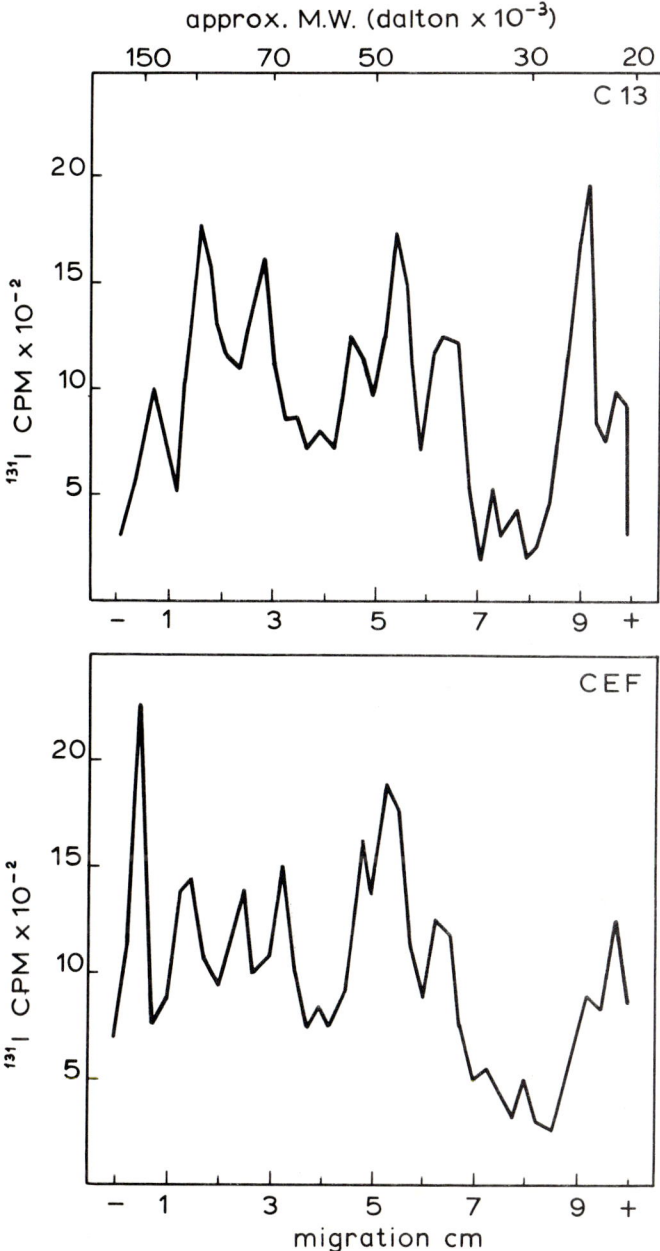

Fig. 3. Electrophoretic profile of TNP-labeled surface proteins purified by affinity chromatigraphy from baby hamster kidney (BHK) fibroblasts (C13) and from chicken embryo fibroblasts (CEF) after removal of membrane-bound serum components of the cutlure medium. Electrophoretic separations shown here and in Fig. 5 were carried out on SDS-9% polyacrylamide tubes under the conditions described in the Materials and Methods section.

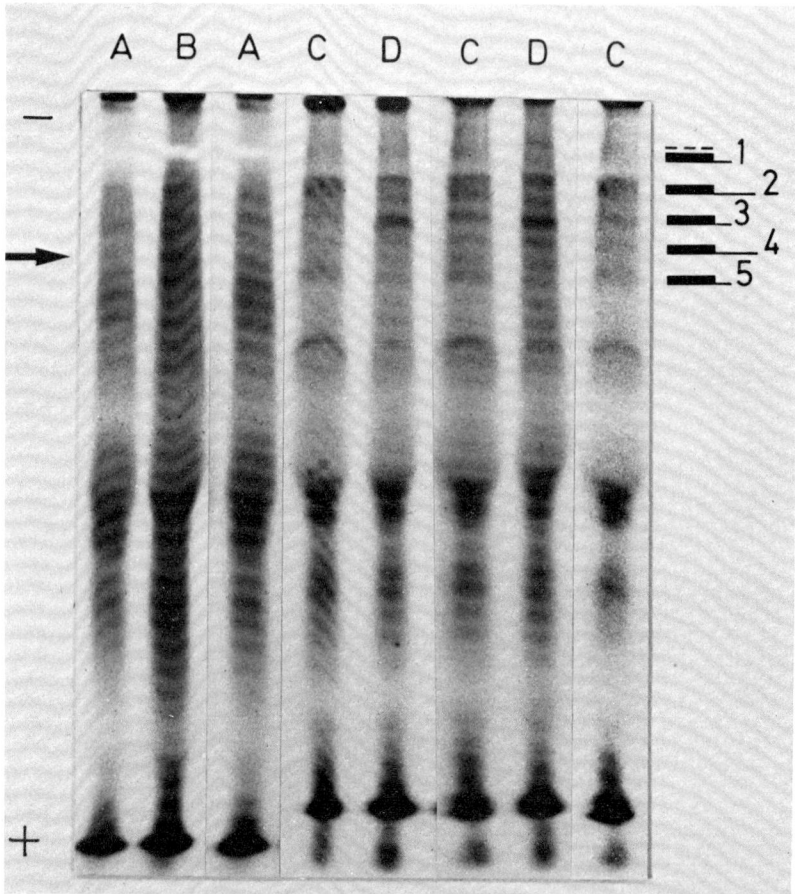

Fig. 4. Autoradiography of purified TNP-labeled surface proteins separated in SDS-9% polyacrylamide slab gel electrophoresis. A) C13 control hamster fibroblasts, B) PY-transformed fibroblasts, C) B4-transformed fibroblasts, D) B4-transformed fibroblasts after Vibrio cholerae neuraminidase digestion of the cell surface (50 IU/5 × 10^6 cells for 1 h at 37°C). Numbers indicates the major high-molecular-weight protein bands; the arrow shows the position of band 4 which is undetectable in control fibroblasts.

one of the peaks (peak C, Fig. 5) was displaced (peak C′) in the patterns of both transformed cell lines. As can be seen in Fig. 5, peak C′ of B4 cells migrates 4 mm behind peak C of C13 cells. This is a difference of 4% in relative mobility. The same result was obtained when surface proteins purified from PY cells were coelectrophoresed with those purified from C13 cells. On the other hand, analysis of transformed B4 and PY cells by co-electrophoresis showed perfect coincidence of the relative mobility of all 7 peaks. Slab gel experiments showed that in both cases the displacement was caused by the presence of an extra band, in the transformed cell pattern of approximately 80–100 k daltons (Fig. 4). Comparison of the relative mobility of tube peaks and slab bands showed that both in control and transformed cells peak B is composed of 2 bands (number 2 and number 3). This is also true for peak D and probably peak F. On the other hand peak C is composed

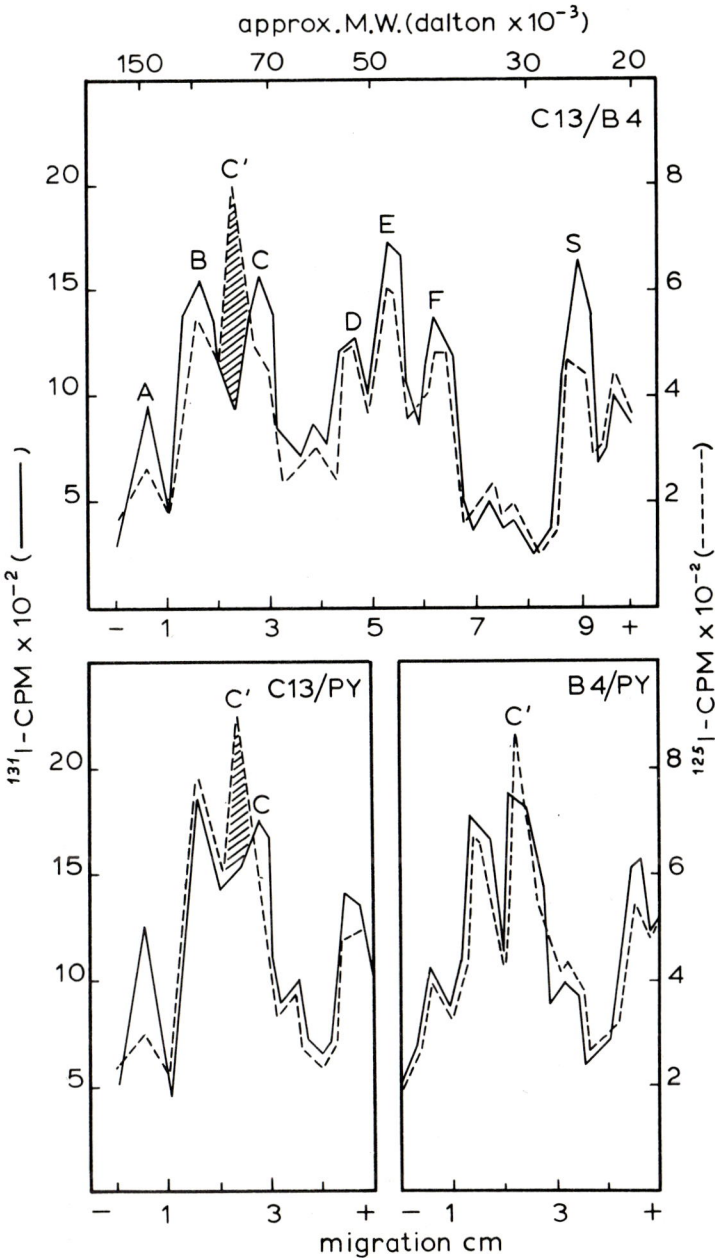

Fig. 5. Electrophoretic pattern of TNP-labeled surface proteins purified from control and virus-transformed baby hamster kidney fibroblasts. Upper plot: ^{131}I-labeled sample purified from control cells (C13) coelectrophoresed in the same gel with ^{125}I-labeled sample purified from RSV-transformed cells (B4). Lower plot on the left: ^{131}I-labeled sample purified from control cells (C13) coelectrophoresed with ^{125}I-labeled sample purified from polyoma-transformed cells (PY). Lower plot on the right: ^{125}I-labeled samples purified from RSV-transformed cells (B4) coelectrophoresed with ^{131}I-labeled samples purified from polyoma-transformed cells (PY). In lower plots only peaks found in the first 5 cm of the gels are shown.

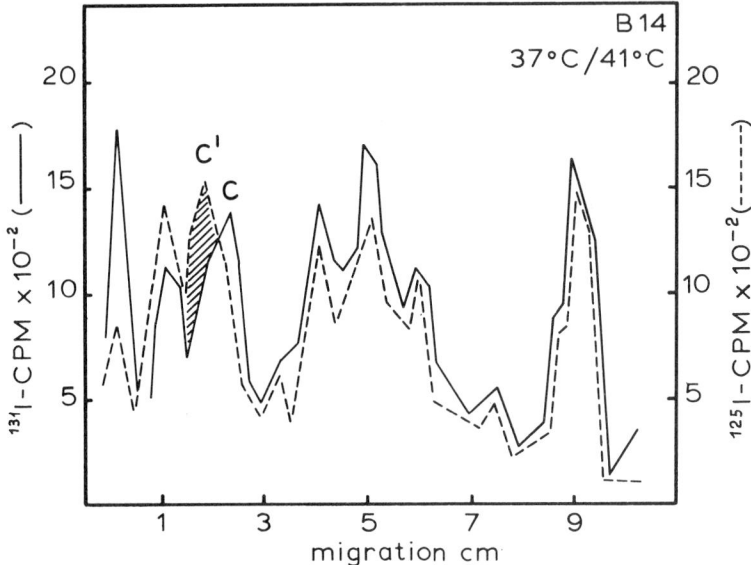

Fig. 6. Electrophoretic pattern of TNP-labeled surface proteins purified from hamster fibroblasts transformed by a temperature-sensitive mutant of RSV. Samples purified from cells grown at the permissive temperature (37°C) were labeled by ^{125}I and coelectrophoresed with samples purified from cells grown at the restrictive temperature (41°C) which were labeled by ^{131}I.

of 2 bands (number 4 and number 5) both in B4 and PY cells, but only one (number 5) in C13 control cells. It thus seemed likely that peak C of control cells exactly coincided with band 5, whereas the extra band 4 in transformed cells — not resolved from the band 5 in tubes — caused displacement of the peak to C'.

To investigate whether the displacement of peak C' in B4 and PY cells was related to transformation or to cell growth, surface proteins purified from C13 and B4 cells in logarithmic phase of growth or in resting condition (confluent monolayers) were analyzed. Peak C was not displaced during the transition of C13 cells from the growing to the resting phase. Peaks C' of growing and resting B4 cells were similarly coincident (7).

Further support for the view that displacement of peak C' was related to malignant transformation came from experiments with the 14B line of BHK fibroblasts transformed by the FU19 temperature-sensitive mutant of RSV. TNP-labeled surface proteins purified from cells grown at 37°C or 41°C were analyzed in SDS-acrylamide gels. At the permissive temperature of 37°C the electrophoretic pattern showed the same displacement of peak C to C' observed in B4- and PY-transformed cells, whereas at 41°C the pattern was similar to that obtained with control C13 cells (Fig. 6 and Ref. 7). Thus in this cell line the extraprotein responsible for the displacement of peak C is exposed at the cell surface at permissive but not at restrictive temperatures, showing correlation with the expression of the transforming gene of RSV.

ACKNOWLEDGMENTS

This work was supported by the Italian National Research Council (C.N.R.). The skillful technical assistance of Ms. Marike Mura-Lagna and Ms. Maria Rosa Amedeo is gratefully acknowledged.

REFERENCES

1. Carraway K: Biochim Biophys Acta 415:379, 1975.
2. Comoglio PM, Tarone G, Prat M, Bertini M: Exp Cell Res 93:402, 1975.
3. Tarone G, Prat M, Comoglio PM: Biochim Biophys Acta 311:214, 1973.
4. Vidal R, Tarone G, Peroni F, Comoglio PM: Febs Lett 47:107, 1974.
5. Comoglio PM, Tarone G, Bertini M: In Bolis L, Hoffman J, Leaf A (eds): "Membranes and Diseases." New York: Raven Press, 1976, p 115.
6. Sartore S, Tarone G, Cantini M, Schiaffino S, Comoglio PM: Dev Biol (Manuscript submitted).
7. Tarone G, Comoglio PM: Exp Cell Res (In press).
8. Chanock B, Hayflyck L, Basile MC: Proc Natl Acad Sci USA 48:41, 1962.
9. Stocker M, Macpherson F: Nature (London) 203:1355, 1964.
10. Pitts J: In Wolstenholme G, Knight J (eds): "Growth Control in Cell Cultures." Edinburgh: Churchill, 1971, p 261.
11. Biquart JM, Vigier P: Virology 47:444, 1972,
12. Aupoix M, Biquart JM, Cochard A: Int J Cancer 14:661, 1974.
13. Axen R, Ernback S: Eur J Biochem 18:351, 1971.
14. Maizel JV: In Maramorosh K, Koprowsky H (eds): "Methods in Virology." New York: Academic Press, 1971, vol 5, p 315.
15. Studier FW: J Mol Biol 79:237, 1973.
16. Talmage D, Claman H: In Williams C, Chase M (eds): "Methods in Immunology." New York: Academic Press 1967, vol 1, p 351.
17. Prat M, Comoglio PM: Immunochemistry 12:9, 1975.
18. Tarone G, Comoglio PM: FEBS Lett 67:364, 1976.
19. Hubbard A, Cohn AA: J Cell Biol 64:438, 1975.
20. Hogg NM: Proc Natl Acad Sci USA 71:489, 1974.
21. Hynes R, Machpherson I: In Schultz J, Block R (eds): "Membrane Transformation in Neoplasia." New York: Academic Press, 1974, p 51.
22. Yamada KM, Weston JA: Proc Natl Acad Sci USA 71:3492, 1974.
23. Ruoslathi E, Vaheri A: J Exp Med 141:497, 1975.
24. Yamada KM, Yamada SS, Pastan I: Proc Natl Acad Sci USA 72:3158, 1975.
25. Yamada KM, Yamada SS, Pastan I: Proc Natl Acad Sci USA 73:1217, 1976.
26. Critcherly DR, Wyke JA, Hynes RO: Biochim Biophys Acta 436:335, 1976

Enzymatic Conversion of Proteins to Glycoproteins By Lipid-Linked Saccharides: A Study of Potential Exogenous Acceptor Proteins

Kathryn E. Kronquist
Veterans Administration, Wadsworth Hospital Center, Wilshire and Sawtelle Boulevards, Los Angeles, California 90073

William J. Lennarz
Department of Physiological Chemistry, The Johns Hopkins University School of Medicine, 725 North Wolfe Street, Baltimore, Maryland 21205

Previous studies have shown that a membrane preparation from hen oviduct catalyzes transfer of oligosaccharide from oligosaccharide-P-P-dolichol to denatured RNase and α-lactalbumin. To gain further insight into the structural requirements of a protein that allow it to serve as a substrate for glycosylation, the acceptor ability of a variety of other modified proteins containing the tripeptide sequence -ASN-X-(SER/THR)- has been investigated. Of 7 proteins tested, 2 (ovine prolactin and rabbit muscle triosephosphate isomerase) could be enzymatically glycosylated by a particulate preparation from hen oviduct. The remaining 5 proteins, assayed as either S-carboxymethylated or S-aminoethylated derivatives, were inactive as carbohydrate acceptors. However, cyanogen bromide treatment of 2 of the inactive proteins, bovine catalase and concanavalin A from jack bean, yielded peptide fragments which served as substrates for glycosylation. These results suggests that for some proteins, disruption of the tertiary structure is sufficient to allow attachment of carbohydrate. Other denatured proteins may possess additional restrictions imposed by their secondary structure. In certain cases, these restrictions are removed when the polypeptide chain is fragmented.

Key words: glycosylation, lipid-linked saccharides, glycoproteins, oligosaccharides

It is now well established that saccharides linked to dolichol pyrophosphate participate as intermediates in the synthesis of the carbohydrate chains of certain N-glycosidically linked membrane (1–3) and secretory proteins (4–7). Relatively little is known about the structural requirements of the protein acceptor, although it has been clear for a number of years that in glycoproteins containing carbohydrate N-glycosidically linked to an ASN residue the sequence following the ASN invariably is -X-SER- or -X-THR- (8).

Recently an approach to gaining further insight into the structural requirement of protein acceptors became available when it was found that 2 proteins of known sequence (RNase A and α-lactalbumin) containing the tripeptide sequences -ASN-X-(SER/THR)-

Received September 13, 1977; accepted September 15, 1977

0091-7419/78/0801-0051$02.90 © 1978 Alan R. Liss, Inc

were shown to serve as acceptors of the oligosaccharide chain of oligosaccharide-P-P-dolichol (9). The reaction is believed to proceed as shown below:

$$(Man)_n(GlcNAc)_2\text{-P-P-Dolichol} + \begin{array}{c}\vdots\\ \text{SER}\\ |\\ \text{X}\\ |\\ \text{ASN}\\ \vdots\end{array} \rightarrow (Man)_n(GlcNAc)_2 - \begin{array}{c}\vdots\\ \text{SER}\\ |\\ \text{X}\\ |\\ \text{ASN}\\ \vdots\end{array} + \text{(Dolichol-P-P)}$$

Of particular interest was the finding that both RNase and α-lactalbumin served as acceptors only after covalent modifications of all cysteine residues under denaturing conditions (9). However, this treatment did not convert all the proteins tested that contained the sequence -ASN-X-(SER/THR)- into active acceptors.

In the present study we have begun to characterize the requirements of the hen oviduct enzyme(s) involved in the transfer of carbohydrate from GDP-Man to exogenous acceptor proteins via oligosaccharide-lipid, and have attempted to gain information about the structural features of proteins that may regulate their ability to serve as carbohydrate acceptors. A variety of proteins of known amino acid sequence which contain 1 or more of the appropriate asparagine glycosylation sites (10–12), but which do not normally exist in a glycosylated form, have been tested as substrates. Of 7 proteins tested, 2 (rabbit muscle triosephosphate isomerase and ovine prolactin) serve to accept oligosaccharide chains from oligosaccharide-lipid after the tertiary structure of the protein has been disrupted by reduction and alkylation. Of 5 proteins that do not accept carbohydrate after such denaturation, 2 (catalase and concanavalin A) have been cleaved by treatment with cyanogen bromide. Experiments with the resulting mixture of polypeptide fragments indicate that, unlike the intact polypeptide, these fragments do serve as oligosaccharide acceptors.

MATERIALS AND METHODS

Catalase (2X crystallized, bovine liver); elastase (Type III, porcine pancreas); glyceraldehyde-3-phosphate dehydrogenase (Baker's yeast); α-lactalbumin (Grade II, bovine milk); prolactin (ovine pituitary); triosephosphate isomerase (Type III, rabbit muscle); trypsinogen (Type I, bovine pancreas); and iodoacetic acid (sodium salt) were all products of the Sigma Chemical Corporation, St. Louis, Missouri. Concanavalin A (2X crystallized, jack bean) was purchased from Miles Laboratories, Inc, Kankakee, Illinois. Ethylene imine was obtained from the Pierce Chemical Company. Ovine prolactin (lot number NIH-P-512), prepared and characterized by the National Institute of Arthritis, Metabolism and Digestive Diseases, was kindly provided by Mr. T. Anderson and Dr. K. E. Ebner, University of Kansas Medical Center, Kansas City, Kansas. Tunicamycin was the gift of Dr. G. Tamura, University of Tokyo, Bunkyo-ku, Tokyo, Japan.

GDP[^{14}C]Mannose (210 mCi/mmole) was obtained from the New England Nuclear Corporation, Boston, Massachusetts and UDP-GlcNAc was purchased from Boehringer-Mannheim Corp. White leghorn laying hens (26–38 weeks old) were purchased from Truslow Farms, Inc., Chestertown, Maryland.

Before derivitization, proteins were dialyzed against distilled water for 12 h to remove salts and stabilizing agents and lyophilized. The residue was dissolved in 6 M guanidine-HCl or in 8 M urea, reduced, and subjected to either S-carboxymethylation (13), or to S-aminoethylation (14), followed by dialysis against distilled water for 3–4 days. The derivatized proteins dissolved readily in water at concentrations of 7–10 mg/ml when the pH was adjusted to 7.2–7.5 by the addition of NaOH. Cyanogen bromide cleavage was performed on lyophilized samples of the carboxymethylated proteins after they were dissolved in 70% formic acid (vol/vol) (14). After incubation for 24 h at room temperature excess reagents were removed by passing the sample over a Biogel P-4 collumn (1 × 19 cm) equilibrated with 50% formic acid (vol/vol). The peptides eluting from the column in the void volume fractions were pooled. The pooled material was diluted to a final concentration of 1% formic acid and lyophilized to dryness.

The oviduct enzyme preparation used in these experiments was a crude membrane fraction isolated by modification of a previously published procedure (5). It was stored at $-20°C$ prior to use. Protein was measured by the method of Lowry et al. (15).

The enzymatic conversion of exogenous native or derivitized proteins to $[^{14}C]$-mannose-labeled glycoproteins was assayed as follows: Standard reaction mixtures contained 600–800 μg of oviduct membrane protein, 100–400 μg of exogenous protein, 20 mM Tris-HCl, pH 7.5, 64 mM sucrose, 64 mM NaCl, 4 mM $MgCl_2$, 250 μM UDP-N-acetylglucosamine and either 8 μM GDP$[^{14}C]$mannose (3×10^6 cpm) or [Man-^{14}C]-oligosaccharide-lipid (2×10^4 cpm) prepared as previously described (16) in a total volume of 120 μl. After incubation at 37°C for 1 h the reaction mixtures were centrifuged at 27,000 × g for 10 min and aliquots of the supernatants and pellets were analyzed by disk gel electrophoresis in the presence of sodium dodecyl sulfate (SDS), urea, and β-mercaptoethanol on 7.5% or 15% polyacryalmide gels (16). After electrophoresis, the gels were stained for protein with Coomassie blue and were then cut into 1.2-mm slices. Gel slices were extracted in capped scintillation vials for 12 h at 65°C with 0.7 ml of 0.1 N NaOH and 1% SDS. Samples were neutralized with HCl and radioactivity was measured after addition of 10 ml of Hydromix (Yorktown Research) using a Packard Tri-Carb liquid scintillation counter. Under these conditions 85–90% of the trichloroacetic acid (TCA) precipitable radioactivity applied to gels was recovered.

A more rapid procedure was also used to assess the conversion of exogenously added protein to $[^{14}C]$mannose-labeled glycoprotein. Analysis by SDS-gel electrophoresis indicated that approximately 90% of the glycosylated exogenous protein was present in the supernatant after the reaction mixture was centrifuged at 27,000 × g, while most of the radioactive endogenous membrane proteins remained in the pellet. Supernatants from centrifuged reaction mixtures were added to test tubes containing 3 ml of 10% TCA and 1 mg of bovine serum albumin as carrier protein. The tubes were boiled for 10 min to hydrolyze and render acid-soluble GDP$[^{14}C]$Man and any [Man-^{14}C]oligosaccharide-lipid present in the assay supernatants. The tubes were chilled on ice, centrifuged, and TCA-insoluble pellets were washed 2 additional times with 5-ml aliquots of cold 10% TCA. The pellets were dissolved in 10 ml of Hydromix and radioactivity was measured. An increase in TCA-precipitable radioactivity in supernatants from assays containing exogenous protein, compared to supernatants from control assays containing only oviduct membrane protein, indicated that exogenously added protein was being enzymatically glycosylated, though it was always necessary to confirm the identity of the radioactive protein product on polyacrylamide gels.

RESULTS

Survey of Ability of Derivitized Proteins to Accept Carbohydrate

A simple TCA precipitation assay (described in Materials and Methods) was initially used to determine the ability of a series of derivitized proteins to serve as oligosaccharide acceptors based on the following observation. When an assay containing denatured α-lactalbumin, a known carbohydrate acceptor (9), was centrifuged at 27,000 × g for 10 min and the supernatant and pellet were analyzed separately by electrophoresis on SDS-polyacrylamide gels, approximately 90% of the total [^{14}C] mannose-labeled α-lactalbumin was recovered in the supernatant, while the majority of the labeled endogenous membrane proteins remained in the pellet. The degree to which the addition of other exogenous proteins resulted in enhanced incorporation of [^{14}C] mannose into the TCA-precipitable fraction in assay supernatants (relative to the background of endogenous incorporation) was used as a preliminary indication of their ability to serve as oligosaccharide acceptors.

A list of the proteins tested, their monomeric molecular weights, and the number of -ASN-X-(SER/THR)- tripeptides each contains, is presented in Table I. The subscript number by each amino acid refers to its position in the published sequence (10–12). Derivitized proteins (200 μg) were incubated in the standard incubation mixture containing GDP[^{14}C]mannose. Assays were centrifuged at 27,000 × g for 10 min and TCA-precipitable radioactivity in supernatants was measured and compared to control assays containing no exogenous protein. A parallel assay containing a known acceptor protein, S-carboxymethylated α-lactalbumin, was done to monitor the activity of the enzyme preparation. The results of such a survey (Table II) showed that, like α-lactalbumin, the addition of 2 other S-carboxymethylated proteins, triosephosphate isomerase and prolactin, to reaction mixtures resulted in a considerable increase in incorporation of radioactivity into TCA-precipitable protein in the supernatant fraction. On the basis of this

TABLE I. Potential Carbohydrate Acceptor Proteins

Protein	Molecular weight (monomers)	Potential site of carbohydrate attachment
Catalase (10)	61,000	-ASX$_{212}$-THR$_{213}$-SER$_{214}$-[a]
		-ASN$_{242}$-LEU$_{243}$-SER$_{244}$-[a]
		-ASN$_{437}$-VAL$_{438}$-THR$_{439}$-[a]
		-ASN$_{479}$-PHE$_{480}$-SER$_{481}$-[a]
Concanavalin A (12)	25,572	-ASN$_{118}$-SER$_{119}$-THR$_{120}$-
		-ASN$_{162}$-GLY$_{163}$-SER$_{164}$-
Elastase (10)	25,906	-ASN$_{66}$-GLY$_{67}$-THR$_{68}$-
		-ASN$_{123}$-ASN$_{124}$-SER$_{125}$
		-ASN$_{215}$-VAL$_{216}$-THR$_{217}$-
Glyceraldehyde-3-phosphate-dehydrogenase (11)	35,549	-ASN$_{146}$-ALA$_{147}$-SER$_{148}$-
		-ASX$_{236}$-VAL$_{237}$-SER$_{238}$-
α-Lactalbumin (10	14,183	-ASN$_{45}$-GLN$_{46}$-SER$_{47}$-
		-ASN$_{74}$-ILE$_{75}$-SER$_{76}$-
Prolactin (10)	22,554	-ASN$_{31}$-LEU$_{32}$-SER$_{33}$-
Triosephosphate isomerase (12)	26,626	-ASN$_{195}$-VAL$_{196}$-SER$_{197}$-
Trypsinogen (10)	23,990	-ASN$_{151}$-SER$_{152}$-SER$_{153}$-

[a]In the case of catalase, for which only fragments have been sequenced, numbers refer to the position of amino acids in the fragment (10), not to the absolute position of these residues in the polypeptide chain.

TABLE II. Incorporation of Mannose From GDP[^{14}C]Mannose Into Exogenous Derivitized Proteins*

	Radioactivity incorporated (cpm)	Incorporation relative to control (no exogenous protein)
Carboxymethylated proteins		
None	9,170	(1.0)
α-Lactalbumin	35,450	3.9
Triosephosphate isomerase	36,500	4.0
Prolactin	17,530	1.9
Catalase	13,500	1.5
Elastase	9,030	1.0
Glyceraldehyde-3-phosphate dehydrogenase	8,230	0.9
Concanavalin A	4,400	0.5
Trypsinogen	5,820	0.6
Aminoethylated proteins		
None	5,860	(1.0)
α-Lactalbumin	26,480	4.5
Prolactin	11,322	1.9
Catalase	2,760	0.5
Elastase	3,420	0.6
Glyceraldehyde-3-phosphate dehydrogenase	6,950	1.2
Concanavalin A	3,110	0.5
Trypsinogen	3,590	0.6

*S-carboxymethylated and S-aminoethylated derivatives of proteins were prepared as described in Materials and Methods. Each derivitized protein (200 μg) was tested for carbohydrate acceptor ability by incubation under standard incubation conditions. After incubation for 60 min glycosylated protein in the supernatant was measured as described in Materials and Methods.

experiment S-carboxymethylated catalase was considered a questionable carbohydrate acceptor since it afforded only a slight increase in radioactivity. In the presence of S-carboxymethylated elastase or glyceraldehyde-3-phosphate dehydrogenase radioactivity was incorporated in supernatants at a level no greater than that of endogenous incorporation, whereas the presence of S-carboxymethylated concanavalin A or trypsinogen suppressed incorporation of [^{14}C] mannose to approximately half that of the endogenous level. These results indicate that the latter 4 proteins were inactive as carbohydrate acceptors under the conditions described.

The S-aminoethylated derivatives of the same proteins were assayed to determine if the introduction of a positively charged aminoethyl group in place of a negatively charged carboxymethyl group would have an effect on the carbohydrate acceptor capacity. The results, also presented in Table II, indicate that the addition of α-lactalbumin or prolactin again resulted in an increase of radioactivity into the supernatant fraction. S-Aminoethylated triosephosphate isomerase was not tested. The aminoethyl derivatives of catalase as well as the aminoethyl derivatives of the other 4 proteins found to be inactive as carboxymethyl derivatives were all inactive.

To confirm the results shown in Table II, the radioactive products formed with the derivitized proteins were analyzed by electrophoresis on SDS-polyacrylamide gels. The supernatant fractions from assays containing S-carboxymethylated α-lactalbumin, prolactin, or triosephosphate isomerase all contained a new radioactive polypeptide which had an electrophoretic mobility slightly slower than that of the respective unlabeled derivatized protein (Fig. 1A, C, and E). As previously observed (9), this slower mobility is consistent with the increased molecular weight expected after labeled oligosaccharide is incorporated into protein. When equivalent amounts of native α-lactalbumin, prolactin, or triosephosphate isomerase were assayed, the new radioactive peaks were absent or greatly reduced (Fig. 1B, D, and F), indicating that these proteins are essentially incapable of serving as carbohydrate acceptors unless their tertiary structure has been disrupted.

SDS-polyacrylamide gel electrophoresis indicated that no new radioactive polypeptides were present in either the supernatants or the pellets from assays containing the S-carboxymethylated forms of catalase, concanavalin A, elastase, glyceraldehyde-3-phosphate dehydrogenase, trypsinogen, or S-aminoethylated glyceraldehyde-3-phosphate dehydrogenase (data not shown). It was concluded that, even though these 5 proteins

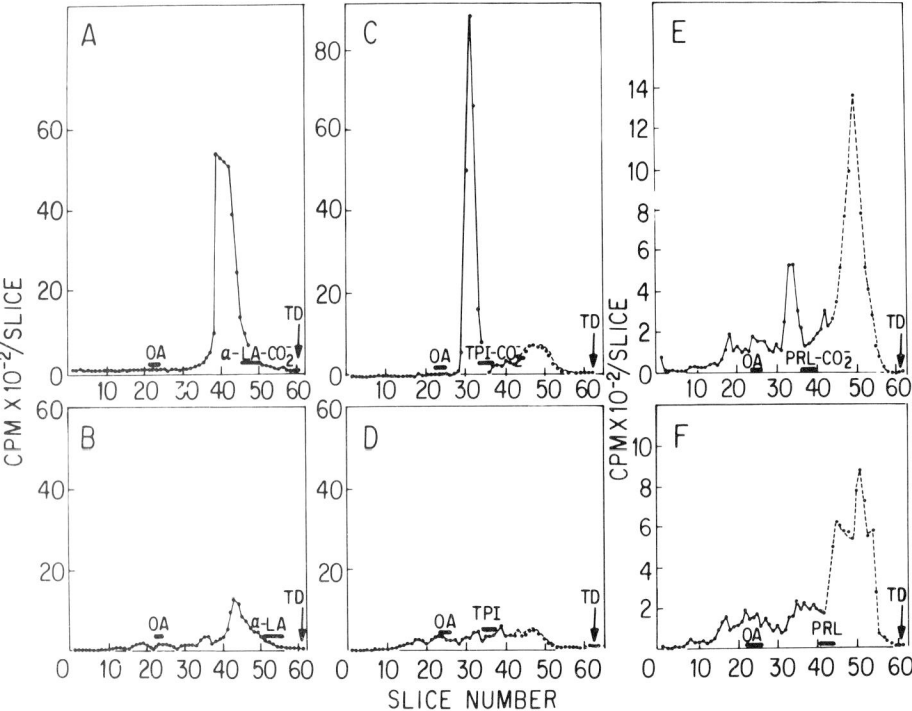

Fig. 1. Electrophoretic analysis of [^{14}C]mannose-labeled polypeptides synthesized from exogenously added native or S-carboxymethylated proteins. Native and derivitized proteins were tested for their ability to accept carbohydrate using standard assay conditions. Proteins added were: A) S-carboxymethylated α-lactalbumin (α-LA-CO_2^-) or B) native α-lactalbumin (α-LA), 200 μg (14 nmoles); C) S-carbosymethylated triosephosphate isomerase (TPI-CO_2^-) or D) native triosephosphate isomerase (TPI), 200 μg (7.5 nmoles); E) S-carboxymethylated prolactin (PRL-CO_2^-) or F) native prolactin (PRL), 200 and 400 μg respectively (9 nmoles and 17 nmoles). After incubation at 37°C for 60 min the reactions were centrifuged at 27,000 × g for 10 min. Aliquots of the supernatant (60 μl) were analyzed by electrophoresis on 7.5% polyacrylamide disk gels in the presence of 1% SDS, 0.5 M urea, and 2% β-mercaptoethanol. The positions of ovalbumin (OA, present endogenously in the oviduct enzyme preparation), native exogenous proteins (α-LA, TPI, PRL) and S-carboxymethylated exogenous proteins (α-LA-CO_2^-, TPI – CO_2^-, PRL – CO_2^-) are shown by the heavy bars. Peaks denoted by a dashed line represent the variable amount of [^{14}C]Man-P-dolichol and [Man-^{14}C]oligosaccharide-lipid recovered in the supernatant. TD marks the front of the tracking dye.

contain at least 1 of the appropriate asparagine acceptor sites, neither the S-carboxymethylated nor the S-aminoethylated derivatives could be glycosylated under the stated assay conditions.

Because of the expense of preparing and utilizing isolated [Man-^{14}C] oligosaccharide-lipid as the radioactive substrate, the above experiments were performed with [Man-^{14}C] oligosaccharide-lipid generated endogenously by addition of GDP[^{14}C] Man and UDP-GlcNAc. Previous studies (17) have shown that the carbohydrate chain transferred to exogenous S-carboxymethylated α-lactalbumin is identical, whether the substrate is preformed, exogenous [Man-^{14}C] oligosaccharide-lipid, or endogenously generated oligosaccharide-lipid. To determine that S-carboxymethylated prolactin and triosephosphate isomerase were indeed glycosylated via saccharide-lipid intermediates, the effect of tunicamycin on glycosylation of these 2 proteins was studied. Work by others (18–21) has shown that tunicamycin specifically inhibits synthesis of the GlcNAc-P-P-dolichol lipid that serves as the acceptor for synthesis of $(Man)_x$ GlcNAcGlcNAc-P-P-dolichol (22). Moreover, using whole cells or slices, it has been shown that tunicamycin blocks glycosylation of a variety of secretory proteins (5–7). As shown in Table III the transfer of mannose from GDP-[^{14}C] mannose to endogenous membrane protein or exogenous derivitized protein was virtually abolished in the presence of the antibiotic tunicamycin. To further corroborate this finding, which indicates that oligosaccharide-lipid does participate in the glycosylation, incubations were carried out in which S-carboxymethylated prolactin or triosephosphate isomerase were incubated with [Man-^{14}C] oligosaccharide-lipid instead of GDP[^{14}C] mannose and UDP-GlcNAc. The incubation supernatants were found to contain labeled proteins having the same mobility as those synthesized using GDP[^{14}C] mannose (data not shown), though the yields were much lower due to the limited amounts of [Man-^{14}C] oligosaccharide-lipid substrate available.

Effect of Membrane Enzyme Concentration, Substrate Concentration and pH on Transfer of Mannose From GDP[^{14}C] Mannose to S-Carboxymethylated Proteins.

Since, with the exception of our early study utilizing sulfitolyzed RNase A as acceptor (9), little was known about the effect of incubation conditions on the extent of glycosylation of exogenous protein acceptors, several variables were examined. As shown

TABLE III. Effect of Tunicamycin on Incorporation of Mannose From GDP[^{14}C] Mannose Into Endogenous and Exogenous Protein*

Addition	Radioactivity incorporated (cpm)		Inhibition of [^{14}C] mannose incorporation (%)
	Minus tunicamycin	Plus tunicamycin	
None	12,730	1,050	92
Carboxymethylated α-lactalbumin	45,500	2,070	96
Carboxymethylated prolactin	22,660	1,330	94
Carboxymethylated triosephosphate isomerase	36,130	1,410	96

*Standard reaction mixtures containing 200 μg of the specified S-carboxymethylated proteins were incubated for 60 min in the absence or presence of tunicamycin at a final concentration of 2 μg/ml. The tunicamycin was dissolved in 0.01 $\underline{N}$ NaOH; samples which were not treated with tunicamycin received an equivalent volume of 0.01 $\underline{N}$ NaOH. Glycosylated protein in the supernatant was measured as described in Materials and Methods

in Fig. 2A, there was a linear dependence between the amount of membrane enzyme added and the incorporation of [^{14}C] mannose from GDP[^{14}C] mannose into exogenous derivitized α-lactalbumin. Using a fixed amount of membrane enzyme a linear dependence between the amount of exogenous protein added and incorporation of TCA-precipitable radioactivity into supernatants was seen using derivitized α-lactalbumin, or triosephosphate isomerase (Fig. 2B). The transfer of mannose from GDP[^{14}C] mannose to S-carboxymethylated α-lactalbumin was stimulated up to eightfold by the addition of unlabeled UDP-N-acetyl-glucosamine (Fig. 2C), which is consistent with our earlier findings that glycosylation was

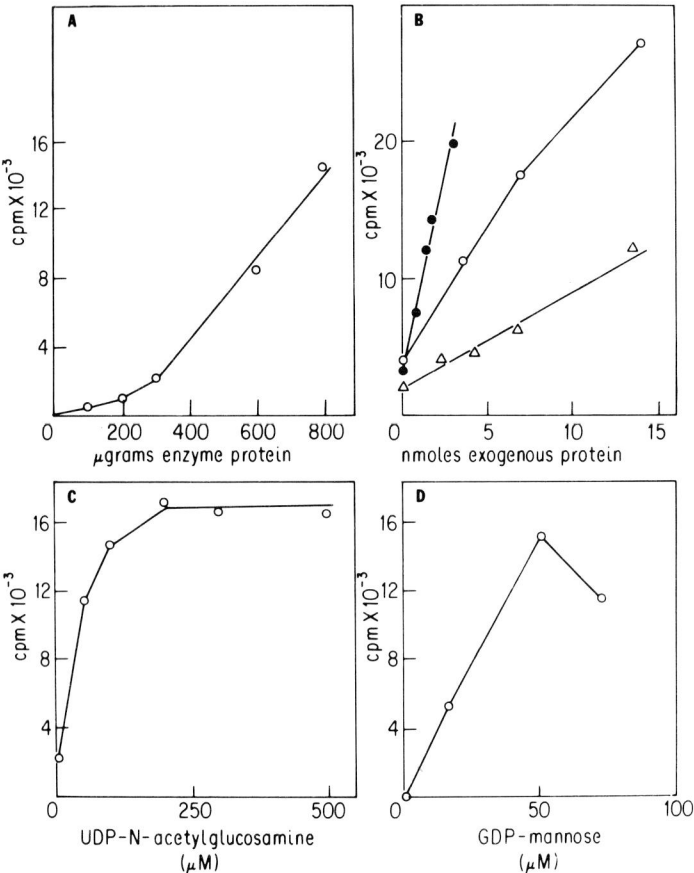

Fig. 2. Effect of enzyme and substrate concentration on glycosylation of S-carboxymethylated proteins. A) Dependence on enzyme protein. Assays contained 20 mM Tris-HCl, pH 7.5, 64 mM sucrose, 64 mM NaCl, 4 mM MgCl$_2$, 180 μg (12 nmoles) of S-carboxymethylated α-lactalbumin, 0-800 μg of oviduct enzyme protein, 250 μM UDP-N-acetylglucosamine, and 8 μM GDP[^{14}C] mannose. B) Dependence on exogenous acceptors: standard assays, described in A, contained 700 μg oviduct membrane enzyme, varying amounts of one of the following S-carboxymethylated proteins: triosephosphate isomerase (-●-●-), α-lactalbumin (-○-○-), prolactin (-△-△-), and 8 μM GDP[^{14}C] mannose. C and D) Sugar nucleotide dependence: standard assays, described in A, contained 700 μg of oviduct enzyme protein, 200 μg (14 nmoles) of S-carboxymethylated α-lactalbumin and varying concentrations of UDP-N-acetylglucosamine (0–600 μM) (C), or GDP[^{14}C] mannose (0–73 μM), specific radioactivity 21 mCi/mmole (D). After incubation for 35 min at 37°C, assays were centrifuged at 27,000 × g for 10 min and TCA-precipitable radioactivity in the supernatants was analyzed as described in Materials and Methods.

blocked by tunicamycin. Maximal sugar incorporation was seen at a UDP-N-acetylglucosamine concentration of 200 μM. As expected, increasing amounts of GDP-mannose (up to 50 μM) also had an increase on mannose incorporation into protein (Fig. 2D). Above 50 μM GDP-mannose, an inhibitory effect was seen, possibly due to end product inhibition by GDP.

As shown in Fig. 3, the optimal pH for glycosylation varies, depending on the exogenous protein used as carbohydrate acceptor. The pH optima for glycosylation of α-lactalbumin (pH 6.4) and triosephosphate isomerase (pH 7.0) are clearly different from the optima for glycosylation of endogenous membrane protein (pH 7.5).

It was clear from this study, as well as an earlier one (9), that a number of proteins containing the necessary -ASN-X-(SER/THR)- tripeptide could not be glycosylated, even after reduction and alkylation. One possible explanation was that, although the acceptor asparagine in the inactive proteins might be accessible to the transferase enzyme(s), additional essential "information" residing in some other region of the polypeptide chain

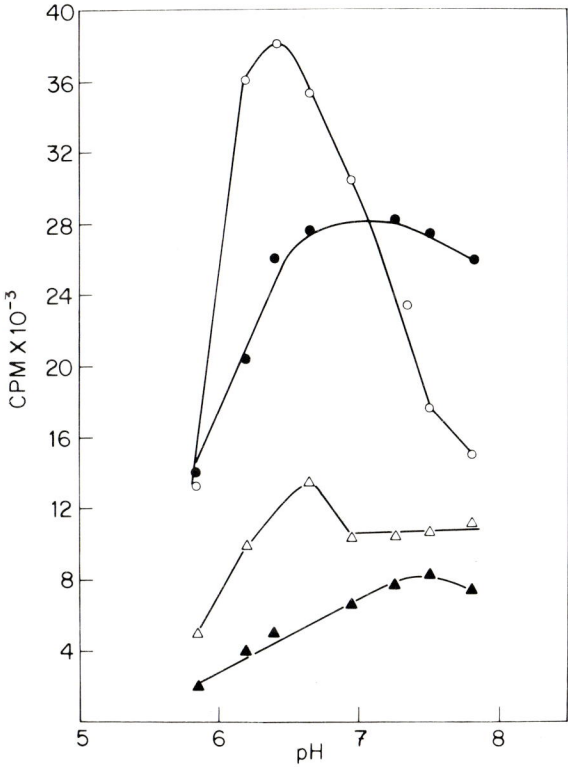

Fig. 3. Effect of pH on glycosylation of S-carboxymethylated proteins. Assays contained 50 mM Tris-maleate buffer at the indicated pH, 64 mM sucrose, 4 mM $MgCl_2$, 250 μM UDP-N-acetylglucosamine, 8 μM GDP[^{14}C] mannose, and either no additions (-▲-▲-) or the following amounts of S-carboxymethylated exogenous protein: α-lactalbumin (-○-○-), 113 μg (8 nmoles); triosephosphate isomerase (-●-●-) 117 μg (4 nmoles); or prolactin (-△-△-), 118 μg (5 nmoles). After incubation for 30 min at 37°C, reaction mixtures were centrifuged at 27,000 × g for 10 min and TCA-precipitable radioactivity in the supernatants was analyzed as described in Materials and Methods.

was lacking. An alternative explanation was that perhaps even though the tertiary structure of all these proteins had been disrupted, in the case of the inactive ones, the appropriate asparagine acceptor residue might still be inaccessible due to restrictions on the protein imposed by remaining domains of ordered structure.

We reasoned that if indeed the latter explanation was valid, cleavage of the protein might yield a smaller peptide which would serve as a substrate for glycosylation. Consequently, cyanogen bromide cleavage fragments from S-carboxymethylated catalase and concanavalin A were prepared and assayed for carbohydrate acceptor ability using the TCA precipitation assay. The results, presented in Table IV, demonstrate that cyanogen bromide cleavage fragments from catalase and concanavalin A significantly stimulated the incorporation of mannose into TCA-precipitable material. In contrast, intact, native or derivitized catalase and concanavalin A did not serve as acceptors. The incorporation of mannose into the cyanogen bromide fragments from catalase and concanavalin A was inhibited 98% by tunicamycin.

The labeled products formed with the mixture of cyanogen bromide cleavage fragments from S-carboxymethylated catalase and concanavalin A, as well as the native and derivitized proteins from which these fragments are prepared were analyzed by electrophoresis on 15% polyacrylamide gels. The results (Fig. 4) show that, relative to the profile of proteins labeled in the absence of exogenous acceptor (Fig. 4D and H), no new mannose-labeled macromolecules were present in assays containing native catalase or concanavalin A (Fig. 4C and G) or the S-carboxymethylated derivatives of these proteins (Fig. 4B and F). However, distinctive new peaks were formed in the presence of cyanogen bromide cleavage fragments from both catalase and concanavalin A (Fig. 4A and E).

TABLE IV. Incorporation of Mannose From GDP[^{14}C] Mannose Into Exogenous Proteins and Polypeptide Fragments*

Addition	Radioactivity incorporated (cpm)	Incorporation relative to control (minus exogenous protein)
Experiment I		
None	5,850	(1.0)
Catalase	5,690	1.0
S-carboxymethylated catalase	7,000	1.2
CNBr fragments from S-carboxymethyl catalase	14,720	2.5
S-carboxymethylated α-lactalbumin	20,000	3.4
Experiment II		
None	3,360	(1.0)
Concanavalin A	1,340	0.4
S-carboxymethyl concanavalin A	1,260	0.4
CNBr fragments from S-carboxymethyl concanavalin A	6,610	2.0
S-carboxymethyl α-lactalbumin	12,870	3.8

*Standard reaction mixtures containing 200 μg of native protein, 200 μg of S-carboxymethylated proteins, or 200 μg of unfractionated cyanogen bromide cleavage fragments prepared from S-carboxymethylated proteins were incubated for 60 min. Glycosylated polypeptides in the supernatant were assayed as described in Materials and Methods.

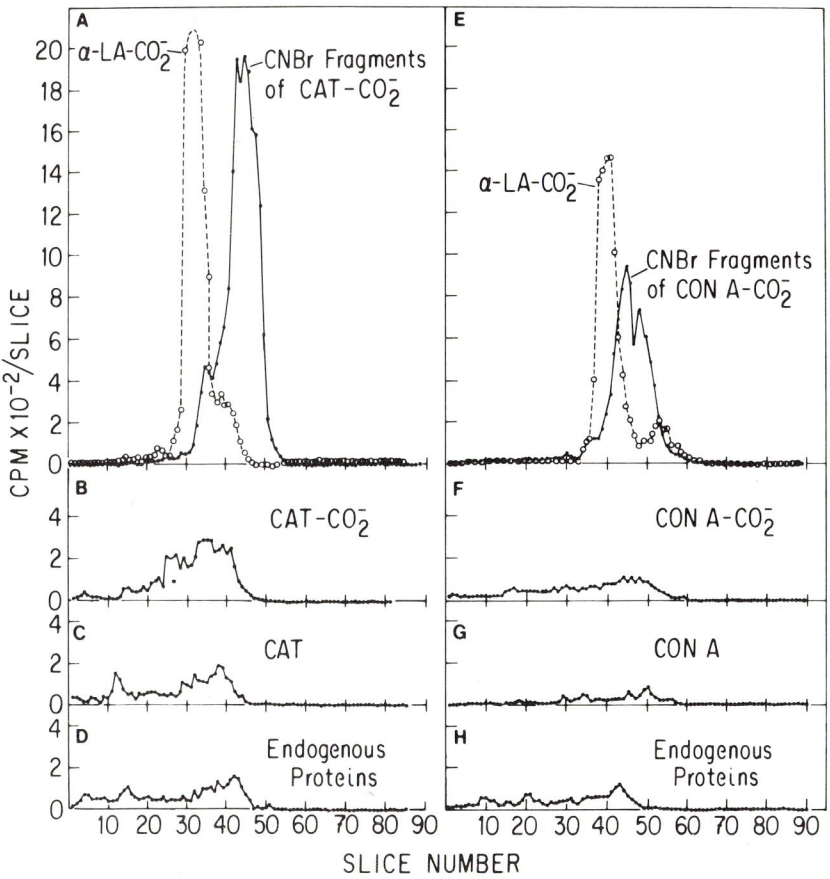

Fig. 4. Electrophoretic analysis of [^{14}C] mannose labeled polypeptides synthesized in the presence of exogenous proteins or polypeptide fragments. The following polypeptides were added to the standard reaction mixture: A) 200 μg of unfractionated cyanogen bromide cleavage fragments from S-carboxymethyl catalase; B) 200 μg of S-carboxymethyl catalase (Cat-CO_2^-); C) 200 μg of native catalase (Cat); D) no exogenous protein added; E) 200 μg of unfractionated cyanogen bromide cleavage fragments from S-carboxymethyl concanavalin A; F) 200 μg of S-carboxymethyl concanavalin A (Con A-CO_2^-); and G) no exogenous protein added. After 60 min at 37°C the reactions were centrifuged at 27,000 × g for 10 min. Aliquots of the supernatants (60 μl) were analyzed on 15% polyacrylamide disk gels in the presence of 1% SDS, 0.5 M urea, and 2% β-mercaptoethanol. The dashed lines in A and E represent the electrophoretic profile of the glycosylated product obtained from an identical assay containing 200 μg of S-carboxymethyl α-lactalbumin (α-LA-CO_2^-), and the dark bars show the position of catalase (Cat) and concanavalin A (Con A). The arrow at the top of the figures marks the front of the tracking dye (TD).

Because of the lack of resolution obtained on 15% gels it was difficult to determine whether these new peaks contained a single or multiple components, but it is clear that they are lower in molecular weight than reference α-lactalbumin (mol. wt. 14,200). According to the published amino acid sequences of these proteins (10–12), tripeptide acceptor sequences are present in 2 cyanogen bromide cleavage fragments from catalase. One fragment (mol. wt. 13,500) has 2 potential acceptor sites; the other fragment (mol. wt. 79,400) contains 1 site or possibly 2, depending on whether amino acid residue 212 is asparagine or aspartate. The potential acceptor sites in concanavalin A are also present in 2 separate cyanogen bromide cleavage fragments of mol. wt. 10,600 and 13,300.

To obtain better resolution of the new glycosylated products formed in the presence of the cyanogen bromide cleavage fragments, gel filtration on Sephadex G-100 was used. The results (Fig. 5) clearly show that at least 2 distinct products are formed in assays containing the cyanogen bromide cleavage fragments from both catalase (Fig. 5A) or concanavalin A (Fig. 5B). No such products were found in the presence of endogenous oviduct proteins (Fig. 5C). The included volume from these columns contained unreacted GDP-[^{14}C]mannose and its breakdown products; the small, partially included peak is free [Man-^{14}C]oligosaccharide. Peaks I and II from both the catalase and concanavalin A fragments were pooled separately and the TCA-precipitable radioactivity in these fractions was determined before and after extensive pronase digestion (24). Pronase rendered over 95% of the radioactivity in the 4 fractions TCA-soluble, indicating that these new products are indeed glycopeptides.

DISCUSSION

Using a simple TCA precipitation assay and a membrane fraction from hen oviduct we have surveyed a variety of denatured proteins and peptides having 1 or more tripeptide sequences of the structure -ASN-X-(SER/THR)- for their ability to accept the oligosaccharide chain of [Man-^{14}C]oligosaccharide- P-P-dolichol. Based on the earlier finding (17) that identical mannose-labeled glycoproteins are synthesized whether GDP[^{14}C]mannose or [Man^{14}C]oligosaccharide-lipid are used as the sugar donor, the assay for glycosylation of exogenous proteins has been simplified and made more economical. In the presence of GDP[^{14}C]mannose and UDP-N-acetylglucosamine, endogenous enzymes in the oviduct membrane preparation catalyze both the synthesis of [Man-^{14}C]oligosaccharide-lipid and the transfer of its oligosaccharide chain to certain exogenous proteins which are readily recovered from the reaction mixture. Inhibition by tunicamycin of mannose incorporation indicates that under the assay conditions used virtually all mannose transfer (95%) into protein is dependent on the synthesis of N-acetylglucosamine-lipid intermediates.

In addition to the previously reported derivatives of α-lactalbumin, ribonuclease A and ovalbumin (9), 2 other exogenous proteins, triosephosphate isomerase and prolactin have been shown to function as oligosaccharide acceptors after reduction and derivitization. In addition, in experiments not shown S-carboxymethylated and S-aminoethylated DNase were found to serve as acceptors under the assay methods used in this study (D. Struck and W.J. Lennarz, unpublished studies). In an earlier study (9) the sulfitolyzed derivatives of elastase, carboxypeptidase, DNase, and alcohol dehydrogenase, all of which contain one or more -ASN-X-(SER/THR)- sequences, were shown to be inactive as carbohydrate acceptors. Based on the current study this list now can be enlarged to include the S-carboxymethyl and S-aminoethyl derivatives of catalase, concanavalin A, elastase, glyceraldehyde-3-phosphate dehydrogenase, and trypsinogen. Thus, of 13 potential carbohydrate acceptor proteins tested so far, 6 have been found to serve as acceptors.

One possible explanation for the inactivity of the remaining 7 proteins is that the charge introduced by derivitization prevents interaction of the protein with membrane-bound oligosaccharide transferase. However, this seems unlikely for several reasons. First, neither the carboxymethyl derivatives (negative charge) nor the aminoethyl derivatives (positive charge) of these 7 proteins served as acceptors. Second, although the carboxymethyl and aminoethyl derivatives of the active proteins quantitatively differed in their ability to accept the oligosaccharide, in all cases both types of derivatives could be shown to be glycosylated.

As shown in the pH dependence studies, pH, and therefore charge, does play a role in the rate of glycosylation. Moreover, the pH affects not only the membrane-bound enzyme system but also the acceptor protein, because the optimal pH for glycosylation

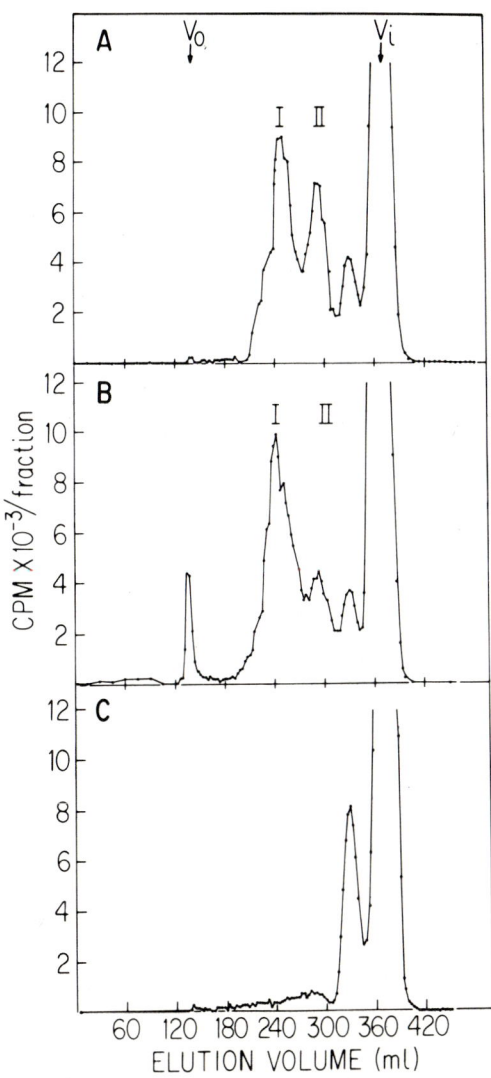

Fig. 5. Analysis of glycosylated polypeptide fragments by gel filtration on Sephadex G-100. Reaction products from standard assays as described in Fig. 4 were applied to a Sephadex G-100 column and eluted with 0.1 M NH_4HCO_3. A) Cyanogen bromide cleavage fragments from S-carboxymethylated catalase; B) cyanogen bromide cleavage fragments from S-carboxymethylated concanavalin A; or C) no exogenous protein. Arrows mark the elution position of the void volume (V_0) and the included volume (V_i).

depends on the particular acceptor protein tested. Nevertheless, under no conditions could the inactive proteins listed above be glycosylated.

Based on these findings, we have considered 2 possible explanations for the finding that certain proteins containing -ASN-X-(SER/THR)- serve as substrates, while others with this tripeptide do not. One possibility is that specific amino acid residues, either in position "X" or in the sequences adjacent to the tripeptide, are required for a protein to be active. Examination of the sequences of both the active and inactive proteins does not reveal any sequence features consistent with this interpretation. The other possibility is that the inactive, denatured proteins still retain regions of order that, in some way, make the acceptor tripeptide sequence unavailable for interaction with the transferase.

If the latter explanation is correct, one would expect that if the size of the polypeptide chain of the inactive proteins was reduced by chemical or enzymatic cleavage, the previously inaccessible tripeptide might become exposed. The results of the current study, showing that a mixture of fragments of catalase, and of concanavalin A, generated by CNBr cleavage are able to accept the oligosaccharide, are consistent with this idea. Moreover, more extensive studies with isolated, purified polypeptide fragments of α-lactalbumin and RNase (D. Struck, G. Hart, K. Brew, G. Grant, R. Bradshaw, and W. J. Lennarz, unpublished studies) show that large portions of the polypeptide chain of active acceptors can be removed without any loss of acceptor ability. Thus, at present it appears the presence of the tripeptide -ASN-X-(SER/THR)- may be sufficient for a protein to be glycosylated, provided that this site is accessible for interaction with the enzyme. We are currently testing this hypothesis by preparing small peptides containing the sequence -ASN-X-SER- or −ASN-X-THR-.

ACKNOWLEDGMENTS

We wish to thank Drs. Douglas Struck, Ursula Czichi, and Gerald Hart for their advice and assistance. The contributions of Ms. Ann Fuhr in preparation of this manuscript, and of Mr. Michael Kluka in handling the chickens are gratefully acknowledged. We are indebted to Mr. T. Anderson and Dr. K. E. Ebner for providing us with samples of prolactin. This work was supported by National Science Foundation Grant PCM-74-11986-A01 and the National Institutes of Health grant 1 R01 GM21451. Dr. Kathryn E. Kronquist was supported by a Postdoctoral Fellowship from the National Institutes of Health (5 F32 GM05325).

REFERENCES

1. Waechter CJ, Lucas JJ, Lennarz WJ: J Biol Chem 248:7570, 1973.
2. Lucas JJ, Waechter CJ, Lennarz WJ: J Biol Chem 250:1992, 1975.
3. Waechter CJ, Lennarz WJ: Annu Rev Biochem 45:95, 1976.
4. Eagon PC, Heath EC: Fed Proc Fed Am Soc Exp Biol 34:678, 1975.
5. Struck DK, Lennarz WJ: J Biol Chem 252:1007, 1977.
6. Duskin D, Bornstein P: J Biol Chem 252:955, 1977.
7. Hickman S, Kulczycki A, Lynch GR, Kornfeld S: J Biol Chem 252:4402, 1977.
8. Marshall RD: Biochem Soc Symp 40:17, 1974.
9. Pless DD, Lennarz WJ: Proc Natl Acad Sci USA 74:134, 1977.
10. Dayhoff MO (ed): "Atlas of Protein Sequence and Structure." Washington, DC: National Biomedical Research Foundation, 1972, vol 5.

11. Dayhoff MO (ed): "Atlas of Protein Sequence and Structure." Washington, DC: National Biomedical Research Foundation, 1973, vol 5, Suppl 1.
12. Dayhoff MO (ed): "Atlas of Protein Sequence and Structure." Washington, DC: National Biomedical Research Foundation, 1976, vol 5, Suppl 2.
13. Castellino F, Vanaman T, Hill RL, Brew K: J Biol Chem 245:4570, 1970.
14. Brew K, Hill RL: J Biol Chem 245:4559, 1970.
15. Lowry OH, Rosebrough NJ, Farr AL, Randall RJ: J Biol Chem 193:265, 1951.
16. Pless DD, Lennarz WJ: J Biol Chem 250:7014, 1975.
17. Struck DK, Chen WW, Lennarz WJ: Fed Proc Fed Am Soc Exp Biol 36:745, 1977.
18. Kuo S-G, Lampen JO: Biochem Biophys Res Commun 58:287, 1974.
19. Tkacz JS, Lampen JO: Biochem Biophys Res Commun 65:248, 1975.
20. Takatsuki A, Kohno K, Tamura G: Agric Biol Chem 39:2089, 1975.
21. Lehle L, Tanner W: FEBS Lett 71:167, 1976.

Sialic Acid: A Specific Role in Hematopoietic Spleen Colony Formation

Quentin Tonelli and Russel H. Meints

Laboratory of Experimental Hematology, School of Life Sciences, University of Nebraska, Lincoln, Nebraska 68588

Vibrio cholerae neuraminidase (VCN) treatment of donor bone marrow cells results in a reduction in the number of hematopoietic colonies (CFUs) formed in the spleens of lethally irradiated mice. Treatment of marrow cells with sodium periodate under mild conditions, known to preferentially oxidize sialic acid, also reduced CFUs while subsequent potassium borohydride reduction restored CFUs to 80% of control levels. Innoculum viability as measured by in vitro incorporation of tritiated precursors into proteins, nucleic acids, and oligosaccharides was unaffected by VCN treatment. The ability of bone marrow cells in culture to respond to the hormone erythropoietin, as measured by the incorporation of ^{59}Fe into cyclohexanone-extractable heme, was also not affected by neuraminidase, making a cytotoxic effect of the VCN preparation unlikely. Incubation of VCN-treated marrow with either β-galactosidase or trypsin had no effect on the VCN-induced reduction in CFUs. These results are consistent with the idea that membrane sialic acid plays a direct and specific role in the implantation and development of CFUs.

Key words: borohydride reduction, spleen colonies, neuraminidase (vibrio cholerae), periodate oxidation, N-acetyl-neuraminic acid, hematopoietic stem cell, erythropoietin

We have been interested in the characterization of the surface properties of the pluripotent hematopoietic stem cell, the precursor of all circulating blood cells. The discrete areas of proliferating hematopoietic tissue found in the spleens of lethally irradiated mice 7–10 days after injection of donor bone marrow (1) are clones derived from a single cell (2, 3), the colony-forming unit (CFUs), believed to represent the hematopoietic stem cell.

Although both bone marrow and spleen will support the proliferation and differentiation of CFUs (4), only a fraction of the total CFUs in the innoculum implant and proliferate in the spleen (5–7). In addition, the spleen has been shown to be a predominantly erythropoietic microenvironment, while bone marrow supports mainly granulopoietic differentiation (4). The inductive events in hematopoietic differentiation are believed to be mediated

Received March 29, 1977; accepted August 2, 1977

Abbreviations: NANA – N-acetylneuraminic acid; VCN – Vibrio cholerae neuraminidase; EPO – erythropoietin; CFU – colony-forming unit; GFH – glucose-free Hanks salt solution

0091-7419/78/0801-0067$02.30 © 1978 Alan R. Liss, Inc

by short-range cellular interactions (8). This differential implantation and subsequent induction of stem cells suggests the existence of a mechanism for CFU site recognition and adhesion, the specificity conferred by cell surface constituents.

Sialic acids occur as the terminal sugar in the oligosaccharide portion of numerous glycoproteins, mucins, and gangliosides (9). Enzymatic removal of sialic acids with neuraminidase (E.C. 1.2.3.18.) has been used to demonstrate an important function for these sugars in numerous biological events. Cleavage of sialic acid destroyed virus receptors (10), and altered both the circulation pattern of lymphocytes (11) and rosette formation (12). Changes in membrane sialic acids have also been shown to alter the velocity of platelet aggregation (13). In addition desialation of serum proteins (14) and erythrocytes (15, 16) results in their clearance from circulation

In an earlier study (17), we examined the role of sialic acid in stem cell function. Treatment of donor marrow with Vibrio cholerae neuraminidase prior to transplantation into lethally-irradiated hosts reduced the number of macroscopic spleen colonies, whereas pretreatment with a variety of proteases and glycosidases had no effect on CFUs. The reduction was independent of enzyme concentration at maximal levels and was lost after heat inactivation of VCN. These results are consistant with an enzymatic rather than cytotoxic effect of VCN. We have therefore proposed a role for sialic acid in the formation of at least some spleen colonies. This study confirms and extends our earlier findings. The VCN-induced reduction in spleen colonies could not be attributed to a decrease in viability of the innoculum, since the ability of the marrow to incorporate precursors of macromolecular synthesis is unaffected after VCN treatment. The response of VCN-treated marrow to the hormone erythropoietin was also unchanged.

Mild periodate oxidation of marrow effected a similar decrease in CFUs as VCN treatment. Periodate oxidation of VCN-treated marrow caused no further reduction in spleen colonies, indicating that both treatments were affecting the same moiety. Borohydride reduction of periodate-treated marrow restored CFU levels to 74–85% that of controls. Sequential incubation of neuraminidase-treated marrow with either β-galactosidase or trypsin did not alter the reduction in spleen colonies. These results suggest a direct and specific role for intact surface sialic acid residues in the formation of at least some hematopoietic spleen colonies.

MATERIALS AND METHODS

Female CF1 mice, 6–10 weeks old were used throughout the study. Cell suspensions were prepared by flushing marrow from tibiofibulae and femora by injection of glucose-free Hanks salt solution (GFH), pH 7.4, containing 0.1 mg/ml deoxyribonuclease I with a syringe and 23 gauge needle. Clumps of cells were dispersed by gentle passage through a pasteur pipette and the suspension filtered through a 200-mesh stainless steel screen. After 10 min at room temperature, cells were washed in a 30-fold dilution of GFH, pelleted by centrifugation, resuspended in GFH, pH 7.4, at 4×10^7 cells/ml and incubated with or without 20 U/ml VCN. GFH contains 1.3 mM $CaCl_2$. One unit of VCN is the enzyme activity required to release 1 μg of sialic acid from its substrate in 15 min at 37°C. VCN was assayed using N-acetylneuraminelactose as a substrate and the released sialic acid estimated by the method of Warren (18), using crystalline N-acetylneuraminic acid (NANA) as the standard.

The culture system has been extensively described elsewhere (19). Briefly, the medium at pH 6.9 contained 65% NCTC 109 made 30 mM with morpholinopropane sulfonic acid, 1 mM with L-glutamine, 30% fetal calf serum, and 5% mouse serum containing 2 μg/ml ferric nitrate. The complete medium was supplemented to 0.05 mg/ml gentamicin. All sera were heat inactivated for 30 min at 56°C. Cells were suspended at 15×10^6 nucleate cells/ml in complete medium and 0.2 ml of suspension were plated in 15×10 mm plastic tissue culture dishes (Linbro FB-16-24-TC). The plates were sealed and incubated at 37°C.

Five replicate cultures were established for each group. For heme synthesis experiments sheep plasma erythropoietin, 50 mU/ml (Connaught, step III, 2.2 units/mg protein, lot 3005-1) in NCTC 109 was added to the appropriate cultures at the start of incubation. After 20 h, 10 μl of ^{59}Fe-labeled mouse serum (0.4 μCi) were added to each plate and incubation continued for an additional 4 h. Cultures were stopped by the addition of 2 ml ice-cold 0.15 M NaCl/5 mM sodium phosphate, pH 7.4, (PBS) and transferred to test tubes. Plates were washed with an additional 1 ml PBS and the cells pelleted by centrifugation. Heme synthesis was measured as previously described (19).

In the remaining cultures, 10 μl of the appropriate isotope were added at the start of incubation. [5-^{3}H] Uridine (29 Ci/mM, 0.2 μCi) containing cultures were incubated for 0.5 h; whereas [methyl-^{3}H] thymidine (41 Ci/mM, 0.2 μCi), L-[4,5-^{3}H] leucine (60 Ci/mM, 0.2 μCi), and D-[1-^{3}H] glucosamine (3 Ci/mM, 1.0 μCi) containing cultures were incubated for 4 h. Cultures were terminated as described above. The cell pellet was washed twice in ice-cold PBS, made to 5% ice-cold trichloroacetic acid (TCA), and allowed to precipitate overnight. The precipitate was centrifuged at $1,000 \times g$, washed twice in 5% TCA, and solubilized in 0.5 ml NCS Tissue Solubilizer. After the addition of 10 ml toluene-based scintillation cocktail and transfer to vials, the samples were counted.

Mice which served as recipients in the CFU assay received 850 rads of ^{60}Co irradiation by rotation in the field of a ^{60}Co therapy source (Picker model, V9M/60). After 24 h, mice received 10^5 nucleate marrow cells via tail vein injection. Seven days later mice were sacrificed and the spleens removed and fixed in alcohol-formalin-acetic acid (750 ml 70% ethanol, 250 ml formaldehyde, 50 ml glacial acetic acid). Following at least 24 h of fixation colonies were counted under a dissecting microscope at $15 \times$.

Marrow cells at 4×10^7 cells/ml which had been incubated previously in either VCN or buffer were incubated with β-galactosidase (20 U/ml, E. coli) or trypsin (Type III, 0.25%) in GFH, pH 7.4, for 1 h at 37°C. After trypsinization, cells were treated with excess soybean trypsin inhibitor. After one wash in a 30-fold dilution of GFH, cells were resuspended in GFH and 10^5 nucleate cells injected into lethally irradiated mice.

The release of material from tritium-labeled cells was used to assay the action of β-galactosidase and trypsin on marrow cells. Galactose and its derivitives were labeled with tritium at the C6 position (20, 21) after galactose oxidase treatment with sodium [^{3}H]-borohydride reduction (22). Galactose oxidase was purified by column chromatography. Two hundred and fifty units of galactose oxidase (Polyporus circinatus) dissolved in 0.5 ml 50 mM sodium phosphate, pH 7.0, was applied to a 1×13 cm Sepharose 6B column and eluted with 50 mM sodium phosphate buffer, pH 7.0. Galactose oxidase activity was measured by the Galactostat procedure (Worthington Biochemicals). The retarded material containing galactose oxidase activity was pooled. Marrow cells in GFH (2×10^7/ml) were incubated for 1 h at 37°C with 1.25 U/ml galactose oxidase. The reaction was terminated by the addition of 30 ml ice-cold PBS. After 2 washes with PBS, cells were treated for 15 min at 25°C with 1 mM freshly-prepared sodium [^{3}H] borohydride (661 mCi/mM). The reaction was terminated, and the cells washed 3 times with ice-cold PBS. Cells were resuspended in GFH, and treated with β-galactosidase or trypsin as described above.

To label sialic acid residues, marrow cell suspensions (2×10^7 cells/ml) were prepared in PBS, pH 7.4. Periodate oxidation was performed in 0.4 mM sodium periodate/PBS for 15 min at 25°C. The reaction was terminated by the addition of 30 ml ice-cold PBS, and the cells centrifuged. The washing procedure was repeated twice. Marrow cells (2×10^7 cells/ml) were reduced in either 1 mM freshly-prepared potassium borohydride or sodium [^{3}H]borohydride (178 mCi/mM)/PBS. After 15 min at 25°, the reaction was terminated and the cells washed as described above.

The products released by VCN hydrolysis of periodate oxidized/[^{3}H]borohydride reduced marrow were chromatographed on a Sephadex G-75 column (1×25 cm) with distilled water as eluant. One milliliter fractions were collected and monitored by liquid scintillation counting. The tritium-labeled, VCN-released material (98%) was present in the retarded peak. These retained fractions were pooled and lyophilized. This material was dissolved in a small quantity of water, and applied to Whatman No. 1 paper.

Descending paper chromatography was run for 18 h at room temperature. The solvent system employed was n-butyl alcohol/acetic acid/water (5/2/2). The paper was dried; a vertical strip containing the standards cut and dipped in $AgNO_3$ (23) or sprayed with thiobarbituric acid (24). The remaining chromatogram was cut into 1-cm horizontal strips, which were placed in a counting vial with 0.1 ml water. After 4 h, 0.5 ml NCS solubilizer was added and incubation continued for an additional 4 h. Ten milliliters of scintillation cocktail was added and radioactivity determined.

VCN was purchased from Calbiochem, La Jolla, California. Trypsin Type III, soybean trypsin inhibitor, DNase I, galactose oxidase, synthetic NANA, and N-acetylneuraminelactose were from Sigma Chemical Company, St. Louis, Missouri. NCTC 109 was from Microbiological Associates, Bethesda, Maryland, fetal calf serum from North American Biologicals, Miami, Florida, NCS and radioisotopes from Amersham/Searle, Arlington Heights, Illinois. CF-1 mice were purchased from Blue Spruce Farms, Altamont, New York.

RESULTS

Viability of Neuraminidase-Treated Marrow

Table I shows the effect of VCN treatment on the ability of bone marrow to incorporate precursors of nucleic acids, protein, and carbohydrate into acid insoluble material. Enzyme treatment of marrow had no effect on the 4-h incorporation of tritiated thymidine, leucine, or glucosamine. In addition, the 0.5-h incorporation of uridine was also unaffected. These results indicate that the reduction in spleen colonies caused by VCN is not mediated through a general decrease in innoculum viability.

The ability of marrow to respond to the hormone erythropoietin (EPO), the specific inducer of erythroid differentiation, was also tested after VCN treatment (Table II). VCN-treated marrow responded to EPO stimulation with the same level of ^{59}Fe incorporation into heme as untreated marrow.

Treatment of Neuraminidase-Treated Marrow With β-Galactosidase or Trypsin

Sialic acid may play a direct or an indirect role in spleen colony formation. A requirement for sialic acid in CFU adhesion to spleen sites would indicate a direct role for NANA; whereas if NANA functions by masking binding sites, an indirect role is indicated. Serum glycoprotein survival times are markedly shortened after their desialation (14). In that system, NANA has been shown to have an indirect role, since cleavage of sialic acid exposes galactose residues, which results in the binding of the glycoprotein to hepatic sites (25, 26). We therefore tested for a similar indirect mechanism to account for the reduction in

TABLE I. Incorporation of Precursors of Macromolecules Following Neuraminidase Treatment*

Treatment	cpm[a]	N[b]	p[c]
	Thymidine[d]		
Control	7,217 ± 434	5	–
Neuraminidase	7,460 ± 237	5	NS[f]
	Uridine[e]		
Control	8,039 ± 829	5	–
Neuraminidase	8,055 ± 75	5	NS
	Leucine[d]		
Control	10,446 ± 705	5	–
Neuraminidase	10,527 ± 820	5	NS
	Glucosamine[d]		
Control	2,214 ± 84	5	–
Neuraminidase	2,171 ± 264	5	NS

*Marrow cell suspensions at 4×10^7 cells/ml were prepared in glucose-free Hanks salt solution (GFH), pH 7.4, and incubated with or without 20 U/ml Vibrio cholerae neuraminidase for 1 h at 37°C. Cells were washed once with a 30-fold dilution of GFH and resuspended at 15×10^6 nucleate cells/ml in 65% NCTC 109, 30% fetal calf serum, and 5% mouse serum; 0.2 ml of the suspension was plated in 15×10-mm plastic tissue culture plates, sealed, and incubated at 37°C. Isotope (10 μl) in NCTC 109 was added at the start of the culture period. [5-^{3}H]Uridine (29 Ci/mM, 0.2 μCi) containing cultures were incubated for 0.5 h; whereas [methyl-^{3}H] thymidine (41 Ci/mM, 0.2 μCi), L-[4,5-^{3}H]leucine (60 Ci/mM, 0.2 μCi), and D-[1-^{3}H]glucosamine (3 Ci/mmol, 1.0 μCi) were incubated for 4 h. Cultures were terminated by the addition of 1 ml ice-cold phosphate buffered saline (PBS). Cells were washed twice in PBS and precipitated overnight in cold 5% trichloroacetic acid (TCA). After 2 washes in 5% TCA, the precipitates were prepared for liquid scintillation counting.
[a]Counts per minute/10^6 nucleate cells ± standard deviation.
[b]Number of cultures.
[c]Probability determined from Student's t test, significance level $p < 0.05$.
[d]4 h pulse, 4 h culture.
[e]0.5 h pulse, 0.5 h culture.
[f]NS: not significant.

colony-forming ability after VCN treatment. Incubation of neuraminidase-treated marrow with β-galactosidase (Table III) had no effect on the reduction in spleen colonies. Similarly, treatment of neuraminidase-treated marrow with trypsin had no effect on the reduction in spleen colonies (Table IV).

To test if β-galactosidase and trypsin were removing material from marrow cells, galactose and its derivitives were labeled with galactose oxidase followed by sodium [^{3}H]-borohydride reduction (Table V). Appreciable incorporation of tritium occurred when marrow cells were treated with VCN and the newly exposed galactosyl residues labeled with tritium. Both β-galactosidase and trypsin released a portion of this radioactivity (Table VI). Treatment of labeled cells with β-galactosidase released only 26% of the total radioactivity, whereas trypsin hydrolysis released 59% of the total radioactivity.

Effect of Periodate Oxidation on Spleen Colonies

Periodate oxidation, under conditions mild enough to be specific for carbohydrates (27, 28) was used to test for a specific role of sialic acid in spleen colony formation (Table VII). In all experiments, periodate oxidation decreased spleen colonies from un-

TABLE II. Incorporation of ^{59}Fe Into Heme Following Neuraminidase (VCN) Treatment*

Group	Percent uptake[1]
Control	0.07 ± 0.04^a
Control + EPO	0.22 ± 0.05^b
VCN	0.12 ± 0.03^c
VCN + EPO	0.21 ± 0.02^b

*Marrow cell suspensions at 4×10^7 cells/ml were prepared in glucose-free Hanks salt solution (GFH), pH 7.4, and incubated with or without 20 units Vibrio cholerae neuraminidase for 1 h at 37°C. Cells were washed once with a 30-fold dilution of GFH, and resuspended at 15×10^6 nucleate cells/ml in 65% NCTC 109, 30% fetal calf serum, 5% mouse serum. The suspension (0.2 ml) was plated in 15 × 10-mm plastic tissue culture plates, and 50 mU/ml erythropoietin (EPO) added to appropriate cultures. Cultures were sealed and incubated at 37°C. After 20 h, 10 µl of ^{59}Fe-labeled mouse sera (0.4 µCi) were added to each plate, and incubation continued for an additional 4 h. Cultures were terminated and the cells washed twice in ice-cold phosphate buffered saline and the incorporation of ^{59}Fe into cyclohexanone-extractable heme measured. Results are expressed as the average percent uptake of 5 replicate cultures ± standard deviation.
[1] Percent uptake of ^{59}Fe per 10^6 nucleate cells ± standard deviation.
Values followed by different letters differ from other experimental values at the significance level of $p < 0.05$, determined from Student's t test.

TABLE III. Treatment of Neuraminidase-Treated Marrow With β-Galactosidase: Effect of Spleen Colonies*

Treatment	Colonies[1]	N[2]
Control	9.0 ± 2.4^a	15
β-Galactosidase	10.4 ± 0.6^a	8
Neuraminidase	4.3 ± 0.9^b	8
Neuraminidase + β-Galactosidase	3.0 ± 1.1^b	8

*Marrow cell suspensions at 4×10^7 cells/ml were prepared in glucose-free Hanks salt solution (GFH), pH 7.4, and incubated with or without 20 U/ml Vibrio cholerae neuraminidase for 1 h at 37°C. Cells were washed once with a 30-fold dilution of GFH and incubated with or without 20 U/ml β-galactosidase (E. coli). After one wash in GFH, 10^5 nucleate marrow cells were transfused via tail vein injection into female CFl mice which had received 850 rads of ^{60}Co irradiation 24 h previously. After 7 days, mice were sacrificed and spleens removed and fixed for at least 24 h in alcohol-formalin-acetic acid. Macroscopic spleen colonies were counted under a dissecting microscope at 15 × magnification.
[1] Colonies/10^5 nucleate marrow cells injected, average ± standard error.
[2] Number of spleens.
Values followed by different letters differ from other experimental values at the significance level of $p < 0.05$, determined from Student's t test.

treated controls (average 51%, range 41–58%), whereas subsequent borohydride reduction restored spleen colonies to 80% of controls (range 74–81%).

Treatment of marrow with VCN, followed by periodate oxidation effected no further reduction in spleen colonies. To test if both treatments were affecting the same surface constituents, periodate-oxidized marrow was reduced with tritiated borohydride, and then incubated with VCN (Table VIII). All of the radioactivity except that which could be attributed to the borohydride treatment alone was released by VCN. The products of VCN cleavage were chromatographically identified (Table IX). The postenzyme incubation

TABLE IV. Treatment of Neuraminidase-Treated Marrow With Trypsin: Effect of Spleen Colonies

Treatment	Colonies[1]	N[2]
Control	16.7 ± 1.3[a]	8
Trypsin	16.9 ± 1.5[a]	8
Neuraminidase	8.8 ± 1.8[b]	8
Neuraminidase + Trypsin	8.5 ± 2.0[b]	8

*Cell suspensions at 4×10^7 marrow cells/ml were prepared in glucose-free Hanks salt solution (GFH), pH 7.4, and incubated with or without 20 U/ml Vibrio cholerae neuraminidase for 1 h at 37°C. Cells were washed once with a 30-fold dilution of GFH and incubated in either 0.25% trypsin/GFH or GFH for 1 h at 37°C. After the addition of excess soybean trypsin inhibitor and one wash in GFH, 10^5 nucleate cells were transfused via tail vein injection into female CF1 mice which had received 850 rads of ^{60}Co irradiation 24 h previously. After 7 days, mice were sacrificed and spleens removed and fixed for at least 24 h in alcohol-formalin-acetic acid. Macroscopic spleen colonies were counted under a dissecting microscope at 15 × magnification.
[1] Colonies/10^5 nucleate marrow cells injected, average ± standard error.
[2] Number of spleens.
Values followed by different letters differ from other experimental values at the significance level of $p < 0.05$, determined from Student's t test.

TABLE V. Products of Galactose Oxidase Treatment and Sodium [^{3}H] Borohydride Reduction of Marrow Cells

Cells reduced with sodium [^{3}H] borohydride	cpm[a]
Cells (untreated)	1,713
Cells incubated with neuraminidase	2,214
Cells incubated with galactose oxidase	2,180
Cells incubated with neuraminidase and galactose oxidase	9,831

*Marrow cell suspensions at 4×10^7 cells/ml were prepared in glucose-free Hanks salt solution (GFH), pH 7.4, and incubated with or without 20 U/ml Vibrio cholerae neuraminidase for 1 h at 37°C. Cells were washed once with a 30-fold dilution of GFH. Cell suspensions (2×10^7 cells/ml) were incubated with or without 1.25 U/ml purified galactose oxidase (Polyporus circinatus) for 1 h at 37°C. Cells were washed 3 times with phosphate buffered saline (PBS). Cell suspensions (2×10^7/ml) in PBS were treated with 1 mm freshly prepared sodium [^{3}H] borohydride for 15 min at 25°C. The reaction was terminated by the addition of a 30-fold dilution of ice-cold PBS. After 2 additional washes with PBS, aliquots were counted for radioactivity.
[a] Counts per minute per 10^6 cells.

supernatant was applied to a 1 × 25-cm Sephadex G-75 column and eluted with distilled water. Ninety-seven percent of the radioactivity was found in the peak of material of < 1,000 molecular weight. This retained material was subjected to descending paper chromatography using n-butyl alcohol/acetic acid/water (5:2:2) as the solvent. Essentially all the radioactivity migrated with either the C^7 analog of sialic acid (NANA$_7$), or N-glycolylneuraminic acid.

Both treatments are therefore affecting the same moiety. These results indicate a direct function for sialic acid in colony formation. The effects of cleavage or modification

TABLE VI. Release of Tritium-Labeled Material From Asialo Marrow Cells by β-Galactosidase and Trypsin

Treatment	cpm[a]
Labeled cells	7,118
Cells after β-galactosidase	5,532
Supernatant after β-galactosidase	1,933
Cells after trypsin	3,074
Supernatant after trypsin	4,491

*Desialated mouse marrow was labeled with tritium by galactose oxidase/sodium [^{3}H]borohydride treatment. Cell suspensions (2×10^7 cell/ml) in glucose-free Hanks salt solution (GFH), pH 7.4, were treated with either 20 U/ml β-galactosidase or 0.25% trypsin for 1 h at 37°C. Cells were pelleted by centrifugation and an aliquot of the supernatant counted for radioactivity. After 3 washes in GFH, cells were counted for radioactivity.
[a]Counts per minute per 10^6 cells.

of NANA are not additive, but result in parallel reductions in spleen colonies. They therefore appear to be exerting their effect through sialic acid.

DISCUSSION

In this study we confirm and extend our earlier findings (17) that VCN treatment of donor bone marrow reduced the number of hematopoietic spleen colonies. The reduction in spleen colonies could not be attributed to a general decrease in the viability of the innoculum, such as that reported by Yuhas (29), since enzyme-treated marrow cells incorporated macromolecular precursors to the same extent as control cells. In addition, VCN treatment has been shown to have no effect on bone marrow oxygen consumption (30). Longer term culture viability was also unaffected since the ability of marrow to respond to erythropoietin was not inhibited after VCN treatment. These results argue against a cytotoxic effect of VCN on mouse bone marrow. Since the action of VCN is directed mainly towards the glycoprotein-bound NANA in intact cells (31), we feel it safe to assume that the reduction in spleen colonies we find after VCN treatment is a result of the cleavage of NANA from cell surface glycoproteins.

It is interesting that only a portion of the total CFUs are lost after VCN treatment. Although in a previous paper (17) we reported that maximal reduction in spleen colonies occurred when marrow cells were incubated with 1 U/ml VCN and that addition of a 50-fold excess of enzyme caused no further reduction, it still is possible that the partial reduction reflects an incomplete accessibility of sialic acid sites on marrow cells to VCN. On the other hand this differential susceptibility to VCN may indicate a heterogeneity within the CFUs pool. For example, VCN lability may reflect the age structure within the CFUs compartment, or it may be indicative of the stage of CFUs within the cell cycle. Edelman (32) has suggested that resting cells in G_1 may actually represent cohorts of cells temporally distributed in a random manner. Since CFUs are normally in a nonproliferative state (33), CFUs may only be susceptible to VCN for a limited period.

The possible indirect role of sialic acid in spleen colony formation was examined. Subsequent incubation of VCN-treated marrow with β-galactosidase or trypsin released a portion of the surface galactosyl residues, but did not alter the reduction in spleen colonies.

TABLE VII. Effect of Periodate Oxidation on Spleen Colonies*

Treatment	Colonies[1]	N[2]
	Experiment 1	
Control	17.9 ± 1.5^a	8
Borohydride	16.0 ± 1.4^a	8
Periodate	9.6 ± 1.4^b	8
Periodate + Borohydride	13.2 ± 1.4^c	8
	Experiment 2	
Control	15.4 ± 1.9^a	6
Neuraminidase	5.9 ± 1.8^b	6
Borohydride	14.2 ± 1.6^a	6
Periodate	9.0 ± 1.8^b	8
Neuraminidase + Periodate	7.2 ± 2.3^b	7
Periodate + Borohydride	13.1 ± 2.3^a	7
	Experiment 3	
Control	14.4 ± 1.1^a	9
Neuraminidase	3.6 ± 1.5^b	10
Borohydride	14.0 ± 1.8^a	10
Periodate	6.0 ± 1.6^c	10
Neuraminidase + Periodate	3.2 ± 0.7^b	10
Periodate + Borohydride	11.8 ± 2.2^a	10

*Marrow cell suspensions at 4×10^7 cells/ml were prepared in glucose-free Hanks salt solution (GFH), pH 7.4, and incubated with or without 20 U/ml Vibrio cholerae neuraminidase for 1 h at 37°C. Cells were washed once with a 30-fold dilution of 0.15 M NaCl/0.005 M sodium phosphate, pH 7.4 (PBS). Cell suspensions (2×10^7 cells/ml) were oxidized in 0.4 mM sodium periodate/PBS for 15 min at 25°C and washed 3 times in PBS. Reduction was performed in freshly prepared 1 mM potassium borohydride/PBS for 15 min at 23°C, and cells washed 3 times in PBS. Marrow cells in GFH (10^5 nucleate cells) were transfused via tail vein injection into female CF1 mice which had received 850 rads of ^{60}Co irradiation 24 h previously. After 7 days, mice were sacrificed and spleens removed and fixed for at least 24 h in alcohol-formalin-acetic acid. Macroscopic spleen colonies were counted under a dissecting microscope at 15 × magnification.
[1]Colonies/10^5 nucleate marrow cells injected ± standard error.
[2]Number of spleens.
Values followed by different letters differ from other experimental values at the significance level of $p < 0.05$, determined from Student's t test.

It seems unlikely that the loss of colony forming ability is mediated through the exposure of penultimate galactose residues following the cleavage of NANA. It is possible, however, that β-galactosidase did not remove all the terminal galactose due to steric hindrance or to different positional specificity. In addition, since sialic acid is often linked to N-acetylgalactosamine, it is possible that this sugar may play a role in colony formation. However, the experiments in which sialic acid was modified in situ suggest that sialic acids play a direct rather than a indirect role in colony formation.

Periodate oxidation of surface sialic acids also reduced spleen colonies. Although it is known that periodate may act at various locations in a carbohydrate or protein polymer, the mild conditions employed in this study suggest that sialic acid residues at the cell surface are preferentially attacked (27, 28).

TABLE VIII. Products of Vibrio Cholerae Neuraminidase Treatment of Modified Marrows Cells

	cpm[a]
Experiment 1	
[^{3}H] Borohydride reduced cells	1,197
Periodate oxidized, [^{3}H] borohydride reduced	
Cells	3,630
Cells after neuraminidase treatment	992
Supernatant after neuraminidase treatment	2,686
Experiment 2	
[^{3}H] Borohydride reduced cells	730
Periodate oxidized, [^{3}H] borohydride reduced	
Cells	3,210
Cells after neuraminidase treatment	763
Supernatant after neuraminidase treatment	2,098

*Marrow cell suspensions were prepared in glucose-free Hanks salt solution (GFH), pH 7.4, and washed twice in PBS. Cell suspensions (2×10^7 cells/ml) were oxidized in 0.4 mM sodium periodate in phosphate buffered saline, pH 7.4 (PBS) for 15 min at 25°C, and washed 3 times in PBS. Reduction was performed in 1 mM freshly prepared sodium [^{3}H] borohydride (178 mCi/mM)/PBS for 15 min at 25°C. After 3 washes in PBS, cells were suspended in GFH, and an aliquot counted for radioactivity, then 20 U/ml Vibrio cholerae neuraminidase was added, and the suspension incubated for 1 h at 37°C. Cells were pelleted by centrifugation and an aliquot of the supernatant counted for radioactivity. After one wash in GFH, cells were counted for radioactivity.
[a]Counts per minute per 10^6 cells.

TABLE IX. Chromatographic Behavior of Neuraminidase-Released Material From Tritium-Labeled Marrow Cells*

Substance applied	Mobility in solvent[a]
N-acetylneuraminic acid[c,d]	1.00
NANA$_7$[c,d]	1.40
N-glycolylneuraminic acid[e]	0.54
Neuraminidase cleavage product[b]	1.38, 0.60

*Mouse marrow cells (2×10^7 cells/ml) labeled with tritium by periodate oxidation/[^{3}H] borohydride reduction were incubated with 20 U/ml VCN in GFH, pH 7.4, for 1 h at 37°C. Cells were pelleted by centrifugation and the supernatant chromatographed on Sephadex G-75. The radioactive fractions were pooled and lyophilized. A sample of this material, along with standards, was applied to Whatman No. 1 paper. Descending chromatography was run for 18 h at room temperature. The paper was dried and cut into 2 vertical strips. One was sprayed with thiobarbituric acid, or dipped in AgNO$_3$; the other was cut into 1-cm horizontal pieces and radioactivity determined.
[a]Mobility relative to N-acetylneuraminic acid, solvent system n-butyl alcohol/acetic acid/water (5:2:2).
[b]These bands contained tritium.
[c]Positive reaction with AgNO$_3$.
[d]Positive reaction with thiobarbituric acid.
[e]Value reported for the mobility of N-glycolylneuraminic acid.

In addition, we have shown that periodate oxidation selectively modified sialic acid residues (Table VI) and that neuraminidase released all of these modified residues (Table V). It has been proposed that the generation of an aldehyde at C^7 of sialic acid by periodate oxidation may result in its interaction with other membrane components through a Schiff's base reaction (34). Presant and Parker (28) have shown that mild periodate oxidation of lymphocytes, at conditions similar to those used in this study, results in specific surface conformational changes. A similar interaction on CFUs could result in a rearrangement of membrane sialic acid-containing components, thereby removing them from participation in colony formation. Such a mechanism has been proposed to account for the reduced adhesion of mycoplasma to periodate-treated epithelial tissue (35). Since periodate oxidation of VCN-treated marrow effected no further reduction in spleen colonies, both treatments appear to exert their reduction through sialic acid.

An interesting finding of this study is the return of colony-forming ability to periodate-treated marrow following subsequent reduction with borohydride, suggesting the importance of the polyhydroxylated side chain of sialic acids in colony formation, Since the anionic sites of sialic acid are not affected by periodate, a charge related role of NANA in colony formation is not supported. Jeanloz and Codington (36) point out the polyhydroxylated side chain at C^7-C^9 of sialic acid appears to be unique among surface components and they suggest that the complex biochemical pathway of its synthesis points to its importance in biological events.

ACKNOWLEDGMENTS

This work was supported by N.I.H. grant AM-174-34-03 and University of Nebraska Research Council NIH Biomedical Support Grant RR-07055. Special thanks to Dr. John McGreer for the generous use of his radiation facilities, and to his staff for their valuable assistance. Thanks to Dr. Gilbert Ashwell for the gift of $NANA_7$.

REFERENCES

1. Till JE, McCulloch EA: Radiat Res 14:213, 1961.
2. Becker AJ, McCulloch EA, Till JE: Nature (London) 197:452, 1963.
3. Wu AM, Till JE, McCulloch EA: J Cell Physiol 69:177, 1967.
4. Trentin JJ: Am J Pathol 65:621, 1971.
5. Kretchmar AL, Conover WR, Transplantation 8: 576, 1969.
6. Siminovitch L, McCulloch EA, Till JE: J Cell Comp Physiol 62:327, 1963.
7. Schooley JC: J Cell Physiol 68:249, 1966.
8. McCulloch EA, Till JE: In Stohlman F (ed): "Hemopoietic Cell Proliferation." New York: Grune and Stratton, 1970, p 15.
9. Gottschalk A: Adv Enzymol 20:135, 1958.
10. McCrae JF: Aust J Exp Biol Med Sci 25:127, 1947.
11. Gesner BM, Woodruff JJ: In Smith RT, Good RA (eds): "Cellular Recognition." New York: Appleton-Century-Crofts, 1969, p 79.
12. Galili V, Schlesinger M: J Immunol 112:1628, 1974.
13. Mester L, Szabados L, Born GNR, Michal F: Nature (London) New Biol 236:213, 1972.
14. Morell AG, Gregoriadis G, Scheinberg IH, Heckman J, Ashwell G: J Biol Chem 246:1461, 1971.
15. Jancik J, Schauer R: Hoppe-Seyler's Z Physiol Chem 355:395, 1974.
16. Aminoff D, Bell WC, Fulton I, Ingebrigtsen N: Am J Hematol 1:419, 1976.
17. Tonelli Q, Meints RH: Science 195:897, 1977.
18. Warren L: J Biol Chem 234:1971, 1959.

19. Goldwasser E, Eliason JF, Sikkema D: Endocrinology 97:315, 1975.
20. Amaral D, Bernstein L, Morse D, Horecker BL: J Biol Chem 238:2281, 1963.
21. Amaral D, Kelly-Falcoz F, Horecker BL: Methods Enzymol 9:87, 1966.
22. Morell AG, Van Den Hamer CJA, Schienberg HI, Ashwell G: J Biol Chem 241:3745, 1966.
23. Wheat RW: Methods Enzymol 8:60, 1966.
24. Warren L: Nature (London) 186:237, 1960.
25. Lenten LV, Ashwell G: J Biol Chem 247:4633, 1972.
26. Rogers JC, Kornfeld S: Biochem Biophys Res Commun 45:622, 1971.
27. Liao TH, Gallop PM, Blumenfeld OO: J Biol Chem 248:8247, 1973.
28. Presant CA, Parker S: J Biol Chem 251:1864, 1976.
29. Yuhas JM, Toya RE, Pazmino NH: J Natl Cancer Inst 53:465, 1974.
30. Olander CP: Personal communication.
31. Barton NW, Rosenberg A: J Biol Chem 248:7353, 1973.
32. Edelman GM, Wang JL: In Mora PT (ed): "Cell Surfaces and Malignancy." DHEW Publication No. NIH 75–796. Washington, DC: US Govt Printing Office, 1976.
33. Lajtha LG: J Cell Comp Physiol (Suppl 1) 67:143, 1963.
34. Novogrodsky A, Katchalski E: Proc Natl Acad Sci USA 70:1824, 1973.
35. Powell DA, Hu PC, Wilson M, Collier AM, Baseman JB: Infect Immun 13:959, 1976.
36. Jeanloz RW, Codington JF: In Rosenberg A, Schengrund C (eds): "Biological Roles of Sialic Acid." New York: Plenum Press, 1976, p 226.

Dimensions and Specificities of Recognition Sites on Lectins and Antibodies

Elvin A. Kabat

Departments of Microbiology, Human Genetics and Development, and Neurology, College of Physicians and Surgeons, Columbia University, The Neurological Institute, Presbyterian Hospital, New York, and the National Cancer Institute, National Institutes of Health, Besthesda, Maryland 20014

A comparison is made of the specific combining sites of a number of lectins and of antibodies with emphasis on those reacting with blood group A, B, and H determinants. The ranges of site sizes and specificities of both groups are similar both from immunochemical studies and from the limited x-ray diffraction data available.

Key words: plant hemagglutinins, carbohydrate binding site

INTRODUCTION

During the past 10 years there has been an extraordinary burst of activity in the study of plant and animal lectins stimulated largely by the findings that they have specific receptor sites for carbohydrate (1–6) and react with glycoproteins in solution or on cell membranes, are involved in the removal of asialoglycoproteins from the circulation (7), and may hold specific nitrogen fixing bacteria in root nodules (8, 9). In many instances they are mitogenic producing blast transformation (2–6) and also react to cause movement of receptors in cell membranes producing patching and capping (2–6).

At the same time there has been extensive interest in antibodies and immunoglobulins and especially in myeloma immunoglobulins for many of these are homogeneous and have specific receptor sites (10, 11, cf.6) so that they may be sequenced and, if crystallized, may be studied by x-ray diffraction. In addition, some antibodies have been obtained in a relatively homogeneous form suitable for sequence and x-ray diffraction studies (12–18, cf.6). To date only very few myeloma antibodies and one lectin but a fair number of enzymes have been studied.

Immunochemical investigations, most frequently by quantitative precipitin and quantitative hapten inhibition assays along the lines developed by the Heidelberger School (19, 6) but more recently by radio (20, 21) and enzyme (22) immunoassay, have made it possible to explore the nature of the specific combining sites of antibodies, myeloma

Received March 4, 1977; accepted August 1, 1977.

0091-7419/78/0801-0079$02.00 © 1978 Alan R. Liss, Inc

globulins, and lectins and to obtain estimates of their dimensions, specificities, and shapes. In the few instances in which x-ray diffraction studies (12–18) are available, these have been in general agreement and the overall range of site sizes of enzyme sites studied crystallographically and of antibody combining sites from immunochemical estimates has been similar (6). For instance the lysozyme site has been shown to accommodate a hexasaccharide composed of alternating N-acetylglucosamine and N-acetylmuramic acid residues (23) while the upper limit for the antidextran site has been shown to be a chain of 6 (34 Å in most extended form) or 7 α1→6 linked D-glucose residues (6). The lysozyme site is a cleft at the surface of the molecule and combining sites of myeloma antidextrans have been shown to be either grooves or cavities in shape (24); human antidextrans although generally restricted in specificity have been shown to be mixtures with both kinds of sites (24).

The lower limit for an antibody combining site has been estimated to be between 1 and 2 glucoses (about 6 Å) (25, 6) and the only lectin studied thus far by high resolution x-ray diffraction, Concanavalin A (Con A), has a carbohydrate binding site comparable in size as determined by quantitative inhibition assays (26, 27), nuclear magnetic resonance (28–30), and x-ray diffraction (31–33). This range of dimensions allowing for shape and conformational differences may well include receptor sites on all antibodies and lectins and a large number of enzymes. Con A also has a hydrophobic pocket located in a cleft between the 2 subunits distinct from the carbohydrate binding site and about 25 Å from it. The combining sites of Bence Jones dimers and Fab fragments studied directly by x-ray diffraction (12–18) fall into the same range of site sizes. Many lectins unlike antibodies are metalloproteins and the metal is necessary for binding activity (1–6).

These 3 classes of substances also show other similarities. Lectins (1–5) and enzymes (34) may exist in multiple forms and antibodies (6) may be mixtures of molecules with combining sites of different sizes even to a single antigenic determinant (6, 19). The former 2 classes are generally paucidisperse being limited to 4 or 5 species while antibodies may be extremely heterogeneous populations but in some instances may show also paucidispersity or even homogeneity (6).

Since specific receptor sites on antibodies and on lectins are directed toward carbohydrate, we are dealing largely with sequential rather than conformational determinants of protein antigens and it becomes possible to explore their nature by using a variety of oligosaccharides to inhibit the precipitin reaction between the glycoprotein or polysaccharide antigens and the antibody or lectin. The site which inhibits at the lowest concentration is considered to be most complementary to the oligosaccharide (19, 6).

A number of lectins, notably from Lens culinaris (lentil) and Robinia pseudoacacia have combining sites of specificity similar to Con A reacting best with mannose but also with fructose, glucose, and DGlcNAc and especially well with glycopeptides of IgG (35–37).

The existence of a large number of lectins with combining sites specific for the blood group A, B, and H antigenic determinants makes possible some comparisons of these receptor sites with antibody combining sites both in size and specificity. The blood group A, B, and H determinants (38–43) have each been found to be of 4 kinds, 2 of which are:

$$\begin{array}{c} \text{LFuc} \\ \alpha\downarrow 1 \\ 2 \end{array}$$

A   DGalNAcα1→3DGalβ1→3DGlcNAcβ1→3

or

B   DGal

```
                    LFuc
                    α↓1
                     2
   A    DGalNAcα1→3DGalβ1→4DGlcNacβ1→6
              or
   B    DGal
```

The other 2 A and B determinants have a second LFuc linked to the DGlcNAc on carbons 4 and 3 respectively. The 4 H determinants have the same structures but lack nonreducing DGalNAc or DGal. Anti-A and anti-B with specificities complementary to each of these structures including the second LFuc and the DGlcNAc have been obtained. The oligosaccharides isolated from blood group substances, the substances themselves, and other structurally similar oligosaccharides and polysaccharides have made possible the gathering of substantial amounts of data on a considerable number of lectin sites.

It has been possible to purify these lectins readily using an insoluble adsorbent, polyleucyl hog A+H substance made by addition of long polyleucyl chains on to the lysines of the blood group A+H glycoproteins from hog gastric mucosa (44, 45). This insoluble adsorbent has blood group A and blood group H determinants as well as determinants with terminal nonreducing DGlcNAcα1→DGalβ1→4 and in addition smaller numbers of determinants from incompletely synthesized chains. It affords considerable versatility for purifying different lectins.

Two lectins which specifically agglutinate human A but not B or O erythrocytes have been shown to vary substantially in the sizes of their specific receptor sites and also in their ability to accommodate related sugars. The receptor site of the lectin of *Helix pomatia* (46–48) appears to be relatively small. Indeed methyl αDGalNAc is the best inhibitor found thus far. Oligosaccharides isolated from blood group A substance are all somewhat less active on a molar basis. Even more surprising in view of its strict specificity for A erythrocytes is the finding that methyl αDGlcNAc is quite a good inhibitor being about one tenth as active and that melibiose (DGalα1→6DGlc) and methyl αDGal also inhibit but are very much less active. The site is thus capable of accommodating portions of these other sugars but the fit is poor. The molecule has a molecular weight of about 79,000 and is composed of 6 subunits of $K^a = 5 \times 10^3$ liter/mole (47). In addition to A substances with which it reacts best, it precipitates with B,H,Le^a and precursor I blood group substances most probably due to the terminal nonreducing DGlcNAc residues and with R_a and R_b rough mutants of *Salmonella typhimurium* having terminal DGalα1→6DGlc determinants (48). Its ability to precipitate is a consequence of its multivalence and the multivalence of the antigen despite the low binding constant of each site. Indeed its high specificity for A erythrocytes is surprising since B erythrocytes have a terminal nonreducing DGalα1→3.

However, methyl αDGal was 1/250 as potent as methyl αDGalNAc in inhibiting precipitation of the lectin by R_a lipopolysaccharide (48). This might not be sufficient to agglutinate B erythrocytes. Since methyl αDGlcNAc is only 1/10 as active as methyl αDGalNAc, it may be that B and O erythrocytes do not have many terminal nonreducing α-linked DGlcNAc accessible at their surface or one might expect agglutination not to be blood group A specific.

By contrast, however, the lectin of *Dolichos biflorus* (49) shows more specificity for the A pentasaccharide,

$$\begin{array}{c} \text{LFuc}\alpha 1 \\ \downarrow \\ 2 \\ \text{DGalNAc}\alpha 1 \rightarrow 3\text{DGal}\beta 1 \rightarrow 4\text{DGlcNAc}\beta 1 \rightarrow 6R, \end{array}$$

being about 1/3 more active than methyl α-DGalNAc, DGalNAcα1→3DGal, and DGalNAcα1→3DGalβ1→3DGlcNAc, all of which were of equal potency. Thus the site is probably specific for DGalNAcα1→3DGal with the LFuc side chain stabilizing a more favorable conformation of the other 2 sugars or involves the DGalNAcα1→3DGal plus some hydrophobic contribution of the fucose to binding. Unlike the *Helix pomatia* site the Dolichos lectin does not react with methyl αDGlcNAc or methyl αDGal and precipitates only with blood group A_1 and A_2 substances and with streptococcal group C polysaccharide which also contains terminal nonreducing α-linked DGalNAc residues. Enzymic de-N-acetylation of the terminal nonreducing DGalNAc of blood group A substances destroys reactivity and re-N-acetylation restores it.

Both the snail and the Dolichos lectins have homogeneous sites as evidenced by the findings that different fractions eluted from the insoluble adsorbent by increasing concentrations of DGalNac all exhibit the same ratio of inhibition of different inhibitors to one another (45, 48); the snail lectin is homogeneous by equilibrium dialysis (46). Although the Dolichos lectin exists in two forms distinguishable by chromatography on Con A-Sepharose these have been shown by Carter and Etzler (50) not to differ in specificity.

Two lectins from the lima bean (51), *Phaseolus lunatus,* have been purified and show blood group A specificity, methyl αDGalNAc being more active than DGalNAc which is better than methyl αDGal. One of the lectins, component III, was more extensively studied, the o- and p-nitrophenyl αDGalNAc being as active as methyl αDGalNAc. Replacement of the N-acetamido group of methyl αDGalNAc by a p-amino or p-nitrobenzamido group gave increased activity; this is probably an indication of a hydrophobic region in the protein capable of reacting with the aromatic moiety on carbon 2.

The soybean agglutinin (52) does not show blood group specificity. It agglutinates all human erythrocytes and also agglutinates mouse, rat, and human cell lines transformed by viral or chemical carcinogens. The purified protein has a K^a of 3×10^4 liter/mole for DGalNAc, a molecular weight of 120,000, and is composed of 4 subunits. It is precipitable by blood group substances with terminal nonreducing α-linked DGalNAc or α- or β-linked DGal residues.

The soybean agglutinin site (52) differs from the others in that both α and β methyl or ethyl DGalNAc inhibit very well, the α being slightly better than the β but both are more potent than DGalNAc itself; the corresponding α- and β-glycosides of DGal are considerably less active but both are more active than DGal. DGalNAcα1→3DGal was identical in inhibiting power to methyl αDGalNAc while the trisaccharide DGalNAcα1→3DGalβ1→3DGlcNAc, the best oligosaccharide inhibitor, was about 2–3 times more potent suggesting that the β1→3-linked DGlcNAc moiety contributes to the binding site. An important finding with soybean lectin is that phenyl αDGalNAc was more potent even than the trisaccharide. This type of finding, namely that phenyl or nitrophenyl glycosides of sugars are often more potent inhibitors than the methyl compounds reported earlier for Con A (27) and *Sophora japonica* (53, 54) lectin, suggests that some of these lectins have a hydrophobic region which may react with the aglycone portion of these glycosides. This may complicate determination of site size. It is of interest that with antibodies to tobacco mosaic virus protein, addition of an octanoyl group to the C terminus of the tripeptide or tetrapeptide making up a portion of the pentapeptide determinant also increased binding substantially (55). An understanding of this finding in terms of site structure would be of great interest. These hydrophobic-binding regions must be reasonably contiguous to the sugar-binding sites. With Con A the binding region contains 2 tyrosines at positions 12 and 100 which are considered to account for the increased activity of the p-nitrophenyl as compared with methyl α-D-mannoside (33).

The soybean agglutinin precipitated poorly with blood group B substances despite the presence of terminal nonreducing DGalα1→3DGal and this was attributable to steric effects of the fucose substituents on the subterminal DGal, the P1 fractions, obtained by mild acid hydrolysis which removes fucose, showing greatly increased activity. Blood group A substances reacted well and thus the reduced activity due to substitution of fucose is less with terminal nonreducing DGalNAc than with DGal.

The *Sophora japonica* lectin which agglutinates A and B cells also resembles soybean in that DGalNAc is a better inhibitor than DGal and methyl α- and β-glycosides of DGal are more active than DGal; methyl glycosides of DGalNAc were not studied. It differs in that β-glycosides are more potent than α-glycosides and in that it does not agglutinate transformed cells.

The anomalous finding that β-linked oligosaccharides of DGal were somewhat more active than α-linked oligosaccharides despite the B specificity was explained (54) when it was found that removal with coffee bean α-galactosidase of terminal αDGal from B substance and periodate oxidation and Smith degradation of B substance to expose I determinants left substantial precipitating capacity of these substances for the *Sophora japonica* lectin. Thus the intact blood group substances were reacting with the lectin by virtue of the terminal α-linked DGal B determinants as well as with βDGal determinants, probably on incomplete chains.

The peanut agglutinin (56), built of subunits of molecular weight 27,500 also agglutinates all human erythrocytes, binds to neuraminidase-treated human, rat, mouse, and guinea pig lymphocytes but stimulates DNA synthesis only in human and rat lymphocytes. It resembles the T agglutinin present normally in human serum which agglutinates neuraminidase-treated erythrocytes. It has a binding site most complementary to DGalβ1→3DGalNAc which is 14, 55, and 90 times as active as DGalβ1→4DGlcNAc, DGal, and DGalβ1→3DGlcNAc, respectively, and is 25 times more active than DGalβ1→3 N-acetyl D-galactosaminitol. Methyl αDGal and methyl βDGal are more active than DGal. The peanut agglutinin precipitates to different extents with various A, B, and H substances and reacts with determinants in the interior of the blood group substances most probably accessible because of incomplete biosynthesis. This is supported by the findings that reactivity is strikingly increased by one stage of periodate oxidation and Smith degradation of the blood group substances exposing DGalβ1→3DGalNAc and DGalβ1→4DGlcNAc determinants and also that fractions of blood group substances more soluble in ethanol precipitated to a greater extent. The best precipitant of peanut agglutinin is antifreeze glycoprotein containing repeating units of DGalβ1→3DGalNAcα1 →O-Thr-Ala-Ala (57). The peanut agglutinin resembles certain human anti-I and anti-i sera of certain groups (58) in many respects and further studies on this relationship are needed.

Ricinus communis seeds contain 2 proteins (59) RCA_I and RCA_{II}, the former being a hemagglutinin, molecular weight 120,000; the latter is highly toxic, molecular weight

60,000. Both react best with lactose and are specific for terminal nonreducing β-linked DGal or with sugars having the DGal conformation on carbons 2,3, and 4 (60). The toxin binds DGalNAc while the hemagglutinin does not (59). With RCA_{II} methyl βDGal and methyl αDGal are better inhibitors than DGal, the β form being slightly more active than the α form. The hemagglutinin is a tetramer with 2 kinds of subunits, called α and β, and has 2 DGal specific binding sites of K^a 3.8×10^3 and 1.2×10^3 liter/mole for lactose and galactose, respectively. The toxin is a dimer of an α and a β' subunit.

Con A precipitates with RCA_I, the mannose-containing carbohydrate of RCA_I being specific for the Con A site. The complex is RCA_I-Con A_4 and can be dissociated by DMan (61).

The H-specific lectins of *Lotus tetragonolobus* (62–64) have proven especially valuable in establishing structures of blood group oligosaccharides. They all have identical sites reacting very strongly with LFucα1→2DGalβ1→4DGlcNAc but not at all

or
DGlc

with LFucα1→2DGalβ1→3DGlcNAc (64) and thus have proven of great value in elucidating structures of compounds in which the DGlcNAc had LFuc on positions 3 or 4 respectively (65), such compounds not being distinguishable by methylation. Substitution of DGalNAcα1→3 or DGalα1→3 on the DGal to form the blood group A or B determinants completely abolishes activity of the active compound.

Bandeiraea simplicifolia seeds contain 2 lectins one of which is blood group B specific (66) while the other is specific for terminal nonreducing DGlcNAc (67). The B-specific lectin is a glycoprotein of molecular weight 114,000 with 4 identical subunits. It agglutinates B and AB erythrocytes strongly, A_1 weakly, and A_2 and O not at all. Blood group A substance precipitates about 20% of the amount precipitated by B substance. Methyl αDGal, DGalα1→6DGlc, and p-nitrophenyl αDGal were equally effective and the best inhibitors; p-nitrophenyl βDGal was slightly less active. Thus with this lectin a hydrophobic contribution of the p-nitrophenyl group is not seen (66). It is of interest that methyl αDGalNac gives an inhibition curve with a slope very different from that found for the methyl αDGal and the other oligosaccharides; if the basis for this difference were understood it might contribute materially to elucidating site structure.

Another lectin, from *Euonymus europeus* (68, 69), shows blood group B and H specificity. It precipitated with B and H substances but not with A_1 substances. B and H specificity are associated with a single molecule since absorption on and elution from an H immunoadsorbent give a purified protein with both B and H specificity. The site is unusual in that it is complementary to the blood group B tetrasaccharide but unlike anti-B does not distinguish between the subterminal DGal linked β1→3 or β1→4 to the DGlcNAc:

LFuc
α↓1
2
DGalα1→3DGalβ1→3 or 4DGlcNAcβ1→

Molecular models show that placing the N-acetamido group on opposite sides in the 1→3 and 1→4 linked compounds produces a striking similarity in contour.

Recently our laboratory has been engaged in a study of lectins produced by the sponges *Axinella polypoides* (70, 71) and *Aaptos papillata* (72). Both of these sponges produce three lectins; two have been purified from each and their sites characterized. Axinella lectin I is a strong mitogen for human peripheral blood T and is less mitogenic for B lymphocytes (71): Axinella lectin II is not mitogenic. Mitogenic response correlated with degree of lymphocyte agglutination and was specifically inhibited by DGal, DFuc, 2 deoxy DGal, and raffinose if added up to 5 h after the lectin (71). Axinella lectin I had a site reacting best and equally with terminal nonreducing p-nitrophenyl βDGal and with DGal-linked β1→6 to DGlc, DGlcNAc, and DGal while lectin II reacted better with p-nitrophenyl βDGal than with these 3 compounds all of which were equally active; methyl βDGal and α-linked compounds showed much lower activity. The site thus involves terminal nonreducing DGal linked β1→6 to a second sugar but some hydrophobic contribution is involved. p-Nitrophenyl αDGal is as active as methyl αDGal. Thus the hydrophobic contribution to binding is intimately associated with the β linkage.

The two Aaptos lectins (72) which have been purified show extraordinary differences in their sites. Like the wheat germ lectin they are specific for β1→4-linked DGlcNAc residues. Aaptos lectin II is inhibited best and equally by N, N', N''-triacetylchitotriose and N,N',N'',N'''-tetraacetylchitotetraose which were 13 times more active than DGlcNAc; it thus has a site complementary to a trisaccharide. With lectin I, N,N',N'',N'''-tetraacetylchitotetraose is better than the trimer establishing the site as complementary to a tetrasaccharide; the tetramer was 2,000 times more active than DGlcNAc.

Hydrophobic contributions to the binding of Aaptos lectin I are very striking. It precipitated with the monofunctional hapten p-nitrophenyl αDGalNAc while the β anomer did not precipitate and was a good inhibitor; both were weak inhibitors of Aaptose lectin II and did not precipitate. Even more interesting was the finding that inhibition of precipitation by a human blood group A substance, MSS 10% 2x, which was reacting because of its content of β-linked DGlcNAc residues, required much larger amounts of the β1→4 DGlcNAc oligomeric inhibitors at 37° than at 4°C.

Wheat germ agglutinin, molecular weight 36,000 with 4 binding subunits, has a site most complementary to the trisaccharide N,N',N'' triacetylchitotriose. It also binds N-acetylneuraminic acid, reacts with a population of mouse leukocytes lacking immunoglobulin and T cell surface markers, which occurs in highest proportion in bone marrow, spleen, and peripheral blood and is found in less than 0.5% of lymph node and thymus cells. It has an insulin-like effect reacting with insulin receptors or fat cells and blocks fertilization (73–76).

This survey of lectin sites as determined by immunochemical methods even for those which react with blood group A,B,H,Le[a],Le[b], I, and i substances shows them to be quite different in site sizes and specificities. As a group they are analogous to antibody-combining sites except that they are distributed more or less individually among different plant and animal species, while among vertebrates each species as far as is known may form the entire repertoire of antibody specificities. It is evident that further understanding of site structure will come only when a sufficient number of sites have been defined crystallographically to make possible understanding of the nature of the bonds by which these sites interact with their specific oligosaccharide determinants and comparison of lectin sites with antibody sites of similar specificities. The mechanisms by which the characteristic biological functions are affected may or may not be related to site structure, but the specific interactions of the site are the *sine qua non* of their action. Table I summarizes the data on various lectin combining sites and their properties.

TABLE I. Specificities and Properties of Sites on Various Lectins

Lectin	Best estimate of site size and complementarity	Hydrophobic forces (H) or conformation (C) important	Relationship of activity of glycosides and to free sugar[a]			Blood group specificity
			α > Free sugar > β β > Free sugar > α	Methylα > Methylβ > Free sugar	p-Nitrophenyl or Phenyl	pNO$_2$ Phenyl = Methyl or Phenyl > Methyl	
Concanavalin A	DManα1→	H		No	Yes[c]		
Helix pomatia	DGalNAcα1→[b]	C	Yes	No			A
Dolichos biflorus	DGalNAcα1→3[LFucα1→2] DGalβ1→	C	Yes	No			A
Lotus tetragonolobus	LFucα1→2DGalβ1→4[LFucα1→3] DGlcNAcβ1→	C or H					H
Aaptos lectin II	DGlcNAcβ1→4DGlcNAcβ1→ 4DGlcNAc	H	Yes	No	Yes		
BS I[d]	DGalα1→	H	Yes	No	No	Yes	B
BS II[d]	DGlcNAc	H	Yes	No	Yes	No	
Axinella lectin I	DGalβ1→6	H	No	Yes?	Yes	No	
Axinella lectin II	DGalβ1→6	H	No	Yes	Yes	No	
Aaptos lectin I	DGlcNAcβ1→4DGlcNAcβ1→ 4DGlcNAcβ1→4DGlcNAc	H	No	Yes	Yes		
Peanut	DGalβ1→3DGalNAc		No	Yes	No	Yes	
RCA 1[e]	DGalβ1→4		Yes[f]	No			
RCA II[e]	DGalβ1→4		No	Yes			
Soybean	DGalNAcα1→3DGalβ1→3DGlcNAc	H	No	Yes	Yes	No	BH
Euonymus europeus	DGalα1→3[LFucα1→2]DGalβ1→3 or 4DGlcNAc						
Lima bean III	DGalNAcα and DGalβ1→	H			Yes	Yes	A
Sophora japonica	DGalNAcα and DGalβ1→			Yes			A and E

[a]With antibodies and myeloma proteins with specificity for carbohydrates for either α or β linkages. The relationship is to my knowledge always
α > free sugar ≥ β or β > free sugar ≥ α.
[b]Concanavalin A also reacts with terminal nonreducing α-linked DGlc, DGlcNAc, DFru to a lesser degree than DMan.
[c]The 4-methylumbelliferyl αDMan is more active than the p-nitrophenyl glycoside (Clegg RM, Loontiens FG, Jovin TM: Biochemistry 16:167, 1977).
[d]Bandeiraea simplicifolia
[e]Ricinus communis
[f]DGal and methyl αDGal equal as inhibitors
[g]DGalNAc also bound

ACKNOWLEDGMENTS

This work was supported in part by National Science Foundation grants BMS-72-02219-AO4 and PCM 76-81029.

REFERENCES

1. Mäkelä O: Ann Med Exp Biol Fenn (Suppl II)35:144, 1957.
2. Sharon N, Lis H: Science 177:949, 1972.
3. Lis H, Sharon N: Annu Rev Biochem 42:541, 1973.
4. Nicolson GL: Int Rev Cytol 39:89, 1974.
5. Cohn E (ed): Ann NY Acad Sci 234:5, 1974.
6. Kabat EA: "Structural Concepts in Immunology and Immunochemistry." 2nd Ed. New York: Holt Rinehart and Winston 1976.
7. Ashwell G, Morell AG: Adv Enzymol 41:99, 1974.
8. Bohlool BB, Schmidt EL: Science 185:269, 1974.
9. Wolpert JS, Albersheim P: Biochem Biophys Res Commun 70:729, 1976.
10. Potter M, Mushinski EB, Rudikoff S, Appella E: Third International Convocation on Immunology," Buffalo, New York. Basel, Switzerland: S. Karger, 1972, p 270.
11. Cohn M: Cold Spring Harbor Symp Quant Biol 32:211, 1967.
12. Poljak RJ, Amzel LM, Avey HP, Chen BL, Phizackerley RP, Saul F: Proc Natl Acad Sci USA 70:3305, 1973.
13. Schiffer M, Girling RL, Ely KR, Edmundson AB: Biochemistry 12:4620, 1973.
14. Segal DM, Padlan EA, Cohen GH, Rudikoff S, Potter M, Davies DR: Proc Natl Acad Sci USA 71:4298, 1974.
15. Epp O, Colman P, Fehlhammer H, Bode W, Schiffer M, Huber R: Eur J Biochem 45:513, 1974.
16. Edmundson AB, Ely KR, Girling RL, Abola EE, Schiffer M, Westholm FA, Fausch MD, Deutsch HF: Biochemistry 13:3816, 1974.
17. Fehlhammer H, Schiffer M, Epp O, Colman PM, Lattman EE, Schwager P, Steigemann W, Schramm HJ: Biophys Struc Mech 1:139, 1975.
18. Edmundson AB, Ely KR, Abola EE, Schiffer M, Panagiotopoulos N: Biochemistry 14:3953, 1975.
19. Kabat EA: "Kabat and Mayer's Experimental Immunochemistry." 2nd Ed. Springfield, Illinois: Charles C. Thomas, 1961.
20. Berson SA, Yalow R: Clin Chem Acta 22:51, 1968.
21. Parker CW: "Radioimmunoassay of Biologically Active Compounds." Englewood Cliffs, New Jersey: Prentice-Hall, Inc., 1976.
22. Engvall E, Perlmann P: Immunochemistry 8:871, 1971.
23. Phillips DC: Sci Am 215: 11, 78, 1966.
24. Cisar J, Kabat EA, Dorner MM, Liao J: J Exp Med 142:435, 1975.
25. Arakatsu Y, Ashwell G, Kabat EA: J Immunol 97:858, 1966.
26. Goldstein IJ, Hollerman GE, Smith EE: Biochemistry 4:876, 1965.
27. Poretz RD, Goldstein IJ: Biochemistry 9:2890, 1970.
28. Brewer CF, Sternlicht J, Marcus DM, Grollman A: Biochemistry 12:4448, 1973.
29. Alter GM, Magnuson JA: Biochemistry 13:4038, 1974.
30. Villafranca JJ, Viola RD: Arch Biochem Biophys 160:465, 1974.
31. Hardman KD, Ainsworth CF: Biochemistry 12:4442, 1973.
32. Becker JW, Reeke GN Jr, Cunningham BA, Edelman GM: Nature (London) 259:406, 1976.
33. Hardman KW, Ainsworth CF: Biochemistry 15:1120, 1976.
34. Markert CL (ed): "Isozymes." New York: Academic Press, 1975, vol 1.
35. Kornfeld S, Rogers J, Gregory W: J Biol Chem 246:6581, 1971.
36. Howard IK, Sage H: Biochemistry 8:2436, 1969.
37. Leseney AM, Bourrillon R, Kornfeld S: Arch Biochem Biophys 153:831, 1972.
38. Kabat EA: "Blood Group Substances." New York: Academic Press, 1956.
39. Marcus DM: N Engl J Med 280:994, 1969.
40. Watkin WM: In Gottschalk A (ed): "Glycoproteins." New York: Elsevier, 1972, p 830.

41. Kabat EA: In Isbell HS (ed): "Carbohydrates in Solution." Washington DC: American Chemical Society, Advances in Chemistry Series No. 117, 1973, p 334.
42. Lloyd KO: In Aspinall GO (ed): "International Review of Science, Organic Chemistry Series Two." London: Butterworths, 1976, vol 7, p 251.
43. Hakomori S, Kobata A: In Sela M (ed): "The Antigens." New York: Academic Press, 1974, vol 2, p 79.
44. Kaplan MH, Kabat EA: J Exp Med 123:1061, 1966.
45. Lloyd KO, Kabat EA, Beychok S: J Immunol 102:1354, 1969.
46. Hammarström S, Kabat EA: Biochemistry 8:2696, 1969.
47. Hammarström S, Kabat EA: Biochemistry 10:1684, 1971.
48. Hammarström S, Lindberg AA, Robertsson ES: Eur J Biochem 25:274, 1972.
49. Etzler ME, Kabat EA: Biochemistry 9:869, 1970.
50. Carter WG, Etzler ME: Biochemistry 14:2685, 1975.
51. Galbraith W, Goldstein IJ, Biochemistry 11:3976, 1972.
52. Pereira MEA, Kabat EA, Sharon N: Carbohydr Res 37:89, 1974.
53. Poretz RD, Riss H, Timberlake JW, Chien S-M: Biochemistry 13:250, 1974.
54. Chien S-M, Lemanski T, Poretz RD: Immunochemistry 11:501, 1974.
55. Benjamini E, Shimizu M, Young JD, Leung CY: Biochemistry 7:1261, 1968.
56. Pereira MEA, Kabat EA, Lotan R, Sharon N: Carbohydr Res 51:107, 1976.
57. DeVries AL, Komatsu S, Feeney RE: J Biol Chem 245:2901, 1970.
58. Feizi T, Kabat EA: J Exp Med 135:1242, 1972.
59. Nicolson GL, Blaustein J, Etzler ME: Biochemistry 13:196, 1974.
60. Drysdale RG, Herrick PR, Franks D: Vox Sang 15:194, 1968.
61. Podder SK, Surolia A, Bachhawat BK: Eur J Biochem 44:151, 1974.
62. Yariv J, Kalb AJ, Katchalski E: Nature (London) 215:890, 1967.
63. Kalb AJ, Biochim Biophys Acta 168:532, 1968.
64. Pereira MEA, Kabat EA: Biochemistry 13:3184, 1974.
65. Rovis L, Kabat EA, Pereira MEA, Feizi T: Biochemistry 12:5355, 1973.
66. Hayes CE, Goldstein IJ: J Biol Chem 249:1904, 1974.
67. Iyer PN, Wilkinson KD, Goldstein IJ: Arch Biochem Biophys 117:330, 1976.
68. Pacák F, Kocourek J: Biochim Biophys Acta 400:374, 1975.
69. Petryniak J, Pereira MEA, Kabat EA: Arch Biophys 78:118, 1977.
70. Bretting H, Kabat EA: Biochemistry 15:3228, 1976.
71. Phillips SG, Bretting H, Kabat EA: J Immunol 117:1226, 1976.
72. Bretting H, Kabat EA, Liao J, Pereira MEA: Biochemistry 15:5029, 1976.
73. Goldstein IJ, Hammarström S, Lundblad G: Biochim Biophys Acta 405:53, 1975.
74. Cuatrecasas P, Tell GPE: Proc Natl Acad Sci USA 70:485, 1973.
75. Oikawa T, Yamanichi R, Nicolson GL: Nature (London) 241:256, 1973.
76. Robinson PJ, Roitt I: Nature (London) 250:517, 1974.

Glycoprotein and Protein Precursors to Plasma Membranes in Vesicular Stomatitis Virus Infected HeLa Cells

Paul H. Atkinson

Departments of Pathology and Developmental Biology and Cancer, Albert Einstein College of Medicine, Bronx, New York 10461

Vesicular stomatitis virus is known to mature at HeLa cell plasma membranes. To study the process, cells, infected with vesicular stomatitis virus, were fractionated after short term labeling studies (1 min pulse, 1 min chase) to determine the assembly kinetics of G protein and M protein into plasma membranes. Newly synthesized M protein was found released in the supernatant from which free polysomes were sedimented during sucrose gradient analysis of these polysomes. If this M protein is particle bound, it must have a density of less than 1.08 g/ml. About 40% of this M protein so labeled was not sedimentable at 165,000 $\times$ g for 16 h. This newly synthesized M protein had not yet assembled into plasma membrane and thus must represent an internal pool. This and previous studies show that it has a subsequent transit time to the plasma membrane of about 2 min. Once associated with plasma membranes, M protein decayed in an approximately logarithmic fashion indicating that newly synthesized M randomly mixes (and turns over) with preexisting M protein. G protein was particle bound in a 1 min pulse, 1 min chase, and was never found released in a soluble form. At the later time when fucose is added to G protein, the oligosaccharide moiety is near to complete, and on completion is about 2,000 in molecular weight. Evidence is presented showing that fucose is probably attached to the N-acetylglucosamine of the protein carbohydrate linkage. G protein to which fucose had just been added was located internally on a membranous fraction of density 1.14 g/ml in sucrose; its subsequent transit time from this pool (which in uninfected cells is between 1–2% of the total cell fucosyl glycoprotein) was about 15 min. Because their densities were different and their transit times were different, internal newly synthesized M and fucosyl G protein which assemble into plasma membranes were not on the same internal membranous component. Association of M protein with the plasma membranes may thus occur from a nonsedimentable soluble cytoplasmic pool by a process of direct adsorption.

Key words: plasma membrane assembly, HeLa cells, purified plasma membranes, intracellular membrane pools, membrane bound and free polysomes, fucose and glucosamine, glycopeptide synthesis, M protein, G protein, vesicular stomatitis virus

Some viruses maturing at the plasma membrane are known to contain as part of their own membrane a glycoprotein (G protein) and in some cases a nonglycosylated species (M protein). In uninfected HeLa cells, glycoproteins and nonglycosylated proteins assemble into membrane by different pathways (1) and in vesicular stomatitis virus in-

Received March 24, 1977; accepted June 24, 1977.

0091-7419/78/0801-0089$03.80 © 1978 Alan R. Liss, Inc

fected HeLa cells G and M proteins associate with the HeLa plasma membrane at different rates after their synthesis (2). For the nonglycosylated matrix proteins similar observations have been made for fowl plague virus (4), Rauscher leukemia virus (5), and Newcastle disease virus (6). In the same context, Meier-Ewert and Compans (7) have concluded that the different polypeptides of influenza virus were assembled into virions by distinct pathways as in Sendai virus assembly (8). Data from our previous publication (Atkinson et al., Ref. 2) is consistent with the hypothesis that there are different routes of assembly of G and M proteins of vesicular stomatitis virus into HeLa cell plasma membranes. Both these proteins have a discernible internal pool (2, 33) precursor to plasma membrane though little else is known about the properties of the M protein pool.

In this paper the state of glycosylation of G protein in transit to the plasma membrane is more fully defined and internal M protein precursor to plasma membrane protein is shown to be in a soluble form or a form of very low sedimentability. It is shown that G protein and M protein are not on the same physical entity prior to the insertion of the former into the plasma membrane. In addition, data are presented showing the turnover characteristics of M and G protein from the plasma membranes, and also detailing the time and position of fucose addition to the carbohydrate chain.

METHODS

Cells, Virus Infection, and Radioactive Labeling

Stock and experimental HeLa S_3 cells were grown at 37°C in Eagle's minimal essential medium (9) in Earle's suspension powder (Grand Island Biological Company, Grand Island, New York, Catalog No. F-14) in the absence of antibiotics. The final glucose concentration was 2 g/liter instead of half this amount as stated under this catalog number. The growth medium was supplemented with 3.5% calf serum, 3.5% fetal calf serum (Grand Island Biological Company), and 1% glutamine. In these conditions, cells grew logarithmically from 10×10^4 cells/ml to 100×10^4 cells/ml, with a generation time of 23 h and were used for experiments after growth to a density of $50-70 \times 10^4$ cells/ml. Cells were tested monthly for Mycoplasma contamination by the culturing procedure of Levine et al. (10) and by Levine's enzymatic assay (11); by both these criteria they were free of such contamination. For long labeling periods (more than 3 h) cells were used in their normal growth density range ($1.5 \times 10^5 - 8.0 \times 10^5$ cells/ml). For infection and short term labeling periods, unless otherwise stated, cells were collected by centrifugation and resuspended in growth medium minus serum supplemented with glutamine (2 mM) at 5–10 times their previous density. Cells were infected with 10 plaque forming units/cell of vesicular stomatitis virus (VSV: Indiana serotype). At 1 h postinfection, actinomycin D (kindly donated by Merck and Company, Inc., Rahway, New Jersey) was added to a concentration of 1 µg/ml and at 1.5 h postinfection fetal calf serum was added to 5% concentration. At times ranging from 3.5 to 5 h after infection when host protein synthesis is maximally inhibited (12), cells were grown in medium containing various radioactively labeled precursors: L-[1-^{14}C]fucose (48.66 mCi/mmole) was used at 0.1 µCi/ml concentration and L-[6-^{3}H]fucose (13.4 Ci/mmole), was used at 1–20 µCi/ml concentration, ^{14}C-labeled amino acids mixture, 15 amino acids (80–400 mCi/mmole) was used at concentrations of 0.1–1.0 µCi/ml, L-[6-^{3}H]fucose (13.4 Ci/mmole) was used at concentrations of 1–20 µCi/ml. D-[1-^{14}C]glucosamine HCl (45–55 mCi/mmole) was used at a concentration of 3–4 µCi/ml. L-[^{35}S]methionine

(157 Ci/mmole) was used at concentrations of 5–50 µCi/ml. These compounds were purchased from New England Nuclear Corporation, Boston, Massachusetts.

Isolation and Purification of HeLa Cell Plasma Membrane Ghosts

After experimental manipulations, cells were harvested from culture by low-speed centrifugation (800 × g · min) and further processed as previously described (2) utilizing Earle's balanced salt solution for washing at pH 6.8 (adjusted on the day of the experiment). Plasma membranes were prepared by one cycle of zonal centrifugation according to the procedure of Atkinson and Summers (13), except for the omission of iodoacetate and azide. Generally, 4 discontinuous gradients were used to isolate plasma membranes from 1×10^8 cells. The ouabain sensitive Na^+, K^+-ATPase recovery and enrichment, RNA and DNA content, morphological appearance, and the presence of actin- and myosin-like proteins in uninfected plasma membranes prepared identically has recently been described (15). Protein was determined by the Lowry method as modified by Ceccarini and Eagle (16).

Sodium Dodecyl Sulfate-Polyacrylamide Gel Electrophoresis

The sodium dodecyl sulfate-polyacrylamide gel (SDS-gel) electrophoresis system of Maizel (17) was used without substantial modification as previously detailed (2): analytical gels were 23–25 cm in length.

Preparation of Polysomes

Membrane bound polysomes and free polysomes were prepared by the method of Grubman, Weinstein, and Shaffritz (18), except sodium deoxycholate (DOC) was substituted for NP40 in the polysome analysis step. Before this, cells were vigorously disrupted in a dounce homogenizer in RSB (19), the nuclei and large plasma membrane fragments removed by brief low speed centrifugation and the remaining material was centrifuged at 16,000 rpm ($\sim$ 30,000 × g) in a Sorvall SS-34 rotor for 30 min. The supernatant was made 1% with respect to DOC and the free polysomes were sedimented and separated on 35 ml 7–52% wt/wt sucrose gradient by centrifugation for 16 h at 16,000 rpm in a SW27 rotor at 0°C. The gradients were buffered with RSB and contained 100 µg/ml heparin. In a control experiment, omission of the DOC made no difference to the sedimentation profile of the polysomes and monosomes. The pellet, obtained above after centrifugation of the homogenate at 30,000 × g, was resuspended in RSB, 100 µg/ml heparin and sedimented through a 15–30 (or 35%) wt/wt sucrose gradient for 30 min at 25,000 rpm in the SW27 rotor at 0–4°C. The pellet so obtained was resuspended in RSB, 100 µg/ml heparin, and 1% DOC and analyzed for polysomes as described above. Polysome profiles of absorbance at 260 nm were determined utilizing a Gilford Spectrophotometer with a flow-cell attachment and fractionated into approximately 40 1.0-ml fractions. Regions containing polysomes and supernatant proteins were separately pooled, precipitated with 10% trichloroacetic acid, washed with 100% acetone, and resuspended in SDS-gel electrophoresis running buffer for analysis on SDS polyacrylamide gels.

Analysis of Pronase-Digested Glycopeptides

Production of glycopeptides by pronase, their subsequent digestion with glycosidases, sizing in gel filtration, glycopeptide standards, high voltage paper electrophoresis, and paper chromatography have been described in detail elsewhere (20, 1, 2) and in general we have used Thyroglobulin Unit A glycopeptide (mol wt 1,800), Thyroglobulin Unit B glycopeptide (mol wt 3,000), [^{14}C]acetyl-Asn-(GlcNAc)$_2$(Man)$_5$ (mol wt 1,393) and [^{14}C]acetyl-Asn-(GlcNAc) (mol wt 379).

RESULTS

Fractionation of HeLa cells into purified plasma membranes allowed observation of the transfer of newly synthesized VSV structural components from a pool inside whole cells to the plasma membranes. In this way "transit times" (1, 2) can be determined, and by extending the time span of the experiments, turnover characteristics can also be observed.

M and G Protein

Our previous paper established the transit times of M and fucosyl G proteins to the plasma membranes from the site of synthesis (2). The present study confirms that M protein continues to accumulate in purified plasma membranes (Fig. 1A) for not more than 5 min after labeling of M protein has ceased in the intact cells (Fig. 1B). This continued accumulation, though of short duration indicates an internal pool of M protein. The maximum transit time of M protein from this pool to the cell surface also, must be 5 min or less. G protein labeled with fucose stopped accumulating in the plasma membranes about 20 min (Fig. 1A) after labeling of fucosyl G protein in the whole cells had ceased (Fig. 1B). This continued accumulation (20 min) is the transit time of fucosyl G protein. G protein labeled with amino acids (i.e., all G protein including the incompletely glycosylated G protein precursor to fucosyl G protein, see below) continued to accumulate in the plasma membranes for the duration of the chase (Fig. 1A). Thus, the transit time of incompletely glycosylated G protein is not less than the duration of the chase in this experiment (75 min) and has not been defined in this study. Chase times were not extended long enough in our previous study (2) to observe turnover of M and G protein and in order to determine the decay or turnover characteristics of these proteins, the chase period was continued over an extended time period. M protein was observed to chase out the purified plasma membranes in near logarithmic decay (Fig. 1A). This turnover behavior is consistent with complete mixing of virus M protein with preexisting molecules in the plasma membrane and random loss from the cell — as has been previously discussed by Witte and Weissman (22) in their observations on oncornavirus glycoprotein. This would imply that M protein does not arrive in the plasma membrane as a complete virion complement of proteins and depart from the cell in the same manner, but rather it is inserted at various multiple sites with a subsequent remixing of newly inserted and previously inserted molecules. There was no obvious logarithimic decay of fucosyl G protein though this may only reflect that the chase period was not long enough or there were insufficient data points to observe such a chase for these species. Newly synthesized N protein in the whole cells stops labeling immediately on the addition of the amino acid chase (Fig. 1C): accumulation of newly synthesized N protein in the plasma membrane does not stop for another 15 min thereafter, which is therefore the N protein transit time. It is of interest to note that this is near the transit time of fucosyl G protein.

To lessen the possibility that cross contamination of subcellular fractions may affect interpretation of the data, the rate of accumulation of polypeptides labeled with ^{14}C-

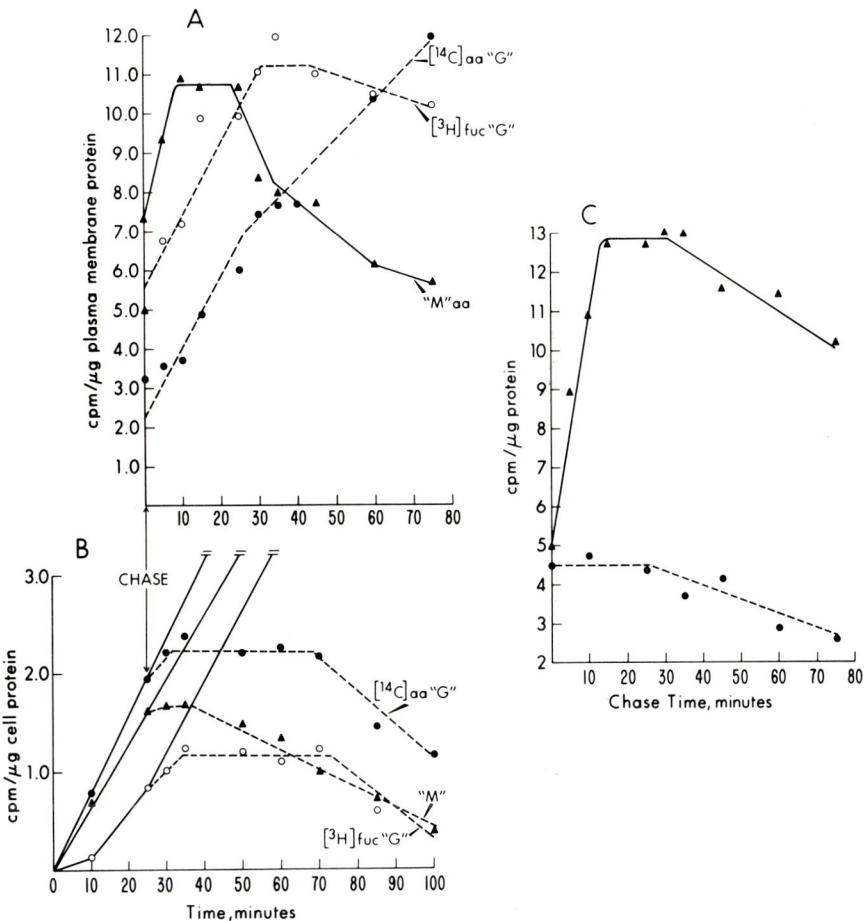

Fig. 1. Chase of 25 min labeled VSV structural proteins into plasma membranes. Cells (1.2 × 10⁹ cells at a density of 3 × 10⁶ cells/ml) were infected 4 h and 25 min with VSV, resuspended in fresh medium at 1.02 × 10⁷ cells/ml then labeled with 34 μCi/ml [³H] fucose and 2.8 μCi/ml ¹⁴C-labeled amino acid mixture. Samples were removed from a portion of this culture 10 and 85 min after the radioactive precursors were added to the main culture and further processed for preparation of plasma membranes, measurement of protein, and radioactivity. "Chase" conditions were obtained in the other culture by adjusting it to approximately 10 × amino acids (essential and nonessential), also, 10 × glutamine, 1 × Earle's balanced salt solution, 0.01 mg/ml phenol red for pH adjustment with 10 N NaOH, and 1.25 g (final concentration 40 mM) fucose. 16.5-ml samples (∼ 1 × 10⁸ cells) were removed and processed as above for plasma maembrane preparation 0, 5, 10, 15, 25, 30, 35, 45, 60, and 75 min after initiation of the chase. VSV structural proteins in homogenates and in the purified plasma membranes were separated and the radioactivity in them quantitated by SDS-polyacrylamide gel electrophoresis. A) The chase of M protein (▲——▲), fucosyl-G protein (○- - -○), and total G protein (●- - -●) into and from the plasma membranes. B) The chase of M protein (▲- - -▲), fucosyl G protein (○- - -○), and total G protein (●- - -●) from the unfractionated cells. The solid lines in this figure show how the respective proteins labeled, without application of the chase. C) The chase of N protein into and from the plasma membranes (▲——▲) and from the cells (●- - -●).

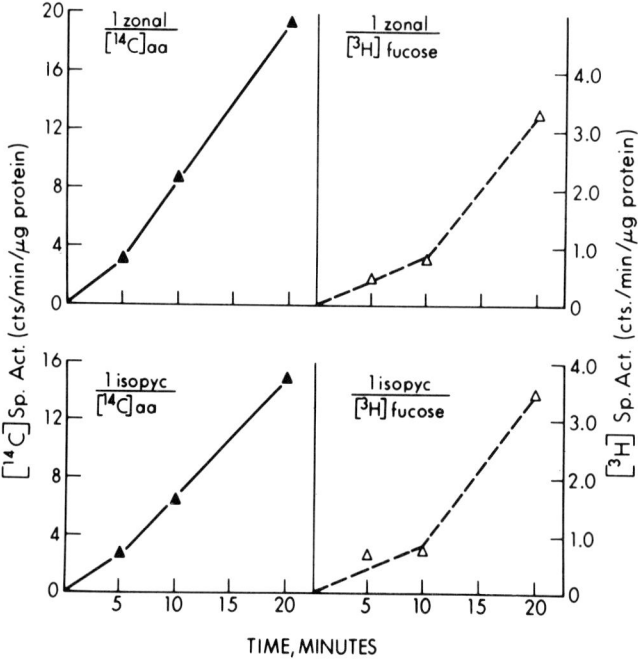

Fig. 2. Effect of more highly purifying plasma membranes on the rapid accumulation of M protein. Cells, 50 ml at 5×10^6 cells/ml, were infected with VSV for 4 h 45 min, at which time 2 μCi/ml ^{14}C-labeled amino acids and 16 μCi/ml [^{3}H]fucose were added. Samples were withdrawn and processed for the plasma membrane preparation at 5, 10, and 20 min of labeling. Protein concentration and radioactivity were determined. Upper Panels. Incorporation of ^{14}C-labeled amino acid-labeled proteins (▲——▲) and [^{3}H]fucosyl glycoproteins (△- - -△) into plasma membranes purified by 1 cycle of zonal centrifugation. Lower Panels. Incorporation of ^{14}C-labeled amino acid-labeled proteins (▲——▲) and [^{3}H]fucosyl glycoproteins (△ - - - △) into plasma membranes purified as above but further purified by banding isopycnically half of the material from the zonal gradient on a 50-ml 0–45% wt/wt sucrose gradient by centrifugation for 16 h, at 45,000 rpm in a SW50 rotor. The band was removed, diluted, and the ghosts collected by centrifugation prior to resuspension in a small volume of 10 mM Tris for analysis of protein and radioactivity.

labeled amino acids was determined in plasma membrane preparations which were more highly purified by centrifugation to their isopycnic density of 1.16–1.17 g/ml in sucrose. Our previous results (2) have shown that the bulk of newly synthesized protein associating with plasma membranes to be M protein when the labeling time with radioactive amino acids is very short (minutes). There was little difference in the lag time (2–3 min) in the rate of accumulation of molecules labeled with radioactive amino acids in plasma membranes purified by rate zonal sedimentation (Fig. 2A) and those purified, in the same experiment, to a higher level by isopycnic banding of the membranes at their density of 1.16–1.17 g/ml (Fig. 2B). The ^{14}C-labeled amino acid specific activity in the membranes purified by isopycnic banding did drop slightly consistent with the removal of soluble molecules such as labeled N and NS proteins. Our previous results (3) had shown that once M protein is bound to plasma membrane ghosts it could not be removed by further sucrose gradient purification, washes, or even with mild detergents known to be capable of removing cytoplasmic tabs from nuclei (19). Hence, loss of labeled molecules is almost certainly the loss of soluble species and not bound M protein. Thus, a procedure known to purify HeLa plasma membranes tenfold (15) with respect to homogenate protein does not significantly alter our transit time estimate for M protein as being 2–3 min. The latter estimate is based on use of the rapid rate zonal purification of plasma membranes (Methods) and the plasma membrane preparations described in the rest of this study were purified only by this method.

Nature of Material 10 Minute Pulse With [³H] Fucose

Our previous results (2) and this study show that the transit time of fucosyl G protein from the site of synthesis to the plasma membranes is 15–20 min. Hence, G protein labeled 12 min with [³H] fucose should all be internal. The nature of this material was explored more thoroughly in this section. VSV-infected HeLa cells, 12-min pulse labeled with [³H] fucose, contain membrane bound material which SDS gel analysis (not shown) demonstrated to be labeled G protein. Pronase digestion of this material followed by high voltage paper electrophoresis at pH 1.9 and pH 6.5 demonstrated that the large proportion of the fucose was attached to glycopeptides. Cellular homogenates from which intact nuclei had been removed, subfractionated on a rate zonal gradient as previously described for the preparation of plasma membranes (14) gave rise to 5 fractions (Table 1): a clear

TABLE I. Distribution of Pulse-Labeled and Long-Term-Labeled Fucosyl G Protein in Cell Fractionation

Cell fraction	Long term labeled [^{14}C] fucose specific activity, cpm/μg protein (relative specific activity)	Recovery %	12 min pulse labeled [^{3}H] fucose specific activity, cpm/μg protein (relative specific activity)	Recovery %
Homogenate	4.5 (1)	100	1.5	100
500 × g · min pellet (nuclei)	4.8 (1.1)	27.7	0.7 (0.4)	11.1
500 × g · min supernatant	4.0 (0.9)	59.7	1.6 (1.1)	72.3
Zonal Fraction I from 500 × g · min supernatant	1.4 (0.3)	9.5	1.0 (0.6)	19.9
Zonal Fraction II	4.3 (1.0)	17.4	3.3 (2.2)	39.0
Zonal Fraction III	3.9 (0.9)	8.0	3.5 (2.3)	21.8
Zonal Fraction IV[a]	22.6 (5.0)	25.9	1.0 (0.7)	3.5
Zonal Fraction V	4.8 (1.1)	1.5	0.6 (0.4)	0.6
Purified plasma membranes	37.9 (9.7)	18.2	0.7 (0.5)	0.8

[a]Contains the partially purified plasma membrane ghosts.

layer on top of the 30% wt/wt sucrose layer containing soluble protein (Fraction I), a white turbid layer containing, among other things, membrane vesicles (Fraction II) which is on top or just penetrating the 30% sucrose layer, a faintly turbid 30% layer containing assorted particles, e.g., vesicles and mitochondria (Fraction III), a turbid band at the 30–45% interface containing plasma membrane ghosts and fragments (Fraction IV) and a clear 45% layer (Fraction V). It can be seen (Table I) that the highest specific activity and the greatest quantity of 12 min ^{3}H pulse-labeled glycoprotein is in Fraction II and III with most of this in Fraction II, the vesicle layer. The plasma membrane layer contains the highest specific activity of the ^{14}C long-term-labeled material which is included as a marker for cell surface fucosyl glycoprotein (35). Glycoprotein pulse labeled 12 min, when removed from Fraction II and centrifuged on a 20–50% wt/wt sucrose gradient banded almost homogeneously at a peak density of 1.14 g/ml (Fig. 3A). Similar observa-

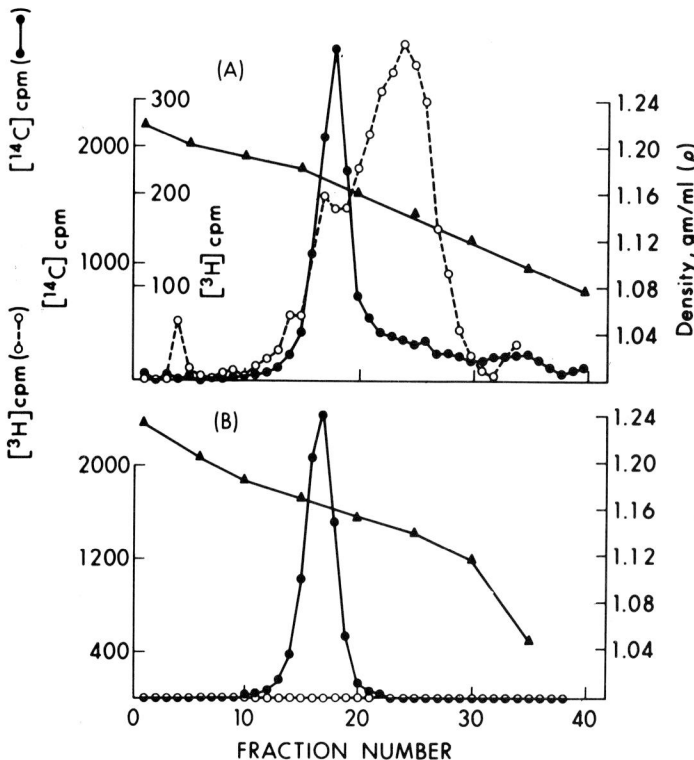

Fig. 3. Comparison of the density of internal fucosyl G protein bearing membrane and plasma membranes. Cells were labeled approximately 16 h with 0.03 μCi/ml [^{14}C] fucose, then infected with VSV (2.25 × 10^8 cells in 50 ml). Approximately 4 h after infection the cells were resuspended at 2 × 10^6 cells/ml and pulse labeled 12 min with 50 μCi/ml [^{3}H] fucose. Plasma membranes are prepared by 1 cycle of zonal centrifugation. In addition, the membranous material located in Zonal Fraction II (Atkinson, 1973) was diluted 25% with 10 mM Tris pH 8, and further analyzed. A) Fraction II material (see also Table I) was mixed with purified plasma membranes and banded isopycnically on a 35-ml 20–50% wt/wt sucrose gradient by centrifugation at 25,000 rpm for 16 h in a SW27 rotor at 4°C. [^{14}C] Fucose-labeled plasma membranes (●—●); [^{3}H] fucose-labeled Fraction II (internal) membranes (○ - - - ○). B) As above except plasma membranes (●—●) were not mixed with Fraction II material. Note the absence of 12 min pulse-labeled fucosyl G protein (○ - - - ○) in the plasma membranes.

tions were made when the unfractionated homogenate was similarly analyzed on a sucrose gradient. The plasma membranes purified from these same cells had little, if any, 12 min pulse-labeled glycoproteins (Fig. 3B). It follows that a 12 min pulse of VSV infected HeLa cells resulted in significant labeling of fucosyl G protein with much of it intracellular as had been suggested from the known transit times. Plasma membranes prepared from infected cells in the same experiment labeled with [^{14}C] fucose were seen to band at a density of 1.17 g/ml (Fig. 3B) at which there is, at most, only a shoulder of 12 min labeled material. This shoulder was almost absent in Fraction II material which had been centrifuged on a similar gradient without the addition of the "marker" plasma membrane material. Thus, the shoulder is probably artifactually adsorbed under these conditions of centrifugation. Hence, material pulse labeled with [^{3}H] fucose for 12 min, was mostly intracellular and was observed at its isopycnic density in a turbid layer at a density typical for vesicles (1.14 g/ml, 13).

State of Completion of the Oligosaccharide Moiety on the Addition of Fucose

In order to more fully assess the meaning of a transit time of fucosyl G protein, the state of completion of the oligosaccharide was analyzed at various labeling times with radioactive fucose. Purified VSV [^{3}H] fucosyl glycopeptides sized on Sephadex G-25 with standard glycopeptides (Fig. 4A) were estimated to be of mol wt 1,950–2,290; when sized on G-50 (Fig. 4B), less ambiguity in the estimate was possible and the glycopeptides appeared to have a mol wt of $\sim$ 2,000. Newly synthesized glycopeptides in infected cell cytoplasm, labeled 5, 10, or 90 min with [^{14}C] fucose appeared (Figs. 5A–C) roughly the same size as the fucosyl glycopeptide in released virus (cf. Fig. 4A) leading to the conclusion that fucose is added near to last in the growing oligosaccharide of fucosyl G protein. [^{3}H] glucosamine in the G glycopeptides formed 3 sizes of substances containing oligosaccharides. The largest size, eluting with blue dextran (Fig. 5C) was either neutral or basic in high voltage paper electrophoresis (HVPE) at pH 1.9 and 6.5 and is probably phosphorylated oligosaccharide arising from a dolichol intermediate (see Discussion). The next size present in the highest molar ratio after 90 min labeling (Fig. 5C) was near to the size of, though distinctly smaller than, the fucosyl glycopeptides. This material was all glycopeptide by HVPE analysis. Glucosamine-labeled glycopeptides in released virus were identical in size to the fucosyl glycopeptides and the apparent lack of this size in infected cells (Figs. 5A–C) may only be a reflection of the fact that the pool of completed G protein within the cell is very small compared with the partially glycosylated pool. The smaller size, mol wt $\sim$ 350–400, appeared in greatest molar ratio in 5-min labeled (Fig. 5A, fractions 52–60) material. This material when [^{14}C] acetylated, was observed to cochromatograph and coelectrophorese (in HVPE) with [^{14}C] acetyl-asparaginyl-N-acetylglucosamine. However, it was also detected in released virus, albeit in very low molar ratio, so it is not known if it is a biosynthetic intermediate. Purified G protein, labeled with [^{14}C] fucose and [^{3}H] glucosamine (Fig. 6A) upon glycosidase digestion gave rise to 2 fucosyl glycopeptides and 3 glucosaminyl products (Fig. 6B). The larger of the 2 fucosyl peaks, also labeled with glucosamine and mol wt $\sim$ 1,600–1,800, was glycopeptide not digested with endo-β-N-acetylglucosaminidase D. Its presence is of interest because it reflects similar endoglycosidase resistance displayed by the uninfected host cells (1) and also quite different to the complete susceptibility of G protein of the same serotype grown in BHK21 cells (23). The smaller of the 2 [^{14}C] fucosyl glycopeptides, colabeled with [^{3}H] glucosamine was pooled and cochromatographed on Sephadex G-25 with standard glycopeptide markers (Fig. 6C). The peak

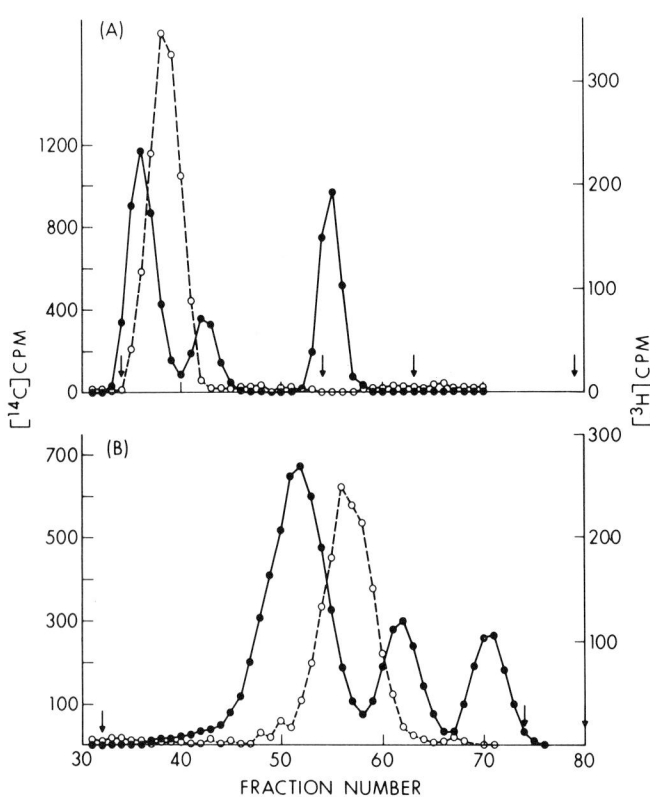

Fig. 4. Molecular weight of VSV fucosyl glycopeptides grown in HeLa cells. VSV labeled with [^{3}H]-fucose was purified, digested with pronase, and chromatographed on Sephadex G-50. The single included peak was pooled, lyophilized, and chromatographed on Sephadex G-25, and the included peak eluting just after the excluded volume mixed with [^{14}C]acetylated Thyroglobulin Unit B glycopeptide (mol wt 3,000), [^{14}C]acetylated Asn-(GlcNAc)$_2$(Man)$_5$ (mol wt 1,393), and [^{14}C]acetylated Asn-GlcNAc (mol wt 379). This mixture was chromatographed on Sephadex G-25. A) VSV glycopeptides (○ - - - ○); [^{14}C]acetylated markers (● - - - ●). The order of appearance of glycopeptide peaks is from left to right: [^{14}C]acetyl Thyroglobulin Unit A, [^{3}H]fucosyl G protein, [^{14}C]acetyl-Asn-(GlcNAc)$_2$(Man)$_5$, [^{14}C]acetyl-Asn-GlcNAc. The arrows show the positions of the standard substances from left to right: blue dextran (excluded), stachyose (mol wt 666) fucose (mol wt 164), and sodium azide. B) The same mixture chromatographed on Sephadex G-50. The peaks are from left to right as listed above. The arrows show the elution positions of blue dextran and fucose. NaN$_3$ elution position is not shown but was at fraction 80.

molecular weight of the [^{14}C]fucosyl glycopeptide was estimated to be 600–660, possibly corresponding to X-Asn-GlcNAc-Fuc, whereas the peak molecular weight of the glucosamine containing material was 500–550, possibly corresponding to X-Asn-GlcNAc. However, there must also have been some glucosamine-labeled oligosaccharide present (an expected product of the glycosidase mixture which contained endo-β-N-acetylglucosaminidase D) probably of the composition GlcNAc-(Man)$_n$ (23) because the bulk of the glucosamine-labeled material was neutral in HVPE at pH 1.9 (Fig. 6D). The bulk of the fucose-labeled material was basic, probably due to the presence of an amino acid such as Asn. Much of this fucose could be removed by mild acid hydrolysis (0.1 N HCl, 85°C,

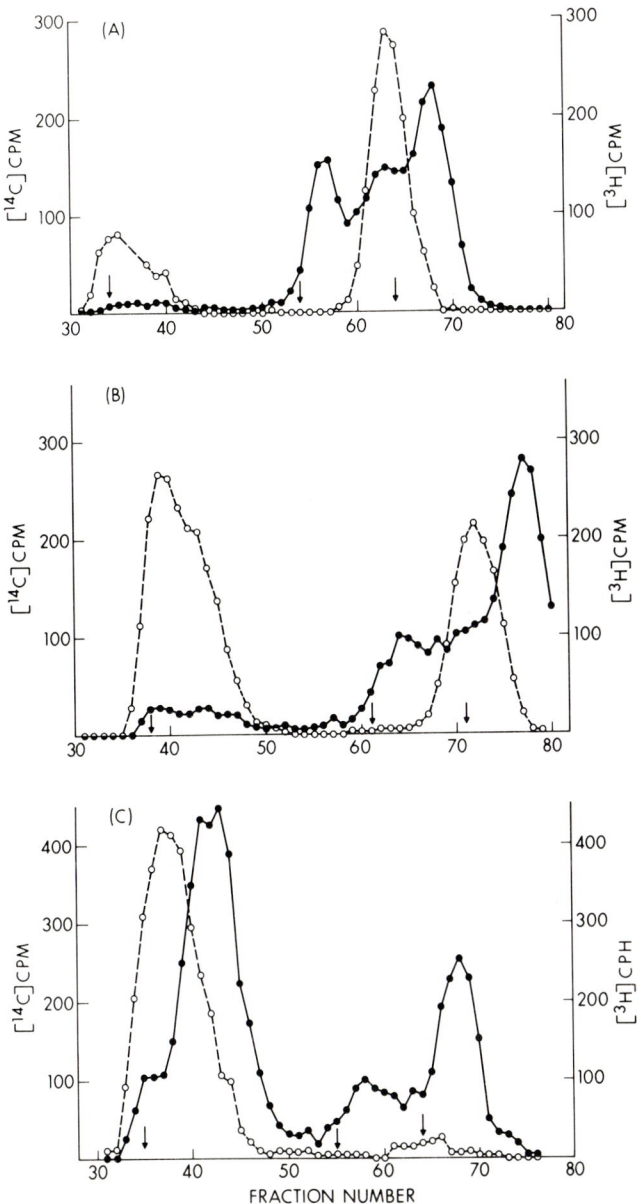

Fig. 5. Time course of synthesis of fucosyl and N-acetylglucosaminyl G protein glycopeptides. Cells, 9.1 × 10^8 cells at a density of 7.6 × 10^6 cells/ml, were infected approximately 5 h with VSV and then pulse labeled with 33 μCi/ml [^{3}H] fucose and 3.3 μCi/ml [^{14}C] glucosamine. At 5, 10, and 90 min of labeling, 20-ml samples were withdrawn and processed as if for plasma membrane preparation. However, after the homogenates had been centrifuged at 500 × g · min the pellet was discarded and the supernatants digested with pronase (10 mg/ml) for 3 days. The digests were chromatographed on Sephadex G-25. A) Five min [^{3}H] fucose-labeled glycopeptides (o - - - o); 5 min [^{14}C] glucosamine-labeled glycopeptides (●——●). The arrows from left to right show the elution positions of the standard substances: blue dextran, stachyose, and fucose. The 2 large peaks to the right of the diagram were found to contain labeled GDP-fucose, glucosamine, and fucose, and possibly other soluble intermediates as well (see Ref. 34). B) Ten min [^{3}H] fucose-labeled glycopeptides (o - - - o) and [^{14}C]-glucosamine-labeled glycopeptides (●——●). The arrows are the same as above. C) Ninety min [^{3}H] fucose-labeled glycopeptides (o - - - o) and [^{14}C] glucosamine-labeled glycopeptides (●——●).

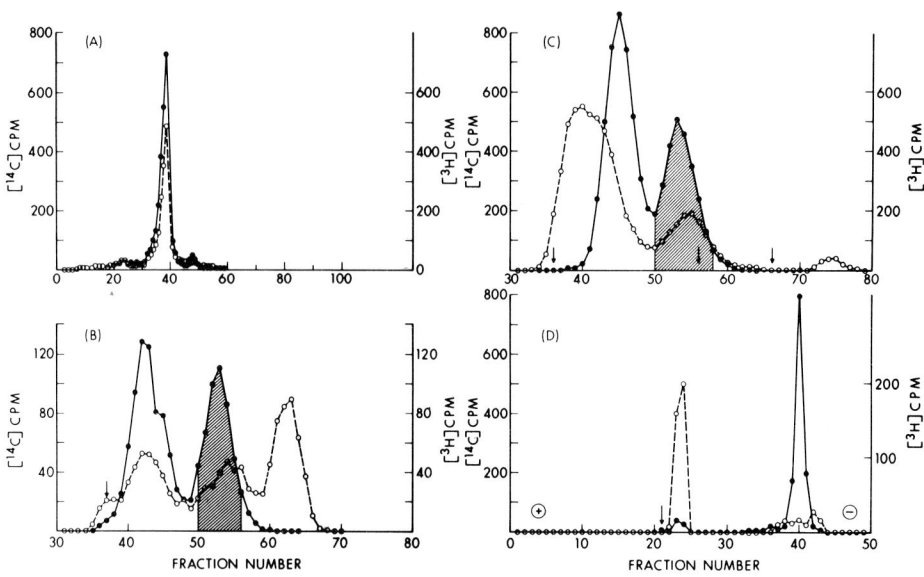

Fig. 6. Fucose in the linkage region of VSV glycopeptides. Cells, 3×10^8 cells at a density of 6×10^6 cells/ml, were infected 1.5 h with VSV then resuspended at 6×10^5 cells/ml. Five hours after infection the cells were labeled with 10 μCi/ml [^{3}H] glucosamine and 0.4 μCi/ml [^{14}C] fucose for 17 h. Virus was then purified from the supernatant medium by the standard methods involving precipitation with polyethylene glycol and banding isopycnically on a 7–52% wt/wt sucrose gradient. The purified virus was then dissociated with 2% SDS-1% mercaptoethanol and 0.01 M phosphate buffer, pH 7.0. A) The dissociated virus was electrophoresed on SDS 7.5% polyacrylamide gels, aliquots of the fractionated gel counted, and the radioactive band pooled for further analysis: [^{14}C] fucose-labeled G protein (●—●); [^{3}H] glucosamine-labeled G protein (○ - - - ○). B) The purified G protein was digested with pronase and then with endo-β-N-acetylglucosaminidase D, β-N-acetylglucosaminidase, β-galactosidase, and neuraminidase. The resultant digest was chromatographed on Sephadex G-25. [^{14}C] Fucose-labeled glycopeptides (●—●); [^{3}H] glucosamine-labeled glycopeptides and oligosaccharides (○ - - - ○). Arrow is elution position of blue dextran; note the double-labeled peak eluting near this which contains endo-β-N-acetylglucosaminidase D-resistant glycopeptides; the shaded area, containing digested material, was pooled for further treatment. C) Pooled material from the previous step was rechromatographed on Sephadex G-25 with standard glycopeptide markers: [^{3}H] acetyl-Thyroglobulin Unit B (mol wt 3,000) eluting in fraction 40; [^{3}H] acetyl-Thyroglobulin Unit A (mol wt 1,800) eluting in fraction 42 (both Thyroglobulin glycopeptides were generous gifts of Dr. A. M. Adamany, Department of Biochemistry) and [^{14}C] acetyl-Asn (GlcNAc)$_2$ (Man)$_5$ (mol wt 1,393) eluting in fraction 46. The peak elution positions of these markers have been determined on the same column when run singly. [^{14}C] Fucose-labeled glycopeptides (●—●, peak eluting at fraction 53); [^{3}H] glucosamine-labeled glycopeptides and oligosaccharides (○ - - - ○, peak eluting at fraction 55). The arrows show the elution position of the standard substances from left to right blue dextran, stachyose, and fucose. The shaded area is the material pooled ("linkage region") from the previous step. D) Pooled material from the column described in C) was subjected to high voltage paper electrophoresis at pH 1.9. The arrow marks the origin. The standard substance galactose runs approximately 3 fractions to the cathode under our conditions: [^{14}C] fucose-labeled glycopeptides (●—●); [^{3}H] glucosamine-labeled material (○ - - - ○).

1 h) indicating its terminal position. These data indicate that fucose is probably attached to the glucosamine of the protein CHO linkage, an assertion which is supported by the known specificity of endo-β-N-acetylglucosaminidase D (41). The existence of a small glycopeptide containing fucose, such as seen in Fig. 6C, argues also in favor of fucose being added much later in the synthesis of the entire oligosaccharide chains, since no rapidly labeled intermediate containing fucose of this size appeared (see Fig. 5A).

Internal Newly Synthesized M Protein

The supernatant fractions obtained at the top of sucrose gradients used to analyze the free and membrane-bound polysomes were prepared from VSV-infected HeLa cells and the distribution of newly synthesized, released G protein and M protein ascertained after a 1 min pulse, 1 min chase of [^{35}S] methionine. This regime of pulse-chasing was adopted for 2 reasons. First, released M and G protein could more readily be identified by their co-migration with VSV structural species; pulse labeling without a chase results in a high heterogenous background obscuring the main components. More importantly, the total time (2 min) is within the known transit time of M protein from its site of synthesis to the plasma membrane and hence M protein labeled in this way should substantially be in an internal pool (2). In addition to the soluble protein component of the sucrose gradient used to prepare the polysomes, the polysome region itself was analyzed on SDS-polyacrylamide gels. Released G protein did not significantly accumulate in the supernatant from which free polysomes were sedimented (Fig. 7A) whereas released M protein did (Fig. 7a) with a specific activity of 27.4 cpm/μg total protein. This experiment was repeated such that the fraction containing the free polysomes was sedimented on a 7–52% sucrose gradient in the absence of DOC to test whether M protein seen in the supernatant actually came from a membranous form contaminating the free polysomes and solubilized by DOC. Identical results were obtained and therefore the appearance of M protein in the supernatant from which the free polysomes were sedimented indicated that it is in a relatively non-sedimentable form and that its density, if particle bound, is less than 1.08 g/ml, i.e., less than that of the sucrose at the top of the gradient. M protein was still found in the supernatant to the free polysomes after a 30 min chase, but with a specific activity of only 13.7 cpm/μg protein. It was observed that no M protein could be detected in the supernatant of the gradient on which the membrane-bound polysomes were analyzed after a 1 min pulse, 1 min chase (Fig. 7C). In several other experiments there were minor amounts of M protein in this supernatant probably a contaminant insufficiently removed in the wash step prior to the polysome gradient. After 30 min chase M protein appeared with a relatively low specific radioactivity of 3.9 cpm/μg total protein (Fig. 7D). After a 1 min pulse, 30 min chase plasma membranes contain substantial amounts of labeled M protein which would mix and dilute with preformed unlabeled M protein. This cannot be so after a 1 min pulse, 1 min chase because little newly synthesized M protein is assembled into plasma membrane within 2 min of its synthesis (Ref. 2 and see above). From this belated appearance of M protein and its relatively low specific activity, it is likely that plasma membrane vesicles are present in both the cell supernatant from which the free polysomes were sedimented and also in the particulate fraction from which the membrane-bound polysomes were sedimented. Vesicles and fragments from other sources containing M protein could also be present. The appearance of released G protein in the supernatant of the gradient containing the membrane-bound polysomes after a 1 min pulse, 1 min chase (Fig. 7C) or in a 1 min pulse, 30 min chase (Fig. 7D) might similarly reflect the presence of intracellular membrane and plasma membrane containing G protein solubilized by DOC.

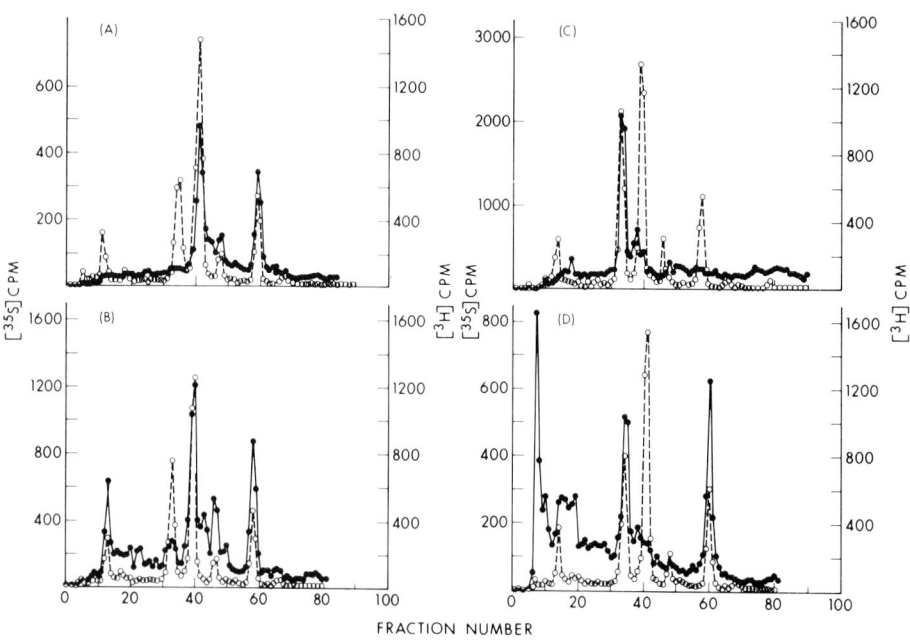

Fig. 7. Analysis of the soluble supernatants of the cell fraction from which free and membrane-bound polysomes were sedimented. Cells, 7.53×10^8 cells at a density of 5.6×10^6 cells/ml were infected 4 h with vesicular stomatitis virus, resuspended in fresh medium lacking methionine at a density of 2.6×10^7 cells/ml, incubated 15 min, and then pulse labeled 1 min with 50 μCi/ml [^{35}S]-methionine (results were essentially the same when these experiments were performed in medium containing complete methionine or at 0.05 normal methionine concentration). The label was then chased by adjusting the culture to 10 × L-methionine. Metabolism was stopped by placing the culture on frozen Earle's solution, after which the culture was washed by centrifugation in Earle's solution. The pellet was resuspended in 8.0 ml RSB and homogenized vigorously in a 0.002-inch clearance dounce homogenizer. Nuclei and the bulk of plasma membrane ghosts and fragments were sedimented at 1,600 rpm for 3 min. The postnuclear supernatant was centrifuged at $31,000 \times g$ for 30 min and this supernatant was saved for analysis of free polysomes. The pellet was resuspended in 6 ml RSB with 100 μg/ml heparin, and sedimented through a 15–30% wt/wt sucrose gradient at 25,000 rpm for 30 min in a SW27 rotor at 0°C. The pellets were resuspended in RSB, 100 μg/ml heparin, and 1% DOC, and sedimented on 7–52% wt/wt sucrose gradients, in RSB, 100 μg/ml heparin, for 16 h at 16,000 rpm at 1°C in a SW27 rotor. A typical free polysome profile was obtained by sedimenting the cell supernatants on 7–52% wt/wt sucrose gradient for 16 h, 16,000 rpm in a SW27 rotor. This profile was substantially similar whether or not 1% DOC was included in the supernatants and whether or not unlabeled methionine was included in the labeling medium of the above composition. The polysome region and the soluble region of the gradient were pooled for further analysis on SDS-polyacrylamide gels. The "membrane bound" polysome region and the soluble region of the gradient were similarly pooled for further analysis. ●—●) Pulse and chase of [^{35}S] methionine; ○ - - - ○) purified VSV marker labeled with [^{3}H] amino acids. The structural species of the marker virus from left to right are respectively: L, G, N, NS, and M proteins. A) Supernatant to the free polysomes 1 min pulse, 1 min chase. B) Supernatant to the free polysomes 1 min pulse, 30 min chase. C) Supernatant to the membrane-bound polysomes 1 min pulse, 1 min chase. D) Supernatant to the membrane-bound polysomes 1 min pulse, 30 min chase.

The overall conclusion from these observations is that the M protein in the supernatant from which free polysomes were sedimented after 1 min pulse, 1 min chase is an intracellular form of M protein and quite probably a soluble form (28). G protein by contrast is particle bound and not found in this fraction. When the cell supernatant (30,000 × g supernatant) containing released M protein pulse labeled 1 min with [^{35}S]-methionine and chased 1 min was centrifuged at 165,000 × g for 16 h, a substantial portion of M protein (40%) was again found to be nonsedimentable. The sedimented and the "soluble" M protein had nearly identical specific activities so it is not known if this represents a subfractionation of M protein. It was concluded from this section of the results that the bulk of M protein after a 1 min pulse, 1 min chase accumulated in the free polysome supernatant, possibly in the cytosol and if particle bound had a different density than the internal G protein labeled for the same period of time.

DISCUSSION

Addition of fucose to VSV G protein in HeLa cells is among the later steps in the synthesis of the molecule, and, on the addition of fucose the glycopeptide is near that of the completed size in the purified virus (mol wt ~ 2,000) as found in this paper. This size is a more accurate estimate than the estimate in our previous paper (mol wt ~ 1,800) (26) and is consistent with that of Robertson et al. (27) for HeLa spinner cells. It is of interest to note that the average mol wt of VSV glycopeptides from virus grown in BHK21 monolayer cultures is 3,150 daltons (23). G protein backbones are made on membrane-bound polysomes (24, 25) where the initial events in glycosylation possibly occur (36); the present data do not pertain to this point. However, fucose is not added to growing oligosaccharides in the polysomes because ^{14}C-labeled amino acid-labeled G protein (i.e., not necessarily glycosylated) continued to chase into the plasma membranes at least 45 min after fucosylated G had ceased to do so. This argues that though the transit time of fucosylated G is ~ 15–20 min, there is another G component which is only partially glycosylated (2). This glycoprotein has a transit time of not less than 75 min and hence G protein is being processed on its way to the plasma membrane at least 55 min prior to the addition of fucose. The mechanism and site of synthesis of protein backbone of this precursor to fucosyl G protein has been studied by others (36). Hunt and Summers (29) observed that core sugars, such as mannose and glucosamine, seemed to be added in the rough endoplasmic reticulum (er) whereas, fucose, a terminal sugar, is added in the smooth er. Their suggestion that the initial glycosylation events may be the en bloc addition of a mannose and glucosamine oligomer receives support from our observation (A. M. Adamany and P. H. Atkinson, unpublished) that a dolichol-bound oligosaccharide could be isolated from VSV-infected HeLa cells labeled for 30 min with [^{3}H]glucosamine. This material was extracted with chloroform:methanol:H_2O (10:10:3) and adsorbed on DEAE-cellulose (DE-52) from which it was eluted with an ammonium acetate in methanol gradient (30, 31). An N-acetylglucosamine-mannose-containing oligosaccharide could be released from the lipid saccharide after hydrolysis in 0.01N HCL/10% methanol at 100°C for 10 min and was found to have a molecular weight of approximately 2,000 as judged by gel filtration on Biogel P-6. Addition of fucose is near to the last events in the glycosylation of G protein and presumably could occur after addition of such a possible glucosamine containing oligosaccharide. There was no evidence of any small glycopeptide intermediates in the biosynthesis of fucosyl glycoprotein. The possibility of an intermediate containing one glucosamine only (see text and Fig. 5A) will have to be further investigated to decide on its validity. G, M, N, NS, and L proteins have been observed to differentially associate with different density membrane fractions. In these studies (34), M protein was found mainly in rough endoplasmic reticulum (density 1.17–1.20 g/ml) and plasma membrane (density 1.15–1.17 g/ml) in a 2–4 min pulse whereas G protein was found mainly in the rough endoplasmic reticulum, but eventually (8 min on) became associated with the light membranes (density 1.12–1.15 g/ml). This latter association is probably similar in time to that posttranslational time when fucose is added to the internal G protein [and uninfected cell glycoprotein, Yurchenco and Atkinson (35)]. The light membrane fraction of Hunt et al. (34) may be similar to the internal membranous component bearing fucosyl glycoprotein of density 1.14 g/ml observed here (Fig. 3).

Our previous paper (2) and that of David (33) showed that there was a small internal pool of M protein precursor to plasma membrane M protein but it was not further characterized. "Soluble" M protein detected in infected cells by us (3, 28) and others could be a variety of forms of M protein including that which was removed from membrane material during cell disruption, and that which never assembles into plasma membranes: we have found in HeLa cells that most of the cellular M protein does not assemble into plasma membranes (unpublished data). It is possible that the transit time of M is time required for some posttranslational modification and that once the modification occurs M enters a pool having 2 compartments that are in rapid equilibrium — a soluble compartment and a membrane-bound compartment. In this interpretation, pulse-labeled M protein would first appear in the soluble pool then transfer to the plasma membrane, mix with unlabeled M protein already in the plasma membrane prior to the onset of labeling, and then transfer back to the soluble pool: The decay of labeled M protein from the plasma membrane seen in Fig. 1 might thus be reequilibration of the plasma membrane M protein back to the soluble pool which is, as a function of chasing the labeled amino acids, getting "cold" M in the chase. There are several points weighing against this interpretation. First, in order for the internal pool to become "cold" in the chase it would have to significantly expand with incoming unlabeled M protein diluting out labeled M protein returning from the plasma membrane. This would imply a steadily expanding internal pool throughout the infectious cycle, including during pulse label accumulation without a chase. In this interpretation, M protein should accumulate either in isolated plasma membranes or in released virus with kinetics consistent with such an expanding precursor pool. Mathematical description of these pools and the expected rates of accumulation of products coming from them was discussed in our previous paper (35): Molecules coming from such an expanding pool would accumulate in a "product" pool with exponential kinetics. There is little evidence for this, and, on the contrary, M protein accumulates in plasma membranes (2) and in released virus (34) with linear kinetics consistent with a small rapidly equilibrated internal pool which does not expand during the term of the experiment. In addition, there is convincing evidence that from about 2 h after infection both the rate of synthesis of M protein and the amount of M protein in the cell steadily decreases in a several hour chase (27a). Secondly, there is evidence that M protein is lost from the cell (Fig. 1b) in amounts more than sufficient to account for the decay process seen from the plasma membranes (Fig. 1a) consistent with accumulation of

M protein in released virus previously observed in other laboratories (27a, 34). Third, the initial rate of loss may seem to resemble the rate at which the M chased into the plasma membrane and therefore seem to be an equilibrium process. However, it can readily be determined that when the last 7 or 8 points on the chase curve are plotted on a logarithmic scale (for specific activity) against time the decay is monophasic, linear, with a half time for turnover of 80–85 min. That is to say, half the newly synthesized M protein inserted into the plasma membrane during the initial 10 min of the chase, is mixed with preexisting M protein then randomly lost from the cell in 80–85 min. Such a rate of turnover would not seem to be inconsistent with virus "bud" formation in the plasma membrane noting at the same time that we have no idea how fast this procedure is. Though in vitro association of M protein with plasma membranes during cell disruption was shown able to occur (3), it does not do so in significant quantities to that already inserted in vivo as determined by mixing-dilution experiments (2). It should be emphasized that the plasma membranes utilized in these studies are ghost structures which can be expected to contain all intermediate forms of budding virus. Plasma membranes purified as vesicles from vigorously disrupted cells may have lost both the intermediate forms as well as bound M protein itself. These studies clearly show an insertion, and random mixing of newly synthesized M protein with preexisting M protein. This mixing may be mixing of individual preexisting M molecules with newly synthesized M or mixing of partially completed aggregates containing newly synthesized and preexisting M protein. The data in this paper do not distinguish these possibilities beyond that mixing of some sort must occur before loss of the M protein from the cell.

The current experiments were designed to further characterize the internal pool of M protein precursor to plasma membrane and the labeling conditions were carefully chosen such that the labeled newly synthesized M protein would mostly be internal as distinct from plasma membrane associated M protein. Furthermore, newly synthesized M protein first appeared in the supernatant from which the free polysomes were sedimented, consistent with previous findings that M protein mRNA is found in the free polysomes of VSV-infected cells (24, 25), and not in the supernatant to the membrane bound polysomes. David (40) has also presented evidence that M protein passes through a soluble phase before being added to the plasma membranes. The current studies thus show that there is an internal "soluble" M protein pool at least some of which must be on its way to assembly in plasma membrane and some of its characteristics described in this study distinguish it from internal fucosyl G protein bearing membrane also on its way to assembly in the plasma membrane. Different routes of assembly for fucosyl G protein and M protein can now be argued from the following facts: that fucosyl glycoprotein, representing a practically completed molecule (the state of completion can be determined from Figs. 4, 5, and 6) has an approximate 20 min transit time to the plasma membrane from the site of addition of fucose and that M protein has a less than 5 min transit time from its site of synthesis (this conclusion is obtainable from Fig. 1); that fucosyl G protein is on an internal membranous component with a density of 1.14 g/ml (derived from Fig. 3) as compared with its final density in plasma membrane (1.16 g/ml); most M protein within its transit time (less than 3 min; see Fig. 2) to the plasma membrane can be observed is soluble form and, if particle bound at all, has a density of not more than 1.08 g/ml. The internal form of M protein precursor to plasma membranes is, therefore, quite different from the component to which fucosyl glycoprotein is bound. These facts together allow for differentiation of assembly routes shown in the model (Fig. 8) for pathways of assembly of these components. The model shows a hypothetical vesicle containing G proteins (knobbed line) with the nonglycosylated end associated with the lipid bilayer [see Mudd (37) for evidence of this orientation and a review of the background evidence and see also Ref. 39]. It also shows M protein (shaded area) in some undefined association with the plasma membrane. A general proviso to Fig. 8 must be that there is no evidence that fucosyl G protein is on a vesicle, as depicted, or that the fucosyl G protein has the orientation shown. In mouse colonic epithelial cells, there is evidence of such a vesicle carrying fucosyl glycoprotein to the cell surface (39). For fucosyl G protein, however, we do know that it is initially on an internal membranous component and that it ends up on the outside of the virus, and therefore, presumably on the outside of the cell prior to virus maturation. Furthermore, though a concentrated "patch" of M and G protein is shown in this set of figures there is no evidence for its initial existence. It is inferred that such a patch must at some stage exist during virus morphogenesis because the virus ends up free of the cell with the equivalent of such a "patch" forming its membranous envelope. However, the mechanisms of patch formation and exclusion of host proteins are not known and are not intended to be the subject of this paper The data in this paper, just summarized above, allow us to distinguish between the 3 alternative pathways of assembly shown in Fig. 8. This study and our previous study (2) would rule out the possibility shown in Fig. 8A because it is inconsistent with the observed kinetics: M and fucosyl G protein do not have the same transit time from the site of synthesis to the plasma membrane as the model implies they do. Neither do they have the same density because G is membrane bound and M appears soluble. The possibility shown in Fig. 8B is unlikely on the basis of the present data: intracellular G protein, once fucosylated and completed, is located on an internal membranous component of density 1.14 g/ml in sucrose and internal (1 min pulse, 1 min chase). Most M protein is in the cell supernatant (also the supernatant from which free polysomes were sedimented) having a peak density in sucrose of 1.08 g/ml. Thus, if M protein is added to a membranous component carrying fucosyl G protein to the plasma membrane prior to assembly into the plasma membrane, the subsequent entry of the conglomerate must be instantaneous, otherwise a sedimentable form of newly synthesized M protein similar to fucosyl G protein should be observed. There is little evidence for this and thus direct addition of M protein from its internal pool to the plasma membrane (Fig. 8C) would seem to explain the observations made in this paper most satisfactorily and also those made on the in vitro binding of M protein to isolated plasma membranes (3, 28). The models are not intended to distinguish the order in which G protein and M protein reach the membrane. However, it is noted (27a) that M and G protein are made at about the same rate from the start of the infectious cycle and thus if M is rapidly synthesized and inserted it may, in fact, be the first VSV protein to modify the plasma membrane. However, the present data do not prove this one way or another but do allow for differentiation of the independent routes G and M proteins take into the plasma membrane. Hay (5) has also proposed the direct addition to plasma membranes of influenza virus matrix protein. Different pathways of assembly have also been proposed for NP protein and M protein in Sendai virus; the latter also being assembled into virions from a small cytoplasmic

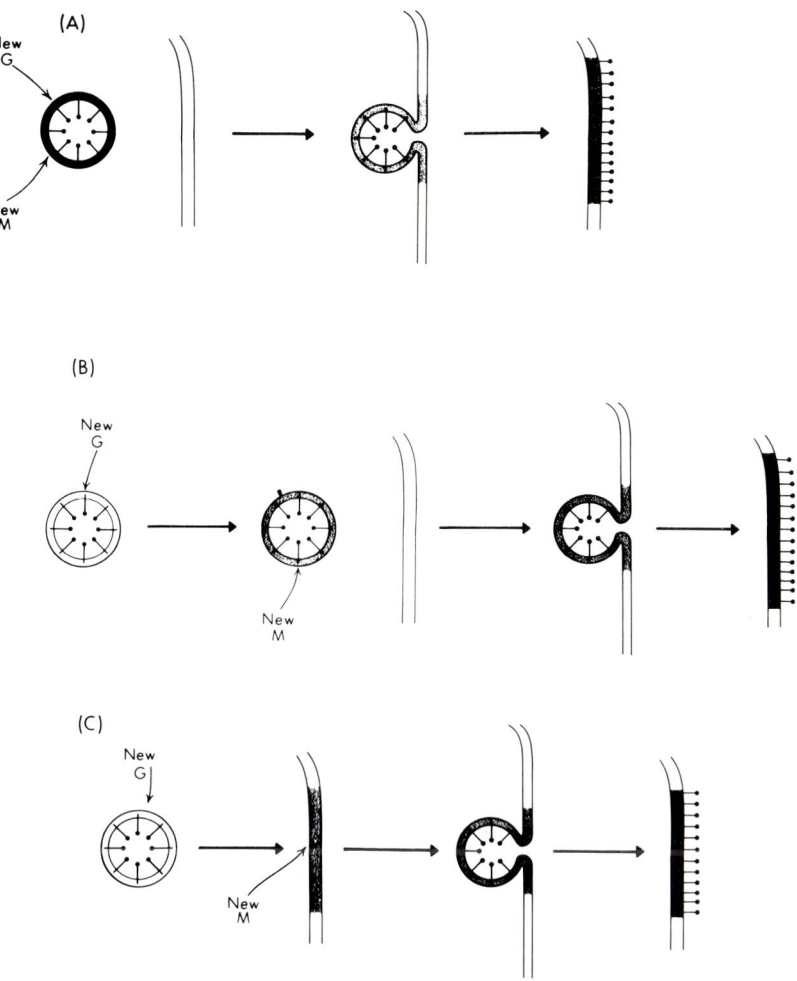

Fig. 8. Three modes of adding newly synthesized fucosyl G protein and newly synthesized M protein to membrane and plasma membrane. In all panels, G protein is represented by a spike and knob, M protein is represented by the shaded area. An internal membranous component is represented by the double membranes circle. However, there is no evidence it is a vesicle; neither is there evidence for the orientation of the G protein. There is also no presumption that the "patch" of M and G protein shown in the plasma membrane (double membranes open structure) assembles in the manner or in the order shown. The M and fucosyl G protein probably present in the plasma membrane prior to the onset of labeling are not shown in these diagrams for sake of clairty. It is possible M and G protein reach the plasma membrane simultaneously or at different times. This model is not intended to distinguish this and the diagram is intended to aid distinction of route only of newly synthesized M protein and newly synthesized fucosyl G protein into the plasma membrane.

pool (32). Hunt et al. have proposed different pathways of assembly for the assembly of VSV components into HeLa cell plasma membranes (34) by different cell fractionation techniques to those used in this study. What sort of interactions M protein would undergo in transition from the relatively nonsedimentable phase to tight binding in the plasma membranes demonstrated (3, 28) is entirely unknown.

ACKNOWLEDGMENTS

This work was supported by U.S. Public Health Service Research Grants CA13402 and CA06576. The author is an Established Investigator of the American Heart Association. I thank Ms. Joyce Tsang for superb technical assistance and Dr. Takashi Muramatsu, Department of Biochemistry, Albert Einstein College of Medicine, Bronx, New York for glycopeptide substrates of known composition.

REFERENCES

1. Atkinson PH: J Biol Chem 250:2123, 1975.
2. Atkinson PH, Moyer SA, Summers DF: J Mol Biol 102:613, 1976.
3. Cohen GH, Atkinson PH, Summers DF: Nature (London) New Biol 231:121, 1971.
4. Hay AJ: Virology 60:398, 1974.
5. Witte ON, Weissman L: Virology 69:464, 1976.
6. Nagai Y, Ogura H, Klenk HD: Virology 69:523, 1976.
7. Meier-Ewert H, Compans RW: J Virol 14:1083, 1974.
8. Portner A, Kingsbury DW: Virology 73:79, 1976.
9. Eagle H: Science 130:432, 1959.
10. Levine EM, McGregor D, Hayflick L, Eagle H: Proc Natl Acad Sci USA 60:583, 1968.
11. Levine EM: Exp Cell Res 74:99, 1972.
12. Mudd JA, Summers DF: Virology 42:328, 1970.
13. Atkinson PH, Summers DF: J Biol Chem 246:5162, 1971.
14. Atkinson PH: In Prescott D (ed): "Methods in Cell Biology." New York: Academic Press, 1973, vol 7, pp 158–188.
15. Costantino-Ceccarini E, Novikoff PM, Atkinson PH, Novikoff AB: J Cell Biol 77:in press, 1978.
16. Ceccarini C, Eagle H: Proc Natl Acad Sci USA 68:229, 1971.
17. Maizel JV Jr: Science 151:988, 1966.
18. Grubman MJ, Weinstein JA, Shafritz DA: J Cell Biol (In press).
19. Penman S: J Mol Biol 17:117, 1966.
20. Muramatsu T, Atkinson PH, Nathenson SG, Ceccarini C: J Mol Biol 80:781, 1973.
21. Wagner RR, Kiley MP, Snyder RM, Schnaitman CA: J Virol 9:672, 1972.
22. Witte ON, Weissman IL: Virology 69:464, 1976.
23. Etchison JR, Robertson JS, Summers DF: J Supramol Struct (Suppl) 1:6, 1977.
24. Grubman MJ, Moyer SA, Banerjee AK, Ehrenfeld E: Biochem Biophys Res Commun 62:531, 1975.
25. Morrison TG, Lodish HF: J Biol Chem 250:6955, 1975.
26. Moyer SA, Tsang JM, Atkinson PH, Summers DF: J Virol 18:167, 1976.
27. Robertson JS, Etchison JR, Summers DF: J Virol 19:871, 1976.
27a. Kang CY, Prevec L: Virology 46:678, 1971.
28. Cohen GH, Summers DF: Virology 57:566, 1974.
29. Hunt LA, Summers DF: J Virol 20:646, 1976.
30. Spiro RG, Spiro MJ, Adamany AM: In Smellie RMS, Beeley JG (eds): "Biochemical Society Symposium No. 40." New York: Academic 1974, pp 37–56.
31. Adamany AM, Spiro RG: J Biol Chem 250:2842, 1975.

32. Famulari NG, Fleissner E: J Virol 17:605, 1976.
33. David AE: J Mol Biol 76:135, 1973.
34. Hunt LA, Summers DF: J Virol 20:637, 1976.
35. Yurchenco PD, Atkinson PH: Biochemistry 16:944, 1977.
36. Katz F, Wirth D, Lodish H: J Supramol Struct (Suppl) 1, 1977.
37. Mudd JA: Virology 62:573, 1974.
38. Schloemer RH, Wagner RR: J Virol 16:237, 1975.
39. Michaels JE, Leblond CP: J Microsc Biol Cell 25:243, 1976.
40. David AE: Virology 76:98, 1977.
41. Koide N, Muramatsu T: J Biol Chem 249:4897, 1974.

Receptor-Mediated Uptake of Lysosomal Enzymes

William S. Sly, Arnold Kaplan, Daniel T. Achord, Frederick E. Brot, and C. Elliott Bell

Departments of Pediatrics, Medicine, and Laboratory Medicine, Washington University School of Medicine, and Division of Medical Genetics, St. Louis Children's Hospital, St. Louis, Missouri 63110

This paper reviews the evidence for two different forms of carbohydrate-mediated recognition of lysosomal enzymes by cell surface receptors. The recognition marker on lysosomal enzymes which are rapidly pinocytosed by fibroblasts contains phosphate, probably linked to D-mannose as a phosphomannosyl moiety. The fraction of lysosomal hydrolases bearing this recognition marker for fibroblasts varies, depending on the tissue source.

Another form of recognition accounts for rapid plasma clearance of infused lysosomal enzymes in the rat. This rapid clearance is mediated by receptors on Kupffer cells and other reticuloendothelial cells that recognize mannosyl residues on the glycoprotein hydrolases.

Key words: receptor, pinocytosis, phosphoglycoprotein, lysosomal enzyme, glycoprotein, mannosyl, recognition

Most mammalian lysosomal hydrolases are glycoproteins. The intracellular role of many of these enzymes in the breakdown of mucopolysaccharides, glycolipids, and glycoproteins has been elucidated through studies of inherited lysosomal storage diseases (1). In these disorders, deficiency of a single enzyme disrupts the catabolic process in which many hydrolases participate in the sequential degradation of macromolecules in lysosomes. The pathological manifestations of the different disorders result from the direct and indirect effects of accumulation of undigested material in the lysosome. Whether these hydrolases also have an extracellular role is unclear. What has become clear, though, is that lysosomal hydrolases do exist in extracellular fluids in some situations, and that the carbohydrate structure of the enzymes is important in their uptake by cells and their proper intracellular localization. In this paper, we review the evidence for two different forms of carbohydrate-dependent recognition and uptake of lysosomal enzymes by mammalian cells. One form of recognition involves pinocytosis receptors on fibroblasts which recognize phosphohexosyl components on some forms of lysosomal enzymes. A second form of recognition accounts for clearance of lysosomal enzymes from the circulation. This clearance is mediated by reticuloendothelial cells including hepatic Kupffer cells and involves recognition of mannosyl components on certain forms of lysosomal hydrolases.

Received May 16, 1977; accepted May 16, 1977.

© 1978 Alan R. Liss, Inc., 150 Fifth Avenue, New York, NY 10011

Like all soluble macromolecules, lysosomal enzymes are internalized at a slow rate through fluid endocytosis, also called bulk-phase endocytosis. No specific form of recognition is involved in this slow form of cellular uptake of macromolecules. The two recognition systems described below mediate uptake of lysosomal enzymes by specific uptake processes which exceed the rate of nonspecific uptake by 100 to 1000-fold.

PHOSPHOHEXOSYL-COMPONENTS ON LYSOSOMAL ENZYMES ARE RECOGNIZED BY PINOCYTOSIS RECEPTORS ON HUMAN FIBROBLASTS

Pinocytosis of lysosomal enzymes by human fibroblasts was initially recognized as uptake of corrective factors for enzyme-deficient human fibroblasts (2). The kinetics of uptake of lysosomal hydrolases has since been studied by a number of investigators with enzymes from different sources, and the uptake process was found to display the selectivity and saturability expected for a receptor-mediated process (3–6). Hickman and Neufeld (7) suggested that many lysosomal hydrolases have a similar recognition marker which is essential to their uptake by human fibroblasts. This suggestion was based on the observation that fibroblasts from patients homozygous for a single gene mutation (I-cell disease) secrete several hydrolases which are not specifically pinocytosed. Later, they reported that periodate treatment of the β-hexosaminidase secreted by normal human fibroblasts destroyed its uptake activity (8). This finding led them to suggest that the recognition marker common to many hydrolases, and presumed defective or masked in enzymes from I-cell disease patients, may be a carbohydrate. Support for this suggestion was provided by Hieber et al (4) who found weak inhibition of pinocytosis of bovine β-galactosidase by certain sugars, and that treatment of this enzyme with partially purified preparations of α-mannosidase reduced its susceptibility to pinocytosis.

We used human β-glucuronidase to study this pinocytosis process. The initial studies indicated that human β-glucuronidase from different tissue sources varied widely in the amount of catalytic activity that was taken up by human fibroblasts (9). The "corrective activity" of β-glucuronidase preparations for the mucopolysaccharide storage abnormality of β-glucuronidase deficient fibroblasts correlated well with the fraction of the catalytic activity taken up from the culture medium by the fibroblasts (9). These observations implied that there were "high-uptake" and "low-uptake" forms of the enzymes in different proportions in β-glucuronidase from different tissue sources and suggested that fibroblasts have a pinocytosis receptor that recognizes high-uptake forms of the enzyme. Extracts of blood platelets and fibroblast secretions were particularly rich in the high-uptake forms of the enzyme (9, 10). When fibroblasts from many mammalian species were examined, most were found to take up high-uptake β-glucuronidase from platelets, implying that the uptake process is a broad biological phenomenon shared by many animal species (11).

Considerable charge heterogeneity was observed in β-glucuronidase preparations from different tissues that were examined by isoelectric focusing (12). However, the high-uptake forms of the enzyme had similar isoelectric properties, regardless of the source of enzyme, and were more acidic than low-uptake, poorly pinocytosed forms of the enzyme. When the acidic, high-uptake fractions of β-glucuronidase from platelet extracts were taken up by cultured fibroblasts, they were converted to less acidic, low-uptake forms (12). Since neuraminidase treatment of high-uptake enzyme did not diminish the uptake activity of high-uptake enzyme, the high-uptake property of acidic enzyme could not be attributed to sialic acid residues on the enzyme.

Recent experiments from this laboratory (5) have implicated phosphohexose in the recognition marker common to lysosomal hydrolases taken up by cultured fibroblasts. This conclusion rests on three lines of evidence. First, alkaline phosphatase treatment of high-uptake forms of β-glucuronidase was shown to destroy the high-uptake property of the enzyme without diminishing its catalytic activity. This treatment also converts some forms of the heterogeneous enzyme to less acidic forms. The effect of alkaline phosphatase on the enzyme was extremely sensitive to inhibition by phosphate. The second line of evidence was provided by the structural requirements of competitive inhibitors of enzyme pinocytosis. Weak inhibition of pinocytosis was observed for D-mannose and other sugars which resemble D-mannose at the 2 and 4 position. However, D-mannose-6-phosphate was 1000-fold more potent an inhibitor of pinocytosis than D-mannose. Yeast mannans which contain phosphomannose were far more potent inhibitors of pinocytosis than yeast mannans which do not contain phosphate. The third line of evidence is that platelet glycoproteins from which β-glucuronidase has been removed are potent inhibitors of pinocytosis of β-glucuronidase uptake, but that their inhibitory activity was destroyed by treatment with alkaline phosphatase.

Thus, although phosphohexose on the enzyme has not yet been demonstrated directly, the evidence is persuasive that phosphate is present on the enzyme in monoester linkage. The information provided by competitive inhibitors of enzyme pinocytosis suggests that the phosphate is linked to D-mannose type carbohydrate on the enzyme. Until recently, no precedent could be cited for phosphomannose containing glycoproteins from mammalian sources, although yeast mannans provided clear examples (13). Phosphate-containing glycoproteins have been isolated from rat brain recently, and the analysis of glycopeptides from these glycoproteins revealed phosphomannose residues in the glycopeptides (14–16).

Is phosphohexose recognition limited to one enzyme and one tissue source? Recent studies suggest that this form of recognition marker is general for many, if not all, high uptake lysosomal hydrolases (27). Inhibition of pincytosis of β-glucuronidase by D-mannose-6-phosphate has been demonstrated for enzyme isolated from liver, spleen, urine, and placenta. Similarly, alkaline phosphatase treatment of β-glucuronidase from each of these sources greatly reduced the susceptibility of the enzyme to pinocytosis by fibroblasts without diminishing the catalytic activity. Pinocytosis of β-hexosaminidase from platelet extracts was also sensitive to inhibition by D-mannose-6-phosphate. That this form of recognition was not limited to enzyme of platelet origin was convincingly shown when β-hexosaminidase was isolated from fibroblast secretions and found to have similar properties. This secretory form of hexosaminidase had very high uptake activity, but its pinocytosis was sensitive to inhibition by D-mannose-6-phosphate, and its susceptibility to pinocytosis by fibroblasts was completely destroyed by treatment with alkaline phosphatase. Neufeld et al (6) have recently extended these observations to another hydrolase, α-L-iduronidase from human urine.

The recognition of a phosphohexose group on a glycoprotein by the pinocytosis receptor on fibroblasts represents a new form of ligand-receptor interaction which is involved in pinocytosis of certain forms of many lysosomal enzymes. The distribution of this receptor on other mammalian cell types, and the physiological significance of this uptake system remains unclear. Fibroblasts from patients with I-cell disease have normal receptor activity for normal human enzymes but secrete recognition defective enzymes (7) in which the phosphohexosyl moiety can be assumed to be masked or defective.

MANNOSYL-DEPENDENT CLEARANCE OF HUMAN β-GLUCURONIDASE BY RAT KUPFFER CELLS

Several systems for clearance of mammalian glycoproteins from the circulation have been described or suggested. These include the well characterized hepatocyte receptor for asialo-glycoproteins (17), a system in avian (18) and rat (19) liver for clearance of agalacto-glycoproteins presumed to terminate in N-acetylglucosamine, and a system for clearance of mannosyl terminal aglycosyl-antibody (20) and RNase B (21). Stahl and co-workers (22, 23) found that a variety of rat lysosomal enzymes was cleared rapidly from rat plasma following infusion, and that periodate treatment of these enzymes markedly delayed their clearance. Simultaneously infused agalacto-orosomucoid dramatically slowed clearance of several rat lysosomal enzymes from rat plasma. These findings led them to attribute clearance of these hydrolases to the system described for agalacto-orosomucoid, and to suggest that N-acetylglucosamine was part of the recognition site on the enzymes which were cleared by this system.

We have developed a rat model system to study the clearance and fate of infused human placental β-glucuronidase (24). The fate of infused human enzyme can be followed in the rat because the human enzyme is completely stable to thermal inactivation conditions which destroy the endogenous rat β-glucuronidase. It is worth emphasizing that human placental β-glucuronidase is predominantly low-uptake enzyme for human and rat fibroblasts, although it is rapidly cleared from rat plasma ($t^{1/2}$ = 3.5 min) following infusion. Enzyme localizes predominantly in rat liver (50–60% of the infused dose) and spleen (3–5%). Subcellular fractionation of liver 18 h after infusion indicated that enzyme cleared by the liver was in lysosomes. Indirect immunofluorescence indicated that most, if not all, of the human enzyme in rat liver was in Kupffer cells (25). The half-life of the heat-stable human β-glucuronidase in liver was 2.6 days and in spleen was 5.8 days. Removal of the liver and spleen prior to the infusion dramatically slowed the clearance of infused enzyme from plasma ($t^{1/2}$ = 60 min) and allowed a large amount of the enzyme to localize in bone. Periodate treatment of the human placental enzyme reduced its binding to concanavalin A Sepharose and converted the enzyme into a very slow clearance form ($t^{1/2}$ > 8 h) (24). These studies suggested that there was a receptor on liver Kupffer cells and other reticuloendothelial cells which recognizes the carbohydrate structure on the enzyme and mediates its clearance from plasma.

The nature of the carbohydrate specificity involved in the clearance was studied using simultaneously infused glycoproteins and sugars as inhibitors of clearance (25, 26 Achord, Brot, Sly and Bell, manuscript in preparation). Asialo-glycoproteins had no effect on clearance. Agalacto-orosomucoid did inhibit clearance, α-methyl mannoside, D-mannose, and L-fucose were all inhibitors of clearance in the nephrectomized rat, while glucose, galactose, and N-acetylglucosamine were without effect. These studies suggested that mannosyl (or fucosyl) recognition was involved in the clearance of human placental β-glucuronidase. The finding that RNase B dramatically inhibited clearance of the enzyme in the nephrectomized rat, while RNase A had no effect on clearance provided further support for mannosyl involvement in the recognition.

If mannosyl-recognition were involved in clearance of infused human placental β-glucuronidase, it was initially puzzling that agalacto-orosomucoid, whether prepared enzymatically or by the periodate treatment method, was also an effective inhibitor of clearance. This led us to study the clearance properties of yeast mannans and agalacto-orosomucoid in the rat. These studies showed that yeast mannans and agalacto-orosomucoid each inhibit the clearance of tracer doses of the other (26). When simple sugars were tested for inhibition of clearance of agalacto-orosomucoid, α-methylmannoside and D-mannose were the most potent inhibitors of clearance of ^{125}I-agalacto-orosomucoid. Thus, even though agalacto-orosomucoid is an N-acetylglucosamine terminal glycoprotein, mannose recognition appears to account for a substantial part of its clearance following infusion into the rat. This explains why agalacto-orosomucoid inhibits clearance of mannyosyl-terminal glycoproteins from rat circulation.

Stahl and co-workers (P. Stahl, personal communication) have recently examined the fate of infused RNase B in the rat and found that ^{125}I-RNase B infused into nephrectomized rats localizes in liver. When parenchymal and Kupffer cells were separated, the RNase B in liver was elevenfold enriched in the Kupffer cell fractions. Their studies support the evidence derived from our studies with infused human β-glucuronidase that there is a glycoprotein clearance system in rat Kupffer cells (and probably other reticuloendothelial cells) that recognizes mannosyl groups on certain glycoproteins and leads to their uptake in vivo.

ACKNOWLEDGMENTS

This work was supported by NIH Grant GM 21096 and a grant from the Ranken Jordan Trust for Crippling Diseases in Children.

REFERENCES

1. Neufeld EF, Lim TW, Shapiro LJ: Annu Rev Biochem 44:357, 1975.
2. Neufeld EF, Cantz M: Ann NY Acad Sci 179:580, 1971.
3. von Figura K, Kresse H: J Clin Invest 53:85, 1973.
4. Hieber V, Distler J, Meyerowitz R, Schmickel RD, Jourdian GW: Biochem Biophys Res Commun 73:710, 1976.
5. Kaplan A, Achord DT, Sly WS: Proc Natl Acad Sci USA 74, 1977.
6. Neufeld EF, Sando GN, Garvin AJ, Rome LH: J Supramol Struc 6:000, 1977.
7. Hickman S, Neufeld EF: Biochem Biophys Res Commun 49:992, 1972.
8. Hickman S, Shapiro LS, Neufeld EF: Biochem Biophys Res Commun 57:55, 1974.
9. Brot FE, Glaser JH, Roozen KJ, Sly WS, Stahl PD: Biochem Biophys Res Commun 57:1, 1974.
10. Hall CW, Cantz M, Neufeld EF: Arch Biochem Biophys 155:32, 1973.
11. Frankel HA, Glaser JH, Sly WS: Pediatr Res 11:811, 1977.
12. Glaser JH, Roozen KJ, Brot FE, Sly WS: Arch Biochem Biophys 166:536, 1975.
13. Ballou CE, Raschke WC: Science 184:127, 1974.
14. Davis L, Javaid JI, Brunngraber EG: FEBS Lett 65:30, 1976.
15. Davis L, Costello JR, Javaid JI, Brunngraber EG, FEBS Lett 65:35, 1976.
16. Davis L, Brettschneider I, Brunngraber EG: Feb Proc Fed Am Soc Exp Biol 36:750, 1977.
17. Ashwell G, Morell AG: In Meister A (ed): "Methods in Enzymology." New York: John Wiley & Sons, 1974, vol 41, pp 99–128.
18. Lunney J, Ashwell G, Proc Natl Acad Sci USA 73:341, 1976.
19. Stockert RJ, Morell AG, Scheinberg IH: Biochem Biophys Res Commun 68:988, 1976.
20. Winkelhake JL, Nicolson GL: J Biol Chem 251:1074, 1976.
21. Baynes JW, Wold F: J Biol Chem 251:6016, 1976.
22. Stahl P, Schlesinger PH, Rodman JS, Doebber T: Nature (London) 264:86, 1976.
23. Stahl P, Six H, Rodman JS, Schlesinger P, Tulsiani DRP, Touster O: Proc Natl Acad Sci USA 73:4045, 1976.
24. Achord D, Brot F, Gonzalez-Noriega A, Sly W, Stahl P: Pediatr Res 11:816, 1977.
25. Achord DT, Brot FE, Bell CE, Sly WS: Fed Proc Fed Am Soc Exp Biol 36:653, 1977.
26. Achord DT, Brot FE, Sly WS: Biochem Biophys Res Commun 77:409, 1977.
27. Kaplan A, Fischer D, Achord D, Sly W: J Clin Invest 1977 (In press).

A New Approach to the Structural Determination of Glycoproteins and Polysaccharides: Anhydrous HF Solvolysis

Andrew J. Mort

MSU/ERDA Plant Research Laboratory, and Department of Biochemistry, Michigan State University, East Lansing, Michigan 48824

From experiments with glycoproteins containing the glycopeptide linkages, arabinose-O-hydroxyproline and galactose-O-serine (plant cell wall glycopeptides), N-acetylgalactosamine-O-serine/threonine (pig submaxillary mucin), and N-acetylglucosamine-N-asparagine (fetuin), it is apparent that anhydrous liquid HF, a reagent commonly used by synthetic peptide chemists for the complete removal of protecting groups from synthetic peptides, cleaves the O-glycosidic linkages of neutral sugars in 1 hr at 0°C, and the O-glycosidic linkages of amino sugars in 3 hr at 23°C. The N-glycosidic linkage of N-acetylglucosamine to asparagine is not cleaved under any conditions that have been tested. Sodium dodecyl sulfate gel electrophoresis of bovine serum albumin treated in HF does not show any degradation of peptide bonds. Some relatively stable enzymes (lysozyme and RNase) have been shown by others to retain most of their enzymic activity after short treatment (1 hr at 0°C) in HF.

With the specificity of HF at 0°C for neutral sugars it should be possible to generate di- or trisaccharides in high yield from polysaccharides containing both neutral and amino sugars with neutral sugars as the reducing termini.

Key words: glycoprotein, deglycosylation, solvolysis, polysaccharide, characterization, HF

Structural studies of the polypeptides of glycoproteins are often hampered by the oligo- or polysaccharide side chains. Deglycosylation of glycoproteins should therefore greatly facilitate the study of the peptide portion of the proteins. A deglycosylated protein could be easier to purify, more susceptible to proteolysis, easier to sequence via Edman degradation, or used as a natural substrate for the study of glycoprotein biosynthesis.

As the known chemical procedures for the degradation of polysaccharides also degrade proteins, and pure enzymes that could be used to deglycosylate proteins are not always readily available, a new method for the specific degradation of polysaccharides is

Received March 28, 1977

Dr. Mort is now at Charles F. Kettering Research Laboratory, 150 East South College Street, Yellow Springs OH 45387

© 1978 Alan R. Liss, Inc., 150 Fifth Avenue, New York, NY 10011

desirable. According to the literature, anhydrous liquid HF might be a reagent that can be used to specifically deglycosylate glycoproteins.

As early as 1869, when Gore (1) tested the reaction of HF with many compounds such as paper and gum arabic, it was known that HF dissolves polysaccharides. Friedenhagen and Cadenbac (2) investigated the reaction of cellulose and wood in liquid HF and found that the cellulose was totally depolymerized into glucosyl fluoride, but the lignin appeared to be undegraded. During the removal of the HF there was a tendency for the glycosyl fluorides to reoligomerize. Friedenhagen (3) and later Katz (4) found that proteins were very soluble in liquid HF. Kock et al. (5) saw no appreciable degradation of ribonuclease in liquid HF at $0°C$ in 2 hr, and only a small increase of amino groups but no disulfide interchange after 24 hr in HF at $30°C$. Ribonuclease (5) and lysozyme (5, 6) retain most of their enzymic activity after 2 hr in HF at $0°C$, but almost none after 2 hr at $30°C$.

The only peptide bond known to be cleaved by anhydrous HF is that of methionine. Lenard et al. (7) found that methionyl glycine was completely cleaved after 36 hr at $30°C$ in HF. In peptides and proteins methionyl bonds are cleaved more slowly. Seryl bonds undergo N to O acyl shift at $30°C$ in HF, but this reaction is also rather slow (8), reaching 65% of completion in peptides after 13 hr. The shift is readily reversible by mild alkali.

When Sakakibara and Shimonishi (9) discovered that HF was a good reagent for the acidolysis of almost all of the protecting groups used to mask sensitive functional groups during peptide synthesis, interest in the use of HF increased. Liquid HF is now one of the most commonly used reagents for final deprotection and cleavage from the solid support in synthetic peptide chemistry. Many biologically active peptides and some functional enzyme segments have been made using HF. Thus HF appears to be also a very good candidate for the deglycosylation of glycoproteins.

MATERIALS AND METHODS

Cell wall glycopeptides were prepared by a 45-min oxidation with sodium chlorite of walls from suspension-cultured cells of tomato as described by Mort and Lamport (10). Fetuin and crab shells were bought from the Sigma Chemical Co., St. Louis, Missouri. Pig submaxillary mucin, prepared according to the method of De Salegui and Polanska (11) but without separating the major and minor fractions, was a gift from Dr. Joseph Sung (Michigan State University).

A Kel-F hydrogen fluoride line was purchased from Peninsula Laboratories, Inc. (P.O. Box 205, San Carlos, California 94070). A description of the line and a review of the use of HF can be found in the literature (12–14).

Before HF treatment the glycoprotein was dried at $60°C$ in a vacuum desiccator overnight; HF was dried in the reservoir over boron trifluoride. With the dry protein (up to 10 mg), 1 ml of anisole,[1] and a Teflon-coated stirring bar in the reaction vessel, the HF line was evacuated and approximately 10 ml of anhydrous HF was distilled from the

---

[1] Anisole was added because sugars are cleaved from the protein as glycosylfluorides, which are very good alkylating agents in HF (15). The large excess of anisole effectively competes for the alkylating species. If one wanted to recover the sugars, the anisole would have to be omitted and some other means used to prevent alkylation of tyrosine, methionine, and other sensitive amino acids. One possibility is to add methanol.

HF reservoir into the reaction vessel immersed in a dry ice-acetone bath. The reaction mixture was then allowed to warm to the desired temperature, as judged by the vapor pressure of the HF, which was measured by a mercury manometer attached to the line.

At the completion of the reaction the HF was evaporated under vacuum as rapidly as possible to keep the time for which the reactants were concentrated as short as possible. The HF was absorbed by a calcium oxide trap between the line and the vacuum pump. Approximately 1 hr after all visible traces of HF had disappeared from the reaction vessel the sample was removed from the HF line, dissolved, depending on its solubility, in either 0.1 M acetic acid or 50% acetic acid in water, and the anisole was extracted into ether. The sample was then dialyzed against distilled water or chromatographed on the appropriate gel filtration column.

Sugars were determined as the trimethylsilyl O-methyl glycosides after methanolysis as described by Bhatti et al. (16). The amounts of polypeptide were calculated from the sum of the amounts of amino acids released after 18-hr hydrolysis in constant boiling HCl as determined by gas chromatography of their heptafluorobutyryl isobutyl esters (17).

For ease of comparison, sugar compositions of the glycoproteins were expressed as the weight of sugar per 100 mg of polypeptide instead of the normally used weight percent of the total protein. Thus the drop in sugar composition is directly proportional to the amount of deglycosylation which has occurred, and the ratio of the sugar composition of treated to untreated protein is the fraction of the original sugar remaining.

Glucosamine and N-acetyl glucosamine were distinguished by forming the pertrimethylsilyl derivatives by reacting with a 1:1 mixture of pyridine and bis (trimethylsilyl) trifluoroacetamide for at least 2 hr and then chromatographing on a 12 foot 3% SP2100 column programed from 165° to 200°C at 1.5 degree/min.

RESULTS

Three glycoproteins and one polysaccharide were chosen to illustrate the effects of HF of glycoproteins. Glycopeptides from tomato cell walls contain many oligosaccharides linked to seryl and hydroxyprolyl hydroxyl groups via O-glycosidic linkages to neutral sugars, galactose and arabinose, respectively (18). Pig submaxillary mucin contains many oligosaccharides attached O-glycosidically via amino sugars, and fetuin contains oligosaccharides linked via both O-glycosidic linkages of amino sugars to serine and threonine, and N-glycosidic linkages of amino sugars to asparagine. Thus all the commonly occurring glycopeptide linkages were represented. Chitin (crab shells) was used to confirm the action of HF on the N-acetylglucosaminyl-N-acetylglucosamine bond and to test for the retention of acetylation of the amino group of N-acetylglucosamine.

Cell Wall Glycopeptides

When glycopeptides from tomato cell walls were treated in anhydrous liquid HF for 1 hr at 0°C, they were almost completely deglycosylated (Table I).

As there is an average of only 3.1 arabinose residues per oligosaccharide on each hydroxyproline in the peptides (19), it is apparent that the hydroxyproline-O-arabinose linkage is cleaved under the conditions listed above. The oligo- or polysaccharide attached to the serine residues has not been characterized, but from the data and knowing that approximately 80% of the serine residues are glycosidically linked, most likely via galactose (18), it is clear that neutral sugar glycosides of serine are also cleaved under these conditions.

TABLE I. Sugar Composition* of Cell Wall Glycopeptides Before and After HF Solvolysis at 0°C for 1 hr

	Before	After	% Sugar remaining
ARA/HYP	5.04	0.14	3
GAL/SER	4.2	0.09	2
GALACTURONIC/SER	7.0	0.04	1.3
RHA/SER	1.7	0.02	1.2
GLC/SER	0.84	0.05	6

*Data expressed as molar ratios. Sugars were determined as their TMS-O-methyl glycosides, and amino acids as their N-heptafluorobutyryl isobutyl esters.

Pig Submaxillary Mucin

Pig submaxillary mucin is a high molecular weight glycoprotein with many short oligosaccharides attached to its serine and threonine hydroxyl groups via N-acetylgalactosamine. The longest has the structure (I) and the others are all the possible combinations of removal of sugars from its nonreducing end (20).

$$\text{I: ser/thr-O} - \text{GalNAc} - \text{Gal} - \text{GalNAc}$$
$$\phantom{\text{I: ser/thr-O - GalNAc - Gal - }}|\phantom{\text{lNAc}}\,|$$
$$\phantom{\text{I: ser/thr-O - GalNAc - Ga}}\text{NGNA}\ \ \text{Fuc}$$

After treatment of pig submaxillary mucin in HF at 0°C for 1 hr, the protein was only partially deglycosylated (Table II).

TABLE II. Sugar Composition of Pig Submaxillary Mucin Before and After HF Solvolysis for 1 hr at 0°C

	Before HF solvolysis mg sugar / 100 mg PSM peptide	After HF solvolysis mg sugar / 100 mg PSM peptide	% Sugar remaining
GalNAc	81	54	67
Gal	32	2.4	7
Fuc	27	0	0
NGNA	42	0	0

From the data of Payza et al. (21) one can calculate that the average number of N-acetylgalactosamine residues per oligosaccharide is 1.5. After the treatment with HF the N-acetylgalactosamine content fell to about two-thirds of the original level. Thus it is easy to conclude that the HF treatment does not cleave the O-glycosidically linked N-acetylgalactosaminyl bond at 0°C.

When the temperature of the HF reaction was raised to room temperature and the reaction time lengthened to 3 hr, pig submaxillary mucin was almost entirely deglycosylated (Table III).

TABLE III. Sugar Composition of Pig Submaxillary Mucin Before and After HF Solvolysis for 3 hr at 23°C

	Before HF solvolysis mg sugar / 100 mg PSM peptide	After HF solvolysis mg sugar / 100 mg PSM peptide	% Sugar remaining
GalNAc	81	5	6
Gal	32	trace	0
Fuc	27	0	0
NGNA	42	0	0

Fetuin

Fetuin contains 2 types of oligosaccharide side chains (II and III), 1 linked O-glycosidically as in pig submaxillary mucin and the other N-glycosidically via N-acetylglucosamine to asparagine. Three of each type are present per mole of fetuin (22).

II: ser/thr — GalNAc — Gal — NANA
                    |
              $(NANA)_{1/2}$

III: AspN — GlcNAc — GlcNAc — Man — Man — Man
                                          | | |
                                       GlcNAc GlcNAc GlcNAc
                                          | | |
                                        Gal   Gal   Gal
                                          | | |
                                       NANA NANA NANA

As with pig submaxillary mucin, HF at 0°C only partially deglycosylated fetuin (Table IV).

Both N-acetylglucosamine and N-acetylgalactosamine remained attached to the protein, but twice as much N-acetylglucosamine as N-acetylgalactosamine remained, indicating that the chitobiose unit of the asparaginyl-linked oligosaccharides was not being cleaved.

After 3 hr in HF at room temperature fetuin was still not completely deglycosylated (Table V).

These results show that the N-glycosidic linkage of N-acetylglucosamine to asparagine is the only sugar linkage commonly found in glycoproteins which is not susceptible to HF solvolysis. This is rather unfortunate as this linkage would have to be broken if HF-deglycosylated proteins were to be used as natural substrates for the assay of the glycosyltransferase involved in addition of the oligosaccharide core region to glycoproteins.

SDS Gels of HF-Treated Proteins

Bovine serum albumin was treated in HF at 0°C for 1 hr, run on a 5.7% gel in 1% SDS, and stained with Coomassie Blue (23). There was no evidence that HF treatment caused degradation of the protein. When fetuin was subjected to the same treatment, its molecular weight fell, as one would expect from the loss of sugars, but it remained as one major band (Fig. 1).

TABLE IV. Sugar Composition of Fetuin Before and After HF Solvolysis for 1 hr at 0°C

	Before HF solvolysis		After HF solvolysis		
	mg sugar	Sugar residues[a]	mg sugar	Sugar residues[b]	% Sugar remaining
	100 mg fetuin peptide	mole fetuin	100 mg fetuin peptide	mole fetuin	
NANA	14.4	13	1.25	1.7	8.7
GlcNAc	7.4	15	3.0	6.0	41
GalNAc	1.34	3	1.26	2.8	94
Man	3.82	9	0	0	0
Gal	4.58	12	0.54	1.4	12

[a]As determined by Spiro (22)
[b]Calculated as: percent of the remaining sugar multiplied by its initial number of residues/mole fetuin.

TABLE V. Sugar Composition of Fetuin Before and After HF Solvolysis for 3 hr at 23°

	Before HF solvolysis		After HF solvolysis		
	mg sugar	Sugar residues	mg sugar	Sugar residues	% Sugar remaining
	100 mg fetuin peptide	mole fetuin	100 mg fetuin peptide	mole fetuin	
NANA	14.4	13	trace	—	—
GlcNAc	7.4	15	1.2	2.4	16
GalNAc	1.34	3	trace	—	—
Man	3.82	9	trace	—	—
Gal	4.58	12	0.34	0.89	7.4

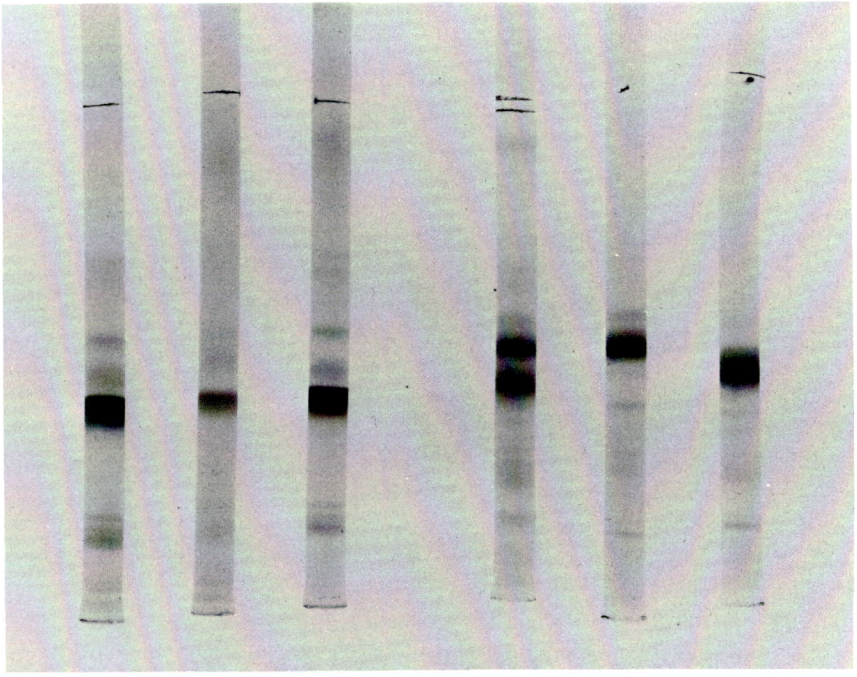

Fig. 1. SDS gels of HF treated and untreated proteins. From left to right: Mixed BSA and HF treated BSA, HF treated BSA, untreated BSA, mixed HF treated fetuin and untreated fetuin, HF treated fetuin, and untreated fetuin. The origin of the gels are at the bottom of the photograph.

Crab Shells

A 1-hr treatment of crab shells in HF at 0°C in the absence of anisole only partially solubilized the chitin. When the soluble portion was chromatographed on a Biogel P2 column all the glucosamine voided the column. However, after 3 hr in HF at room temperature by far the majority of the glucosamine was retarded by the column and eluted where one would expect free N-acetylglucosamine. At the void of the column was a small peak of glucosamine, presumably still linked to the protein component of the shells, and between the void and included volume was another small peak of glucosamine containing an unidentified compound, possibly a pigment.

The trimethylsilyl derivative of material from the major peak off the P2 column was made with no prior hydrolysis and subjected to gas chromatography. Only a single peak was observed eluting with the same retention time as N-acetylglucosamine well separated from what would have been the elution time of glucosamine.

DISCUSSION

From the experiments described here I conclude that anhydrous liquid HF at 0°C can be used to totally deglycosylate glycoproteins containing only neutral sugars. In proteins containing amino sugars in a direct glycopeptide linkage there are 2 possibilities. If the glycopeptide linkage involves a hydroxy amino acid, it can be cleaved by HF at room temperature in 3 hr, but if it is an N-glycosidic link to asparagine, it cannot be cleaved using HF under any of the conditions that I have tested.

As HF at 0°C causes no peptide bond cleavage and at room temperature (from the results of others) none or very little, it can be used effectively to assist in the determination of the structure of the peptide portion of glycoproteins by complete or almost complete removal of the oligosaccharide side chains of the proteins.

It may be possible to take advantage of what appears to be a high selectivity of HF towards neutral and acidic sugar linkages at 0°C. Polysaccharides such as hyaluronic acid should be readily decomposed to disaccharides of N-acetylglucosaminyl glucuronic acid, the inverse of what is normally found with partial acid hydrolysis (glucuronyl N-acetylglucosamine). Of course care must be taken to prevent the reoligomerization observed by Friedenhagen and Cadenbac (2). HF treatment at 0°C may allow an easier determination of the structure of the mannosyl core region present in many glycoproteins by generating N-acetylglucosaminyl-mannosyl disaccharides or trisaccharides.

The specificity of HF to different glycosidic linkages could also be used to distinguish glycopeptide linkages. HF treatment of glycopeptides with O-glycosidically linked oligosaccharides via neutral sugars should completely remove the sugars from the peptide. In the commonly occurring N-acetylgalactosaminyl-linked glycopeptides HF treatment at 0°C will leave a single N-acetylgalactosamine residue per oligosaccharide, immediately proving the involvement of N-acetylgalactosamine in the glycopeptide linkage. In the case of asparaginyl-linked oligosaccharides a chitobiose unit will remain attached to protein for each oligosaccharide side chain after HF treatment at 0°. Confirmation of the N-asparaginyl link could be made by a 3-hr treatment at room temperature, which should leave only a single N-acetylglucosamine attached to the protein per oligosaccharide.

ACKNOWLEDGMENTS

These experiments were performed in the laboratory of Dr. D.T.A. Lamport in partial fulfillment of the requirements for my PhD. I thank him for the freedom and encouragement to investigate the effects of HF. I also thank Dr. Shirley Rodaway for running the SDS gels, and Marlee Pierce for assistance with the manuscript.

This research was supported under ERDA contract No. EY-76-C-02-1338*000.

REFERENCES

1. Gore J: J Chem Soc 22:396, 1869.
2. Fredenhagen K, Cadenbach G: Angew Chem 46:113, 1933.
3. Fredenhagen K: Z Phys Chem, Abst A 164:176, 1933.
4. Katz JJ: Arch Biochem Biophys 51:293, 1954.
5. Koch AL, Lamont WA, Katz JJ: Arch Biochem Biophys 63:106, 1956.
6. Aimoto S, Shimonishi Y: Bull Chem Soc Jp 48:3293, 1975.
7. Lenard J, Schally AV, Hess GP: Biochem Biophys Res Commun 14:498, 1964.
8. Lenard J, Hess GP: J Biol Chem 239:3275, 1964.
9. Sakakibara S, Shimonishi Y, Kishida Y, Okada M, Sugihora H: Bull Chem Soc Jp 40:2164, 1967.

10. Mort A, Lamport DTA: Plant Physiol (Suppl) (Abstract 80) 56:16, 1975.
11. De Salegui M, Polonska H: Arch Biochem Biophys 129:49, 1969.
12. Stewart JM, Young JD: "Solid Phase Peptide Synthesis." San Francisco: WH Freeman and Company, 1969.
13. Lenard J: Chem Rev 69:625, 1969.
14. Sakakibara S: In Weinstein B (ed): "The Chemistry and Biochemistry of Amino Acids, Peptides and Proteins." New York: Marcel Dekker 1971, vol 1, p 51.
15. Wagner A: In Olah GA (ed): "Friedel-Crafts and Related Reactions." New York: Wiley-Interscience, 1965, vol IV, p 235.
16. Bhatti T, Chambers RE, Clamp JR: Biochem Biophys Acta 222:339, 1970.
17. MacKenzie SL, Teneschuk D: J Chromatogr 111:413, 1975.
18. Lamport DTA, Katona L, Roerig S: Biochem J 133:125, 1973.
19. Lamport DTA, Miller DH: Plant Physiol 48:455, 1971.
20. Carlson D: J Biol Chem 243:616, 1968.
21. Payza N, Rizui S, Pigman W: Arch Biochem Biophys 124:68, 1969.
22. Spiro RG, Bhoyroo VD: J Biol Chem 249:5704, 1974.
23. Fairbanks G, Steck J, Allach DW: Biochemistry 10:2606, 1971.

Toward a Mechanism of Myoblast Fusion

K. A. Knudsen and A. F. Horwitz

Department of Biochemistry and Biophysics, University of Pennsylvania Medical School, Philadelphia, Pennsylvania 19174

Myoblasts derived from chick pectoral muscle explants and grown in vitro in low calcium medium are harvested with EDTA. These cells, when agitated in suspension, show a calcium mediated, reversible aggregation which is relevant to myoblast fusion. With time, progressively harsher methods are required to disrupt the aggregates until, finally, dispersion resistant aggregates and multinucleate cells appear.

From these kinetic data we conclude that myoblast fusion results from a sequence of stages including recognition, adhesion, membrane union, and further morphologic changes. Various agents and manipulations, known to inhibit myotube formation, affect these stages differently.

Our observations are compatible with protein mediated recognition and adhesion-stages, and with a role for the cytoskeleton, Ca^{2+}, and fluid membrane lipids at a stage at, or just prior to, membrane union. One model consistent with our observations is an adhesive step, possibly via a gap junction-like structure, followed by directed movement of protein, which results in closely apposed or exposed regions of lipid bilayer, which then fuse.

Key words: myogenesis, fusion, adhesion, recognition

The early development of muscle includes many interesting and well-described phenotypic changes. The fusion of mononucleate myoblasts into myotubes is one such change (1, 2). This process is of major importance since it is a key event in the production of muscle fibers. Like that of other processes involving membrane fusion, e.g., exocytosis, endocytosis, secretion, etc., the mechanism of myoblast fusion remains a mystery. However, several models have been proposed (3–5). In one class of possibilities, fusion is postulated to result from the union of exposed regions of membrane lipid. Other possibilities envision fusion as resulting from a complicated sequence of events. The first alternative has its precedent in studies of model lipid vesicles, where vesicle fusion is sensitive to the fluid properties of the membrane and the presence of calcium ions (6, 7). The observation that myoblast fusion requires both Ca^{2+} and temperatures above $30°C$ can be interpreted to be consistent with this mechanism if the temperature dependence is explained in terms of changes in the membrane's fluidity (8, 9).

Received April 21, 1977; accepted May 16, 1977.

© 1978 Alan R. Liss, Inc., 150 Fifth Avenue, New York, NY 10011

Data derived from microscopic observations of fusing myoblast cultures, however, suggest the second, more complex, fusion mechanism (10, 11). Myoblasts are observed to sort out from other cell types, to align, and to fuse only with each other (12, 13). The time course of this process is slow, requiring several hours. This sorting out could be apparent, i.e., the result of fusion itself, or it could be due to recognition and adhesion phenomena that are distinct from membrane fusion. In order to further investigate the mechanism of fusion the effects of various agents and experimental manipulations on myotube formation have been assayed (14–19). Such studies also implicate a complex, multistep mechanism for fusion. However, the conclusions to be drawn from such microscopic observations are ambiguous due, in part, to cell attachment to the substratum and to cell migration which are necessary for myoblast fusion in culture (20).

We have eliminated some of these ambiguities by studying myoblast fusion in suspension. The details of the procedures and a more complete discussion of the results have been published (21). A summary of more recent results are included in Table I (Knudsen, K.A. and Horwitz, A. F., unpublished observations). In brief, we harvest (with EDTA) chick pectoral myoblasts grown in culture for $\sim$ 52 h in low calcium medium (to reversibly inhibit fusion) and gently agitate the washed cells in suspension. The cells aggregate appreciably after short times (5–20 min) only in the presence of calcium. Fibroblasts are excluded from these aggregates. To provide evidence that this aggregation is relevant to fusion, we assayed the effects on aggregation of various manipulations reported to affect myotube formation. We found that the influence of variations in pH, $[Ca^{2+}]$, temperature, and cell culture age on the rate of aggregation quantitatively parallels that for myotube formation. Although many biochemical processes are sensitive to pH, $[Ca^{2+}]$, and temperature, they often show different optima. This is readily apparent in comparing the effects of these variables on myoblast aggregation to those published on other cell types (22, 23). Therefore, we conclude that the quantitative correlations of the effects of several different variables on aggregation with those for myotube formation provide reasonable evidence that the aggregation phenomenon is relevant to fusion. In addition, the specificity implied by the exclusion of fibroblasts, which do not fuse with myoblasts, along with the influence of age in culture on aggregation and on myotube formation, supports this conclusion.

MULTINUCLEATE CELLS RESULT FROM A SEQUENCE OF STAGES

We return to the original question concerning the mechanism of fusion. In our suspension assay we find that after short incubation times, when specificity and the effects of various manipulations ($[Ca^{2+}]$, pH, etc.) are apparent, the aggregates are readily dispersed by repeated pipetting or by incubation with EDTA or trypsin. Furthermore, following fixation of the cells and staining of the nuclei, we find that there is no increase in multinucleate cells. These observations support the conclusion that this early interaction of myoblasts precedes fusion, i.e., membrane union and cytoplasmic continuity. Since the reversible aggregation shows specificity at this time, we have termed the initial interaction recognition.

If the aggregates are incubated in suspension for longer times, progressively more harsh methods are required for their dispersal. After $\sim$ 30 min EDTA is no longer effective, and by 1–2 h the aggregates begin to show trypsin resistance. An increase in multinucleate cells is not seen before $\sim$ 2 h. We interpret these kinetic observations as evidence support-

TABLE I. Fusion Competent Myoblasts Aggregate in Suspension and Proceed to Form Multinucleate Cells. Various Agents and Manipulations Block This Process at Different Stages

Morphology	Single cells	Aggregates			Rosettes	Multinucleate Cells
Stages	Single cells	Recognition	Reversible adhesion	Irreversible adhesion	Membrane union	Morphologic changes
Dispersion		EDTA / EGTA	Trypsin (EDTA Resistant)	— (Trypsin Resistant)	—	—
Time	0	~ 5–20 min	~ 30–60 min	~ 1–2 h		> 2 h
Blocks		↛	↛	↛	↛	
		low Ca	low Ca	low Ca (?)	ND	
		energy poisons	energy poisons (?)	energy poisons (?)	ND	
		trypsin	trypsin	trypsin	NE	
		glutaraldehyde	glutaraldehyde (?)	glutaraldehyde (?)	ND	
		temp < 30°C	temp < 30°C	temp < 30°C (?)	ND	
		NE	20 mM Mg^{++}	ND	ND	
		NE	NE	Cytochalasin b	ND	
		NE	NE	Colchicine	ND	
		25OH Chol	ND	ND	ND	
		NE	NE	elaidate	ND	

NE – No effect.
(?) – Block occurs at one or both steps indicated.
ND – Not determined.

ing a fusion mechanism comprised of a sequence of distinct stages. Recognition is followed by adhesion, defined operationally as that stage in which the aggregates are resistant to EDTA dispersion but still sensitive to trypsin. The appearance of trypsin-resistant aggregation signals yet another change in cellular interactions. It seems most likely that membrane union, and the morphologic changes subsequent to it, parallels the onset of trypsin-resistant aggregation or occurs shortly after it.

We further support the notion that fusion proceeds via a sequence of distinct stages by our observation of the effects of chemical agents and other manipulations on myoblast interaction in suspension (Table I). As stated above, the effects of Ca^{2+}, pH, temperature, and culture age are evident at the initial stage of aggregation. (They might, of course, influence subsequent stages as well.) Furthermore, myoblasts treated with trypsin, glutaraldehyde, energy poisons, or grown with their sterol synthesis inhibited neither aggregate nor fuse (Knudsen, K. A., Wight, A. A., and Horwitz, A. F., unpublished observations). These observations point to a critical, and perhaps central, role for recognition in the process of fusion. These adhesive interactions that lead to (but precede) myoblast fusion appear to be unique. For example, while cells aggregate in the presence of 20 mM Mg^{2+} or Con A, they do not fuse (18, 24, Table I). In addition, myoblasts grown for 52 h on plates treated to inhibit cell attachment (Sylgard coated) (25) and in the presence of low Ca^{2+} medium, tend to reside in aggregates. However, these aggregates do not form multinucleate cells, i.e., myoballs.

Other manipulations inhibit fusion differently (Table I). While aggregates form in the presence of 20 mM Mg^{2+}, they do not become resistant to EDTA dispersion. [It is important to query whether this aggregation is indeed relevant to fusion, i.e., analogous to that mediated by calcium. Evidence implying this has been published (26).] This implicates another distinguishable stage occurring before fusion. Furthermore, the addition of cytochalasin b or colchicine to cells in suspension inhibits both the appearance of trypsin-resistant aggregates and the subsequent formation of multinucleate cells. However, the myoblasts treated with these agents show calcium dependent aggregation which, with time, becomes EDTA resistant. Cells grown under conditions which cause enrichment of the membrane lipids with elaidate block fusion similarly. These studies point to yet another critical and distinguishable stage, which occurs either at or slightly before membrane union.

MECHANISTIC CONSIDERATIONS IN MULTINUCLEATE CELL FORMATION

Recognition

Inhibitor studies of the type discussed can, with reservation, provide insights into the molecular processes occurring during cell fusion. Several different experimental manipulations appear to influence the onset of aggregation, i.e., recognition. In sum, this process is sensitive to trypsin, and requires calcium, glycolytic or redox energy, temperatures above $30°C$, and normal synthesis of cholesterol. The cholesterol requirement could arise either as part of a recognition assembly mechanism or due to changes in the membrane properties resulting from the inhibition. The slow time course of reversal of the effect suggests the former (Wight, A. A. and Horwitz, A. F., unpublished observations). The role, if any, of other lipids at the recognition step is not clear. Studies with cells enriched in elaidate, which produces a less fluid membrane, demonstrate that although fusion is inhibited, recognition still occurs.

The nature of the calcium requirement has been probed further using the inophore A23187. At levels of ionophore catalyzing a net transfer of calcium across the cellular membrane, myoblasts do not form myotubes in medium containing slightly suboptimal calcium concentrations. In addition, the presence of ionophore does not inhibit myotube formation in medium containing calcium concentrations that normally support fusion (24; Holtzer, S., Wight, A., Scarpa, A., Horwitz, A., and Holtzer, H., unpublished observations). Other experiments show that ruthenium red, an impermeable reagent that competes for calcium-binding sites (27, 28) also inhibits myotube formation (Holtzer, S., Wight, A., Scarpa, A., Horwitz, A., and Holtzer, H., unpublished observations). The implication is that Ca^{2+} acts on the outside of the cell in promoting fusion. We do not exclude other, possibly intracellular, roles for calcium occurring at subsequent stages. The most straightforward interpretation of an external role for calcium is that the ion functions in recognition as an intercellular bridge between anionic groups on different cells or as a bridge of anionic sites on a given cell to induce an adhesive membrane conformation.

A recurrent theme in recognition studies is the participation of cell surface carbohydrates (23, 29–31). Although it seems likely that sugars might participate in myoblast recognition as well, we have little evidence supporting this. We find that while periodate can inhibit calcium mediated aggregation, its effect is partially or entirely blocked by inhibitors of proteolysis. In addition, efforts to find a monosaccharide or disaccharide that blocks either aggregation or fusion have not yet been successful. While we and others (19) note that neuraminidase can block myotube formation, we find that the enzyme has little effect on recognition or EDTA resistant adhesion.

A final point concerning recognition follows from the older literature. In mixed cultures of young and old myoblasts or of myoblasts and other cell types, the fusion competent myoblasts sort out (12, 13, 24). In our suspension assay optimal calcium-mediated aggregation is not seen until the cells are also optimally competent to form myotubes. Furthermore, we note that myoblasts treated for a short time with low concentrations of trypsin and washed thoroughly do not aggregate with each other or with untreated cells. These observations imply that both cells in a bicellular aggregate must possess appropriate surface molecules for recognition to occur.

Adhesion and Fusion

As more harsh methods are required to disperse the aggregates, it seems likely that the cell-cell interactions have changed and that altered, different, or additional molecules are involved. This adhesion could be comprised of reversible (trypsin-dissociable) and irreversible (trypsin-resistant) stages. Alternatively, the appearance of trypsin-resistant aggregation could signal the onset of membrane union.

Several agents and manipulations prevent the formation of trypsin-resistant aggregates as well as fusion. For example, the addition of colchicine, colcemid, or cytochalasin b, or the growth of cells under conditions that alter their fatty acyl chains all produce Ca^{2+} dependent aggregates that are dissociable with trypsin but not with EDTA. In addition, continued energy metabolism, presence of calcium, and temperatures above $30°C$ also appear required to form multinucleate cells. These observations are open to many interpretations. However, the involvement of the cytoskeleton does suggest that a particular orientation and/or directed movement of membrane molecules is necessary for the formation of multinucleate cells. The requirement for fluid lipid suggests that fusion proceeds via the union of lipid bilayers since recognition, and, hence, the postulated clustering of membrane protein (see below) are not inhibited.

CONCLUSION

Returning to the models for myoblast fusion discussed in the introduction, we conclude that our observations clearly favor the notion that myotubes form as a result of traversal through a sequence of distinct stages. The recognition and adhesion stages are trypsin sensitive and hence most likely (glyco-) protein mediated. The clustering of intramembrane particles (IMP) and formation of gap junction-like structures correlates with these early events in multinucleate cell formation (26, 33, 34). Since the IMP clustering also appears in the presence of Mg^{2+} (no Ca^{2+}), which mediates formation of aggregates that are sensitive both to EDTA and trypsin, we suggest that they form during the recognition process and might be relevant to it (26).

The mechanism of membrane union, since it is unique to fusion, is of particular interest. The influence of membrane lipid alterations at what might be the stage at or immediately preceding fusion is consistent with the suggestion that membrane union results from fusion of lipid bilayers. The continued calcium requirement is also consistent with this suggestion (7, 35, 36). The observed role for the cytoskeleton at this stage could be to direct movements of membrane protein away from the fusion site, thus generating an exposed region of lipid bilayer. Alternatively, the cytoskeleton could mediate, without lateral movements of protein, apposition of lipid bilayers. Freeze fracture studies implicate a movement of membrane protein away from the site of membrane union, supporting the former alternative (37, 38).

ACKNOWLEDGMENTS

This work was supported by grants from the NIH (GM 23244 and HL 18708) and the Cystic Fibrosis Foundation and benefitted from facilities made available from NIH grant GM 20138. This work was performed during the tenure of the W.D. Stroud Established Investigatorship of the American Heart Association (AFH). KAK is a NIH predoctoral trainee (IT32 GM 07229-01). We thank Ms. Alice Ann Wight for her useful discussions and valuable contributions.

REFERENCES

1. Bischoff R, Holtzer H: J Cell Biol 41:188, 1969.
2. Yaffe D, Dym H: Cold Spring Harbor Symp Quant Biol 37:543, 1972.
3. Poste G, Allison AC: Biochem Biophys Acta 300:421, 1973.
4. Lucy JA: Nature (London) 227:815, 1970.
5. Ahkong QF, Fisher D, Tampion W, Lucy JA: Nature (London) 253:194, 1975.
6. Prestegard JH, Fellmeth B: Biochemistry 13:1122, 1974.
7. Papahadjopoulos D, Poste G, Schaeffer BE, Vail WJ: Biochim Biophys Acta 352:10, 1974.
8. Shainberg A, Yagil G, Yaffe D: Exp Cell Res 58:163, 1969.
9. van der Bosch J, Schudt C, Pette D: Exp Cell Res 82:433, 1973.
10. Holtzer H, Rubinstein N, Fellini S, Yeoh G, Chi J, Birnbaum J, Okayama M: Quart Rev Biophys 8:523, 1975.
11. Holtzer H, Jones KW, Yaffe D: J Neurol Sci 26:115, 1975.
12. Yaffe D, Feldman M: Dev Biol 11:300, 1965.
13. Okazaki K, Holtzer H: J Histochem Cytochem 13:726, 1965.
14. Bischoff R, Holtzer H: J Cell Biol 36:111, 1968.
15. Sanger JW, Holtzer S, Holtzer H: Nature (London) New Biol 229:121, 1971.
16. van der Bosch J, Schudt C, Pette D: Biochem Biophys Res Commun 48:326, 1972.
17. Nameroff M, Trotter JA, Keller JM, Munar E: J Cell Biol 58:107, 1973.
18. Den H, Malinzak DA, Keating HJ, Rosenberg A: J Cell Biol 67:826, 1975.
19. Schudt C, Pette D: Cytobiologie 13:74, 1976.
20. Holtzer H, Rubinstein N, Dienstman S, Chi J, Biehl J, Somlyo A: Biochimie 56:1575, 1974.
21. Knudsen K, Horwitz A: Dev Biol 58:328, 1977.
22. Umbreit J, Roseman S: J Biol Chem 250:9360, 1975.
23. Marchese RB, Vosbeck K, Roth S: Biochim Biophys Acta 457:385, 1976.
24. Schudt C, van der Bosch J, Pette D: FEBS Lett 32:296, 1973.
25. Lass Y, Fischbach GD: Nature (London) 263:150, 1976.
26. Schudt C, Dahl G, Gratzl M: Cytobiologie 13:211, 1976.
27. Rahamimoff R, Alnaes E: Proc Natl Acad Sci USA 70:3613, 1973.
28. Szubinska B, Luft JH: Anat Rec 171:417, 1971.
29. Ashwell G, Morell AG: Adv Enzymol 41:99, 1974.
30. Roseman S: In Lee EYC, Smith EE (eds): "Biology and Chemistry of Eucaryotic Cell Surfaces." New York: Academic Press, 1974, pp 317–354.
31. Talmich KW, Burger, MM: In Whelan W (ed): "Biochemistry of Carbohydrates." Baltimore: University Park Press, MTP International Review of Science, Series 1, 1975, vol 5, pp 43–93.
32. Yaffe D: Exp Cell Res 66:33, 1971.
33. Rash JE, Fambrough D: Dev Biol 30:168, 1973.
34. Rash JE, Staehelin, LA: Dev Biol 36:455, 1974.
35. Papahadjopoulos D, Vail WJ, Newton C, Nir S, Jacobson K, Poste G, Lazo R: Biochim Biophys Acta 465:479, 1977.
36. Poste G, Papahadjopoulos D: Proc Natl Acad Sci USA 73:1603, 1976.
37. Lawson D, Raff MC, Gomperts B, Fewtrell C, Gilula NB: J Cell Biol 72:242, 1977.
38. Satir B, Schooley C, Satir P: J Cell Biol 56:153, 1973.

Cell Surface Carbohydrate Recognition and the Viability of Erythrocytes in Circulation

David Aminoff, William C. Bell, and William G. VorderBruegge

Departments of Internal Medicine (Simpson Memorial Institute) and Biological Chemistry, The University of Michigan, Ann Arbor, Michigan, 48109

The presence of sialic acid on the cell surface is crucial for the survival of mammalian erythrocytes in circulation. In contrast, sialidase-treated chicken erythrocytes retain their viability in circulation. Galactose oxidase treatment of chicken red blood cells has no effect on their viability. However, after sialidase treatment, galactose oxidase treatment results in the rapid elimination of the chicken erythrocytes from circulation. This is compatible with the interpretation that consecutive treatment with the 2 enzymes abolishes the ability of the chicken erythrocytes to regenerate the sialic acid on the cell surface.

Mammalian asialo-erythrocytes are sequestered in the liver and spleen. We have shown that at the cellular level there is a preferential recognition of sialidase-treated as compared to normal erythrocytes by mononuclear spleen cells and Kupffer cells of the liver. This recognition manifests itself in both autologous and homologous systems by rosette-like adhesions. These adhesions may represent the normal physiological mechanism for the removal of senescent erythrocytes from circulation by liver and spleen since it has been previously reported that older erythrocytes contain decreased amounts of sialic acid.

The mechanisms in mammals for the elimination from circulation of asialo-erythrocytes and asialo-glycoproteins, while analogous in many respects, are definitely not identical.

Key words: erythrocyte aging; erythrocyte analysis; erythrocyte viability; galactose oxidase; hepatocytes; Kupffer cells; liver; neuraminic acids; rosettes; sialic acids; sialidase; spleen

Sialic acid is an indigenous component of the erythrocytes of most animal species (1–3). The sialic acid (N-acetyl- or N-glycolylneuraminic acid) residues are predominantly attached to D-galactose or N-acetyl-D-galactosamine at the nonreducing end of the oligosaccharide chains or glycoproteins and glycolipids (4). These sialyl residues on the erythro-

---

Abbreviations: HBSS-CF — Hank's balanced salt solution, calcium free; HBSS-CF-BSA — Hank's balanced salt solution, calcium free with 1% bovine serum albumin (wt/vol); TBA — thiobarbituric acid assay.

Received May 2, 1977; accepted July 15, 1977.

© 1978 Alan R. Liss, Inc., 150 Fifth Avenue, New York, NY 10011

cyte surface appear to play a passive role as receptor sites for the hemagglutinins of the myxovirus group of viruses (5) and in man as part of the serological blood group M and N determinants (4). Otherwise, little is known regarding the biological significance of sialic acid on the red cell surface.

The correlation between decreased sialic acid content, electrophoretic mobility, and the age of the erythrocytes in circulation (6–9) has given rise to an intriguing hypothesis. Could the decrease in sialic acid content be correlated with the physiological mechanism of erythrocyte senescence and therby be the responsible signal for the removal of the older erythrocytes from circulation?

A series of studies were therefore initiated (9–11), and this communication is a summary of the findings to date.

MATERIALS AND METHODS

The animals used for these experiments were White Leghorn chickens, 1–1.6 kg; mongrel dogs, 13–18 kg; goats, 17–20 kg; New Zealand White rabbits, 3–4 kg; and male Sprague-Dawley rats, 200–250 g.

Sialidase was prepared as previously described (9, 11). Galactose oxidase (68 Worthington units/mg) and the horseradish peroxidase (698 Worthington units /mg) were obtained from Worthington Biochemicals, Freehold, New Jersey.

Determination of sialic acid content of erythrocytes was made by 3 different methods (9): a) direct determination of total bound sialic acid in stroma by the modified resorcinol method (12, 13), b) hydrolysis of stroma with 0.1 N HCl at 80°C and the sialic acid released into the supernatant determined as total sialic acid (12, 13) and as the free sialic acid by the TBA assay (14), and finally c) the total sialic acid released from intact erythrocytes upon treatment with sialidase was again determined by the resorcinol and TBA procedures.

Bleeding and Treatment of Erythrocytes

Dogs and goats were bled and transfused via the jugular vein. Rabbits were bled from the central ear artery and transfused through the marginal ear veins. Chickens were bled and transfused via the wing veins. Anesthetized rats were bled by sterile cardiac puncture.

Erythrocytes were removed from plasma and buffy coat and were washed twice with isotonic saline followed by a final wash of isotonic saline containing 0.01 M $CaCl_2$. These cells were then used for one or more of the following experiments. a) Reagent blank: labeled directly with $Na_2{}^{51}CrO_4$ (1.5 μCi per 1 kg body weight) for 30 min at 37°C with occasional gentle swirling. b) Sialidase treatment: simultaneously with the same amount of radioactive $Na_2{}^{51}CrO_4$ solution previously used, the erythrocytes were treated with the equivalent of 0.2 units of sialidase in 1 ml of solution per 1 ml of packed red blood cells (RBC). For these studies we define one unit of enzyme as that amount of enzyme that will release 1 μmole of sialic acid per minute at 37°C from ovine submaxillary mucin at pH 5.5, acetate buffer, and in the presence of 1 mM $CaCl_2$. c) Galactose oxidase treatment: Simultaneous with the $Na_2{}^{51}CrO_4$ labeling, the cells can also be treated with 5 units of galactose oxidase per 1 ml of packed cells for 30 min at 37°C either directly on the untreated erythrocytes, or d) after sialidase treatment as in (b) above to give cells that have been treated with both sialidase and galactose oxidase.

After these various incubations the cells were sedimented at 500 × g for 5 min, washed 3 times with sterile bacteriostatic saline, and resuspended in fresh sterile solution and injected autologously.

Radioactive Counting

Blood samples were obtained from the transfused animals 30 min after transfusion and every 24 h thereafter until the radioactivity of the samples reached 50% (the half-life, expressed in days) of the counts exhibited by the 24-h sample. The counts, using 0.5-ml samples, were corrected for [^{51}Cr] decay by the inclusion of the 24-h sample at each counting interval. The radioactivity of subsequent samples was also corrected for any change in hematocrit reading, which varied from day to day. The percent of cells surviving in the first 24 h were derived from the counts obtained in the 30-min sample ("zero" time) and at 24 h, respectively.

Labeling of Erythrocytes With NaB^{3}H$_4$ After Treatment With Galactose Oxidase.

The procedure used was essentially that previously described by Gahmberg and Hakomori (15). The technique was applied to erythrocytes with or without prior treatment with sialidase (11).

Isolation of Hepatocytes and Kupffer Cells From Rat Liver.

The method used is based on that of Wincek et al. (16). Using an ether anesthetized rat, the abdominal vena cava was ligated above the right renal vein and cannulae were inserted, one into the portal vein and the second through the right atrium into the thoracic vena cava. Residual blood in the liver was removed by perfusion via the portal vein with 400 ml of HBSS-CF at flow rate of 20 ml/min. A suspension of colloidal iron (0.3 g/ml) in 2.2 M sucrose was slowly injected (approximately 20 sec) into the portal vein cannula. Ten minutes were allowed for the phagocytosis of the injected iron after which the liver was flushed through by perfusion with 100 ml of HBSS-CF to remove the excess iron. The liver was then perfused with 0.05% (vol/vol) solution of collagenase in HBSS-CF at a flow rate of 20 ml/min for 8–10 min. The perfusion solutions were oxygenated with 100% O_2 and maintained at 37°C. The liver was then excised and placed in 20 ml HBSS-CF-BSA at 4°C and gently teased until the liver cells were dissociated. These were filtered through Dacron gauze and the Kupffer cells separated from the hepatocytes by the use of a magnet (16). The cells were used immediately and had a viability of 80–90% on the basis of the Trypan blue exclusion test. The Kupffer cells containing iron filings represented 5–10% of the total number of liver cells in the suspension.

Isolation of Spleen Cells

A rat was injected with 200 USP units of sodium heparinate into the abdominal vena cava and the spleen excised after 3 min. It was minced with scissors in HBSS-CF-BSA at 4°C and the suspension of cells swirled for 3 min and filtered through Dacron gauze. The filtrate was layered onto 4 ml of Ficoll-Paque (Pharmacia Fine Chemicals, Uppsala, Sweden) and centrifuged at 2,000 × g for 15 min at room temperature on a swing-out head. The layer of spleen mononuclear cells was aspirated from the Ficoll/HBSS interface, washed twice in HBSS-CF-BSA at 4°C to remove the Ficoll and then used as a 1% suspension (vol/vol) in HBSS-CF-BSA. The viability of spleen cells was 80–90% on the basis of the Trypan blue exclusion test.

"Rosette" Assay

The method used was based on the procedure developed for T cells (17). One drop of the liver or spleen cell suspension (1%, vol/vol) was added to one drop of RBC suspension (1%, vol/vol) in a 7 × 75 mm plastic tube and this was centrifuged at room tempera-

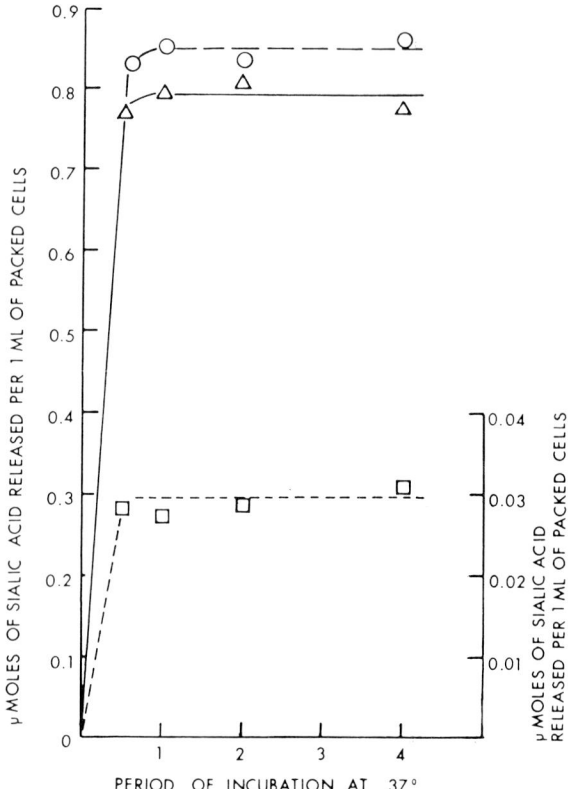

Fig. 1. Rate of release of free sialic acid from red cells on treatment with sialidase at 37°C in $CaCl_2$/saline as determined by the thiobarbituric acid assay (see text for experimental conditions): the scale to the right refers to the amount of sialic acid released from rabbit erythrocytes; △—△) goat, ○—○) dog, □---□) rabbit.

ture at 2,000 × g using a swing-out head. The pellet of cells was then gently aspirated with a Pasteur pipette and a wet slide prepared for direct microscopic examination (magnification 430 ×) and for staining with Wright's stain. A Spencer "Bright-line" hemocytometer was used to quantitate observations. Three or more erythrocytes adhering to a liver or spleen cell are considered to constitute a "rosette."

RESULTS

Effect of Sialidase on Erythrocytes

The rate and extent of release of sialic acid residues from the erythrocyte surface by sialidase is shown in Fig. 1. Table I summarizes the data on the types of sialic acid and the total amount present on the erythrocytes of different species examined. Only sialic acid was released. There was no indication of any other sugar released simultaneously. Moreover, there were no polypeptides or amino acids released from the erythrocytes (9).

Simultaneous Labeling With $Na_2\,^{51}CrO_4$ and Incubation With Enzymes

It was found (9) that, wherever feasible, it is preferable to carry out the labeling step at the same time as the incubation with enzyme, thereby minimizing the physical trauma to the erythrocytes.

TABLE I. Type and Amount of Sialic Acid on Erythrocytes of Different Species

Species	Number of animals tested	Type of sialic acid	μmoles/ml of packed cells	Number of sialic acid residues per erythrocyte × 10^{-7}
Dog	5	NAN	0.93 ± 0.17	7.2
Goat	3	NGN	0.89 ± 0.14	1.1
Rabbit	6	NAN	0.057 ± 0.02	0.32
Chicken	6	NAN	0.29 ± 0.02	1.37

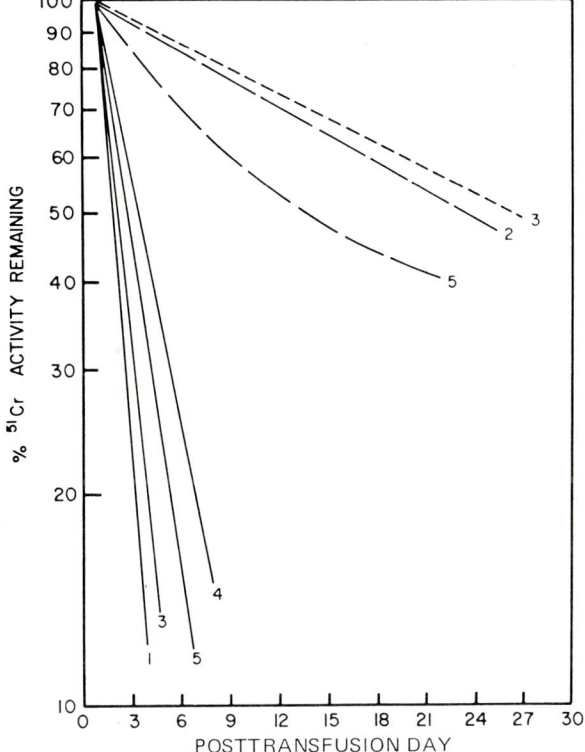

Fig. 2. Posttransfusion survival of dog erythrocytes as determined by the ^{51}Cr technique, plotted on semilogarithmic scale. The radioactivity at 24 h after transfusion is taken as 100% standard for comparison; ———) cells incubated with sialidase in CaCl$_2$/saline; — —) cells incubated in CaCl$_2$/saline; and – – –) untreated cells. The numbers identify the dogs.

Survival of Sialidase-Treated Erythrocytes in Circulation

A typical set of results for the survival of erythrocytes in circulation of the dog are shown in Fig. 2, where the effect of sialidase on the decreased "viability" of the erythrocytes is readily apparent from the decreased half-life (in days) of cells in circulation. This decreased viability is also very readily demonstrated by the percentile of RBC surviving the first 24 h, Table II. Table II also summarizes the similarities and differences in behavior shown by erythrocytes from different species. The results indicate that while all the mammalian erythrocytes lose their viability on treatment with sialidase, chicken erythrocytes appear to be fully viable even though they have lost all their sialic acid.

TABLE II. Survival of Enzyme-Treated and -Untreated Erythrocytes in Circulation

Animal	Number	Untreated B-days	CaCl$_2$/Saline A%	CaCl$_2$/Saline B-days	Sialidase A%	Sialidase B-days
Dog	5	26	97	22	11	3
Goat	5	11	72	10.5	6	4
Rabbit	5	12.5	92.5	13.3	57.5	4.3
Chicken	6	10.5	93	9.5	93	10

A equals the % survival, 24 h postinjection.
B equals the half-life (in days) of those RBC that survive the first 24 h.

TABLE III. Relation Between Amount of Sialic Acid Removed With Sialidase and the Viability of Dog Erythrocytes (see text for experimental details)

Animal number	Amount of sialic acid released, %	Survival of erythrocytes after 24 h, %
1	100	0
2	36	17
3	12	14
4	4	91
5	0	81

Correlation Between the Amount of Sialic Acid Cleaved and the Viability of Dog Erythrocytes.

Washed erythrocytes from a number of dogs were incubated separately with different amounts of sialidase under otherwise identical conditions for the same period of time. Each sample of erythrocytes was labeled simultaneously with the same amount of Na$_2$ ^{51}CrO$_4$ and the cells recovered by centrifugation, washed several times with isotonic saline, resuspended in saline, and transfused autologously. The sialic acid content of the supernatants and washes were then determined by the TBA procedure (14). Table III summarizes the data obtained.

Optimal Concentration of Galactose Oxidase for Erythrocyte Treatment

The rate and extent of oxidation of asialo-erythrocyte stroma from dog and rabbit with varying amounts of galactose oxidase is shown over a period of time in Fig. 3. On the basis of these results, it was decided to use 5 units of galactose oxidase per ml of packed cells for 30 min at 37°C in all subsequent experiments utilizing galactose oxidase. Under these conditions there was less than 0.1% hemolysis.

Effect of Galactose Oxidase With and Without Prior Sialidase Treatment on the Viability of Erythrocytes in Circulation

Table IV summarizes the data obtained with the different animals on the viability of erythrocytes in circulation after various treatments with enzymes. The important facts to emerge were: a) galactose oxidase treatment alone has no deleterious effect on the viability of dog and chicken erythrocytes; b) galactose oxidase treatment alone does, however, have an effect on the viability of rabbit erythrocytes in circulation; c) sialidase followed by galactose oxidase treatment of erythrocytes does not restore the viability of

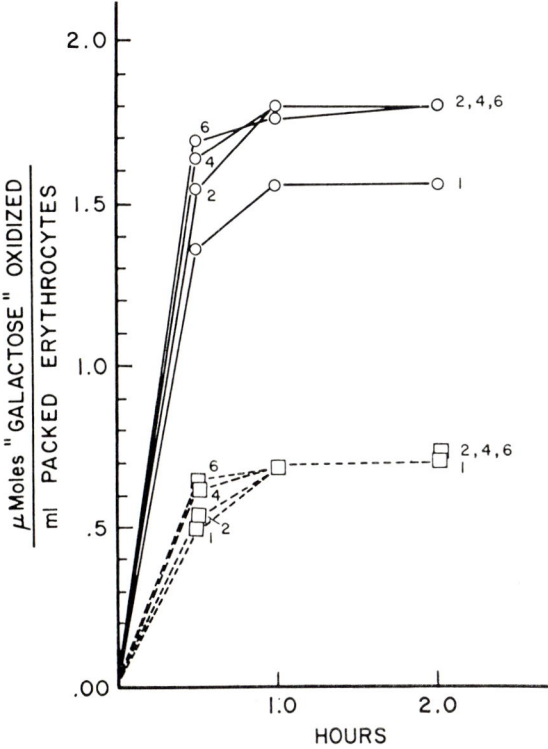

Fig. 3. Rate and extent of oxidation of asialo-erythrocyte stroma from dog (□) and rabbit (○) with varying amounts of galactose oxidase, 1, 2, 4 and 6 units respectively. See text for experimental details.

TABLE IV. The Effect of Sialidase and Galactose Oxidase on the Erythrocytes of Various Species

Animal	Treatment	% Survival after 24 h	Half-life of survivors, days
Dog	Control	83	25
	Sialidase	12	2
	Galactose oxidase	70	25
	Sialidase + galactose oxidase	18	3
Rabbit	Control	86	12
	Sialidase	34	3
	Galactose oxidase	15	4
	Sialidase + galactose oxidase	22	3.5
Chicken	Control	89	8.5
	Sialidase	100	7.5
	Galactose oxidase	100	8.5
	Sialidase + galactose oxidase	2	2

dog and rabbit erythrocytes (the significance of this will be discussed later); d) sialidase followed by galactose oxidase treatment of chicken erythrocytes now renders the erythrocytes nonviable, as compared to the sialidase-treated chicken erythrocytes which are fully viable (Table II).

TABLE V. Incorporation of Tritium From Tritiated Sodium Borohydride Into Sialidase and Galactose Oxidase-Treated Erythrocytes

Animal	Treatment	cpm × 10^4
Dog	Control	2.5
	Sialidase	2.0
	Galactose oxidase	3.9
	Sialidase + galactose oxidase	20.0
Rabbit	Control	3.0
	Sialidase	2.3
	Galactose oxidase	28.0
	Sialidase + galactose oxidase	36.1

Chemical Evidence for the Susceptibility of Rabbit Erythrocytes to Oxidation With Galactose Oxidase

The above data indicated quite clearly the difference of rabbit erythrocytes from those of the other animals examined in terms of their susceptibility to oxidation with galactose oxidase. This difference in biological response was readily reflected by the chemical test for the formation of carbonyl residues using tritiated borohydride (Refs. 15, 18–21, Table V).

Rosettes Obtained With Asialo-Erythrocytes and Liver and Spleen Cells.

Rosettes were obtained under the conditions of assay with asialo-erythrocytes and Kupffer cells of liver and mononuclear cells of the spleen, but not with hepatocytes (Fig. 4). Preliminary attempts at visual quantitation of rosette-formation is summarized in Table VI and indicate that sialidase-treated rat erythrocytes formed rosettes at least 10 times as often with either rat Kupffer cells or spleen mononuclear cells as compared to the autologous rat erythrocytes not treated with sialidase.

DISCUSSION

Methods have been described for the determination of the sialic acid in erythrocytes using either the intact erythrocytes or stroma obtained therefrom. The greatest amounts of sialic acid were released by the action of sialidase on erythrocytes. Complete removal of the sialic acid and its effect on the viability of the resultant erythrocytes in circulation was monitored by tagging the erythrocytes with $Na_2{}^{51}CrO_4$. Two criteria were used to ascertain viability of the erythrocytes in circulation: a) percent survival 24 h after injection, and b) half-life of those erythrocytes that survive the first 24 h after injection (Table II, Fig. 2).

A decreased viability in circulation of sialidase-treated erythrocytes was observed in the case of mammalian but not in chicken erythrocytes. This decreased viability of sialidase-treated erythrocytes could be attributed to increased fragility, to development of autoimmune response, or to panagglutinin-like interactions. For reasons outlined elsewhere (9), however, we believe that the sialidase treatment has a unique action on the mammalian nonnucleated erythrocytes. This phenomenon is reminiscent of the behavior of plasma glycoproteins (22), where the removal of sialic acid residues is a signal for the clearance of the glycoprotein molecule from the circulation. In our experiments, the amount of sialic acid removed also appears to be critical. At least 12% of the total sialic acid has to be removed from dog erythrocytes to markedly affect the 24-h survival of erythrocytes. (Table III). These findings correlate with the observations of others who noted that older erythrocytes contain ~ 10–15% less sialic acid than younger cells (8, 23). The older cells apparently cannot resialate their surfaces and this despite the presence of sialyltransferases in both mammalian plasma (24, 25) and erythrocyte stroma (25).

In contrast to the generalized behavior of mammalian erythrocytes, it would appear that chicken erythrocytes retain their viability in circulation. This could be attributed to a number of reasons, three of which were previously enumerated (9) and can be briefly summarized as follows: a) presence of a nucleus in the chick erythrocyte and therefore of CMP-NAN synthetase, a crucial enzyme in the pathway of resialation, b) a different microcirculation in chicken and mammals, and c) the chemotactic recognition site for senescent avian erythrocyte is a sugar other than galactose, e.g., N-acetylglucosamine (cf. Ref 26). We initially favored (a), but this was ruled out by our inability to demonstrate the presence of CMP-NAN synthetase in either mammalian or chicken erythrocytes. At present we have no data to enable us to choose between the two alternative hypotheses.

The experiments with galactose oxidase were initiated with two objectives in mind: a) to ascertain whether the mechanism for removal of asialo-erythrocytes from circulation is similar to that of asialo-glycoproteins (22), and b) to elucidate the reason for the observed difference in the viability of chicken and mammalian asialo-erythrocytes.

In the system studied so elegantly by Ashwell et al. (22), oxidation of the mammalian asialo-glycoprotein with galactose oxidase restores their viability in the circulation. In this respect the asialo-erythrocytes behave quite differently. Oxidation with galactose oxidase does not restore their viability. The reason for this difference in behavior is not known but we are presently investigating this problem. The preliminary indications would be that the carbonyl group engendered in position C6 of galactose of N-acetylgalactosamine by the treatment with galactose oxidase is the critical structure since it can be detected biologically on a) the terminal sugar of the oligosaccharide chain of mammalian erythrocyte after removal of sialic acid, b) the avian asialo-erythrocyte, and c) the terminal nonreducing sugar in rabbit erythrocytes.

Recently it has been shown that the mammalian asialo-erythrocytes were sequestered in the liver and spleen, thereby accounting for the loss of "viability" in circulation (27, 28). We were interested in the mechanism by which the sialidase-treated erythrocytes were sequestered in these organs and examined the phenomenon at the cellular level.

The observations hereby reported support the following conclusions: a) the apparent lack of reactivity of the hepatocytes is an illustration of the specificity of the interaction, b) the Kupffer cells of liver and mononuclear cells of spleen preferentially react with sialidase-treated as compared to normal erythrocytes, be they of autologous or homologous origin, c) the adhesion phenomenon observed in the formation of rosettes could explain the selective retardation of sialidase-treated erythrocytes in the sinusoids of the liver and spleen, and thereby their sequestration from the circulation.

Normal erythrocytes, not treated with sialidase, also give rosette-like interactions, albeit at a much smaller frequency. We are, therefore, tempted to propose that the consequences of desialation on the viability of erythrocytes in circulation is but an extension of the natural or physiological process of removal of senescent erythrocytes by the liver and spleen.

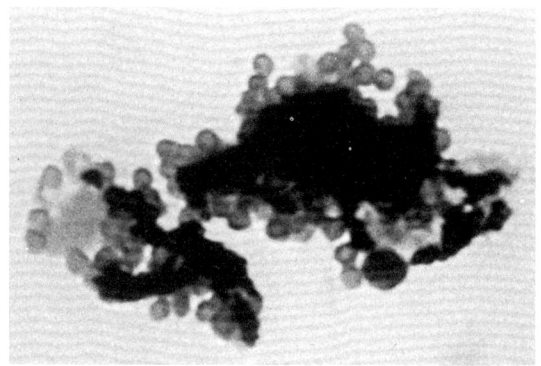
4Aa

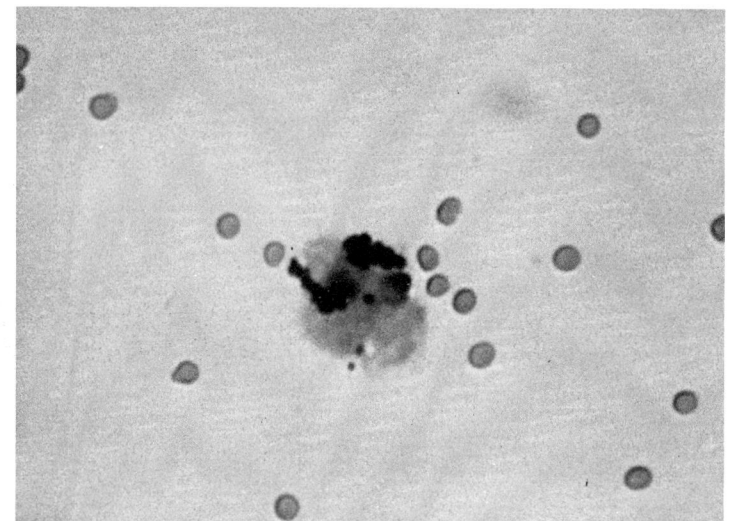
4Ab

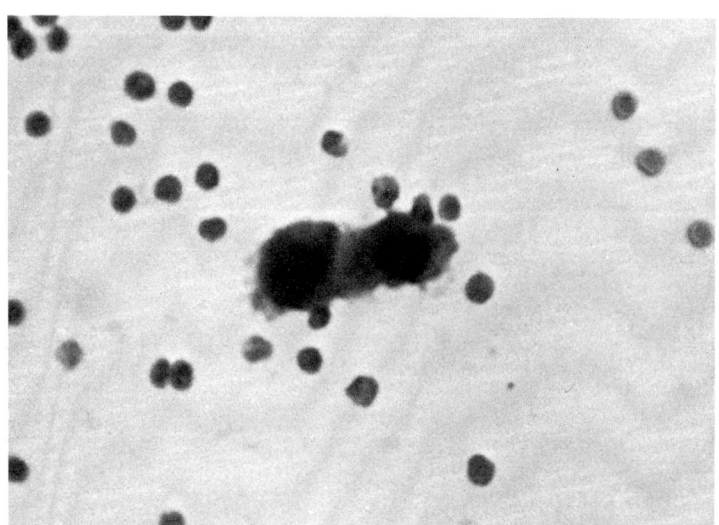
4Ac

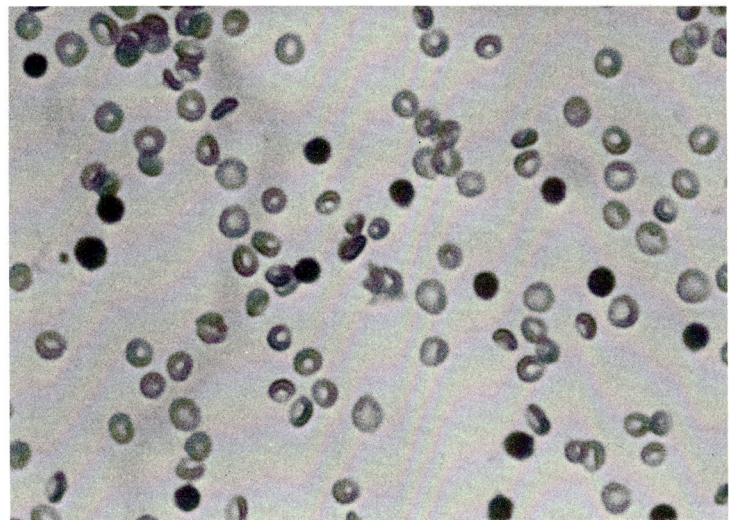

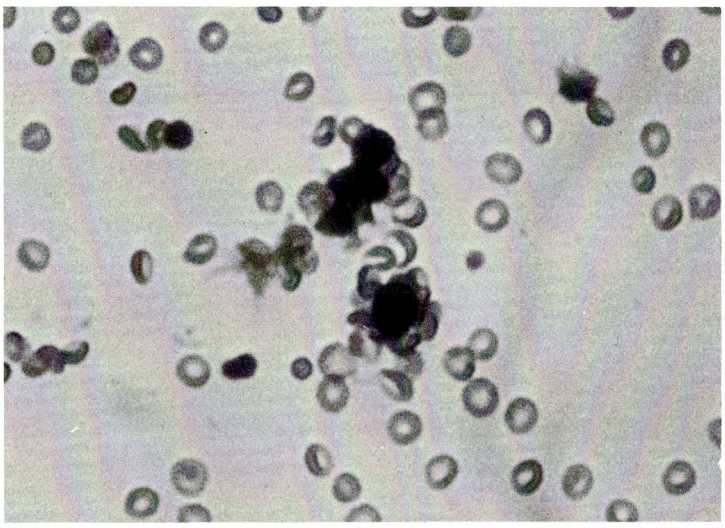

Fig. 4. A) Rat Kupffer cells, identified by partial phagocytosis of colloidal iron filings, form rosettes with sialidase-treated rat erythrocytes (a), but not with untreated erythrocytes (b). Rat hepatocytes with sialidase-treated rat erythrocytes did not form rosettes (c). B) Rat spleen mononuclear cells form rosettes with sialidase-treated rat erythrocytes (a) but not with untreated erythrocytes (b). Original magnification 200 ×, Wright's stain.

TABLE VI. Rosette Formation of Sialidase-Treated and -Untreated Rat Erythrocytes With Rat Kupffer, Hepatocytes, and Spleen Cells*

Cell type	Erythrocytes	
	Treated	Untreated
Kupffer	20–25	2–5
Hepatocytes	1–2	1
Spleen	10–15	1

*The values are given as percentages of viable liver or spleen cells which form rosettes, and represent the mean of at least 3 animals tested.

TABLE VII. Comparison of the Two Systems for Removal of Asialo-Complexes

	Glycoprotein	RBC
1. Site of sequestration	Liver only	Liver and spleen
2. Cells involved	Hepatocytes	Kupffer and mononuclear spleen
3. Galactose oxidase	Restores viability	No effect

This system for the elimination of mammalian asialo-erythrocytes shows a remarkable analogy to the system of Ashwell et al. (22). However, there are important differences (Table VII) which suggest that indeed two distinct systems are involved.

ACKNOWLEDGMENTS

This project was supported in part by grant HL AM 17881 from the National Institutes of Health, and in part by Office of Naval Research contract N-00014-76-C-0269.

REFERENCES

1. Yamakawa T, Suzuki S: J Biochem (Tokyo) 38:199, 1951.
2. Yamakawa T, Suzuki S: J Biochem (Tokyo) 40:7, 1953.
3. Klenk E, Lauenstein K: Hoppe Seyler's Z Physiol Chem 291:249, 1952.
4. Zahler P: Vox Sang 15:81, 1968.
5. Gottschalk A, Belyavin G, Biddle F: In Gottschalk A (ed): "Glycoproteins." 2nd ed. Amsterdam: Elsevier Publishing Company, 1972, p 1082.
6. Danon D: Bibl Haematol (Pavia) 29:178, 1968.
7. Yaari A: Blood 33:159, 1969.
8. Greenwalt TJ, Steane EA, Pine NE: In Jamieson GA, Greenwalt TJ (eds): "Glycoproteins of Blood Cells and Plasma." Philadelphia: JB Lippincott Company, 1971, p 235.
9. Aminoff D, Bell WC, Fulton I, Ingebrigtsen N: Am J Hematol 1:419, 1976.
10. Aminoff D, VorderBruegge WF, Bell WC, Sarpolis K, Williams R: Proc Natl Acad Sci USA 74:1521, 1977.
11. Bell WC, Williams R, Aminoff D: Proc Natl Acad Sci USA Acad Sci USA 74:4205. 1977.
12. Svennerholm L: Biochim Biophys Acta 24:604, 1957.
13. Cassidy JT, Jourdian GW, Roseman S: J Biol Chem 240:3501, 1965.
14. Aminoff D: Biochem J 81:384, 1961.
15. Gahmberg CG, Hakomori S: J Biol Chem 48:4311, 1973.
16. Wincek TJ, Hupka AL, Sweat FW: J Biol Chem 250:8863, 1975.
17. Brain P, Gordon J, Willets W: Clin Exp Immunol 6:681, 1970.
18. McLean C, Werner DA, Aminoff D: Anal Biochem 55:72, 1973.

19. Steck TL, Dawson G: J Biol Chem 249:2135, 1974.
20. Carraway KL, Colton DG, Shin BC, Triplett RB: Biochim Biophys Acta 382:181, 1975.
21. Gahmberg CG: J Biol Chem 251:510, 1976.
22. Ashwell G, Morell AG: Adv Enzymol 41:99, 1974.
23. Baxter A, Beeley JG: Biochem Soc Trans 3:134, 1975.
24. Hudgin RL, Schachter H: Can J Biochem 49:829, 1971.
25. Kim YS, Perdomo J, Bella A, Nordberg J: Biochim Biophys Acta 244:505, 1971.
26. Lunney J, Ashwell G: Proc Natl Acad Sci USA 73:341, 1976.
27. Gregoriadis G, Putman D, Louis L, Neerunjun D: Biochem J 140:323, 1974.
28. Durocher JR, Payne RC, Conrad ME: Blood 45:11, 1975.

Proteins Containing Reductively Aminated Disaccharides: Chemical and Immunochemical Characterization

Gary R. Gray, Barbara A. Schwartz, and Barbara J. Kamicker

Department of Chemistry, University of Minnesota, Minneapolis, Minnesota 55455

Synthetic glycoproteins can be prepared by reductive amination of proteins and reducing carbohydrates in the presence of sodium cyanoborohydride. The reaction proceeds readily in aqueous solution at pH 6–9 to give high degrees of substitution. The degree of substitution can be determined by amino acid analysis, as the 2° amine linkage formed with the ϵ-amino groups of lysine is stable to acid-catalyzed protein hydrolysis conditions.

Antisera have been obtained to bovine serum albumin conjugates containing reductively aminated cellobiose, lactose, and maltose. Preliminary experiments demonstrate that antiserum to the cellobiose-BSA conjugate is hapten-specific, and the structural features of the hapten recognized by the antibodies were established by hapten inhibition experiments. These studies demonstrate that antibodies recognize both the terminal β-glucosyl and acyclic reduced glucosyl residues.

Key words: cyanoborohydride, synthetic glycoproteins, carbohydrate antigens

Increasing interest in the synthesis of carbohydrate antigens has led to the development of several methods of covalently attaching carbohydrate haptens to carrier proteins (1–7). Most of these methods, however, either require prior chemical modification of the carbohydrate, result in low degrees of substitution, or require nonphysiological conditions for the coupling reaction. Recently, the direct reductive amination of reducing carbohydrates to proteins with cyanoborohydride anion has been shown to overcome these difficulties (8–9). This method relies on the ability of cyanoborohydride anion to selec-

Received March 31, 1977.

© 1978 Alan R. Liss, Inc., 150 Fifth Avenue, New York, NY 10011

tively reduce a Schiff base in the presence of a carbonyl at pH > 5 (10). This paper describes the experimental procedure for accomplishing this type of covalent modification, demonstrates that 2° amine linkages are in fact formed with free lysyl amino groups of the protein, and presents preliminary experiments designed to determine the specificities of antibodies formed in rabbits after immunization with a bovine serum albumin conjugate of cellobiose (11).

MATERIALS AND METHODS

Materials

Sodium cyanoborohydride (Alfa Inorganics Division, Ventron Corporation, Danvers, Mass.) was freshly recrystallized as its dioxane complex as described by Borch et al. (10). ϵ-N-1-(1-deoxyglucitol)lysine, α-N-1-(1-deoxyglucitol)lysine, and ϵ-N-di-[1-(1-deoxyglucitol)]lysine were synthesized as described by Schwartz and Gray (9), and α-N-acetyl-ϵ-N-1-(1-deoxyglucitol)lysine, 1-deoxy-1-aminocellobiitol, α-N-acetyl-ϵ-N-1-(1-deoxycellobiitol)-lysine, and α-N-acetyl-ϵ-N-di[1-(1-deoxycellobiitol)]lysine were prepared as described by Kamicker et al. (11).

Coupling of Disaccharides to BSA

Bovine serum albumin (68 mg, 1 μmol), disaccharide (100 mg, 292 μmol), and sodium cyanoborohydride (100 mg, 1.59 mmol) were dissolved in 5.0 ml of 0.2 M potassium phosphate (pH 6–9) and incubated at 37°C. Aliquots (0.5 ml) of the reaction mixture were withdrawn at various times and applied to a 2.5 × 45 cm Bio-Gel P-2 column in 0.1 M potassium phosphate (pH 7.0) containing 0.02% sodium azide (Fig. 1). Fractions (3 ml) were assayed for total carbohydrate by the phenol-sulfuric acid procedure (12) with glucose or galactose as standard, where appropriate, and for total protein by absorbance at 280 nm, using a molar extinction coefficient of 4.57×10^4 for BSA. Fractions containing the conjugates were pooled, dialyzed extensively against distilled water, and lyophilized.

RESULTS

Synthesis of Disaccharide-BSA Conjugates

The effect of pH on the rate of coupling of maltose to BSA is shown on Fig. 2. The rate of reductive amination is essentially the same at pH 6 and 7, but increases at higher pH values. In our hands, however, the most satisfactory results have been obtained at pH 7–8, where denaturation and alkaline degradation are minimized. Not all the disaccharides examined were coupled at the same rate however (Fig 3). Melibiose, lactose, and cellobiose were consistently coupled at a faster rate than maltose.

Amino Acid Analysis of BSA Conjugates

In order to demonstrate that coupling occurs to lysine residues, the conjugates were compared with synthetic α-N-1-(1-deoxyglucitol)lysine (1), ϵ-N-1-(1-deoxyglucitol)-lysine (2a) and ϵ-N-di[1-(1-deoxyglucitol)]lysine (3a) by amino acid analysis. The results of amino acid analysis of the synthetic lysine conjugates are summarized in Table I. Actual retention times were slightly variable in different runs, but they were constant as a fraction of the retention time of an internal lysine standard, and are therefore expressed as

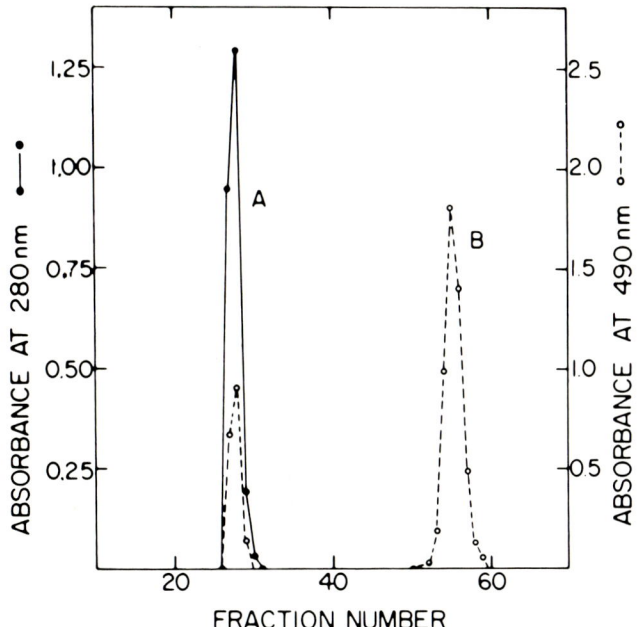

Fig. 1. Purification of disaccharide-BSA conjugates by chromatography on Bio-Gel P-2 (2.5 × 45 cm). In this experiment a 0.5 ml aliquot of a cellobiose-BSA reaction mixture was applied to the column after 72 h. Total carbohydrate was determined by the phenol-sulfuric acid method (12) using 2.0 ml aliquots (A) and 0.1 ml aliquots (B). The cellobiose-BSA conjugate (A) contains 11 mol glucose/mol BSA. B) Unreacted cellobiose.

R_{lys} values. The α- and ε-glucitol lysine derivatives (1 and 2a, respectively) are well separated from lysine under the standard conditions of analysis on a Beckman Model 119 Analyzer, with R_{lys} values of 0.82 and 0.89, respectively. Both 1 and 2a have retention times different from other amino acids under these conditions. The 3° amine, ε-N-di[1-(1-deoxyglucitol)]lysine (3a) was also synthesized by the cyanoborohydride procedure (under conditions of excess reducing sugar) in order to provide a standard for assessing the importance of 3° amine formation in the reductive amination reaction. Compound 3a was also readily observed by amino acid analysis, at a retention time of $R_{lys} = 0.61$.

Both the 2° amine (2a) and 3° amine (3a), the major lysine derivatives expected to be formed in the reductive amination reaction, are therefore readily observed by amino acid analysis. In order to examine the stability of these derivatives to the conditions of acid-catalyzed protein hydrolysis, each was again analyzed after exposure to 6N HCl at 105°C for 21 or 96 hr.

The stability of 2a to protein hydrolysis conditions was determined utilizing a standard solution containing α-N-acetyl-ε-N-1-(1-deoxylactitol)lysine (2b) and valine. The valine concentration (1.93 mM) was established by amino acid analysis and the concentration of 2b was established by the phenol-sulfuric acid method (12) with galactose as standard. Amino acid analysis of the mixture before hydrolysis (Table II) gave valine and a trace of lysine, but after hydrolysis, 2 other components were observed. The major component in the 21-h hydrolysis (R_{lys} = 0.89) is ε-N-1-(1-deoxyglucitol)lysine (2a), the expected product formed by hydrolysis of the N-acetyl and terminal galactosyl residues. The other component (R_{lys} = 0.94) must arise from the degradation of 2a, as its increase in concentration with a longer hydrolysis time exactly corresponds to the decrease in the concentration of 2a. The combined area of the R_{lys} = 0.89 and 0.94 components remains constant under the conditions of protein hydrolysis, indicating that they give exactly the same color yeild. From the known concentration of 2a before hydrolysis, the color yield of the R_{lys} = 0.89 and 0.94 components was found to be 90.0% of the lysine color yield. Free lysine was not formed during the hydrolysis, indicating that, as expected, the 2° amine linkage is stable to protein hydrolysis conditions.

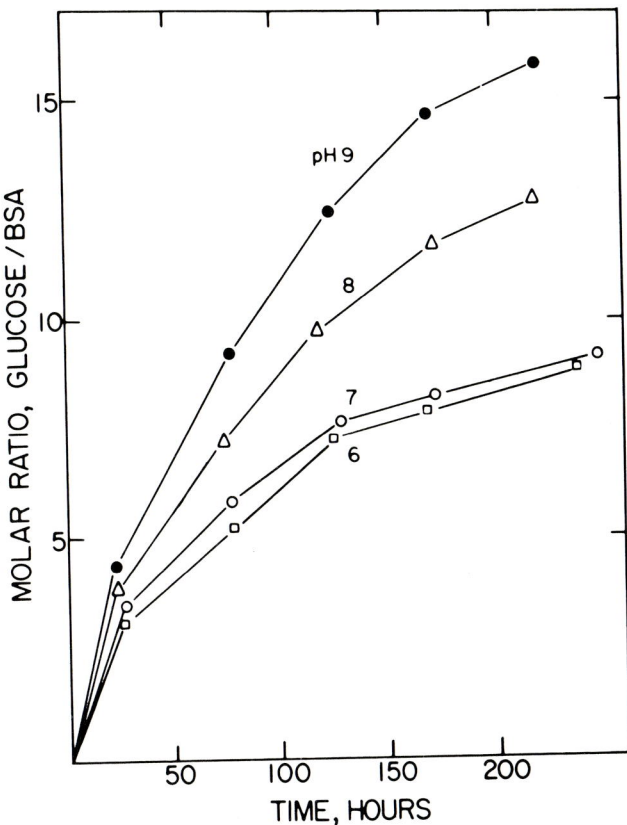

Fig. 2. The effect of pH on the rate of coupling of maltose to bovine serum albumin. Each reaction mixture contained maltose (100 mg), BSA (68 mg), and NaBH$_3$CN (100 mg) in 5 ml of 0.2 M potassium phosphate (pH 6–9) at 37°C. The degree of substitution was determined as described for Fig. 1.

The stability of 3a to protein hydrolysis conditions (Table III) was determined in a similar manner, utilizing a standard solution containing α-N-acetyl-ε-N-di[1-(1-deoxylactitol)]lysine (3b) and valine. The valine concentration of this mixture (2.10 mM) was established by amino acid analysis, and the concentration of 3a and its degradation products were established using the known color yield of 2a (90.0% of the lysine color yield). Under the conditions of acid-catalyzed protein hydrolysis, compound 3a was converted to two other components having R_{lys} values of 0.71 and 0.85 (Table III). Only 3a (R_{lys} = 0.61) and the unknown component with R_{lys} = 0.71 were observed to a significant extent in the 21-h hydrolysis, but the components with R_{lys} = 0.71 and 0.85 were predominant in the 96-h hydrolysis. The total concentration of 3a and its degradation products (0.75 mM) did not decrease with the time of hydrolysis, and the concentration of lysine did not increase. These results indicate that the 3° amine linkage of 3a is stable to acid hydrolysis, and that the degradation products of 3a have the same color yield as 3a.

Selected disaccharide-BSA conjugates, for which the degrees of substitution were known from direct analysis, were hydrolyzed in 6 N HCl at 105°C for 21 hr, then subjected

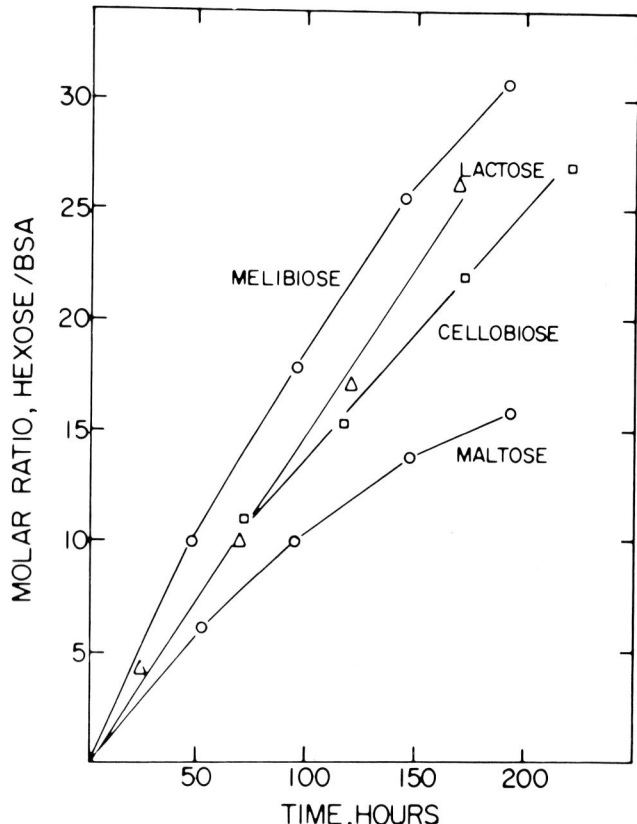

Fig. 3. The rate of coupling of maltose, cellobiose, lactose, and melibiose to bovine serum albumin. Each reaction mixture contained disaccharide (100 mg), BSA (68 mg), and NaBH$_3$CN (100 mg) in 0.5 ml of 0.2 M potassium phosphate (pH 8.0), 37°C. The degree of substitution was determined as described for Fig. 1.

TABLE I. Amino Acid Analysis of Synthetic Lysine Conjugates

Compound	R_{lys}[a]
α-N-1-(1-Deoxyglucitol)lysine (1)	0.82
ε-N-1-(1-Deoxyglucitol)lysine (2a)	0.89
ε-N-Di-[1-(1-deoxyglucitol)]lysine (3a)	0.61

[a]Expressed as a fraction of the retention time of lysine

to amino acid analysis. The amounts of ε-N-1-(1-deoxyglucitol)lysine (2a) and lysine found by amino acid analysis were used to calculate the degree of disaccharide substitution (Table IV). In each case, the molar proportions of 2a and lysine were calculated relative to the known leucine content of BSA (63 leu/BSA; 13), and the observed molar proportion of lysine was subtracted from that of a control sample (59 lys/BSA) to give the molar

TABLE II. Stability of ε-N-1-(1-Deoxyglucitol)lysine (2a) to Protein Hydrolysis Conditions

Amino acid	R_{lys}	Concentration (mM)		
		Standard solution	21-h Hydrolysis	96-h Hydrolysis
Valine	0.51	1.93	1.93	1.93
2a	0.89	1.80	1.26	0.78
Unknown	0.94	0	0.54	1.02
Lysine		trace	trace	trace

TABLE III. Stability of ε-N-di-[1-(1-Deoxyglucitol)] Lysine (3a) to Protein Hydrolysis Conditions

Amino acid	R_{lys}	Concentration (mM)		
		Standard solution	21-h Hydrolysis	96-h Hydrolysis
Valine	0.51	2.10	2.10	2.10
3a	0.61	[a]	0.40	0.08
Unknown	0.71	0	0.31	0.47
Unknown	0.85	0	0.04	0.21
Lysine		trace	0.53[b]	0.49[b]

[a] Contamination of 3b by an unknown which gave lysine and free sugar after hydrolysis prevented the use of direct analysis for carbohydrate.
[b] Formed by hydrolysis of an unknown contaminant, not 3a; see Ref. 9

TABLE IV. The Degree of Substitution of Synthetic Disaccharide-BSA Conjugates

BSA conjugate	Degree of substitution		
	Direct analysis	Amino acid analysis	
		2a formed[b]	Lysine lost
Cellobiose	26.8[a]	18.3	31.2
Lactose	34.1	35.2	45.8
Maltose	15.8	17.5	26.7
Melibiose	17.5	14.7	21.8

[a] Protein concentration from absorbance at 280 nm; carbohydrate concentration by the phenol-sulfuric acid method
[b] Combined areas of R_{lys} = 0.89 and R_{lys} = 0.94 components

proportion of lysine missing. In all conjugates examined, the amount of lysine lost is greater than the amount of 2a formed, and in fact greater than the degree of disaccharide substitution established by direct analysis of protein and total carbohydrate. In most cases, however, the amount of 2a formed is fairly close to the amount expected from direct analysis. The difference between the amount of lysine lost and the amount of 2a formed was not due to degradation of lysine during hydrolysis, as control samples containing BSA and varying amounts of free sugar and alditol were hydrolyzed without loss of lysine.

Compound 3a and its degradation products were not observed in hydrolyzates of the BSA conjugates, but the presence of small amounts of these components cannot be ruled out. The retention times of 3a (R_{lys} = 0.61) and its degradation products (R_{lys} = 0.71 and 0.85) are close enough to those of leucine (R_{lys} = 0.58), tyrosine (R_{lys} = 0.69), and glucosamine (R_{lys} = 0.85) that accurate quantitation is difficult when the latter amines are present.

Immunochemical Properties of the Disaccharide-BSA Conjugates

Antisera to the cellobiose-BSA, lactose-BSA, and maltose-BSA conjugates, obtained from rabbits by the procedure of Vaitukaitis et al. (14), were examined for the presence of hapten-specific antibodies. The precipitin curves for the reaction of the conjugates with their homologous antisera (Figs. 4–6) demonstrate that they are effective precipitating antigens over a wide range of concentration. The cellobiose-BSA and lactose-BSA antisera appear to be primarily hapten-specific, in that there is little cross-reaction with BSA, but the maltose-BSA antiserum appears to be less specific for the conjugate.

Antiserum to the cellobiose-BSA conjugate, which appeared to contain the highest percentage of hapten-specific antibodies, was used to explore in detail the structural features of the hapten recognized by the antibodies. Inhibitors containing all or part of the structural features of the hapten, or those of similar structure, were preincubated with the antiserum over a wide range of concentration, and the optimal amount of the cellobiose-BSA conjugate (6 µg, Fig. 4) was then added. The percent inhibition was determined from the decrease in the amount of protein in the precipitin (15). These studies (Fig. 7) demonstrate that the antiserum is predominantly specific for the reductively aminated cellobiose hapten. Compounds containing the cellobiose residue reductively aminated to the ε-amino group of α-N-acetyl lysine (4a; 6) or to ammonia (4b) are found to inhibit the agglutination reaction to the greatest extent, giving maximal inhibitions of 86–92%. Cellobiitol (4c) is also an effective inhibitor, giving a maximal inhibition of 80%. These inhibitors have in common a β-glucosyl residue linked 1 → 4 to an acyclic reduced glucose residue. Compounds which lack these structural features are ineffective as inhibitors at comparable concentrations. Especially noteworthy are the lack of inhibition by α-N-acetyl-ε-N-1-(1-deoxyglucitol)lysine (5), which structurally represents the linkage region of the hapten, and cellobiose, which contains the reducing glucose residue in its cyclic form.

DISCUSSION

The cyanoborohydride anion catalyzed reductive amination of reducing disaccharides to BSA yields conjugates containing high degrees of substitution over a broad pH range. The reaction proceeds somewhat slowly at neutral pH, but is much faster under more alkaline conditions. This pH effect has also been observed in the reductive amination of reducing disaccharides to aminoethyl polyacrylamide gels by cyanoborohydride (16), but the mechanism was not established.

The products formed in the reductive amination reaction, however, are more thoroughly characterized. That secondary amines are formed with the lysine ε-amino

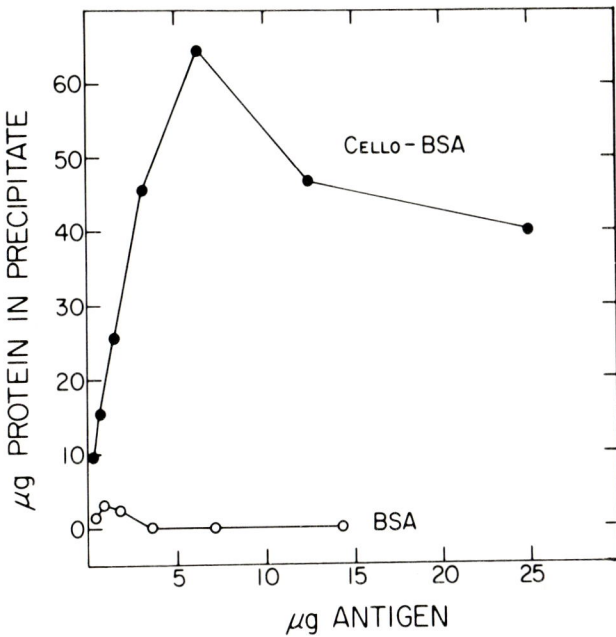

Fig. 4. Quantitative precipitin reactions of immune rabbit serum with BSA and the reductively aminated cellobiose-BSA conjugate (23.7 mol reduced disaccharide/mol BSA). Each reaction mixture contained 50 µl of 4-fold-diluted serum from day 120 and the indicated amount of antigen in a total volume of 500 µl.

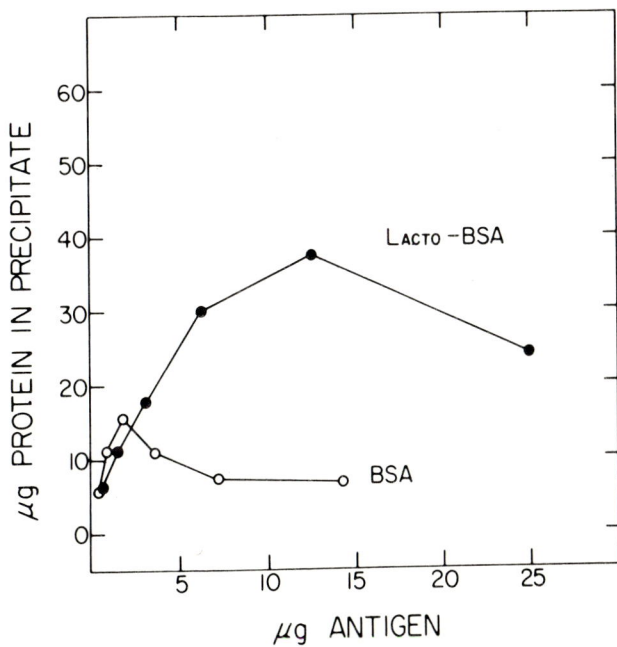

Fig. 5. Quantitative precipitin reactions of immune rabbit serum with BSA and the reductively aminated lactose-BSA conjugate (17.7 mol reduced disaccharide/mol BSA). Each reaction mixture contained 50 µl of 4-fold-diluted serum from day 120 and the indicated amount of antigen in a total volume of 500 µl.

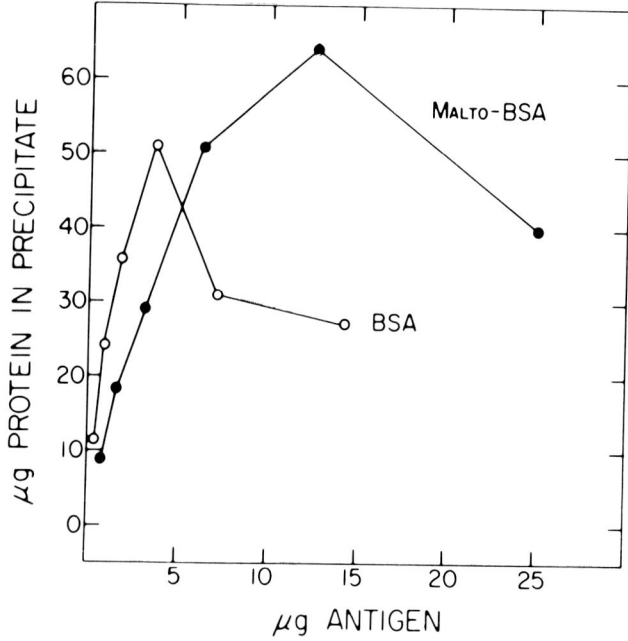

Fig. 6. Quantitative precipitin reactions of immune rabbit serum with BSA and the reductively aminated maltose-BSA conjugate (20.4 mol reduced disaccharide/mol BSA). Each reaction mixture contained 50 μl of 4-fold-diluted serum from day 120 and the indicated amount of antigen in a total volume of 500 μl.

groups of BSA was demonstrated by the identification and quantitative estimation of the amount of ϵ-N-1-(1-deoxyglucitol)lysine (2a) formed by complete acid hydrolysis of the conjugates. The 3° amine, ϵ-N-di-[1-(1-deoxyglucitol)]lysine (3a), was not observed in hydrolyzates of the conjugates examined, but the formation of 3° amines has been observed in the reductive amination of excess disaccharides to α-N-acetyl lysine (9, 11).

Under acid-catalyzed protein hydrolysis conditions, both 2a and 3a are converted to derivatives which give the same color yield but have longer retention times on amino acid analysis. Compound 2a (R_{lys} = 0.89) is converted to a component with R_{lys} = 0.94, and compound 3a (R_{lys} = 0.61) is converted to 2 components with R_{lys} = 0.71 and 0.85. The formation of these derivatives is not unexpected, since polyols are known to be readily dehydrated to anhydroalditols when heated with strong acids (17). This type of structure readily accounts for the facts that the unknowns have the same color yield, but longer retention times than their parent derivatives. The fact that 3a is converted to 2 other products under these conditions again supports the proposed formation of anhydroalditols, i.e., 2 anhydroalditols are expected, resulting from anhydrization of one or both glucitol residues.

Bovine serum albumin conjugates containing reductively aminated disaccharides appear to be very effective antigens in rabbits. Antisera obtained against the lactose-BSA and cellobiose-BSA conjugates were highly hapten-specific, but antiserum to the maltose-

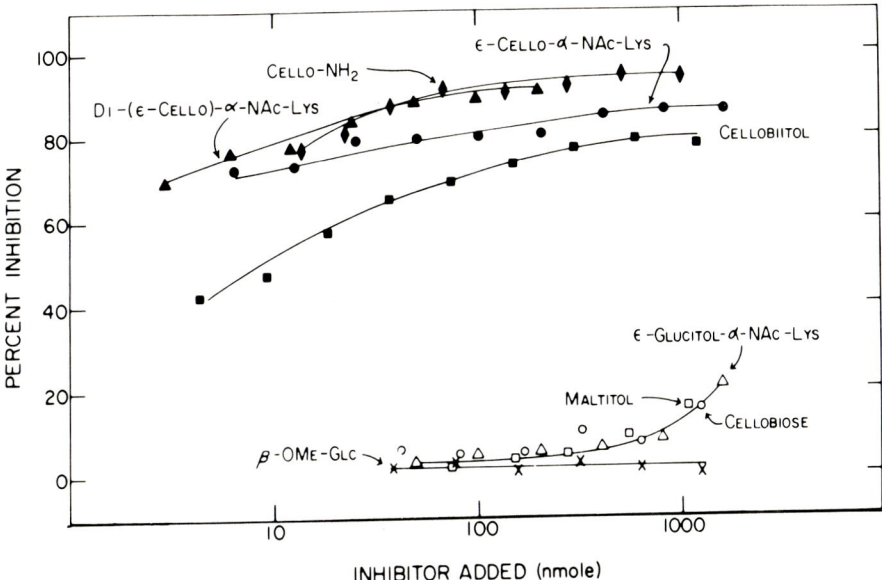

Fig. 7. Inhibition of the precipitin reaction between immune rabbit serum and the cellobiose-BSA conjugate by 1-deoxy-1-aminocellobiitol (♦——♦), α-N-acetyl ε -N-di[1-(1-deoxycellobiitol)] lysine (▲——▲), α-N-acetyl-ε -N-1-(1-deoxycellobiitol)lysine (●——●), cellobiitol (■——■), α-N-acetyl-ε -N-1-(1-deoxyglucitol)lysine (△——△), cellobiose (○——○), maltitol (□——□), and methyl β-D-glucopyranoside (X——X). Fourfold-diluted immune rabbit serum (50 μl) and varying amounts of inhibitor were combined and diluted to a final volume of 460 μl with 0.15 M NaCl. After incubation for 2 hr at 37°C, 50 μl of antigen solution (containing 6 μg Cello-BSA) was added, and the reaction mixtures were thoroughly mixed and incubated for 48 hr at 4°C. The amount of precipitin formed was determined as described in the text.

BSA conjugates appears to be less specific for the hapten. In order to establish the precise structural features of the cellobiose-BSA conjugate recognized by the antibodies, synthetic analogs incorporating various structural features of the hapten were prepared and examined for their ability to inhibit the agglutination reaction between conjugate and homologous serum. These studies demonstrate that antibodies to the conjugate recognize the non-reducing terminal sugar residue, the anomeric configuration of that residue, and the reducing sugar residue in its acyclic form.

These results again demonstrate the importance of the linking "arm" as a determinant for antibodies formed against the hapten of synthetically derived carbohydrate-protein antigens. Avery and Goebel (18) have shown that proteins containing diazotized p-aminophenyl glycosides, for example, p-aminophenyl β-D-glucopyranoside, elicit antibodies complementary to both the β-D-glucopyranosyl and p-aminophenyl residues. Proteins containing disaccharide aldonic acid haptens also elicit antibodies specific for both the terminal carbohydrate moiety and the acyclic aldonic acid arm (2, 6). The observed importance of the linking arm of reductively aminated disaccharide-BSA antigens as an antigenic determinant may not, however, always occur, since the results reported herein are based on antiserum obtained against only 1 conjugate. Based on the work of others cited above, however, it is probable that proteins containing reductively aminated disaccharides cannot be used to obtain antibodies against a single, specific nonreducing monosaccharide determinant. It has been demonstrated with conjugates prepared by the aldonic acid procedure (3, 6) that the importance of the "arm" as a determinant in antibody specificity decreases in conjugates prepared from larger oligosaccharides. The same would be expected to be true for conjugates prepared by the cyanoborohydride procedure, and experiments to determine the minimal size of an oligosaccharide necessary to overcome the influence of the acyclic arm are in progress.

ACKNOWLEDGMENTS

This work was supported by grant CA15325 from the National Institutes of Health, and by a grant from the University of Minnesota Graduate School. Gary R. Gray is the recipient of Faculty Research Award 143 from the American Cancer Society. Barbara J. Kamicker is the recipient of a 1976 Lando Summer Research Fellowship from the Department of Chemistry, University of Minnesota.

REFERENCES

1. Lee YC, Stowell CP, Krantz MJ: Biochemistry 15:3956, 1976.
2. Lönngren J, Goldstein IJ, Niederhuber JE: Arch Biochem Biophys 175:661, 1976.
3. Zopf DA, Ginsburg V: Arch Biochem Biophys 167: 345, 1975.
4. Lemieux RU, Bundle DR, Baker DA: J Am Chem Soc 97:4076, 1975.
5. Himmelspach K, Westphal O, Teichmann B: Eur J Biochem 1:106, 1971.
6. Arakatsu Y, Ashwell G, Kabat EA: J Immunol 97:858, 1966.
7. Westphal O, Feier H: Chem Ber 89:582, 1956.
8. Gray GR: Arch Biochem Biophys 163:426, 1974.
9. Schwartz BA, Gray GR: Arch Biochem Biophys 181:542, 1977.
10. Borch RF, Bernstein MD, Durst HD: J Am Chem Soc 93:2897, 1971.
11. Kamicker BJ, Olson RM, Drinkwitz DC, Gray GR: Arch Biochem Biophys (In press).
12. Dubois M, Gilles KA, Hamilton JK, Rebers PA, Smith F: Anal Chem 28:350, 1966.
13. Stein WH, Moore S: J Biol Chem 178:79, 1949.
14. Vaitukaitis JL, Robbins JB, Nieschlag E, Ross GT: J Clin Endocrinol 33:988, 1971.
15. Lowry OH, Rosebrough NJ, Farr AL, Randall RJ: J Biol Chem 193:265, 1951.
16. Baues RJ, Gray GR: J Biol Chem 252:57, 1977.
17. Soltzberg S: Adv Carbohydr Chem Biochem 25:229, 1970.
18. Avery OT, Goebel WF: J Exp Med 50:533, 1929.

Comparative Biochemistry of Nucleotide-Linked Sugars

Victor Ginsburg

National Institutes of Health, National Institute of Arthritis, Metabolism, and Digestive Diseases, Bethesda, Maryland 20014

Nucleotide-linked sugars have 2 general biochemical functions: they are i) intermediates in the formation of monosaccharides found in complex carbohydrates and ii) glycosyl donors of these monosaccharides. Few sugars arise by reactions not involving nucleotide-linked intermediates. Of these few, glucose, mannose, and N-acetylglucosamine are important in that they are transformed after attachment to nucleotides into most other monosaccharides. Several different nucleotides are involved in these transformations. What factor governs the choice of a particular nucleotide carrier for a given reaction is not apparent, but the use of different nucleotides separates pathways of synthesis and offers a means for their independent control by creating reactions unique to the synthesis of certain products and therefore suitable for regulation. Carrying sugars on different nucleotides may also be advantageous by increasing the accuracy of synthesis of complex carbohydrates. For example, a transferase responsible for the transfer of fucose from GDP-fucose is less likely to transfer galactose from UDP-galactose by mistake than galactose from GDP-galactose. The role of nucleotides in the synthesis of complex carbohydrates thus appears related to the specificity of enzymes that catalyze the modification and transfer of nucleotide-linked sugars.

Key words: pyrophosphorylase, glycosyltransferase

Nucleotide-linked sugars are generally formed by pyrophosphorylases catalyzing reactions between nucleoside triphosphates and a sugar phosphate, as shown in Fig. 1.[1] Although these reactions are freely reversible in vitro they are rendered irreversible in vivo by the rapid hydrolysis of inorganic pyrophosphate. In many cases the sugar incorporated into the nucleotide is modified before it is transferred. In the example shown, glucose is converted into the pentose, L-arabinose, by modifying enzymes (via the intermediate formation of glucuronic acid and xylose). Finally, the modified sugar is transferred to acceptors to form complex carbohydrates. In the synthesis of complex carbohydrates, then, nucleotide-linked sugars have 2 roles. In addition to being glycosyl donors, they are also intermediates in the synthesis of many monosaccharides. In fact, most sugars arise as nucleotide-linked derivatives from nucleotide-linked precursors; relatively few are formed directly from metabolic intermediates by pathways not involving nucleotides. Of these few "primary" sugars, glucose, mannose, and N-acetylglucosamine are especially important in that they are transformed into most other sugars after attachment to nucleotides. Some reactions involved in these transformations are shown in Fig. 2. The first 4 reactions occur in pathways leading to 9 out of the 10 monosaccharides commonly found in the complex carbohydrates of animal cells, as shown in Fig. 3. The tenth monosaccharide, L-iduronic acid, is unusual in that it is formed from glucuronic acid after the incor-

[1] For references to most reactions mentioned in this paper, see Refs. 1 and 2.

Manuscript received May 10, 1977; Accepted for publication May 25, 1977.

© 1978 Alan R. Liss, Inc., 150 Fifth Avenue, New York, NY 10011

1. Activation of sugars:

   glucose−1−P + UTP $\xrightarrow{\textit{pyrophosphorylase}}$ UDP−glucose + PP$_i$

2. Modification of nucleotide-linked sugars:

   UDP−glucose $\xrightarrow{\textit{modifying enzymes}}$ UDP−L−arabinose

3. Transfer of nucleotide-linked sugars:

   UDP−L−arabinose + acceptor $\xrightarrow{\textit{glycosyltransferase}}$ arabinosyl−acceptor

Fig. 1. Nucleotide-linked sugars and the formation of complex carbohydrates.

1. Epimerization:
   UDP−glucose ⟶ UDP−galactose

2. Oxidation:
   UDP−glucose ⟶ UDP−glucuronic acid

3. Decarboxylation:
   UDP−glucuronic acid ⟶ UDP−xylose

4. Reduction:
   GDP−mannose ⟶ GDP−L−fucose

5. Rearrangement:
   TDP−glucuronic acid ⟶ TDP−apiose

Fig. 2. Examples of some modifications of nucleotide-linked sugars.

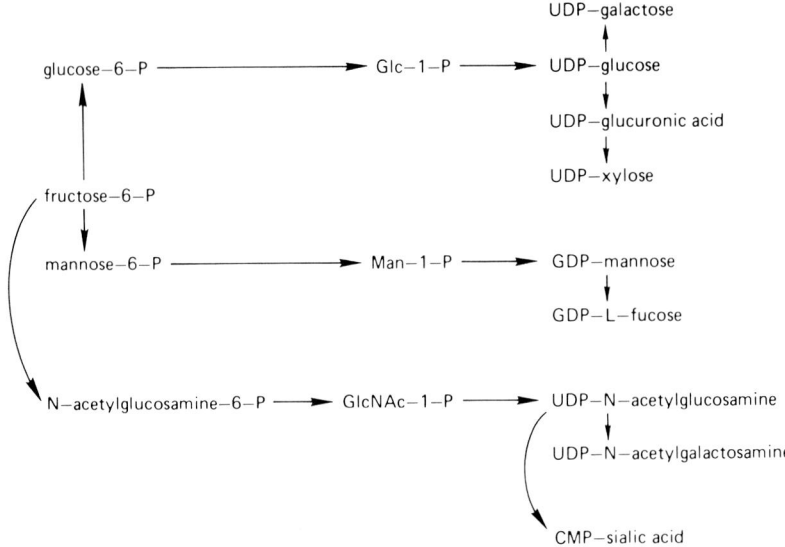

Fig. 3. Formation of sugars found in the complex carbohydrates of animal cells.

poration of glucuronic acid into mucopolysaccharides (4), and its synthesis will be discussed later.

As far as is known, in animals a given sugar is carried by only one base. Fucose, for example, is always carried by GDP and glucose by UDP. In contrast, plants and bacteria utilize several bases to carry the same sugar. In plants, glucose is carried by UDP, ADP, and GDP, as shown in Fig. 4, and glucosyltransferases specific for each nucleotide form glucans with different linkages. Similarly, as shown in Fig. 5, glucose is carried by UDP, TDP, and CDP in bacteria, and modifying enzymes specific for each nucleotide transform glucose into other sugars.

What factors govern the choice of a particular base for a given reaction are not apparent, but clearly the use of different bases sharply separates pathways of synthesis in plants and bacteria. One advantage to the organisms of such a separation might be the possibility for the independent control of various pathways by creating pyrophosphorylase reactions unique to the synthesis of certain products and therefore suitable for regulatory control. CDP-Paratose, for example, is a specific and potent inhibitor of CDP-glucose pyrophosphorylase and presumably regulates its own synthesis by feedback inhibition of this enzyme. CDP-Glucose gives rise to different deoxysugars in different bacteria (Fig. 6), and in each case the nucleotide-linked deoxysugar specifically and nearly completely inhibits CDP-glucose pyrophosphorylase. This type of inhibition implies that CDP-glucose is

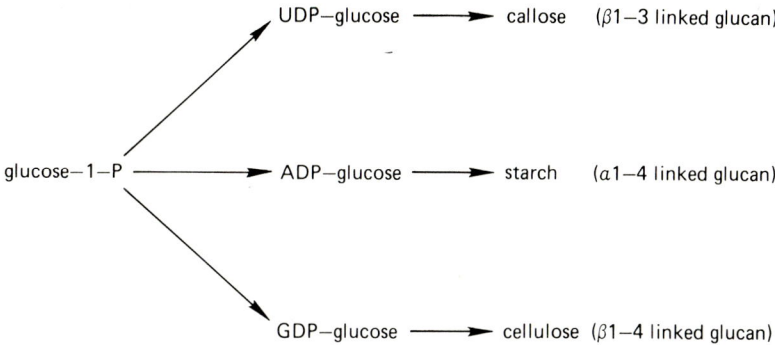

Fig. 4. Formation of plant glucans.

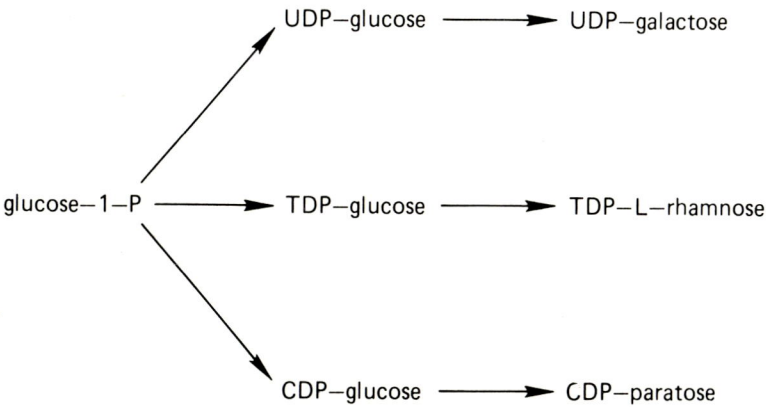

Fig. 5. Reactions of Salmonella paratyphi.

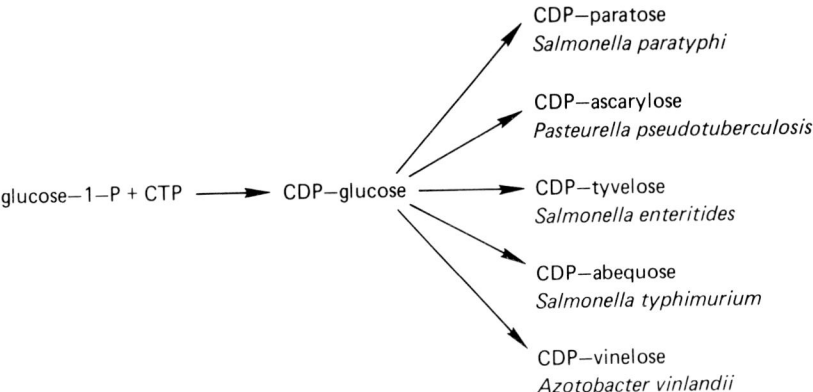

Fig. 6. Formation of deoxysugars in bacteria.

utilized by these bacteria only as a precursor of the different deoxysugars, does not give rise to other sugars, and is not a glycosyl donor itself. That this actually is the case is indicated by the occurrence of CDP-glucose pyrophosphorylase in various bacteria; only those bacteria which make deoxysugars contain the pyrophosphorylase (3), suggesting that in these bacteria, glucose is carried by CDP so as to create a synthetic reaction, which is the first irreversible step in an unbranched pathway and therefore eminently suited for regulation.

Although carrying sugars on different bases may be advantageous for independent regulation, it does not appear to be absolutely necessary. As an example, let us consider the regulation of glycogen synthesis in animals as compared to the regulation of glycogen and starch synthesis in bacteria and plants. As shown in Fig. 7, ADP-glucose is the glucosyl donor in the synthesis of glycogen and starch in bacteria and plants, and regulation of the synthesis of these polymers is at the level of ADP-glucose pyrophosphorylase (5) as indicated by the arrow. Regulation at this step in the sequence implies not only that the pyrophosphorylase step is rate-limiting, but also that the ADP-glucose produced is used only in the synthesis of glycogen and starch and is neither an intermediate in the synthesis of other sugars nor a glucosyl donor in the synthesis of other complex saccharides. Use of ADP as a carrier allows the pyrophosphorylase step to be unique to the synthesis of glycogen and starch in bacteria and plants. However, this does not appear to be necessary.

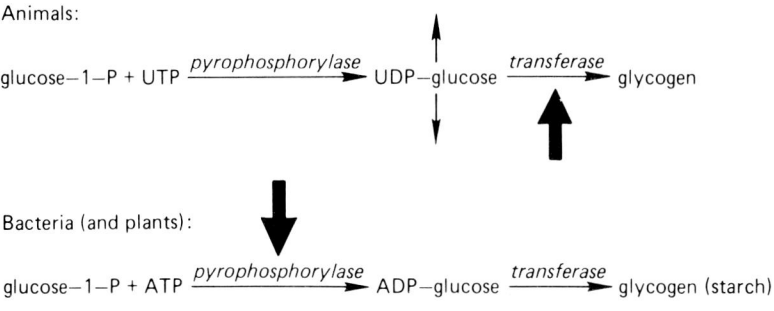

Fig. 7. Regulation of glycogen synthesis.

Animals control the synthesis of glycogen quite well by regulation of the glycosyltransferase rather than the pyrophosphorylase (Fig. 7), presumably because UDP-glucose, which gives rise to glycogen in animals, has many other functions. UDP-Glucose is the precursor of other nucleotide-linked sugars (Fig. 3) and also is a glycosyl donor in the synthesis of complex carbohydrates such as glycolipids and collagen.

Separate pathways for the synthesis of deoxysugars also are not necessary. L-Fucose arises as GDP-L-fucose by reduction of GDP-mannose in both animals and bacteria. In those organisms which also use GDP-mannose as mannosyl donors, the level of GDP-L-fucose is controlled not by inhibition of GDP-mannose pyrophosphorylase, but rather by inhibition of GDP-mannose oxidoreductase, while GDP-mannose regulates its own synthesis by inhibition of GDP-mannose pyrophosphorylase (6). Thus, the levels of both nucleotide-liked sugars are independently regulated, even though they are attached to the same base.

The role of nucleotide-linked sugars as glycosyl donors suggests another possible advantage of using different bases. In the stepwise synthesis of oligosaccharide chains, the structures produced are determined by the inherent specificities of the glycosyltransferases that catalyze each step. However, enzyme specificity is a relative concept, and glycosyltransferases can make mistakes. Carrying monosaccharides on different nucleotides may insure a more accurate synthesis of complex carbohydrates by decreasing the probability of error. Glycosyltransferases have a specificity for both base and sugar, so that an enzyme responsible for the transfer of L-fucose from GDP-L-fucose, for example, is less likely to make a mistake and transfer galactose from UDP-galactose than from GDP-galactose.

As previously mentioned, L-iduronic acid is unusual among the sugars of animals in that it is formed from glucuronic acid by epimerization after glucuronic acid is incorporated into polysaccharides, not while the glucuronic acid is nucleotide-linked. L-Iduronic acid is the C-5 epimer of glucuronic acid (Fig. 8) and occurs, along with glucuronic acid,

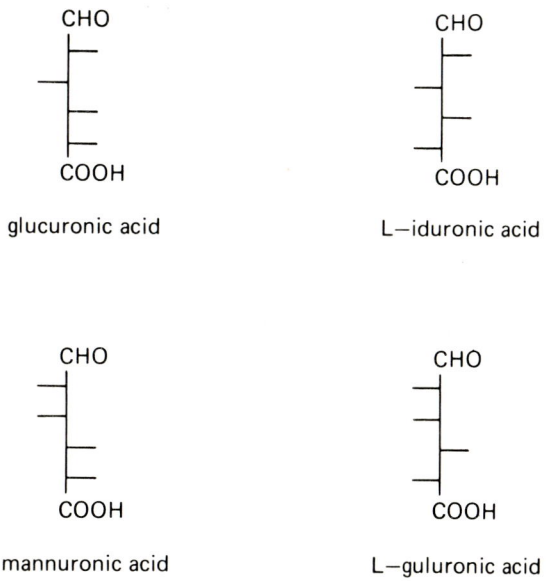

Fig. 8. Some uronic acids.

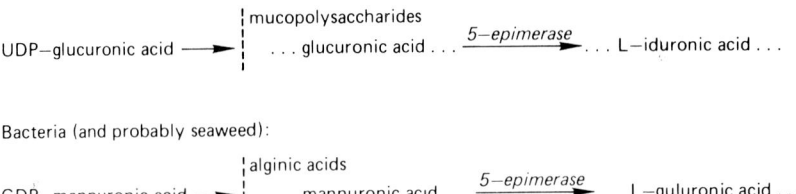

Fig. 9. Formation of L-iduronic acid and L-glucuronic acid.

in mucopolysaccharides such as heparin, heparan sulfate, and dermatan sulfate. An analogous pair of sugars is mannuronic acid and its C-5 epimer, L-guluronic acid (Fig. 8). L-Guluronic acid occurs, along with mannuronic acid, in the alginic acids of seaweeds and bacteria. The analogy between the pairs of uronic acids extends to their synthesis. Like the formation of L-iduronic acid from glucuronic acid, L-guluronic acid is formed from mannuronic acid after the incorporation of mannuronic acid into polysaccharides (7) (Fig. 9). Why is the synthesis of L-iduronic acid and L-guluronic acid different from other sugars? A possible reason is suggested by investigations on the solubility of alginic acids in the presence of calcium ion. Addition of calcium ions to solutions of sodium alginate leads to increased viscosity, gel formation, and precipitation of the alginate. The level of calcium ion required to produce these changes depends on the relative amounts of L-guluronic acid and mannuronic acid in the sample of alginic acid. Samples rich in L-guluronic acid have a higher affinity for calcium ion than samples low in L-guluronic acid (8). Alginic acid covers the exterior surface of seaweeds, and it has been proposed by Larsen that alginic acid may be synthesized in a soluble form (i.e., without L-guluronic acid) and transported to where it is to be laid down. Then some of its mannuronic acid is epimerized to L-guluronic acid, causing the polymer to precipitate with calcium ion. By analogy, the same thing may happen with L-iduronic acid. For example, heparan sulfate, which is a component of cell membranes, might be synthesized as a soluble polymer, transported to where it is to be deposited, and some of its glucuronic acid then epimerized to L-iduronic acid.

REFERENCES

1. Ginsburg V, Kobata A: In Rothfield L (ed): "Structures and Function of Biological Membranes." New York: Academic Press, 1971, p 439.
2. Nikaido H, Hassid WZ: Adv Carbohydr Chem Biochem 26:351, 1971.
3. Nikaido H, Nikaido K: J Biol Chem 241:1376, 1966.
4. Malmström A, Fransson L, Höök M, Lindahl U: J Biol Chem 250:3419, 1975.
5. Furlong CE, Preiss J: J Biol Chem 244:2539, 1969.
6. Kornfeld R, Ginsburg V: Biochim Biophys Acta 117:79, 1966.
7. Haug A, Larsen B: Carbohydr Res 17:297, 1971.
8. Smidsrød O, Haug A: Acta Chem Scand 19:329, 1969.

Immunochemistry of Streptococcal Group C Polysaccharide and the Nature of its Crossreaction With the Forssman Glycolipid

John E. Coligan, Blair A. Fraser, and Thomas J. Kindt

The Rockefeller University, New York, New York 10021

Acid hydrolysis of streptococal Group C polysaccharide yields a disaccharide, 3-O-α-N-acetylgalactosaminosyl-N-acetylgalactosamine (3-O-α-GalNAc-GalNAc) which expresses Group C antigenic activity. This disaccharide, which exists as a side chain in the intact polysaccharide, can completely inhibit the binding between Group C polysaccharide and most Group C antibodies, indicating that this unit is the immunodominant feature of the intact polysaccharide. The α anomeric configuration and N-acetylation are required for the expression of the antigenic activity by the haptenic disaccharide. Also obtained from the acid hydrolysis of the Group C polysaccharide are rhamnose oligosaccharides with structural identity to the Group A-variant polysaccharide and with Group A-variant antigenic activity. It is inferred from these data that the Group A-variant polysaccharide structure is the core unit of the Group C polysaccharide.

The nature of the immunologic crossreactivity between the Forssman glycolipid and Group C polysaccharide, which possess identical nonreducing terminal digalactosamine units, was investigated. Rabbit anti-Group C antibodies bound the Forssman glycolipid with approximately the same affinity as 3-O-α-GalNAc-GalNAc and were capable of mediating lysis of sheep red blood cells (SRBC). Antibody fractions isolated from anti-sheep hemolysin were likewise able to bind Group C polysaccharide. The heterologous reactions were in most assay systems weaker than reactions with the immunizing antigen.

Key words: streptococcus, polysaccharide, Forssman, crossreaction, Group C

Abbreviations: GalNAc — N-acetyl-D-galactosamine; GlcNAc — N-acetyl-D-glucosamine; SRBC — sheep red blood cells; PBSA — 0.15 M NaCl buffered to pH 7.0 with 0.05 M phosphate containing 0.005% NaN_3; α-GalNAc-Sepharose — para-aminophenyl-O-α-D-N-acetylgalactosamine attached to Sepharose 2B; stach — stachyose; raff — raffinose; ManNAc — N-acetyl-D-mannosamine; lact — lactose; 3-O-α-GalNAc-GalNAc — 3-O-α-N-acetylgalactosaminosyl-N-acetylgalactosamine; $GalNH_2$ — galactosamine.

All authors are now located at Building 5, Room B2-31, NIH-NIAID, Bethesda, MD 20014

Received March 23, 1977; accepted June 21, 1977.

© 1978 Alan R. Liss, Inc., 150 Fifth Avenue, New York, NY 10011

The cell wall polysaccharide antigens of the β-hemolytic streptococci provide a means of serologically classifying these organisms into defined groups. Recent studies have been directed at determining the structures and defining the chemical basis of serologic specificity for these antigens.

Group A-variant carbohydrate consists of a linear homopolymer of rhamnose which contains equal amounts of 1,2- and 1,3-linked residues (1). The Group C polysaccharides are postulated to contain the A-variant polysaccharide as a core structure to which immunodominant amino sugar side chains are linked (1). The side chains of the Group A polysaccharide consist of single β-linked N-acetylglucosamine (GlcNAc) residues, which can be removed by enzymatic (2) or chemical (1) means. Structural studies (1) suggest that a disaccharide, 3-O-α-N-acetylgalactosaminosyl-N-acetylgalactosamine (3-O-α-GalNAc-GalNAc), is the side chain moiety of the Group C polysaccharide.

This report describes the isolation, structure, and antigenic nature of this disaccharide and provides evidence that the Group A-variant polysaccharide and the Group C polyrhamnose core are very similar or identical. In addition, it was recognized that the 3-O-α-GalNAc-GalNAc side chains of Group C polysaccharide are identical in structure to the nonreducing terminus of the Forssman glycolipid. Therefore, the extent of serologic cross-reactivity between these antigens of bacterial and mammalian origin was investigated. Rabbit anti-sheep hemolysin was used as a source of antibodies to Forssman glycolipid.

MATERIALS AND METHODS

Streptococcal Group C polysaccharide and A-variant polysaccharide were isolated from strains C74 and A486, respectively, as previously described (3). Antisera directed against these polysaccharides were prepared by immunization of rabbits with heat-killed, pepsin-digested vaccine (4). The Forssman glycolipid was isolated from fresh-frozen horse spleen (Pel-Freeze Bio-Animals, Inc., Rogers, Ark.) as previously described (5, 6). Rabbit anti-sheep hemolysin was purchased from BBL (Div. Becton, Dickinson, and Co., Cockeysville, Md.).

Binding assays utilized the modified Farr procedure as described by Bernstein et al. (7). Labeled carbohydrates were prepared by tyrosylation as described by Gotschlich et al. (8) followed by radioiodination with ^{125}I (9). Inhibition of binding studies were performed as previously described (3). Hemagglutination assays were performed in microtiter plates using 50 μl of an antibody dilution and 50 μl of 2% SRBC per well. Hemolytic assays were performed as described by Papirmeister and Mallette (10) in which 1.25×10^8 SRBC are allowed to stand in the presence of antibody and 24 $C'H_{50}$ complement units for 30 min at 37°C.

Haptens possessing streptococcal Group C and Group A-variant activity were isolated from streptococcal Group C polysaccharide after hydrolysis (0.5 N HCl, 100°C, 30 min) and subsequent chromatography on Bio-Gel P-2 (3). Bovine cerebrosides (Supelco, Inc., Bellefonte, Pa), para-nitrophenyl-N-acetyl-D-α-galactosaminide (Research Products International Corp., Elk Grove Village, Ill.), and para-nitrophenyl-N-acetyl-β-D-galactosaminide (Sigma Chemical Co., St. Louis, Mo.), were used as inhibitors in various assays.

Compounds were assayed for reducing power by the Park-Johnson ferricyanide method (11), for neutral sugars by the phenol-H_2SO_4 assay (12), for $GalNH_2$ by a Fluram assay (3), and for reducing GalNAc by the Benson modification (13) of the Morgan-Elson assay (14).

Galactosamine-containing disaccharides were N-acetylated as previously described (3).

The IgG fractions of rabbit antisera, obtained by DEAE-cellulose chromatography, to Group C carbohydrate or SRBC were further fractionated on a α-GalNAc-Sepharose immunoadsorbent column as described in the Results. The column was prepared commercially (Photo Research Associates, Berkeley, Ca.) by coupling para-aminophenyl-O-α-D-N-acetylgalactosamine to Sepharose 2B using cyanogen bromide.

Samples were permethylated according to the method of Hakomori (15). The masses of the molecular ions of permethylated oligosaccharide derivatives were determined in a Dupont Model 21-490 spectrometer modified for chemical ionization. The samples were introduced by the direct probe and examined at low resolution using isobutane as the reactant gas. In order to determine and quantitate the linkage of the constituent monosaccharides in the polyrhamnose oligosaccharides isolated from streptococccal Group C polysacchride, the permethylated oligosaccharides were analyzed as described by Stellner et al. (16). The partially methylated alditol acetates were identified by comparison of retention times using known standards on a Varian Model 3700 gas chromatograph equipped with a CDS-III integrator. A 10-foot column of 3% ECNSS-M coated on Gas Chrom Q (Applied Science Laboratories, State College, Pa.) was used to separate the sugar derivatives.

Protein sequence analyses were carried out using a Beckman 890B sequencer with a modified DMAA program (111374), as previously described (17).

RESULTS

As reported previously (1), the streptococcal Group C polysaccharide used in the following studies consisted of rhamnose (41%, wt/wt) and GalNAc (43%, wt/wt) with small amounts of GlcNAc (2%, wt/wt) and amino acids (6%, wt/wt). Figure 1 depicts the fractionation by gel filtration of an acid hydrolysate of this polysaccharide. The dot-shaded area (Fractions 101–109) designates fractions possessing Group C antigenic activity; this pool contained 85% wt/wt GalNAc. This active compound was isolated utilizing Bio-Rex 70 chromatography and preparative thin-layer chromatography as previously described (3).

The antigenically active compound was permethylated and its molecular weight determined by chemical ionization mass spectrometry. The mass peak at 536 M/e, identified as the molecular ion, was consistent with a structure of GalNAc-GalNAc. The disaccharide yielded color equivalent to an equimolar amount of GalNAc in both the Park-Johnson ferricyanide reducing sugar test (11) and a modified Morgan-Elson assay (13). These results agree with previous methylation data (1) and support a structure of 3-O-α-GalNAc-GalNAc for the disaccharide. The α-linkage was determined by immunochemical studies discussed later in this report.

It was of interest to examine the immunochemical relationship between the Group C polyrhamnose backbone and the A-variant polysaccharide. Oligosaccharide fragments of the Group C polyrhamnose backbone, devoid of GalNAc, were obtained from the limited acid hydrolysis (Fig. 1, hatched area, Fractions 60–71). These rhamnose oligosaccharides possessed no Group C antigenic activity. However, as shown in Table I, material derived from the hatched area of Fig. 1 was 13 times more effective on a weight basis than intact Group C polysaccharide in inhibiting the binding of ^{125}I-A-variant polysaccharide to anti-A-variant antiserum. Rhamnose oligosaccharides isolated from Fractions 72–80 of Fig. 1 were 5 times more effective than intact Group C polysaccharide in inhibiting the A-variant binding reaction.

The rhamnose oligosaccharides (hatched area, Fig. 1) were analyzed by the methylation procedure of Stellner et al. (16) and were found to contain nearly equal amounts of

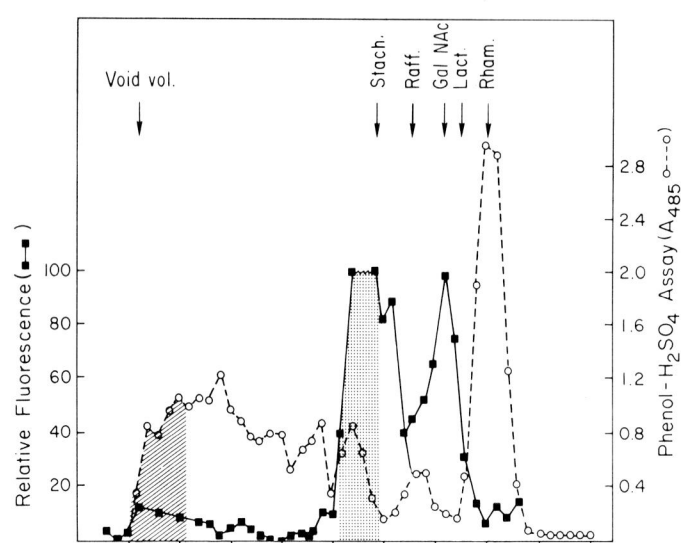

Fig. 1. Bio-Gel P-2 chromatography of hydrolyzed (0.5 N HCl, 30 min, 100° C) streptococcal Group C polysaccharide in 5% acetic acid. Fractions were analyzed for rhamnose by phenol-H_2SO_4 assay (circles) and galactosamine by a Fluram assay (squares). The Fluram assay was performed after additional hydrolysis in 4 N HCl at 100°C for 1 h. The dot-shaded area represents Group C antigenic activity and the hatched area represents Group A-variant activity.

TABLE I. Inhibition of the Binding of ^{125}I-A-Variant Polysaccharide to Rabbit Anti-A-Variant Serum by Group A-Variant Polysaccharide, Group C Polysaccharide, and Group C-Rhamnose Oligosaccharides*

Carbohydrate inhibitor	Amount required to give 50% inhibition	
	µg/ml	ratio
A-variant polysaccharide	9	1
Group C polysaccharide	2,500	278
Group C-rhamnose oligosaccharides	200	22

*Group C-rhamnose oligosaccharides refer to material isolated from Group C polysaccharide by limited acid hydrolysis (Fig. 1, hatched area). These oligosaccharides are essentially devoid of GalNAc.

1,2- and 1,3-linked rhamnose moieties. This result is identical to that found for the A-variant polysaccharide (1). In addition, ^{13}C-NMR spectroscopy showed the Group C-rhamnose oligosaccharides and A-variant polysaccharide to be indistinguishable (Live, Coligan, and Kindt, to be published). These results indicate that the Group C polysaccharide possesses a polyrhamnose unit which is very similar, if not identical, to the Group A-variant polysaccharide.

Specificity of Rabbit Anti-Streptococcal Group C Polysaccharide Antisera

The ability of 3-O-α-GalNAc-GalNAc to inhibit the binding of ^{125}I-streptococcal

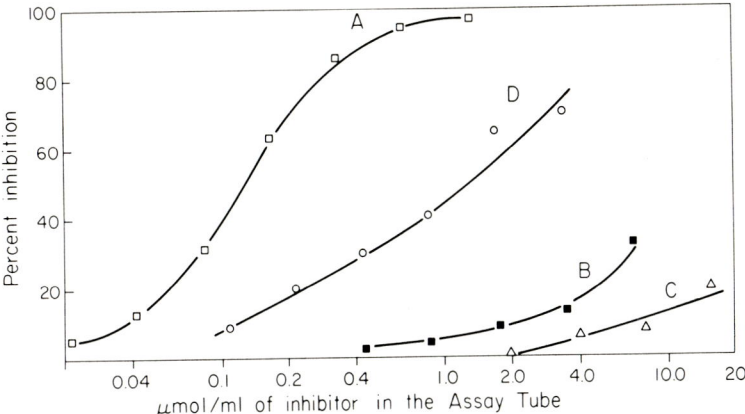

Fig. 2. Inhibition of binding of ^{125}I-streptococcal Group C polysaccharide to rabbit 4588 anti-Group C antiserum by 3-O-α-GalNAc-GalNAc (curve A); de-N-acetylated galactosamine dimer (curve B); GalNAc (curve C); and p-nitrophenyl-N-acetyl-α-D-galactosamide (curve D).

Group C polysaccharide to an anti-Group C antiserum (4588) is shown in Fig. 2 (curve A). In order to investigate whether this inhibition was a general phenomenon, eight additional antisera obtained from various stages of the immunization procedure described by Krause (4) were examined in the inhibition assay. The binding of 7 antisera was inhibited from 82–100% and the binding of one was inhibitied 38% when the assay was performed at a concentration of 3.6 μmol/ml of 3-O-αGalNAc-GalNAc.

Examination of Fig. 2 (curve C) shows that monomeric GalNAc possesses only slight Group C antigenic activity. In comparison (Fig. 2, curve D), para-nitrophenyl-N-acetyl-α-D-galactosaminide is considerably more antigenically reactive than GalNAc, being approximately 10-fold less effective than 3-O-α-GalNAc-GalNAc as an inhibitor. Under the same assay conditions, the β para-nitrophenyl derivative was ineffective as an inhibitor. This illustrates the importance of fixing the nonreducing GalNAc end of the disaccharide in the α anomeric position and possibly of increasing the size of the hapten.

The importance of N-acetylation to Group C antigenic activity is shown by the following results. A compound isolated from the acid hydrolysate of Group C polysaccharide (Fig. 1, Fractions 110–115) was shown to be a de-N-acetylated form of 3-O-α-GalNAc-GalNAc (3). This disaccharide possessed little antigenic activity (Fig. 2, curve B). However, re-N-acetylation yielded an inhibition curve essentially superimposable on curve A in Fig. 2.

One instance in which the binding of a homogeneous antibody (4136) to Group C polysaccharide was especially sensitive to inhibition by GalNAc (compare Table II to Fig. 2, curve C) was examined in greater detail. As shown in Table II, GlcNAc is nearly as effective an inhibitor as GalNAc, whereas N-acetylmannosamine (ManNAc) is ineffective as an inhibitor; the nonacetylated amino sugars were ineffective as inhibitors.

In order to examine whether all antibodies in a particular anti-Group C polysaccharide antiserum were reactive with 3-O-α-GalNAc-GalNAc, antiserum 4588 was fractionated on a column of α-GalNAc-Sepharose (3). As shown in Table III, several populations of antibodies were obtained, some of which (pools 2–4) possess restricted heterogeneity as determined by N-terminal L chain sequence analysis and cellulose acetate electrophoresis.

TABLE II. Inhibition of the Binding of Homogeneous Antibody 4136 to ^{125}I-Streptococcal Group C Carbohydrate by various Amino Sugars

Amino sugars	Assay tube concentration (μmol/ml)	% inhibition of binding
Galactosamine·HCl	7.4	0
N-acetylgalactosamine	6.5	30.2
Glucosamine·HCl	7.5	0
N-acetylglucosamine	7.1	20.3
Mannosamine·HCl	7.5	0
N-acetylmannosamine	6.7	0
p-Nitrophenyl-N-acetyl-α-D-galactosaminide	1.6	32.0
3-O-α-GalNAc-GalNAc	0.25	68.0

Pools 1 and 5 were nonspecifically eluted and are heterogeneous. Pool 1 differs from pool 5 in that it contains IgG other than that specific for Group C polysaccharide.

The amounts of 3-O-α-GalNAc-GalNAc required to give 50% inhibition of binding of ^{125}I-Group C polysaccharide to the 5 antibody pools are indicated in Table III. With the exception of pool 5, there is a direct correlation between avidity for the α-GalNAc-Sepharose immunoadsorbent and the ability of 3-O-α-GalNAc-GalNAc to inhibit binding of the antibodies to labeled Group C. The binding of Group C polysaccharide by antibody pools 2 through 5 was 100% inhibited when 3-O-α-GalNAc-GalNAc was added in sufficient excess over the labeled Group C polysaccharide.

Crossreactivity of Streptococcal Group C Polysaccharide and the Forssman Glycolipid

The Forssman glycosphingolipid is known to have a nonreducing terminus identical to the 3-O-α-GalNAc-GalNAc isolated from streptococcal Group C polysaccharide (5, 18). The capacity of the Forssman antigen to inhibit the binding of the Group C antibody pools of rabbit 4588 is shown in Table III. Notice that the inhibitory capacity of the Forssman hapten on a molar basis is similar to that of 3-O-α-GalNAc-GalNAc for all the antibody pools. In addition, the ability of the Forssman hapten to inhibit binding of streptococcal ^{125}I-Group C polysaccharide to 8 other Group C antisera (mentioned previously) was comparable on a molar basis to 3-O-α-GalNAc-GalNAc.

As shown in Table III, there is a correlation between avidity for the α-GalNAc-Sepharose column and the ability of the anti-Group C antibody pools to lyse SRBC. This agrees with the fact that the Forssman glycolipid is a major antigenic grouping on SRBC (18). All hemolysis was inhibitable by intact Group C polysaccharide.

The IgG fraction from rabbit anti-sheep hemolysin was fractionated on an α-GalNAc-Sepharose column as shown in Fig. 3. The elution of antibodies in pool 4 was slightly retarded by the column compared to pools 1, 2, and 3, indicating weak binding to the α-GalNAc-Sepharose column. Antibodies in pools 5 and 6 more tightly bound to the column and were eluted by the solutions indicated.

A direct correlation was observed between the hemolytic efficiency of antibody pools 4, 5, and 6, and their avidities for the α-GalNAc-Sepharose column (Table IV). These pools when used in sufficient quantity were capable of greater than 80% hemolysis; this hemolysis could be inhibited completely by Group C polysaccharide. Hemolysis titers for pools 1, 2, and 3 could not be determined because all dilutions gave extensive hemagglutination. Hemagglutination of SRBC by the anti-sheep hemolysin fractions are also given in

TABLE III. Characteristics of Antibodies Obtained by Fractionation of Rabbit 4588 Anti-Streptococcal Group C Polysaccharides on a Column of α-GalNAc-Sepharose

Antibody pool	Elution conditions[a]	N-terminal L chain sequence	Amount required to give 50% inhibition of binding to ^{125}I-Group C polysaccharide		% hemolysis[b]
			3-O-α-GalNAc-GalNAc (nmol/ml)	Forssman hapten (nmol/ml)	
1	PBSA[c]	heterogeneous[d]	2,955[e]	>2,000[f]	<1
2	0.33% GalNAc	A-D-V-V-M-T-Q[h]	384	113	<1
3	0.33% GalNAc in PBSA[g]	D-V-V-M-T-Q-T-P-A-S-V-E-A-A-V-G-G-T-V-T-I-K-X-Q-A	68	76	10
4	0.33% GalNAc in 2 M NaCl	D-V-V-M-T-Q-T-P-A-S-V-E-A-A-V-G-G-X-V-X-I-K-C-Q-A	48	73	32
5	5 M guanidine·HCl, pH 8.0	heterogeneous[d]	82	78	44

[a]The column was eluted sequentially with the solutions listed from top to bottom. The A_{280} was allowed to reach baseline between elutants.
[b]The extent of lysis was determined for a 5 mg/ml solution of antibody in the presence of 1.25×10^8 erythrocytes for 30 min at 37°C and in the presence of 24 C'H$_{50}$ complement units.
[c]This fraction was not specifically bound to the α-GalNAc-Sepharose column and contained IgG as well as antibodies to Group C carbohydrate.
[d]Because heterogeneity was detected by cellulose acetate electrophoresis, no attempt was made to determine the sequence of this fraction.
[e]Fraction 1 value was obtained by extrapolation of the inhibition curve through the 50% point. Maximum inhibition obtained with the concentration of inhibitor used was 30%.
[f]At 2,000 nmol/ml, 35% inhibition of binding was achieved.
[g]Antibody fraction 3 eluted in a distinct peak after fraction 2.
[h]Amino acids are represented by the single letter code. Unidentified residues are represented by the letter X.

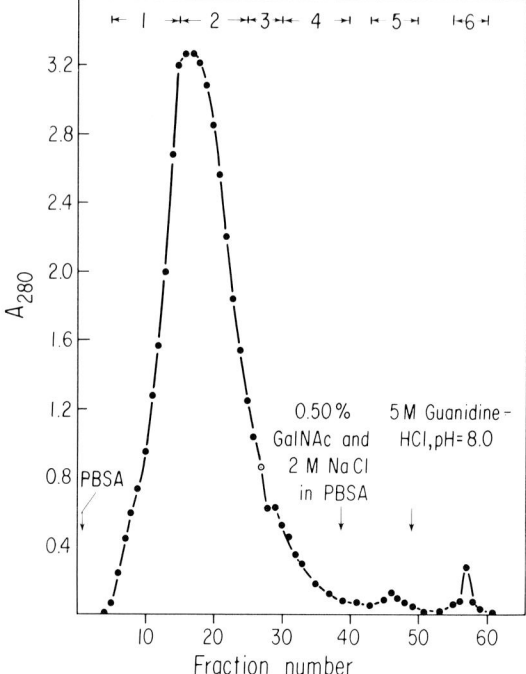

Fig. 3. Fractionation of rabbit anti-sheep hemolysin on a column of α-GalNAc-Sepharose. As shown by the arrows, the column was eluted sequentially with PBSA, 0.5% GalNAc, and 2 M NaCl in PBSA, and 5 M guanidine·HCl (pH 8.0). Six pools were made as indicated.

TABLE IV. Reactivity With Sheep Red Blood Cells of Antibody Fractions Obtained by Fractionation of Anti-Sheep Hemolysin on α-GalNAc-Sepharose

Antibody pool[a]	% hemolysis[b]	% inhibition of hemolysis by Group C carbohydrate[c]	Hemagglutination titer[d]
1	ND[e]	NA[f]	>2,000
2	ND	NA	>2,000
3	ND	NA	>4,000
4	44	100	16
5	51	100	16
6	79	100	32

[a]Refer to Fig. 3.
[b]Extent of hemolysis for a 1 mg/ml solution of antibody in the presence of 1.25×10^8 SRBC for 30 min at 37°C and in presence of 24 C'H$_{50}$ complement units.
[c]Determined at the 20% hemolysis point using 5 mg/ml of streptococcal Group C carbohydrate as inhibitor.
[d]Initial IgG concentration for each fraction was 0.8 mg/ml.
[e]ND = hemolysis of this fraction could not be determined due to extensive hemagglutination.
[f]NA = not applicable.

Table IV. Pools 1, 2, and 3 were much more efficient at hemagglutination than pools 4, 5, and 6 (Table IV).

Figure 4 shows that all the antibody pools obtained from the fractionation of the anti-sheep hemolysin were capable of binding ^{125}I-Group C polysaccharide. There is direct correlation between avidity for the α-GalNAc-Sepharose column and the ability of the IgG to bind ^{125}I-Group C polysaccharide. Pools 1, 2, and 3 are not directly comparable to pools 4, 5, and 6 in their binding efficiency for Group C polysaccharide since the former contain nonreactive IgG in addition to antibody.

For comparison, the binding curves for antibody pools 2, 4, and 6 isolated from 4588 anti-Group C carbohydrate antiserum on α-GalNAc-Sepharose (Table III) are also depicted in Fig. 4. These pools are essentially equivalent to pools 5 and 6 isolated from the fractionation of anti-sheep hemolysin in their avidity for the α-GalNAc-Sepharose column. However, the binding efficiency of the anti-sheep hemolysin pools for ^{125}I-Group C polysaccharide is considerably less than the pools isolated from rabbit 4588 anti-Group C polysaccharide antiserum. The binding to ^{125}I-Group C polysaccharide of all fractions shown in Fig. 4 could be completely inhibited by Forssman hapten, 3-O-α-GalNAc-GalNAc and Group C polysaccharide.

DISCUSSION

A disaccharide, 3-O-α-GalNAc-GalNAc, can be isolated in yields which represent 20–30% of the initial weight of streptococcal Group C polysaccharide. Studies using this disaccharide have provided information on the structure of Group C polysaccharide, on the specificity of anti-Group C antibodies, and on the nature of the crossreaction between the Group C polysaccharide and the Forssman glycolipid.

The isolation of 3-O-α-GalNAc-GalNAc in good yield, coupled with previous chemical studies on Group C polysaccharide (1), has led to proposal of the following structure for this polysaccharide antigen:

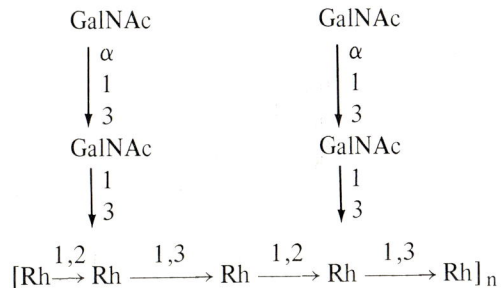

This structure is consistent with the apparent chemical identity of the Group C polyrhamnose backbone and the Group A-variant polysaccharide. The Group A-variant polysaccharide was previously shown to consist of equal amounts of 1,2- and 1,3-linked rhamnose units (1) which were shown by Smith degradation to be alternating linkages (Kindt, unpublished). It had previously been shown that the Group A polysaccharide possesses this same alternating 1,2- and 1,3-linked rhamnose backbone (1). Studies are currently underway to determine if other streptococcal group-specific carbohydrates may possess an A-variant-like core structure.

Antibodies directed against streptococcal Group C polysaccharide recognize 3-O-α-

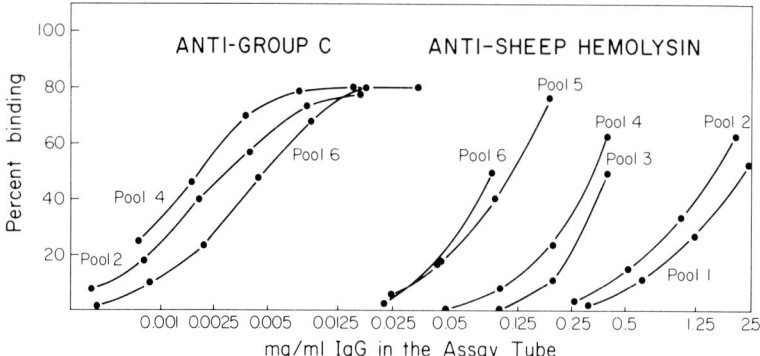

Fig. 4. Binding of ^{125}I-streptococcal Group C polysaccharide by fractions of rabbit anti-Group C polysaccharide and rabbit anti-sheep hemolysin obtained by fractionation on an α-GalNAc-Sepharose column.

GalNAc-GalNAc as an immunodominan group. However, antibodies that did not bind to the α-GalNAc-Sepharose column required relatively large amounts of this hapten to inhibit their binding to the Group C polysaccharide. This result suggests that Group C antibodies of this type either have low affinity for 3-O-α-GalNAc-GalNAc or that they are directed against a larger determinant in the Group C polysaccharide which may include a portion of the rhamnose backbone. It is unlikely that the rhamnose backbone itself can function as a Group C antigenic determinant since polyrhamnose containing fractions isolated from the Group C acid hydrolysate (Fig. 1, Fractions 60–100), which were essentially devoid of amino sugar, possessed no Group C antigenic activity. The immunodominance of the 3-O-α-GalNAc-GalNAc is not surprising since side chains appear to be the major immunodeterminants in antigenic glycans isolated from other microorganisms (18–20).

N-acetylation of the 3-O-α-GalNAc-GalNAc is required for full antigenic activity (3). In one instance, reported here, a homogeneous antibody (4136) appears to have specificity for the N-acetyl group of GalNAc. This was indicated by the almost equal effectiveness of GlcNAc, as compared to GalNAc, in inhibiting the binding of 4136 antibody to Group C polysaccharides whereas ManNAc was ineffective as an inhibitor. This difference in inhibitory capacity of the amino sugars can be explained on the basis of the steric position of the N-acetyl groups. The N-acetyl groups on the C-2 position of GalNAc and GlcNAc occupy sterically identical positions in relation to the substituents on C-1, C-3, C-5, and C-6, whereas the N-acetyl on the C-2 of ManNAc occupies a sterically opposite position in relation to these positions.

As was predicted, due to the presence in both of a common 3-O-α-GalNAc-GalNAc nonreducing terminus, the Forssman glycolipid and streptococcal Group C polysaccharide have been demonstrated to be crossreactive antigens. Antisera and their component antibodies directed aganist Group C polysaccharide show similar reactivity with 3-O-α-GalNAc-GalNAc and the Forssman glycolipid. In addition, antibodies directed against Group C carbohydrate as would be expected, are capable of lysing SRBC. However, as can be seen when values in Table III and Table IV are compared, 5 times more Group C antibody compared to antibody isolated from anti-sheep hemolysin is required to achieve the same level of hemolysis. This is true even with the antibody fractions which bound the α-GalNAc-Sepharose column with approximately the same affinity. This difference in hemolytic efficiency may be due to differences in the antigen binding sites of the antibodies or it may also be possible that these homogeneous Group C antibodies do not efficiently fix complement.

When anti-sheep hemolysin was passaged over α-GalNAc-Sepharose, 6 antibody fractions were obtained, all of which were capable of reacting with SRBC (Table IV) and binding ^{125}I-streptococcal Group C polysaccharide (Fig. 4). However, even though these antibodies were able to bind Group C polysaccharide, they were significantly less efficient than antibodies isolated from anti-Group C polysacchararide antisera. These data indicate that antibodies present in anti-sheep hemolysin reactive with 3-O-α-GalNAc-GalNAc, which are presumably anti-Forssman glycolipid antibodies, possess antibody combining sites significantly different from anti-Group C antibodies even though they recognize the same determinant. It is likely that anti-Forssman glycolipid antibodies require more than the nonreducing terminal 3-O-α-GalNAc-GalNAc of the glycan portion of the Forssman antigen for efficient binding. Kabat (22) has shown that anti-dextran antibodies, while being reactive with smaller dextran-derived oligosaccharides, do not bind with highest affinity until penta- or hexasaccharides are used.

Pools 1, 2, and 3, isolated from anti-sheep hemolysin, were more efficient at hemagglutination than pools 4, 5, and 6. The former pools may contain antibodies other than anti-Forssman glycolipid antibodies which are more efficient in agglutination. Alternatively, anti-Forssman glycolipid antibodies which bind to α-GalNAc-Sepharose may recognize a smaller portion of the glycan moiety of the Forssman glycolipid than anti-Forssman antibodies which do not bind to α-GalNAc-Sepharose; these latter antibodies may agglutinate SRBC more efficiently.

The crossreaction between the Forssman glycolipid and streptococcal Group C polysaccharide has several biological implications. First, one might expect that animals which are positive for Forssman antigen would be poor antibody responders to Group C polysaccharide. Although no systematic studies have been carried out, mice which are thought to be Forssman-positive (23), do not respond well to Group C polysaccharide but produce high concentrations of antibody to Group A polysaccharide. Second, although there are other antigens present in both the Group C streptococci and SRBC, this crossreaction of major antigenic determinants could cause complications in experimental studies in which SRBC are employed as immunogens, especially if the immunized animals have been previously exposed to Group C streptococci. Further aspects and implications of crossreactions between eukaryotic cell surface structures and microbial antigens have been discussed in detail by Springer (24, 25).

Detection and measurement of various antigenic glycosphingolipids isolated from cell surfaces have presented various technical problems (26). Radioimmunoassays using purified bacterial polysaccharides as antigens and detergent-solubilized glycolipid preparations may alleviate difficulties often encountered when using complement fixation and hemagglutination techniques. Assay for Forssman hapten using the Group C streptococcal radioimmunoassay system can perhaps be extended to the assay of other antigenic glycolipids like lactosylceramide and blood group glycolipids using other purified bacterial carbohydrates (26, 27). In addition, it may be possible to use these chemically defined bacterial polysaccharides as immunoadsorbents in further purification of antibodies specific for these glycolipids. These purified antibodies could be made into immunoadsorbents and utilized for isolating glycolipid components of solubilized membrane preparations.

ACKNOWLEDGMENTS

This work was supported by Grant PCM75-21310 from the National Science Foundation and by grant AI08429 from the National Institute of Allergy and Infectious Diseases, National Institutes of Health. Blair A. Fraser is a recipient of a Postdoctoral Research Fellowship (AI05181) from the National Institute of Allergy and Infectious Diseases, NIH.

The authors thank Drs. M. Mudgett, R.M. Krause, and M. McCarty for helpful ideas and discussions concerning portions of this work. The authors also gratefully acknowledge the excellent technical assistance provided by Ms. Maryellen Feeney.

REFERENCES

1. Coligan JE, Schnute WC, Jr, Kindt TJ: J Immunol 114:1654, 1975.
2. McCarty M: The Harvey Lectures, Series 65:73, 1973.
3. Coligan JE, Fraser B, Kindt TJ: J Immunol 118:6, 1977.
4. Krause RM: Adv Immunol 12:1, 1970.
5. Siddiqui B, Hakomori S: J Biol Chem 246:5766, 1971.
6. Sung SJ, Esselman WJ, Sweeley CC: J Biol Chem 248: 6258, 1973.
7. Bernstein D, Klapper DG, Krause RM: J Immunol 114:59, 1975.
8. Gotschlich E, Rey M, Trian R, Sparks KJ: J Clin Invest 51:89, 1972.
9. Greenwood FC, Hunter WM, Glover JS: Biochem J 89:114, 1963.
10. Papirmeister B, Mallette MF: Arch Biochem Biophys 57:94, 1955.
11. Park JT, Johnson MJ: J Biol Chem 181:149, 1949.
12. Dubois M, Gilles KA, Hamilton JK, Rebers PA, Smith F: Anal Chem 28:350, 1956.
13. Benson RL: Carbohydrate Res 42:192, 1975.
14. Morgan WTJ, Elson LA: Biochem J 28:988, 1934.
15. Hakomori S: J Biochem 55:205, 1964.
16. Stellner K, Saito H, Hakomori S: Arch Biochem Biophys 155:464, 1973.
17. Thunberg AL, Kindt TJ: Biochemistry 15:1381, 1976.
18. Fraser BA, Mallette MF: Immunochemistry 11:581, 1974.
19. Luderitz O, Staub AM, Westphal O: Bacteriol Rev 30: 192, 1966.
20. Lipke PN, Raschke WC, Ballou C: Carbohydrate Res 37:23, 1974.
21. Robbins PW, Uchida T: Fed Proc 21:702, 1962.
22. Kabat EA, In "Structural Concepts in Immunology and Immunochemistry." New York: Holt, Rinehart, and Winston, 1968, p 82.
23. Stern K, Davidson I: J Immunol 77:305, 1956.
24. Springer GF: In : Proceedings of the Symposium on New Approaches for Inducing Natural Immunity to Pyogenic Organisms. DHEW Publication No. (NIH) 74-553, p 23, 1974.
25. Springer GF: Prog Allergy 15:9, 1971.
26. Marcus DM, Schwarting GA: Adv Immunol 23:203, 1976.
27. Pazur JH, Cepure A, Kane JA, Hellerquist CG: J Biol Chem 248:279, 1973.

Use of Common Plant Lectins for Isolation and Characterization of Constitutive and Developmentally Regulated Cell Surface Associated Glycoproteins of Dictyostelium discoideum

John E. Geltosky, Jasodhara Ray, and Richard A. Lerner

Department of Immunopathology, Scripps Clinic and Research Foundation, La Jolla, California 92037

Glycoproteins as a class of molecules have been implicated as serving crucial roles in cell recognition events. Using 3 common plant lectins, we have isolated and identified a number of cell surface associated glycoproteins. The appearance of at least 5 of these proteins is under developmental regulation.

Key words: cell surface glycoproteins, Dictyostelium discoideum, lectin affinity chromatography, developmental regulation

In the presence of an adequate supply of nutrients, Dictyostelium discoideum exist as independent amoebae; upon deprivation of the nutrients, the cells stream toward one another, become mutually cohesive, and form tight multicellular aggregates. These structures then undergo further morphogenetic changes, eventually giving rise to the fruiting body, which consists of 2 distinct cell types: stalk cells and spores.

When the protein content of the plasma membrane of cells derived during various stages of the developmental sequence is analyzed, only very minor changes are noted (1, 2). One therefore needs selective probes with which to focus upon molecules of potential interest. One such class of molecules, the cell surface associated glycoproteins, has been implicated as being involved in mediating cell recognition events. We have decided to utilize a number of common plant lectins to select these molecules from the population for further study. Herein, we describe the developmentally regulated appearance on the cell surface of a number of glycoproteins.

MATERIALS AND METHODS

Cell Growth and Development

NC4, the wild type strain, was used for all experiments. Conditions for cell growth, development on pads, and harvesting protocol were as previously described (3).

Jasodhara Ray is on leave of absence from Bose Institute, Calcutta, India.

Abbreviations: Con A – Concanavalin A; AP lectin – Abrus precatorius lectin; RCA_1 – Ricinus communis agglutinin, mol wt = 120,000 daltons; αmm – α-methyl-D-mannoside.

© 1978 Alan R. Liss, Inc., 150 Fifth Avenue, New York, NY 10011

Radioiodination of Cell Surfaces and Subsequent Solubilization

Cells were harvested and plasma membrane associated proteins selectively radioiodinated according to published procedures. Adherence to these labeling conditions results in over 90% of the isotope being bound to membranous material (4). Labeled cells were lysed with 0.5% NP40 in the presence of 0.01% phenylmethylsulfonyl fluoride and then centrifuged at 100,000 $\times$ g of 60'. The resulting supernatant was then treated with the lectin columns.

Preparation of Lectins

Con A (5), AP lectin (6) and RCA_1 (7) were all prepared according to published procedures. The latter was a kind gift of Dr. C. R. Birdwell. Each lectin was coupled to ECD-Sepharose following activation with CNBr (8). Con A was coupled at a concentration of 15 mg/ml, whereas AP lectin and RCA_1 were each coupled at a final concentration of 6 mg/ml.

Lectin Affinity Chromatography

In all experiments, a 1-ml column equilibrated with PBS was utilized. The 100,000 $\times$ g supernatant (usually from 2×10^8 cells) was added to the column, and the unbound material was eluted with PBS at a flow rate of approximately 1 ml/min. When the effluent radioactivity was < 0.01% of the input amount, selective elution with the appropriate sugar was initiated. α-Methyl-D mannose (0.1 M) was used for Con A and D-galactose (0.1 M) for both the AP lectin and RCA_1. The resulting peak of radioactivity was exhaustively dialyzed versus distilled water and then lyophilized.

SDS Gel Electrophoresis

The lyophilized samples were solubilized and electrophoresed on slab gels, the resultant gel being dried down and subjected to autoradiography as previously described (9).

RESULTS

Establishment of Conditions for Chromatography

In order to define the capacity for each lectin column, a glycoprotein with well defined sugar content was treated with 1 ml of packed lectin-Sepharose (Table I, lines 1, 4, 6). In each case, the columns displayed an adequate capacity. A possible explanation for the relatively low capacity of the AP lectin and RCA_1 columns is that incomplete desialiation occurred; this possibility was not further investigated because we are adding very small amounts of Dicyostelium protein and the capacity would be adequate.

Chromatography of Radioiodinated Plasma Membrane Proteins and Their Subsequent Identification

When the labeled material was washed through a 1-ml Con A column at an approximate flow rate of 1 ml/min, 0.25% of the input material was retained and subsequently eluted with 0.1 M αmm. Under the same conditions when both AP lectin and RCA_1 columns were used, approximately 1.5% of the input material was specifically eluted with 0.1 M D-galactose. When untreated Sepharose columns were used instead, only 0.09% of the material was eluted with galactose.

TABLE I. Capacity and Specificity of Lectin Columns

Column material[a]	Sample description		Eluants (% radioactivity eluted)			
		PBS	αmm	D-galactose	L-fucose	N-acetyl-D-glucosamine
Con A-Sepharose	horseradish peroxidase[b]	10	90	NT[d]	NT	NT
Con A-Sepharose	125I-NC4 0-h sonicate	99	0.25	<0.01	<0.01	0.02
BSA-Sepharose	125I-NC4 6-h sonicate	99	<0.01	NT	NT	NT
AP lectin-Sepharose	desialiated fetuin[c] (1 mg/ml)	87	0.5	12	NT	NT
AP lectin-Sepharose	NC4 0-h sonicate	98	0.05	1.5	NT	NT
RCA$_1$-Sepharose	desialiated fetuin (1 mg/ml)	88	0.7	11	NT	NT
RCA$_1$-Sepharose	NC4 0-h sonicate	98	NT	1.6	NT	NT

[a]One-milliliter lectin-Sepharose columns contained one of the following: Con A at 10–15 mg/ml packed Sepharose; AP lectin and RCA$_1$ at 6 mg/ml packed Sepharose; BSA at 3 mg/ml of packed Sepharose.
[b]Measured at 403 nm.
[c]Fetuin desialiated by published procedure (10). The material was then radioiodinated with chloramine T (11).
[d]Not tested.

Developmentally Regulated Appearance of Cell Surface Glycoproteins

Con A. As we have previously described (9), the appearance of 2 cell surface associated Con A binding proteins is under developmental regulation (Fig. 1). By 6 h in development, roughly the time at which the cells begin to display a mutual cohesiveness, a new protein, the 150 k molecule, appears on the surface of the cell. Coincidental with this occurrence, a molecule of approximate mol wt of 180 k is no longer detected on the cell surface.

AP lectin. There are approximately 12 readily distinguishable polypeptides associated with the cell surface that binds to the galactose specific AP lectin (Fig. 2). The appearance of at least 3 of these glycoproteins is under developmental regulation. Vegetative cells display a protein of approximate mol wt of 34,000 daltons that disappears by 6 h into development and never reappears. At 9 h of development, a molecule of approximate mol wt of 38,000 daltons appears on the cell surface and is retained until at least the 18-h stage. An interesting event takes place at the 6-h point in development. A molecule of approximate mol wt of 24,000 daltons appears and by 9 h has disappeared from the cell surface.

RCA$_1$. RCA$_1$ recognizes many of the same proteins as does AP lectin. The procedure with RCA$_1$, however, does not detect the 3 developmentally regulated proteins as does the AP lectin (Fig. 3). The most heavily labeled bands are recognized by both lectins.

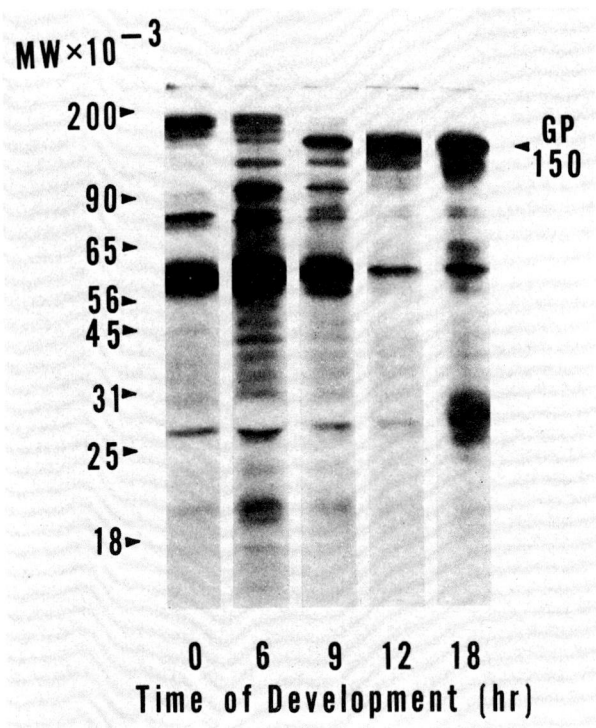

Fig. 1. Con A binding proteins on the surface of cells during development. Cells were harvested during various periods of development and subjected to the procedures described in Materials and Methods. This figure represents an autoradiogram derived from a 7–15% acrylamide slab gel on the αmm eluted material. Equivalent amounts of radioactivity were used for each sample. Gp150 refers to the developmentally regulated glycoprotein of mw of 150,000 daltons.

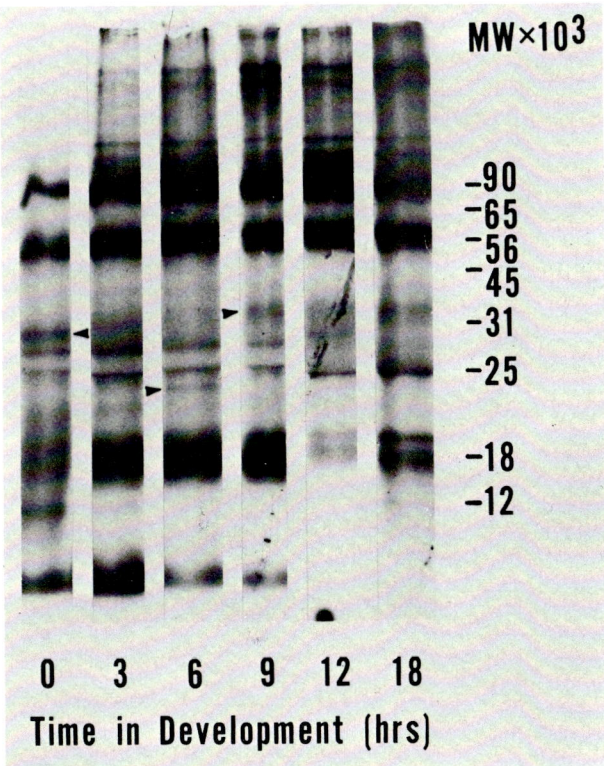

Fig. 2. AP lectin binding proteins on the surface of cells during development. The conditions here are identical to those described in Fig. 1 except that AP lectin columns were used.

DISCUSSION

The combined techniques of selective radioiodination of plasma membrane associated proteins followed by lectin affinity chromatography provide a powerful means of isolating molecules of potential interest for further study. In this case, cell surface associated glycoproteins are focused upon. Treatment of solubilized plasma membrane protein with Con A results in at least a 400-fold enrichment of glycoprotein from the original population. In the cases with both the AP lectin and RCA_1, a 70-fold enrichment for these particular glycoproteins results. These figures are based on radioiodination data, so that they are admittedly spurious. When a greater quantity of protein is passed over a Con A column and optical density at 280 nm monitored, a 100-fold enrichment of Con A binding proteins from the population results (JEG, unpublished results).

A number of cell surface associated glycoproteins are recognized by both Con A and the AP lectin: the 29, 56, 90, and 130 k proteins. Of interest is that none of the developmentally regulated proteins is apparently recognized by more than 1 lectin. This situation can obviously be exploited for purification purposes. It is curious that RCA_1 does not recognize the 3 developmentally regulated proteins that AP lectin does. Possibly the binding properties of the 2 lectins are diverse enough to create this situation.

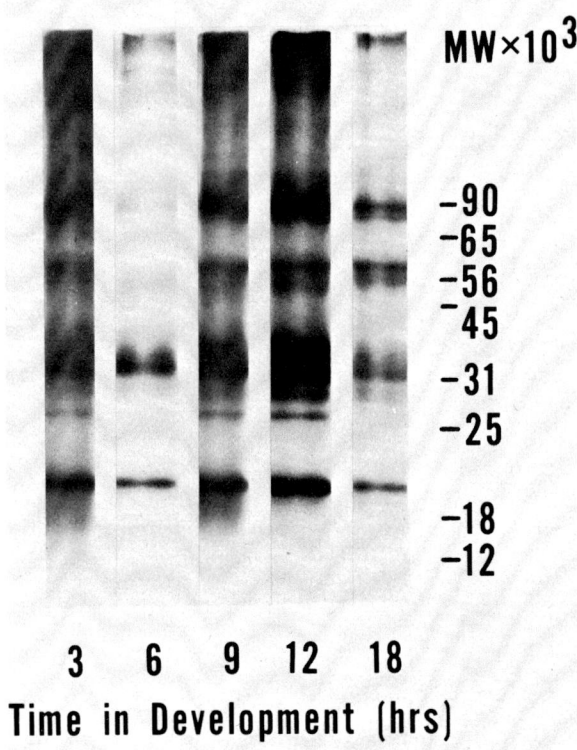

Fig. 3. RCA_1 binding proteins on the surface of cells during development. The conditions here are identical to those described in Fig. 1 except that RCA_1 columns were used.

The rapid disappearance from the cell surface of the 180k, the 34k, and the 24k proteins serves to illustrate the principle that selective loss of certain molecules from the cell surface does occur. Such events may, indeed, play integral roles in cell cohesion events. Experiments are currently in progress to assess this possibility.

The AP lectin and RCA_1 binding proteins have potential additional interest, since galactose binding lectin is associated with the cell surface of Dictyostelium discoideum (4, 12). If, in fact, this lectin activity is used in the normal sequence of developmental events, its receptor could be a galactose containing glycoprotein. These investigations are also in progress.

ACKNOWLEDGMENTS

This research was supported by National Institutes of Health grant GM-22500-01. This is publication number 1312 from the Department of Immunopathology, Scripps Clinic and Research Foundation, La Jolla, California 92037.

REFERENCES

1. Smart JE, Hynes RO: Nature 251:319, 1974.
2. Siu C-H, Lerner RA: (In press).
3. Sussman M: In Prescott D (ed): "Methods in Cell Physiology." New York: Academic Press, 1966 vol 2, pp 397–410.
4. Siu C-H, Lerner RA, Ma G, Firtel RA, Loomis WF: J Mol Biol 100:157, 1976.
5. Agrawal BBL, Goldstein IJ: Biochem Biophys Acta 147:262, 1967.
6. Ray J, Som S, Sen A: Arch Biochem Biophys 174:359, 1976.
7. Nicolson GL, Blaustein J: Biochim Biophys Acta 226:543, 1972.
8. Porath J, Aspberg K, Drevin H, Axen R: J Chromatog 86:53, 1973.
9. Geltosky JE, Siu C-H, Lerner RA: Cell 8:391, 1976.
10. Spiro RG: J Biol Chem 239:567, 1964.
11. McConahey PJ, Dixon FJ: Int Arch Allergy Appl Immunol 29:185, 1966.
12. Rosen SD, Kafka JA, Simpson DL, Barondes SH: Proc Natl Acad Sci USA 73:577, 1973.

The Initial Glycosylation of the Sindbis Virus Membrane Proteins

Bartholomew M. Sefton

Tumor Virology Laboratory, The Salk Institute, San Diego, California 92112

The mechanism by which the membrane proteins of Sindbis virus are initially glycosylated during growth of the virus in chick cells was studied. The experiments suggest strongly that the 2 viral glycoproteins are glycosylated before release from the polysome and that this glycosylation involves transfer of a large, 1,800 dalton oligosaccharide to the polypeptide chains. The donor of the oligosaccharide is most probably a lipid.

Key words: lipid-linked oligosaccharides, Sindbis virus, nascent polypeptides

It is of interest to know when, where, and how membrane proteins are glycosylated. A series of experiments was therefore carried out to characterize the initial addition of carbohydrate to the membrane proteins of Sindbis virus during growth of the virus in chick embryo cells. This system has 2 particular advantages. First, the structures of the oligosaccharide side chains of the 2 virion membrane proteins, E1 and E2, are now known with some certainty (1, 2). Second, infection of chick cells with Sindbis virus causes sufficient inhibition of host-cell protein synthesis that several hours after infection the only membrane proteins being synthesized by these cells are those of the virus (3).

It is now clear that each glycoprotein of Sindbis virus contains 2 oligosaccharides which are linked to the polypeptide by an N-glycosidic bond between asparagine and N-acetyl glucosamine. One site of attachment on each polypeptide contains a complex A-type oligosaccharide and the other site on each polypeptide contains a simpler B-type oligosaccharide. The 2 A-type oligosaccharides and the 2 B-type oligosaccharides are very similar, if not identical in structure. Two details of the structure of these oligosaccharides must be kept in mind to properly interpret the experiments to be described. 1) The cores of all the viral oligosaccharides contain mannose. 2) If these oligosaccharides were synthesized by polymerization of monosaccharides on the polypeptide, the third sugar to be incorporated would be mannose.

Received March 29, 1977

© 1978 Alan R. Liss, Inc., 150 Fifth Avenue, New York, NY 10011

All the experiments described below involve pulse-labeling of infected chick cells late after infection. At this time in the infection, the cells are both synthesizing new viral proteins and assembling and releasing mature virions. As a result, the cells, at the time of the addition of the labeled compound, already contain the full spectrum of possible forms of the viral proteins, from only partially synthesized proteins to fully modified proteins. The experiments therefore examine the entry of label into an ongoing system.

MATERIALS AND METHODS

Cells

All experiments were performed with secondary chick embryo cells. In general, the cells were seeded at a density of 1.5×10^6 cells per 35 mm plastic petri dish (Falcon) and infected with wild-type Sindbis virus at a high multiplicity 18–24 hr after subculturing. In experiments which involved labeling the cells with radioactive sugars, the cells were incubated before infection in glucose-free Eagle's minimal essential medium (MEM) supplemented with 2% tryptose phosphate broth, 4% calf serum, 1% heat-inactivated chicken serum, 5 mM sodium pyruvate, and the nonessential amino acids. Once infected, the cells were incubated in the above medium supplemented in addition such that it contained glucose at a concentration of 10% of normal (100 µg/ml).

In most other experiments, cells were incubated, both before and after infection, in MEM supplemented with 2% tryptose phosphate broth and 1% calf serum. In most experiments, virus-infected cells were treated with actinomycin D (1 or 2 µg/ml) throughout the infection.

Glucose Starvation

Glucose-starved cells were produced by incubation of the cells, before and after infection, in medium containing a very low concentration of glucose and during labeling, in medium containing no glucose. The actual concentration of glucose in the incubation medium was not measured. The MEM contained no glucose, but glucose was present in the tryptose phosphate broth and in the serum, which together constituted 6% of the medium volume. Actual labeling of the cells was done in glucose and serum-free MEM. In some experiments, the cells were incubated just prior to labeling in completely glucose-free medium for 5–30 min.

Radioactive Labeling of Infected Cells

The details of the individual experiments are presented in the figure legends. Pulse-labeling was generally carried out at 37°C, in air, in medium containing only enough sodium bicarbonate to maintain in appropriate pH. Labeling for longer than 30 min was done in a CO_2 incubator.

SDS Polyacrylamide Gel Electrophoresis

(SDS) polyacrylamide gel electrophoresis was performed in 1-mm thick slab gels essentially as described by Laemmli (4), except that the buffer in the resolving gel had a pH of 8.6. Samples were initially dissolved in 2% SDS, 0.1 M dithiothreitol, 5 or 10 mM sodium phosphate, pH 7.0, 20% glycerol, 1% mercaptoethanol, and 2 mM phenylmethansulfonylfluoride (PMSF). All samples were then mixed with fresh mercaptoethanol, usually 10% (vol/vol) just prior to electrophoresis. Following electrophoresis, the gels were stained and fixed and then impregnated with diphenyloxazole (5). Fluorography was as described by Laskey and Mills (6), using Kodak RP/R54 (now XR-5) film. Exposures varied in length from 8 h to 6 weeks.

Extraction of Lipids

Redistilled chloroform and methanol were used in all the extractions. Chloroform/methanol (2:1, vol/vol), usually 5 ml, was added to a frozen cell pellet, $2-6 \times 10^7$ cells, and the solution was mixed vigorously with a vortex mixer. After at least 15 min at room temperature, the suspension was centrifuged and the resulting supernatant removed. The cellular debris was then reextracted with 2.5–3 ml of chloroform/methanol (2:1), centrifuged, and the 2 supernatants combined. This supernatant was then filtered through a Whatman GF/C filter or a HAWP millipore filter to remove insoluble cellular material and then washed extensively with 5 mM $CaCl_2$, as described by Folch, Lees, and Sloane-Stanley (7).

The remaining cellular material was washed once or twice more with chloroform/methanol (2:1), dried under N_2, and then washed 2 or 3 more times with approximately 8 ml of H_2O. Methanol (usually 0.3–0.6 ml) was than added to the insoluble material and the suspension dried with N_2. The cellular material was then extracted twice with chloroform/methanol/water (10:10:3, vol/vol) as just described for chloroform/methanol (2:1) (8). These 2 supernatants were combined and filtered as above. Finally, the resulting insoluble material was washed once or twice with approximately 3 ml of chloroform/methanol/water (10:10:3, vol/vol) and dried with N_2.

Mild Acid Hydrolysis

Lipids to be hydrolyzed were dried with N_2 and then dissolved in 0.4 ml of tetrahydrofuran. To this was added 0.1 ml of 0.5 normal HCl and the mixture was incubated for 60 min at 50°C (9). The solution was then neutralized with 0.025 ml of 2 M tris-(hydroxymethyl)aminomethane base (Tris) and dried with N_2.

Pronase Digestion

Digestion of samples with pronase for gel filtration was as described previously (1).

Gel Filtration

Gel filtration was exactly as described before except that Bio-Gel P-4 (200/400 mesh, Bio-Rad) was used instead of Bio-Gel P-6 (1).

RESULTS

When does the initial addition of carbohydrate occur? When is mannose added to E2?

The virion glycoprotein E2 is released from the polysome as a precursor polypeptide, PE2, which is approximately 10,000 daltons larger than E2. The precursor is cleaved to yield E2 only after an obligatory lag of 20 min (10). This cleavage is necessary for the incorporation of the polypeptide into virions.

One knows that this cleavage occurs 20 min after synthesis because radioactive amino acids are detected in E2 no sooner than 20 min after the onset of labeling. Similarly, one can estimate the time of incorporation of mannose into this polypeptide by measuring the lag with which radioactive mannose appears in E2. Such an experiment is shown in

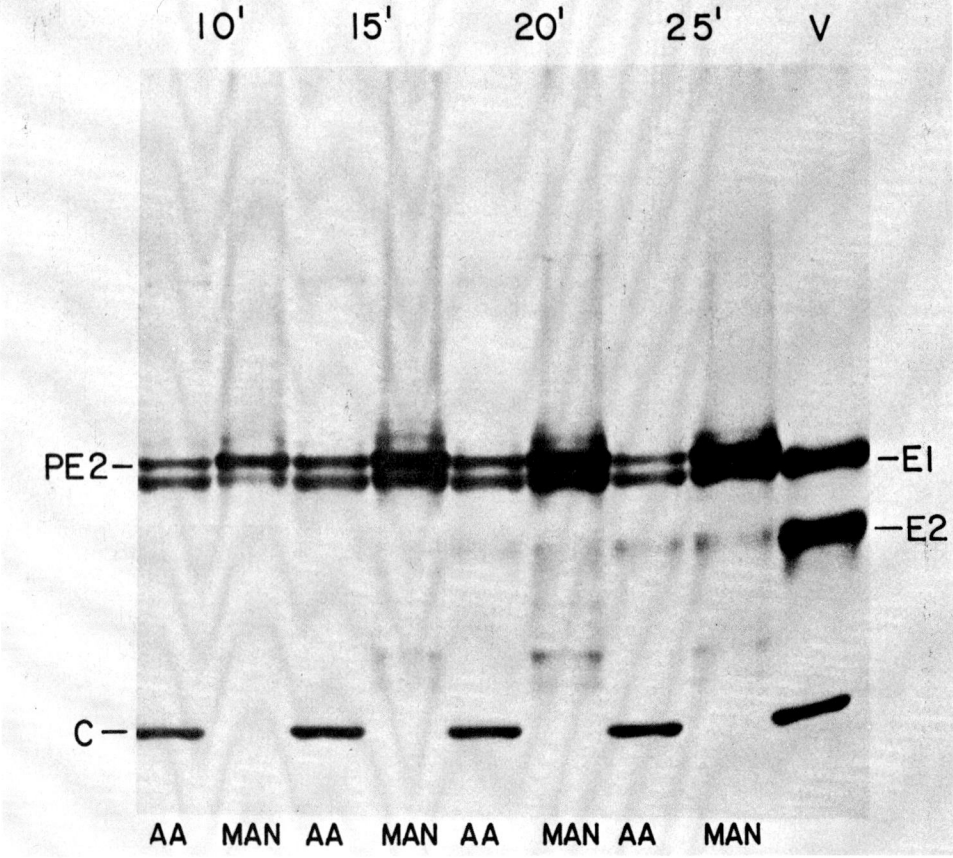

Fig. 1. It takes 20 min to label E2 with either [^{3}H]mannose or [^{3}H]amino acids. Infected cells were labeled with mixed [^{3}H]amino acids (50 μCi/ml, Amersham) or with [^{3}H]-2-mannose (200 μCi/ml, 2 Ci/ml, Amersham) for 10 min, 7 hr after infection with Sindbis virus. The labeling medium was MEM containing no nonradioactive amino acids, a reduced concentration of glucose (100 μg/ml), sodium pyruvate (5 mM), and 1% chicken serum. At the end of the 10 min labeling period, the cells were washed 3 times with, and then further incubated in, enriched Eagle's medium supplemented with 5 mM sodium pyruvate, the nonessential amino acids, and 1% calf serum. At the appropriate intervals, the cells were washed once with Tris-buffered saline, dissolved in SDS, and subjected to SDS polyacrylamide gel electrophoresis. The figure is a fluorogram of such a gel. The length of time between the addition of the label and the dissolving of the cells in SDS is designated at the top. The labeled precursor is designated at the bottom. Reprinted with permission from Cell.

Fig. 1. The figure is an autoradiogram of an SDS polyacrylamide gel of Sindbis-infected chick cells which had been dissolved in SDS at various times after the addition of either radioactive amino acids or radioactive mannose to the cells. It can be seen that no radioactivity is detectable in E2 15 min after the start of labeling but that E2 is labeled with both amino acids and mannose by 20 min. This indicates that the final addition of both amino acids and mannose to PE2 occurred at the same time, between 15 and 20 min before cleavage. Interpretation of this experiment is not complicated by differential dilution of the labels by the intracellular pools because PE2 could be labeled with both amino acids and mannose within 1 min of addition of the label.

Do Cells Contain Unglycosylated Membrane Proteins?

This result suggested that glycosylation occurred before synthesis of the protein was complete. To examine this question in another way, an effort was made to detect, within the infected cell, fully synthesized but not yet glycosylated viral proteins. In order to do this, it was necessary to know how to identify an unglycosylated protein.

It has been observed by Kaluza that the glycoproteins of Semliki Forest virus which are synthesized by infected cells growing in glucose-free medium migrate more rapidly than usual in an SDS polyacrylamide gel (11). A similar phenomenon was observed in glucose-starved Sindbis-infected chick cells. The proteins synthesized by infected cells growing in either complete or glucose-free medium are compared in Fig. 2. During glucose starvation little PE2 or E1 were synthesized. Instead, several new polypeptides which migrated more rapidly than E1 were seen. As is described below, it was possible to demonstrate that these abberrant polypeptides were forms of the viral membrane porteins which were less glycosylated than usual. The new polypeptide with the greatest electrophoretic mobility appeared to contain no carbohydrate at all, while the other apparently larger polypeptides appeared to be glycosylated, but only partially. It is therefore possible to ask whether unglycosylated or partially glycosylated viral glycoproteins are ever present in cells growing in complete medium by comparison with the polypeptides synthesized in glucose-starved cells.

The demonstration that the aberrant polypeptides synthesized during glucose starvation were indeed under-glycosylated viral membrane proteins came from 3 observations. 1) The methionine-containing tryptic peptides of the aberrant polypeptides were identical to those of authentic PE2 and E1. 2) The aberrant polypeptide with the greatest electrophoretic mobility could not be labeled with either glucosamine or mannose. 3) This same polypeptide would not bind to 2 lectins which bind PE2, E1, and other mannose-containing glycoproteins.

Sindbis-infected cells, growing in complete medium, were labeled very briefly with [^{35}S] methionine and the labeled proteins were compared with those synthesized during glucose starvation by SDS polyacrylamide gel electrophoresis (Fig. 3). Full-sized E1 was detectable in cells labeled for 30 sec and full-sized PE2 was detectable in cells labeled for 120 sec. No discrete polypeptides similar to those made during glucose starvation were seen in any of the samples from cells growing in complete medium. Literally, this result says that the lifetime of an unglycosylated E1 molecule is much shorter than 30 sec.

How Are These Proteins Glycosylated?

Clearly these proteins are glycosylated very early in their lifetime. An effort was made to determine whether this initial addition of carbohydrate occurred by the polymerization of monosaccharides on the polypeptide, with nucleotide sugars as the immediate donor, or by transfer of a preassembled oligosaccharide, with a lipid-oligosaccharide as the immediate donor. In theory, these 2 mechanisms can be distinguished if infected cells are labeled with a radioactive sugar for only an instant and the sizes of the protein-bound radioactive oligosaccharides then examined. One would detect small, unfinished protein-bound oligosaccharides if growth occurred by the sequential addition of monosaccharides.

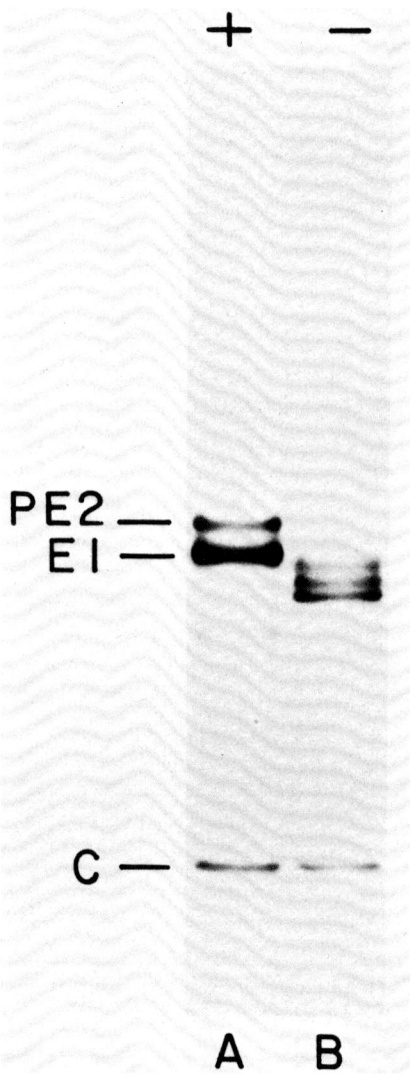

Fig. 2. Effect of glucose starvation on the synthesis of the Sindbis virus glycoproteins. Infected cells were labeled with [^{35}S]methionine (25 μCi/ml, 250 Ci/mmole, New England Nuclear) for approximately 30 min. Glucose-starved cells, prepared as described in Materials and Methods, were labeled 7.5 hr after infection in glucose-free and methionine-free medium. The control cells were labeled 6 hr after infection in methionine-free medium. The labeled cells were then dissolved for 20 min at 4°C in 1.0 ml of 1% NP40, 0.15 M NaCl, and 0.01 M sodium phosphate, pH 7.0 (glucose-starved cells) or of 1% NP 40, 0.2 M NaCl, 0.05 M Tris-HCl, pH 8.0, and 2 mM PMSF (control cells). Nuclei were removed by centrifugation and the samples prepared for electrophoresis. The figure is a fluorogram of an SDS polyacrylamide gel containing these 2 samples. a) Cells in complete medium; b) Glucose-starved cells. Reprinted with permission from Cell.

Sindbus Virus Glycosylation 627

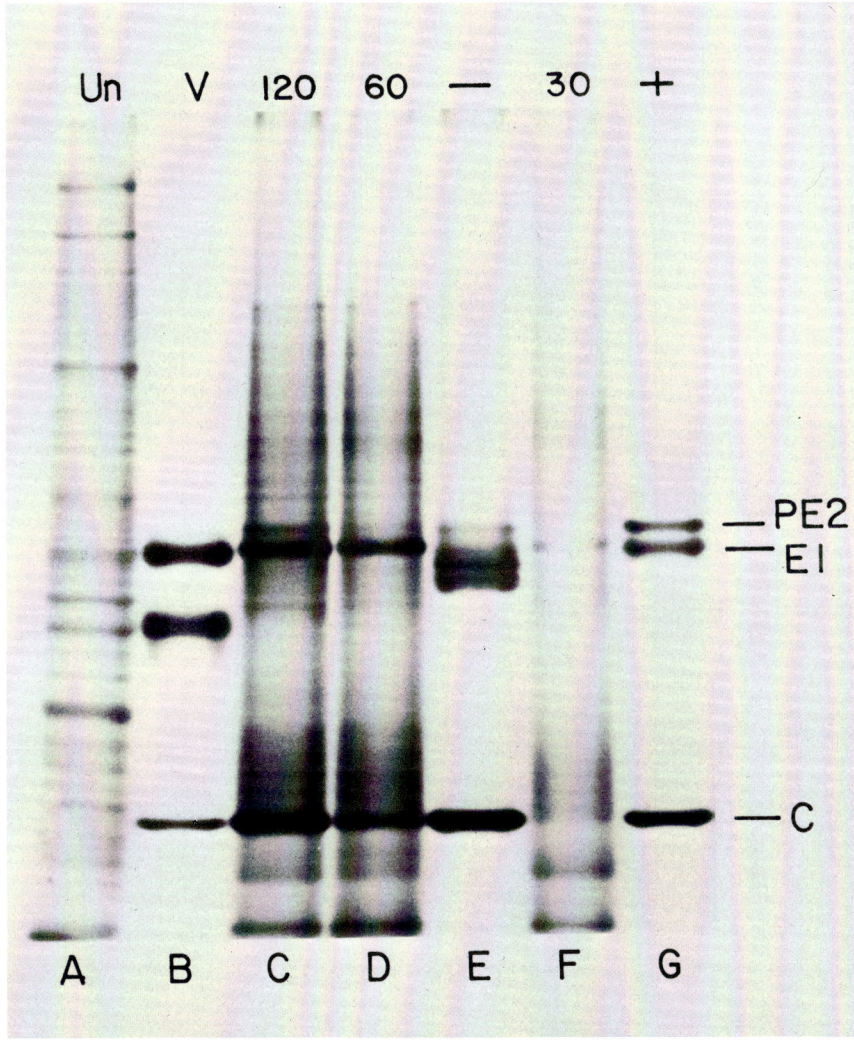

Fig. 3. Size of the glycoproteins of Sindbis virus labeled during a very brief exposure of Sindbis-infected cells to [^{35}S]methionine. Sindbis-infected cells, incubated in complete MEM, were labeled with [^{35}S] methionine (180 μCi/ml) for 30, 60, or 120 sec, 7.25 hr after infection. The cells were then washed with Tris-buffered saline, dissolved in SDS, and subjected to SDS polyacrylamide gel electrophoresis. Washing of the cells took approximately 12 sec. The figure is a fluorogram of the gel.
a) Uninfected chick cells labeled with mixed [^{3}H] amino acids for 10 min; b) Purified Sindbis virions labeled with mixed [^{14}C] amino acids; c) Infected cells labeled for 120 sec with [^{35}S] methionine; d) Infected cells labeled for 60 sec with [^{35}S] methionine; e) Glucose-starved infected cells labeled for 20 min with [^{35}S] methionine; f) Infected cells labeled for 30 sec with [^{35}S] methionine; g) Infected cells labeled for 10 min with [^{35}S] methionine. Reprinted with permission from Cell.

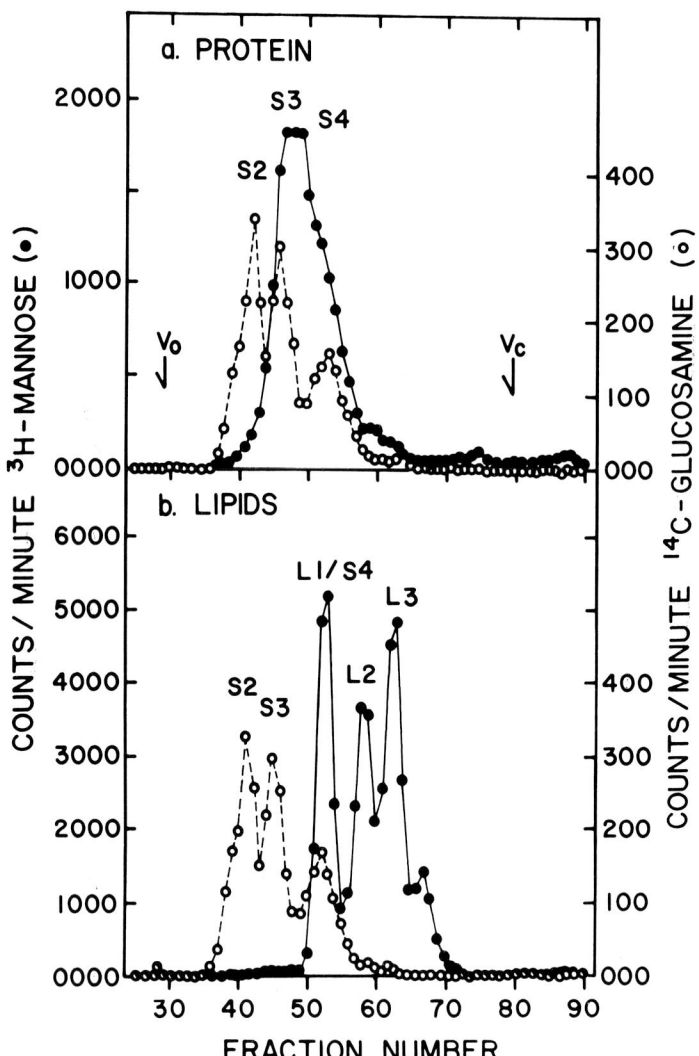

Fig. 4. Comparison of the size of the protein-bound mannose-labeled glycopeptides with that of the lipid-bound mannose-labeled oligosaccharides in Sindbis-infected cells labeled with [^{3}H]mannose for 50 sec. Sindbis-infected cells were labeled with [^{3}H]mannose (500 μCi/ml) for 50 sec and harvested and extracted as described in the legend to Fig. 5 and in Materials and Methods. The extracted lipids were hydrolyzed in mild acid, and the remaining cellular debris was digested with pronase. Both samples were then mixed with [^{14}C]glucosamine-labeled pronase-digested Sindbis virions and applied to a column of Bio-Gel P-4. •) [^{3}H]mannose; ○) [^{14}C]glucosamine; a) pronase-digested protein; b) acid-hydrolyzed lipids. Reprinted with permission from Cell.

On the other hand, one would detect only large protein-bound oligosaccharides if glycosylation occurred by the transfer of a preassembled moiety.

Sindbis-infected cells were therefore labeled with [^{3}H]mannose for 50 sec and the size of the radioactive, protein-bound oligosaccharides was determined. The lipids of these cells were extracted, and the size of the lipid-linked mannosyl oligosaccharides was also examined. Figure 4 compares the sizes of the pulse-labeled protein-bound oligosaccharides (Fig. 4a), the pulse-labeled lipid-bound oligosaccharides (Fig. 4b) and the oligosaccharides of the mature viral glycoproteins. Size determination was done by gel filtration of Bio-Gel P4. All the pulse-labeled oligosaccharides linked to protein were as large or larger than the smallest oligosaccharide of the virion, S4, which is 1,800 daltons in size (12). In contrast, a majority of the pulse-labeled, lipid-linked oligosaccharides were smaller than the smallest glycopeptide of the virion. It is important to note that the protein-bound oligosaccharides from both the cells and the virions were solubilized by digestion with pronase. As a result these molecules are glycopeptides which contain a few amino acids, rather than true oligosaccharides. The lipid-linked oligosaccharides, however, were solubilized by hydrolysis with mild acid and thus contain no noncarbohydrate material. Three distinct species of lipid-linked oligosaccharides were detected and have been labeled L1, L2, and L3. L2 and L3 are smaller than L1 by an amount consistent with their containing at least 1 and 2 fewer carbohydrate residues, respectively.

Since no protein-bound incomplete oligosaccharides were detected after labeling for 50 sec, it seemed unlikely that these oligosaccharides were being polymerized on the polypeptide from monosaccharides. Rather it appeared that transfer of a preassembled moiety must be occurring. The presence of large amounts of lipid-linked oligosaccharides in these cells is strong circumstantial evidence that lipid is the donor in this process. This idea is supported by the observed metabolic instability of oligosaccharides L2 and L3. Figure 5 shows the size of the labeled lipid-linked oligosaccharides as a function of time that the cells were labeled. Size determination is again by gel filtration with the glycopeptides of the assembled virion as markers. It can be seen that L2 and L3 comprise a significant fraction of the lipid-linked oligosaccharides only if the labeling period is much shorter than 5 min.

DISCUSSION

These experiments argue strongly that the 2 glycoproteins of Sindbis virus are glycosylated before they are released as identifiable polypeptides from the polysome, and that the glycosylation occurs by way of the transfer of a preassembled oligosaccharide. Further, the circumstantial evidence is strong that the donor of the oligosaccharide is a lipid.

The idea that glycosylation is coincident with synthesis comes from the observation that the intervals before which radioactive amino acids and radioactive mannose enter E2 are indistinguishable, and from the observation that unglycosylated viral glycoproteins were undetectable, even in infected cells which had been labeled with [^{35}S] methionine for only 30 sec.

The idea that glycosylation occurs by way of transfer of an oligosaccharide comes from the observation that no obviously incomplete protein-bound oligosaccharides could be found, even in cells labeled with [^{3}H] mannose for only 50 sec. That this labeling interval is indeed sufficiently short to have detected oligosaccharide polymerization is shown by the fact that it was possible to detect large amounts of incomplete lipid-linked oligosaccharides in the same cells. The presence of these lipid-linked oligosaccharides makes it very likely that lipid is the donor in the glycosylation of the viral glycoproteins.

Two specific points are worth emphasizing. Sindbis-infected chick cells attach both A-type and B-type oligosaccharides to the viral glycoproteins. Since no protein-bound

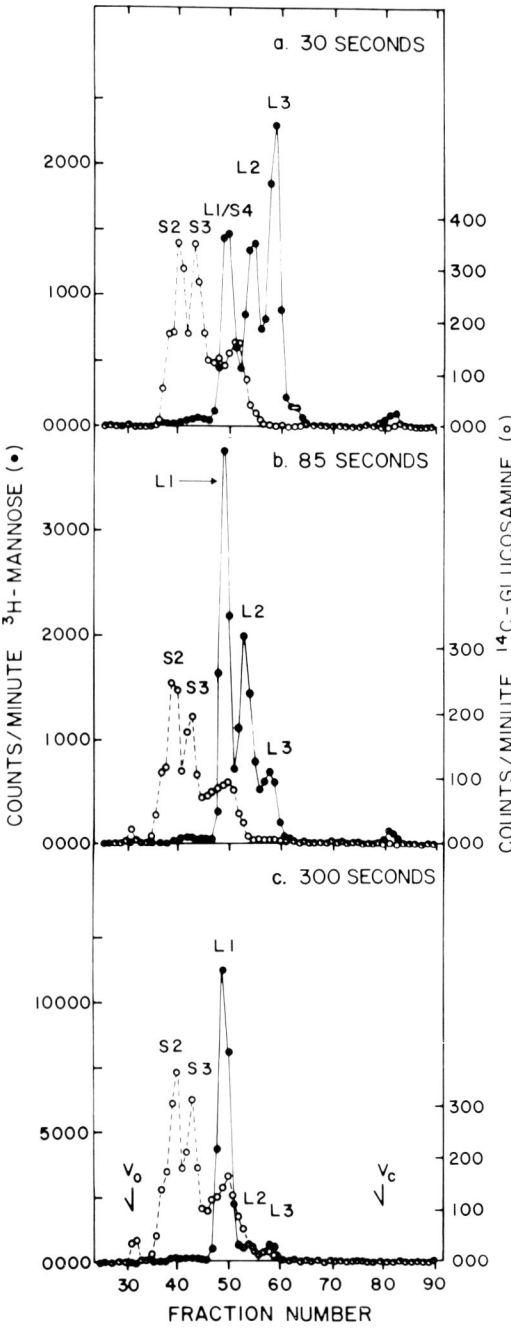

Fig. 5. Size of the mannose-labeled lipid-bound oligosaccharides in Sindbis-infected chick cells. Sindbis-infected chick cells were labeled with [^{3}H]-2-mannose (500 μCi/ml) for the indicated times, 6.5 to 7.5 hr after infection. At the end of the labeling period the cells were chilled on ice, washed once with cold Tris-buffered saline, scraped into cold Tris-buffered saline, centrifuged into a pellet, and frozen at $-70°$C until extracted. The lipids were extracted, subjected to mild acid hydrolysis, mixed with [^{14}C] glucosamine-labeled pronase-digested Sindbis virions and applied to a column of Bio-Gel P-4. ●) [^{3}H] mannose-labeled acid-hydrolyzed oligosaccharides; ○) [^{14}C] glucosamine-labeled pronase-digested virion glycopeptides; a) cells labeled for 30 sec; b) cells labeled for 85 sec; c) cells labeled for 300 sec. Reprinted with permission from Cell.

intermediates were detected, it appears that both types of asparagine-linked oligosaccharides are first assembled on lipid intermediates. Further, while the oligosaccharides attached to these newly synthesized proteins are large, at least 1,800 daltons in size, they are not as large as the largest A-type oligosaccharides found on the mature proteins. Completion of these oligosaccharides, which involves at least the addition of fucose and sialic acid, occurs no sooner than 20 min after the synthesis of the proteins (1).

It is becoming clear that most, if not all, asparagine-linked oligosaccharides are attached to nascent polypeptides by transfer of a preformed moiety from a lipid donor. The work described here makes it very likely that this is the mechanism of glycosylation of membrane proteins in vivo. In addition, there is a large body of work which has delineated the pathways by which lipid-linked mannosyl-oligosaccharides are synthesized in vitro, and there is very good evidence that these molecules function as donors in the glycosylation of endogenous proteins in vitro (9, 13–15). It has proved more difficult, however, to demonstrate the glycosylation of specific secretory proteins using a purified oligosaccharide lipid as donor. It seems very likely now that these problems are due not to the fact that immunoglobulin or ovalbumin are normally not glycosylated by transfer of an oligosaccharide but are instead due to the fact that the normal site of asparagine-linked oligosaccharide attachment is already occupied on postpolysomal proteins. This has been shown directly in the case of ovalbumin. Kiely et al. observed that nascent ovalbumin polypeptides already contain some glucosamine and mannose (16). Finally, it now appears that the site of oligosaccharide attachment may be accessible only during the synthesis of a protein. This is suggested by the observation that RNase A, a protein which is not glycosylated but which contains a bona fide site for carbohydrate attachment, can be glycosylated in vitro only after severe denaturation (17).

ACKNOWLEDGMENTS

I thank Claudie Berdot for help with the experiments. This work was supported by Grants CA-17289 and 14195 from the National Cancer Institute.

REFERENCES

1. Sefton BM, Keegstra K: J Virol 14:522, 1974.
2. Burke D: "Studies on the Sindbis Virus Glycoproteins." PhD Thesis, State University of New York at Stony Brook, 1976.
3. Strauss JH Jr, Burge BW, Darnell JE: Virology 37:367, 1969.
4. Laemmli UK: Nature (London) 227:689, 1970.
5. Bonner WM, Laskey RA: Eur J Biochem 46:83, 1974.
6. Laskey RA, Mills AD: Eur J Biochem 56:335, 1976.
7. Folch J, Lees H, Sloane-Stanley GH: J Biol Chem 226:497, 1957.
8. Behrens NH, Parodi AJ, Leloir LF: Proc Natl Acad Sci USA 68:2857, 1971.
9. Lucas JJ, Waechter CJ, Lennarz WJ: J Biol Chem 250:1992, 1975.
10. Schlesinger S, Schlesinger MJ: J Virol 10:925, 1972.
11. Kaluza G: J Virol 16:602, 1975.
12. Burge BW, Strauss JH: J Mol Biol 47:449, 1970.
13. Parodi AJ, Behrens NH, Leloir LF, Carminatti H: Proc Natl Acad Sci USA 69:3268, 1972.
14. Richards JB, Hemming FW: Biochem J 130:77, 1972.
15. Hsu A-F, Baynes JW, Heath EC: Proc Natl Acad Sci USA 71:2391, 1974.
16. Kiely ML, McKnight GS, Schimke RT: J Biol Chem 251:5490, 1976.
17. Pless DD, Lennarz WJ: Proc Natl Acad Sci USA 74:134, 1977.

Developmentally Regulated Lectins in Cellular Slime Molds and Embryonic Chick Tissues

Samuel H. Barondes

Department of Psychiatry, School of Medicine, University of California, San Diego, La Jolla, California 92093

Several species of cellular slime mold (including D. discoideum and P. pallidum) and a number of embryonic chick tissues (including muscle, heart, brain, and liver) contain lectin activities that can be extracted and assayed as hemagglutinins. In all cases studied the lectin activities show significant changes with differentiation. The studies with cellular slime molds are more advanced; and suggest that lectins play a role in developmentally regulated cell cohesion. The function of the embryonic chick lectins in differentiation is presently under investigation.

Key words: lectins in slime molds, cellular slime molds, dictyostelium discoideum, polysphondylium pallidum, lectins, lectin activity and development

Lectins may be defined as polyvalent, carbohydrate-binding proteins. They are generally assayed as cell agglutinins since oligosaccharides on the surface of some cell types (e.g., erythrocytes) may react with the carbohydrate-binding site of a specific lectin. The polyvalency of the lectins leads to cross-linking of the cells which is observed as agglutination.

Lectins were first identified in extracts of plant material. Recently, however, they have been turning up in a variety of other tissues. Despite the wide distribution of these proteins, their role in cell function has not yet been unequivocally established. Indeed, lectins may have different functions in different tissues and under different situations. In this view, the common property that unites these substances under the heading lectins may be more operational than biologically significant.

Work in our laboratory has suggested that one important role of lectins is in certain differentiative processes. This was first indicated from studies of the relationship between lectin activity in cellular slime molds and the differentiation of these simple eukaryotic organisms (1, 2). More recently, however, striking changes in lectin activity in a number of developing embryonic chick tissues have been observed. The purpose of this report is to briefly summarize work with the cellular slime molds which indicates that developmentally

Received April 22, 1977.

© 1978 Alan R. Liss. Inc., 150 Fifth Avenue, New York, NY 10011

regulated lectin activity plays a role in the cohesiveness which slime mold cells display with differentiation; and also to document that there are striking changes in lectin activity in a number of embryonic chick tissues whose function is presently being investigated.

STUDIES WITH SLIME MOLDS

Much of the work with developmentally regulated slime mold lectins has been summarized recently (1). In essence these studies show the following:

1. When the cellular slime molds Dictyostelium discoideum (2) or Polysphondylium pallidum (3) differentiate from a vegetative noncohesive form to a social cohesive form, upon food deprivation, there is concomitant appearance of lectin activity which can be extracted upon appropriate homogenization of these eukaryotic cells. At the time that the cells are vegetative and show no cohesiveness, little or no lectin activity is detectable. As cohesiveness develops and the cells form aggregates, lectin activity rises such that about 1% of the protein in soluble extracts of these cells is lectin.

2. The lectin activity is detectable on the surface of these slime mold cells. A number of different types of studies have indicated this. In one series of studies antisera to purified lectins from D. discoideum (4) and P. pallidum (5) have been used with standard immunofluorescent and immunoferritin techniques to demonstrate cell surface location of the lectin.

3. Lectin activities from the different species of cellular slime molds are discriminable. For example, the lectins from D. discoideum (2 discrete lectins have been purified) are discriminable from the lectin thus far purified from P. pallidum in that: a) there are marked differences in the relative potency of a number of saccharide inhibitors of the hemagglutination activity of these lectins (3); b) the lectins are different physicochemically (6, 7).

4. Substances appear on the surfaces of developing D. discoideum and P. pallidum cells with a high affinity for slime mold lectins (association constants in the range of 10^9 M^{-1}) (8). Surface receptors on cohesive cells have a higher affinity for the lectin than surface receptors on vegetative cells; and there is some degree of species specificity observed (8).

5. The inference from these studies is that cohesion of slime mold cells is mediated by interaction between cell surface lectin and complementary oligosaccharide-containing receptors. Both of these substances are detectable on the cell surface and appear with the development of cohesiveness. Since there is some degree of species specificity not only in the lectins but also in the oligosaccharide receptors, the system could be used for species specific sorting out which has been documented in cellular slime molds. Some direct evidence in support of this inference has been provided. For example, cohesiveness of P. pallidum cells can be blocked by high concentrations of sugars that interact with the purified lectin but only poorly by identical concentrations of sugars which interact poorly with this lectin (3). Likewise, monovalent antibody (Fab) fragments raised against the purified lectin inhibit cell cohesion whereas control Fab fragments do not (9).

The inference from all of this work is that lectins play a specific role in the differentiation of cellular slime molds. This role includes mediation of developmentally regulated cell cohesion and may also include species specificity in the cohesion process.

STUDIES WITH EMBRYONIC CHICK TISSUES

Because of the specific role of developmentally regulated lectins in slime mold cell differentiation, it seemed possible that similar molecules might play a role in differentiation and cohesion of vertebrate cells. Chick embryo tissues were chosen for investigation since there is a great deal of information about chick embryology and since chick embryos are readily available and inexpensive sources of reasonably large quantities of immature tissue. Our initial studies used chick embryo pectoral muscle since Teichberg et al. (10) had found evidence of lectin activity in this tissue. We found that lectin activity in this tissue was strikingly developmentally regulated (11). Extracts of pectoral muscle tissue from 8-day-old chick embryo showed very low levels of lectin activity; but the levels rose at least 10-fold in the ensuing 4–8 days and declined thereafter (11). Lactose and thiodigalactoside were the best inhibitors of hemagglutination activity.

Lectin protein is fairly abundant in embryonic chick muscle. We have recently purified a lectin from this tissue by affinity chromatography on a Sepharose column derivatized with p-aminophenyl lactoside (12). It represents 0.1% of the extractable soluble protein of 12–16-day-old embryonic chick pectoral muscle. An "impurity" possibly representing a second lectin has also been detected in this tissue. Given the purified protein (separated from the "impurity" by preparative isoelectric focusing) we have raised antibodies to it in rabbits. In preliminary studies we have demonstrated that myoblasts in primary chick muscle cultures specifically bind this antibody on their surface. This indicates that the lectin is present on the myoblast surface. Although studies with the chick muscle system are much less advanced than those with the slime mold system, our preliminary evidence suggests that there are considerable similarities. In particular the cell surface location of embryonic chick muscle lectin suggests that, as in the case of slime molds, some specific role in cell cohesion may be subserved by this material. There is already some controversy about this in the literature based on earlier studies (13, 14).

More recently we have studied other embryonic chick tissues. The other tissues examined, heart, liver, and brain, all have detectable lectin activity (15). As with the chick muscle lectin, lactose is the best inhibitor of hemagglutination of the sugars studied. This suggests that the lectin activity from the other tissues may be similar to the lectin activity isolated from embryonic chick muscle. However, the developmental time course of these lectin activities is strikingly different (Fig. 1). In none of the tissues studied is there as striking a change in activity as with the chick pectoral muscle. Over the time period studied, heart muscle activity shows either little change or a progressive decline. Brain shows a specific increase to around 12 days of embryonic development with decline thereafter (this decline has been substantiated by further experimentation, Ref. 15). Liver shows a decline and then a secondary rise after hatching. The lectin activity that appears posthatching is discriminable from that in developing liver, based on the relative potency of a number of sugars in inhibiting hemagglutination activity (15). Attempts to purify the lectins from these tissues are now underway in the hope of determining their relationship.

One point already established by these preliminary studies is that developmentally regulated lectins are characteristic not only of differentiating cellular slime molds but also of all the embryonic tissues that we have examined. Whereas the relationship of the lectin activities from these various tissues is presently unclear, it seems likely that a specific role of lectins in cellular differentiation will be recognized as a general phenomenon.

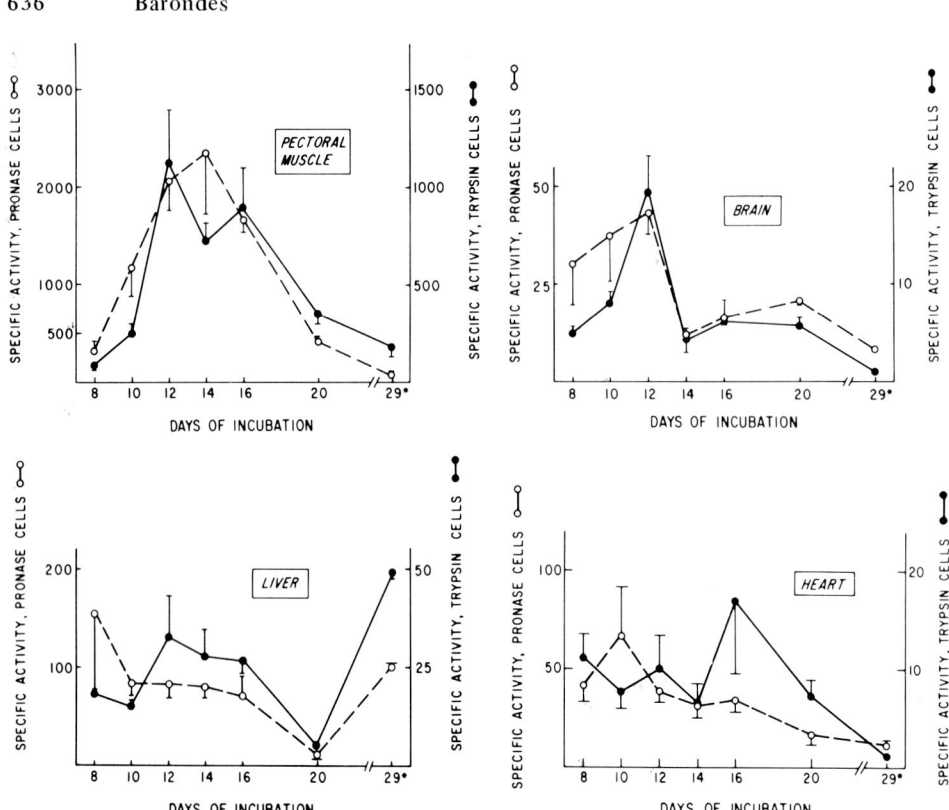

Fig. 1. Developmentally regulated lectin activity in extracts of 4 embryonic chick tissues. Lectins were assayed as agglutinins of treated rabbit erythrocytes. The rabbit erythrocytes were either treated with trypsin or pronase, then fixed with glutaraldehyde. Extracts of the indicated tissues were serially diluted and mixed with the appropriate erythrocytes in microtiter V-plates to determine the hemagglutination titer of each extract (for details see Refs. 11 and 15). Specific activity of each extract is the reciprocal of the titer divided by the protein in the extract in milligrams. Results are the average of studies with a number of extracts, generally at least 5. The bars represent standard error of the mean. Twenty-nine* days indicates one week after hatching which occurs at day 22. Further details of these experiments are presented in Ref. 15.

REFERENCES

1. Barondes SH, Rosen SD: In Barondes SH (ed): "Neuronal Recognition." New York: Plenum Press, 1976, pp 331–356.
2. Rosen SD, Kafka J, Simpson DL, Barondes SH: Proc Natl Acad Sci USA 70:2554, 1973.
3. Rosen SD, Simpson DL, Rose JE, Barondes SH: Nature (London) 252:128, 1974.
4. Chang C-M, Reitherman RW, Rosen SD, Barondes SH: Exp Cell Res 95:136, 1975.
5. Chang C-M, Rosen SD, Barondes SH: Exp Cell Res 104:101, 1977.
6. Frazier WA, Rosen SD, Reitherman RW, Barondes SH: J Biol Chem 250:7714, 1975.
7. Simpson DL, Rosen SD, Barondes SH: Biochim Biophys Acta 412:109, 1975.
8. Reitherman RW, Rosen SD, Frazier WA, Barondes SH: Proc Natl Acad Sci USA 72:3541, 1975.
9. Rosen SD, Haywood PL, Barondes SH: Nature (London) 263:425, 1976.
10. Teichberg VI, Silman I, Beutsch DD, Resheff G: Proc Natl Acad Sci USA 72:1383, 1975.
11. Nowak TP, Haywood PL, Barondes SH: Biochem Biophys Res Commun 68:650, 1976.
12. Nowak TP, Kobiler D, Roel LE, Barondes SH: J Biol Chem, in press.
13. Den H, Malinzak DA, Rosenberg A: Biochem Biophys Res Commun 69:621, 1976.
14. Gartner TK, Podleski TR: Biochem Biophys Res Commun 67:972, 1975.
15. Kobiler D, Barondes SH: Dev Biol, in press.

Properties of a Double Gradient Model for Retinotectal Specificity

Richard B. Marchase and Stephen Roth

Department of Biology, The Johns Hopkins University, Baltimore, Maryland 21218

The properties of a double gradient model for retinotectal specificity are discussed. The model utilizes only two complementary molecules, each located on both retina and tectum, to determine position along the dorsoventral axis. Two possible modes of interaction between these molecules are assumed. One of these allows all possible bonds between a retinal cell and tectal loci to be formed. This results in a rigid retinotectal projection in which the adhesion of each retinal cell to its normal tectal locus is maximal. The other assumes stochastic interactions between the molecules, and results in retinotectal specificity only if additional constraints are imposed on the system. Either of these modes of interaction predicts adhesive preferences for retinal cells to tectal halves similar to those observed experimentally.

Key words: retinotectal specificity, adhesive selectivity

In this formulation of neuronal specificity, Sperry (1, 2) attributed the selectivity that characterizes the circuitry of the brain to interactions between cell surface, recognition molecules. The axonal tip of each retinal ganglion cell, for example, was thought to possess a unique battery of cytochemical markers. Interactions between these molecules and similar recognition molecules at potential termini in the tectum were proposed to result in differential affinities between the axons and the various tectal loci. Sperry reasoned that each axon would adhere preferentially to, and ultimately synapse with, the area of tectum possessing a particular set of surface molecules, while bypassing sites with less appropriate displays.

Attardi and Sperry (3) obtained evidence for such rigid specificity from experiments with adult goldfish in which optic nerves were severed and portions of neural retina were ablated. The fibers that regenerated from the remaining retina formed connections solely with their appropriate tectal sites, at times crossing empty tectal areas to do so. Subsequently, this experimental strategy of creating size disparities between retina and tectum was exploited imaginatively and zealously and produced two opposing interpretations. One view, propounded by Gaze and Keating (4, 5), suggested that the relationship between retinal and tectal cells was far less rigid than that implied by the hypothesis of neuronal

Accepted for publication May 5, 1977

specificity. This view was supported by experiments in which retinal fibers failed to project solely to their normal tectal loci following experimental manipulation, but rather, seemed to adjust their connections depending on the complement of retinal fibers present and the extent of tectum available for innervation. They termed this type of mechanism "systems matching."

The other point of view, reviewed recently by Meyer and Sperry (6), emphasized the original hypothesis of neuronal specificity. They argued that the adjustments that occurred in some experiments were the results of regulative phenomena common to a number of developmental systems (7). The neural cells were proposed to lose their old cytochemical identities and acquired new ones that were appropriate for their positions in the regulated field.

Our approach to retinotectal specificity (8, 9) has utilized an assay capable of directly measuring the adhesion of dissociated, retinal cell bodies to limited areas of tectal surface. Either dorsal or ventral halves of neural retina were labeled with ^{32}P and dissociated to form a single cell suspension. Tecta were split into equal-sized dorsal and ventral halves and fixed to the bottom of a petri dish. The labeled retinal cells were added to the dish and it was placed on a reciprocating shaker. After a collection period of about 1 h, the tectal halves were washed and counted in a scintillation counter to determine the numbers of radioactive retinal cells that adhered. Since the numbers of collisions between the labeled retinal cells in suspension and each of the tectal halves was equal, comparisons between the numbers of cells adhering to dorsal and to ventral halves provided an operational definition of adhesive specificity.

With this assay, cell bodies from dorsal retina were shown to adhere preferentially to ventral half-tecta, while cell bodies from ventral retina preferred dorsal half-tecta. Tecta that had never been innervated were also used as collecting surfaces and displayed selectivity similar to that shown by innervated tecta. The preferential adhesion could not therefore, be attributed solely to the adhesion of retinal cell bodies to retinal axons on the surface of the tecta. Thus, an adhesive selectivity between neural retina and optic tectum that mimicked the innervation pattern of the tectum by the retina was demonstrated.

A MODEL FOR RETINOTECTAL SPECIFICITY

Experiments investigating the biochemical basis for this adhesive specificity (10) implicated a model for retinotectal specificity presented previously (11, 12) that is an outgrowth of the ideas of sperry (1, 2). The model considers retinotectal specificity to be determined independently along two orthogonal axes. Two molecules, which possess lock-and-key complementarity toward each other, are proposed as determining retinotectal specificity in the dorsoventral axis, with two other complementary molecules functioning in a similar fashion but independently in the anterioposterior axis. Specificity in each axis is achieved through a double-gradient distribution of the respective molecules, as depicted for the dorsoventral axis in Fig. 1. Ventral retina, rich in locks, would adhere preferentially to dorsal tectum, which is rich in keys. Similarly, dorsal retina, rich in keys, would prefer ventral tectum, rich in locks. Distribution along the anterioposterior axis would be determined through interactions of two independent molecules.

In order to determine the distribution of retinal axons along the tectal surface that such a model would establish, it is necessary to define the manner in which a retinal axon's battery of recognition molecules will interact with the molecules at any point on the tectal surface. Two modes of interaction seem physically feasible. In the first of these, the strength of an interaction between retinal axon and tectal locus would be determined by

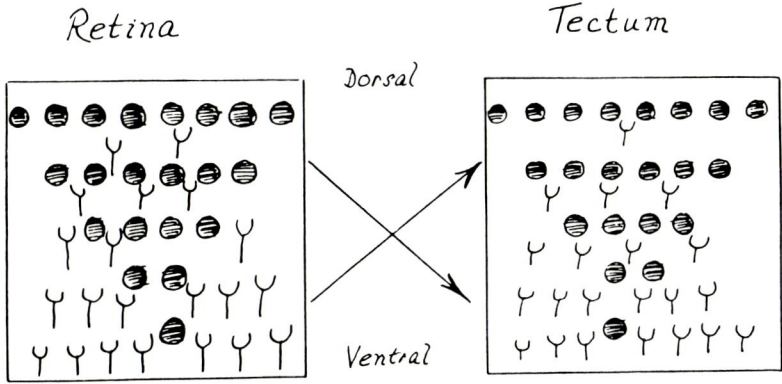

Fig. 1. A simple model of gradients of complementary molecules in the dorsoventral axis that could provide adhesive connections corresponding to the observed retinotectal map. Retinal cells from the dorsal region have many ●'s but few Y's. The ventral part of the tectum has cells with few ●'s but many Y's, thereby allowing a maximum number of bonds (Ỷ) to be formed. In the same fashion, ventral retinal retina cells, rich in Y's and poor in ●'s, would form the most stable connections with the dorsal tectum since it is rich in ●'s but poor in Y's. The gradients are pictured as linear, with the concentration of each adhesive molecule being zero at the appropriate extreme.

the total number of lock and key pairs that could possibly be formed. For instance, in an array of 10 cells in which both the number of locks and the number of keys varied linearly from 0 to 9, a retinal cell that possessed 6 locks would also possess 3 keys (Fig. 2). Its interaction would be maximal with the tectal cell possessing 6 keys and 3 locks since 9 pairs could be formed. Its interaction with the tectal site possessing 5 keys and 4 locks would result in 8 bonds being formed. A graph of this retinal cell's interactions with all the potential tectal loci results in a linear profile that peaks at the tectal locus possessing numbers of locks and keys exactly complementary to its own (Fig. 2). It is clear that this manner of interaction results in retinotectal specificity and that even in a mismatch situation every retinal cell would prefer its normal tectal locus.

Another possible mode of interaction is more stochastic in nature. In this case, the strength of interaction between a retinal cell and a tectal locus would be determined by the numbers of collisions between keys on retina and locks on tecta, and vice versa. The formula for a relative strength of interaction would be:

$$\text{Strength} \propto [\text{locks}_\text{retina}] \cdot [\text{keys}_\text{tectum}] + [\text{keys}_\text{retina}] \cdot [\text{locks}_\text{tectum}].$$

Using this assumption, profiles of interaction strength for retinal cells as functions of tectal position can be determined and are graphed in Fig. 3. This mode of interaction also results in linear profiles, but in this case all retinal cells maximize their binding strength at either one extreme or the other. In order for normal retinotectal specificity to result, it is necessary to invoke additional rules, such as limiting the number of retinal cells that can synapse

	Retinal Cells		Tectal Loci	
	No of locks	No. of keys	No. of locks	No. of keys
a)	0	9	a') 9	0
b)	1	8	b') 8	1
c)	2	7	c') 7	2
d)	3	6	d') 6	3
⋮	⋮	⋮	⋮	⋮
j)	9	0	j') 0	9

Possible no. of bonds (d, d') = [3_{locks} retina ↔ 3_{keys} tectum] + [6_{keys} retina ↔ 6_{locks} tectum] = 9

Possible no. of bonds (d, c') = [2_{locks} retina ↔ 2_{keys} tectum] + [6_{keys} retina ↔ 6_{locks} tectum] = 8

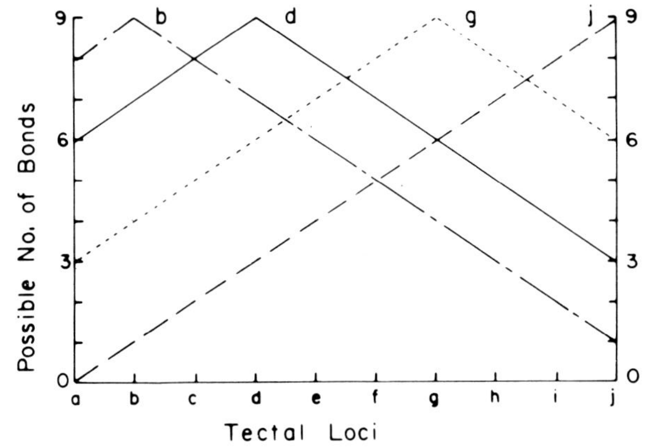

Fig. 2. A double gradient model for retinotectal specificity along one axis. An array of 10 retinal cells, a–j, and 10 corresponding tectal loci, a'–j', is depicted, each possessing locks and keys in the numbers indicated. The strength of the interaction between a retinal cell and a tectal locus is determined by the maximum number of lock-and-key bonds possible, given each loci's numbers of locks and keys. Profiles for 4 retinal cells are shown. In this model every retinal cell adheres maximally to its ultimate tectal locus, so that retinotectal specificity results if each retinal cell maximizes its strength of interaction. This arrangement also provides guidance for a retinal cell migrating on the surface of the tectum, for the strength of interaction increases monotonically as a retinal cell approaches its final locus.

with any tectal loci. With some such limiting constraints, competition among the retinal cells would result in a normal retinotectal projection. Such competition has been suggested to play a role in determining synaptic specificity previously (13, 14). The behavior of such a double gradient model to experimentally created size disparities would depend on the additional constraints imposed, but it is clear that this model could display aspects of "systems matching."

One might interpret the first rules as arising from molecular interactions that were irreversible or at least long-lived enough so that all possible pairs could be formed at any one time. The second set of rules would arise from relatively short-lived interactions that were being formed and broken continually. The numbers of pairs at any time would then be functions of the concentrations of the components.

The proposed, double gradient model might thus be consistent with either a rigid or a "systems matching" interpretation. It seems worth noting that such totally different behavior might result from the same molecular configuration and depend primarily on a difference in the interactions assumed among the proposed recognition molecules. Whether these phenomena may play a role in the species or developmental differences found physiologically is, of course, unknown.

THE MAGNITUDE OF THE ADHESIVE SELECTIVITY

Since the profiles for the strengths of interactions between retinal cells and tectal loci in the double gradient model have been calculated (Figs. 2 and 3), the probability of

	Retinal Cells			Tectal Loci	
	No. of locks	No. of keys		No. of locks	No. of keys
a)	0	9	a')	9	0
b)	1	8	b')	8	1
c)	2	7	c')	7	2
d)	3	6	d')	6	3
⋮	⋮	⋮	⋮	⋮	⋮
j)	9	0	j')	0	9

Relative strength of interaction $(d, d') = [3_{\text{locks retina}} \times 3_{\text{keys tectum}}] + [6_{\text{keys retina}} \times 6_{\text{locks tectum}}] = 45$

Relative strength of interaction $(d, c') = [3_{\text{locks retina}} \times 2_{\text{keys tectum}}] + [6_{\text{keys retina}} \times 7_{\text{locks tectum}}] = 48$

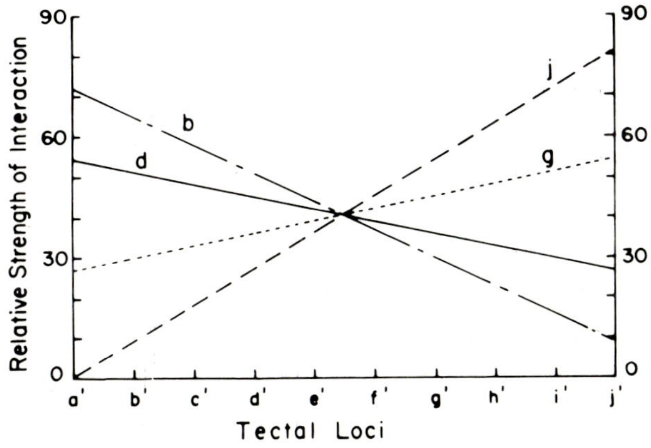

Fig. 3. A double gradient model for retinotectal specificity along one axis in which the interactions between locks and keys are stochastic. The strength of the interaction between a retinal cell and a tectal locus is dependent on the concentrations of locks and keys according to the formula:
$$\text{strength} \propto [\text{locks}_{\text{retinal}}] \cdot [\text{keys}_{\text{tectum}}] + [\text{keys}_{\text{retina}}] \cdot [\text{locks}_{\text{tectum}}]$$
Profiles for 4 retinal cells are shown. In this model, all retinal cells prefer either one extreme of tectum or the other. Retinotectal specificity could result if the number of spaces available at each tectal loci were limited. Competition among retinal fibers would then establish the normal projection.

adhesion for any retinal cell to the respective halves of tectum can also be determined if one assumes that adhesion probability is proportional to strength of interaction. By expressing this probability adhesion as a function of position in the retina and then integrating over half a retina, one can simulate a collection experiment for retinotectal specificity and determine the hypothetical adhesion of the cells of a half-retina to appropriate and inappropriate tectal halves. The calculations for this exercise are given in the Appendix. The results are that in both a rigid model (Fig. 2) and a plastic model (Fig. 3), 1.8 times as many retinal cells would adhere to the appropriate half-tecta as to the inappropriate one. These results agree well with the overall experimental ratio for actual collection assays of 1.7 (8, 9, 10). This experimental ratio seems to be about constant for both dorsal and ventral suspensions and for both neural and pigmented retina. The true significance of this correspondence is unknown. It certainly lends support to a double gradient model for retinotectal specificity, but further experimentation is required to show that the similarity is not coincidental.

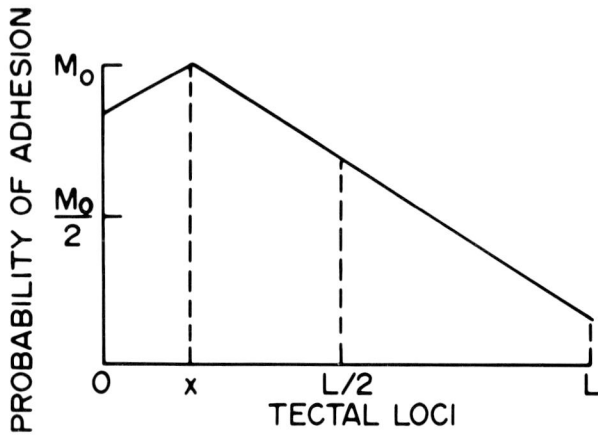

Fig. 4. Profile of adhesion probability for retinal cell x in the rigid model of retinotectal specificity in which the maximum numbers of bonds between retinal cell and tectal loci are assumed.

APPENDIX
Simulations of Collection Experiments

Rigid model (see Fig. 2 and Fig. 4)

O and L are the limits of the tectal field.

M_O = maximal probability of adhesion.

$\pm M_o/L$ = slope of every profile if the concentration of each of the recognition molecules is 0 at one of the extremes.

$p(x,y)$ = probability of a retinal cell that adheres maximally at tectal locus x adhering to tectal locus y.

$$R(x) = \frac{\text{probability at a retinal cell that adheres maximally at x adhering to appropriate tectal half}}{\text{probability of a retinal cell that adheres maximally at x adhering to inappropriate tectal half}}$$

for $x = 0$ to $x = \frac{L}{2}$,

$$p(x,x) = M_0 \qquad p(x, \frac{L}{2}) = M_0 - \frac{M_C}{L}(\frac{L}{2} - x) = M_0(\frac{1}{2} + \frac{x}{L})$$

$$p(x,0) = M_0 - \frac{M_0}{L}x \qquad p(x,L) = M_0 - \frac{M_0}{L}(L-x) = \frac{M_0 x}{L}$$

By computing areas of trapezoids:

$$R(x) = \frac{1/2\,[p(x,0) + p(x,x)]\,x + 1/2\,[p(x,x) + p(x,L/2)]\,(L/2 - x)}{1/2\,[p(x,L/2) + p(x,L)]\,L/2}$$

$$R(x) = \frac{1/2\,[M_0 - \frac{M_0}{L}x + M_0]\,x + 1/2\,[M_0 + M_0(1/2 + \frac{x}{L})]\,(L/2 - x)}{1/2\,[M_0(1/2 + \frac{x}{L}) + M_0\frac{x}{L}]\,\frac{L}{2}}$$

$$R(x) = \frac{\frac{-2}{L^2}x^2 + \frac{1}{L}X + \frac{3}{4}}{\frac{1}{L}X + \frac{1}{4}}$$

Averaging over a half retina, $X = 0$ to $x = \frac{L}{2}$:

$$R_{ave} = \frac{\int_0^{L/2} R(x)dx}{L/2}$$

$$R_{ave} = 1 + \frac{3 \ln 3}{4}$$

$$R_{ave} = 1.82$$

Plastic model (see Fig. 3 and Fig. 5)

O and L are the limits of the tectal field.

M_0 = the maximum probability of adhesion, i.e., the adhesion of the most dorsal retinal cell to the most ventral tectal locus, and vice versa.

$M(x)$ = maximum probability for retinal cell whose normal position in the retinotectal map is x.

$$M(x) = M_0 - \frac{M_0 x}{L} \qquad \text{for } x = 0 \text{ to } x = \frac{L}{2}$$

$R(x) = \dfrac{\text{probability of retinal cell whose normal position is x adhering to appropriate tectal half}}{\text{probability at retinal cell whose normal position is x adhering to inappropriate tectal half}}$

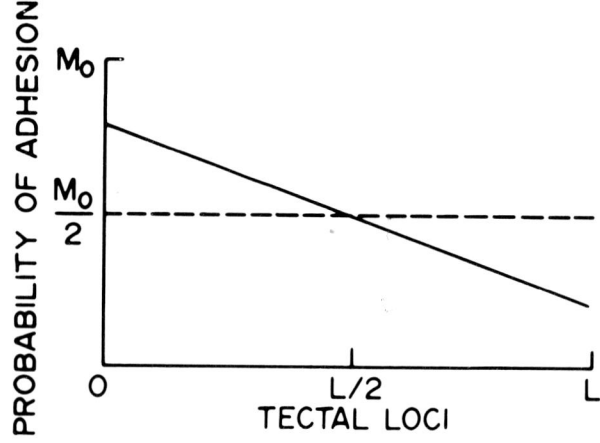

Fig. 5. Profile of adhesion probability in the stochastic model in which the strengths of interaction between retinal cell and tectal loci are assumed proportional to the concentrations of the molecules involved.

By computing areas of rectangles and triangles:

$$R(x) = \frac{\frac{M_0}{2}\frac{L}{2} + \frac{1}{2}(\frac{M_0 - M_0 x}{L} - \frac{M_0}{2})\frac{L}{2}}{\frac{M_0}{2}\frac{L}{2} - \frac{1}{2}(M_0 - \frac{M_0 x}{L} - \frac{M_0}{2})\frac{L}{2}}$$

$$R(x) = \frac{3/2 - x/L}{1/2 + x/L}$$

Averaging over half a retina, $x = 0$ to $x = \frac{L}{2}$:

$$R_{ave} = \frac{\int_0^{L/2} R(x)dx}{L/2}$$

$R_{ave} = 4 \ln 2 - 1 = 1.77$

ACKNOWLEDGMENTS

This work was supported by a graduate fellowship to Richard B. Marchase from the Danforth Foundation and grants to Stephen Roth from the National Institute of Child Health and Human Development. This is contribution No. 916 from the McCollum–Pratt Institute.

REFERENCES

1. Sperry RW: J Comp Neurol 79:33, 1943.
2. Sperry RW: Proc Natl Acad Sci USA 50:703, 1963.
3. Attardi DG, Sperry RW: Exp Neurol 7:46, 1963.
4. Gaze RM, Keating MJ: Nature (London) 237:375, 1972.
5. Keating MJ: In Gottlieb G (ed): "Neural and Behavioral Specificity." New York: Academic Press, 1976, p 59.
6. Meyer RL, Sperry RW: In Gottlieb G (ed): "Neural and Behavioral Specificity." New York: Academic Press, 1976, p 111.
7. Weiss P: "Principles of Development." New York: Holt, 1939.
8. Barbera AJ, Marchase RB, Roth S: Proc Natl Acad Sci USA 70:2482, 1973.
9. Barbera AJ: Dev Biol 46:167, 1975.
10. Marchase RB: J Cell Biol, in press.
11. Marchase RB, Barbera AJ, Roth S: In: "Cell Patterning." (CIBA Foundation Symposium 15), Amsterdam: Associated Scientific Publishers, 1974, p 315.
12. Roth S, Marchase RB: In Barondes S (ed): "Neural Recognition." New York: Plenum Press, p 227.
13. Ramon y Cajal S: "Studies on Vertebrate Neurogenesis (Translated by Guth L) Springfield, Illinois: CC Thomas, 1960.
14. Prestige MC, Willshaw DJ: Proc R Soc London Ser B 190:77, 1975.

Affinity Labeling of a Cell Surface Receptor for Epidermal Growth Factor

Manjusri Das, Tokichi Miyakawa, and C. Fred Fox

Department of Bacteriology and the Parvin Cancer Research Laboratories, Molecular Biology Institute, University of California, Los Angeles, California 90024

The membrane receptor for epidermal growth factor (EGF) on 3T3 cells has been identified and specifically labeled radiochemically using a photoreactive derivative of EGF. Photoreactive EGF, labeled with ^{125}I, was incubated with 3T3 cells and then photolyzed in situ to generate a nitrene capable of reacting with a wide variety of chemical bonds. Analysis of the system by sodium dodecyl sulfate/polyacrylamide gel electrophoresis revealed, besides the band of EGF, only 1 other major radioactive band which migrated at approximately 190,000 daltons. This band was absent when a nonresponsive and nonbinding variant of 3T3 was used. A direct proportionality between binding activity and crosslinked complex formation was demonstrated using a variety of binding conditions. The crosslinked complex in intact cells is accessible to the action of a macromolecule like trypsin at 4°, suggesting a cell surface location for this complex. Upon incubation of cells at 37°, radioactivity from previously formed EGF-receptor crosslinked complex is converted by cellular action to 3 forms of mol wt $\leq$ 58,000 daltons. These are not accessible to trypsin action upon intact cells.

Key words: mitogen receptor, heterobifunctional crosslinker, photoreactive reagent

Epidermal growth factor (EGF), a single chain polypeptide of mol wt $\sim$ 6,000 stimulates the proliferation of various epidermal and epithelial tissues (1) and is a potent mitogen for human fibroblasts (2, 3), human glial cells (4), and for mouse cell lines such as 3T3 (5, 6). The factor has been isolated from the submaxillary glands of male mice (7–9), and also from urine of pregnant women (10). Both human and murine EGF have identical biological activities and compete for the membrane receptor site(s) in a wide variety of vertebrate cells (10). The striking similarities in properties of EGF and its receptors from 2 widely different mammalian species suggest a vitally important function for this growth regulatory system.

Binding of EGF to responsive cells is mediated through highly specific cell surface receptors (2, 11, 12). There are about 40,000–100,000 EGF binding sites per cell (2, 11), and it may be possible to purify this receptor protein in quantities sufficient for biochemical characterization. Identification of hormone receptors by conventional techniques involves solubilization and affinity purification (13–17). This widely used procedure may not be applicable in certain cases, where solubilization from membranes results in a loss of hormone binding activity. However, it may be possible to identify such labile receptor proteins by affinity labeling using modified hormone (18–22). In this technique, the binding of hormone to the receptor takes place under more physiological conditions, and the

Accepted for publication June 10, 1977.

© 1978 Alan R. Liss, Inc., 150 Fifth Avenue, New York, NY 10011

hormone-receptor complex is covalently crosslinked before solubilization or disruption of membranes. This property can also be exploited to study the physiological fate of the hormone-receptor complex. In this communication, we report the synthesis of a novel photoreactive derivative of EGF and the use of this modified hormone in identification and specific radiolabeling of a membrane receptor for EGF on 3T3 cells.

MATERIALS AND METHODS

Synthesis of Methyl 3-[(p-Azidophenyl)dithio] propionimidate (PAPDIP)

The heterobifunctional crosslinking reagent (structure shown below) was synthesized by a procedure to be published elsewhere (Miyakawa and Fox, to be published).

$$N_3-\langle\rangle-S-S-CH_2-CH_2-C\underset{OCH_3}{\overset{NH}{\diagup\diagdown}}$$

Preparation of ^{125}I-Labeled EGF and PAPDIP-^{125}I-Labeled EGF

EGF from male mouse submaxillary glands was purified to electrophoretic homogeneity using the procedure of Savage and Cohen (23). Radioiodinated EGF was prepared using Na-^{125}I and chloramine-T. Carrier-free Na-^{125}I (40 mCi) and EGF (100 μg) in 0.2 M potassium phosphate, pH 7.5 (total volume, 300 μl), were mixed vigorously with 40 μl of chloramine-T (50 mg/ml) for 1 min at 4°. The reaction was terminated by the addition of 40 μl of sodium metabisulfite (100 mg/ml) and 40 μl of potassium iodine (0.7 g/ml). The labeled protein was separated from unreacted Na-^{125}I by gel filtration through Sephadex G-10 with a buffer containing 20 mM potassium phosphate, pH 7.5, and 0.2 M NaCl. The specific activity of ^{125}I-labeled EGF was approximately 5×10^5 cpm/ng of protein. A photoreactive derivative of radioiodinated EGF was prepared by reacting ^{125}I-labeled EGF (50 μg) with 4 mg of PAPDIP in 0.5 ml of 0.1 M triethanolamine-HCl/0.2 M NaCl, pH 8.5, for 1 h in the dark, at 23° with stirring. The remaining unreacted imidoester groups in PAPDIP were then inactivated by the addition of 30 μl of 1 M ammonium acetate and incubation at 23° for 10 min. The resulting suspension, containing some insolubilized PAPDIP, was filtered through a Sephadex G-10 column (eluted with 20 mM potassium phosphate/0.2 M NaCl, pH 7.5) to separate PAPDIP-^{125}I-labeled EGF from excess reagents. The imidoester group in PAPDIP can react readily with primary amino groups, but reacts very slowly with sulfhydryls, amides, and hydroxyls. Murine EGF has only 1 primary amino group, and therefore the extent of derivation is not expected to proceed beyond 1 equivalent of crosslinker per mole of EGF. Since this derivative was prepared using small quantities of ^{125}I-labeled EGF, the extent of derivatization of ^{125}I-labeled EGF by PAPDIP was not determined.

Cell Culture

Monolayer cultures of Swiss 3T3 cells were grown and maintained at 37° in a 10% CO_2 atmosphere in Dulbecco's modified Eagle medium (DME) containing 10% fetal calf serum. Cells were subcultured by trypsinization. To prepare experimental cultures for binding and crosslinking, approximately 10^5 cells were inoculated into 35-mm Falcon culture dishes containing 3 ml of DME medium with 5% fetal calf serum. The cultures were incubated at 37° in a humidified 10% CO_2 atmosphere until a confluent monolayer of cells was formed. A variant 3T3 cell (NR-6), which neither responds to nor binds EGF, was a gift from Dr. Harvey R. Herschman of this institution. These NR-6 cells were grown and maintained under the same conditions as the responsive 3T3 cells, with the exception that the medium was supplemented with newborn-calf serum instead of fetal-calf serum.

Binding of [^{125}I]-EGF

Confluent monolayers in 35-mm culture dishes, containing about 5×10^5 cells, were washed 3 times with Earle's balanced salt solution containing 5 mM HEPES, pH 7.4, and 0.1% bovine serum albumin (EBSS-0.1% BSA), and then 1 ml of EBSS-0.1% BSA was added to each plate. ^{125}I-labeled EGF was added to initiate binding. After the incubation periods described in the figure legends, unbound radioactivity was removed by rapidly washing the cell monolayers 5 times with 3 ml of EBSS-0.1% BSA for each wash. The washed monolayers were solubilized in 1 ml of 0.5 N NaOH and counted in a gamma counter. Specific binding was determined by measuring the difference in cell bound radioactivity in the presence and absence of 1 μg of unlabeled EGF. The cell bound radioactivity in the presence of an excess amount of unlabeled EGF was considered to be "nonspecific." In the experiments reported here, the nonspecific binding amounted to less than 10% of the total binding.

Binding and Photolysis of PAPDIP-^{125}I-Labeled EGF

Cells grown to confluence in 35-mm culture dishes were washed 3 times with Earle's balanced salt solution containing 5 mM HEPES, pH 7.4, and 0.02% bovine serum albumin (EBSS-0.02% BSA). Then 1 ml of EBSS-0.02% BSA was added to each plate. Binding was initiated by the addition of PAPDIP-^{125}I-labeled EGF in the dark. After incubation for periods described in the legend, unbound radioactivity and proteins were removed by thoroughly washing each monolayer 6 times in the dark with a total of 18 ml of Earle's balanced salt solution buffered with 5 mM HEPES, pH 7.4 (EBSS), and containing no BSA. Finally, 1 ml of EBSS was added to each monolayer and the photolysis was performed at 4° for 5 min with a uv lamp (UVL.21, Ultraviolet Products Inc.) set at a distance of 5 cm from the sample. After removal of EBSS from the plates, the cells were solubilized with 100 μl of 3% sodium dodecyl sulfate (SDS) and 0.6% N-ethylmaleimide in 0.1 M Tris-HCl, pH 6.8, counted in a gamma counter for measurement of binding, and then analyzed by gel electrophoresis for detection of crosslinked complexes.

Gel Electrophoresis

Electrophoresis was performed in a slab gel apparatus on 5–20% acrylamide gradients, with electrode buffer containing 0.1% SDS in 0.025 M Tris/0.192 M glycine, pH 8.3 (24). After staining with Coomassie blue, the slabs were dried down onto paper and autoradiograms were made on x-ray films (No-Screen film, NS-5T, Eastman Kodak Company).

RESULTS

Characteristics of ^{125}I-Labeled EGF Binding

The time dependence of binding of ^{125}I-labeled EGF to 3T3 cells at 37°, 23° and 4° is shown in Fig. 1. At 37°, the amount of cell bound radioactivity decreased after 40 min of incubation, suggesting that there may be extensive internalization and cellular degradation of the hormone at this temperature, similar to that shown to occur with

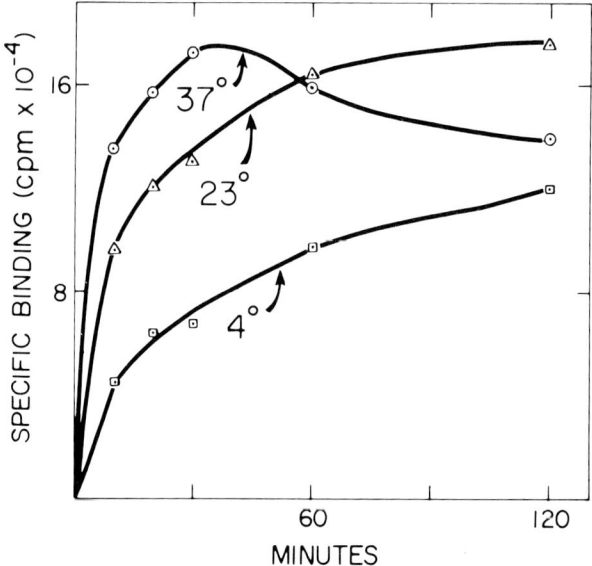

Fig. 1. Time course of ^{125}I-labeled EGF binding to 3T3 cells at 37°, 23°, and 4°. ^{125}I-labeled EGF (30 ng, 5 × 10^5 cpm/ng) was added to each culture dish in 1 ml of Earle's balanced salt solution containing 5 mM HEPES, pH 7.4, and 0.1% bovine serum albumin. At the indicated times, cell-bound radioactivity was determined as described under "Materials and Methods." (Reproduced from: Ref. 32).

human fibroblasts (25). At 23° and below, there was apparently no net loss of cell-bound radioactivity with prolonged incubation suggesting that at 23°, the rate of endocytosis of membrane bound hormone is significantly lower (26).

The effect of concentration on binding of ^{125}I-labeled EGF to 3T3 cells at 37°, 23°, and 4° was also studied. A Scatchard plot of the binding data at 37° showed that there are about 94,000 EGF receptor sites per cell. The apparent dissociation constant of the binding reaction was calculated to be about 1.1×10^{-9} M. Native unlabeled EGF competed effectively with ^{125}I-labeled EGF for binding to 3T3 cells, suggesting that iodination did not appreciably alter the binding characteristics of EGF.

A study on the effect of gangliosides on binding showed that the binding was unaffected by incubation of ^{125}I-labeled EGF with ganglioside GM$_1$ (10 µg/ml), ganglioside GM$_2$ (10 µg/ml), or crude bovine brain gangliosides (150 µg/ml), suggesting that the receptor is probably not a ganglioside. Brief incubation (10 min) of the cells with lectins such as concanavalin A, ricin agglutinin, or phytohemagglutinin at concentrations ranging from 0.05–1 mg/ml resulted in only partial inhibition of binding. The maximal inhibition was about 50% in each case. However, treatment of the cells with trypsin (50 µg/ml) at 4° for 30 min resulted in a complete loss of binding activity.

Binding and Crosslinking of PAPDIP-^{125}I-Labeled EGF to 3T3 Cells

The characteristics of binding of PAPDIP-^{125}I-labeled EGF to 3T3 cells were very similar to those of ^{125}I-labeled EGF binding. Photolysis of PAPDIP-^{125}I-labeled EGF bound to 3T3 cells resulted in the formation of a single crosslinked complex, as detected by electrophoretic and radiochemical analysis of the system (Fig. 2). The crosslinked

Receptor for Epidermal Growth Factor

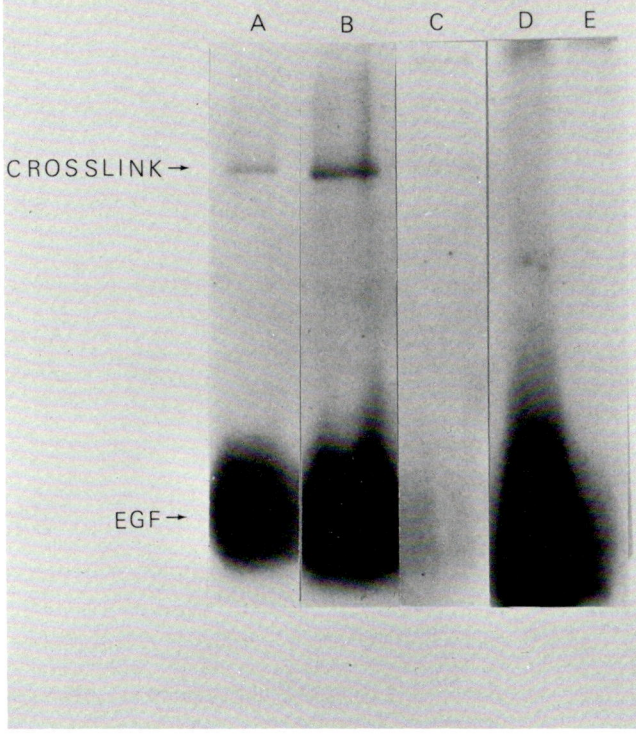

Fig. 2. Autoradiography of 5–20% polyacrylamide gradient gels. Cells were incubated with PAPDIP-^{125}I-labeled EGF in 1 ml of Earle's balanced salt solution containing 5 mM HEPES, pH 7.4, and 0.02% bovine serum albumin at 23° for 10 min in the dark. After incubation, each monolayer was washed 6 times with 3 ml each time of Earle's balanced salt solution containing 5 mM HEPES, pH 7.4, and then photolyzed and subjected to electrophoresis as described under "Materials and Methods." The direction of migration was from top to bottom. A, B) 3T3 cells incubated with 2 and 25 ng of PAPDIP-^{125}I-labeled EGF; C) 3T3 cells incubated with 25 ng of PAPDIP-^{125}I-labeled EGF and 500 ng of unlabeled EGF; D) PAPDIP-^{125}I-labeled EGF, 0.5 ng (no cells); E) NR-6 cells incubated with 25 ng of PAPDIP-^{125}I-labeled EGF. (Reproduced from: Ref. 32).

complex had a mol wt of approximately 190,000. This high-molecular-weight band was not observed when cells were incubated with underivatized ^{125}I-labeled EGF or when the PAPDIP-^{125}I-labeled EGF bound to cells was not photolyzed. The band was also absent when the nonresponsive and nonbinding variant NR-6 was used. Determination of radioactivity on the crosslinked complex band formed with increasing concentration of PAPDIP-^{125}I-labeled EGF showed a direct proportionality between binding activity and covalent complex formation (Fig. 3). Incubation of cells with PAPDIP-^{125}I-labeled EGF in the presence of varying amounts of competing unlabeled EGF resulted in a decrease in crosslinking, concomitant with the reduction in binding of radioactive EGF (data not shown). The linear relationship between binding and crosslinking was further confirmed using "down regulated" cells. Incubation of 3T3 cells with high concentrations of EGF, followed by several washes with EGF-free medium resulted in a severe reduction of binding activity with either ^{125}I-labeled EGF or PAPDIP-^{125}I-labeled EGF. This "down regulation" of receptor activity was also found to be accompanied by a parallel reduction in covalent complex formation (Fig. 4). Time dependence of complex formation at 23° is

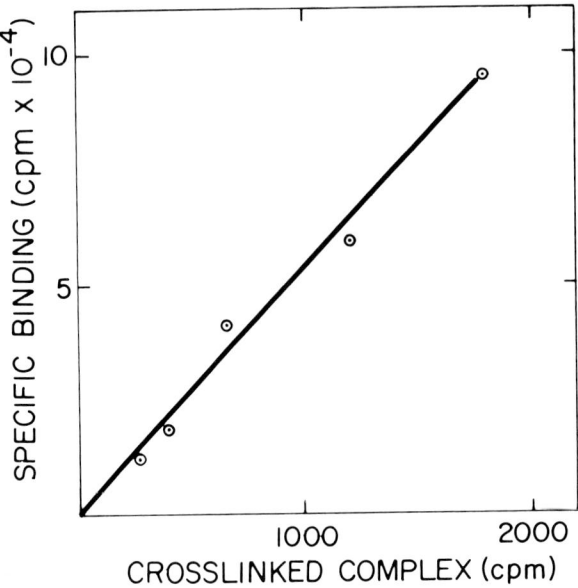

Fig. 3. Relationship between binding activity and crosslinked complex formation. 3T3 cells were incubated with 1, 2, 5, 10, and 25 ng of PAPDIP-^{125}I-labeled EGF at 23° for 10 min, and the cell-bound PAPDIP-^{125}I-labeled EGF was photolyzed. Binding activity was determined as described under "Materials and Methods." The amount of crosslinked complex formed was determined after electrophoretic analysis of the system by counting the covalent complex-containing region of the dried gel slab in a gamma counter. Dried gel strips of similar dimensions from adjacent regions were also counted to correct for background radioactivity. (Reproduced from: Ref. 32).

shown in Fig. 5. The results confirmed the close correspondence between binding activity and crosslinking. However no other new crosslinked species was observed with increasing periods of incubation.

When cells containing bound EGF and the crosslinked complex were incubated at 37°, there was loss of total cell-bound EGF with time and also a decrease in radioactivity in the covalent complex band, accompanied by appearance of new radioactive bands of molecular weights 58,000, 43,000, and 33,000 (Fig. 6). In contrast to the high-molecular-weight band (mol wt 190,000), which was degraded by trypsin treatment of intact cells at 4°, the lower-molecular-weight bands were apparently not accessible to trypsin (Fig. 6). This suggests that the high-molecular-weight crosslinked complex on the cell surface may be internalized and degraded to smaller products during incubation at 37°. Although trypsin could degrade the receptor and the EGF-receptor complex present on the cell surface, it could not release any significant portion of the cell-bound radioactivity into the medium. Murine EGF is known to be relatively insensitive to trypsin (9), and it appears that the EGF binding site on the receptor is likewise insensitive to trypsin, and degradation of a part of the receptor molecule does not result in dissociation of the bound EGF into the medium. Also, with increasing time of incubation at 37°, the cell surface-bound EGF enters into the cell, and thereby becomes totally insensitive to cold trypsin treatment.

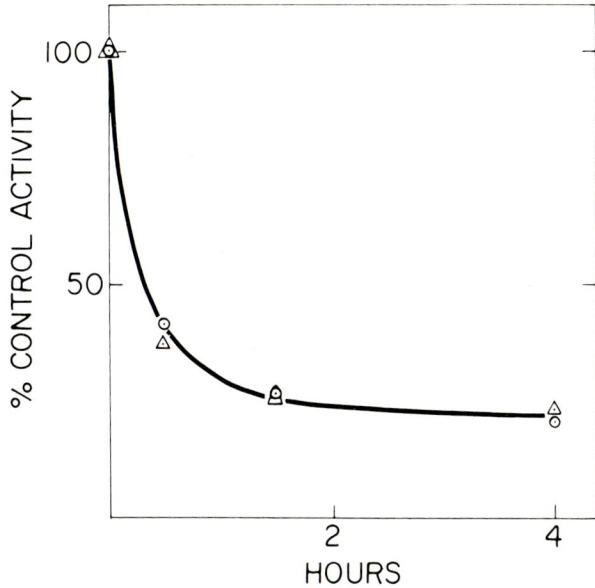

Fig. 4. Effect of "down regulation" of receptor activity on crosslinked complex formation. Monolayers of 3T3 cells in 35-mm culture dishes were incubated at 37° for the indicated time periods, as described under "Materials and Methods," with unlabeled EGF (50 ng/ml) in Dulbecco's modified Eagle medium (DME) containing 5% fetal calf serum. Each monolayer was then washed for 2 h with a total of 18 ml of DME plus 5% fetal calf serum. The washed cells were then incubated for 10 min at 23° with PAPDIP-^{125}I-labeled EGF (12 ng) in 1 ml of binding medium. After photolysis of the cell bound PAPDIP-^{125}I-labeled EGF, the cells were solubilized and subjected to gel electrophoresis. Specific binding (○—○) was determined, as described under "Materials and Methods," and the degree of crosslinking (△—△) was measured, as described in the legend for Fig. 3. [Reproduced from (32)]

DISCUSSION

In attempting to identify a hormone receptor using an affinity labeling procedure, the possibility of nonspecific associations should be excluded. The crosslink reported here is unlikely to be an artifact, since it meets the following criteria: 1) Only 1 protein in 3T3 membrane has been crosslinked to EGF; 2) no crosslinking is observed with the nonbinding variant clone NR-6; 3) there is a decrease in crosslinked complex formation with increasing concentrations of unlabeled EGF; 4) a direct proportionality between binding activity and crosslinked complex formation has been shown to exist under a variety of binding conditions. The photoreactive derivative of EGF used here thus acts as a typical affinity label for its receptor, and there appears to be only 1 protein involved in specific recognition and binding of EGF to 3T3 cells.

The photoreactive reagent (PAPDIP) used in this study for derivatizing EGF and crosslinking it to receptor has a photosensitive arylazide group. This group is readily photolyzed to an arylnitrene which can attack C-H bond, replacing hydrogen (27, 28), thereby allowing the crosslinking to take place even in the absence of any particular functional group on the binding protein. However, the amount of radioactivity in the EGF-receptor crosslinked complex was a small fraction, 1.5–2%, of the total bound radioactivity.

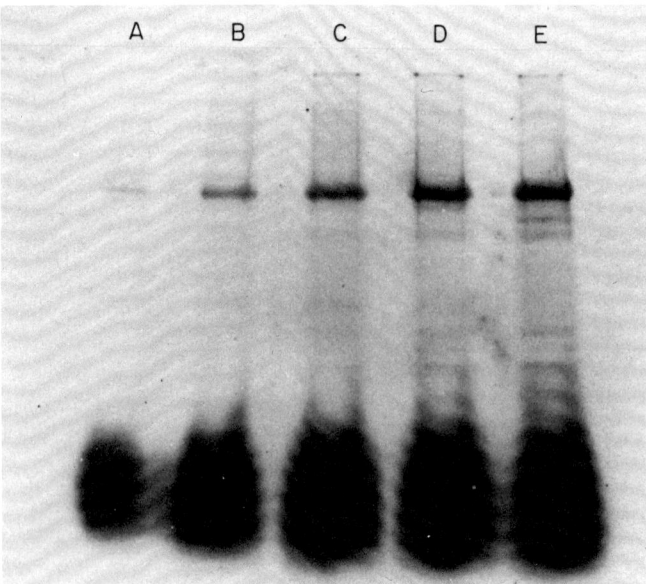

Fig. 5. Time dependence of complex formation. 3T3 cells in 35-mm culture dishes were incubated with 12 ng of PAPDIP-^{125}I-labeled EGF at 23° in the dark for varying time periods. After incubation, each monolayer was washed to remove unbound radioactivity, and then photolyzed and subjected to electrophoresis as described under "Materials and Methods." Total radioactivity bound at 5 min (A), 10 min (B), 20 min (C), 40 min (D), and 30 min (E) was 35,100, 55,910, 71,640, 91,450, and 102,100 cpm per monolayer, respectively. (Reproduced from: Ref. 32).

A likely explanation for this is that the receptor binding site in EGF is situated far from its N terminus containing the photoreactive group. This would reduce the probability of intermolecular crosslinking between EGF and its receptor, but would increase the chances of intramolecular crosslinking within the molecule of EGF. The stokes radius of EGF is about 12 Å, calculated on the assumption that it is a globular protein, whereas the heterobifunctional reagent used in this study to derivatize EGF has a length of about 11 Å. Therefore, despite the small size of EGF as a protein, the crosslinker cannot span the whole molecule of EGF. Also, the arylazide protion of the crosslinker is lipophilic, allowing for possible penetration of the hydrocarbon core and its resultant crosslinking with lipid rather than receptor.

The heterobifunctional reagent is cleavable under reducing conditions. Although such a reagent is ideal for specifically labeling the receptor without permanent attachment to EGF, the cleavability of the crosslink does not allow the use of reducing SDS gels for receptor subunit characterization. The high-molecular-weight protein that has been complexed with EGF could thus be an oligomer composed of smaller subunits held together by disulfide bridges, but studies in progress utilizing noncleavable crosslinkers to determine the mol wt of the EGF receptor in reducing gels indicate that the receptor has no apparent subunit structure.

Cell-associated EGF is rapidly degraded at 37°. However the rate of release of radioactivity from covalently bound EGF is slow compared to the rate of degradation of noncovalently bound EGF (Fig. 6). A possible explanation for this is selective iodination of

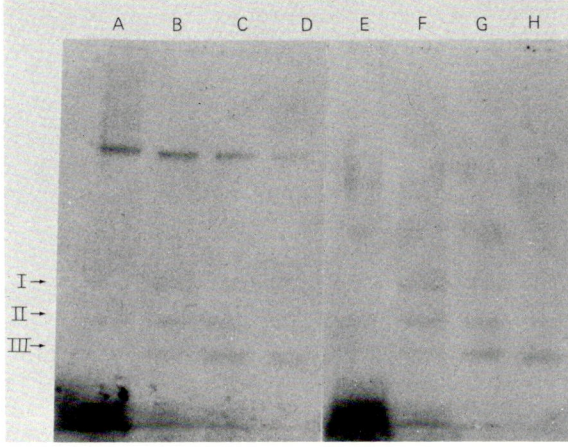

Fig. 6. Degradation of cell-bound EGF at 37°. Cells were incubated with PAPDIP-^{125}I-labeled EGF in 1 ml of Earle's balanced salt solution containing 5 mM HEPES, pH 7.4, and 0.02% bovine serum albumin at 23° for 10 min in the dark. After incubation, each monolayer was washed 6 times with 3 ml each time in Earle's balanced salt solution containing 5 mM HEPES, pH 7.4 (EBSS), and then photolyzed. Then 1 ml of EBSS was added to each plate and the monolayers were incubated at 37° for 0 min (A), 30 min (B), 60 min (C), and 120 min (D). The supernatant fraction was removed, and the cells were solubilized and analyzed by electrophoresis and subsequent autoradiography as described in "Materials and Methods." Total radioactivity in lanes A, B, C, and D was 60,586, 32,596, 18,614, and 14,043 cpm, respectively. Another set of cell monolayers was treated identically, except that after incubation at 37° for 0 min (E), 30 min (F), 60 min (G), and 120 min (H), the cells were treated with trypsin (50 μg/ml) in 1 ml of EBSS at 4° for 30 min. The supernatant was carefully aspirated, and the cell-bound radioactivity was analyzed as described above. Total radioactivity in lanes E, F, G, and H was 59,756, 35,604, 20,366, and 14,166 cpm, respectively.

the tyrosine residues near the N-terminal region of EGF, the part of the molecule that becomes covalently linked to receptor. This would reduce the accessibility of the radioactive portion of the EGF molecule to proteolytic attack. It may also be necessary for EGF to be dissociated from the EGF-receptor complex in order to become hydrolyzed efficiently.

Recent studies have led to a greater recognition of the role of EGF as a physiological effector involved in the control of growth, proliferation, and senescence of cells (29, 30). EGF acts as a mitogen at very low concentrations (1–10 ng/ml), and the apparent dissociation constant for EGF binding to cells correlates well with the estimates of concentration of this peptide in mouse serum (31). The EGF receptor system is therefore an excellent model to explore the mechanism of action of mitogens in general. Cells carrying an in situ radiolabeled receptor may be particularly useful in identifying the cellular components, other than the receptor, which are involved in the expression of mitogenic response.

REFERENCES

1. Cohen S: Natl Cancer Inst Monogr 13:13, 1964.
2. Hollenberg MD, Cuatrecasas P: J Biol Chem 250:3845, 1975.
3. Carpenter G, Cohen S: J Cell Physiol 88:227, 1976.
4. Westermark B: Biochem Biophys Res Commun 69:304, 1976.

5. Armelin HA: Proc Natl Acad Sci USA 70:2702, 1973.
6. Rose SP, Pruss RM, Herschman HR: J Cell Physiol 86:593, 1975.
7. Cohen S: J Biol Chem 237:1555, 1962.
8. Taylor JM, Mitchell WM, Cohen S: J Biol Chem 247:5928, 1972.
9. Savage CR, Inegami T, Cohen S: J Biol Chem 247:7612, 1972.
10. Cohen S, Carpenter G: Proc Natl Acad Sci USA 72:1317, 1975.
11. Carpenter G, Lembach KJ, Morrison MM, Cohen S: J Biol Chem 250:4297, 1975.
12. O'Keefe E, Hollenberg MD, Cuatrecasas P: Arch Biochem Biophys 164:518, 1974.
13. Olsen RW, Meunier JC, Changeux JP: FEBS Lett 28:96, 1972.
14. Cuatrecasas P: Proc Natl Acad Sci USA 69:1277, 1972.
15. Lefkowitz RJ, O'Hare D, Haber E: Proc Natl Acad Sci USA 69:2828, 1972.
16. Pricer WE, Hudgin RL, Ashwell G, Stockert RJ, Morell AG: Methods Enzymol 34:688, 1974.
17. Tate RL, Winard RJ, Kohn LD: Methods Enzymol 34:692, 1972.
18. Galardy RE, Greig LC, Jamieson JD, Printz MP: J Biol Chem 249:3510, 1974.
19. Ji TH: J Biol Chem 252:1566, 1977.
20. Katzenellenbogen JA, Johnson HF, Myers HN: Biochemistry 12:4085, 1973.
21. Marver D, Chiu WH, Wolff ME, Edelman IS: Proc Natl Acad Sci USA 73:4462, 1976.
22. Ruoho AE, Kiefer H, Roeder PE, Singer SJ: Proc Natl Acad Sci USA 70:2567, 1973.
23. Savage CR, Cohen S: J Biol Chem 247:7609, 1972.
24. Baum SG, Horwitz MS, Maizel JV: J Virol 10:211, 1972.
25. Carpenter G, Cohen S: J Cell Biol 71:159, 1976.
26. Huang YO, Hoff S, Wisnieski B: Fed Proc 35:1703, 1976.
27. Knowles JR: Acc Chem Res 5:155, 1972.
28. Fleet GWJ, Knowles JR, Porter RR: Biochem J 128:499, 1972.
29. Todaro GJ, DeLarco JE, Cohen S: Nature 264:26, 1976.
30. Rheinwald JG, Green H: Nature 265:421, 1977.
31. Cohen S: J Invest Dermatol 59:13, 1972.
32. Das M, Miyakawa T, Fox CF, Aheronov A, Pruss RM, Herschman HR: Proc Natl Acad Sci 74:2790, 1977.

Lectin Receptors and Cell Surface Recognition

R. Colin Hughes

National Institute for Medical Research, Mill Hill, London NW7 1AA, England

Ricin-resistant (Ric^R) baby hamster kidney (BHK) cell lines have been classified into a small group (3 lines) showing a minimal surface change from wild type and a larger group (20 lines) which exhibit more extreme alterations in surface properties. Glycopeptides released by pronase from some Ric^R lines in the second group show a lower content of sialic acid, galactose, and N-acetylglucosamine and greatly reduced binding activity for ricin. Treatment of receptor deficient glycopeptides or Ric^R cells with neuraminidase reveals new ricin receptors and renders the cells very sensitive to ricin. Several classes of ricin receptors are postulated for BHK cells, some of which are cryptic and under independent genetic control from receptors selected against with ricin.

The cell lines showing greatly altered surface properties in general adhere poorly to a substratum and also aggregate poorly compared to wild type or Ric^R cells showing minimal surface change. The lactoperoxidase iodinateable 250K glycoprotein of normal BHK cells is lacking in all but one of these Ric^R cell lines. The role of 250K glycoprotein in normal cell adhesion is considered and a hypothesis proposed relating changes in surface organization of the 250K glycoprotein to alterations in receptors induced by ricin selection.

Key words: ricin, glycoprotein receptors, mutation, adhesion

The biological functions of cell surface carbohydrates are poorly understood. However, surface glycolipids and glycoproteins are widely believed to play important roles in cellular adhesiveness and in the recognition of target cells by effector substances such as polypeptide hormones, plant and bacterial toxins, and the antiviral protein interferon (reviewed in Ref. 1).

Although the isolation and structural analysis of surface carbohydrate receptors for hormones and toxins are well advanced and some progress has been made in characterization of aggregation factors, little is known of the genetic control and specificity of these receptors. As one approach to these outstanding problems several groups (2-12) have attempted to modulate surface carbohydrate composition by selection of cells resistant to toxic lectins having well-defined specificities for particular carbohydrate sequences. Like many of the effector substances listed above, the initial events in mediation of lectin toxicity presumably require binding to cell surface carbohydrates (12-16). Therefore, loss of surface carbohydrate receptors would be expected as a common phenotype among

Received April 11, 1977.

lectin resistant cells. The availability of such surface mutants or variants affords a unique opportunity to analyze the structure, biosynthesis, and functions of cell surface carbohydrates in mammalian cells.

Most of the discussion in this paper will concern ricin-resistant (Ric^R) baby hamster kidney (BHK) cells isolated in our laboratory and described in detail elsewhere (7, 8). Recent studies of the response of these cells to toxic lectins have suggested an unexpected complexity of surface receptors for ricin. A minor proportion of these receptors appear to be critically involved in mediating ricin toxicity as suggested earlier by Nicolson (15). It is argued on the basis of other evidence that these same receptors may also be involved in surface membrane organization and in cellular adhesiveness, and simple models are presented incorporating these hypotheses.

METHODS

Cell Lines

Ricin-resistant clones of baby hamster kidney cells (BHK21/C13 cells) have been isolated by treating monolayers of mutagenized (methyl-N-nitro-N-nitrosoguanidine treated) cells with 1 $\mu g/ml^{-1}$ ricin in modified Eagle's medium plus 10% fetal calf serum as described (7, 8). After several days at 38.5°C a few single cells remained attached to the culture plate and these were grown into reasonably sized colonies, removed by trypsinization, and propagated in ricin containing medium to high densities. Usually each cell line has been recloned at least once and subsequently passaged in medium minus ricin for periods up to 2 years. Routinely during continuous passage, cultures of each Ric^R cell line have been tested as described below to detect phenotypic reversion to ricin sensitivity. So far no detectable reversion has occurred in any of the cell lines discussed here.

Resistance to Lectins

Trypsinized cells in Ham's F10 medium plus 10% fetal calf serum are plated on 60 mM Falcon dishes at approximately 1,000 cells per dish. After 1–2 hr at 39°C ricin is added at various known final concentrations and the colonies formed after 5–7 days in the presence of ricin counted and compared to the number of colonies formed in control dishes containing no ricin. The effects of Phaseolus vulgaris lectin (PHA) and concanavalin A (con A) are tested similarly.

In other experiments the inhibitory effect of ricin on protein synthesis has been measured by pretreating cells with ricin at various concentrations for 1–4 hr at 39°C, to allow penetration into the cells, followed by pulsing the cells with [^{3}H] leucine. The incorporation of radioactivity into a total acid precipitable fraction is then measured as described (17) and compared with the incorporation taking place in untreated cells over the same pulse period.

Binding of ^{125}I-Lectins

Lactoperoxidase iodinated ricin and concanavalin A were purified after extensive dialysis by affinity chromatography on Sepharose 6B and Sephadex G50, respectively. Radioiodinated peanut (Arachis hypogaea) agglutinin was purified on Biogel P150 as described fully elsewhere (S. Rosen and R. C. Hughes, manuscript in preparation). All binding procedures were carried out at 4°C and in the case of ricin and con A suitably controlled for nonspecific binding in the presence of hapten inhibitors (lactose and α-methylmannoside, respectively).

Neuraminidase Treatments of Cells

Cell monolayers were washed with phosphate buffered saline and treated at 36°C for 30 min with 1 ml saline containing neuraminidase (Hoechst Pharmaceuticals, Kew Bridge, England) (50 units/ml). After further washing with buffered saline the treated cells were used for binding experiments with ^{125}I-lectins or for cytotoxicity tests with ricin. Full details of the experiments will be reported elsewhere (S. Rosen and R. C. Hughes, in preparation).

Surface Labeling

Cells growing in monolayer cultures on 60 mM dishes were labeled by the lactoperoxidase-glucose oxidase iodination procedure modified as described (18; R. Nairn and R. C. Hughes, in preparation). After labeling, the cells were solubilized in 0.5% NP40 and solubilized glycoproteins run through affinity columns (10 ml; 1 mg/ml) of ricin-Sepharose 4B. Initial elution with 0.5% NP40 was followed by 0.1 M lactose in 0.5% NP40. Iodinated cells were also treated with 1% sodium dodecyl sulphate-1% β-mercaptoethanol in electrophoresis buffer, pH 6.7, and analyzed by slab polyacrylamide gel electrophoresis. Radioactive bands were detected by radioautography or gel slicing and radioactive counting in a gamma spectrometer.

Adhesion Assays

Cells were dispersed by brief trypsinization in EDTA containing buffers and aggregation in trypsin-free medium was followed by Drs. J. G. Edwards and J. McK. Dysart, Department of Cell Biology, University of Glasgow, Scotland, using established procedures described elsewhere (19–22). In some experiments cells dispersed with EDTA alone were used with essentially similar results. Adhesion of lightly trypsinized cells to protein covered coverslips followed the procedure of Rabinovitch and DeStefano (23) except that the cells were prelabeled by incorporation of ^{32}P-phosphate as described previously (22).

RESULTS AND DISCUSSION

Phenotypic Classes of Ricin Resistant (Ric^R) Cells

Preliminary studies from our laboratory and others have shown a considerable degree of phenotypic variation among lectin-resistant cell lines (8, 10, 11). This is shown by an analysis of the cross-resistance of cell lines to lectins other than the selective lectin and by lectin-binding experiments (Table I). By this means 4 main classes of Ric^R BHK cell variants are obtained. Two of these classes, II and IV, are distinguished by cross-resistance to PHA which binds to closely similar carbohydrate sequences to ricin on sensitive cells. The simplest explanation for lectin resistance is a loss of surface receptors and mutants of class III and IV show this phenotype, while class I and II mutants bind ricin normally. It is already apparent, however, that several different phenotypes exist within each class of Ric^R mutants. There are almost certainly, for example, several mechanisms leading to loss of surface receptors.

TABLE I. Properties of Ric^R Variants

Class	Cells	Resistance[a] to			Ricin binding	Phenotype name
		Ricin	PHA	Con A		
I	Ric^R 1	+	WT		WT	Ric^R PHA^S Con A^S bin$^+$
	2	+	WT		WT	
	22	+	WT	WT	WT	
II	12	+	+	−	WT	Ric^R PHA^R Con A^S bin$^+$
	16	+	+	−	WT	
	19	+	+	WT	WT	
III	17	+	WT		−	Ric^R PHA^S Con A^S bin$^-$
IV	9	+	+		−	Ric^R PHA^R Con A^S bin$^-$
	10	+	+		−	
	20	+	+		−	
	7	+	+	−	−	
	14	+	+	−	−	
	15	+	+	−	−	
	18	+	+	−	−	
	21	+	+	−	−	

[a]Resistance related to wild type (WT): +) at least 5-fold greater than WT; −) at least 5-fold less than WT

Modification of Ricin Receptors in Ric^R Cells

Direct evidence for a block in carbohydrate chain assembly which could contribute to loss of these lectin receptors has been found in lectin-resistant cells. Several of these lines show low activity in a specific N-acetylglucosaminyl transferase, e.g., in PHA and Ric^R CHO cells (24, 25) and in the Ric^R 14 clone of Ric^R BHK cells (7, 26) (Table I). The enzyme transfers N-acetylglucosamine (GlcNAc) in β-glycosidic linkage to a mannose rich core of glycosidase-treated soluble glycoproteins, e.g., α_1-acid glycoprotein and immunoglobulin. This enzyme presumably is needed for cell surface glycoprotein biosynthesis and complete assembly of ricin receptors. Thus, the absence of a GlcNAc transferase and GlcNAc transfer to core sequences would prevent the subsequent addition of peripheral sequences containing galactose sialic acid and fucose residues, even though the specific glycosyl transferases building up these peripheral sequences may be fully active in mutant cells. Presumably assembly of the carbohydrate sequences containing binding sites for ricin could be affected at several enzymatic steps accounting for the many Ric^R cell lines containing normal levels of GlcNAc transferase which have, nevertheless, lost ricin binding sites (Table I). Evidently, however, synthesis of the mannose containing core sequences is not affected in any of our lines, since the cells are at least as sensitive as wild type cells to con A. The recently described "double mutants" selected for resistance to PHA and Con A (9) may show defective core assembly.

In Ric^R 17 cells (Table I) loss of ricin receptors does not deplete surface receptors for PHA, showing that ricin and PHA receptors, although similar structurally, are nevertheless distinct. The PHA-resistant cells described by Stanley (9), some of which are as sensitive as wild type to ricin, underline these differences in receptor sites for the 2 lectins. In general, however, mutational loss of ricin and PHA receptors occurs simultaneously.

Recently, Dr. Saul Rosen, Mr. Lionel Wilson, and I have been able to identify in BHK wild type cells the glycopeptides binding to a ricin-Sepharose affinity column. The glycopeptides were produced by pronase digestion of [^{3}H]glucosamine labeled cells and separated into 5 fractions of different molecular weights by chromatography on Sephadex G50. When each glycopeptide fraction was passed through a ricin-Sepharose affinity column about 22% of the label in one of these fractions, or less than 5% of the total labeled material, adsorbed to the column and was eluted with lactose, while the other glycopeptide fractions had essentially no binding activity. The specific ricin-binding fraction contains all the common monosaccharide constituents of glycoproteins including as major components galactose, glucosamine, mannose, and sialic acid, lesser amounts of galactosamine, and a trace of fucose. From the apparent molecular weight of 3,800 we calculate a carbohydrate chain of about 15–20 sugars. When we analyzed glycopeptide fractions of Ric^{R}14 and Ric^{R}21 cells in the same way there was a marked shift of [^{3}H]glucosamine-labeled material to glycopeptides eluting as components of lower molecular weight (about 3,100) on Sephadex G50. These fractions contained less galactose, sialic acid, and glucosamine and were enriched in mannose. The results are consistent, therefore, with glycosylation defects due to GlcNAc transferase in Ric^{R}14 and unknown causes in Ric^{R}21 and a structural change in many cell surface glycopeptides, including a small fraction of ricin-binding glycopeptides. The shift in molecular weight suggests deletion of about 3–6 sugar residues per carbohydrate chain. Ogata et al. (27) have recently reported on the carbohydrate structure of BHK glycoproteins. A mannose-rich core region is substituted with a branched peripheral sequence containing 2 sialic acid-galactose-glucosamine sequences. Deletion of these sequences in the Ric^R mutants would reduce the size of the branched carbohydrate chain by close to the amount estimated from gel filtration on Sephadex G50.

Cryptic Receptors

Although the enzymic defect differs in Ric^{R}14 and 21, the end result is the same, i.e., loss of functional receptors (Fig. 1). Recently Dr. Saul Rosen has shown that new receptors for ricin are exposed on these resistant cells by treatment with neuraminidase. Neuraminidase-treated cells bind as much ricin as wild type cells and revert to a completely sensitive state (LD$_{10}$ shifts from > 20 to 0.5 μg/ml ricin). Wild type cells also show a 2-fold increase in ricin binding after neuraminidase and interestingly become measurably more sensitive to ricin (LD$_{10}$ shifts from 1 μg/ml to 0.25 μg/ml ricin). The cryptic ricin receptors which are revealed by removal of sialic acid bind a highly purified preparation of peanut agglutinin (28), kindly provided by Dr. N. Sharon, Weizmann Institute, Rehovot, Israel. Very similar binding of peanut agglutinin is found for the Ric^R and wild type cells, indicating that assembly of the cryptic receptors does not require the enzymes defective in Ric^{R}14 and, presumably Ric^{R}21 cells and the carbohydrate structures of these ricin receptors are probably very different. From the specificity of peanut agglutinin (28) we conclude that the cryptic receptors include the sequence galactosyl β1,3 N-acetylgalactosamine or possibly galactosyl β1,3 N-acetylglucosamine. The former sequence is part of the smaller carbohydrate chain of glycophorin (31). Earlier, Nicolson (29) and Kornfeld's group (30) had reported a greatly increased binding of ricin to glycophorin after neuraminidase treatment.

When ^{125}I-labeled glycoproteins from lactoperoxidase surface-labeled Ric^{R}14 or 21 cells were solubilized in 0.5% NP40, treated with neuraminidase and passed through a

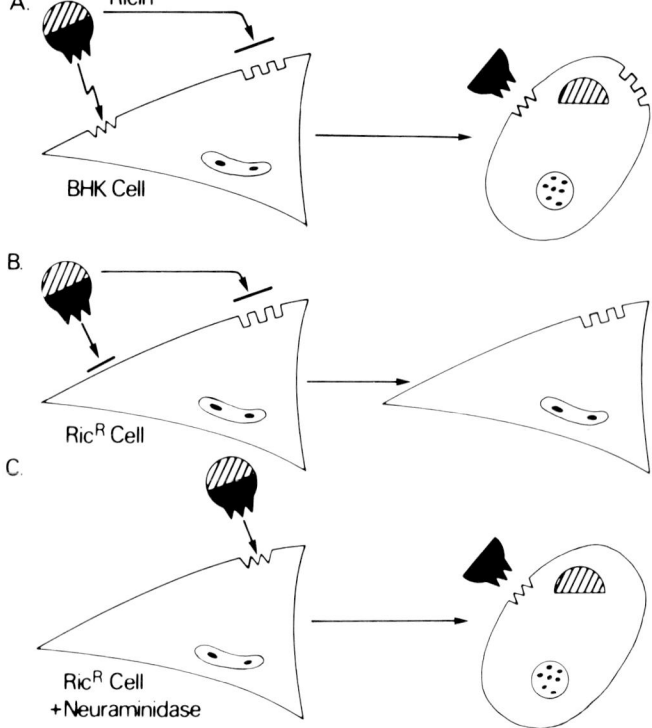

Fig. 1. Receptor control of susceptibility to ricin cytotoxicity. a) Ricin or a ricin subunit enters a susceptible cell through a specific cell receptor. As a consequence, cytoplasmic protein synthesis is inhibited and the cell dies. b) Ricin cannot enter a resistant cell line because there is no compatible receptor. c) Ricin can enter a resistant cell line pretreated with neuraminadase which exposes free ricin receptors on the cell surface.

ricin-Sepharose affinity column, essentially all of the glycoproteins adsorbed to the column and were eluted with lactose. All of the ^{125}I-labeled components of wild type cells adsorbed without neuraminidase treatment.

Although our evidence is so far indirect, we therefore picture the BHK cell surface ricin receptors as curiously similar in organization to glycophorin (31). Thus, in normal BHK cells at least 10 different polypeptides are substituted with probably several complex carbohydrate chains of the following average sequence: sialic acid, galactose, N-acetylglucosamine, mannose core, peptide, and several perhaps simpler oligosaccharides containing a sialic acid, galactosyl β1, 3 N-acetylhexosamine sequence. The former chains bind ricin and mediate toxicity in wild type cells and their synthesis is defective in several resistant variants. The latter chains bind ricin only after neuraminidase treatment, but then mediate ricin toxicity (Fig. 1).

Surface Organization of Ricin Receptors

As pointed out by Nicolson (15), lectin-resistant cell lines showing essentially normal binding of the selective lectin (Table I) indicate that the majority of lectin binding to susceptible cells may be irrelevant for mediation of lectin cytotoxicity. There must be 2 types of exposed ricin receptors, R1 and R3 (Fig. 2), on sensitive cells which are similar structurally, since loss of GlcNAc transferase ($Ric^R 14$) affects assembly of both sets of receptors but which can be regulated independently as shown by loss of only R1 receptors in some Ric^R mutants (Table I, class II). By analogy, neuraminidase reveals 2 sets of cryptic ricin receptors (Fig. 2, R2 and R4) of which only R2 receptors mediate cytotoxicity. Some properties of the various types of ricin receptors in BHK cells indicated by these experiments are given in Table II.

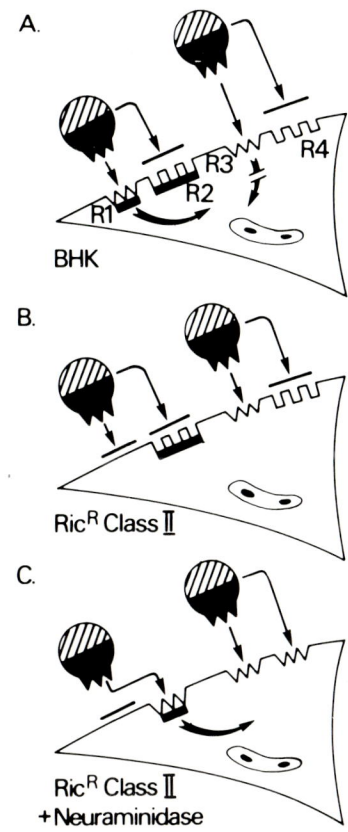

Fig. 2. Classes of ricin receptors on BHK cells. Four main types of receptors are identified: R1 receptors bind ricin before neuraminidase treatment and are under transmembrane control. R2 receptors are under transmembrane control but bind ricin only after neuraminidase treatment. R3 and R4 receptors bind ricin before or after neuraminidase, respectively, but do not interact transmembranally. a) Ricin bound to R1 receptors enters the cell to inhibit cytoplasmic protein synthesis. Ricin bound to R3 receptors remains at the cell surface and is without effect. b) In Ric^R cell R1 receptors are missing while R2 receptors are unchanged and R3 receptors may or may not be expressed normally. Ricin does not internalize. c) After neuraminidase R2 receptors convert to productive receptors and ricin enters the cell. Ricin binding may be enhanced by conversion of R4 receptors.

TABLE II. Properties of BHK Ricin Receptors

1.	R1, R3.	Mutation can affect both sites. Some GTs involved in assembly of both R1 and R3 (GlcNAc-GT deficient clone lacks R1 and R3). Common CHO structures for R1, R3.
2.	R1, R3.	Mutational loss of R1 sites occurs when assembly of R3 sites is normal. Some GTs unique for R1 Differences in CHO structure for R1, R3
3.	R3 > R1.	In WT cells. Since class II mutants bind almost normal amounts of ricin.
4.	[R2 + R4] ⩾ [R1 + R3].	Since > 100% increase in ricin binding by WT cells after neuraminidase.
5.	R1, R2.	Assembled independently No mutation affecting both R1 and R2 yet found Different CHO structures R2 binds peanut agglutinin after neuraminidase treatment
6.	R1/R3 and R2/R4 type receptors carried on same polypeptides	

By lactoperoxidase labeling Nicolson (15) was able to show the deletion of 2 surface membrane polypeptides of molecular weight 80,000 and 35,000 from Ric^R BW5147 lymphoma. He suggested that these components facilitate transport of ricin-receptor complexes into susceptible BW5147 cells. Nicolson further suggested that the glycoprotein receptors involved in ricin uptake may be connected to cytoplasmic contractile elements, whereas the majority of surface receptor glycoproteins are not (see Fig. 2, R1 and R3 respectively). Similar analysis of Ric^R clones 12, 16, and 19 shows no discrete deletions of polypeptides of this size (R. Nairn and R. C. Hughes, in preparation). However, in each case the 250K glycoprotein or LETS glycoprotein of Hynes (32) is missing from the labeled profiles. We know that 250K glycoprotein binds ricin and following Nicolson's reasoning it could be argued that ricin toxicity is mediated by binding to R1 receptors of the 250K glycoprotein, which in turn is linked to cytoplasmic microfilaments either directly or by an intermediate integral membrane component (baseplate) since the 250K glycoprotein itself seems not to be integrated into the surface membrane (37, 41). However, the 250K glycoprotein is not obligatory for a cell to be sensitive to ricin; for example, virus-transformed BHK cells or neuraminidase-treated class II mutants (Table I) are fully sensitive, which suggests that the baseplate or other integral transmembrane (glyco) proteins also carry exposed (BHK py) or cryptic (class II Ric^R mutants) ricin receptors that function in the absence of 250K glycoprotein.

The concurrent loss of the 250K glycoprotein and ricin R1 receptors (Fig. 2) in most BHK Ric^R mutants, only class I mutants and the Ric^R 14 clone (Table I) express normally the 250K glycoprotein (R. Nairn and R. C. Hughes, in preparation), raises the interesting possibility that in some way functional R1 receptors (Fig. 2) are involved in interaction between 250K glycoprotein and the baseplate of the surface membrane (Fig. 3). In this way selection against R1 receptors with ricin would be expected to interfere with normal 250K glycoprotein-baseplate binding, and perhaps other 250K glycoprotein functions, e.g., as an aggregation factor.

Adhesive Properties of Ric^R Cells

In collaboration with Drs. J. G. Edwards and J. McK. Dysart, Department of Cell Biology, University of Glasgow, Scotland, we have found that all of the Ric^R BHK cell lines so far tested are defective to a greater or lesser degree in short term or long term aggregation experiments (Table III). Further, most of the Ric^R cell lines show slower-than-wild-type rates of attachment of labeled single cells to collagen or serum coats (22). In general, a particular cell line shows similar adhesive properties when tested in cellular aggregation or substrate adhesion assays, suggesting some common steps in these very complex and diverse processes. The fact that one cell line, Ric^R 19, is apparently normal in substrate attachment while forming cellular aggregates less well than wild type (although there is some day-to-day variation), does indicate that these processes are separable, however.

TABLE III. Adhesive Properties of Ric^R Variants (with J. G. Edwards)

Cells	Class	Gelatin	Cell – Substratum		Cell – Cell	
			Calf serum	Fetal calf serum	Short term	Long term
Ric^R 7	IV	–	–	–	–	–
12	II	–	–	–	–	–
14	IV	–	–	–	–	–
16	II	–	–	–	–	–
17	III	–	–	–	–	–
19	II	+	WT	WT	–	WT
21	IV	–	–	–	–	–
22	I	+	WT	WT	WT	WT

+ denotes greater than wild type (WT) adhesiveness
– denotes less than wild type adhesiveness

Our general conclusions so far are that Ric^R cells bearing wild type amounts of lectin receptors and showing a normal surface labeling pattern are closest to wild type behavior in adhesion assays, e.g., Ric^R 22 (Table III). On the other hand, a correlation seems to exist between loss of ricin receptors, loss of 250K glycoprotein in all but one case (Ric^R 14), and cellular adhesivity. However, a simple correlation is excluded by the poor adhesiveness of clones Ric^R 12 and 16 (Table III) which bind ricin well (Table I). This implies that R1 glycoprotein receptors, which are modified in all our Ric^R mutants, as shown by neuraminidase-induced reversion to ricin sensitivity, may be involved both in mediating ricin toxicity and in promoting aggregation. Whether this is because of a unique carbohydrate sequence present on R1 but absent on R3 receptors or due to the putative transmembrane control of the R1 glycoprotein receptors, e.g., a role in some cis reorganization of the cell surface during intercellular aggregation or substrate attachment, remains to be determined.

Model

A simple model incorporating these points is shown in Fig. 3. In this model the 250K glycoprotein-baseplate interactions are mediated by binding of R1-type receptors on the 250K glycoprotein by a lectin (agglutinin)-like activity on baseplate or baseplate

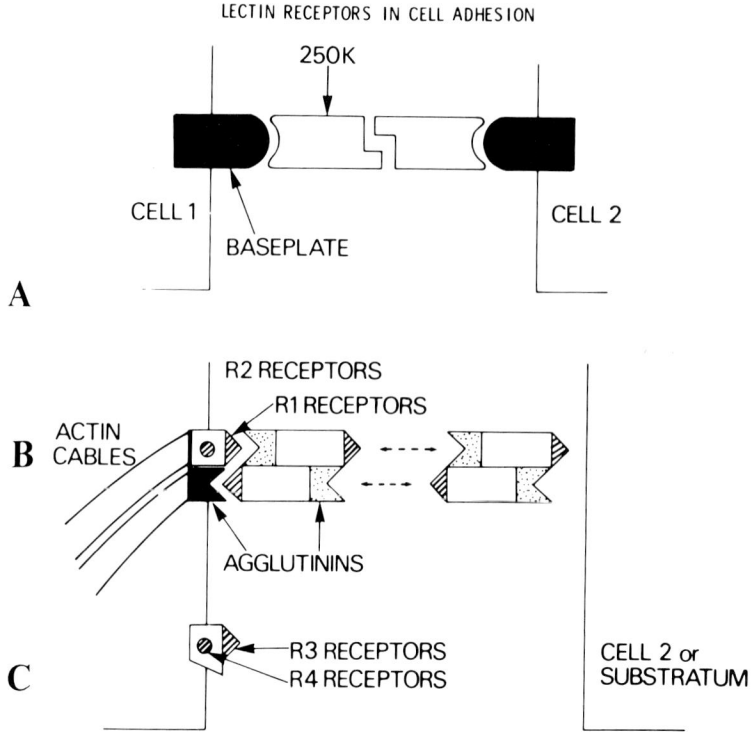

Fig. 3. Models of cell recognition. A) The large external glycoprotein (LETS or 250K glycoprotein) is anchored to an integral membrane protein or glycoprotein (baseplate) and interacts to form adhesive bonds between neighboring cells. As a consequence of ricin selection, interaction between 250K glycoproteins is usually greatly reduced (cellular adhesiveness is reduced) and in most but not all cases in interaction between the 250K glycoprotein and baseplate is also prevented (loss of surface exposed 250K glycoprotein). B) Lectin receptors (R1) mediating ricin cytotoxicity are present on both the baseplate glycoprotein and the 250K glycoprotein. R1 receptors play a role in baseplate-250K glycoprotein interactions via an agglutinin subunit of the baseplate glycoprotein and an agglutinin of different specificity present on the 250K glycoprotein molecule. The intractions between R1 receptors and the agglutinin site of the 250K glycoprotein are postulated to be responsible for the formation of adhesive bonds between cells or onto a substratum. Structural changes in R1 sites leading to loss of binding by ricin similarly may weaken or prevent interactions of R1 sites with baseplate and 250 K glycoprotein agglutinin sites. Definitions of ricin receptors R2, R3, and R4 are given in Fig. 2. See text for further discussion.

subunit. Recently the Glasgow group has described (33) a membrane-bound hemagglutinin in wild-type BHK cells which has many of the proposed properties of the baseplate (Fig. 3). Thus, the hemagglutinin is solubilized by deoxycholate, indicating an integral membrane protein, and hemagglutination is inhibited by N-acetylgalactosamine (a ricin-binding sugar) or thio-di-galactoside. This factor (membrane-HA) therefore is clearly distinguishable from the 250K hemagglutinating (250K-HA) activity, discovered by Yamada et al. (34, 35). We propose, however, that the 250K-HA also plays a part in 250K-baseplate interactions although its main role in the simple model is to mediate intercellullar attachment and possibly attachment to substrate (Fig. 3). The Yamada et al.

hemagglutinin from chick cells is not inhibited by simple sugars, except free amino sugars or other amines. However, interaction may require an extended oligosaccharide sequence or perhaps a basic peptide, the conformation of which is maintained by interaction with an intact R1 oligosaccharide unit similar to human blood group MN antigenic determinants, for instance (36). Alternatively, the peptide sequence differences between the chick component and the analogous BHK 250K glycoproteins (37) may imply a different recognition site for hemagglutination in BHK cells.

The model (Fig. 3) can account at least in part for the pleiotropic changes seen in the Ric^R mutants. The R1 receptors are recognized by at least 4 ligands, namely baseplate and 250K agglutinins, ricin, and PHA. We have already seen in $Ric^R 17$ cells how R1 binding of ricin may be decreased without affecting the PHA-binding. Taking this point further, therefore, other mutations affecting ricin-binding may leave recognition of modified R1 receptors by baseplate-HA unimpaired, e.g., in the $Ric^R 14$ clone which expresses surface 250K glycoprotein normally. Since adhesiveness of $Ric^R 14$ cells is low, however, the mutation does affect recognition of R1 receptors by 250K-HA. In $Ric^R 22$ cells only ricin-binding to modified R1 receptors is affected, since these cells are sensitive to PHA, are wild type in adhesiveness, and express the surface-bound 250K glycoprotein. The scheme does not account satisfactorily for the properties of the $Ric^R 19$ clone. In these cells presumably 250K-HA interactions with modified R1 receptors are broken since the cells express no surface-bound 250K glycoprotein. Perhaps these cells synthesize and secrete the glycoprotein normally however, as do some virus-transformed cells (38), giving essentially wild type adhesion of $Ric^R 19$ cells to substrate coats. Future characterization of the R1 receptors in wild type and Ric^R mutants and their interactions with BHK cell agglutinins will be an interesting test of the models presented here.

CONCLUSIONS

The wide range of phenotypes evident among the lectin-resistant cells so far isolated should provide valuable aids at many different levels. Such cells have obvious value for study of the control of synthesis and secretion of glycoproteins, as well as for defining more precisely the roles of carbohydrates in membrane functions, both at the cell surface as well as intracellularly, e.g., the mitochondrial calcium-binding glycoprotein which appears to require a full complement of sialic acid for cation binding. Similarly, the complex relationships between lysosomal function and complete glycosylation of lysosomal enzymes (39, 40), and the breakdown of normal relationships in rare human genetic diseases, e.g., I-cell disease may conceivably be approached using readily available lectin-resistant cell lines as models. Finally, the possibility of correlating changes in surface structure with adhesiveness may ultimately define more precisely some of the surface molecules necessary for this important biological function.

ACKNOWLEDGMENTS

I thank colleagues at Mill Hill, A. Meager, R. Nairn, S. Rosen, A. Ungkitchanukit, and L. Wilson and also J. G. Edwards and J. McK. Dysart of Glasgow University for active collaboration and valuable discussion.

REFERENCES

1. Hughes RC: Essays Biochem 11:1, 1975.
2. Ozanne B: J Virol 12:79, 1973.
3. Culp L, Black PH: J Virol 9:611, 1972.
4. Wright JA: J Cell Biol 56:666, 1973.
5. Gottlieb C, Skinner AM, Kornfeld S: Proc Natl Acad Sci USA 71:1078, 1974.
6. Zachowski A, Prigent B, Monsigny M, Paraf A: Biochimie 56:1621, 1974.
7. Meager A, Ungkitchanukit A, Nairn R, Hughes RC: Nature 257:137, 1975.
8. Meager A, Ungkitchanukit A, Hughes RC: Biochem J 154:113, 1976.
9. Stanley P, Caillibot V, Siminovitch L: Somat Cell Genet 1:3, 1975.
10. Stanley P, Caillibot V, Siminovitch L: Cell 6:121, 1976.
11. Stanley P, Siminovitch L: In Vitro 12:208, 1976.
12. Hyman R, Lacorbiere M, Stavarek S, Nicolson G: J Natl Cancer Inst 52:963, 1974.
13. Nicolson GL, Lacorbiere M, Hunter TR: Cancer Res 35:144, 1975.
14. Refsnes K, Olsnes S, Pihl A: J Biol Chem 249:3557, 1974.
15. Nicolson GL, Robbins JC, Hyman R: J Supramol Struct 4:15, 1976.
16. Olsnes S, Pihl A: In Cuatrecasas P (ed): "Receptors and Recognition: The Specificity and Action of Animal, Bacterial and Plant Toxins." London: Chapman and Hall, 1976, p 129.
17. Hughes RC, Gardas A: Nature 264:63, 1976,
18. Nairn R: PhD Thesis, University of London, 1976.
19. Edwards JG, Campbell JA: J Cell Sci 8:53, 1971.
20. Vicker MG, Edwards JG: J Cell Sci 10:759, 1972.
21. Edwards JG: In Pain RH, Smith BT(eds): "New Techniques in Biophysics and Cell Biology." vol 1, London and New York: John Wiley & Sons, 1973, p1.
22. Edwards JG, Dysart JMcK, Hughes RC: Nature 264:66, 1976.
23. Rabinovitch M, DeStefano MJ: J Cell Biol 59:165, 1973.
24. Gottlieb C, Baenziger J, Kornfeld S: J Biol Chem 250:3303, 1975.
25. Stanley P, Narasimhan S, Siminovitch L, Schachter H: Proc Natl Acad Sci USA 72:3323, 1975.
26. Hughes RC: In "Specificity in Plant Disease." NATO Advanced Study Institutes Series A: Life Sciences. London and New York: Plenum Press, 1976.
27. Ogata S, Muramatsu T, Kobata A: Nature 259:580, 1976.
28. Latan R, Skutelsky E, Danon D, Sharon N: J Biol Chem 250:8518, 1975.
29. Nicolson GL: J Natl Cancer Inst 50:1443, 1973.
30. Adair WL, Kornfeld S: J Biol Chem 249:4696, 1974.
31. Marchesi VT, Furthmayer H, Tomita M: Annu Rev Biochem 45:667, 1976.
32. Hynes RO: Biochim Biophys Acta 458:73, 1976.
33. Dysart J McK, Edwards JG: FEBS Lett 75:96, 1977.
34. Yamada KM, Yamada SS, Pastan I: Proc Natl Acad Sci USA 72:3158, 1975.
35. Yamada KM, Yamada SS, Pastan I: Proc Natl Acad Sci USA 73:1217, 1976.
36. Lisowska E, Morawiecki A: Eur J Biochem 3:237, 1967.
37. Nairn R, Hughes RC: Biochem Soc Trans 4:165, 1976.
38. Burridge K: Proc Natl Acad Sci USA 73:4457, 1976.
39. Hickman S, Shapiro LJ, Neufeld E: Biochem Biophys Res Commun 57:55, 1974.
40. Hieber V, Distler J, Myerowitz R, Schmikel R, Jourdian GW: Biochem Biophys Res Commun 73:710, 1976.
41. Graham JM, Hynes RO, Davidson EA, Bainton DF: Cell 4:353, 1975.

Author Index

A

Achord, Daniel T., 547
Albersheim, Peter, 139
Ali, Iqbal U., 369
Aminoff, David, 569
Atkinson, Paul H., 525
Ayers, Arthur R., Jr., 139

B

Barker, R., 75
Barondes, Samuel H., 633
Bell, C. Elliott, 547
Bell, William C., 569
Bernacki, R.J., 237
Bertini, Marilena, 475
Brot, Frederick E., 547
Brownell, Anna G., 225
Burke, David, 343

C

Carlson, Russell, 139
Coligan, John E., 601
Comoglio, Paolo M., 475
Cooper, Sheldon M., 131

D

Das, Manjusri, 647
Deppert, Wolfgang, 455
Dermer, Gerald B., 391
Destree, Antonia T., 369
Dreyer, W.J., 407

E

Ebel, Jürgen, 139

F

Fang, Faye, 119
Fischman, Donald A., 295
Fleischer, Becca, 159
Fong, Shao-Ling, 99
Fox, C. Fred, 647

Fram, Steven R., 253
Fraser, Blair A., 601
Friedlander, Martin, 295
Fuentes, R., 75
Furcht, L.T., 279
Furthmayr, Heinz, 201

G

Gander, J.E., 119
Geltosky, John E., 613
Ginsburg, Victor, 595
Grant, Chris, W.M., 1
Gray, Gary R., 583
Grimes, W.J., 51

H

Hahn, Michael, 139
Hargrave, Paul A., 99
Hascall, Vincent C., 181
Hatten, Mary E., 269
Hirschberg, Carlos B., 111
Hood, L., 407
Horwitz, A.F., 563
Huang, H.V., 407
Hughes, R. Colin, 657
Humphreys, Susie, 311
Humphreys, Tom, 311
Hunt, Lawrence A., 215
Hynes, Richard O., 369

K

Kabat, Elvin A., 515
Kamicker, Barbara J., 583
Kamm, Arthur R., 51
Kano, K., 27
Kaplan, Arnold, 547
Katz, Flora N., 325
Kean, Edward L., 353
Keegstra, Kenneth, 343
Keski-Oja, Jorma, 91
Kindt, Thomas J., 601
Knipe, David M., 325
Knudsen, K.A., 563

Korytnyk, W., 237
Kronquist, Kathryn E., 487

L

Lampson, Lois A., 43
Leeden, R.W., 437
Lennarz, William J., 487
Lerner, Richard A., 613
Levy, Ronald, 43
Lilien, Jack, 381
Lodish, Harvey F., 325

M

McDonough, James, 381
Marchase, Richard B., 637
Mautner, Vivien M., 369
Meints, Russel H., 503
Merrick, J.M., 27
Milgrom, F., 27
Miyakawa, Tokichi, 647
Mora, P.T., 171
Morris, Edwin R., 11
Mort, Andrew J., 553
Mosher, Deane F., 91

N

Nunez, H.A., 75

O

O'Connor, J., 75

P

Paul, B., 237
Porter, N.K., 237

R

Ray, Jasodhara, 613
Rees, David A., 11
Rothman, James E., 325
Roth, Stephen, 637
Royston, Ivor, 43
Rustum, Y., 237

S

Saksela, Olli, 91
Sambray, Yugalkishore, 131
Sano, James, 311
Sarkar, Siddhartha, 67
Schifferle, R., 27
Schwartz, Barbara A., 583
Sefton, Batholomew, 621
Serianni, A., 75
Sharma, M., 237
Sharom, Frances J., 1
Sidman, Richard L., 269
Sly, William S., 547
Summers, Donald F., 215

T

Tarone, Guido, 475
Thom, David, 11
Tökés, Zoltán A., 391
Tonelli, Quentin, 503
Trenkner, Ekkhart, 67

V

Vaheri, Antti, 91
Valent, Barbara S., 139
Van Nest, Gary A., 51
Vijay, Inder K., 253
VorderBruegge, William G., 569

W

Walker, T.E., 75
Walter, Gernot, 455
Welsh, E. Jane, 11
Wendelschafer-Crabb, G., 279
Winterbourne, D.J., 171
Wolpert, Jack, 139
Woodbridge, P.A., 279

Y, Z

Yeh, Mary, 111
Zadarlik, K., 27

Subject Index

A

Actin cytoskeleton, 369
Adhesion, 369, 563, 657
Adhesive selectivity, 637
Affinity chromatography, 475
Aggregation factor, 311
Agglutination, lectin-induced, 269
Alignment of cells, 369
α-D-glucose, 225
α-D-mannose, 225
Amino acid sequence, 201
Amino-terminal, 99
Antibodies, 43, 67
Area-code hypothesis, 408
Attachment of cells, 369

B

Bacterial polysaccharides, 139
β-D-galactose, 225
Binding, 91
Biosynthesis, by the retina, 353
Blastocyst, 225
Borohydride reduction, 503
Bovine erythrocytes, 27
Breast, 391

C

Carbohydrate(s), 225
    antigens, 583
    attachment, 99
    binding site, 515
    conformations, 75
Carboxyl-terminal, 99
Carcinoma, 391
Cell
    membranes, 225
    migration, 269
    viability, 455
Cell surface, 51
    antigens, 295
    embryonic, 269
    glycoproteins, 613
    recognition molecules, 407

Cellular slime molds, 633
Characterization, 553
Chromosomal modifications, 532
Circular dichroism, 11
Collagen, 225
Concanavalin A, 225
Conformational analysis, 11
Cooperative cation binding, 11
Cooperative interactions, 11
Crossreaction, 601
Cyanoborohydride, 583
Cystic fibrosis fibroblasts, 215

D

Deglycosylation, 553
Development, 67
Developmental regulation, 613
Dimeric lectins, 269
Dictyostelium discoideum, 613, 633
DNA translocation, 531
Dolichyl-phosphate-mannose, 353

E

Electron microscopy, 201
Elicitor(s), 139
Embryology, 225
Embryonic cell surface, 269
Endoplasmic reticulum, 325
Envelope glycoprotein, 215
Erythrocyte
    aging, 569
    analysis, 569
    viability, 569
Erythropoietin, 503

F

Fc receptors, 131
Fibroblasts, 93
Fibronectin, 93
Fluorescense microscopy, 225
Forssman, 601
Fucose & glucosamine, 525
Fungus, 139
Fusion, 563

G

Galactofuranosyl, 119
Galactose oxidase, 569
Gangliosides, 1, 437
GDP-mannose, 119
Glucan, 139
Glycoconjugate, 171, 311
Glycolipid, 1, 51
Glycopeptides, 119, 343
   synthesis, 525
Glycoproteins, 51, 311, 325, 343, 353, 391, 547, 553
   envelope, 215
   LETS, 369
   receptors, 657
Glycosaminoglycan, 171
Glycosphingolipids 1, 437
Glycosylation, 325, 487
Glycosyl phosphates, 75
Glycosyltransferase, 215, 595
G protein, 525
Group C, 601

H

HeLa cells, 525
Hematopoietic stem cell, 503
Hepatocytes, 569
Heterobifunctional crosslinker, 647
Heterophile antigen, 27
HF, 553
Host-pathogen interaction(s), 139
Host-symbiont interaction(s), 139
Hyaluronic acid, 171

I

Immune precipitation, 369
Immune system, 407
Immunochemistry, 201
Immunofluorescence, 93, 369
Implantation, 223
Infectious mononucleosis, 27
Intracellular glycosylation, 455
Intracellular membrane pools, 525

K

Kupffer cells, 569

L

Lectin's, 139, 225, 269, 633
   activity and development, 633
   affinity chromatography, 613
   in chicks, 633
   -induced agglutination, 269
   in slime molds, 633
LETS glycoprotein, 369
L-fucose, 225
Lipid-linked oligosaccharides, 621
Lipid-linked saccharides, 487
Lipopolysaccharide(s) 141
Liver, 569
Lysosomal enzyme, 547

M

Mannolipids, 353
Mannosyl, 547
Mannosyltransferase, 119
   of the retina, 353
Matrix protein, 325
Membrane, 27
   asymmetry, 325
   biogenesis, 325
   glycoproteins, 131
   structure, 1
Membrane bound & free polysomes, 525
Microvilli, 225
Mitogen receptor, 647
Monosaccharides, 225
Morula, 225
Mouse
   cerebellum, 67
   leukemia, 131
M protein, 525
Multigene families, 407
Mutation, 657
Mycelial wall polysaccharides, 139
Myoblast differentiation, 295
Myogenesis, 563

N

N-acetyl-D-galactosamine, 225
N-acetyl-D-glucosamine, 225
N-acetyl-neuraminic acid, 503
Nascent polypeptides, 621

Neoplastic transformation, 475
Neuraminic acids, 569
Neuraminidase, 295
Neuraminidase (vibrio cholerae), 503
Neurons, 437
Nervous system, 437
Nitrogen fixation, 139
Nuclear magnetic resonance, 11
Nucleotide, 353
    sugar hydrolysis, 455

O

Oligosaccharide, 487
    lipids, 353
    structures, 437
Optical rotation, 11
Organ culture, 391

P

Pathogenesis, 139
Paul-Bunnell antigen, 27
Penicillium, 119
Periodate oxidation, 503
Phosphoglycoprotein, 547
Phosphomannan, 119
Phosphorylation site, 99
Photoactivable reagent, 647
Phytoalexin(s), 139
Pinocytosis, 547
Plant(s), 139
    cell cultures, 139
    hemagglutinins, 515
Plasma membrane, 325, 475
    assembly, 525
Polyprenyl-phosphates, 353
Polysaccharides, 11, 311, 553, 601
Polysphondylium pallidum, 633
Preimplantation embryos, 225
Primary neuronal cultures, 269
Prostate, 391
Protease, 295
Protein(s)
    fucose-binding, 225
    matrix, 325
    synthesis inhibition, 369
Proteoglycans, 311
Proteolysis, limited, 99

Purified plasma membranes, 525
Pyrophosphorylase, 595

R

Receptor, 93, 547
Recognition, 547, 563
Retina, 353
Retinotectal specificity, 637
Reversion of transformed
    morphology, 369
Rhodopsin, 99
Ricin, 657
Ricinus communis agglutinin$_I$, 225
Ricinus communis agglutinin$_{II}$, 225
Rod cell membrane, 99
Rosettes, 569

S

Scanning electron microscopy, 225
Secretion, 93
Sialic acids, 569
    uptake, 111
Sialidase, 569
Sialoglycolipids, 111
Sialoglycoprotein, 111, 201
Sindbis virus, 343, 621
Solvolysis, 553
Soybean(s), 139
Spectroscopic techniques, 11
Spleen, 569
    colonies, 503
Sponges, 311
Spreading of cells, 369
Streptococcus, 601
Subcellular distribution, 437
Sugar linkages, 343
Surface carbohydrates, 67
Symbiosis, 139
Synergistic interactions, 11
Synthesis by transformed cells, 369
Synthetic glycoproteins, 583

T

T & B cells, 43
^{13}C-enriched carbohydrates, 75
^{13}C-NMR, 75

Tissue culture, 225
Transformation, 171
Transformed cells, 51
Trophoblast, 225
Tumourigenicity, 171

V

Vesicular stomatitis virus, 215, 525

W

Wheat germ agglutinin, 225

Z

Zona pellucida, 225

QP
701
C34

SEP 19 1978